FROM FIRST FLIGHT TO PRIVATE CERTIFICATE

THE STUDENT PILOT'S FLIGHT MANUAL

Including FAA Knowledge Test Questions, Answers and Explanations, and Practical (Flight) Test for Airplanes

EIGHTH EDITION

William K. Kershner

IOWA STATE UNIVERSITY PRESS / AMES

To Betty, Cindy, and Bill

WILLIAM K. KERSHNER began flying in 1945 at fifteen, washing and propping airplanes to earn flying time. He obtained the private, commercial, and flight instructor certificates, becoming a flight instructor at nineteen. He spent four years as a naval aviator, mostly as a pilot in a night fighter squadron, both shore- and carrier-based. He flew nearly three years as a corporation pilot and for four years worked for Piper Aircraft Corporation, demonstrating airplanes to the military, doing experimental flight testing, and acting as special assistant to William T. Piper, Sr., company president. Kershner holds a degree in technical journalism from Iowa State University. While there he studied Air Science courses such as aerodynamics, performance, and stability and control. He holds airline transport pilot, commercial, and flight and ground instructor certificates and has flown airplanes ranging from 40-HP Cubs to jet fighters. He also is the author of *The Advanced Pilot's Flight Manual, The Flight Instructor's Manual,* and *The Basic Aerobatic Manual.* For several years he has operated a one-airplane, one-instructor aerobatic school using a Cessna 152 *Aerobat.* Kershner received the General Aviation Flight Instructor of the Year Award, 1992, at the state, regional, and national levels and the Elder Statesman of Aviation Award for 1997 from the National Aeronautic Association, and was inducted into the Flight Instructor Hall of Fame in 1998.

ILLUSTRATED BY THE AUTHOR

© 1998 William K. Kershner. Previous editions and printings © 1993, 1990, 1979, 1973, 1968, 1964, 1960 Iowa State University Press, Ames, Iowa 50014. All rights reserved.

♾ Printed on acid-free paper in the United States of America

Authorization to photocopy items for internal or personal use, or the internal or personal use of specific clients, is granted by Iowa State University Press, provided that the base fee of $.10 per copy is paid directly to the Copyright Clearance Center, 222 Rosewood Drive, Danvers, MA 10923. For those organizations that have been granted a photocopy license by CCC, a separate system of payments has been arranged. The fee code for users of the Transactional Reporting Service is 0-8138-1609-2/98 $.10.

First, second, third, fourth, fifth, sixth, and seventh editions, © 1960, 1964, 1968, 1973, 1979, 1990, 1993 carried through forty-five printings.

Eighth edition, 1998

Library of Congress Cataloging-in-Publication Data

Kershner, William K.
    The student pilot's flight manual : including FAA knowledge test questions, answers, and explanations and practical (flight) test for airplanes / William K. Kershner.—8th ed.
        p.   cm.
    "From first flight to private certificate."
    ISBN 0-8138-1609-2
    1. Airplanes—Piloting—Handbooks, manuals, etc. 2. Airplanes—Piloting—Examinations—Study guides. I. Title.
    TL710.K43   1998
    629.132'52—dc21   98–2725

Last digit is the print number: 9 8 7 6 5 4 3 2

Contents

A note for private pilots

This book was written basically as a step-by-step ground and flight reference for the person starting to fly and working toward the private certificate, but with additions it is also intended for use as a reference for the Flight Review for the private pilot, single-engine, land airplane.

FAR 61.56 describes the general requirements for the Flight Review and notes that it consists of a review of the current general operating and flight rules of FAR 91 (selected sections are also included there) plus a review of maneuvers and procedures that, with the discretion of the person giving the review, are necessary to demonstrate that you can safely exercise the privileges of your pilot certificate.

The HOME STUDY section in Chapter 27 can be used as a general reference for the oral part of the review. Subject areas and chapter references are listed for points you feel should be looked at again. Then read the suggested chapter or chapters to get up-to-date on *details* of a particular maneuver or procedure. Also a good look at FAR Part 91—*General Operating and Flight Rules*—and a recheck of the included Part 61—*Certification: Pilots, Flight Instructors, and Ground Instructors*—would be in order to note your responsibilities and limitations. It's suggested that you take another look at the Knowledge Test in the chapter as a check of your knowledge in preparing for the Flight Review.

After you've had a good review of your general knowledge, use Chapter 28 as an initial quick reference for maneuvers and then go back to the chapters recommended for more detailed study of the theory and procedures. Chapter 28 covers the knowledge and skill requirements for the private pilot *practical test* and in some instances may repeat areas of knowledge covered in Chapter 27 (such as certificates and documents, weight and balance, and weather information) but will likely bring up some new ideas for you to use on the oral part of the review as well.

Consult the *Pilot's Operating Handbook* for the airplane you're flying to check and fill in the numbers in the form at the end of Chapter 28. If there is a question, procedures recommended by the *Pilot's Operating Handbook* and your instructor for your airplane will naturally take precedence over those in this book, since as a text it must take a general look at maneuvers and operational procedures.

Remember that you'll be taking a *review,* not a practical test, and you can't "fail" it. If you're more rusty than you thought, the flight instructor won't make any notations in your logbook or send anything to the Flight

preparing for a Flight Review . . .

Standards District Office. Your instructor might recommend some practice or more dual and then, when you've done this and straightened out any problems, will fly with you and endorse your logbook to the effect that the Flight Review is completed.

One problem you may face is "checkitis" as described in Chapter 28. It may have been several years since you had someone overseeing your procedures and flying technique or asking questions about FAR 91, so you can expect to be a little tense. Give yourself a good review at home and be familiar with the various maneuvers as listed in Chapter 28 and you'll do fine.

One question that came to mind when adapting this book to be used also as a Flight Review guide was whether the name "Student Pilot's Flight Manual" would turn off a private pilot who is "beyond all that." The answer is, of course, that if the needed material is available the name doesn't matter; all pilots, whatever certificates or ratings they hold, will be student pilots as long as they fly, in the broad sense of the term. That's the appeal of flying; the fact that you'll never learn it all and are constantly challenged to improve.

After over 50 years of flying this writer is still a "student pilot" and hopefully will remain so. So forgive the title.

Bill Kershner

PREFACE

This manual is a gathering of material used in pre-flight and postflight briefings and in-flight instruction. The maneuvers are written in the probable order of introduction to the student. The spin was included because the writer felt that students should have some idea of just what this maneuver entailed, even if they would never be required to practice one. No attempt was made to set up a rigid schedule of maneuvers because such a schedule is precluded by variation in student ability.

This book is not intended to just help the reader "get past" the FAA written and practical (flight) tests but includes information for use in the day-to-day process of learning to fly airplanes.

There is emphasis on making judgment calls in various areas of flying (formally known as aeronautical decision making) to help the student pilot start setting up individual methods of coming to the right decision for a go or no-go situation.

At various points in the book, judgment factors are included to achieve this end.

I would like to thank the following for help on this manual for the first five editions:

Colonel E. F. Quinn, Professor of Air Science at Iowa State University, whose encouragement and backing got the book started.

Major J. W. Carothers and Captain H. E. Fischer also of the Iowa State University Air Science Department, who reviewed the first manuscript and made suggestions and comments.

Dr. M. L. Millett of the Department of Aeronautical Engineering at Iowa State University, who reviewed both the first and final drafts of the first manuscript, with particular attention to the aerodynamics and instrument theory, and who made comments and suggestions that improved the book.

The Army Aviation School personnel at Fort Rucker, Ala., who reviewed and made many valuable suggestions.

The Air Force ROTC Headquarters personnel at Maxwell Field, Ala., who reviewed and encouraged the completion of the manual.

A. J. Prokop, who checked the first manuscript; Eugene Anderson, also of the Des Moines Aviation Safety District Office, who checked the manuscript with particular emphasis on emergency recovery by instruments; and Glen Bower, flight instructor of Ames, Iowa, who reviewed and commented on the manual.

The FAA General Operations Branch Office personnel in Washington, D.C., who made several pertinent corrections and suggestions.

J. M. Bennett of Morgantown, W.Va., who furnished reports on the instrument training program at the University of West Virginia.

Particular thanks must go to K. R. Marvin, former head of the Department of Technical Journalism at Iowa State University, who backed the effort.

John Murray, of the Weather Bureau Airport Station at Williamsport, Pa., for his review and suggestion for revision of the chapter on weather.

Jerome Bosworth and John Leck of the Williamsport, Pa., Flight Service Station for their help in obtaining the latest material about the functions of the FSS.

Delbert W. Robertson of the Chattanooga, Tenn., Weather Bureau Airport Station for answering questions and furnishing weather data.

James D. Lee of the Crossville, Tenn., Flight Service Station for furnishing FSS information.

Colonel Leslie McLaurin, flight instructor and manager of the airport here at Sewanee, for suggestions and ideas that were used in the book.

W. D. Thompson, Chief, Flight Test and Aerodynamics, Commercial Aircraft Division of Cessna Aircraft Company, for furnishing data on Cessna airplanes.

I've had many good suggestions for improving the book from student and private pilots and flight instructors. I would especially like to mention Allen Hayes of Chartair, Inc., Ithaca, N.Y., and Robert S. Woodbury of Massachusetts Institute of Technology.

Again, I've been lucky in having knowledgeable individuals helping with this book over the years. Any errors are mine, though. I would especially like to thank the following people:

William Thompson, who sent me data on Cessna airplanes with dispatch with valuable personal comments on the aerodynamics of flight.

Thanks must also go to Cessna and Grumman for allowing me to use information on their airplanes as examples.

Appreciation is expressed to Dr. A. V. Ray of Burlington, Ontario, for suggestions on improving the section in Chapter 26 on the physiology of the eye.

Special thanks must go to Elizabeth Motlow, who flew the example cross-country with me and took pictures of the checkpoints.

The people at the Chattanooga Weather Service Office, as always, were most cooperative and came through with actual material and, more importantly, simple explanations for the computerized methods of weather reporting and forecasting. In particular, I would mention Ed Higdon and W. R. Wright. Others who helped are Ray R. Casada, Hugh Pritchard, Jr., Delbert Robertson, M. H. Smith, Jack Phillips, and Sam DeLay.

Thanks to Jules Bernard of Tullahoma, Tenn., for help.

David F. Shaw, of Penn Yan, N.Y., gave good suggestions for later printings.

Thanks to Larry Crawford, of the Nashville Flight Service Station, for furnishing actual weather information.

Kermit Anderson, formerly of the FAA at OKC, gave much appreciated and valuable help.

My wife encouraged me, as always, and gave practical help by typing the rough copies of the manuscript in the first four editions. Mrs. Barbara Hart typed the smooth copy for the fourth edition.

For the sixth edition, I would add:

Thanks to Margaret Puckette of Corvallis, Ore., and Tim Keith-Lucas of Sewanee for suggestions that would make the book more useful to the student and low-time private pilot.

Eleanor Ulton of Sewanee did her usual outstanding job of translating my scribbling into understandable English.

Nancy Bohlen, my astute editor at ISU Press, kept a sharp eye on my grammar and helped assure that female pilots were given proper recognition. Her efforts have made my books much better over the years of our association.

Thanks to Lynne Bishop, also an editor at ISU Press, for her patience, humor, and ability to catch my mistakes in this edition.

My son, Bill, currently flying for a major airline, gave much help, particularly with the answers and explanations in Chapter 27.

Special thanks to Kathleen Schlachter, meteorology instructor at the FAA Academy at OKC, for translating weather presentations to the METAR/TAF format.

To Hap's Air Service, Ames, Iowa, for donating the airplane, fuel, and air time used during the photo session for the cover. Thanks also are expressed to James Kurtenbach for piloting the plane, to Kristen Krumhart for being the student pilot, and to Peter Krumhart for taking the photograph.

This book is aimed at the individual who is working directly for the private pilot (airplane, single-engine, land) certificate; therefore, other certificates and ratings (recreational pilot or private pilot, seaplane, etc.) are not emphasized here. The basics of flying an airplane *are* covered, so people working on those other certificates/ratings will still find it useful.

Thanks to Jim Oliver, training specialist at the Nashville Flight Service Station, for answering my questions and furnishing material to update the weather services and NOTAM materials.

As a textbook, a general approach must be taken here in discussing maneuvers and flight areas such as weather services and other rapidly changing information services. Some airplanes may require specific techniques in, for instance, use of flaps, carburetor heat, or recommended spin entries and recoveries. The procedures for a particular airplane as outlined in the *Pilot's Operating Handbook*, Airplane Flight Manual, or the equivalent information for your airplane will naturally be the final guide for operations. And, as noted in Chapter 3, your flight instructor will also have suggestions for your airplane and operating conditions.

William K. Kershner
Sewanee, Tennessee

PART ONE

BEFORE THE FLIGHT

1

STARTING TO FLY

There are many reasons why people want to start flying. Maybe you are a younger person who wants to make it a lifetime career or maybe you are a slightly more senior citizen who always wanted to fly but until now haven't had the money. Whether a man or woman, young or old, you still may have a few butterflies in your stomach while worrying about how you will like it or whether you can do it. That's a natural reaction.

What can you expect as you go through the private pilot training course? You can expect on most flights to work hard and to come down from some flights very tired and wet with perspiration, but with a feeling of having done something worthwhile. After others, you may consider forgetting the whole idea.

Okay, so there will be flights that don't go so well, no matter how well you get along with your instructor. The airplane will seem to have decided that it doesn't want to do what you want it to. The situation gets worse as the flight progresses and you end the session with a feeling that maybe you just aren't cut out to be a pilot. If you have a couple of these in a row, you should consider changing your schedule to early morning instead of later afternoon flights, or vice versa. You may have the idea that everybody but you is going through the course with no strain at all, but every person who's gone through a pilot training course has suffered some "learning plateaus" or has setbacks that can be discouraging.

After you start flying, you may at some point decide that it would be better for your learning process if you changed instructors. This happens with some people and is usually a no-fault situation, so don't worry about a change or two.

It's best if you get your FAA (Federal Aviation Administration) medical examination out of the way very shortly after you begin to fly or, if you think that you might have a problem, get it done before you start the lessons. The local flight instructors can give you names of nearby FAA aviation medical examiners.

How do you choose a flight school? You might visit a few in your area and see which one suits you best. Watch the instructors and students come and go to the airplanes. Are the instructors friendly, showing real interest in the students? There should be pre- and postflight briefings of students. You may not hear any of the details but can see that such briefings are happening. Talk to students currently flying at the various schools and get their opinions of the learning situation. One good hint about the quality of maintenance of the training airplanes is how clean they are. Usually an airplane that is clean externally is maintained well internally, though certainly there are exceptions to this.

What about the cost? It's a good idea to have money ahead so that you don't have to lay off and require a lot of reviewing from time to time. Some flight schools give a discount if you pay for several hours ahead of time.

You are about to set out on a very rewarding experience; for an overall look at flight training and future flying you might read the following.

THE BIG THREE

As you go through any flight program, particularly in a military flight program, you will hear three terms used many times: *headwork, air discipline,* and *attitude toward flying.*

You may be the smoothest pilot since airplanes were invented, but without having a good grip on the above requirements, you'll last about as long as Simon Legree at an Abolitionist meeting.

Fig. 1-1. Headwork is remembering to put the landing gear down.

■ Headwork

For any pilot, private or professional, the most important thing is good headwork. Nobody cares if you slip or skid a little or maybe every once in a while land a mite harder than usual. But if you don't use your head—if you fly into bad weather or forget to check the fuel and have to land in a plowed field—you'll find people avoiding you. Later, as you progress in aviation and lead flights or fly passengers, it's a lot more comfortable for all concerned if they know you are a person who uses your head in flying. So as the sign says—THILK, er, think.

■ Air Discipline

This is a broad term but generally means having control of the aircraft and yourself at all times. Are you a precise pilot or do you wander around during maneuvers? Do you see a sports car and decide to buzz it? Air discipline is difficult at times. It's mighty tough not to fly over that good-looking member of the opposite sex who happens to be sunbathing right where you are doing S-turns across the road—but be firm!

More seriously, air discipline is knowing, and flying by, your own limitations. This means, for instance, holding down for bad weather and not risking your life and your passengers' lives. It also means honestly analyzing your flying faults and doing something about them. In short, air discipline means a mature approach to flying.

■ Attitude

A good attitude toward flying is important. Most instructors will go all out to help someone who's really trying, even a complete bungler. Many an instructor's favorite story is about ol' Joe Blow who was pretty terrible at first, but who kept at it until he got the word, and is now flying rockets for Trans-Galaxy Airlines. With a good attitude you will get plenty of help from everybody. More students have failed in flying because of poor headwork and attitude than for any other reason. This doesn't imply "apple polishing." It does mean that you are interested in flying and study more about it than is required by law.

2

THE AIRPLANE AND
HOW IT FLIES

Four forces act on an airplane in flight: *lift, thrust, drag,* and *weight* (Fig. 2-1).

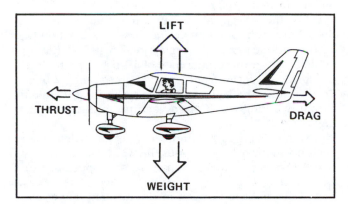

Fig. 2-1. The four forces. When the airplane is in equilibrium in straight and level cruising flight, the forces acting fore and aft (thrust and drag) are equal, as are those acting at 90° to the flight path (lift and weight, or its components).

THE FOUR FORCES

■ Lift

Lift is a force exerted by the wings. (Lift may also be exerted by the fuselage or other components, but at this point it would be best just to discuss the major source of the airplane's lift, the wings.) It is a force created by the "airfoil," the cross-sectional shape of the wing being moved through the air or, as in a wind tunnel, the air being moved past the wing. The result is the same in both cases. The "relative wind" (wind moving in relation to the wing and airplane) is a big factor in producing lift, although not the only one (Fig. 2-2).

Lift is always considered to be acting perpendicularly both to the wingspan and to the relative wind (Fig. 2-3). The reason for this consideration will be shown later as you are introduced to the various maneuvers.

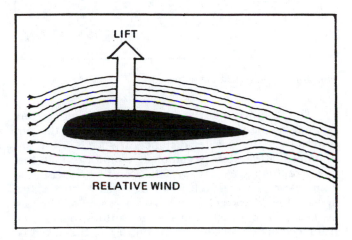

Fig. 2-2. The airfoil.

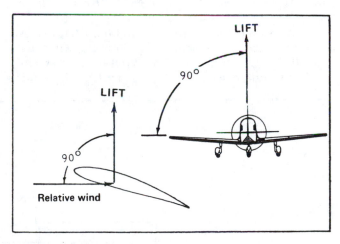

Fig. 2-3. Lift acts perpendicular to the relative wind and wingspan.

As the wing moves through the air, either in gliding or powered flight, lift is produced. How lift is produced can probably be explained most simply by Bernoulli's theorem, which briefly puts it this way: "The faster a fluid moves past an object the less sidewise pressure is exerted on the body by the fluid.'' The fluid in this case is air; the body is an airfoil. Take a look at Figure 2-4, which shows the relative wind approaching an airfoil, all neatly lined up in position 1. As it moves past the airfoil (or as the airfoil moves past it—take your choice), things begin to happen, as shown by the subsequent numbers.

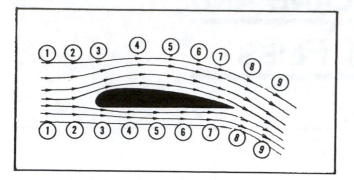

Fig. 2-4. Airflow past the airfoil.

The distance that the air must travel over the top is greater than that under the bottom. As the air moves over this greater distance, it speeds up in an apparent attempt to reestablish equilibrium at the rear (trailing edge) of the airfoil. (Don't worry, equilibrium *won't* be reestablished.) Because of this extra speed, the air exerts less sidewise pressure on the top surface of the airfoil than on the bottom, and lift is produced. The pressure on the bottom of the airfoil is normally increased also and you can think that, as an average, this contributes about 25 percent of the lift; this percentage varies with "angle of attack" (Fig. 2-5).

Some people say, "Sure, I understand what makes a plane fly. There's a vacuum on top of the wing that holds the airplane up.'' Let's see about that statement.

The standard sea level air pressure is 14.7 pounds

per square inch (psi), or 2116 pounds per square foot (psf). As an example, suppose an airplane weighs 2000 pounds, has a wing area of 200 square feet, and is in level flight at sea level. (The wing area is that area you would see by looking directly down on the wing.) This means that for it to fly level (lift = weight) each square foot of wing must support 10 pounds of weight, or the wing loading is 10 pounds psf (2000 divided by 200). Better expressed: There would have to be a difference in pressure of 10 pounds psf between the upper surface and the lower surface. This 10 psf figure is an average; on some portions of the wing the difference will be greater, on others, less. Both surfaces of the wing can have a reduced sidewise pressure under certain conditions. However, the pressure on top still must average 10 psf less than that on the bottom to meet our requirements of level flight for the airplane mentioned. The sea level pressure is 2116 pounds psf, and all that is needed is an average difference of 10 psf for the airplane to fly.

Assume for the sake of argument that in this case the 10 psf is obtained by an *increase* of 2.5 psf on the bottom surface and a *decrease* of 7.5 psf on the top (which gives a total difference of 10 psf). The top surface pressure varies from sea level pressure by 7.5 psf. Compared to the 2116 psf of the air around it, this is certainly a long way from a vacuum, but it produces flight!

Note in Figures 2-2 and 2-4 that the airflow is deflected downward as it passes the wing. Newton's law, "For every action there is an equal and opposite reaction,'' also applies here. The wing deflects the airflow downward with a reaction of the airplane being sustained in flight. This can be easily seen by examining how a helicopter flies. Some engineers prefer Newton's theory over Bernoulli's theory. But the air *does* increase its velocity over the top of the wing (lowering the pressure), and the downwash also occurs. The downwash idea and how it affects the forces on the horizontal tail will be covered in Chapters 9 and 23.

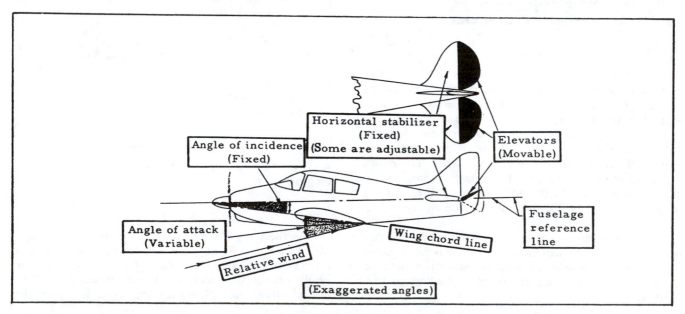

Fig. 2-5. Nomenclature.

ANGLE OF ATTACK. *The angle of attack is the angle between the relative wind and the chord line of the airfoil.* Don't confuse the angle of attack with the angle of *incidence.* The angle of *incidence* is the *fixed* angle between the wing chord line and the reference line of the fuselage. You'd better take a look at Figure 2-5 before this gets too confusing.

The pilot controls angle of attack with the elevators (Fig. 2-5). By easing back on the control wheel (or stick) the elevator is moved "up" (assuming the airplane is right side up). The force of the relative wind moves the tail down, and because the wings are rigidly attached to the fuselage (you hope) they are rotated to a new angle with respect to the relative wind, or new *angle of attack*. At this new angle of attack the apparent curvature of the airfoil is greater, and for a very short period lift is increased. But because of the higher angle of attack more drag is produced, the airplane slows, and equilibrium exists again. (More about drag later.)

If you get too eager to climb and *mistakenly* believe that the reason an airplane climbs is because of an "excess" of lift (and so keep increasing the angle of attack), you could find that you have made a mistake. As you increase the angle of attack the airplane slows and attempts to reestablish equilibrium, so you continue to increase it in hopes of getting an "excess" of lift for more climb. You may make the angle of attack so great that the air can no longer flow smoothly over the wing, and the airplane "stalls" (Fig. 2-6).

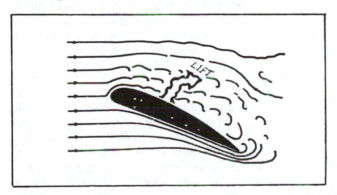

Fig. 2-6. The stall.

It's not like a car stalling, in which case the engine stops; the airplane stall is a situation where the lift has broken down and the wing, in effect, is no longer doing its job of supporting the airplane in the usual manner. (The engine may be humming like a top throughout the stall.) There is still some lift, but not enough to support the airplane. You have forced the airplane away from the balanced situation you (and the airplane) want to maintain. For the airplane to recover from a stall, you must decrease the angle of attack so that smooth flow again occurs. In other words, point the plane where it's going! This is done with the elevators, the angle of attack (and speed) control (Fig. 2-5). For most lightplane airfoils the stalling angle of attack is in the neighborhood of 15°. Stalls will be covered more thoroughly in Chapters 12 and 14.

At first, the student is also confused concerning the *angle of attack* and airplane *attitude.* The attitude is how the plane looks in relation to the horizon. In Figure 2-7 the plane's attitude is 15° nose up, but it's climbing at an angle of 5°, so the angle of attack is only 10°.

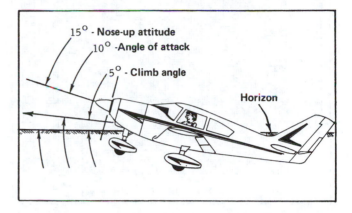

Fig. 2-7. Pitch attitude, climb angle (flight path), and angle of attack.

In a slow glide the nose attitude may be approximately level and the angle of attack close to that of the stall. Later in your flying you'll be introduced to the attitude of the wings (wing-down attitude, etc.), but for now only nose attitudes are of interest.

The coefficient of lift is a term used to denote the relative amounts of lift at various angles of attack for an airfoil. The plot of the coefficient of lift versus the angle of attack is a straight line, increasing with an increase in the angle of attack until the stalling angle is reached (Fig. 2-8).

Lift depends on a combination of several factors. The equation for lift is:

$$L = C_L S \frac{\varrho}{2} V^2, \quad \text{or} \quad L = C_L \times S \times \frac{\varrho}{2} \times V^2$$

where L = lift, in pounds

C_L = coefficient of lift (varies with the type of airfoil used and the angle of attack). The coefficient of lift, C_L, is a dimensionless product and gives a *relative* look at the wings' action. The statement may be made in groundschool that, "At this angle of attack, the coefficient of lift is point five (0.5)." Point five what? "Just point five, and it's one-half of the C_L at one point zero (1.0)." Just take it as the relative effectiveness of the airfoils at a given angle of attack. Later, the coefficient of *drag* will be discussed.

S = wing area in square feet

$\frac{\varrho}{2}$ = air density (ϱ) divided by 2. Rho (ϱ) is air density, which for standard sea level conditions is 0.002378 slugs per cubic foot. If you want to know the mass of an object in slugs, divide the weight by the acceler-

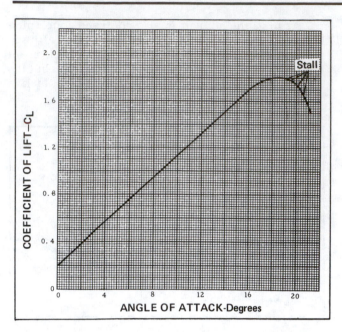

Fig. 2-8. Relative lift increases with angle of attack until the stall angle is reached.

ation of gravity, or 32.2. (The acceleration caused by gravity is 32.2 feet per second per second at the earth's surface.)

V^2 = velocity in feet per second squared

When you fly an airplane, you'll be working with a combination of C_L and velocity; but let's talk in pilot terms and say that you'll be working with a combination of angle of attack and airspeed. So lift depends on angle of attack, airspeed, wing area, and air density. For straight and level flight, lift equals weight. Assuming that your airplane weighs 2000 pounds, 2000 pounds of lift is required to maintain level flight. This means that the combination of the above factors must equal that value. The wing area (S) is fixed, and the air density (ϱ) is fixed for any one time and altitude. Then C_L (angle of attack) and velocity (airspeed) can be worked in various combinations to maintain the 2000 pounds of lift required. Flying at a low airspeed requires a high angle of attack, and vice versa. As the pilot you will control angle of attack and, by doing so, control the airspeed. You'll use power (or lack of power) with your chosen airspeed to obtain the desired performance.

While the factors of lift are being discussed, it might be well to say a little more about air density (ϱ). The air density decreases with increased altitude and/or temperature increase. Airplanes require more runway to take off on hot days or at airports of high elevation because of decreased air density. You can see in the lift equation that if the air is less dense the airplane will have to move faster through the air in order to get the required value of lift for flight—and this takes more runway. (The airspeed mentioned is called "true airspeed" and will be discussed in more detail in

Chapter 3.) Not only is the lift of the wing affected, but the less dense air results in less power developed within the engine. Since the propeller is nothing more than a rotating airfoil, it also loses "lift" (or, more properly, "thrust"). Taking off at high elevations or high temperatures can be a losing proposition, as some pilots have discovered after ignoring these factors and running out of runway.

Interestingly enough, you will find that lift tends to remain at an almost constant value during climbs, glides, or straight and level flight. *Don't* start off by thinking that the airplane glides because of decreased lift or climbs because of excess lift. *It just isn't so.*

■ **Thrust**

Thrust is the second of the four forces and is furnished by a propeller or jet. The propeller is of principal interest to you at this point, however.

The theory of propellers is quite complicated, but Newton's "equal and opposite reaction" idea can be stated here. *The propeller takes a large mass of air and accelerates it rearward, resulting in the equal and opposite reaction of the plane moving forward.*

Maybe it's time a few terms such as "force" and "power" should be cleared up. Thrust is a *force* and like the other three forces is measured in pounds. A *force* can be defined as a tension, pressure, or weight. You don't necessarily have to move anything; you can exert force against a very heavy object and nothing moves. Or you can exert a force against a smaller object and it moves. When an object having force exerted upon it moves, *work* has been done.

Work, from an engineering point of view, is simply a measure of *force* times *distance.* And while at the end of a day of pushing against a brick wall or trying to lift a safe that won't budge, you feel tired, actually you've done no *work* at all. If you lift a 550-pound safe 1 foot off the floor, you'll have done 550 foot-pounds of *work* (and no doubt strained yourself in the bargain). If you lift a 50-pound weight to a height of 11 feet, you'll have done the same *work* whether you take all day or 1 second to do it—but you won't be developing as much *power* by taking all day. So the *power* used in lifting that 50 pounds up 11 feet, or 550 pounds up 1 foot, in 1 second would be expressed as:

Power = 550 foot-pounds per second

Obviously, this is leading somewhere, and you know that the most common measurement for power is the term "horsepower." One horsepower is equal to a power of 550 foot-pounds per second, or 33,000 foot-pounds per minute (60 seconds × 550). Whether the average horse of today can actually do this is not known, and unfortunately nobody really seems to care.

The airplane engine develops horsepower within its cylinders and, by rotating a propeller, exerts thrust. In straight and level, unaccelerated, cruising flight the thrust exerted (pounds) is considered to equal the drag (pounds) of the airplane.

You will hear a couple of terms concerning horsepower:

Brake horsepower—the horsepower developed at the crankshaft. In earlier times this was measured at the crankshaft by a braking system or absorption dynamometer known as a "prony brake." *Shaft horsepower* means the same thing. Your airplane engine is always rated in brake horsepower, or the power produced at the crankshaft. Brake horsepower and engine ratings will be covered more thoroughly in Chapter 23.

Thrust horsepower—the horsepower developed by the propeller in moving the airplane through the air. Some power is lost because the propeller is not 100 percent efficient, and for round figures you can say that the propeller is *at best* about 85 percent efficient (the efficiency of the fixed-pitch propeller varies with airspeed). The thrust horsepower developed, for instance, will be only up to about 85 percent of the brake horsepower.

If you are mathematically minded you might be interested in knowing that the equation for thrust horsepower (THP) is: THP = TV ÷ 550, where T is thrust (pounds) and V is velocity (feet per second) of the airplane. Remember that a *force* times a *distance* equals *work*, and when this is divided by time, *power* is found. In the equation above, *thrust* is the force, and velocity can be considered as being distance divided by time, so that TV (T × V) is power in foot-pounds per second. Knowing that 1 horsepower is 550 foot-pounds per second, the power (TV) is divided by 550 and the result would give the horsepower being developed—in this case, *thrust horsepower*. (See Chapter 9.) THP = TV mph ÷ 375, or TV knots ÷ 325 (more about "knots" later).

For light trainers with fixed-pitch propellers a measure of the power (brake horsepower) being used is indicated on the airplane's tachometer in rpm's

(revolutions per minute). The engine power is controlled by the throttle. For more power the throttle is pushed forward or "opened"; for less power it is moved back or "closed." You'll use the throttle to establish certain rpm (power) settings for cruise, climb, and other flight requirements.

TORQUE. Because the propeller is a rotating airfoil, certain side effects are encountered. The "lift" force of the propeller is the thrust used by the airplane. The propeller also has a drag force. This force acts in a sidewise direction (parallel to the wing span or perpendicular to the fuselage reference line).

The propeller rotates clockwise as seen from the cockpit, causing a rotating mass of air to be accelerated toward the tail of the airplane. This air mass strikes the left side of the vertical stabilizer and rudder. This air mass, called "slipstream" or "propwash," causes the airplane to veer or "yaw" to the left. Right rudder must therefore be applied to hold the airplane on a straight track (Fig. 2-9). This reaction increases with power, so it is most critical during the takeoff and climb portion of the flight. The slipstream effect is the biggest factor of torque for the single-engine airplane.

An offset fin may be used to counteract this reaction. The fin setting is usually built in for maximum effectiveness at the rated cruising speed of the airplane, since the airplane will be flying most of the time at cruising speed (Fig. 2-10).

The balance of forces at this point results in no yawing force at all, and the plane flies straight with no right rudder being held.

Sometimes the fin may not be offset correctly due to tolerances of manufacturing, and a slight left yaw is present at cruising speed, making a constant use of right rudder necessary to hold the airplane straight. To take care of this, a small metal tab is attached to the trailing edge of the rudder and is bent to the left. The relative air pressure against the tab forces the rudder to the right (Fig. 2-11).

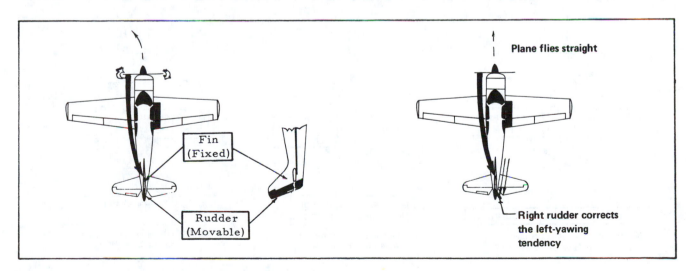

Fig. 2-9. The slipstream effect makes the airplane want to yaw to the left.

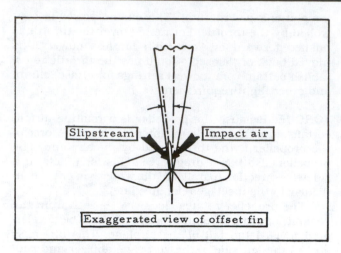

Fig. 2-10. The fin of this example airplane is offset to balance the yawing forces at cruise.

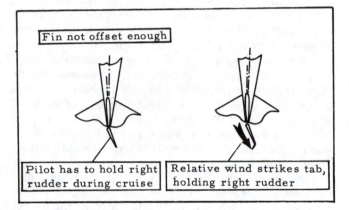

Fig. 2-11. The bendable rudder tab corrects for minor yaw problems at cruise. Some airplanes have a rudder tab that is controllable from the cockpit. (See Chapter 8.)

On many lightplanes this adjustment can be accomplished only on the ground. The tab is bent and the plane is test flown. This is done until the plane has no tendency to yaw in either direction at cruising speed.

Assuming that you have the tab bent correctly and the plane is balanced directionally, what happens if you vary from cruising speed? If the same power setting of 2400 rpm, for instance, is used, the arrows in Figure 2-12 show a simplified approach to the relative forces at various airspeeds.

In a climb, right rudder is necessary to keep the plane straight. In a dive, left rudder is necessary to keep it straight.

In a glide there is considered to be no yawing effect. Although the engine is at idle and the torque effect is less, the airspeed is lower and the effect of the offset fin is also less.

Some manufacturers use the idea of "canting" the engine slightly so that the thrust line of the propeller (or crankshaft) is pointing slightly to the right. The correction for a left-yawing tendency at cruise is taken care of in this way rather than by offsetting the fin. In such installations, however, right rudder is still necessary in the climb and left rudder in a dive under the conditions shown in Figure 2-12.

Another less important contributor to the "torque effect" is the tendency of the airplane to rotate in an opposite direction to that of the propeller (for every action there is an opposite and equal reaction). The manufacturer flies each airplane and "rigs" it, making sure that any such rolling tendency is minimized. They may "wash in" the left wing so that it has a greater angle of incidence (which results in a higher angle of attack for a particular nose attitude) than that of the right wing. This is the usual procedure for fabric-covered airplanes, resulting in more lift and more drag on that side. This may also contribute very

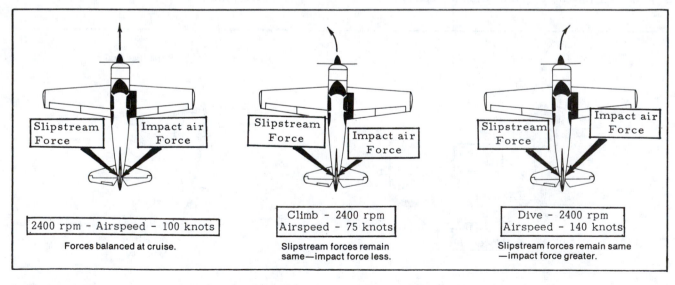

Fig. 2-12. A comparison of forces at various airspeeds (constant rpm).

slightly to the left yaw effect. In some airplanes, a small metal tab on one or both of the ailerons can be bent to deflect the ailerons as necessary, using the same principle described for the rudder tab (the tab makes the control surface move). The controls and their effects will be discussed in Chapter 8, and you may want to review this section again after reading that chapter. Figure 2-13 shows the ailerons and aileron tab.

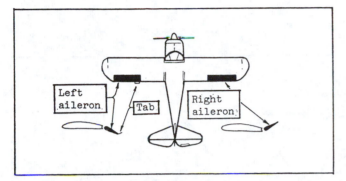

Fig. 2-13. Your trainer may have a bendable tab on the aileron(s).

Two additional factors that under certain conditions can contribute to the torque effect are gyroscopic precession and what is termed ''propeller disk asymmetric loading'' or ''P factor.'' Gyro precession acts *during* attitude changes of the plane, such as those that occur in moving the nose up or down or yawing it from side to side. Gyro precession will be discussed in Chapters 3 and 13. Asymmetric loading is a condition usually encountered when the plane is being flown at a constant, positive angle of attack, such as in a climb or the tail-down part of the tailwheel airplane takeoff roll (Chapter 13). The downward-moving blade, which is on the right side of the propeller arc as seen from the cockpit, has a higher angle of attack and higher thrust than the upward-moving blade on the left. This results in a left-turning moment.

Actually the problem is not as simple as it might at first appear. To be completely accurate, a vector system including the propeller angles and rotational velocity and the airplane's forward speed must be drawn to get a picture of the exact angle of attack difference for each blade. In other words, if the plane is flying at an angle of attack of 10° this *does not* mean that the downward-moving blade has an effective angle of attack 10° greater than normal and that the upward-moving blade has an effective angle 10° less than normal, as might be expected.

From a pilot's standpoint, you are only interested in what must be done to keep the plane straight. When speaking of ''torque,'' the instructor is including such things as the rotating slipstream, gyroscopic effects, asymmetric disk loading (P factor), and any other power-induced forces or couples that tend to turn the plane to the left.

■ Drag

Anytime a body is moved through a fluid (such as air), drag is produced. Drag acts parallel to and in the same direction as the relative wind. The ''total'' drag of an airplane is composed of two main types of drag, as shown by Figure 2-14.

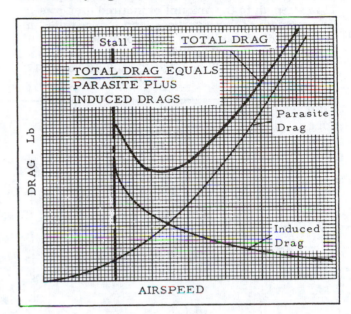

Fig. 2-14. The total drag of the airplane is made up of a combination of parasite and induced drag.

Parasite drag (Fig. 2-15)—the drag composed of (1) ''form drag'' (the landing gear and radio antennas, the shape of the wings, fuselage, etc.), (2) skin friction, and (3) airflow interference between components (such as would be found at the junction of the wing and fuselage or fuselage and tail). As the word ''parasite'' implies, this type of drag is of no use to anybody and is about as welcome as any other parasite. However, parasite drag exists and it's the engineers' problem to make it as small as possible. Parasite drag increases as the square of the airspeed increases. Double the airspeed and parasite drag increases *four* times. Triple the airspeed and parasite drag increases *nine* times.

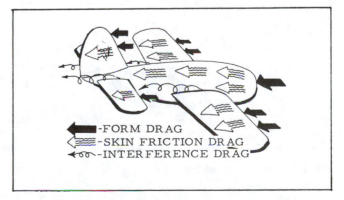

Fig. 2-15. Parasite drag.

Induced drag—the drag that results from lift being produced. The relative wind is deflected downward by the wing, giving a rearward component to the lift vector called induced drag (the lift vector is tilted rearward—see Fig. 2-16). The air moves over each wing tip toward the low pressure on the top of the wing and vortices are formed that are proportional in strength to the amount of induced drag present. The strength of these vortices (and induced drag) increases radically at higher angles of attack so that the *slower* the airplane flies, the *much greater* the induced drag *and* vortices will be (Figs. 2-14, 2-16, and 2-17).

■ Weight

Gravity is like the common cold, always around and not much that can be done about it. This can be said, however: Gravity always acts "downward" (toward the center of the earth). Lift does not always act opposite to weight, as you will see in Chapter 9.

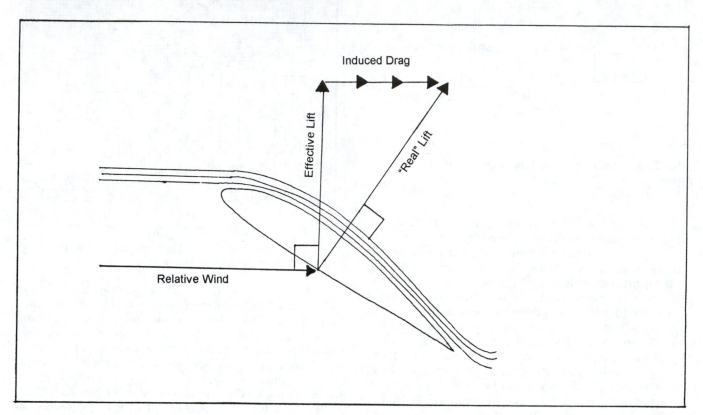

Fig. 2-16. Induced drag.

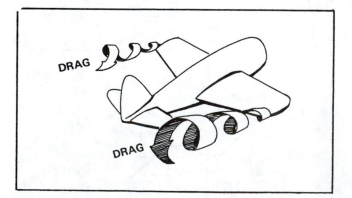

Fig. 2-17. Wing tip vortices.

3

COCKPIT—INSTRUMENTS
AND SYSTEMS

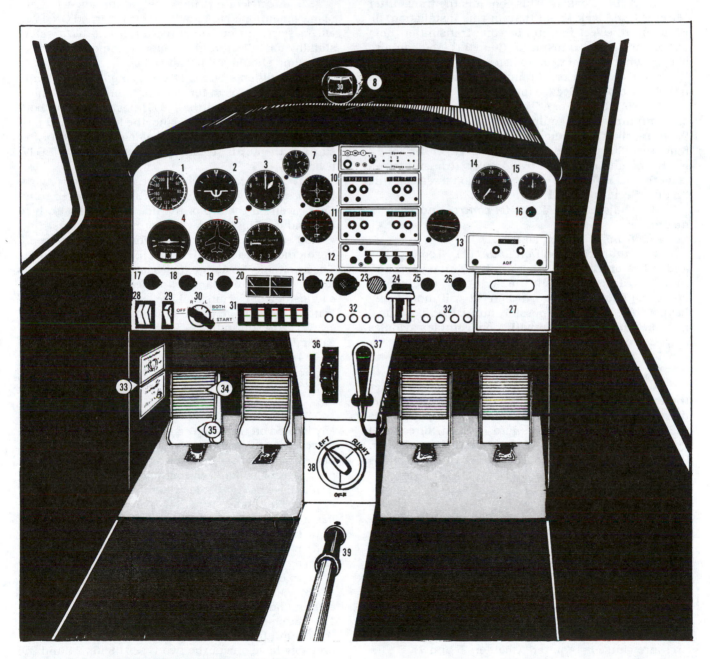

Fig. 3-1. The controls, radios, and instruments for a "composite" general aviation trainer.

Figure 3-1 shows controls and instruments for a "composite" airplane. The seats and control wheels have been "removed" for clarity. Chapter references indicate where more information may be found concerning each item. If there is no reference, the item is covered in this chapter.

1. *Airspeed indicator* (ASI).
2. *Attitude indicator.*
3. *Altimeter.*
4. *Turn coordinator.*
5. *Heading indicator.*
6. *Vertical speed indicator* (VSI).
7. *Clock.*
8. *Magnetic compass.*
9. *Audio Console.* This contains the transmitter selector (the black knob) that, in this installation, allows you to select the number one transmitter (item 10) or the second transmitter (item 11). You can also choose whether you want to listen to a selected set on the cabin speaker or through the earphones (also noted in item 37). See Chapter 21.
10. *NAV/COMM* or *COMM/NAV set*, a combination of communications equipment and VHF omnirange (VOR) receiver and indicator (also called a "one-and-a-half" set). The left side of the equipment, or the "one," is the communications *transceiver* (transmitter/receiver); the right side is a navigation *receiver* (the "half"). See Chapter 21.
11. The second *NAV/COMM control* and *VOR indicator.*
12. *Transponder* (Chapter 21).
13. *Automatic direction finder* (ADF) control box and indicator (Chapter 21). It's unlikely that your "primary" trainer has all the aviation electronics (avionics) equipment shown here; it will more likely have a NAV/COMM and possibly an ADF and/or transponder. You'll get familiar with more complex avionics as you progress to higher certificates and ratings.
14. *Tachometer* (indicates engine rpm).
15. *Ammeter.*
16. *Suction gauge*, the instrument that indicates the drop in inches of mercury created by the engine-driven vacuum pump. The vacuum pump provides the suction to operate the gyro horizon, directional gyro, and in some cases, the turn and slip indicator. (In most cases the turn coordinator or turn and slip is electrically driven as a backup.)
17. *Primer* (Chapter 5).
18. *Parking brake.*
19. *Rheostat* to control the brightness of the instrument panel and radio dial lights (Chapter 26).
20. *Engine instrument cluster*, which may include fuel quantity, fuel pressure, oil pressure, and oil temperature gauges. High-wing trainers that depend on gravity for fuel feeding do not have a fuel pressure gauge. Low-wing airplanes have an engine-driven fuel pump (something like that in your car) and must have an electrically driven pump as a standby, so a fuel pressure gauge is required (Chapters 5 and 7).
21. *Carburetor heat* (Chapter 4).
22. *Throttle* (Chapters 2, 5, and 7).
23. *Mixture control* (Chapter 5).
24. *Electric flap control.* You would move the handle down to indentations (notches) to preselect, for instance, 10°, 20°, or 30° of flap deflection (Chapter 13).
25. *Cabin heat control (Chapter 4).*
26. *Cabin air control.* A combination of cabin air and cabin heat may be used to control temperature and flow rate in the cabin.
27. *Map compartment* (used to hold most everything except maps).
28. *Master switch*, which, as will be discussed later in this chapter, is a "split" switch to control both the alternator and battery.
29. *Electrically driven fuel pump* (low-wing airplanes) mentioned in item 20. This is turned ON as an aid in starting (for some airplanes) and as a safety standby for takeoff and landings if the engine-driven fuel pump should fail (Chapter 7).
30. *Ignition switch.* The airplane has left (L) and right (R) magnetos to furnish ignition. The airplane is flown with the key in the BOTH position, using both magnetos. To start the airplane, the key is turned past the BOTH position to the spring-loaded START. After starting, the key is released and moves back to BOTH (Chapter 4).
31. *Electrical switches.* These will be for navigation (position) lights, wing tip strobes, and a red rotating beacon on the top of the fin. There will also be a taxi and landing light switch for most airplanes.
32. *Circuit breaker panel* for the items depending on the electrical system, such as radios and lights. These circuit breakers (fuses are used for various items in some airplanes) are designed to stop any overloading of circuits and are "safety cutoffs" to prevent fire or damage to the system.
33. *Aircraft papers.* The Airworthiness Certificate must be *displayed* (see end of chapter).
34. *Toe brake* (left) (Chapter 6).
35. *Rudder pedal* (Chapters 6 and 8).
36. *Elevator or stabilator trim wheel and indicator.*
37. *Microphone.* For receiving, the speaker (usually located on the cabin ceiling) is used in most cases; however, your trainer may have earphones. Many planes have both a speaker and earphones, and the pilot selects whichever method of receiving is preferred. See the right side of item 9.
38. *Fuel selector valve.* This example airplane has a tank in each wing and is selected to use fuel from the left wing tank. The OFF selection is very seldom used, but the selector valve should *always* be checked before starting to make sure the fuel selector is on an operating (and the fullest) tank (Chapter 23).
39. *Manual flap control.* The flap handle is operated in "notches," giving various preset flap settings. The button on the end of the handle is used to unlock the control from a previous setting. Your trainer will probably have one of the two types (items 24 and 39) shown here, but not both. Or your trainer may not have flaps (Chapter 13).

REQUIRED INSTRUMENTS

■ Altimeter

The altimeter is an aneroid barometer with the face calibrated in feet instead of inches of mercury. As the altitude increases, the pressure decreases at the rate of about one inch of mercury for each thousand feet of altitude gain. The altimeter contains a sealed flexible diaphragm, correct only at sea level standard pressure (29.92 inches of mercury) and sea level standard temperature (59°F or 15°C). Indicating hands are connected to the diaphragm and geared so that a small change in the diaphragm results in the proper altitude indications. The altimeter has three hands. The longest indicates hundreds of feet and the medium-sized hand, thousands of feet. Another pointer indicates tens of thousands of feet, but it is doubtful if you will have a chance to use this one for a while.

A small window at the side of the face allows the altimeter to be set to the latest barometric pressure as corrected to sea level. As you know, the atmospheric pressure is continually changing and the altimeter, being a barometer, is affected by this.

Suppose that while you're flying, the pressure drops at the airport (maybe a low-pressure area is moving in). After you land, the altimeter will still be "in the air" because of the lower pressure. The instrument only measures the pressure and has no idea where the ground is unless you correct for pressure changes. The altimeter is usually set at field elevation so that a standard may be set among all planes. In other words, pilots fly their altitude with respect to sea level. The elevations of the airports around the country vary, and if the altimeters were set at zero for each field there would be no altitude standardization at all. The altitude you read when the altimeter has been set to the present corrected station barometric pressure is called "mean sea level" (MSL) or indicated altitude. Remember that this does not tell you how high you are above the terrain at a particular time. As an example, assume your field elevation is 720 feet. Set the altimeter at 720 feet with the setting knob, as shown in Figure 3-2.

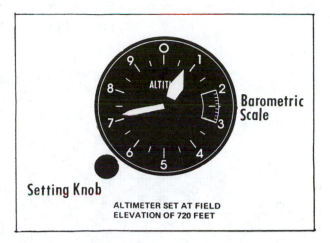

Fig. 3-2. Altimeter setting.

If you want to fly at 2000 feet above the local terrain, your indicated altitude will be 2720, or 2720 feet above sea level (Fig. 3-3). If you were asked your altitude by a ground station (or another plane) you'd answer, "2720 feet MSL."

With the altimeter set to sea level you can call FAA Flight Service Stations or towers in the vicinity of the destination airport and get the altimeter setting. They'll give their barometric pressure in inches of mercury, corrected to sea level. You'll set this in the small barometric scale on the altimeter and this will correct for any pressure difference between your home field and the destination. The altimeter should read the destination field's elevation after you land. If not, there's an altimeter error.

Airplanes used for instrument flying must have had their altimeter(s) and static pressure system (see Fig. 3-5) tested and inspected to meet certain minimums, within the past 24 calendar months by the manufacturer or a certificated repair facility.

When you fly from a high- to a low-pressure area, the altimeter reads too high: H to L = H. When you fly from a low to a high the altimeter reads too low: L to H = L.

In other words, when you fly from a high-pressure region into one of lower pressure, the altimeter "thinks" the plane has climbed and registers accordingly, when actually your altitude above sea level may

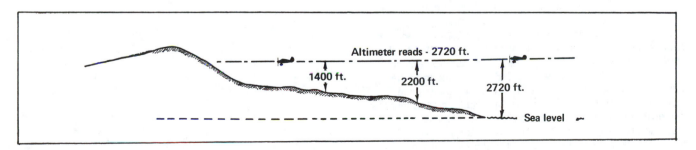

Fig. 3-3. *Absolute altitude* is the height above the surface. *True altitude* is the height above mean sea level. *True altitude* is a constant 2720 feet here.

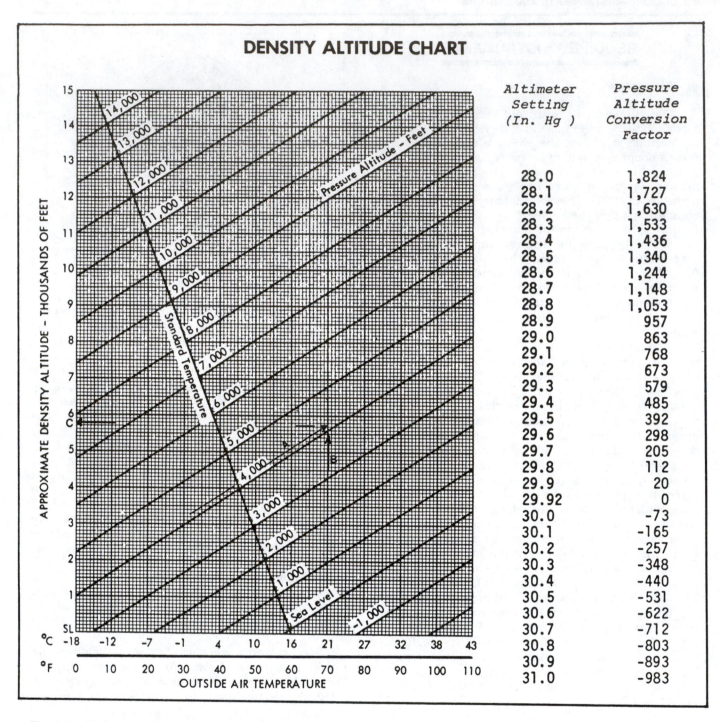

DENSITY ALTITUDE CHART

Altimeter Setting (In. Hg)	Pressure Altitude Conversion Factor
28.0	1,824
28.1	1,727
28.2	1,630
28.3	1,533
28.4	1,436
28.5	1,340
28.6	1,244
28.7	1,148
28.8	1,053
28.9	957
29.0	863
29.1	768
29.2	673
29.3	579
29.4	485
29.5	392
29.6	298
29.7	205
29.8	112
29.9	20
29.92	0
30.0	-73
30.1	-165
30.2	-257
30.3	-348
30.4	-440
30.5	-531
30.6	-622
30.7	-712
30.8	-803
30.9	-893
31.0	-983

Fig. 3-4. Finding the density-altitude when altimeter setting and temperature are known. Another example:

Assume an altimeter setting of 29.80 and an outside air temperature of 21°C at an indicated altitude of 4000 feet. What is the density-altitude as found by this chart?

First, you'd correct the indicated altitude to the pressure altitude by *adding* 112 feet (see the conversion factor above for 29.80). The pressure-altitude is 4112 feet (it's higher than indicated altitude because the pressure is *lower* than the standard of 29.92 inches of mercury).

You would move up and to the right on the pressure altitude at 4112 feet (A) until intersecting the 21°C outside air temperature (B) to get a density-altitude of approximately 5800 feet—sure, it's hard to read that closely on the graph; a calculator gave an answer of (C) 5734 feet. Try reading *that* on the graph! (FAA-T-8080-15)

not have changed at all. Since you will be flying in reference to the altimeter, this means that you will fly the plane down to what appears to be the correct altitude and will be low. If the outside temperature at a particular altitude is lower than standard, the altimeter will read higher than the airplane's actual altitude if not corrected. HALT (*H*igh *A*ltimeter because of *L*ow *T*emperature). Higher than standard temperature means an altimeter that reads low. Figure 19-12 shows how to correct for a nonstandard temperature.

As an example of flying from a high- to a low-pressure area, suppose you take off from an airport where the altimeter setting is 30.10 inches (corrected to sea level) and fly to one with an actual altimeter setting of 29.85 (corrected to sea level). If you forget to reset the altimeter before you land, the altimeter would read approximately 250 feet higher than the actual field elevation. In other words, sitting parked at the new airport's field elevation of 1000 feet, the altimeter would read 1250 feet, and "thinks" you're still 250 feet above the airport.

At lower altitudes, for approximately each 1.0 inch of mercury change the altimeter (if not reset for the new pressure) would be 1000 feet in error. The actual atmospheric pressure change was from 30.10 down to 29.85, a value of 0.25 inches, or the altimeter "climbed" 250 feet during your flight. The barometric scale in the setting window follows the needle movement as you use the setting knob, and vice versa. If you use the knob to *decrease* the 1250-foot reading to the field elevation of 1000 feet, the window will indicate 29.85. Or, if when approaching the airport, you were given the correct altimeter setting by radio, you would *decrease* the window setting to 29.85 and thereby *decrease* the altimeter indication by 250 feet and be right on the field elevation when you land.

Pressure-altitude is shown on the altimeter when the pressure in the setting window is set at 29.92, meaning that this is your altitude as far as a standard-pressure day is concerned. Unless the altimeter setting on the ground is exactly 29.92 and the pressure drop per thousand feet is standard, the pressure altitude and indicated altitude will be different. Pressure altitude computed with temperature gives the density-altitude, or the standard altitude where the density of the air at your altitude is normally found.

Density-altitude is used for computing aircraft performance. In your flying, you will use indicated altitude or the altitude as given by the actual barometric pressure corrected to sea level. Figure 3-4 is a chart for computing density-altitude if pressure altitude and temperature are known.

As an example, the altimeter setting is 29.92, (which is the standard), the pressure altitude is 2000 feet, and the outside air temperature is 70°F (21°C). From the temperature, move up until the 2000-foot pressure altitude line is intersected. Running from that point horizontally, a density-altitude of about 3200 feet is found. In other words, the warmer than standard temperature has lowered the air density so that the airplane would operate as if it was at 3200

feet, when it's actually 2000 feet above sea level.

Note that there is a pressure altitude conversion factor in Figure 3-4 to be used if the altimeter setting is *not* the standard 29.92. You would calculate the density-altitude by correcting the pressure altitude with the conversion factor and *then* finding the density-altitude as before.

For instance, the altimeter setting is 29.5, (29.50) the field elevation is 2900 feet, and the outside temperature is 80°F (27°C). First, you'd correct the pressure altitude by adding 392 feet (the altitude correction at 29.5 inches of mercury) to get 3292 of feet pressure altitude (call it 3300). Draw a line up the vertical line from 80°F to intersect this pressure altitude value and then move horizontally from this point to read a density-altitude of about *5300 feet*. Although the airplane is sitting at an airport at 2900 feet above sea level, because of the lower atmospheric pressure and higher temperature (making the air "thin") it will only perform as if it's at 5300 feet above sea level. *The runway might not be long enough for a safe takeoff under these conditions.* The computations seem to be a little complicated right now, but come back and look at Figure 3-4 again after you've read Chapter 17.

■ Airspeed Indicator

The airspeed system is composed of the pitot and static tubes and the airspeed indicator instrument (Fig. 3-5). The pitot-static system measures the dynamic pressure of the air. As the airplane moves through the air, the relative wind exerts a "ram" pressure in the pitot tube. This pressure expands a diaphragm that is geared to an indicating hand. This pressure is read as airspeed (in miles per hour, or knots) rather than pressure. In short, the pitot-static system measures the pressure of the relative wind approaching the wings (and the entire plane, for that matter). A pitot tube alone would not tell the pilot how much of this pressure was the dynamic pressure giving the wings their lift. The static tube equalizes the static pressure within and without the system, leaving only the dynamic pressure or airspeed being measured. You will check the pitot tube and static vent openings during the preflight inspection, as will be covered in Chapter 4.

Let's discuss *airspeed* and *groundspeed* while we're on the subject. Once a plane is airborne, it is a part of the air. The airplane moves through the air, but the air itself may move over the ground. The plane's performance is not affected by the fact that the entire mass is moving. It is only affected by its relative motion to the mass. A boat may drift downstream with the current at the rate of 7 miles an hour, but to have control, or steerageway, it must move relative to the water. That is, it must also have water moving past the rudder.

At a certain power setting for a certain attitude, a plane will move through the air at a certain rate, this rate being measured by the airspeed indicator. Figure

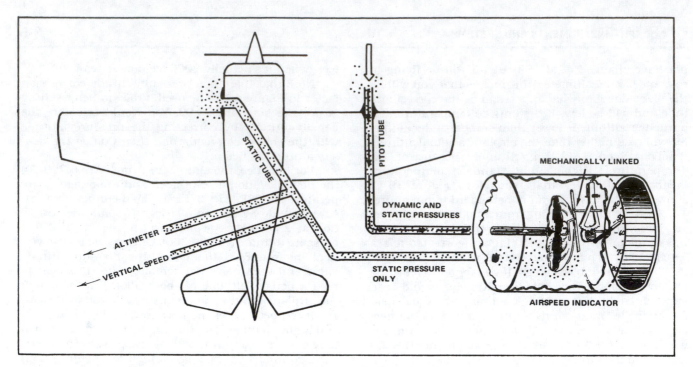

Fig. 3-5. Typical general aviation trainer airspeed indicator and pitot-static system. The airspeed, altimeter, and vertical speed indicators rely on the static tube for the ambient static pressure. In addition, the airspeed needs the pitot tube for the entrance of dynamic pressure into the diaphragm. As shown, the pitot tube also contains static pressure, which is "cancelled" by the static pressure in the static tube and *inside the case.*

3-6 shows the idea, using "true" airspeed or its real speed in relation to the air moving past the airplane.

Groundspeed is like walking on an escalator. If you're walking against the escalator at its exact speed, you make no progress. By walking *with* the escalator, your speed is *added* to its speed. Chapter 10 goes into more detail on this.

Lightplane airspeed indicators are calibrated for standard sea level operation (59°F and 29.92 inches of mercury). As you go to higher altitudes, the air will be less dense and therefore there will be less drag. The plane will move faster, but because of the lesser density, airspeed will register less than your actual speed. This error can be corrected roughly by the rule of thumb: "Add 2 percent per thousand feet." This means that if your "calibrated" airspeed is 100 mph at 5000 feet, your "true" airspeed is 5 × 2 = an additional 10 percent = (.10)(100) = 10 mph + 100 mph

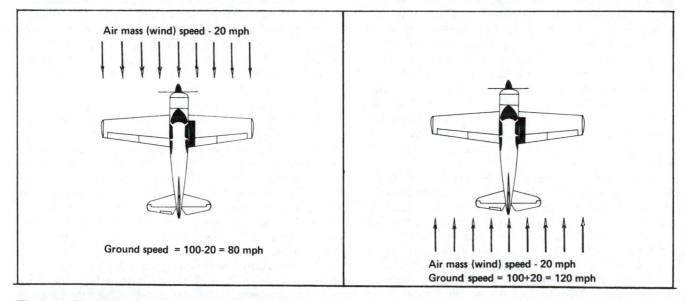

Fig. 3-6. The airplane's true airspeed is 100 mph; its groundspeeds are 80 and 120 mph, as shown.

= 110 mph. This works generally up to about 10,000 feet, but for closer tolerances and altitudes above this a computer should be used.

The computer takes into consideration any deviation of temperature and pressure from the "normal lapse rate." *The standard temperature at sea level is 59°F (15°C). The normal lapse rate, or temperature drop, is 3½°F (2°C) per thousand feet.* The "2 percent per thousand feet" rule of thumb uses this fact and does not take into consideration any change from the normal lapse rate or from standard conditions.

The calibrated airspeed mentioned above is the corrected indicated airspeed, so in order to find the true airspeed the following steps would apply:

1. *Indicated airspeed (IAS) + instrument and position error = calibrated airspeed (CAS).*

2. *Calibrated airspeed plus pressure altitude and temperature correction (2 percent per thousand feet, or use a computer) gives the true airspeed (TAS).* (Most lightplanes have IAS correction information, but for others, assume IAS to equal CAS.)

Looking back at the lift equation in Chapter 2 you'll see in it the expression ($\rho \div 2$) × V². This is the dynamic pressure that contributes to the wing's lift and is measured by the airspeed indicator, as discussed earlier. For instance, at an indicated 100 mph the dynamic pressure is about 25.5 psf. Your airspeed indicator is made so that when dynamic pressure of that value enters the pitot tube the hand will point to 100 mph. At sea level air density your motion relative to the air will actually *be* 100 mph when the airspeed indicates this amount (assuming that the instrument is completely accurate).

As you know, the air density decreases with altitude. When you are at 5000 feet and are indicating 100 mph (25.5 psf), the airspeed indicator doesn't reason that the plane is actually moving faster in relation to the air and making up for the lower density in the dynamic pressure equation. All the indicator knows is that 100 mph worth of dynamic pressure is being routed into it and leaves it up to you to find the "true" airspeed, or the plane's actual speed in relation to the air mass at the higher altitude.

If your plane indicates 50 mph at the stall at sea level, it will indicate 50 mph at the stall at 5000 or 10,000 feet (assuming the same airplane weight and angle of bank; in Chapter 9 you'll see why such a condition is made). This is because a certain minimum dynamic pressure is required to support the airplane, and for your plane it happens to be an indicated 50 mph (or about 6.4 psf).

Another airspeed you may hear about is equivalent airspeed (EAS). At higher airspeeds and altitudes, compressibility errors are induced into the system. Looking back at Figure 3-5, you see that there is static pressure in the *instrument case*, which (theoretically, anyway) cancels out the static pressure in the *diaphragm* (let in through the pitot tube). At high speeds where compressibility is a factor, the effect is that of "packing" and raising the pressure of the *static* air in the pitot tube and diaphragm. The static air in the *case* doesn't get this effect, so that indicated (and calibrated) airspeeds are shown as *higher* than

the actual value. A correction is made to obtain the "correct" CAS; this is called the *equivalent* airspeed. EAS would then be used to find the TAS. For high-speed airplanes the correction would be:

1. IAS plus (or minus) instrument and position error = CAS.

2. CAS minus compressibility error = EAS.

3. EAS corrected for pressure altitude and temperature = TAS.

You won't have to cope with compressibility errors for the airplane you are training in, but you should be aware that such factors do exist.

The calibrated airspeed is the "real" airspeed affecting the flying of the airplane and is the dynamic pressure [($\rho/2$)V²] in pounds per square foot (psf). If you want to find the dynamic pressure working on the airplane at, say, 100 knots (more about knots shortly) you use the equation V² knots/295. At 100 knots, the result is 33.9 psf (call it 34 psf).

You'll note that examples of true airspeed in this section used statute miles per hour. While flying, you'll use the nautical mile (6080 feet) and knots (1 knot equals 1 nautical mile per hour). Also, 1 nautical mile equals 1.152 statute mile. This book will follow the data as produced in the *Pilot's Operating Handbooks* and use knots (K) for speed and nautical miles (NM) for distance, with a few exceptions to be noted in later chapters. While this will seem strange at first, you'll soon get accustomed to thinking in these terms in flying.

■ Tachometer

The lightplane tachometer is similar in many ways to a car speedometer. One end of a flexible cable is attached to the engine and the other is connected to the instrument. The rate of turning of the cable, through mechanical or magnetic means, is transmitted as revolutions per minute (rpm) on the instrument face. The average lightplane propeller is connected directly to the engine crankshaft so that the tachometer registers both the engine and propeller rpm. The propeller is usually geared down in larger planes, and in that case the ratio of engine speed to propeller speed can be found in the Airplane Flight Manual. At any rate, you always use the engine rpm for setting power because this is what is indicated on the tachometer in the direct drive *or* geared engine.

Most tachometers have a method of recording flight hours, based on a particular rpm setting (say, 2300 rpm). This means that the engine is not "building up time" as fast when it's at idle or at lower power settings. The time indicated on the recording tachometer is used as a basis for required inspections and overhauls, and the reading is noted in the logbooks whenever such work is done on the airplane and/or engine.

■ Oil Pressure Gauge

Every airplane is equipped with an oil pressure gauge (Fig. 3-7), and to the majority of pilots it is the most important engine instrument (the fuel gauge

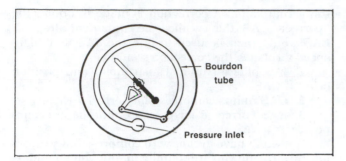

Fig. 3-7. Oil pressure gauge.

runs a close second). Some oil pressure gauges have a curved Bourdon tube in the instrument. Oil pressure tends to straighten the tube, and through mechanical linkage a hand registers the pressure in pounds per square inch. The move now is toward using electrical transmitters and indicators, particularly for more complex engine installations.

Oil pressure should reach the normal operating value within 30 seconds after the engine starts in temperate conditions. The instrument should be checked about every 5 minutes in flight, since it is one of the first indicators of oil starvation, which means engine failure and a forced landing. If the oil pressure starts dropping, land at an airport as soon as safely possible.

There have been many cases of the instrument itself giving bum information. After you notice the falling (or low) oil pressure, keep an eye on the oil temperature gauge for a rising temperature as you turn toward an airport or good landing area. Pilots have depended solely on the oil pressure gauge and landed in bad places when the engine was getting plenty of oil (and the oil temperature stayed normal).

■ Oil Temperature Gauge

The vapor-type temperature gauge is commonly used for lightplanes. The indicating head or instrument contains a Bourdon tube and is connected by a fine tube to a bulb that contains a volatile liquid. Vapor expansion exerts pressure, which is read as temperature on the instrument. Most trainers do not have a cylinder head temperature gauge, so the oil temperature gauge is the only means of telling if the engine is running hot. Other models of airplanes use the principle of electrical resistance change for measuring oil temperature. Figure 3-8 is a typical general aviation oil temperature gauge.

Section 2 of the *Pilot's Operating Handbook* for your airplane will contain the limits of operation and markings for the engine instruments (see Fig. 3-30).

■ Compass

The airplane compass is a magnet with an attached face or "card" that enables you to read directions from 0° to 360°.

The magnet aligns itself with the Magnetic North Pole, and the plane turns around it.

Every 30°, the compass has the number of that heading minus the 0. For instance, 60° is 6 and 240° is 24 on the compass card. It's broken down further into 10° and 5° increments. The compass card appears to be backward in the diagram, but as the plane actually turns around the compass, the correct reading will be given.

In a shallow turn, the compass is reasonably accurate on headings of East and West but leads ahead of the turn as the plane's heading approaches South and lags behind as it approaches North. If you are on

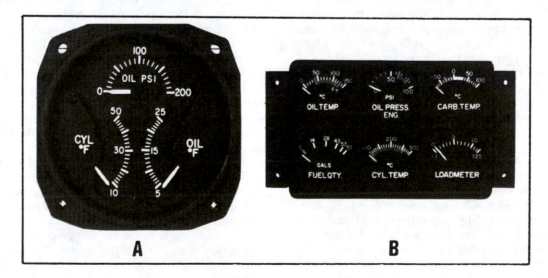

Fig. 3-8. (A) Engine instrument cluster with cylinder head and oil pressure and temperature gauges. (B) Another engine instrument cluster arrangement (*Castleberry Instruments and Avionics, Inc.*).

a heading of *North* and make a turn in either direction, the compass will initially roll to a heading *opposite* to the bank. On a heading of *South* when the turn is started, the compass will initially roll to a heading indication *exceeding* that actually turned (North-Opposite, South-Exceeds, or NOSE). The amount of lead or lag depends on the amount of bank and the latitude at which you are flying. Knowing the amount of lead or lag in the compass you are using will be helpful in rolling out on a heading.

Figure 3-9 is a simplified drawing of the magnetic compass. The magnets tend to point parallel to the earth's lines of magnetic force. As the compass is brought nearer to the Magnetic North Pole, the magnets will have a tendency to point toward the earth's surface.

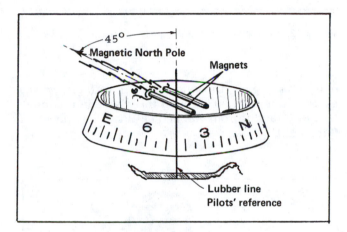

Fig. 3-9. Magnetic compass.

If directly over the Magnetic North Pole, the compass would theoretically point straight down, causing a great deal of confusion to all concerned. The magnetic compass is of little use close to the pole, so other methods of navigation are used in this area.

The suspension of the compass card causes certain problems. For instance, if you are flying on a generally easterly heading and accelerate, the compass will turn toward a northerly heading even though the plane has not turned. This occurs on a westerly heading as well.

If on these headings the plane decelerates, a southerly indication is given under these conditions, as can be seen by referring to Figure 3-9. The following can be noted:

Acceleration—The compass indicates a more northerly heading.

Deceleration—The compass indicates a more southerly heading.

One way of remembering what happens is ANDS (Acceleration-North, Deceleration-South).

Acceleration/deceleration errors are not considered a factor in North and South headings.

When the plane is climbed, dived, or banked (particularly a bank of more than 15°–20°), inaccuracies

result. Notice in a climb or glide (on an east or west heading) that the compass inches off its original indication because of the airplane's attitude (assuming that you have kept the nose from wandering). The compass card and magnets have a one-point suspension so that the assembly acts as a pendulum and also will tilt in reaction to various forces and attitudes of flight (acceleration, banks, nose attitudes, etc.). This tilting action is also a major contributing factor to the reactions of the compass. The "floating" compass as described here is used for reference rather than precision flying (Fig. 3-10). In turbulence the compass swings so badly that it's very hard to fly a course by it. It's a good instrument for getting on course or keeping a general heading between checkpoints on cross-country, but never blindly rely on it. There will be more about this instrument in later chapters.

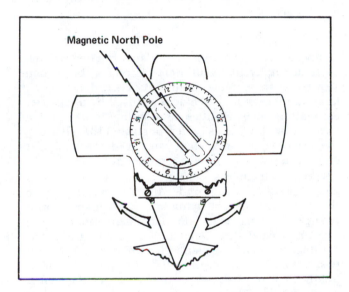

Fig. 3-10. The lubber line and instrument case are fastened to the aircraft, which turns around the free-floating compass card and magnets.

Figure 3-11 is a newer type of magnetic compass with a vertical card. The small airplane is fixed; the card turns in a more easily readable way and is like the new faces on the heading indicator in Figure 3-22. Chapter 15 goes into more detail on use of the magnetic compass.

ADDITIONAL INSTRUMENTS

The instruments discussed in the earlier part of this chapter are all that are required for daytime VFR flight, but the instruments discussed below will also be used in your training.

Federal Aviation Regulations require that emer-

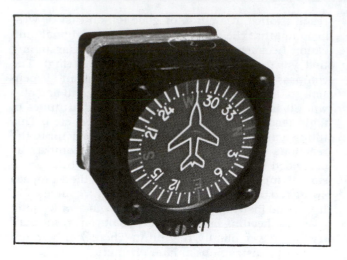

Fig. 3-11. Vertical card magnetic compass. (*Hamilton Instruments, Inc.*)

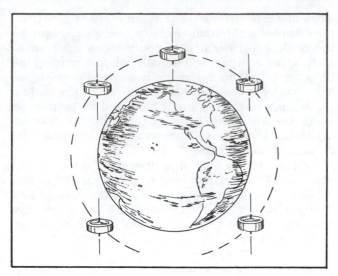

Fig. 3-12. The gyro maintains rigidity in space.

gency recovery from a loss of visual references (such as accidentally flying into clouds or fog) be demonstrated on the private flight test. There have been a large number of fatal accidents caused by pilots overestimating their abilities and flying into marginal weather. It can be easily demonstrated that flying by the seat of your pants when you have lost visual references is a one-way proposition—*down*.

The instruments discussed earlier are used for your day-to-day flying under the VFR (visual) type of flying; for instrument flying (and your instruction in emergency flying using the instruments) more equipment is needed. The FAA requires that on the practical (flight) test the airplane be equipped with "appropriate flight instruments" for checking the pilot's ability to control the airplane by use of the instruments. It's very likely that your trainer will be equipped with all of the instruments covered in this chapter, but techniques are given in Chapter 15 for using only the turn coordinator or turn and slip (plus airspeed and altimeter) to maintain control the airplane under simulated or actual instrument conditions.

■ Gyro Instruments in General

Gyro instruments work because of the gyroscopic properties of "rigidity in space" and "precession."

If you owned a toy gyroscope or top in your younger days, the property of "rigidity in space" is well known to you. While the top was spinning, it could not be pushed over but would move parallel to its plane of rotation (on the floor) in answer to such a nudge. The gyroscope resists any effort to tilt its axis (or its plane of rotation) (Fig. 3-12). This property is used in the attitude indicator and heading indicator.

The property of "precession" is used in the turn and slip indicator. If a force is exerted against the side

of a rotating gyro, the gyro reacts as if the force was exerted at a point 90° around the wheel (in the direction of rotation) from the actual place of application (Fig. 3-13).

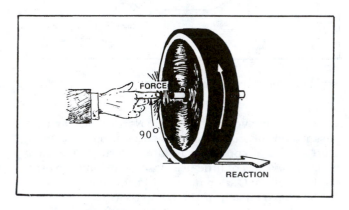

Fig. 3-13. Precession.

■ Turn and Slip Indicator (Needle and Ball)

The turn and slip indicator is actually two instruments. The slip indicator is merely a liquid-filled, curved glass tube containing an agate or steel ball. The liquid acts as a shock dampener. In a balanced turn the ball will remain in the center, since centrifugal force offsets the pull of gravity (Fig. 3-14).

In a slip there is not enough rate of turn for the amount of bank. The centrifugal force will be weak, and this imbalance will be shown by the ball's falling down toward the inside of the turn (Fig. 3-15). (You experience centrifugal force any time you turn a car—

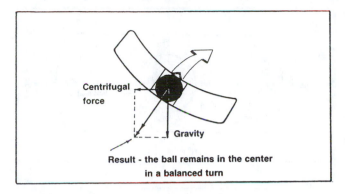

Fig. 3-14. A balanced right turn.

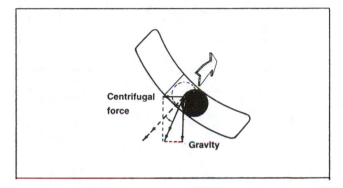

Fig. 3-15. A slipping right turn.

particularly in an abrupt turn at high speed as you tend to move to the outside of the turn.)

The skid is a condition in which there is too high a rate of turn for the amount of bank. The centrifugal force is too strong, and this is indicated by the ball's sliding toward the outside of the turn (Fig. 3-16).

The slip and skid will be discussed further in "The Turn," in Chapter 9.

The turn part of the turn and slip indicator, or "needle" as it is sometimes called, uses precession to

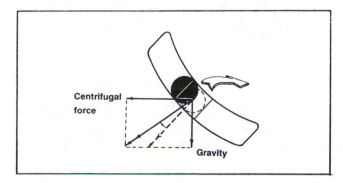

Fig. 3-16. A skidding right turn.

indicate the direction and approximate rate of turn of the airplane.

Figure 3-17 shows the reaction of the turn and slip indicator to a right turn. Naturally, the case is rigidly attached to the instrument panel and turns as the airplane turns (1). The gyro wheel (2) reacts by trying to rotate in the direction shown by (3), moving the needle in proportion to the rate of turn (which controls the amount of precession). As soon as the turn stops, the precession is no longer in effect and the spring (4) returns the needle to the center. The spring resists the precession and acts as a dampener, so the nose must actually be moving for the needle to move. The ball may be off to one side even though the needle is centered. For instance, in a side or forward slip (to be covered later) the airplane may be in a bank with the pilot holding opposite rudder. The ball will be in a more extreme position than that shown in Figure 3-15.

Some turn and slip indicators are calibrated so that a "standard-rate turn" of 3° per second will be indicated by the needle's being off center by one needle width (Fig. 3-17). This means that by setting up a standard-rate turn it is possible to roll out on a predetermined heading by the use of a watch or clock. It requires 120 seconds (2 minutes) to complete a 360° turn. There are types of turn and slip indicators calibrated so that a double-needle-width indication indicates a standard-rate turn. These have a "doghouse" to indicate a standard-rate turn (Fig. 3-1, item 2). If your heading is 070° and you want to roll out on a heading of 240°, first decide which way you should turn (to the right in this case). The amount to be turned is 240 − 70 = 170°. The number of seconds required at standard rate is 170 ÷ 3 = 57. If you set up a standard-rate turn, hold it for 57 seconds, and roll out until the needle and ball are centered, the heading should be very close to 240°. You will have a chance to practice these timed turns "under the hood" (by instruments alone) during your training.

One of the most valuable maneuvers in coping with bad weather is the 180° turn, or "getting the hell out of there." It is always best to do the turn *before* you lose visual contact, but if visual references are lost inadvertently, the 180° turn may be done by reference to the instruments.

The advantage of the turn and slip over other gyro instruments is that the gyro does not "tumble" or become erratic as certain bank or pitch limits are exceeded (see "Attitude Indicator").

A disadvantage of the turn and slip is that it is a rate instrument, and a certain amount of training is required before the student is able to quickly transfer the indications of the instrument into a visual picture of the airplane's attitudes and actions.

The gyro of the turn and slip and other gyro instruments may be driven electrically or by air, using an engine-driven vacuum pump or an outside-mounted venturi. (Some airplanes use an engine-driven *pressure* pump.) The venturi is generally used

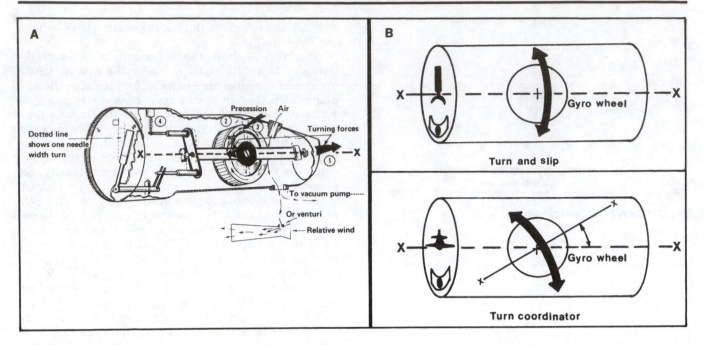

Fig. 3-17. (A) Turn and slip indicator (vacuum-driven). Newer instruments are electrically driven. (B) The gyro wheel in the turn and slip is attached in the instrument so that when it reacts to yaw it can only tilt around the long axis of the instrument (X-X). It is limited in its tilt action by stops. The gyro wheel in the turn coordinator moves (in reaction to precession) around an axis 30° from the long axis of the instrument and so reacts to both roll and yaw inputs. Since precession acts while a force is being *applied,* the turn coordinator only reacts to yaw (rate of turn) once the bank is established.

on older lightplanes because of its simplicity; it is nothing more than a venturi tube causing a drop in pressure and drawing air through the instrument past the vanes on the gyro wheel as the plane moves through the air (Fig. 3-17).

■ Turn Coordinator

The turn coordinator also uses precession and is similar to the turn and slip in that a standard rate may be set up but, unlike that instrument, it is designed to respond to *roll* as well.

Note that the gyro wheel in the turn and slip (Fig. 3-17) is vertical and lined up with the axis of the instrument case when yaw forces aren't acting on it. In the turn coordinator the gyro wheel is set at approximately 30° from the instrument's long axis, allowing it to precess in response to both yaw and roll forces (Fig. 3-17). After the roll-in (or roll-out) is complete, the instrument indicates the rate of turn. Figure 3-18 compares the indications of the two instruments in a balanced standard-rate turn (3° per second) to the left. Note that the turn coordinator also makes use of a slip indicator (ball).

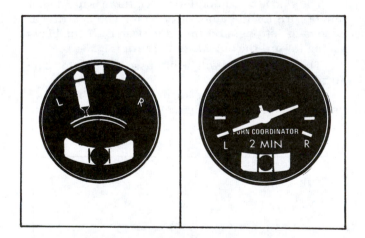

Fig. 3-18. A comparison of a standard-rate turn as indicated by the turn and slip and the turn coordinator. The "2 MIN" indicates that if the wing of the miniature airplane is placed on the reference mark, it will take 2 minutes to complete a 360° turn (3° per second).

■ Attitude Indicator

The attitude indicator (Fig. 3-19) or gyro horizon (artificial horizon) depends on the "rigidity in space" idea and is a true attitude instrument. The gyro plane of rotation is horizontal and remains in this position, with the aircraft (and instrument case) being moved about it. Attached to the gyro is a face with a white horizon line and other reference lines on it. When the instrument is operating correctly, this line will always represent the actual horizon. A miniature airplane, attached to the case, moves with respect to this artificial horizon precisely as the real airplane moves with respect to the real horizon. This instrument is particularly easy for students to use because they are able to "fly" the small airplane as they would the large airplane itself.

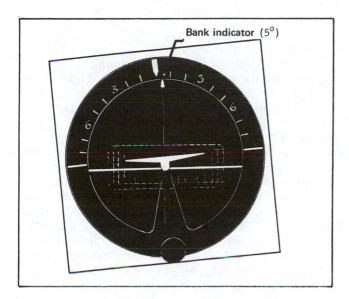

Fig. 3-19. Attitude indicator. Shown is a nose-up, left-wing down (5° bank) attitude. Although it appears to indicate a shallow climbing turn to the left, a check with other instruments is needed to confirm this.

There are limits of operation on the less expensive attitude indicators and these are, in most cases, 70° of pitch (nose up or down) and 100° of bank. The gyro will "tumble" above these limit stops and will give false information as the gyro is forced from its rotational plane. The instrument will also give false information during the several minutes required for it to return to the normal position after resuming straight and level flight.

"Caging" is done by a knob located on the instrument front and is useful in locking the instrument before doing acrobatics, which include the spin demonstration discussed later. Because it is possible to damage the instrument through repeated tumbling, this caging is a must before you do deliberate aerobatics. Most of the later attitude indicators do not have a caging knob. Figure 3-20 compares two types of attitude indicators.

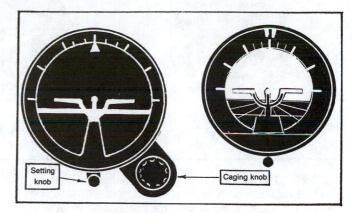

Fig. 3-20. Attitude indicators (caging and noncaging types).

The more expensive attitude indicators have no limits of pitch or bank, and aerobatics such as rolls or loops can be done by reference to the instrument. These are used in jet fighters and are generally electrically driven rather than air driven, as in the case of the older type gyros, because of the loss of efficiency of the air-driven type at high altitudes.

Attitude indicators have a knob that allows you to move the miniature airplane up or down to compensate for small deviations in the horizontal-line position.

The attitude indicator is more expensive than the turn and slip but allows the pilot to get an immediate picture of the plane's attitude. It can be used to establish a standard-rate turn, if necessary, without reference to the turn coordinator or turn and slip indicator, as will be shown.

An airplane's rate of turn depends on its velocity and amount of bank. For any airplane, the slower the velocity and the greater the angle of bank, the greater the rate of turn. This will be evident as soon as you begin flying and is probably so even now.

A good rule of thumb to find the amount of bank needed for a standard-rate turn at various airspeeds (mph) is to divide your airspeed by 10 and add 5 to the answer. For instance, airspeed = 150 ÷ 10 = 15; 15 + 5 = 20° bank required. This thumb rule is particularly accurate in the 100–200 mph range.

Figure 3-21 compares the indications of the turn and slip, turn coordinator, and attitude indicator in a standard-rate turn to the right at 100 K *true airspeed*. (True airspeed would be the proper one to use in the rule of thumb, but indicated or calibrated airspeed is okay for a quick result.)

A rule of thumb for finding the required bank for a standard-rate turn when you are working with knots is to divide the airspeed by 10 and multiply the result by 1½. As an example, at an airspeed of 156 K you'd get 15.6° (call it 16°). One and one-half times 16 is 24° required.

Fig. 3-21. The turn and slip, turn coordinator, and attitude indicator indications in a standard-rate turn at 100 K. Setting up the thumb rule for the bank angle only works for the balanced turn (no slip or skid).

■ Heading Indicator

The heading indicator functions because of the gyro principle of "rigidity in space" as did the attitude indicator. In this case, however, the plane of rotation is vertical. The older heading indicator has a compass card or azimuth scale that is attached to the gyro gimbal and wheel. The wheel and card are fixed and, as in the case of the magnetic compass, the plane turns about them.

The heading indicator has no "brain" (magnet) that causes it to point to the Magnetic North Pole; it must be set to the heading indicated by the magnetic compass. The instrument should be set when the magnetic compass is reading correctly, and this is done in straight and level flight when the magnetic

compass has "settled down."

You should check the heading indicator against the compass about every 15 minutes in flight. More about this in Chapter 25.

The advantage of the heading indicator is that it does not oscillate in rough weather and it gives a true reading during turns when the magnetic compass is erratic. A caging/setting knob is used to cage the instrument for aerobatics or to set the proper heading.

A disadvantage of the older types of heading indicators is that they tumble when the limits of 55° nose up or down or 55° bank are reached. Newer heading indicators have higher limits of pitch and bank and have a resetting knob rather than a caging knob. Figure 3-22 compares the faces of the old and new types.

More expensive gyros, such as are used by the

Fig. 3-22. Three types of heading indicator presentations.

military and airlines, are connected with a magnetic compass in such a way that this creep is automatically compensated for.

The greatest advantage of the heading indicator is that it allows you to turn directly to a heading without the allowance for lead or lag that is necessary with a magnetic compass.

Figure 3-23 shows the schematic of the engine-driven vacuum system of a popular trainer. The air filter is on the firewall under the instrument panel.

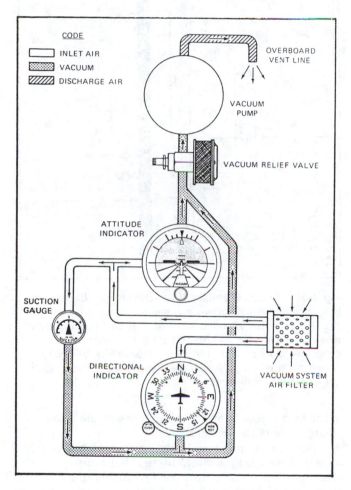

Fig. 3-23. Engine-driven vacuum pump system.

■ Vertical Speed Indicator

This is an instrument that is useful in maintaining a constant rate of climb or glide.

It is a rate instrument and depends on a diaphragm for its operation, as does the altimeter. In the rate of climb indicator the diaphragm measures the change of pressure rather than the pressure itself, as the diaphragm in the altimeter does.

The diaphragm has a tube connecting it to the static port (see Fig. 3-5), or the tube may just have access to the outside air pressure in the case of cheaper or lighter installations. This means that the

inside of the diaphragm has the same pressure as the air surrounding the plane. Opening into the otherwise sealed instrument case is a capillary, or very small tube. This difference in the diaphragm and capillary tube sizes means a lag in the equalization of pressure in the air within the instrument case as the altitude (outside pressure) changes.

Figure 3-24 is a schematic diagram of a vertical speed indicator. As an example, suppose the plane is flying at a constant altitude. The pressure within the diaphragm is the same as that of the air surrounding it in the instrument case. The rate of climb is indicated as zero.

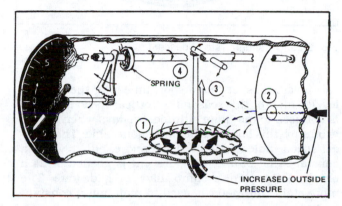

Fig. 3-24. A simplified view of a vertical speed indicator. As the plane descends, outside pressure increases. The diaphragm expands immediately (1). Because of the small size of the capillary tube (2), pressure in the case is not increased at the same rate. The link pushes upward (3) rotating the shaft (4), which causes the needle to indicate 400 fpm down as shown. The spring helps return the needle to zero when pressures are equal and also acts as a dampener. The instrument here is shown proportionally longer than the actual rate of climb indicator so that the mechanism may be clearly seen.

The plane is put into a glide or dive. Air pressure inside the diaphragm increases at the same rate as that of the surrounding air. However, because of the small size of the capillary tube, the pressure in the instrument case does not change at the same rate. In the case of a glide or dive, the diaphragm would expand—the amount of expansion depending on the difference of pressures. The diaphragm is mechanically linked to a hand, and the appropriate rate of descent, in hundreds (or thousands) of feet per minute, is read on the instrument face.

In a climb the pressure in the diaphragm decreases faster than that within the instrument case, and the needle will indicate an appropriate rate of climb. The standard rate of descent in instrument work is 500 fpm (Fig. 3-25).

Because in a climb or dive the pressure in the case is always "behind" the diaphragm pressure, a certain amount of lag results. The instrument will still indi-

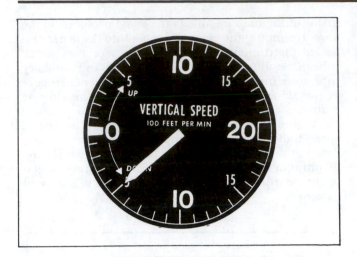

Fig. 3-25. Standard rate of descent.

cate a vertical speed for a short time after the plane is leveled off. For this reason the vertical speed indicator is not used to maintain altitude. On days when the air is bumpy, this lag is particularly noticeable. The vertical speed indicator is used, therefore, either when a constant rate of ascent or descent is needed or as a check of the plane's climb, dive, or glide rate. The more stable altimeter is used to maintain a constant altitude.

■ Summary of the Additional Instruments

Gyro instruments and the vertical speed indicator are not covered in detail as would be done in a training manual for an instrument rating. The calibrations and procedures used by instrument pilots are not covered for obvious reasons. For the noninstrument-qualified private pilot or the student these instruments are to be used in an emergency only as a means of survival. Even the most expensive instruments cannot compensate for lack of preflight planning or poor headwork in flight.

ELECTRICAL SYSTEM

Since most trainers these days have electrical systems for starting and for lights and radios, Figure 3-26 is included to give an idea of a simple system.

The battery stores electrical energy, and the engine-driven alternator (or generator) creates current and replenishes the battery, as necessary, as directed by the voltage regulator.

The ammeter (item 15, Fig. 3-1) indicates the flow of current, in amperes, from the alternator to the battery or from the battery to the electrical system. When the engine is operating and the master switch (item 28, Fig. 3-1) is ON, the ammeter indicates the charging rate of the battery. If the alternator isn't working or

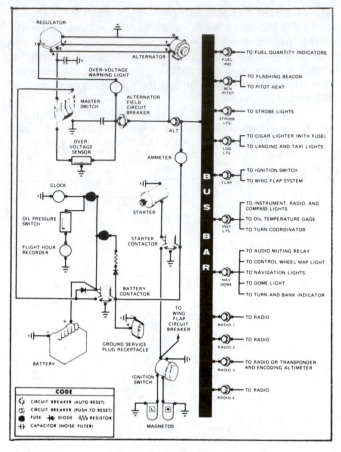

Fig. 3-26. Electrical system. If you don't have a background in electricity, these diagrams can seem very complicated. Your job as a pilot will be to know what system is affected by a loss of electrical power. You'll need to know what components are protected by circuit breakers or fuses and where the breakers (or fuses) are in the cockpit. This diagram shows the electrical items.

if the load is too great, the ammeter will show the discharge rate of the battery.

Note in Figure 3-26 that the system has a split master (electrical) switch (upper left-hand corner). The right half of the switch (BAT) controls all electrical power in the airplane. The left half (ALT) controls the alternator. For normal operations both switches should be ON so that the system (BAT) is working and the alternator is "replenishing" the electricity used by the various components, as is the case for your automobile (except the airplane doesn't depend on the battery/alternator for ignition, but more about this shortly). If the alternator started acting up, you'd turn it off and rely on the battery (for a while, anyway), conserving electricity by turning off all unnecessary electrical equipment. You should not, in this particular system, turn off the battery switch when the alternator is operating.

If the battery is completely dead, hand starting (propping) the airplane (Chapter 5) may get the engine running, but the alternator needs some electrical power to energize it to start charging. It won't be building up the battery, even after the engine is running. An outside source of electricity (with jumper cables, for instance) is necessary for beginning the charging process by the alternator. Let the pros at the airport do it.

Circuit breakers or fuses are installed to protect the electrical circuits (item 32, Fig. 3-1). The circuit breakers "pop out" or the fuses burn out when an overload occurs, breaking the connection between the item causing the problem and the alternator-battery system. The circuit breakers may be reset, but a 2-minute cooling period is usually recommended. Most pilots figure that if a breaker pops three times in a row, it's best to leave it out. In the case of a system using fuses, spare fuses are sometimes carried in clips in the map compartment.

The ignition system is shown in Figure 3-26, but it is self-sustaining and independent of the alternator-battery system. An airplane engine does *not* need an alternator-battery system to operate. In fact, for many years most did not have one; the engine was started by hand. The magnetos are run by the engine and in turn furnish the spark for combustion—a sort of "you rub my back and I'll rub yours" arrangement that works very well. The magnetos will be covered again in following chapters.

AIRCRAFT DOCUMENTS

Every aircraft must carry three documents at all times in flight: (1) Certificate of Aircraft Registration or ownership, (2) Airworthiness Certificate (this must be displayed), and (3) Airplane Flight Manual or equivalent form containing an equipment list and limitations. Look in your plane and know where these documents are.

■ Certificate of Aircraft Registration

The Certificate of Aircraft Registration (Fig. 3-27) contains the name and address of the owner, the aircraft manufacturer, model, registration number, and manufacturer's serial number. When a plane changes owners, a new registration certificate must be obtained.

■ Airworthiness Certificate

The Airworthiness Certificate (Fig. 3-27) is a document showing that the airplane has met the safety requirements of the Federal Aviation Administration and is "airworthy." It remains in effect as long as the aircraft is maintained in accordance with the Federal Aviation Regulations. This means that required inspections or repairs are done and notice of this is entered in the aircraft and/or engine logbook.

AIRPLANE FLIGHT MANUAL OR

PILOT'S OPERATING HANDBOOK

The *Pilot's Operating Handbook*s for general aviation airplanes are laid out as follows for standardization purposes:

Section 1: General—This is a general section and contains a three-view of the airplane and descriptive data on engine, propeller, fuel and oil, dimensions, and weights. It contains symbols, abbreviations, and terminology used in airplane performance.

Section 2: Limitations—This section includes airspeed limitations such as shown in Figure 3-28.

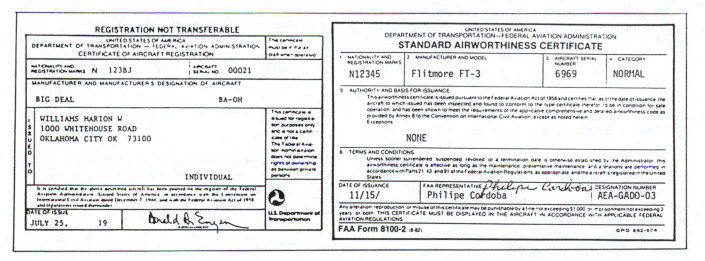

Fig. 3-27. The Certificate of Aircraft Registration and Airworthiness Certificate.

AIRSPEED LIMITATIONS

	SPEED	KCAS	KIAS	REMARKS
V_{NE}	Never Exceed Speed	141	141	Do not exceed this speed in any operation.
V_{NO}	Maximum Structural Cruising Speed	104	107	Do not exceed this speed except in smooth air, and then only with caution.
V_A	Maneuvering Speed: 1600 Pounds 1450 Pounds 1300 Pounds	95 90 85	97 93 88	Do not make full or abrupt control movements above this speed.
V_{FE}	Maximum Flap Extended Speed	89	85	Do not exceed this speed with flaps down.
	Maximum Window Open Speed	141	141	Do not exceed this speed with windows open.

AIRSPEED INDICATOR MARKINGS

MARKING	KIAS VALUE OR RANGE	SIGNIFICANCE
White Arc	42 - 85	Full Flap Operating Range. Lower limit is maximum weight V_{So} in landing configuration. Upper limit is maximum speed permissible with flaps extended.
Green Arc	47 - 107	Normal Operating Range. Lower limit is maximum weight V_S at most forward C.G. with flaps retracted. Upper limit is maximum structural cruising speed.
Yellow Arc	107 - 141	Operations must be conducted with caution and only in smooth air.
Red Line	141	Maximum speed for all operations.

Fig. 3-28. Airspeed limitations and airspeed markings as given in a *Pilot's Operating Handbook*. Note that the stall speeds (V_{s1} for flaps up, and V_{so} for landing configuration) are given for maximum certificated weight. Compare the airspeed limitations with the indicator markings here. V_{NO} is the maximum structural cruising speed (or max speed for *Normal Operations*).

The primary airspeed limitations are given in knots and in both indicated and calibrated airspeeds (although some manufacturers furnish mph data as well), and the markings are in knots and indicated airspeeds.

The maneuvering speed, which is the maximum indicated airspeed at which the controls may be abruptly and fully deflected without overstressing the airplane, depends on the stall speed, which decreases as the airplane gets lighter. More about this in Chapter 23. Figure 3-29 shows some markings for a fictitious airplane.

Included here also are power plant limitations (engine rpm limits, maximum and minimum oil pressures, and maximum oil temperature and rpm range to be used) (Fig. 3-30). Weight and center of gravity limits are included for the particular airplane. Maneuver and flight load factor limits (also to be discussed in Chapter 23) are listed, and operation limits (day and night, VFR, and IFR) are found here. Fuel limitations (usable and unusable fuel) and minimum fuel grades are covered in this section. An air-

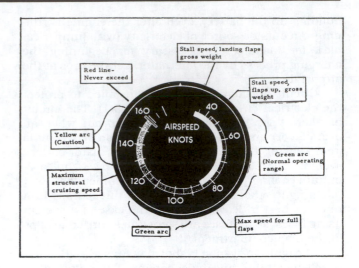

Fig. 3-29. Important airspeed markings. For most airplanes manufactured before the 1976 models the values were given in miles per hour and calibrated airspeed; 1976 and later models will have the airspeed markings as *indicated* airspeed in knots. (1 K = 1.152 statute mph—see Chapter 19.) Your instructor will give you the word about this for the airplane you are using.

plane may not be able to use some of the fuel in flight and this is listed as "unusable fuel." For example, one airplane has a total fuel capacity of 24 U.S. gallons with 22 U.S. gallons being available in flight (2 gallons unusable).

Section 3: Emergency procedures—Here are checklists for such things as engine failures at various portions of the flight; forced landings, including ditching; fires during start and in flight; icing; electrical power supply malfunctions; and airspeeds for safe operations.

There are amplified procedures for these "problems" including how to recover from inadvertently flying into instrument conditions and recovering after a vacuum system failure, spin recoveries, rough engine operations, and electrical problems.

Section 4: Normal procedures—This section also covers checklist procedures from preflight through climb, cruise, landing, and securing the airplane. It's followed by amplified procedures of the same material. Recommended speeds for normal operation are summarized, as well as being given as part of the am-

POWER PLANT INSTRUMENT MARKINGS

INSTRUMENT	RED LINE MINIMUM LIMIT	GREEN ARC NORMAL OPERATING	RED LINE MAXIMUM LIMIT
Tachometer	- - -	2000 - 2750 RPM	2750 RPM
Oil Temperature	- - -	100° - 240°F	240°F
Oil Pressure	10 psi	30 - 60 psi	100 psi

Fig. 3-30. Power plant limitations.

plified procedures. Environmental systems for the airplane and other normal procedures are also included.

Section 5: Performance—Performance charts (takeoff, cruise, landing) with sample problems are included with range and endurance information, airspeed calibration, and stall speed charts.

Section 6: Weight and balance and equipment list—Airplane weighing procedures are given here, with a loading graph and an equipment list with weights and arms of the various airplane components.

Section 7: Airplane and systems descriptions—The airframe, with its control systems, is described and diagrammed. The landing gear, engine and engine controls, etc., are covered in this section. Also, fuel, brake, electrical, hydraulic, and instrument systems are laid out in detail. Additional systems such as oxygen, anti-icing, ventilation, heating, and avionics are covered here.

Section 8: Airplane handling, service, and maintenance—Here is the information you need for preventive maintenance, ground handling, servicing and cleaning, and care for that particular airplane.

Section 9: Supplements—This is devoted to covering optional systems with descriptions and operating procedures of the electronics, oxygen, and other nonrequired equipment. You should be well familiar with the *Pilot's Operating Handbook* of your airplane.

■ Logbooks

The Federal Aviation Regulations require that the registered owner or operator keep a maintenance record.

Airplanes must have had an annual inspection within the preceding 12 calendar months to be legal. There are some exceptions for special flight permits or a progressive (continuing) inspection procedure, but the vast majority of airplanes fall under the requirements of an annual inspection. You might check with your instructor about your airplane. The annual inspection must be done, even if the airplane has not been flown during that time.

In addition, if the airplane is operated carrying persons for hire (charter, rental, flight instruction, etc.), it must have had an inspection within the past 100 hours of flying (tachometer time or other approved method of noting the time). The 100-hour limitation may be exceeded by not more than 10 hours if necessary to reach a place at which the inspection can be done. The excess time must be included in the next 100 hours of time in service.

If your airplane has had any major repairs or alterations since its manufacture, a copy or copies of the major repair and alteration form must be aboard the airplane (normally it's attached to the weight and balance information for convenience). This lists the airplane make and model, registration number, manufacturer's serial number, and the owner's name and address; it may also have that particular airplane's latest empty weight, the empty center of gravity location

in inches from the datum, and the useful load of the plane. The useful load includes the weight of usable fuel, occupants, and baggage. The maximum weight minus the empty weight equals the useful load. (Use actual passenger weights.) The gas weight is 6 pounds per gallon, and the oil, 7½ pounds per gallon. When you are loading the baggage compartment, do not exceed the placard weight, even if you have plenty of weight left before reaching the maximum for the airplane. You may overstress the compartment as well as adversely affect the position of the center of gravity of the plane. (Chapter 23 covers this subject in more detail.) Any major repairs or additions of equipment such as radios, etc., that would affect the position of the center of gravity are noted on this form.

On the back of the major repair and alteration form is a description of the work accomplished. The person authorized for such work signs the form and gives the date that the repair or alteration was completed.

The 100-hour inspection can be completed and signed by a certificated airframe and power plant mechanic, but the annual inspection must be supervised by an A and P mechanic or facility holding inspection authorization. If you get the 100 hours done by such mechanics or facilities, it can count as an annual inspection. If an annual inspection is done anytime in, say, the month of May (which means that it is good until the end of the following May) and it's necessary to get a 100-hour inspection in August, this 100-hour inspection, if done by the same people who did the annual, can count as a new annual inspection—which means that the next annual is due by the end of the following *August*, not the next May. The inspectors will note the date and recorded tach time (or other flight time), and the type of inspection will be noted in the logbooks.

The logbooks are not required to be in the plane at all times, as are the first three documents discussed, but they must be available to an FAA inspector if requested. This means that if you're operating away from the plane's normal base, it would be wise to take the logbooks with you.

RADIO EQUIPMENT

The instructor may explain the use of the radio equipment before the first flight, particularly if you are flying from an airport with a control tower or a Flight Service Station (FAA radio station). You will have a chance to watch and listen to the instructor for the first few flights and will soon be asking for taxi instructions and takeoff and landing clearances yourself, under supervision. If you are flying from an airport where a radio is required, you might also look at Chapter 21 soon after you start flying.

Most small airports have a "Unicom" or aeronautical advisory station for giving traffic information and wind and runway conditions to pilots operating in the vicinity of the airport. You may get an early start in the use of radio while operating out of an airport so equipped.

SUMMARY

Don't expect to have all the information in this chapter and the following four chapters down cold before you start flying. Read them over and then use them for review purposes after you've been introduced to the various subjects (the cockpit, preflight check, starting, and taxiing) by the instructor. This is a workbook and is intended to be used both before and *as* you fly. Remember, too, that your instructor and the *Pilot's Operating Handbook* will have recommendations as to what is best for your particular airplane and operating conditions.

For more-detailed information on the systems discussed in this chapter, read Appendix D in the back of this book.

4

PREFLIGHT CHECK

An organized check is the best way to start a flight. The best way to proceed is to make a clockwise (or counterclockwise, if you like) check starting at the cockpit and getting each item as you come to it. Don't be like Henry Schmotz, student pilot, who checked things as he thought of them. He'd check the prop, then the tailwheel and run back up front to look at the oil, etc. One day Henry was so confused that he forgot to check the wings (they were gone—the mechanic had taken them off for recovering) and tried to take off. He wound up in dire straits in the middle of a chicken farm right off the end of the runway. After he cleaned himself up, the first thing he did was to set up an organized preflight check and has had no trouble since.

One of the biggest problems the new student (and the more experienced pilot too) has is knowing what to look for in the preflight check. The first few times you look into the engine compartment and see that maze of wires, pipes, plates, etc., it seems that you wouldn't know if anything was out of place or not. The airplane engine assembly is compact and sealed. What's meant by this is: (1) There should be no evidence of excess *fluid leaks* (oil, brake fluid, or other fluids) in the compartment or on the engine. You may expect a certain small amount of oil, etc., to be present; but if in doubt about the "allowed amount," ask someone. (2) There should not be any loose wires or cables. Every wire or cable should be attached at both ends. If you find one dangling loosely, ask your instructor or a mechanic about it. Don't feel shy about asking questions.

Look at the inside of the bottom cowling to check for indications of fluid leaks. Some airplanes have cowlings that require a great deal of effort to remove—it's a bit impractical to remove what appears to be several hundred fine-threaded screws to check the engine before each flight. This is a bad design by the manufacturers, since the preflight check is definitely more than just checking the oil.

Discussed here (and shown in Fig. 4-1) is a clockwise check of a typical high-wing general aviation airplane with side-by-side seating. The *Pilot's Operating Handbook* for your airplane has a suggested preflight check and you and your instructor may want to use it, referring to the *Handbook* or a checklist as you do the check. As you walk out to the plane, look at its general appearance; sometimes discrepancies can be seen better at a distance. Start the actual check at the cockpit.

Your instructor will point out the following items and note the things to look for on your particular airplane:

1. *Switch off.* You'll be checking the propeller so make sure the magneto switch is in the OFF position. This does not guarantee that the ignition is off, but you will double check the magnetos themselves later (in some airplanes).

Lower the flaps fully so that the hinges and operating rods may be examined more closely. If the flaps are electrical, before you turn the master switch ON make sure that the propeller area is clear. In rare cases, if the starter solenoid or other components are malfunctioning the starter might automatically engage of its own accord (the key is not even in the starter switch) and could be a real hazard to people or animals near the propeller. Be sure that the master switch is OFF after the flaps are down. Remove control locks.

For some planes with an electric stall warning horn you may want to turn the master (electrical) switch on and move the tab on the wing to hear the horn. For other airplanes with a stall warning light (no horn) and the stall warning tab well out on the wing, have somebody else move it while you watch the light on the instrument panel. Make sure the master switch if OFF after either type of these checks is completed. Chapter 12 goes into more detail on the principle of stall warning devices.

Nearly every nut on the airplane is safetied by safety wire, an elastic stop nut, or a cotter pin; there *are* a few exceptions, so if in doubt about a particular item, ask your instructor.

2. *Tires and brakes.* The tires should be inspected for fabric showing or for deep cuts. Roll the airplane a foot or so to see all sides of the tire. (Obviously, if this was a jet airliner it would be quite a job to roll it back and forth, but this book was written for lightplanes.) Check the brakes and brake lines for evidence of hydraulic leaks. Check the landing gear struts for damage and for proper inflation of the oleo strut if the plane has this type of shock absorber.

3. *Look for wrinkles* or any signs of strain or fracture along the side of the fuselage near the landing gear. If the plane has been landed hard, this will be the first place to show it. Check for pulled or distorted rivets in a metal airplane.

4. If the aircraft has *wing tanks*, check the one

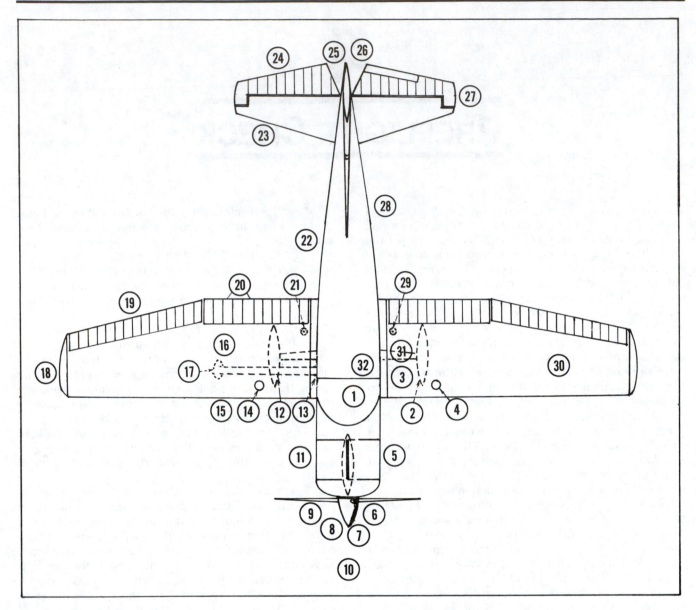

Fig. 4-1. The preflight check.

on this side *visually* at this point; don't rely on tank gauges. You should drain fuel from the tank into a clear container (commercially available) to check for dirt and/or water. Hold the container up to the light (or sun) to get a better look. If contaminants are found, the fuel should be repeatedly drained and checked until it is clear. As a student pilot, however, the presence of foreign matter in the fuel should call for a discussion with your instructor and an extra check before flying. One disadvantage of holding the fuel up to the sky is that a full container of water or jet fuel (!) will appear blue, or the color of 100 Low Lead aviation fuel. Some instructors suggest that the container be held against the side of a white area of the fueselage to check its true color. Jet fuel, which can cause a fatal crash if used in a reciprocating engine, smells like kerosene and a quick sniff of the fuel could avoid a bad problem.

It's best to fill the fuel tanks in the evening after the last flight; this keeps moisture from condensing on the sides of the tank during the cooler night temperatures and so decreases the possibility of water in the fuel.

Some pilots make up a crude tank dipstick from a 12-inch wooden ruler or wooden dowel so that they can get a look at the relative amount of fuel in the tank; they calibrate the sticks by filling the tanks a gallon or two at a time and marking them. One caution is that rubberized tanks may be punctured by a *sharp* stick. Calibrated fuel dipsticks are available for various airplanes through aviation supply houses.

5. *Engine compartment.* Open the cowling on the pilot's side and check:

a. *Ignition lead ("leed") wires* for looseness or fraying. There are two magnetos, each firing its own set of plugs (two separate leads and plugs for each cylinder). The dual ignition system is for safety (the engine will run on either magneto alone), and each magneto will be checked during the pretakeoff check at the end of the runway. The magnetos are self-supporting ignition and have nothing to do with the battery. The battery-alternator system only furnishes power to the starter and other electrical equipment.

b. *Each magneto* has a ground wire on it. When the ignition switch is turned to RIGHT or LEFT, the ground wire is electrically disconnected on that magneto, and it is no longer "grounded out" but is ON. You may do the same by a physical disconnection of the ground wire on a magneto. It may also vibrate loose, and that magneto will be ON even though the switch inside says OFF. It may vibrate off inside the magneto where it can't be seen, so it is best to always consider the prop as being "hot" even after checking the switch and ground wires. *Check both magneto ground wires at this point* if possible.

c. *Look at the carburetor heat and cabin heat muffs for cracks.* These are shrouds around the exhaust that collect the heated air to be routed to the carburetor or cabin heater.

Carburetor heat is necessary because float-type carburetors are very effective refrigerators. The temperature of the air entering the carburetor is lowered for two reasons: (1) The pressure of the mixture is lowered as it goes through the venturi (narrow part of the throat), which causes a drop in temperature. (2) The evaporation of the fuel causes a further drop in temperature. The result is that there may be a drop of up to 60°F below the outside temperature in the carburetor. If there is moisture in the air, these droplets may form ice on the sides of the venturi and on the jets, resulting in loss of power. If the situation gets bad enough, this can cause complete stoppage of the engine. That's where carburetor heat comes in. In a plane with a fixed-pitch prop (no manifold pressure gauge) the usual indication of icing is a dropping back of rpm with no change in throttle setting. There may be only a small amount of ice in the venturi and the effects may not be noticeable at full or cruising throttle, but when the throttle is closed to the idle position, all outside air may be shut out and the engine literally strangles. So . . . *before you close the throttle in the air, pull the carburetor heat;* give it about 10 seconds to clear out any ice before pulling the throttle back. (See item 21, Fig. 3-1.)

Carburetor icing boils down to this: It isn't the heat, or the lack of it, it's the humidity. At very low temperatures (say about 0°F) the air is so dry that carburetor icing is no

problem. A wet cloudy day with a temperature of 50–70°F is the best condition for carburetor icing to occur.

As you pull the carburetor heat control knob in the cockpit, a butterfly valve is opened, allowing the prewarmed air from around the exhaust to mix with the outside air coming into the carburetor. This prevents ice from forming or melts ice already formed. (Fig. 4-2 shows a *full* hot setting.)

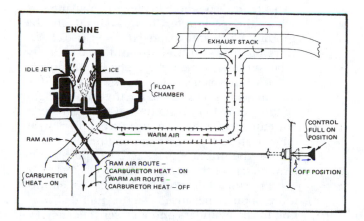

Fig. 4-2. Carburetor heat system.

Here are some of the indications of carburetor icing. The rpm starts creeping back slowly. Being a normal pilot, you'll probably open the throttle more to keep a constant rpm setting. A few minutes later (or sooner if icing conditions are bad) you'll have to open the throttle farther. Finally, realizing there is ice about, you'll pull the carburetor heat and get a shock. There'll be about a 100 rpm drop plus some soul-shaking reactions from the engine for a few seconds as the ice melts.

If there's a lot of ice, here's what happens. The ice melts and becomes a deluge of water into the engine. A little water helps compression (World War II fighters used water injection for bursts of power), but you can get too much of a good thing. Some pilots recommend partial heat in this situation, but the best answer is to use full heat as soon as signs of icing appear and continue using it as long as necessary, realizing that some power is lost because of the warmer air. When using full carb heat in cruise, leaning the mixture will regain most of the lost power. Check with your instructor about the best procedure for your airplane.

As part of the pretakeoff check, pull on the carburetor heat with the engine running at 1800 rpm (or whatever rpm the manufacturer may ask for) and watch for a drop-off of about 100 rpm. It will vary from plane to plane, but this is about the normal drop. The heated air going into the engine is less dense than an equal volume of the outside cold air, and the engine will suffer a slight loss in power. In other words, if there's no drop-off, you're not getting heat to the carburetor, and it would be wise to find out what's up before flying.

The air coming through the carburetor heat system is not filtered, as the normal outside air coming to the carburetor is (see item 8), so that it's possible to pick up dust and dirt when the carburetor heat is used on the ground. It's best to avoid extensive use of carburetor heat on the ground, particularly when taxiing, which may "stir up the dust."

Carburetor heat richens the mixture (the warmer air is less dense for the same weight of fuel going to the engine), so that using carburetor heat with the mixture full rich at higher altitudes makes the engine run rough unless the mixture is leaned. Chapter 5 covers the use of the mixture control.

The fuel injection system is becoming more popular because of better fuel distribution to the cylinders and a lesser tendency to icing problems, but the carburetor will still be used for quite a while.

Most lightplane cabin heaters operate on the same principle of using hot air from around the exhausts. Pulling the cabin heat control opens a valve and allows the hot air to come into the cockpit. The effectiveness of the carburetor (and cabin) heat depends on the temperature of the air in the shroud(s), and this depends on the power being produced. If on a cold day you've been descending for some time at a low power setting with carb and cabin heats on and suddenly find that your feet are getting cold, you can be sure that the carburetor heat is becoming as ineffective as the cabin heat. Get power back on to assure effectiveness of the carb heat and to also warm the engine, or you shortly may have trouble getting the engine to respond. Larger planes will have the more efficient and more expensive gasoline cabin heaters.

d. The fuel strainer drain (not just for the fuel tanks) is located at the bottom of the airplane near the firewall. (The firewall separates the engine compartment from the passenger compartment.) Drain the fuel into a clear container, as noted in item 4, and check for contaminants.

e. While you're browsing in the engine compartment, your instructor will show you how the *engine accessories* should look. (Is that wire that's dangling there normal?) Since you probably don't have experience as an airplane mechanic, many things will just have

to *look* right. Know the preflight check so thoroughly that anything not in its place will stick out like a sore thumb.

f. If the oil tank cap is on this side, *check the oil.* In most planes of this type, the oil is on the right, but don't count on it. The minimum allowable amount will vary from plane to plane. The instructor will give you this information. Most lightplanes of the 100-hp range hold 6 quarts and have a minimum flyable level of 4 quarts. Make sure the cap is secured after you've checked the oil. If you add oil, be sure it is the recommended grade or viscosity for the season and your airplane. Know whether your airplane uses ashless dispersant oil or straight mineral oil (Chapter 23).

g. *Check the alternator* for good wiring and connections (if you can see it; some airplanes have alternators that are pretty inaccessible without taking the cowling off). To repeat from Chapter 3, the alternator serves the same purpose as the one in your car; it keeps the system up for starting and operating the electrical accessories such as lights and radios and some engine and flight instruments as noted earlier. The battery box (and battery) is located in the engine compartment in some airplanes and should be checked for security here.

Figure 4-3 shows the components of a current trainer using a Lycoming 0-235 L2C engine that develops 110 brake horsepower at 2550 rpm. (You might review Chapter 2 about brake and thrust horsepower.) The propeller has been removed.

Looking at the various parts: AB—alternator drive belt. AL—alternator. BB—battery box. CB—carburetor. CH—carburetor heat hose from the muffler shroud (CM). CS—carburetor intake screen. CX—the common exhaust outlet or stack. CY—one of the four cylinders. EP—external electric power plug (for use on those cold mornings when the battery is dead; this is optional equipment). FC—filler cap (oil). FS—fuel strainer. HT—hose from the muffler shroud for cabin heat. IM—intake manifold (route of the fuel-air mixture from the carburetor to the cylinder; there are four of them here). ML—magneto (left). MR—magneto (right). OB—oil breather line. This releases pressure created in the crankcase by the hot oil vapors (when you go out to the airplane after it has just come down, you may see a small patch of light brown froth of oil or water on the ground under the outlet pipe. You can trace the route here from the front of the crankcase to the outlet opening). In the winter, the breather hose (sticking down below the cowling) should be checked to assure that the moisture, condensed to water, has

Fig. 4-3. Components of the engine and systems of a current trainer as discussed in this chapter.

not frozen over and sealed the outlet. If the outlet is sealed by ice, the pressures could build up in the crankcase and damage it. OC—oil cooler. The oil is routed from the engine-driven oil pump through this small radiator to the engine and finally to the sump (it's slightly more complicated than that). OF—oil filter. OL—oil line from the oil cooler. OS—the oil sump or oil storage tank. SP—spark plugs. Each cylinder has two plugs. SW—spark plug (ignition) wire or lead. VV—vacuum system overboard vent line. (See Fig. 3-23.) XM—exhaust manifold.

It's unlikely that you'll get the cowling this much off of your airplane during any preflight check, but you might quietly watch the mechanics working on an engine in the hangar or look at an engine in the process of being overhauled. (Don't move *any* propellers; the mechanic might not appreciate it and it could be dangerous.)

h. Make sure the *cowling* is secured.
i. On some trainers the *vent for the static system* is located on the left side of the fuselage just behind the cowling. The static system instruments are, as you recall, the altimeter, rate of climb indicator, and airspeed indicator (which also needs a source of ram pressure, the pitot tube). The designers have located the static pressure orifice at a point where the actual outside static pressure exists (without any ram or eddying effects), and the hole must be clear to ensure accurate readings of the three instruments. Other locations for the static opening are on the pitot tube (which would then make it a *pitot-static tube*) and on the side (or sides) of the fuselage between the wing and the tail assembly. Check with your instructor about the particular system for your airplane. (See Fig. 3-5.)

6. *Check the propeller* for spinner security and for nicks. Propellers are bad about sucking up gravel and then batting it back against the tail. A sharp nick is called a "stress raiser" by engineers because stresses can be concentrated at the point of the nick, possibly causing structural failure and the loss of a propeller tip. (You may have used this same principle by notching a piece of wood to break it more easily.) Since the propeller is turning at 40 revolutions per second, an unbalanced propeller could literally pull the engine from its mount. If you have any doubts about prop damage when making the preflight check, you'd better have a mechanic or a flight instructor take a look at it. Figure 4-4 shows a method that mechanics use to eradicate comparatively small stress raisers. *Don't move the prop. The ground wire might be off inside the mag where you can't see it—or if the en-*

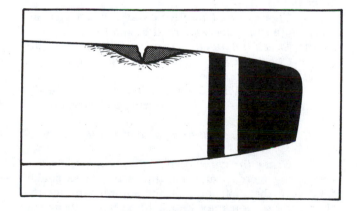

Fig. 4-4. An exaggerated view of a "stress raiser" and the method of "dressing" the nick to eliminate this problem. A nick of this relative size usually means that the propeller is sent back to the manufacturer or discarded. (Enough nicks—and dressing—on one blade could cause a balance problem.) The shaded area shows removed material.

gine is hot, it may kick over if the prop is moved, even if the mags are working normally.

7. *Look in the engine-cooling openings* for objects that might hinder air flow. Planes have hit small birds, and these have been jammed into the engine fins causing a hot spot, followed by the ruining of a cylinder. The fins are finely engineered and any obstruction cuts down the cooling efficiency. You may be able to check the alternator and its drive belt at this point in some airplanes. Birds like to build nests in there, too, and can do it overnight.

8. *Look at the carburetor intake screen.* The screen is there to stop bugs, dirt, and other foreign matter from going into the carburetor. Check for security and see if the screen is clogged.

9. *Check the oil radiator* for obstructions or damage.

10. *Step back to check alignment of the landing gear.* You'll never make a hard landing, of course, but that character who just flew the plane might have. Look behind before you step back. (One student stepped back and sat in a 5-gallon can of freshly drained oil.) This would be the point where you'd check the nosewheel tire and oleo for signs of wear or damage.

11. *Open the right side of the cowling.*
   a. *Check the ignition leads and the right magneto.* If the leads are frayed, bring this to the attention of the instructor.
   b. *Check the oil* if it's on this side—probably it will be.
   c. *Generally take a good look at the items that were hard to check from the other side.* Don't be like Oswald Zilch who checked only the side the oil was on. One day a mechanic took off two of the cylinders, and since Oswald never checked that side, he tried to take off and ran through the same chicken farm that Henry Schmotz, of no-wing fame, had recently vacated.
   d. *Secure the cowling.*

12. *Check the tire and landing gear strut. Check the brake assembly.*

13. *Check the strut fittings* at the fuselage for safetying (if they aren't covered by a fairing).

14. *Check the fuel quantity visually on this side.*

15. *Check the pitot-static tube,* if it's on this side, for obstructions.

16. *Look under the wing for tears or wrinkles* in the skin structure. A wrinkle means a strained member.

17. *Check the strut wing attachments* for safetying.

18. *Grasp the wing* in line with the two spars, if it is a fabric-covered trainer, and give it a brisk shake. Make sure the instructor is not standing under the other wing or you may knock him or her cockeyed. This is a bad way to start a flight. (On the other hand you may *want* to. Use your own judgment.)

The reason for holding the wings in this manner is that you are putting a direct stress on the spars; if there has been any damage to these members or there are broken drag wires, the damage will show up by sound and/or by wrinkling in the skin.

The metal trainers with plastic wingtips are somewhat harder to check for damage in this manner because pressure exerted on these plastic tips may crack them. In that case, grasp the wing on the leading edge just inboard of the plastic tip. The main wing structure may consist of a single box spar with the metal airfoil shape built around attached ribs (with the ailerons and flaps added) rather than a two-spar and rib system. You should try to move it to check for wrinkles, but you also should look at the rivets in the wing to check for signs of overstress. Sometimes, but not always, this will be indicated by the paint being cracked around the rivet, or it may appear that certain rivets have moved slightly or seem elongated.

19. *Check the ailerons* for excessive play or slackness between the two. Look at the aileron actuating rod and hinge assembly for security. Check the aileron counterweights for security; these are weights placed ahead of the aileron hinge line to balance it for aerodynamic reasons.

20. *Check the flap hinges and tracks* for cracks or excessive wear. Look at the general condition of the flaps (no stretched rivets or cracks).

21. *Drain some fuel* from the right tank drain to check for water or other contamination. Some flight school operators advocate the check for water once a day rather than every flight. However, it's better to drain them every flight so that *you* can see that it's clear, rather than taking someone's word for it. (A $25,000 airplane and your neck are a pretty high price to pay just to avoid draining 15 cents worth of gasoline.)

22. *Look alongside the fuselage for tears or wrinkles or pulled rivets.* If the static vent is in this area, make sure it's clear of dirt. (Look under the belly, also.)

23. If the *stabilizer* is externally braced, look at the brace wires or struts and check for slack wire or loose attachments. Most stabilizers now are of cantilever construction and have no external bracing. Grasp the tip of the stabilizer and check for internal damage by exerting up and down forces on it. *Don't try to pick the tail up by the stabilizer tip. It's quite possible that you might damage it.* Many lightplanes have signs painted on the stabilizers warning against lifting the tail this way, but you can get a pretty good idea if things aren't up to snuff by shaking it. Better to find out now than at 500 feet. Some planes have an adjustable stabilizer instead of a trim tab. Some trainers have a "stabilator" (stabilizer-elevator combination), and the entire horizontal tail surface acts as the elevator.

24. *Move the elevators* up and down to check for full movement. Note any binding of the hinges and check for safetying on all hinge bolts. Look under the elevator and stabilizer for tears and wrinkles. If your

plane has a stabilator (a one-piece "flying tail", see Chapter 8), check its movement and condition.

25. *Check the rudder* for binding and limits of operation. Look for fraying of the cables and safetying of the cable attachments at the rudder horn.

26. *Tailwheel.* If your airplane is the tailwheel type, naturally this would be the place to check the tailwheel. The rudder is connected to the tailwheel by coiled springs that absorb ground shocks to the rudder and also make for smoother directional control on the ground. Check these springs. Check the tailwheel attachment bolts for signs of wear or shear breakage. The last student may have landed slightly sideways, and this puts a large shear stress on the attaching bolts.

27. *Check the other stabilizer and elevator and look at the trim tab.* Is it set at nose up? Nose down? Neutral? Check the trim tab cables for wear if they are accessible. Make sure the tab does not have excessive play.

28. *Check this side for tears or wrinkles.*

29. *Check this tank for water and other contamination.*

30. *Check this wing, aileron, flap, and strut. Check the pitot tube* if it is on this side.

31. *Check the strut fuselage attachments* for safety.

32. *You should now be back at the door, ready to get in.* If for some reason you are not back at the door, ask somebody how to get there.

A suggestion for the preflight check—make sure the switches are OFF and then check all of the fuel items at the same time. Note that there are *five* items that pertain to the fuel or fuel systems to be checked: the *two* tanks (should be checked visually) and the *three* drains (two wing tank drains and the main system drain). You might be less likely to overlook one of these items if you get them at the same time and then

proceed with the clockwise check (starting from the cockpit as illustrated). If you use this system, make sure you don't pull a Henry Schmotz.

It's a good idea, for high wing airplanes in particular, to drain the wing tank drains first to get water or other contaminants that have settled to the tank bottom, then drain the main (lowest) drain.

It could happen that draining the main drain first would pull contaminants down into the line between the tank and the carburetor so that they would not show up, as in the wing tank draining. The garbage might only make itself known at a very bad spot just after lift-off.

Some systems require that each tank be drained through the fuel strainer (lowest) drain. Your instructor will issue you a *Pilot's Operating Handbook* shortly after you start to fly, and discussions and your reading will make things much clearer about the systems of your airplane.

When you are flying solo, make sure that the other seat belt and shoulder harness is fastened so that there are no "loose ends" to hang out of the airplane. A seat belt and/or shoulder harness banging in the slipstream against the fuselage sounds like an engine malfunction. Pilots have put airplanes down in strange places and damaged them because it sounded as if the engine was blowing up. As will be said many times during your flight training, always maintain control of the airplane, *then* fix the problem.

After the complete check, the airplane is okay, and you are ready to get in, fasten your belt and shoulder harness and start the engine.

Your instructor will, on the first flight, have you adjust the seat height and position for best visibility and operation of the controls. You should use this setting for following flights so that your visual perception or control travel efforts aren't changed.

5

STARTING THE
AIRPLANE

The plane with a starter has one disadvantage against it's many advantages. There is nobody out front cranking the prop and keeping people from getting hit by it. A dog or small child can easily get under the nose unseen and be killed or seriously injured when you start the engine. *Before you turn the master switch ON to start the engine shout, "Clear?" and receive the acknowledgment, "Clear!" from somebody on the ramp who is in a position to see.* If you don't get this reply but go ahead, turn the master switch ON, and push the starter button anyway, you may have reason to regret it the rest of your life.

Sure, if you holler, all of the airport personnel will get out of the way, but the visitors or kids won't know what the word means. Some pilots are criminal in their starting procedures; they shout, "Clear!" and before the word is out of their mouths are starting the engine. This is as bad as saying nothing. Get set up to start, get acknowledgment of your shout, then start the engine. Don't check that everything is clear and then dawdle in the cockpit until somebody has a chance to stick a sawhorse or something under the prop.

GENERAL PROCEDURE

(Use your *Pilot's Operating Handbook*)

1. *Preflight check complete.*
2. *Seat belts and shoulder harness fastened.*
3. *Controls move freely.*
4. *Fuel ON fullest tank.*
5. *Mixture rich.*
6. *Prime as needed.*
7. *Throttle cracked ¼ inch.*
8. *Brakes ON.*
9. *Radios and other unnecessary electrical equipment OFF.*
10. *Make sure that the propeller area is clear, turn the master switch ON, and then turn the electrically driven fuel pump ON* (if the airplane is so equipped) before starting to see if it's working. You can hear it and see the fuel pressure come up. You'll turn the electric pump OFF after the engine starts to check that the engine-driven pump is working. (See Chapter 7.)

11. *"Clear?"* After receiving acknowledgment, ensure that the master switch is ON and engage the starter (airplanes with no electrical fuel pump).

COLD WEATHER STARTING

When the mercury is nudging down toward zero, the airplane's engine, like the automobile's, can get cantankerous to start at times. The principle for starting on a cold day is generally the same for both. In the car you'll pump the accelerator, or the automatic choke will be helping. You want a rich mixture, or a high ratio of gasoline to air, during the start. The same thing applies to starting the plane. Prime the engine well and leave the primer out, so you can give it a shot if needed after the engine starts. It's possible that you may have to keep it running with the primer until the engine starts to warm up. Once the engine is running on its own, push the primer in and lock it, or the engine will run rough, particularly at low rpm, because raw gas will be pulled into the engine, making the mixture too rich. Take a look ahead at Figure 23-17. There you'll see that the primer system is a separate source of fuel and injects the fuel directly into the cylinder(s).

HOT WEATHER STARTING

The engine will run hot and stay hot a long time after being shut down in the summer. A hot engine has a tendency to "load up," or flood itself, and the fuel-air ratio should be comparatively low for best starting.

Suppose you come out to the airplane just after it has come down on a hot day. You run a preflight check and the prestart check is completed. The guys on the ramp give you a thumbs up and you push the starter. The prop turns over, and over, and over. If you've done the prestart procedure correctly, the chances are that the engine is loaded. One way to clear it out is to turn off all the switches, open the throttle wide, and have somebody pull the prop through several times. Then close the throttle and start the prestart procedure again. The reason for the open throttle is that as the engine turns over, the added air through the carburetor helps to clear the cylinders of excess fuel.

Another procedure is to open the throttle at least halfway as soon as you realize the engine is loaded and then continue the start. The engine will clear itself out and start. Of course, you'll have to pull the throttle back as soon as it starts, or everybody in the hangar will hear the noise and figure that you'll be coming through there any second.

If the engine is obviously loaded you can open the throttle and have the mixture at idle cutoff as the start is continued. This is a more complicated approach for the inexperienced pilot, since the throttle must be retarded and the mixture control moved out of idle cutoff as the engine starts. This procedure is mentioned for later use.

Sometimes during a start the engine will backfire, causing the fuel in the carburetor to be ignited. *If you suspect that there is a fire in the carburetor intake, keep the engine turning over. The fire will be sucked into the cylinders where it belongs.* A fire extinguisher may have to be used outside the airplane in some conditions. You may put the mixture to idle cutoff, turn the fuel off, and open the throttle as you keep the engine turning over with the starter. Review the *Pilot's Operating Handbook* for specific procedures.

After the fire is out, shut down the engine and have a mechanic inspect the carburetor and engine compartment for damage.

ACCELERATING PUMP

While discussing the subject of starting, it might be well to mention that some trainers have an accelerating pump on the carburetor whereby a quick forward movement of the throttle results in an added spurt of fuel being injected into the carburetor throat. This is done so that if the pilot suddenly opens the throttle the engine gets enough fuel to accelerate quickly and smoothly. The accelerating pump is also useful in starting the airplane and can often be used in lieu of the primer. By pumping the throttle, raw fuel is injected into the carburetor throat and an extra rich mixture is pulled into the cylinders when the engine starts turning over. (You have a similar accelerating pump in your car.) The primer is normally set up to send fuel *directly* to the cylinders rather than injecting fuel into the carburetor throat, as the accelerating pump does. *Because the accelerating pump puts fuel into the carburetor, overzealous use of it can mean*

fuel is pooled in the carburetor and carburetor air box (see Fig. 4-2), and the chances of a fire can be greater than when using the primer.

When the temperature is very low, using the primer is the most effective means of ensuring that sufficient fuel is available for starting. (Excessive priming tends to wash oil from the cylinder walls with resulting increased engine wear, but it's a necessary evil.) A method you may try in cold weather when you become more proficient is to use the primer as the engine is turning over. This tends to decrease the chances of loading up because the engine will usually start when conditions are "right" for it.

MIXTURE CONTROL

The purpose of the mixture control is to allow the pilot to set the ratio of fuel-to-air mixture going to the engine. As a quick and dirty figure, a mixture of 1 pound of fuel to 15 pounds of air is about right to support combustion.

Assume that your airplane doesn't have a mixture control, but the carburetor has been set at sea level by a mechanic to give the amount of fuel to the engine to get this 1:15 ratio. As the airplane climbs, the air becomes less dense and weighs less per cubic foot. The carburetor, though, is sitting there fat, dumb, and happy, letting the engine get the same amount (weight) of fuel as it did at sea level, but now it is getting less than the correct amount (weight) of air. The mixture, which was just right for sea level, is now overly "rich" (too much fuel for the air available), combustion efficiency drops, and power is lost. You are also using more fuel than necessary. You may have run into this problem in driving to high elevations in your car.

Now suppose you do all of your flying from an airport high in the Andes, where the air is "thin." There still being no mixture control in the cockpit, the mechanic has adjusted the carburetor to get the magic 1:15 ratio of fuel to air at that high elevation. When you fly down to a sea level airport, you'll find that as you descend the engine starts to run rough, coughs a couple of times, and quits. Why? Because as you descend, the air becomes denser and the fuel that the carburetor had been metering to the engine at the high altitude is not enough to support a combustion mixture in the denser air. The engine is running too "lean"; in fact, it doesn't want to run at all.

Lest you think that the 1:15 ratio (or 15:1 whichever way you prefer to think of it) is the only mixture for an airplane to operate with, it might be added that combustion can occur in mixtures as widely divergent as 1:7 (very rich) to 1:20 (very lean), but these are the extremes. The mixture control in the airplane allows you to take care of such variations as necessary.

When the mixture control is full-forward, the setting is full-rich. This full-rich position is always set slightly richer than would give the best fuel-to-air ratio for full power. This is done to be on the safe side because a rich mixture aids in engine cooling and only a slight amount of power is lost. As the control is pulled out or moved aft, the mixture becomes progressively leaner until the "idle cutoff" position is reached at the full-aft or full-out position. The fuel is cut off completely in the idle cutoff position. This is the way the engine with a mixture control is shut down.

The mixture control is normally in the full-rich position for takeoff and other high-power conditions and for starting. However, you can see that for takeoffs from airports at high elevations this could be too rich and power would be lost. Later you'll learn the techniques for leaning the mixture for takeoffs under these conditions, but for now use full-rich unless the instructor says otherwise. You may never take off from an airport of an elevation (or density-altitude) where this leaning is necessary. In Chapter 23 more details will be given on cruise control (setting power and mixture) in flight.

STOPPING THE ENGINE

Occasionally, in hot weather the engine of older type trainers won't stop when the ignition is turned off (no mixture control). The engine is so hot that it's "preigniting," or setting off the fuel mixture without benefit of spark plugs. After you cut the switch, open the throttle wide as the engine runs down. This will send fuel and cool air into the cylinders, lowering the temperature and keeping preignition from occurring.

In some cases, the engine will kick backward after the switch is cut. This is a form of preignition, and opening the throttle will usually stop any tendencies toward this engine-punishing action.

The throttle is opened on the shutdown in the winter to send fuel into the cylinders to help the next start. This is particularly important if your primer isn't as good as it should be.

For airplanes equipped with mixture controls, the shutdown technique is to pull the throttle back to idle and pull the mixture all the way aft (to the idle cutoff position). The engine will quit smoothly. The beauty of using the mixture control rather than the ignition switch to shut down the engine is that the possibility of preignition is eliminated (there's no fuel left in the cylinders to be ignited). If the idle cutoff is improperly set, you might find that the engine tends to continue running at idle because some fuel is still getting through the carburetor. One method of shutting down in this case is to leave the control in idle cutoff and advance the throttle until the incoming air leans the mixture to a point where combustion can no longer be supported. (In extreme conditions you may have to turn the fuel selector off and wait it out.)

As will be noted in Chapter 7, the instructor may check the ignition switch for proper operation before stopping the engine normally with the mixture control.

PROPPING THE PLANE

Nearly all planes have starters these days, but the following is presented for your possible use. If you plan on doing any propping (cranking the propeller by hand), you should receive instruction from someone with experience; it's an extremely risky business. Another rule is to *never* prop an airplane without a competent pilot or mechanic, *who is familiar with the particular airplane,* at the controls. Propping a plane without a competent operator inside is asking for great excitement, tire tracks on your sports jacket, and loss of an airplane.

The idea of starting is the same as with an electric starter except you have manpower instead of electricity turning the prop. There'll be a little more conversation in this case. (Women *can* and *do* prop planes too.)

Man out front: "OFF AND CRACKED!" (Meaning switch off and throttle cracked.)

You: (After checking to make sure that the fuel valve is ON, the mags are OFF, and the throttle is cracked) "OFF AND CRACKED."

Man out front: (After pulling the prop through several times to get the engine ready for start) "BRAKES AND CONTACT!"

You: Check the brakes and say, "BRAKES AND CONTACT," *before* turning the switch on. Notice that you're giving the benefit of the doubt in all cases.

He pulls the prop again and the engine starts. If in hot weather the engine loads up and doesn't start, the propper will have to clear the cylinders of the excess fuel. If he trusts you, he may say, "Keep it hot and give me half," meaning for you to leave the switch on and open the throttle halfway, adding that you are expected to hold brakes firmly and pull the throttle back as soon as the engine starts. He may push against the prop hub to see if you are holding the brakes, will then pull the prop through until the engine starts, and will be very, very unhappy if you forget to pull the throttle back and are not holding any brakes. One student chased a mechanic with an airplane for 50 yards one day. Of course, soon afterward the mechanic chased the student with a wrench.

In most cases with students, the person propping will say, "SWITCH OFF AND THROTTLE OPEN," and you will make sure the switch is off and the throttle is open. When the plane is ready to start, the propper will say, "THROTTLE CLOSED, BRAKES AND CONTACT." You'll close the throttle (he can hear it close from outside), say, "THROTTLE CLOSED, BRAKES AND CONTACT," hold the brakes, and turn the switch on.

The whole idea in starting an airplane is safety. That prop is a meat cleaver just itching to go to work on somebody. Don't you be the one who causes it to happen, and don't be the one that it happens to.

"CONTACT!" may sound dramatic, but unlike "SWITCH ON!" which, on a noisy ramp may sound like "SWITCH OFF!", there is only one meaning to the word (the magnetos are, or are to be, *hot*). Besides, "CONTACT!" evokes memories of biplanes and barnstorming, helmets and goggles. (Great.)

■ When You're Swinging the Prop

Always figure the switch is on, no matter what the person in the cockpit says. Push against the prop hub to see if the pilot is holding the brakes. He (or she) may be an old buddy but having no brakes can do you a lot of damage here, so don't take any chances.

Before you start to prop the plane, look at the ground under the prop. Is there oil, water, or gravel that might cause your feet to slip out from under you? If the ground doesn't look right, then move the plane—better a tired back than a broken one.

You: "OFF AND CRACKED."

Cockpit: (After checking) "OFF AND CRACKED."

A lot of nonpilots think that the prop is turned backward and "wound up" and released, but this is wrong. You will turn the prop in its normal direction—that is, clockwise as seen from the cockpit, counterclockwise as seen by you when standing in front of the plane. You're doing the same thing a starter does—turning the prop over until the engine starts. (Your car starter doesn't turn the engine backward and then let go to make it start.)

Put both hands close together on the prop about halfway between the hub and the tip (Fig. 5-1). Don't wrap your fingers around the trailing edge; just have enough of the tips there to pull the prop through, because if the engine should kick back, your fingers would suffer. (You may also be fired into orbit if you hang on tight enough.) Stand at about a 45° angle to the prop; this should have your right shoulder pointing in the general direction of the hub. Keep both feet on the ground but have most of your weight on the right foot. As you pull the propeller through, step back on the left foot. This moves you back away from the prop each time. Listen for a sucking sound that tells you the engine is getting fuel. This will be learned from experience; an open or half-open throttle does not have this sound.

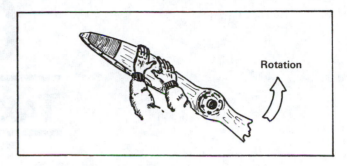

Fig. 5-1. Propping the airplane.

When you think the engine is ready, step back and say, "BRAKES AND CONTACT." After the acknowledgment from the cockpit, step forward and give the prop a sharp snap. Then step backward and to one side as the engine starts.

The starting problems you may encounter have already been discussed.

Don't stand too far from the prop. This causes you to lean into it each time you pull it through.

If the prop is not at the right position for you to get a good snap, have the person in the cockpit turn the switch off. Then move the prop carefully to the position required. *As far as you are concerned, the switch is always on.*

AFTER STARTING

The oil pressure should come up to normal within 30 seconds in warmer weather. Any undue delay should cause you to shut the engine down. Your instructor will have advice for cold weather starts.

After the engine is running well, you should retract the flaps (they were down for the preflight) and (if electrically driven) you will want to wait until the alternator is in operation and helping the system.

6

TAXIING

Some students at first have more trouble taxiing the airplane than flying it, so you might want to take special note of this chapter.

COMMUNICATIONS

At uncontrolled airports you'll set one of the communications transceivers to the Unicom, or local advisory, frequency to coordinate your takeoff, pattern, and landing with other traffic in the area.

For instance, at small airports where you have to taxi on the runway (no taxiways) to get to the point of takeoff, it's customary to announce in the blind on Unicom such statements as "SEWANEE–FRANKLIN COUNTY TRAFFIC, CESSNA SEVEN FIVE FIVE SEVEN LIMA, BACK TAXIING ON RUNWAY TWO-FOUR." This lets incoming traffic know that you're going to use Runway 24 and are taxiing on the runway itself the "wrong way" to get to the end for takeoff. Your airport may have different procedures, but your instructor will brief you on them.

At most controlled airports (with a tower) you will listen to ATIS (Automatic Terminal Information Service)—a continual broadcast of recorded noncontrol information concerning winds, ceiling, visibility, and runway(s) in use—and when you contact ground control for taxi instructions you'll let them know that you have "Information Bravo" or "Charlie," the latest information (the recording is updated periodically and a new alphabet symbol assigned to it).

Your instructor will probably let you handle the radio after the first couple of flights. There is a condition known as "mike fright," in which a normally intelligent person forgets everything he or she was going to say on the radio as soon as a microphone is raised to their lips. Just about every pilot has had it, and even experienced pilots have had it when they did not think through what they were going to say. Those smooth-talking tower controllers also have run into it a time or two in their careers.

There's more detail on communications in Chapter 21.

WIND INDICATORS

Get in the habit of checking the wind indicators before getting into the plane so that you'll know which runway is in use. There are three main types of wind indicators (Fig. 6-1): (1) the windsock, a cloth sleeve that has the advantage of telling both wind direction and velocity; (2) the wind tee, a free-swinging T-shaped marker that resembles an airplane; and (3) the tetrahedron (a Greek word meaning "four bases or sides"), a four-sided object that resembles an arrowhead when seen from the air.

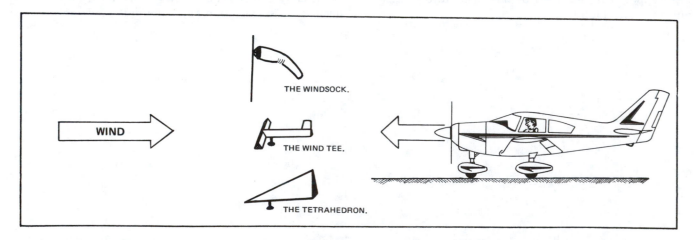

Fig. 6-1. Airport wind indicators.

Airplanes normally take off and land into the wind because the lower groundspeed shortens the ground run.

Some students remember the correct way to use the windsock at first by thinking of the plane as being blown from a trumpet. Think of the tee as an airplane. The way it's heading is the way you want to take off and land. The tetrahedron is an arrowhead pointing the way for you to take off and land.

■ Segmented Circle

The wind indicator is usually in a segmented circle to draw the pilot's attention to it from the air (Fig. 6-2). The runways are shown as extensions from the circle. If one or more of the runways has a right-hand traffic pattern (the normal pattern is left-hand), this will be indicated by the extensions. Perhaps there is a hospital or town on one side of the airport and it's best to keep the air traffic away from that area.

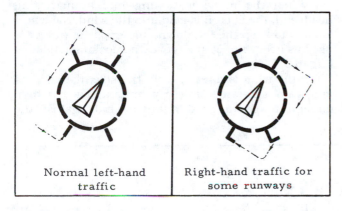

| Normal left-hand traffic | Right-hand traffic for some runways |

Fig. 6-2. The segmented circle.

Take off and land with the other traffic even though this does not agree with the wind indicator. However, remember one thing: When solo, *you* are the pilot in command of your airplane. If some character is trying to break his neck by landing downwind (with the wind) in strong wind, or other such foolishness, you'd be better off to continue to circle until that mission is accomplished, and then you can land in the proper direction.

Incidentally, hard-surfaced runways are numbered by their magnetic headings to the nearest 10°. Runway 22 means that the plane's magnetic heading on this runway will be about 220° when taking off or landing. The actual heading may be 224°, but it is called 22. If the actual heading was 226°, then the runway would be 23, or considered to be 230°. The opposite direction on the same runway (22) would be called 4 (40°). (See Fig. 6-3.) These numbers are painted on the runways. This enables transient pilots to pick out the active runway more easily from the air after the tower has given them landing instructions or Unicom has given them an advisory.

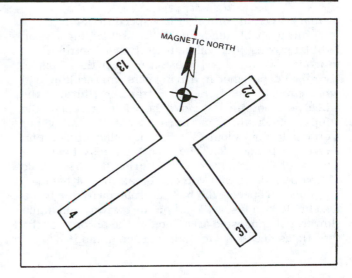

Fig. 6-3. Runway numbering system.

The wind is not always right down the runway, so you'll use the runway that mostly coincides with the wind direction and the crosswind takeoff and landing techniques your instructor will show you. (See Chapter 12.)

TAXIING

The majority of the trainers used today are of the tricycle-gear type, which, if nothing else, has cut down on the number of taxiing accidents. In the older tailwheel types the nose was high and obscured the view of the pilot, so continual "S-turning" was required to keep the taxi area ahead in sight. Sometimes the nose was swung just in time for the pilot to realize that the airplane was bearing down fast on a large, immovable and/or expensive object. The lower nose of the nosewheel type makes it easier to see ahead at all times.

Also, the nosewheel steering is usually more positive and gives better control in strong wind conditions.

The airplane on the ground is out of its element. It lumbers and shakes and moves awkwardly. But in order to fly, you first must taxi, and you'll find that a certain degree of skill is required to move the airplane on the ground under varying wind conditions.

Most tricycle-gear airplanes are steered on the ground with the rudder pedals, which are connected to the nosewheel *and* rudder. Some airplanes have stiff connections between the rudder pedals and the nosewheel, while others have a softer-feeling spring-loaded bungee to transmit the pedal pressure to the nosewheel. In any type, push the rudder pedal in the desired direction of turn. As will be noted later, brakes may be used in some cases to tighten the turn. Other airplanes have a free-castering type of nosewheel (not connected to the rudder pedals), with brakes being the primary steering control at low-speed taxiing.

Your probable tendency in the first taxiing session (and later ones, too) will be to grab that control wheel in front of you and try to use it to steer the airplane. Your past experience in driving cars can result here in what educators call "negative transfer"; turning the wheel does nothing but move those ailerons out on the wings, which are sometimes an aid in taxiing but won't help you avoid that million-dollar corporation jet you're headed for. The instructor may have you keep your hands off the wheel during the first couple of taxi periods so that you can break the car habit.

The taxi speed, which is controlled with the throttle, should be a fast walk out on the taxiway and much slower in a congested area such as the apron. Use the throttle as necessary to maintain the taxi speed.

■ Flight Controls in Taxiing

When taxiing downwind and the wind is strong, hold the wheel forward so that the wind strikes the down elevator and keeps the tail from rising.

When taxiing upwind (into the wind), hold the wheel back so that the wind force and propwash hold the tail down. At any rate, always hold the wheel so that forces acting on the elevator will hold the tail down (Fig. 6-4).

As noted earlier, the tricycle-gear airplane generally has more positive reactions in taxiing and doesn't

tend to weathercock as easily as the tailwheel type. One problem with the latter is that a too-fast taxi, moving downwind, can be a hazard if sudden braking is required. The airplane may tend to nose over and, if the wheel or stick is pulled back (which is a usual reaction to "keep the tail down"), the tailwind gets under the elevators and magnifies the problem, sometimes at the price of a propeller or worse.

Under certain wind conditions the ailerons are useful in taxiing.

In Figure 6-5 the wind, coming from the side and behind, wants to make the plane weathercock or turn into the wind and is also trying to push the plane over. The elevators are down, and opposite rudder is needed to keep the plane from weathercocking. By turning the wheel to the right, the right aileron is up and the left is down as shown. The impact of the wind on the down left aileron helps fight the overturning tendency, and the added area of the down aileron works against the weathercocking force. The aileron control (wheel or stick) is held away from the wind in a quartering tailwind.

When the plane is taxiing into a quartering headwind, the aileron is held into the wind, the wheel is back to keep the tail down, and opposite rudder is used as needed to stop any weathercocking tendency (Fig. 6-6).

If the wind is directly from the side, the ailerons are of no assistance unless your taxi speed is high enough to get a good relative wind against the

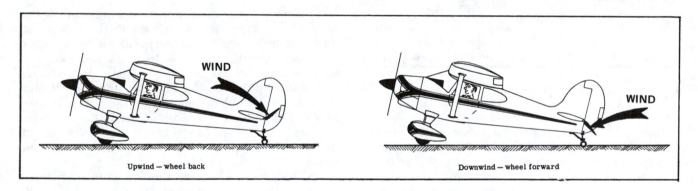

Fig. 6-4. Elevator handling—upwind and downwind taxiing (tailwheel type).

Fig. 6-5. This applies to tricycle-gear airplanes also.

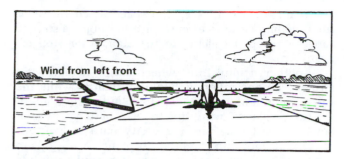

Fig. 6-6. Control positions for a crosswind (tailwheel airplane).

ailerons; if you are taxiing that fast, *slow down.*

"In days of old, when men were bold" and planes were light and ailerons large, ailerons were used opposite to the turn. The drag of the down aileron helped

the plane to turn. The planes are now heavier and the ailerons smaller, so ailerons have a negligible effect unless you happen to be doing at least 25 K or so and, again, this is taxiing too fast.

If there is a strong, quartering tailwind, it is possible to nose over a tricycle-gear airplane, especially if brakes are abruptly applied. What happens is that the plane "bows." The wind is able to get under the upwind wing and the stabilizer, and the plane may rotate around an imaginary line drawn between the nosewheel and the downwind main wheel. This reaction is aggravated if the elevators are in the *up* position. The same tendency is present, particularly in high-wing airplanes (where the center of gravity is farther above the wheels), when an abrupt turn is made while taxiing downwind in a strong wind. One memory aid is that when the wind is strong and behind the airplane (straight or quartering), hold the wheel as far away

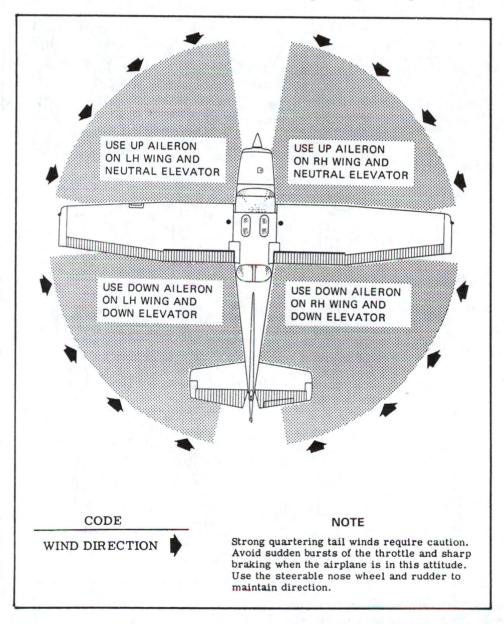

Fig. 6-7. Taxiing diagram.

from the wind as possible. (Push and turn it away from the wind.)

Although the tricycle landing gear has made taxiing much safer and easier, you will still need to exercise caution at all times. Figures 6-5 and 6-6 show control deflections as they would apply to either landing gear type.

This matter of using controls in taxiing may be quite confusing at first, but just visualize what you want the wind to do when it strikes a particular control surface and go on from there. After a while it'll become second nature.

Figure 6-7 gives the control deflections as shown in a *Pilot's Operating Handbook*.

TAXI SIGNALS. Figure 6-8 shows the signals you might get when being directed by a line attendant.

The main thing is to keep the speed down very low when taxiing close to aircraft or other objects so that you can respond quickly to the signals (or stop on your own).

Another helpful hint, when taxiing a high-wing airplane among other high-wing airplanes while the sun is out and high, is to look down at the *shadows* of the two airplanes' wing tips to get a better picture of how far apart they are. It's pretty hard to look out directly at your and the other airplane's wing tips to judge the distance between them. If in doubt anytime before taxiing in tight quarters, shut the engine down and get a taxi director and/or people to "walk your wing" (walk with each wing tip as you start back up and taxi slowly, to watch for possible collision hazards). Sometimes, after a lot of car driving experience,

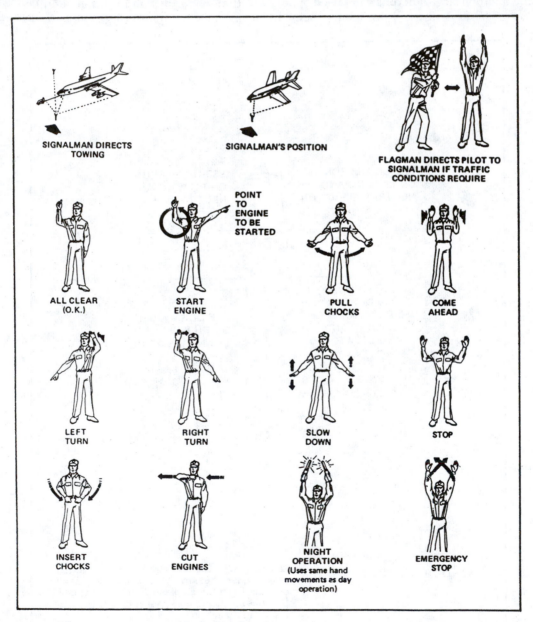

Fig. 6-8. Taxi signals. You should be well familiar with these signals, both as a director or directee. (*Aeronautical Information Manual*)

people who are just starting to fly forget that there are 18- or 20-foot "projections" on each side of the cabin. You may decide to shut down and manually pull or push the airplane past the danger spot.

BRAKES

Most airplanes have separately controlled brakes for each main wheel. Your plane may have either heel brakes or toe brakes; that is, the pedal is operated by your heel or toe. The *toe brake* is a part of each of the rudder pedals and is applied by bending your foot so that the ball or your toes apply the pressure (Fig. 6-9). You will taxi with your heels on the floor until the brake is needed. Then slide your foot up and apply brakes along with the rudder. Toe brakes are hydraulic, similar in action to those in your car, except you aren't able to brake one wheel at a time in the car.

The *heel brake* is usually a cheaper, mechanical brake and is found only on older airplanes. The pedals are located under the rudder pedal and slightly toward the inside of the pedal. Instead of using hydraulic pressure to do the braking, the heel brake usually has a steel cable running from the pedal to its wheel. It is used by placing your heels on the pedals as shown in Figure 6-10 and applying pressure.

Don't use brakes for taxiing unless you have to. There will be days when you have to taxi crosswind in

a tailwheel type for a long distance with the wind so strong that opposite rudder won't be enough (Fig. 6-11). In many cases you can help steering by applying short bursts of power to help the rudder and tailwheel do the job. Obviously, any blasting of the rudder must be of short duration. You don't want to take off except on the runway. If none of these work, you may have to use downwind braking from time to time.

After continued usage, the brake may get hot and fade out completely, and you're worse off than when you started. If it's that windy, it would be best to use brakes or whatever else is needed to get back to the hangar and lock the plane in it.

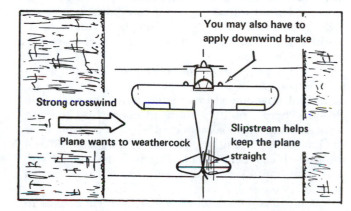

Fig. 6-11. Taxiing in a strong direct crosswind (tailwheel type).

Fig. 6-9. The toe brakes.

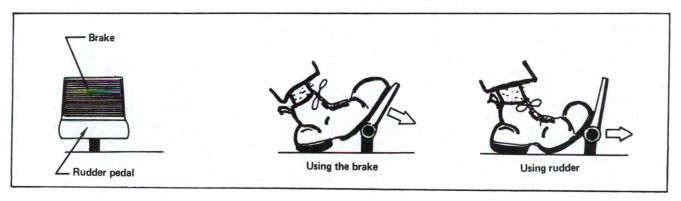

Fig. 6-10. Heel brakes are being phased out.

The brakes are used for stopping the plane at slow speeds, holding it at a standstill, and making a tight turn.

If you have to make a tight turn, use full rudder and then apply brake. Never lock a wheel to turn the plane. This is hard on tires. If you get in a spot requiring a locked-wheel turn, stop the engine, get out, and push the plane around. An easy way to lose friends is to start the engine on the ramp, jam on full throttle, and brake to make a jazzy turn. The pilot whose plane is being blasted by your propwash and flying gravel is going to take a dim view of your taxi technique.

Don't jam on both brakes quickly at the same time! You don't have to be an aeronautical engineer to figure out what would happen if you did. You may get away with it in a tricycle-gear plane, but even in this it could throw a great stress on the nosewheel assembly and slide the tires.

Don't depend on the brakes—they may fail when least expected. In a congested area, taxi so that the plane will slow down immediately when the throttle is closed. There's nothing more disconcerting than to come roaring up to somebody's $100,000 airplane at the gas pit, applying brakes, and feeling that sickening mushiness as they fail. This is a time when you wish you had a ceremonial sword to fall on—and as quickly as possible.

Don't taxi with a lot of power and use brakes to keep the speed down. The brakes will burn up very quickly. This seems to be one of the most common faults of students' taxiing. Many people flew for years before brakes became universal for airplanes, so don't become dependent on them.

Your plane will probably have a parking brake. The instructor will show you how to use it and will warn you that it is sometimes unreliable and that it is best to chock the wheels or tie the plane down if you have to leave it for any period of time.

A type of brake used on some tricycle-gear airplanes consists of a hand brake that operates both brakes simultaneously when pressure is applied. This system, when used in conjunction with the steerable nosewheel, also allows small-radius turns. This system requires a short period of transition for pilots not accustomed to using a hand brake, but it proves to be quite satisfactory.

■ Some Important Notes

You should check the brake effectiveness immediately as power is added and the airplane starts to move out of the parking spot. You can see if they are working *before* picking up speed and boring into something expensive.

Some pilots, when approaching the airport for landing, briefly press the brake pedals to check for firmness. If the pedals feel mushy or one or both "go to the floor," they are warned about possible braking problems *before* they are on the ground and running out of runway.

Figure 6-12 is from the *Aeronautical Information Manual* (FAA) and shows holding position markings (sometimes called "holdback lines"). These mark the boundaries between the taxiway and the runway and shouldn't be crossed until you've ascertained that there is no traffic on the final approach (uncontrolled airport) or the tower has cleared you onto the active runway for the takeoff phase (controlled airport).

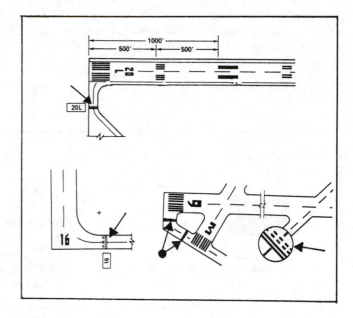

Fig. 6-12. Taxiway holding position markings (arrows). (*Aeronautical Information Manual*)

7

PRETAKEOFF OR COCKPIT CHECK

The pretakeoff check is the last chance you'll have of finding something wrong with the airplane before the flight, so it should be a thorough one. In the preflight check on the line it was found that the airplane *looked* okay; here you will find out if the airplane *operates* okay (Fig. 7-1). Following is a *general* look at the procedure and you should *use the pretakeoff checklist for your particular airplane.*

PROCEDURE

1. *Taxi to the warm-up spot by the runway and turn the plane into the wind.* The modern air-cooled engine is designed to operate in the air, and the less it runs on the ground the better off it is. It's better to turn into the wind when you operate it on the ground. Don't have your tail pointed across the runway when you make the run-up. Your propwash could give landing planes some problems.

2. *Run the engine up to about 1000 rpm and leave it there while you check the controls.* This gives the engine a chance to be warming and smoothing while you run through the rest of the cockpit check. If you are flying a plane with an electrical system, it's a good idea to run the rpm up to where the alternator is effective, usually at about 800–1200 rpm. The days are gone when a pilot sat on the end of the runway and took a lot of time warming up the engine. It's possible to damage an engine by prolonged ground runs. When you finish the cockpit check and the engine will take the throttle smoothly, take off.

3. *Check the ailerons by moving the wheel or stick to the left and say to yourself, "Wheel to the left, left aileron up, right aileron down."* Look at *both* ailerons. Move the wheel in the opposite direction and watch how the ailerons move. *DON'T JUST WIGGLE 'EM TO SEE IF THEY MOVE FREELY—MAKE SURE THEY MOVE IN THE RIGHT DIRECTION.* Aileron cable arrangements may have been reversed by mechanics who replaced them. If you take off with the ailerons reversed (turning the wheel to the left makes the plane bank to the right), a crash is almost certain to occur before you realize what's happening.

4. *Move the wheel through its full backward and forward range.* (Wheel back, elevator up, etc.) Look back at the elevators and check the elevator trim

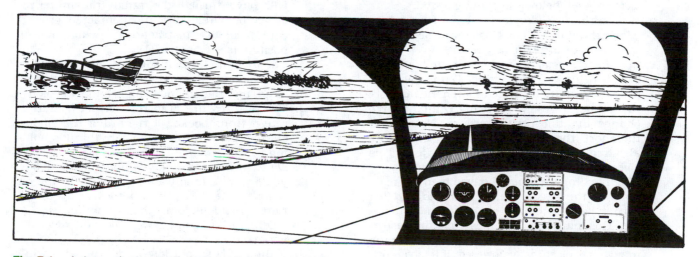

Fig. 7-1. A thorough pretakeoff check is vital to a safe flight. (Sounds rather pompous, eh?) Do it anyway!

tab to see that it is neutral. Check the trim tab control setting in the cockpit to see that it coincides with the actual setting of the tab. If your plane has an adjustable stabilizer, look back to check its actual setting with the stabilizer indication in the cockpit. For future reference (in case you are at a strange field and the indicator is broken) you might check the number of turns of the cockpit trim control from either full-up or full-down that is required to get the neutral position. Obviously this technique should be necessary only on a temporary basis; the indicator should be repaired at the earliest possible time. It's considered good practice by many pilots to roll the elevator tab or adjustable stabilizer control through its full travel (up and down) to check for possible binding; at any rate make sure it's in the correct setting for takeoff.

5. *You checked the rudder control while taxiing, but look back and move it again for full play.* In some tricycle-gear airplanes you will not be able (and shouldn't try) to move the rudder while the plane is stationary. But you should have some idea of rudder control reaction during the process of taxiing. For airplanes with controllable rudder trim this would be the time to check the setting for takeoff.

6. *Check the flaps for operation and equal deflection if you haven't done so during the preflight check.* Of course, the trainer you're flying could easily be landed without them, but it's better to find out *now* that they don't work than to suddenly discover it when trying to land on the shortest runway in three counties. If your flaps don't pass the test, taxi back and get them fixed. If you wait to try them in the air for landing, you might find that only one will extend. (And no student pilot should be doing slow rolls on final approach.)

7. *Start on the left side of the instrument panel and check each instrument as you move from left to right.* There may be a slight variation of this order of checking for your particular airplane, and you may not have some of the instruments mentioned.

   a. The *airspeed indicator* should be reading zero, or well below the stalling speed.

   b. The *attitude indicator* small reference airplane should be set as close to the actual attitude of your airplane as possible.

   c. Set the *altimeter* and check *both hands* for proper field elevation setting. Too many pilots look only at the hundred-foot hand and set the altimeter wrong—particularly after a large change in atmosperic pressure. Notice that the short (thousand-foot) hand in Figure 3-2 also registers approximately 720 feet. This gives you a double check.

   d. When taxiing you should have noted the reactions of the *needle in the turn and slip* (left turn, needle to the left, etc.) or turn coordinator. This naturally assumes that the instrument is powered by an engine-driven vacuum pump or is electrically powered. If it is venturi-driven, it shouldn't be operating when you taxi because if it does you've been taxiing quite a

Fig. 7-2. Use care in setting the heading indicator to the magnetic compass.

few knots too fast.

   e. The heading *indicator* is set with the *magnetic compass.* (Check that the compass is reading right for your approximate heading.) Some instructors recommend setting the heading indicator after the airplane is lined up with the runway on known runway headings; this can give a double check of the magnetic compass. Take a good look at Figure 7-2. The heading indicator has been set incorrectly because the pilot automatically set its lubber, or reference, line to relatively match that of the magnetic compass. The pilot thinks that the heading indicator was set to about 260° when actually it was about 280°. Say the compass reading to yourself and set *that* number on the heading indicator.

   f. The *vertical speed indicator* should be indicating a zero rate of climb.

   g. Check and set the *airplane* clock.

   h. The instructor will show you how to check the *radio* if the plane has one and if you are in a position to receive. (In a controlled field you will have turned the radio transmitter-receiver on after start and contacted ground control for taxi instructions to your present position (see Chapter 21).

   i. The *tachometer* is holding steady. There are no excessive fluctuations of the instrument hand, and the engine sounds and feels okay.

   j. The *oil pressure* is in the normal operating range. (You checked it right after the engine started and want to make sure that it's holding.)

   k. *Oil temperature* is okay. On a really cold day you'll have a long wait if you wait for the oil temperature to come up to the operating range. Just look to see that it's not *too* high or whether it's beginning to *move* if the day isn't very cold.

   l. If the airplane is a low-wing type with wing tanks, check to see if the *engine-driven fuel pump* is working properly with the *electric boost pump OFF.* The fuel pressure should stay in the normal range. Turn the electric pump ON again as an emergency standby for takeoff unless the *Pilot's Operating Handbook* specifically says otherwise. Its job is to aid in starting (for some airplanes) and to be as an emergency standby for takeoffs and landings in case the engine-driven pump fails (it, or course, would be turned on if the engine-driven pump failed anytime in flight). *After takeoff* you will normally turn it off after reaching a safe altitude. Watch the fuel pressure as you turn it off; the engine-driven pump may have failed and the electric pump may have been carrying the load. If, after shutting it off, the fuel pressure falls and the engine starts cutting out, turn the pump on and enter the traffic pattern for landing.

m. Check the *ammeter* for proper charge if the plane has an electrical system.

n. Note the *suction gauge* for normal value. (This instrument indicates the drop in atmospheric pressure—in inches of mercury—established for the vacuum-type gyro instruments by the engine-driven vacuum pump.) Most vacuum gauges in normal operation will indicate a drop in pressure of between 4 and 5 inches, but the instructor and/or *Pilot's Operating Handbook* will furnish the specifics for your airplane.

8. *Fuel* on fullest tank. Check the fuel gauges to see if the indications jibe approximately with the fuel amount shown in your visual inspection of the fuel quantity.

9. Check *again* that you won't blast another plane behind you, then run the engine up to 1800 rpm or that recommended by the manufacturer and check *each magneto.* Turn the switch from BOTH to R (right magneto only). Watch the tachometer for an rpm drop. You are allowed a 75 to 125 rpm drop, depending on the airplane. (If the rpm drops more than is allowable, it's best to take the plane back to the hangar.) Turn the switch back to BOTH and watch it pick up. This gives you a double check on the rpm loss.

Turn the switch from BOTH to L, or left, and watch for the drop. Now turn back to BOTH and watch the rise. Make sure the switch is on BOTH after you check the mags. Compare the rpm drops.

*Pilot's Operating Handbook*s give both a maximum allowable drop on each magneto and a maximum allowable difference between mag drops. For instance, one airplane cites a maximum drop of 150 rpm for each magneto or a maximum of 75 rpm drop differential between the two.

When you turn the switch to LEFT, the right magneto is grounded out and the engine is running on the left magneto only. Since one spark plug does not give the smooth burning of the mixture in a cylinder as that which occurs with both plugs firing, a drop in rpm is expected (and the limits are given by the manufacturer). If you switch to one of the mags and the engine does as well (no drop) as it did on BOTH, you are probably still running on both mags (a ground wire is off or the ignition switch is bad). Make sure this is reported so that no one can be hurt if they move the prop with the "hot" mag still in action. Remember, at these earlier stages anything unusual on the preflight or pretakeoff check should be reported to an instructor or mechanic immediately. If you should inadvertently go to the OFF position while checking the mags, close the throttle before turning the switch back on. You'll usually have time to close the throttle *and* turn the switch on before the engine dies and you won't get a backfire.

Occasionally, you'll have an excessive drop in rpm that is caused by fouled spark plugs. On some makes of engines the left magneto is more susceptible to this problem than the right because it feeds the bottom plugs on *all* of the cylinders and the bottom plugs are more likely to be fouled by oil. (The right magneto fires all of the top plugs in these engines.) This plug problem is most likely to be present if the plane has been sitting idle for a long period or if the piston rings are worn. The scarcity of grade 80 fuel, necessitating the use of grade 100, has caused plug fouling by lead deposits. On other makes of engines, the right magneto fires the *left top* and *right bottom* plugs. (And, of course, the left magneto fires the *right top* and *left bottom.*) You might check with your instructor on the ignition setup for your particular airplane.

If your plane has a mixture control, it is often possible to clear a fouled plug by leaning the mixture during the run-up. This causes the temperature in the cylinder to rise enough to "burn out" the trouble. Needless to say, this treatment can be overdone from the engine's viewpoint. You are better off at your point of experience to taxi back to have a mechanic look at the problem.

Don't make your run-up over loose gravel because the propeller will pick it up and will be damaged. (If you doubt the "picking up" ability of a prop, watch when an airplane is sitting over a puddle of water with the engine running; a minor waterspout will result.)

When the flight is over, just before shutting down the engine, the instructor may pull the *throttle back to idle* and check the integrity of the ground wires (or P-leads as they are sometimes called) by turning the ignition switch OFF and then back ON. If the engine starts to quit, the switch and ground wires are operating normally and there's less chance of an inadvertent engine start if someone moves the propeller while the airplane is sitting tied down or in the hangar. If the engine continues to run with the ignition switch OFF, shut it down with the mixture control and notify a mechanic immediately and/or hang a warning sign on the propeller if you have to leave the airplane unattended.

After the ignition check is done, it's a good idea to run up the engine to 1200–1500 rpm for several seconds to help assure that the spark plugs are "cleared out" (not fouled by lead).

Another good idea is to remove the ignition key

from the switch after the engine is dead. This would assure that it's not inadvertently turned on as could happen if the key was left in the switch. There may be a designated place in the airplane or a clipboard for it.

While we're on the subject of securing the airplane, sometimes the master (electrical system) switch is left ON, which, of course, may result in a dead battery and the requirement for battery and jumper cables or other outside sources of electrical power. The best way to avoid such a hassle is to use a postflight checklist as furnished by the manufacturer. One problem is that sometimes the checklist has items to be checked or done *while the engine is still running and your head is down* while you are reading. There have been cases of airplanes easing into other airplanes and narrow escapes of small children running toward the idling (and invisible) propeller "to see mommy or daddy back from a flight." Keep an eye out as you go through any shutdown procedures.

Okay, now back to the pretakeoff check.

10. While at 1800 rpm, pull the *carburetor heat* ON. Watch for a drop of about 100 rpm, showing that the warm air is getting into the engine, cutting down its power, and that the carburetor heat is working. Leave the carburetor heat on for 10 seconds. If the rpm suddenly increases during that time, it means that you picked up carb ice since starting up and warns you that icing conditions exist for this flight. (The heat cleared it up—temporarily.) Push the heat OFF for the takeoff, since it will cut down on the power available. If carburetor icing conditions are bad, you can leave it ON until the throttle is full open for the takeoff. At fields of high elevation with the mixture in full rich and carburetor heat full ON you'll find that takeoff power is practically nonexistent. The actual high density-altitude (the air is less dense), plus a full-rich mixture setting, *plus* the even more lowered density of the warm air from the carburetor heat can cut down power drastically, as noted in Chapter 4.

11. Close the *throttle* and check the *idle rpm* with the carburetor heat ON to see if the engine operates properly in this condition, which may be the same as found on the latter part of the approach when the runway is "made." If the engine wants to quit or runs roughly, you should have it checked by a mechanic because the propeller will windmill in flight even when the engine is dead and would mask the problem. It would be a shock to need to add power to avoid that cow that just sauntered out on the runway and find that the engine had already quit when you closed the throttle and you have no power available. If the recommended landing procedure for your airplane is *not* to use carb heat, check the idle with it OFF. Idle rpm should be between 500 and 700 and the engine should idle without loping. If you didn't lock the primer after starting, the engine will draw raw gas through the primer system and will tend to load up and quit at idle. Check the primer control for security.

Push the carb heat for takeoff for the same reasons noted in item 10.

12. If the field is short or if you so desire, hold *brakes* and make a *full-power run-up*. With experience (and word from the instructor) you'll be able to see what is normal for a "static" rpm indication (no windmill effects because of forward speed).

13. *Make sure the door is closed and latched.* Quite a stir can be created by the sudden opening of a door just after lift-off. *Check with your instructor about possible procedures if this should occur.* (In addition to specifics for your airplane, the instructor will tell you that maintaining control of the airplane is *first*, *last* and *always* when a door comes open.) Assure yourself that your seat is latched securely; you don't want to slide away from the controls at a critical point during the takeoff.

14. Switch to *tower frequency*, ask for *takeoff clearance*, and follow the tower's instructions when you are fully ready to go. Usually the tower will give one of three basic instructions in such a case:

a. *Hold position* (or *hold short*), meaning you are not to go onto the runway.

b. *Taxi into position and hold,* meaning taxi onto the runway and line up. But remember you are *not* cleared for takeoff; this is done usually because of other traffic or possible wake turbulence.

c. *Cleared for takeoff.* Just what it says. If your condition was (a) above, you'd check for landing traffic (tower personnel are human, too) and taxi onto the runway and take off. In (b) this instruction would clear you to open the throttle and go. You would acknowledge the tower in each case to show that you understood. On uncontrolled fields after the check is complete turn the airplane, and if there is no landing traffic taxi onto the runway for takeoff. Read Chapter 21 for more details in this regard.

PART TWO

PRESOLO

8

EFFECTS OF
CONTROLS

Most of your first flight will consist of finding out how the various controls affect the airplane. The instructor will climb the plane to a safe altitude (at least 1500 feet above the ground) and demonstrate the use of each of the flight controls. You will then use the controls until the plane's reaction to each becomes familiar to you.

The instructor will also trim the airplane for straight and level flight and show you how the plane flies by itself, demonstrating its natural stability. During the flight the instructor will also show you the practice areas and point out outstanding landmarks and their reference to the airport.

Bumpy air on this first flight may cause you to be tense at first. The "bumps" are caused by uneven heating of the earth's surface. A plowed field reflects heat; a forest or water area absorbs it. As you fly over the terrain, the rising air currents are of different velocities, so the plane rises and drops slightly. As the air rises, it is cooled and soon becomes stable and smooth. The altitude at which this occurs varies from day to day, but on a hot summer day you may climb to 6000 or 8000 feet before finding smooth air.

ELEVATORS

The elevators control the pitching motion of the plane and are your airspeed control (Fig. 8-1). Under normal flying conditions (the plane is right side up), pressing back on the wheel moves the elevators "up."

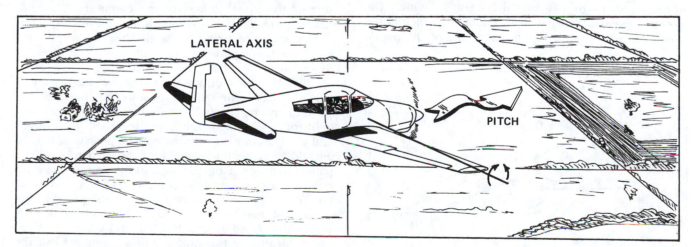

Fig. 8-1. Elevators control movement about the lateral axis (pitch).

8-1

The relative wind forces the tail downward, the nose moves up, and *if you started with sufficient power and airspeed* the plane starts to climb. Conversely, if the wheel is pressed forward, the nose goes down and the plane dives. If the plane is inverted and you press the wheel back, obviously the plane will not climb but will dive.

The best way to look at elevator movement, however, is to think of it as follows: wheel (or stick) back, the nose moves toward you; wheel forward, the nose moves away from you. This works for all attitudes.

At a low airspeed you might still think you can climb, or climb at a greater rate, by pulling the nose up, but you find that the up-elevators don't "elevate" the airplane but just decrease the airspeed. The nose moves up, *but* the increased drag causes the airplane to climb less or even start sinking! Power makes the airplane climb.

Let's take a situation such as the one you'll face when you're first introduced to the normal climb. You're flying along at cruise power and airspeed and want to climb. Back pressure is gently exerted on the wheel until the nose moves to the normal climb position relative to the horizon, as shown by your instructor. (You will also open the throttle to climb power.)

Okay, when you applied back pressure on the wheel, the elevators moved up. The relative wind moving past the elevators results in the tail moving downward (Fig. 8-2). Since the wings are rigidly attached to the fuselage, they are rotated as well. This results in an increase in angle of attack (and drag), the plane starts slowing up to the speed for best rate of climb, you apply climbing power, and the plane climbs. *This is the technique for a steady climb.* You'll find you can "zoom" the airplane, that is, get a short-term climb by trading airspeed for altitude, using the momentum of cruise speed. In fact, because of this excess energy available at cruise speed, the airplane will start to climb as soon as you start to ease the nose up. (You will see that the airspeed will start decreasing immediately.)

The job of the elevators in the climb is that of airspeed control—to get to the speed at which the greatest excess horsepower is available, which is the speed for best rate of climb. As a rule of thumb for

most trainers, this will be found at about 1.4–1.5 times the flaps-up, power-off stall speed. (This is for calibrated airspeed, or CAS. You might review this in Chapter 3.)

To repeat: The elevators control the angle of attack (and indirectly the airspeed) of the airplane. Because most lightplanes don't have angle of attack indicators, you'll be watching the airspeed indicator for the plane's reactions to your use of the elevators. Therefore, the elevators are considered to directly control the airspeed.

Power (or lack of it) gives the required performance. Suppose a light trainer uses the same airspeed for best rate of climb speed and recommended glide speed. You would maintain with the elevators the same airspeed for both maneuvers; the difference in performance (climb or glide) would be in the amount of power used (climb power or idle). Or you could fly straight and level at that same speed by using *some* power. So one of the things you'll learn is that the "elevators" are really misnamed; everybody uses the term, though, so it's a little late to change it. Probably it started back in the early days of flying when it was believed that this was their function—to "elevate" the airplane.

The trainer you will be flying may have a "stabilator" or "flying tail" instead of a stabilizer-elevator combination. The entire horizontal tail of the plane moves and acts as a stabilizer or elevators, as required. The principle of operation is the same—the "stabilator" is an angle of attack and airspeed control and, like the elevator, has a trim tab to help the pilot correct for various airplane loadings and airspeeds.

■ Elevator Trim Tab

Elevator or stabilator trim tabs are used to hold elevator or stabilator pressure for the pilot. If an unusually heavy load was placed in a back baggage compartment, the tail would be heavy, the nose would want to rise, and you would have to hold forward pressure on the wheel to keep from slowing up. The trim tab is controlled from the cockpit, and during flight the plane can be trimmed for climb, glide, or straight and level. Some planes use an adjustable stabilizer for this purpose.

Larger planes require use of the trim tab or adjustable stabilizer when speed or power is changed, but your instructor probably will have you fly all maneuvers *at first* with the plane trimmed for straight and level so that the varying control pressures can be felt. Some students get in the habit of trimming the plane full nose-up for landing, but this can lead to difficulties if a go-around is required. The application of power may cause the nose to rise so sharply that the plane is stalled before the nose can be lowered. This will be discussed further in Chapter 14.

Returning to the example of a heavy load in the baggage compartment (Fig. 8-3): The nose wants to go up; the pilot holds the nose at the proper position and

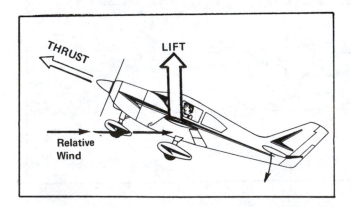

Fig. 8-2. Relative wind at the instant of the plane's rotation.

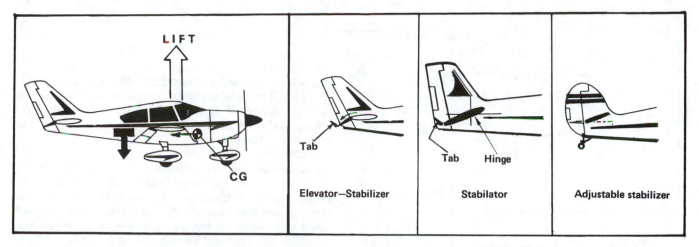

LIFT

Tab

Elevator—Stabilizer

Tab Hinge

Stabilator

Adjustable stabilizer

CG

Fig. 8-3. Pilot has to hold the wheel forward because the plane wants to nose upward. (Center of gravity is moved aft.) The trim tab or the adjustable stabilizer does it for you. "Nose-down" trim is set in all cases here.

trims the plane "nose down" until there is no pressure against the wheel. When the tab control is moved to the "nose-down" position, the tab on the elevator moves up. The relative wind strikes the tab, forcing the elevator down for the pilot. (This is strictly an example; you'll read later that the baggage compartment is *not* to be overloaded.)

RUDDER

The rudder controls the yawing motion of the airplane (Fig. 8-4). Push the left rudder pedal and the nose yaws to the left; push the right and the opposite occurs. The prime purpose of the rudder is to overcome the adverse yaw of the aileron. In certain maneuvers such as the slip and crosswind landing, ailerons and rudder are used opposite to each other, but 99.99 percent of the time they are used together in the air.

On tailwheel-type airplanes the rudder is connected to the tailwheel by coil springs, and this is the main factor in ground steering. The tricycle-gear types have a mechanical linkage between the rudder pedals and the nosewheel so that ground steering depends even less on the rudder. (In the tailwheel airplane, sometimes a blast of slipstream striking the rudder can help in turning; for tricycle-gear types it has little effect.) "Rudder" is another term that is misleading. A rudder of a boat is the main tool for turning. An airplane's rudder is not the primary control used for turning; in normal flight it is an auxiliary to the ailerons.

However, you'll find that by application of rudder in the air it is possible to turn. By pushing the rudder pedal you skid the airplane, causing one wing to move faster than the other (Fig. 8-5). This added lift causes the airplane to bank and the rest of the turn is normal. It can be rolled out in the same manner, as will be demonstrated to you.

■ Rudder Trim

In Chapter 2 the "bendable" rudder tab was discussed; some of the later model trainers have a rudder bungee or tab that is controllable from the cockpit. You can adjust the rudder trim as necessary in flight to get the desired reaction. In these airplanes, one of your jobs during the pretakeoff check at the end of the runway will be to set the trim properly. (It's set to a "nose-right" indication to help you on that right rudder to overcome "torque" on takeoff and climb.)

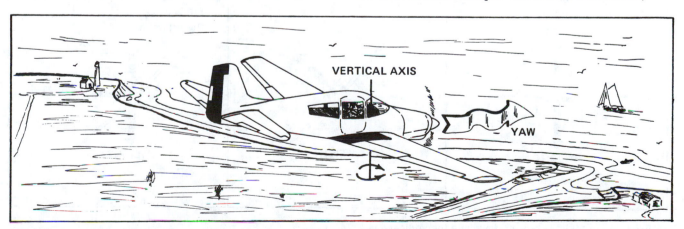

VERTICAL AXIS

YAW

Fig. 8-4. The rudder controls movement about the vertical axis (yaw).

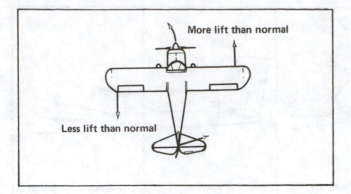

Fig. 8-5. Banking the airplane by rudder alone.

AILERONS

The ailerons roll the airplane. If the stick is moved to the left (in an airplane with a wheel, the wheel is turned to the left), the left aileron moves up and the right aileron moves down. The relative wind strikes the control surfaces as shown and the airplane rolls to the left (Fig. 8-6).

The aileron tab(s) for trimming the airplane later-ally was discussed in Chapter 2 (Fig. 2-13).

The plane will continue to roll as long as the ailerons are deflected. In fact, this generally would be the way you'd do a slow roll—plus the use of other controls to aid the process.

Ailerons are used to bank the airplane. The airplane banks in order to turn. There's nothing in the world that gives less traction than air. When a rudder pedal is pushed, the nose yaws around, but the plane tries to continue in a straight line. True, it would gradually turn because the yawing would speed up one wing, giving it more lift and banking it, but it would be an inefficient and uncomfortable process. When the airplane is banked, the lift force acts as a centripetal, or holding, force as well and can be considered to be broken down into two parts—this is a legal assumption according to engineering mechanics (Fig. 8-7).

The centripetal force (or horizontal component of lift) then helps the plane to stay in the turn. It is the force acting against the tendency of the airplane to continue in a straight line. *So in order to turn properly, the airplane must be banked.*

Fig. 8-6. The ailerons control movement about the longitudinal axis (roll).

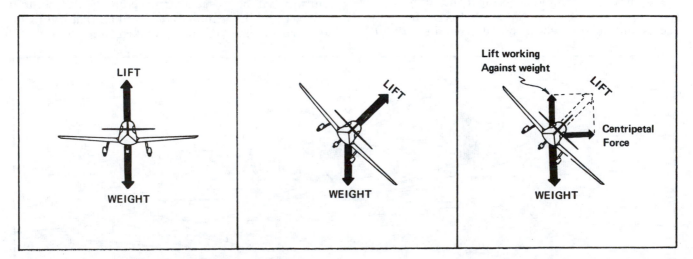

Fig. 8-7. In a turn, lift is broken down into the vertical component (lift working against weight) and the horizontal component (which turns the airplane).

9

THE FOUR FUNDAMENTALS

During the first flight, after you are familiar with the effects of the controls, the instructor will introduce you to one or more of the Four Fundamentals: *the turn, the normal climb, the normal glide,* and *straight and level.* These are the basic maneuvers of flying. You'll be using them or a combination of them for the rest of your flying career, so get them down pat now.

THE TURN

In the discussion of ailerons it was found that the plane must be banked in order to turn in the air. Fine—all you have to do is jam the wheel over in the direction you want and you turn that way. *Not so fast!* There's more to it than that. Let's see what happens when you give it left aileron (that is, you turn the wheel or move the stick to the left).

The lift and drag vectors (Fig. 9-1) show that although the wheel is turned to the left there's more drag on the right because of the down aileron. The plane rolls to the left but the nose yaws to the right. The plane will slip to the left and finally set up a balanced turn. The right yaw you got is called "adverse aileron yaw," and the rudder is used to correct for this.

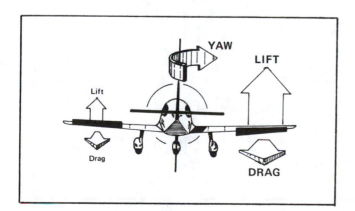

Fig. 9-1. Adverse aileron yaw.

The rudder is used any time the ailerons are used in a turn. Most planes have a differential aileron movement (the aileron moves farther up than down) and aileron designs that decrease this initial yaw effect.

To make a turn to the left, left aileron and left rudder are used smoothly together. As soon as the desired amount of bank is reached, the controls are neutralized. The wheel is moved smoothly back to the neutral position, the pressure on the control wheel is eased off, and rudder pressure is no longer needed. This is called "neutralizing" the controls—the relaxing of exerted pressure on them. This term will be used throughout this chapter. Air pressure should streamline the control surfaces when the control wheel and rudder pedal pressures are relaxed. If the controls are not neutralized, the turn will become steeper and steeper.

You're in the turn. If you let go of the controls the plane will turn indefinitely (or at least as long as the fuel lasts) *except for one small detail.*

Lift is considered to act perpendicularly to the wingspan. Consider an airplane in straight and level flight where lift equals weight. Assume the airplane's weight to be 2000 pounds.

In Figure 9-2A everything's great—lift equals weight. If the plane is banked 60° as in Figure 9-2B, things are not so rosy. The weight's value or direction does not change. It's still 2000 pounds downward. The lift, however, is acting at a different angle. The vertical component of lift is only 1000 pounds, since the cosine of 60° is 0.500 (60° is used for convenience here; you certainly won't be doing a bank that steep in the earlier part of your training). With such an unbalance the plane loses altitude. The answer, Figure 9-2C, is to increase the lift vector to 4000 pounds so that the vertical component is 2000 pounds. This is done by increasing the angle of attack. Back pressure is applied to the elevators to keep the nose up. To establish a turn to the left:

1. Apply left aileron and left rudder smoothly.

2. As the bank increases, start applying back pressure smoothly.

3. When the desired amount of bank is reached, the ailerons and rudder are neutralized, but the back

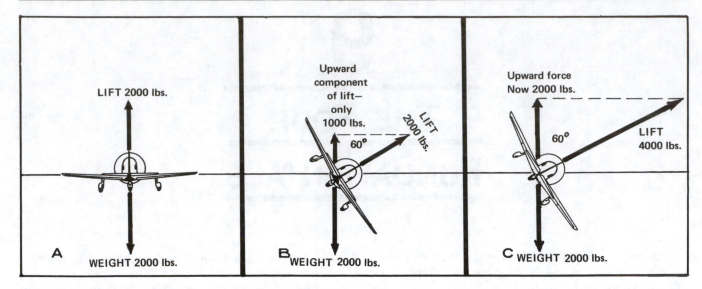

Fig. 9-2. The vertical component of lift must be equal to weight in a constant-altitude turn.

pressure is held as long as you are in the bank. Now you can turn indefinitely.

To roll out of the left turn:

1. Apply right aileron and right rudder.

2. As the bank decreases, ease off the back pressure.

3. When the plane is level, the ailerons and rudder are neutralized and there should be no more back pressure.

If the rudder and aileron are not coordinated properly, the plane will skid or slip. The skid occurs when *too much* rudder is used. You'll try to slide toward the outside of the turn as you do when a car turns a corner. The slip is a result of *too little* rudder, and you'll feel that you are sliding toward the inside of the turn. These errors can be felt and give you an idea of the expression "seat of the pants flying." Students have

been known to wear out the seats of perfectly good pants, sliding around during the presolo phase.

If in our example of the 60° bank, you roll out level but don't get rid of the extra 2000 pounds of lift you have, obviously the airplane will accelerate upward (the "up" and "down" forces will no longer be in balance). Maybe your idea of coordination is that demonstrated by a go-go dancer, but it's necessary in flying too. Think of control pressures rather than movement. The smoother your pressures, the better your control of the airplane.

If you aren't particularly interested in the trigonometric approach to back pressure, Figure 9-3 gives another approach.

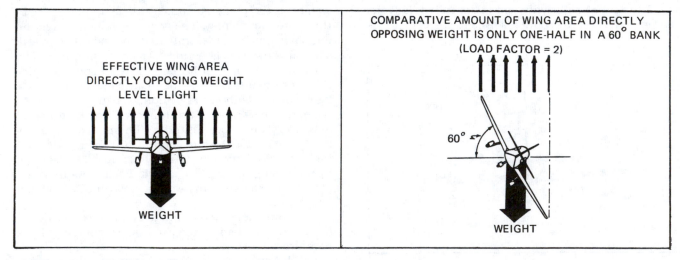

Fig. 9-3. The effects of bank angle on the amount of lift that is directly opposing weight.

■ Load Factors in the Turn

The turn introduces a new idea. It is possible to bank so steeply that the wings cannot support the airplane. The lift must be increased so drastically that the critical angle of attack is exceeded, and the plane stalls. In the previous example at 60° of bank it was found that our effective wing area was halved; therefore, each square foot of wing area had to support twice its normal load. This is called a *load factor of 2.* A plane in normal, straight and level flight has a load factor of 1, or 1 "g." You have this same 1 g load on your body at that time. Mathematically speaking, the load factor in the turn is a function of the secant of the angle of bank. The secant varies from 1 at 0° to infinity at 90°; so to maintain altitude indefinitely in a constant 90° bank, an infinite amount of *lift* is required—and this is not available.

For those interested in the mathematics of the problem, the following is presented.

It is interesting to note that the stalling speed of the aircraft can be figured for each degree of bank. In the airplane weighing 2000 pounds and having a wing area of 200 square feet, the wing loading is 10 pounds per square foot. With a load factor of 2, the wing loading is

$$20 \text{ pounds per square foot} = \frac{\text{weight}}{\text{wing area}} \text{ or } \left(\frac{W}{S}\right)$$

Using the equation

$$L = C_L S \frac{\varrho}{2} V^2$$

and solving for V (stall), you find that the stall is at $C_{L_{max}}$ so that value would be used in the equation:

$$V_{stall} = \sqrt{\frac{L}{C_{L_{max}} S \frac{\varrho}{2}}}$$

But in level flight L = W so that

$$V_{stall} = \sqrt{\frac{W}{C_{L_{max}} S \frac{\varrho}{2}}}$$

And since $C_{L_{max}}$ (the maxium coefficient of lift), S and ϱ are considered to be constants for a given situation, the new stall speed is directly proportional to the square root of the change in the wing loading W ÷ S. Then V_{stall} is a function of the square root of the load factor. So the stall speed in a 60° bank is $\sqrt{2}$ times that of the stall speed in level flight. If the plane stalls at 50 K in level flight, its stall speed in a 60° bank is $50 \times \sqrt{2} = 50 \times 1.414 = 70.7$ K. *All of this is assuming that you hold enough back pressure to maintain level flight* (Fig. 9-4). If no back pressure is held, the airplane will be in a diving turn, there is still a load

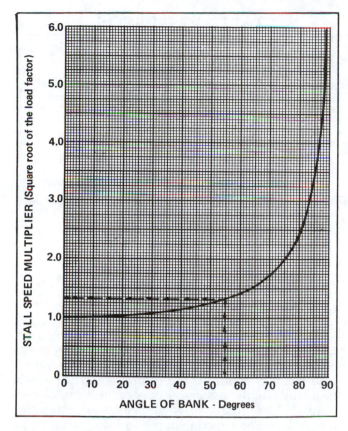

Fig. 9-4. Choose an angle of bank (in this case, 55°). Multiply the corresponding number (1.32) times the wings-level (normal) stall speed to get the stall speed at this angle of bank.

factor of 1, and the stall speed is not increased. If you make this kind of turn, it is wise to have a lot of altitude. Needless to say, the Federal Aviation Administration frowns on such capers at low altitudes because you are likely to leave pock marks on surface structures when you hit.

Simply stated—if you want to find the stalling speed of your plane in a level turn, either of the following will apply:

$$V \text{ stall in the turn } = V \text{ stall (level) } \sqrt{\text{load factor}}$$

or

V stall in the turn
= V stall (level) $\sqrt{\text{secant of the angle of bank}}$

Figure 9-5 is a diagram of the increase of stall speeds with bank. The numbers have been rounded off, but note that the stall speed increase with bank for any flap setting increases at the ratio indicated by Figure 9-4. Gross weight (maximum certificated weight) is used, but for lighter weights the multiplier for bank increase would be the same. (The airplane will always stall in a 60° bank at a speed of 1.414 times the wings-level stall speed whether it's given in mph or knots.)

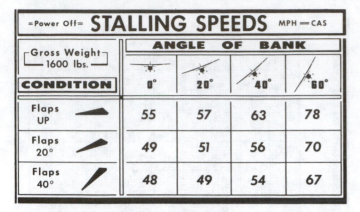

=Power Off= STALLING SPEEDS MPH—CAS				
Gross Weight 1600 lbs.	ANGLE OF BANK			
CONDITION	0°	20°	40°	60°
Flaps UP	55	57	63	78
Flaps 20°	49	51	56	70
Flaps 40°	48	49	54	67

Fig. 9-5. Stall speed increase with bank.

Now back to the important, practical side of flying. Putting it simply: The stall speed goes up in the turn; and the steeper the bank, the faster the stall speed jumps up, as can be seen in Figures 9-4 and 9-5. The load factors just discussed are "positive" load factors and are attained by pulling the wheel back, causing you to be pressed down in the seat. A negative load factor is applied if the control wheel is pushed forward abruptly, and in this case you feel "light" and tend to leave the seat. Lightplanes are generally stressed to take a maximum positive load factor varying from 3.8 to 6, and a negative load factor of between 1.52 and 3, depending on the make and model. Both you and the plane are able to stand more positive g's than negative. Positive or negative load factors can be imposed on the plane by sharp up or down gusts as well as by the pilot's handling of the elevators. (See Chapter 23.)

■ Review of the Turn

1. Apply aileron and rudder pressure in the desired direction of turn.
2. Begin back pressure.
3. When desired steepness of bank is reached, neutralize the ailerons and rudder but hold enough back pressure to keep the nose at the correct place on the horizon; turn until you get ready to roll out. Check the nose, wings, and altimeter for a smooth level turn.

■ The Rollout

1. Apply opposite ailerons and rudder. Gradually ease off the back pressure as the wings become level.
2. As the wings become level, neutralize the ailerons and rudder (the back pressure will be completely off at this point).

Ride with the turn. You'll have a tendency at first to lean against the turn. Remember that in a proper turn the resulting forces will hold you upright in the seat.

If you are in a side-by-side airplane, the position of the nose seems different in left and right turns (Fig. 9-6). You will be sitting to the left of the center line of the fuselage so that when the airplane is turning left the nose will seem high and the tendency is to lose altitude as you "correct" for this. In a right turn, the nose will appear too low and the tendency will be to gain altitude. Your instructor will demonstrate the proper nose position for climbs, glides, turns, and straight and level and may also demonstrate that elevator trim can be used to maintain back pressure in an extended turn situation (retrimming as the airplane rolls out). Figure 9-6 shows the views from the left seat during left and right constant altitude turns.

Fig. 9-6. Perspective in left and right constant attitude turns. Note that your reference (which is a point on the cowling *directly in front of you,* and not the center of the nose) is at the same position in relation to the horizon. Don't use the center of the cowling as a reference.

■ Coordination Exercises

After you are familiar with the idea of the turn, the instructor may have you do what is known as a coordination exercise. This consists of heading the plane at a point on the horizon and rolling from left to right bank, and so forth, without letting the nose wander from the point. It is a very good way to discover how much control pressure is needed to get the desired effect with the plane.

The beginning of the maneuver is like that of the beginning of the turn. Ailerons and rudder are used together. Then, before the nose has a chance to move, hold it on the point with opposite rudder. Apply opposite aileron and more opposite rudder and roll it to the other side, then repeat the process. You will find it difficult at first but will soon get the knack and be able to roll smoothly from one side to the other without the nose moving from the point.

Although in parts of the maneuver the controls are crossed and not coordinated as they are in the turn, you will find that your turns will profit by a short period of this maneuver.

THE NORMAL CLIMB

Another of the Four Fundamentals is the normal climb, and at first you will practice starting it from normal cruising flight. To climb, the nose is eased up to the proper position relative to the horizon, the throttle is opened smoothly to the recommended climb power, and, as the airspeed starts to decrease, right rudder is applied to correct for torque. The recommended airspeed for best rate of climb is maintained.

You had a brief introduction to the climb in Chapter 8 when the elevators were discussed. There you noted that the elevators do not make the airplane climb; they are used to attain and maintain the *airspeed* at which the best rate of climb is found. That airspeed, engineers have found, is the one at which the greatest amount of excess horsepower is available at climb power for your airplane. (The rate of a steady climb depends on the amount of excess horsepower available.) An equation for determining rate of climb is:

$$\text{Rate of climb (fpm)} = \frac{33{,}000 \times \text{excess horsepower}}{\text{weight of plane}}$$

The horsepower in that equation is *thrust* horsepower, or the power being effectively developed through the propeller, as mentioned in Chapter 2. You also remember from that chapter that *power* is *force* times *distance per unit of time,* and 1 horsepower is equal to 550 foot-pounds per second, or 33,000 foot-pounds per minute. That's where the 33,000 in the equation comes in; it's set up for a rate of climb in *feet per minute.* By illustration, suppose your plane weighs 1500 pounds, has an engine capable of developing 100 *brake* (or shaft) horsepower at the recommended climb power setting at a particular altitude, and the propeller is 85 percent efficient at the climb speed. This means there will be 85 *thrust* horsepower available at the climb speed. Suppose, also, that at that speed 55 *thrust* horsepower is necessary to maintain a constant altitude. This means that 30 thrust horsepower is available for use in climbing. From the above equation we find the rate of climb to be 660 feet per minute:

$$\text{R/C (fpm)} = \frac{33{,}000 \times 30}{1500} = 660 \ \text{fpm}$$

So the excess thrust horsepower is working to raise a weight (the airplane) a certain vertical distance in a certain period of time (rate of climb). If no excess horsepower was available at the climb speed, the rate of climb would be zero. If you throttled back until the power being developed was *less* than that needed just to keep the airplane flying at a particular airspeed, a "negative" rate of climb would result; the airplane would descend at a rate proportional to your "deficit" power.

As a pilot you won't worry about excess horsepower, but you will use the recommended climb airspeed and power setting and let the rate of climb take care of itself.

The revolutions per minute of a fixed-pitch prop such as on a light trainer are affected by the airspeed. Assume you set the throttle for cruising at 2400 rpm. If the airplane is dived, the added airspeed will give the propeller a windmilling effect and the rpm will speed up to, say, 2500 or 2600 rpm. If you pull the nose up for a climb, the airspeed decreases and the rpm drops below the cruising setting. Therefore, it is necessary to apply more throttle to obtain climb power (usually full open [forward] throttle for most trainers). Normal climb speed is about 1.4 to 1.5 times the stall speed and results in the best *rate of climb.* Your instructor will show you the correct climb attitude and airspeed for your particular plane (Fig. 9-7).

■ To Climb

1. Ease the nose up to normal climb position and maintain back pressure to keep it there.
2. Increase the power to the climb value.
3. As speed drops, apply right rudder to correct for torque.
4. Don't let the nose wander during the climb or transition to the climb.

■ To Level Off from the Climb

1. Ease the nose down to level flight position.
2. As speed picks up, ease off right rudder.
3. Throttle back to maintain cruise rpm.
4. Don't let the nose wander during the transition.

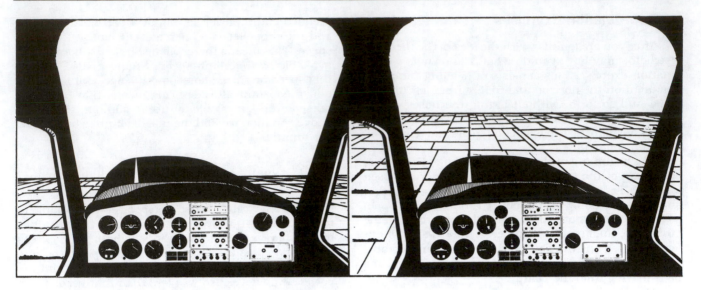

Fig. 9-7. The climb and level-flight attitudes. This doesn't necessarily mean that you are climbing or flying level. These are attitudes only.

■ Probable Errors

1. Climbing too steeply—overheating the engine and actually decreasing the rate of climb.

2. Too shallow a climb—easy on the engine but not a good way to gain altitude.

3. Over- or undercorrecting for torque during the climb transition.

4. Rough transition to the climb or straight and level flight.

5. Not keeping the nose lined up on a reference point—letting it wander.

After you've gotten the feel of the controls for the climb well in hand, your instructor will have you trim the airplane for it. As you level off, the trim will have to be changed for straight and level cruising flight.

■ The Climbing Turn

This is a combination of the two fundamentals covered so far. It is, as the name implies, a turn while climbing.

For practice at first, the climbing turn will be executed from a straight climb. That is, establish a normal climb and then start your turn.

Make all your climbing turns very shallow (Fig. 9-8). Steep banks require added back pressure (angle of attack) to keep the nose up, which sharply reduces the rate of climb because of greater drag (induced drag; see Chapter 2). *A 10° bank is plenty.*

Torque must be considered in making a climbing turn. *Anytime the plane is climbing, right rudder pressure or right rudder trim must be used.*

Suppose you are climbing and want to turn to the right. You are already holding right rudder, but as

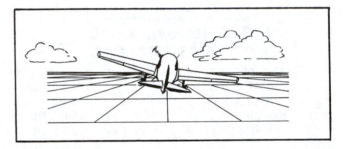

Fig. 9-8. Keep the bank shallow in the climbing turn or else the climb will suffer.

right aileron is applied more right rudder is needed to take care of aileron yaw, so the turn is something like the following.

Climb:

1. Back pressure.

2. Right rudder.

Climbing right turn:

1. Right aileron and more right rudder—make the turn shallow.

2. More back pressure.

3. Neutralize the ailerons and return to enough right rudder to correct for torque.

Rolling out to resume the straight climb:

1. Left aileron and very little left rudder (in some planes, easing off the correction for torque may be all that is needed).

2. As the wings become level, neutralize the ailerons and get back on that right rudder.

Again, any time you make a turn it requires added back pressure.

The torque effect is as if slight left rudder was being held. In a left-climbing turn, torque will tend to skid you into a steeper bank. In a right-climbing turn, it will tend to skid you out of the turn.

When you have the climb and climbing turn idea well in mind and are able to control the aircraft in a reasonable manner, you should start the habit in your climbing turns of maintaining a constant bank and making definite 90° turns as you climb. (This will be your introduction to precision flying.)

It's best not to climb straight ahead for long periods because you can't see other airplanes over the nose. Make occasional shallow turns to each side of your "base course" or lower the nose briefly to check ahead.

THE NORMAL GLIDE

The normal glide is a third fundamental and the least understood by the beginning student. The nose is not pushed *down* to glide as commonly believed, but, on the contrary, back pressure is needed to keep the glide from getting too steep.

In Figure 9-9A the airplane is trimmed for straight and level flight. The force caused by the slipstream and relative wind acting on the stabilizer just balances the force of the weight trying to pull the nose down. The plane is in balance. Pulling the throttle back to idle causes a lessening of the tail force by two factors: (1) The slipstream effect has dropped to nil because the propeller of an idling engine produces a very weak slipstream, and (2) the plane starts slowing immediately now that the drag is greater than the thrust—hence the relative wind speed decreases. (The decreased slipstream is by far the biggest factor in this nosing-down process, however. Item 2 mentioned above can be neglected for all practical purposes.)

The result is shown in Figure 9-9B. The plane noses down until the forces are balanced again. Your

speed now may be higher than desired for the glide. If you want to glide at 70 K, you must slow the plane to that speed by holding back pressure so that the plane's glide attitude is something like that in Figure 9-10. Normal glide attitude in most lightplanes is only slightly more nose-down than the straight and level position (Fig. 9-11).

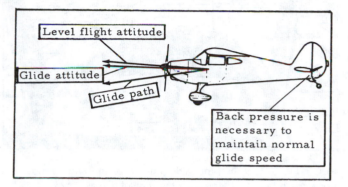

Fig. 9-10. Level flight and glide attitudes.

Tailwheel-type airplanes are shown in Figure 9-9 and 9-10, but the tricycle-gear trainer reacts the same way.

The recommended normal glide speed gives you the most feet forward for altitude lost. For a fixed-gear trainer the normal glide ratio is about 9:1, that is, 9 feet forward for each foot of altitude lost.

If you glide at a too-fast airspeed, the 9:1 ratio drops off to maybe only 5:1. You can tell the steep glide by sight (nose position, airspeed increase), sound (increase in pitch of sound of wind), and feel (the controls become firmer).

Trying to "stretch the glide" by pulling the nose up also causes the glide ratio to suffer. You can recog-

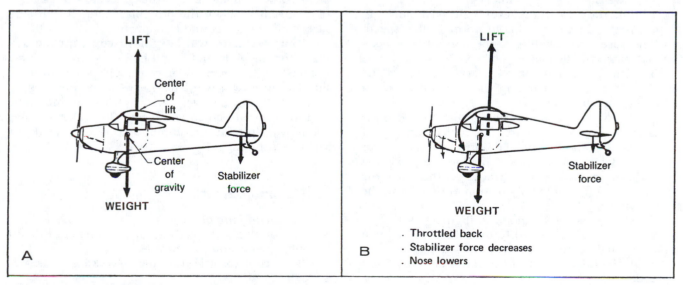

Fig. 9-9. The airplane is designed to have a nosing down tendency when power is reduced.

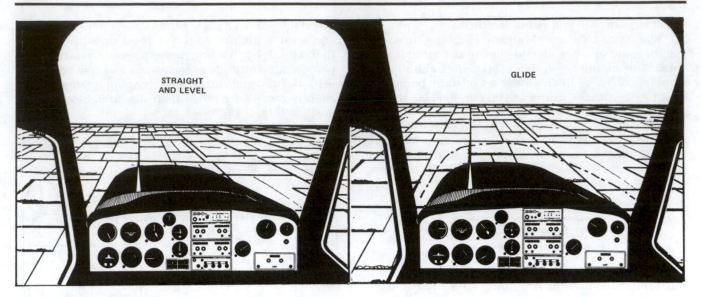

Fig. 9-11. Notice that the glide attitude is only slightly more nose down than that for straight and level flight.

nize the too-slow glide by sight (nose high), sound (decrease in wind noise), and feel (the elevator pressure feels "mushy"). You've increased the angle of attack to such a point that drag holds the plane back, but gravity is still pulling the plane down with the same force, resulting in a lower glide ratio. *So stick to that recommended glide speed.*

Pull the carburetor heat ON before closing the throttle in flight unless the Pilot's Operating Handbook *recommends otherwise.* Its use will be mentioned here each time the glide is covered, but your trainer and conditions may not always require it.

Assume that you are flying straight and level and want to glide:

1. Carburetor heat ON.
2. Close the throttle (to idle).
3. Hold the nose in the level flight position.
4. As the airspeed drops to normal glide speed, ease the nose down slightly to the normal glide position as shown by the instructor.
5. Clear the engine about every 200 feet of the descent. This means running the rpm up to about 1500 and then back to idle.

You'll notice that quite a bit of back pressure is necessary to hold the nose up. The instructor may let you trim the plane for the glide. In that case, roll or move the trim until you feel no force against the wheel. Most instructors want you to get the feel of the plane and will probably only demonstrate trim in the glide at first.

To return to straight and level flight:

1. Apply power smoothly to cruise rpm as you simultaneously ease off the back pressure.
2. Push carburetor heat OFF after cruising flight is established.

With more experience you'll open the throttle up to about 100 rpm less than the cruise setting so that when the carburetor heat is pushed OFF the rpm will be at the cruise setting.

The altimeter appears to "lag" in the climb or glide, particularly in the glide. If you want to level off at a specific altitude, say 1500 feet, it is wise to start the above procedure about 50 feet above that, or 1550 feet. The plane has momentum downward that requires altitude to change and maybe your reactions aren't perfect yet, so give yourself about 50 feet for recovery from the glide. Of course, the bigger the airplane and the faster you are descending, the more margin you'll use.

If the back pressure is not eased off as you open the throttle, the nose will rise sharply as the slipstream strikes the up-elevators. The elevators are comparatively ineffective in the glide, so you use a lot of up-elevator. When the prop blast hits all that surface, something's bound to happen.

Your instructor will later have you trim the airplane for the glide and you'll be pretty busy at first, readjusting the trim when you apply power to level off. The nose will want to rise with the addition of power, being helped very well by the nose-up trim, and in some airplanes you'll tend to underestimate the forward pressure required to keep the nose down while retrimming.

■ Common Errors

In starting the glide:

1. Not giving the carburetor heat enough time to work before closing the throttle.
2. Abrupt throttle movement—jerking it back to idle.
3. Holding the nose up too long, resulting in too slow a glide speed.

4. Letting the nose wander as the glide is entered.
5. Forgetting to clear the engine in the glide.
In recovery from the glide:
1. Opening the throttle abruptly.
2. Letting the nose wander.
3. Not releasing back pressure or changing trim in time—nose goes up and plane climbs.

The carburetor heat is the first thing ON before, and the last thing OFF after, the glide.

If icing conditions are bad, the carburetor may ice enough to cause engine stoppage between the time you push the heat off and get around to opening the throttle, particularly if you are the fumbling-fingers type. Granted, this would be an unusual occurrence, but why take the chance? Do it correctly now until it becomes a habit.

Clear that engine! In the winter, the engine may cool so much after an extended glide without clearing that it won't take throttle. In the summer the engine may "load up" and quit. *If you want to see what an actual emergency landing is like, don't clear the engine.*

Your instructor will have you glide at various flap settings at the recommended landing-approach airspeed(s).

GLIDING TURN

Like the climbing turn, the gliding turn is a combination of two of the Four Fundamentals. Combine the glide and the turn and you get a gliding turn. An extended steep gliding turn is called a spiral.

Probably the first thing you'll notice about the gliding turn is that the rudder seems to have lost much of its effect. Although the ailerons are ineffective in comparison with level flight, they still feel the same as they did in the climbing turn because the airspeed is approximately the same for climb and glide. The rudder, however, is suffering from lack of slipstream (Fig. 9-12). Concentrate on coordination again.

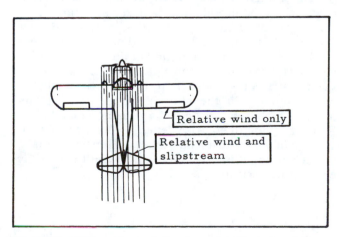

Fig. 9-12. The rudder and elevator get a slipstream during cruise and climb.

Use the controls so that the airplane reacts correctly.

Any time a turn is made, back pressure is required. If you have back pressure in the glide and make a turn, it will require more pressure. The principle is the same for climbing, gliding, or level turns. When the plane banks, the lift vector must be increased. If you use insufficient back pressure:
1. In the climbing turn—no climb, only turn.
2. In the level turn—turn plus a shallow dive.
3. In the gliding turn—turn plus a steeper dive.

In rolling into or out of a gliding turn, the rudder is not as effective as you've been used to because of the weaker slipstream.

■ Common Errors

1. Not enough back pressure in the turn—plane dives.
2. Not enough rudder on roll-in and roll-out.
3. Not easing off back pressure when rolling out—the nose comes up and the plane slows up.

Your instructor will have you make 90° gliding turns in each direction after you have a good idea of the procedure.

STRAIGHT AND LEVEL FLYING

One of the most important of the Four Fundamentals is straight and level flying; some pilots have trouble even after years of experience. Nearly every student has difficulty keeping the wings level. The average student tends to stare out over the nose with the glazed gaze of a poleaxed steer. Don't try to keep the wings level by watching for the cant of the nose, because the nose line is so short that one of the wings can be down quite a way before the nose shows it. Correct straight and level flying means that you are directionally straight and longitudinally and laterally level (Fig. 9-13).

The plane, if trimmed properly and left alone, will do a better job of flying than you are able to do. Many a student flies a "Chinese Cross-Country" (WUN WING LO) and wonders about that slightly uncomfortable feeling.

This is what's happening. You are unconsciously holding the aileron one way or the other. A plane with a stick is flown with the right hand and sometimes you allow the weight of your arm to pull the stick in that direction, resulting in a right slip. As the wheel is flown with the left hand, sometimes left aileron is slipped in.

The plane is banked and wants to turn, so you helpfully hold opposite rudder to keep the nose from moving and are now in a slip. In Figure 9-14, the resultant direction is downward and the plane starts losing altitude gradually. To stop this, you use back pressure to maintain altitude, and the controls are in the position shown in Figure 9-14. You are holding right

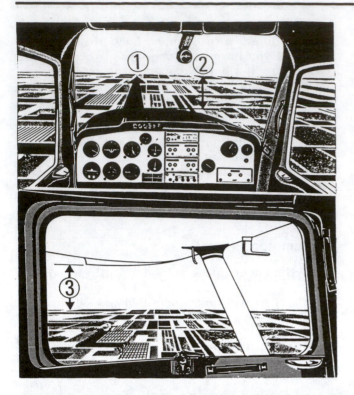

Fig. 9-13. In straight and level flight the airplane should be kept (1) directionally straight (no yaw), (2) longitudinally level (nose at the proper position with respect to the horizon), and (3) laterally level (the wings the same distance *above* the horizon for the high-wing craft, or the same distance *below* the horizon for the low-wing airplane).

aileron, left rudder, and up-elevator when you should be able to fly "hands off."

The instructor will show you how to trim the airplane for straight and level flight. Generally this is done as soon as you reach the assigned practice altitude. It consists of placing the nose at the correct attitude, leaving the climb power on until the expected cruise airspeed is reached (then set up cruise rpm), and trimming until the stick or wheel force against your hand is zero, (the same technique as described in the normal glide).

Larger planes not only have controllable tabs for the elevators but have tabs for the rudder and ailerons as well. The pilot can trim the plane for the attitude or speed desired.

Most lightplanes have a bendable aileron tab similar to the one on the rudder as discussed in Chapter 2. If the plane has a tendency to fly wing-low at cruise, the pilot is able to bend the aileron tab so that this is corrected. The procedure is the same as used for the rudder tab. Bend the tab and fly. If further correction is needed, land, make the correction, and fly again.

Once you have established good straight and level flight, there's nothing to do but watch the plane to see that it keeps this attitude. Some students feel that they should do *something* and consequently the plane receives quite a workout.

Try not to be tense, but if you are, don't worry about it. Flying is a strange experience at this point, and if you are completely relaxed you're one student in a million. However, if you are *too* tense the learning process slows down.

If a wing goes down momentarily, as may happen on a hot, bumpy day, bring it up with aileron and rudder together. As soon as the wings are level, neutralize the controls. If you try to use aileron alone, the plane will yaw from its heading.

Usually another "bump" will raise that wing, but this return to wings-level flight is not always immediate, so you can help the process. But don't overdo it because on bumpy days you'll be worn out in 30 minutes.

All through this book so far it has been stated that power controls altitude and that elevators are to be used to establish the proper airspeed for the maneuver involved. You will find, however, that in straight and level *cruising* flight, for minor altitude adjustments it's a lot simpler to use the elevators. For instance, if you are 50 feet low you will exert back pressure on the wheel and regain the altitude rather than adding power, which from previous discussions would be the "proper" thing to do. (Use "energy effect.") However, you might also take a look at the tachometer to make sure that the altitude loss (or gain) wasn't caused by an improper setting.

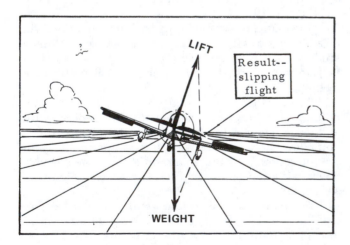

Fig. 9-14. Exaggerated view of wing-low flying.

■ **Common Errors**

1. Flying with one wing low.
2. Holding back pressure unconsciously, causing the plane to climb. (This is usually a sign of tenseness.)
3. Fighting the bumps. Every once in awhile release the wheel or stick to see if you have been working at it too hard.

NOTE

These are the Four Fundamentals. These and stalls will be the backbone of your presolo flying.

THE FOUR FUNDAMENTALS AND
INSTRUMENT INDICATIONS

The FAA requires that on the practical (flight) test you demonstrate the ability to recover from an emergency situation such as accidentally flying into clouds or fog. This means that you must be able to get out of such a situation through use of the flight instruments.

Tests have shown that students got better grades on the flight test when this training started early, that is, when flight with reference to instruments was integrated with the pre- and postsolo maneuvers. The FAA recommends that this integrated method be used in training all student pilots.

As in all of your flight training, the steps must be taken logically. When you are able to do the Four Fundamentals and feel at ease in the airplane, the instructor will have you go through these same maneuvers, directing attention to the instrument indications during the process of each. Later, you will use an extended visor cap, or "hood," that will restrict your vision to the instrument panel. At this point, however, it will be better to see the plane's attitude as well as the instruments so that you will be able to tie in the plane's actions with the instrument indications (and vice versa). In order to fly without outside visual references, you must be able to "see" what the plane is doing through the instruments.

Figures 9-15 through 9-20 show the Four Fundamentals (plus climbing and descending turns) as seen from the cockpit and by the instrument indications.

It will likely be at this stage that you will first note the idiosyncrasies of the magnetic compass. You will see for yourself that in actual flight the compass is affected by steep banks, attitude changes, and acceleration or deceleration of the airplane. In many cases the compass may initially show a change of direction of up to 30° as soon as the plane is banked for a turn;

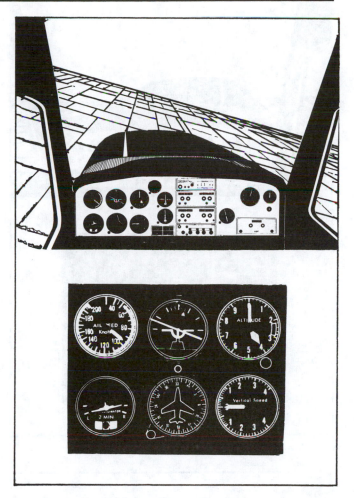

Fig. 9-15. A balanced constant-altitude turn (standard-rate). The airspeed is slightly less than cruise because of bank (and back pressure). The heading indicator shows a left turn, and the turn coordinator shows a balanced standard-rate turn. Altitude is constant.

in some cases it may show a turn in the opposite direction at first. In any event, the various actions of the compass for each maneuver cannot be shown here; in Chapter 15 more information will be given about it. The heading indicator gives a true indication of the amount of turn, but the compass is accurate only when the plane is flying straight and level in balanced, unaccelerated flight.

During these maneuvers your attention will be on the instrument panel a large part of the time and the instructor will maintain a close watch for other airplanes.

There is no specific amount of dual required on the emergency maneuvers for the private certificate; it will rest on the instructor's judgment of your ability in this area.

However, the instructor will require that you be able to safely control the airplane by use of the instruments *before* you go on a solo cross-country. Precision

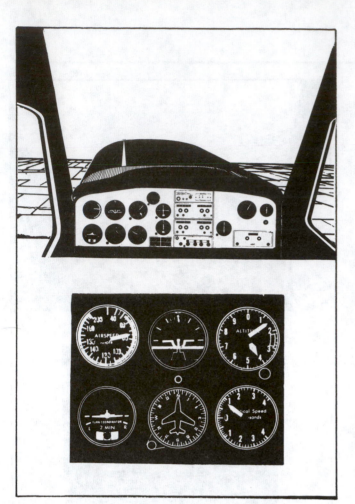

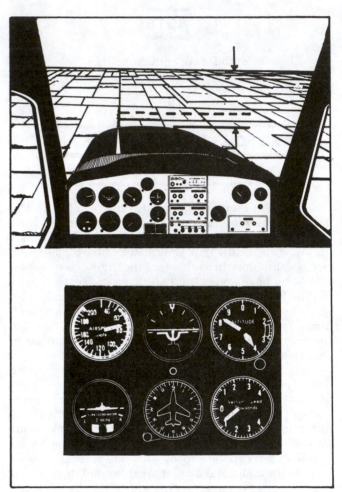

Fig. 9-16. Normal straight climb. The airspeed is steady at climb speed. The attitude indicator shows that the attitude is nose-up, wings-level. The heading indicator shows a constant heading; the turn coordinator indicates balanced straight flight. The altitude is increasing as shown by the altimeter and vertical speed indicator.

Fig. 9-18. Normal power-off glide (straight descent). The airspeed is steady at the proper glide speed. The attitude indicator shows a slight nose-low, wings-level attitude. The heading indicator shows a constant heading; the turn coordinator indicates straight balanced flight. The altitude is decreasing as indicated by the altimeter and vertical speed indicator.

Fig. 9-17. Climbing turn, standard-rate. The airspeed is steady at climb speed. The attitude indicator shows 11° bank, climb attitude. The heading indicator shows a left turn. The turn coordinator shows a standard-rate balanced turn. Altitude is increasing as indicated by the altimeter and vertical speed indicator.

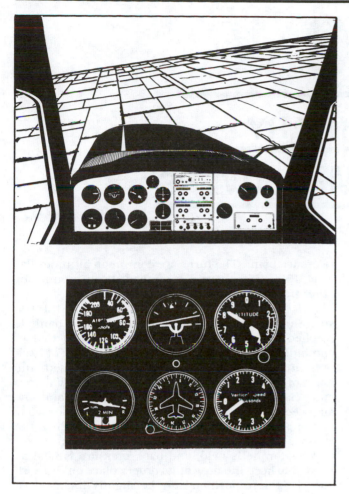

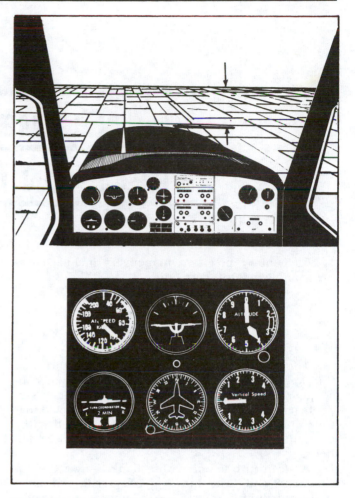

Fig. 9-19. Balanced, standard-rate, gliding turn. The airspeed is steady at glide speed. The attitude indicator shows a slight nose-low, 10° bank. The heading indicator shows a right turn. The turn coordinator indicates a balanced, standard-rate right turn. The altitude is decreasing as shown by the altimeter and vertical speed indicator.

Fig. 9-20. Straight and level flight. Airspeed steady at expected cruise value. Attitude indicator shows nose- and wings-level flight. Heading indicator shows a constant heading. Turn coordinator, straight and balanced flight. Altitude constant as shown by altimeter and vertical speed indicator.

instrument flight will not be required, but you will be expected to stay ahead of the airplane in this area.

T-TAIL AIRPLANES

The current trend is toward building airplanes with T-tails (the horizontal tail surfaces are on top of the fin and rudder). The horizontal tail is out of the slipstream (Fig. 9-21) and changes in power don't cause the pitch changes discussed earlier in the chapter (Figs. 9-9 and 9-10), or at least such changes are smaller. You might watch for the differences if you go from one type to the other (T-tail or standard) during your flying career.

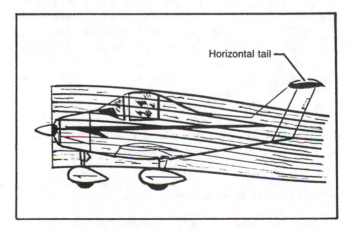

Fig. 9-21. The T-tail is usually out of the slipstream so that the pitch change with power is decreased.

10

ELEMENTARY

PRECISION MANEUVERS

Elementary precision maneuvers include steep turns and wind drift correction.

STEEP TURNS

The only difference between the steep turn and the normal turn is the steepness of bank. Many instructors have made this statement, and many students have nodded agreement but mentally noted that they didn't believe a word of it. The fact is that the techniques are the same with one or two minor additions.

A steep turn is one with a bank of over 30°. In Chapter 8 and "The Turn," in Chapter 9, you noted that in medium turns (15°–30° bank) you were able to take care of the need for increased lift by increasing back pressure. As the turn gets progressively steeper, you will find that the lift must be increased so much that you run out of angle of attack. In other words, the airplane stalls before you get enough lift to maintain altitude.

In the steep turn you may need increased power to maintain altitude because two things are happening: (1) The required added back pressure means a higher angle of attack and greater induced drag (see Chapter 2) and (2) the added load factor in the turn causes an increase in the stall speed. The added power serves to compensate for the added drag (which might otherwise finally result in an altitude loss), and it also helps lower the stall speed. You'll note in doing stalls that the stall speed is lower when power is used (Chapter 12).

■ The 720° Power Turn

The 720° power turn is a confidence-building precision maneuver and consists of 720° of turn with a coordinated roll-in and roll-out and a bank of 45°–60°. You will use added power and vary your bank as needed to maintain altitude (Fig. 10-1).

Pick a point on the horizon or a road below as a reference point. The turns are done at an altitude of at least 1500 feet above the surface. Look around for other traffic before starting the turn.

As you roll into the turn, open the throttle smoothly so that as the desired angle of bank is reached the power setting is approximately that of climbing power. Because of the great amount of back pressure needed after the turn is established, the angle of attack will be such that the airspeed will drop appreciably. You have a high power setting and low airspeed, so you will have to correct for torque effect.

Once the desired amount of bank is reached, the ailerons are neutralized, sufficient right rudder to correct for torque is held, and back pressure is held as needed to keep the nose at its proper place on the horizon. This nose position may be slightly higher than that in medium turns because of the required increase in angle of attack (added back pressure). A good rule of thumb is to have your bank and power established before you have made over 45° of turn.

You will find that the back pressure required is somewhat greater than you had anticipated.

If you see that the plane is losing altitude, the best method of bringing it back is to shallow the bank slightly. This will increase the vertical component of lift and the nose will return to the correct position. Then resume your steeper bank. Or if the plane is climbing, steepen up slightly and/or relax back pressure slightly.

The reason for not making the back pressure take care of all of the altitude variations is this: If you are in a 60° bank, the stall speed is increased 1.414 times that in level flight. If you have to climb or increase the lift at this bank, obviously you will run out of angle of attack. At 60° it takes 2 pounds of lift to get 1 pound of the vertical component. This is a losing proposition, and with steeper banks these odds jump sharply (see Figs. 9-4 and 9-5).

Check your wings, nose, and altimeter as you turn. Watch for the checkpoint. Some students make three or four complete turns instead of two because they aren't watching for the point. After all, this is a precision maneuver.

After you've completed the first 360° of turn, if

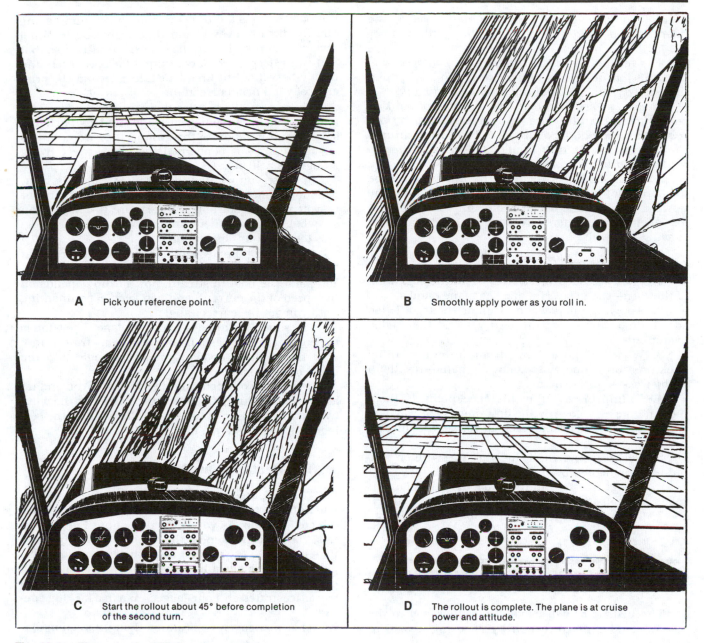

A — Pick your reference point.

B — Smoothly apply power as you roll in.

C — Start the rollout about 45° before completion of the second turn.

D — The rollout is complete. The plane is at cruise power and attitude.

Fig. 10-1. Elements of a 720-degree power turn.

you held altitude, the plane may wallow or bump slightly as it flies through its wake turbulence. The more slipstream you hit on the second 360° the better your turn (though this could be argued). It's possible, however, to hit a bump even if the altitude ran wild.

The idea is not to start out, say at 3000 feet, lolly-gag around from 2800 to 3200, and then feel proud because you happen to be back at 3000 feet when you roll out.

While your instructor won't be too critical about your performance at this stage of the game, it's a good idea to learn the technique now. Then you won't have to sweat 720s later when you should be practicing advanced stalls or other maneuvers.

The second 360° is a continuation of the first;

make corrections as needed. Watch for the checkpoint and start rolling out and throttling back about 45° before you are lined up with it. The roll-out should be smooth and timed so that the plane is in a straight and level attitude and at cruising power when the checkpoint is reached.

The hardest part about the roll-out is keeping the nose from coming up. It takes a lot of back pressure to hold it up in the turn, and it will take concentration in easing off that back pressure. During the first few times it will seem that you don't relax back pressure—you press forward to get the nose down where it belongs. To review the 720° turn:

The 720° power turn uses the same fundamentals as the shallow or medium turn. A steep turn requires only a longer period of deflection of the ailerons and rudder in order to obtain the steeper bank.

The reason slight right rudder may be needed is that the torque effects are caused by added power and slower airspeed, not the steepness of bank.

Power is used because (1) the angle of bank is such that the stall speed is increased, and (2) the required added angle of attack is increasing drag to the point where altitude may no longer be efficiently maintained by back pressure alone.

Some students get the idea that the nose can be held up by top rudder. The infinitesimal amount of the upward thrust angle doesn't compare with the great loss of efficiency suffered by slipping the airplane and slowing it down.

PROBABLE ERRORS

1. Back pressure added too soon at the beginning of the turn—the nose rises and the plane climbs.

2. Applying full power or opening the throttle too soon—airspeed has not dropped yet and the engine overspeeds.

3. When the nose drops, trying to pull it up by back pressure alone, not shallowing bank—results in high stresses on the plane.

4. Failure to watch for the checkpoint. (Make a mental note as it goes by the first time. That way you won't make the instructor sick as you go around and around.)

5. Not releasing all of the back pressure on roll-out—the plane climbs.

6. Forgetting to throttle back as the plane is rolled out—also causes the airplane to climb.

7. The usual coordination problems of slipping and skidding.

WIND DRIFT CORRECTION MANEUVERS

The instructor will usually introduce you to the idea of wind drift correction by choosing a road or railroad that has a crosswind component and having you fly directly over it and then alongside it. You'll reverse course and fly in the opposite direction so that you can get practice in correcting for both a left and right crosswind. But first, let's take a look at the principle of wind drift correction.

You've seen lightplanes flying on windy days and noticed that they sometimes appeared to be flying sideways. In relation to the ground they were, but as far as the air was concerned the planes were flying straight as a string.

Suppose you wanted to cross a stream in a boat, paddling or motoring briskly from X to Y. The current is swift, so if you start out, as in Figure 10-2A, you can see what happens—there's much sweat but little progress toward your destination.

By pointing the nose directly across the stream at Y, you wind up at Z. In Figure 10-2B you play it smart. You point the boat upstream, how far up depends on the speed of the current, and you keep experimenting until you get the correct angle.

The same idea applies to the airplane. The wind is the current in this case. If you want to fly from X to Y with the wind as shown, you must angle into the wind, or set up a "crab."

If you start out pointing the nose at Y, you end up as shown in Figure 10-3A. So you correct for the wind by making a balanced turn and rolling out at the correction angle as shown in Figure 10-3B.

■ The Rectangular Course

The purpose of the rectangular course is to give you a chance to fly the airplane while your attention is directed outside. As you learned the Four Fundamentals, you concentrated on the airplane and only noted its reference to the ground as a whole. Now you'll begin to use the plane and make it follow a definite path over the ground. During the first few minutes you'll feel as frustrated as a one-legged man in a kicking contest.

This maneuver consists of flying a rectangular

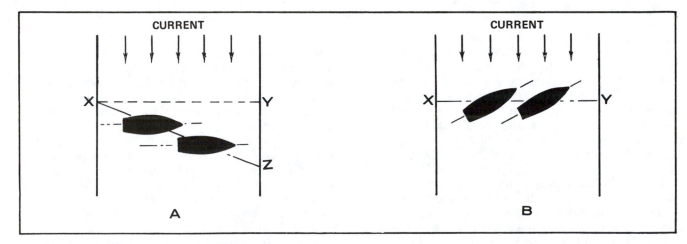

Fig. 10-2. Boats and river currents.

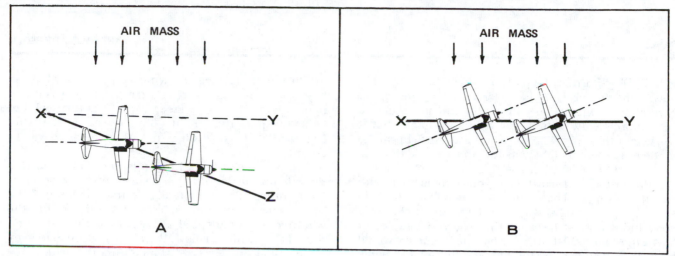

Fig. 10-3.   The principle applied to the airplane and air mass.

pattern around a large field or fields and is done for three reasons:

1. To get into the habit of flying the plane while dividing your attention between the cockpit and objects outside.

2. To learn to correct for wind drift in flying a straight course in preparation for cross-country flying

3. To get practice in low precision flying in prepa-

ration for traffic pattern flying.

The rectangular course is done at 600 feet because mistakes are easily seen at that altitude and it builds confidence to fly at a lower altitude.

The instructor will try to pick a field lying so that the wind is blowing diagonally across it. This will give you practice in wind correction on all four legs (Fig. 10-4).

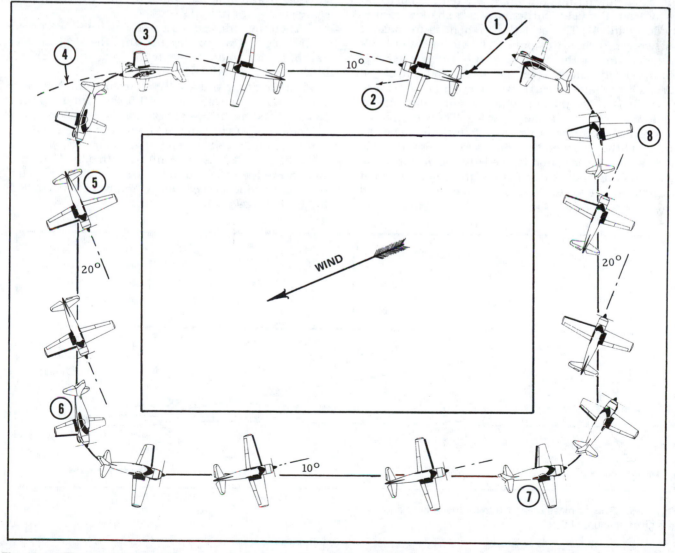

Fig. 10-4.   The rectangular course.

PROCEDURE. Enter the pattern at a 45° angle, flying downwind (point 1). It is easier to set up your first correction downwind because drift can generally be seen more easily. This is just a human foible, but it works out that way. The turns around the field can be made either left or right. Assume in this case that the turns are to the left (Fig. 10-4).

Roll out of the turn at what you think is the proper correction angle. The plane's distance from the field should be far enough away that the boundary is easily seen and yet not so far away that your mistakes can't be seen. About 600–800 feet from the boundary would be a good distance for the average lightplane. Some students roll out parallel to the boundary and then make the correction, but it's advisable for you to get into the habit of making a correction as soon as possible.

Assume that at point 2 you undercorrected and the plane is drifting into the field. Make a balanced turn of a few degrees into the wind and see how things fare. If you are now holding your own and are not too close to the field to see the boundary without straining, fly the plane to the end of the field.

At point 3 you'll have to make a left turn. Since there is a quartering tailwind, it will tend to push you away from the field, particularly if a shallow turn is made (point 4). The best thing, then, is to make a fairly steep turn at this point. The steepness of bank will depend on the wind velocity—the greater the wind velocity, the steeper the bank. Roll out of the turn with whatever correction you think necessary (point 5). Don't count on having the same angle of correction as you had on the other leg. The wind probably won't be crossing this leg at the same angle. As shown in the diagram, the wind is more from the side here, so more correction is needed. The vector of the wind pushing you into the field on the first leg was not as great as the vector you're fighting on this second leg.

In the second leg the wind is still somewhat behind the plane, so when you get to point 6 you can figure on another fairly steep turn. This turn will not be as steep as the one at point 3 because the wind component on your tail is not as great as it was at point 3.

Your angle of bank (the steepness of the turn) is directly proportional to your speed over the ground.

Continue the rectangular course. At point 7 the bank must be shallow, since you are flying into the wind (your groundspeed is low). At point 8 the first part of the turn is shallow, then it is steepened to keep the plane at the correct distance from the field.

PROBABLE ERRORS

1. Poor wind drift correction—not setting up correction soon enough.

2. Not recognizing drift or not making a firm correction after recognizing it.

3. Not maintaining altitude. Most students tend to climb in a rectangular course—there are a few rugged individualists who lose altitude.

4. Coordination problems. Some students who make perfect turns at higher altitudes get so engrossed in watching the field that their turns are awesome spectacles indeed.

When the airspeed indicator was being discussed in Chapter 3, it was mentioned that the airplane was part of the air when it was flying. Let's go into that a little deeper.

Suppose that you are flying from point A toward point B, as shown in Figure 10-5. The wind is 20 K. You are to point the nose at B and make no correction for the wind. At the same time you leave A, a balloon is released there. At the end of an hour both you and the balloon are 20 nautical miles south of the A-B line. You covered a lot of ground but still ended up as far south of the line as the balloon. You were both carried by the air mass itself.

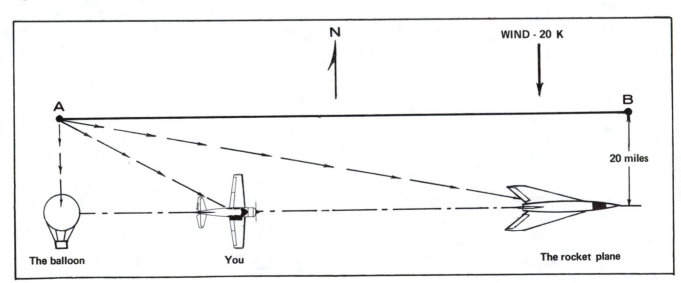

Fig. 10-5. The air mass will move the three aircraft equally in a given amount of time.

A jet or rocket plane, flying the same heading, would be 20 NM south of the A-B line at the end of an hour, even if it cruised at 1500 K.

Old-time pilots used to say, "I was flying along and all of a sudden the wind got under the wing and almost flipped the plane over." They hit turbulence, or disturbances within the air mass. The air mass itself didn't suddenly "blow" against just a part of the plane.

Another idea that died a hard death is that holding rudder corrects for wind drift. Stories are still told of pilots flying the mail in the 1930s and landing with one leg numb from holding rudder for "that blank crosswind." If you hold rudder, you defeat your purpose. When you skid a plane, it tends to continue in a straight line, and because of the added drag its airspeed will drop off. You'll end up in cross-controlled flight.

Skidding the plane defeats your purpose. The must comfortable and efficient means of compensating for drift is to maintain balanced flight (no slip or skid). You will be shown later that there is one time, and one time only, when this does not apply (see Chapter 13).

■ S-Turns Across the Road

S-turns across a road, like the rectangular course, are good maneuvers for getting you used to dividing your attention between the airplane and the ground. The primary purpose, however, is to show you how to correct for wind in a turn. There was a brief introduction to this idea in the rectangular course, but S-turns give you a chance to acquire finesse.

S-turns are a series of 180° turns of about a quarter-mile radius using a road as nearly perpendicular to the wind as possible. (This is for planes of 90–95 K cruise. Faster planes will use a greater radius.)

The reasons for S-turns are:

1. To fly the airplane while dividing your attention between the cockpit and the ground.

2. To learn to correct for wind drift by varying the bank. This will come in handy later in circling and remaining near the same spot in a strong wind.

3. To get practice in precision flying in preparation for advanced maneuvers later.

S-turns are also done at 600 feet for the same reasons as given for the rectangular course.

The object is to fly a series of semicircles of the same size, making smooth balanced turns and correcting for wind drift by varying the steepness of the bank. The plane should cross the road in a level attitude and with the wings parallel to the road.

PROCEDURE. A road that runs as nearly 90° to the wind as possible is picked. It is best to enter the S-turns downwind, because drift is more easily detected and corrected by the student if correction requires a steepening of the bank rather than shallowing it out. The first turn can be made in either direction.

As shown in Figure 10-6, the initial bank must be steep; otherwise, the wind would "push" the plane too far from the road before the turn is completed (point 1).

Assuming that you have the correct steepness of bank set in, when point 2 is reached the bank must be shallowed or the plane would be turning at the same rate of turn in degrees per minute but, as it began to head into the wind, the groundspeed would drop. It would appear to "pivot" and would not follow the smooth curve of the semicircle but would end up at point 3.

The shallowing of the turn should be such that the

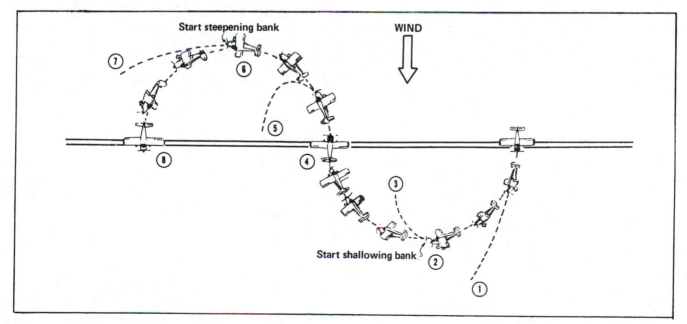

Fig. 10-6. S-turns across the road.

wings are level as the plane crosses the road, still at 600 feet (point 4).

After you cross the road, the bank should be a shallow one in the opposite direction. If a steep bank is used, a path such as at point 5 would result.

When point 6 is reached, the bank must be steepened in order to have the turn completed as the plane crosses the road again. If the bank was not steepened at point 6, a path like that of point 7 would occur. The plane would not cross the road at the correct place nor would the wings be level or parallel to the road.

Continue the series until you get tired or run out of road. Remember: *The angle of bank is directly proportional to the groundspeed.* (The greater the groundspeed, the greater the angle of bank.

The angle of bank in S-turns must be constantly changing. The hardest part of trying to keep the maneuver smooth is at point 8, where you have to roll from a steep bank one way to a steep bank in the opposite direction.

If the wind is strong, many students do not have a shallow enough bank on the upwind side of the road. Strangely enough they are able to correct on the downwind side without much trouble. Many times in this situation the plane's path over the ground looks like Figure 10-7.

PROBABLE ERRORS

1. Failure to properly correct for drift.

2. Rolling out of the turn too soon or too late, resulting in crossing the road with wings not parallel to it.

3. Gaining or losing altitude.

4. Coordination problems in the turns (jerky or slipping and skidding).

In the S-turn you are interested in a definite path over the ground and therefore must correct for the movement of the air mass of which you are a part.

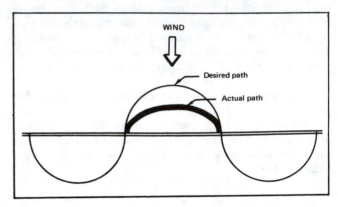

Fig. 10-7. A common error in S-turns across the road.

11

ELEMENTARY EMERGENCY

LANDINGS

At some time during the period of rectangular courses or S-turns, your instructor will demonstrate an elementary emergency landing or "simulated emergency," as you may hear it called.

Engines nowadays are very reliable. Cases of engine failure are rare, but you will still be given emergency-landing practice because pilots still run out of fuel and leave oil caps off. It is still possible, too, that the engine could fail structurally, so it's better to be prepared.

The elementary emergency is given below 1000 feet and requires a 90° gliding turn at most. The instructor will pull the carburetor heat and close the throttle at some point during the maneuvers and demonstrate the procedure.

If it is at all possible, you want to land into the wind. If the plane stalls at 50 K and you have a 15 K headwind, your groundspeed (or relative velocity to the ground) at touchdown will be 35 K. The advantages of this are obvious. However, sometimes a field may not be available for an upwind landing. Or if the field is steeply sloped, it's better to land uphill even if it's downwind. Obstacles may require that you land downwind.

Sometimes students who've been doing the rectangular course or S-turns for several minutes and have been correcting for a stiff wind forget the wind direction as soon as the throttle is closed for the simulated emergency.

Picking the right kind of field (at low altitude you don't have much time to shop around) comes with experience. If you've lived on a farm you can spot a good field without much trouble.

Table 11.1 is a list and description of various types of fields with comments about each type.

Obviously there will be times when some of the landing spots labeled "poor" and "very poor" will be the only ones available. It then becomes the case of making the best of a bad situation. As one instructor put it, "Hit the softest, cheapest thing in the area as slowly as possible"—which pretty well covers it.

WHAT TO DO

In the case of an actual forced landing below 1000 feet:

1. Establish the max distance glide airspeed.

2. Pull on the carburetor heat. You can do this simultaneously as you ease the nose up to conserve altitude and establish the proper glide airspeed. The carburetor may have iced up and will clear out when heat is applied. Normally you will have some warning

TABLE 11.1   Types of fields for emergency landings

Type of Field	Description	Comments
Good—pasture	Brown-green; may have livestock or cow paths across it	You can land in a pasture, but land in the part well away from the stock. Cows like to eat airplane fabric so don't be gone too long to telephone (unless you have a metal plane).
Good—freshly cut wheat, barley, oat fields, etc.	Brown grass but mower marks can usually be seen	Very good for emergency landing, if any place could be considered good for an emergency landing.
Fair—field of high grass, wheat, etc.	Rich green or golden	Not much landing roll, but field may be soft under camouflage of grass. Possibly plane will nose over.
Poor—high corn	Green, or later brown	Field itself usually rough. Plane will probably be damaged.
Poor—freshly plowed field	Dark brown	If the field is soft the plane will nose over. If you must land in a plowed field, land parallel to the rows.
Poor—terraced field	Contour lines	Countour lines show that it is sloping. Never land across the contours.
Very poor—swamps, woods, etc.		Land as slowly as possible, but don't stall it in.

beforehand, such as rpm dropping, but you may not have noticed. The carburetor heat should be pulled ON immediately because if the engine is dead it will be cooling and any residual heat in the system will be dissipating rapidly. If you wait too long, there may not be enough heat left to clear out any ice. (If that was the problem.)

3. Pick your field and start the approach.

4. Switch to another fuel tank (if your airplane has more than one).

5. Electric boost pump ON (if equipped).

6. Mixture RICH.

Do not waste time trying other methods of starting. Do items 4–6 after you have the field selected and the approach set up. Then get back to your approach. Don't try to use the starter.

The propeller usually windmills after an engine failure and will start again if the fault is remedied. If it does, climb to an altitude of 2000 or 3000 feet while circling the field. If you are *sure* of the reason for the failure (tank run dry, carburetor icing, etc.) and have remedied it, continue your flight. Otherwise keep your high altitude and pick a route of good terrain back to the airport.

Once you have selected a field, you have very little time to change your mind. Pilots have been killed trying to change fields in an emergency because they made turns too steep and stalled at low altitudes.

A normal glide is a must for an emergency landing. If you are high and dive at the field, the airspeed goes so high that the plane "floats" and you still overshoot, or go past the field.

If you try to "stretch a glide," the rate of descent actually increases as the angle of attack gets in the critical range. It's better to hit something when you have control than to stall and drop in (Fig. 11-1).

If you are going to land in trees, don't stall the plane above and drop in. Get as slow as possible, still having control, and fly it into the trees.

It's impossible to list exact procedures for every situation, but you'll try to keep the airplane under control without stalling and will try to pick a landing point that will allow the lowest deceleration (something soft like shrubs or weeds, rather than a stone wall or solid tree trunk). As for when to turn off all switches before impact, your *Pilot's Operating Handbook* will have such information. Your primary goal is to keep the cockpit intact.

Sometime after you have had the simulated emergency demonstrated, the instructor will close the throttle and say, "Emergency landing." You will then go through the procedure. The instructor will let you glide to a reasonable low altitude or an altitude where you could see whether the field could be made, then will say, "I got it," open the throttle, take the plane up, and point out any mistakes you may have made.

Review the *Pilot's Operating Handbook* for information on these procedures.

■ Probable Errors

1. Failure to establish a normal glide promptly.

2. Indecisiveness in selecting a field.

3. Trying to stretch the glide if low and diving if high.

4. Making steep turns close to the ground.

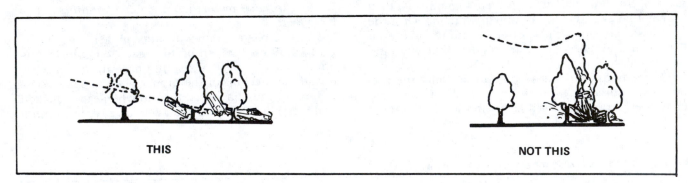

THIS   NOT THIS

Fig. 11-1. It's better to have control at impact than to stall or spin the airplane in.

12

STALLS AND
SLOW FLIGHT

A stall is a condition in which the angle of attack becomes so great that the flow over the airfoil breaks down and the wings can no longer support the airplane. An airplane can be stalled at any speed, attitude, or power setting—as many a dive bomber pilot found out when he tried to pull out of a dive too quickly. He may have been doing 400 K but the plane was stalled as completely as if he had climbed too steeply. What happens is shown in Figure 12-1.

With the knowledge that a stall is caused by too great an angle of attack, the proper recovery is always to decrease that angle of attack and get the air flowing smoothly again whether you are doing 40, 400, or 4000 K.

You won't be doing accelerated stalls like that just discussed until later, but if you know what a stall is the accelerated stall will be no different from the first elementary ones. The stalls in this phase will be done by gradually pulling the nose up and slowing the plane to the stalled condition.

In the earlier days of aviation, keeping the wings level during stalls was a problem. In the older planes, control effectiveness in the stall was lost in alphabetical order—AER. The Ailerons were the first to go, followed by rapidly decreasing Elevator effectiveness, and last of all by the Rudder. Generally, if power was used during the stall the Rudder was effective throughout.

In many cases, use of ailerons alone to raise a wing during a stall resulted in an opposite effect to that desired. When the aileron was applied, the down aileron on the low wing caused adverse yaw effects, resulting in a further slowing of that wing and causing it to stall first. This was disconcerting, to say the least, and led to some interesting wing-leveling techniques. Pilots found that it was possible, for instance, to raise the low wing during the stall by applying opposite rudder—which yawed the plane in that direction, speeded up the low wing, gave it added lift, and caused it to rise. Sometimes this opposite rudder was applied so enthusiastically that the high wing was slowed to the point that it abruptly exceeded the criti-

Fig. 12-1. Pilot releases bomb, discovers that altitude is very low; pulls back stick abruptly. Plane changes attitude but momentum carries it down. Result—a high-speed stall. Lacking altitude for a smooth pull-out, plane strikes the ground.

cal angle of attack and stalled first, thereby causing a great deal of adrenalin to flow in occupants of the plane.

Planes type-certificated under the Federal Aviation Regulations (as all U.S general aviation planes are now) must meet certain rolling (ailerons) and yawing (rudder) criteria throughout the stall. The Federal Aviation Administration, therefore, now encourages the use of coordinated controls to keep the wings level during the stall. While keeping the wings level is not critical at the altitudes where stalls will be practiced, it will be very important when you start landings, as will be shown later.

There is one very important point: Get the nose down (decrease the angle of attack) *first* before trying to level the wings. Once the angle of attack is below the stall value, the ailerons and rudder are less likely to get you into trouble in *any* airplane.

Several methods of design are used to aid lateral control throughout the stall. The idea in each case is to have the wing tips stall last. If a wing tip stalled first (and it is unlikely that both will stall at exactly the same time), a dangerous rolling tendency may occur as the stall breaks.

The tips may be made to stall last by "washout." The manufacturer builds a twist into the wing so that the tips always have a lower angle of incidence (and a resulting lower angle of attack) and will still be flying when the root area has stalled (Fig. 12-2).

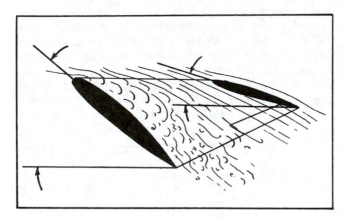

Fig. 12-2. Exaggerated view of wing tip washout. The ailerons are still effective even though the wing root is in the stalled condition.

Another method of having the root stall first is the use of slots near the tip. The opening (Fig. 12-3) near the leading edge allows the air to maintain a smooth flow at angles of attack that would result in a stall for the unslotted portion of the wing.

Still another method is that of stall strips (Fig. 12-4) or spoiler strips being placed on the leading edge of the root area of the wing, breaking the airflow and resulting in an earlier stall for this portion. Some airplane wings have higher lift-type airfoils at the tip, causing the wing tip to stall last. In some cases several of these design techniques may be combined.

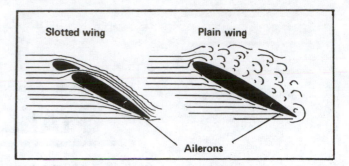

Fig. 12-3. Plain and slotted wings at equal angles of attack.

Fig. 12-4. The stall strip as a means of ensuring that the root section stalls first.

STALLS AS AN AID TO LANDINGS

As you'll see, a normal landing is nothing more than a stall. The average lightplane landing is begun at an altitude of about 15 or 20 feet from a normal glide. At this point you begin the landing transition, or start "breaking the glide." From here it becomes a matter of your judgment in trying to have the airplane completely stalled just as it touches. This means that you have the wheel or stick full back and the plane stops flying at the instant it contacts the ground.

The instructor will show you a series of straight-ahead stalls that will be good preparation for landing practice later.

In practicing stalls, it's important not to stare blindly over the nose. Get an idea of what the plane is doing by looking out of the corners of your eyes. The instructor will probably have you practice landings from a normal glide at 2000 or 3000 feet above the ground. It's impossible to see the ground by looking over the nose of the average trainer when it's in the landing attitude; so as soon as you start the transition, switch your scan to along the left side of the nose, noting the plane's attitude out of the corner of your eye and keeping the wings level through coordinated use of the controls.

Once the nose is at the landing attitude (the instructor will show you the proper nose position), it becomes a matter of pinning it at that position by continued back pressure.

APPROACH TO STALLS—POWER-ON
OR POWER-OFF

You'll practice these stalls in takeoff or climb configuration (power-on) or approach configuration (power-off), straight ahead, or in 20° banked turns. For the straight ahead stalls be sure to "clear the area" by making two 90° turns in opposite directions because the nose will be in one place and form a blind area. Recover at least 1500 feet above the ground.

Approach to stalls is so named because the airplane is not allowed to stall, but recovery is effected as soon as indications of a stall occur (buffeting or decay of control effectiveness).

■ Approach to a Stall—Power-Off

PROCEDURE (STRAIGHT AHEAD)

1. Clear the area.
2. Carburetor heat ON—throttle back to idle.
3. Apply gentle back pressure to raise the nose to about the landing touchdown attitude—don't let the nose wander.
4. Hold the nose at this point by continued back pressure—keep the nose directionally straight by looking alongside it at a reference on the horizon. The use of the elevators in approaching the stall can be likened to a vicious cycle. As the plane is slowed up by the elevators, the nose tends to drop—requiring more up-elevator, which slows the plane further, requiring more elevator, etc. The point where the wheel is all the way back and the nose drops is the stall "break." In the approach to a stall the process does not go as far as the break (Fig. 12-5).
5. Indications of the approaching stall are sensed by:
   a. Sight—check the attitude and airspeed indicator.
   b. Feel—the controls are ineffective and mushy. The airplane may shudder or vibrate as the stall approaches.
   c. Sound—the wind and engine noise level drops as the plane slows.

Lower the nose to level flight and apply full power simultaneously. (Easy with that throttle—don't ram it open.)

6. Carburetor heat OFF.

The idea of this maneuver is for you to learn to recognize the approaching stall. You could recover easily by leaving the throttle back at idle; it just takes a little more altitude, that's all. Don't start thinking that power is the big thing in the recovery. The change in the plane's attitude is the most important part. Your instructor will also have you practice these stalls in 20° banked turns in each direction.

■ Approach to a Stall—Power-On

Use cruising power on this one (Figs. 12-6 and 12-7). The technique is fundamentally the same except you'll have to raise the nose higher in order to speed up the approach to the stall. The plane wants to "hang on" longer when power is used. A thing to remember: At higher rpm you have torque effects as the plane slows and the plane is less laterally stable because of gyroscopic effects. That is, you may have more trouble keeping the wings level.

PROCEDURE

1. Clear the area.
2. Slow the airplane to approximately the climb speed or slightly below by throttling back and maintaining a constant altitude, then set the rpm to cruise.
3. Ease the nose slightly higher than for the power-off version, maybe 5° higher. (Your instructor will demonstrate the nose position.) Keep the wings level.
4. Keep the nose in line with a point on the horizon. Remember that torque will be acting on the plane.
5. As the approaching stall is noted (sight, sound, feel), lower the nose and open the throttle fully (Fig. 12-7).

Granted, you'll be practicing approaches to stalls at an altitude where the recovery will be no problem. Still, you might as well get in the habit of using full power and recovering with as little altitude loss as possible because someday you may be low and won't have time to reason, "Well, I'm pretty low so maybe I'd

A Ease nose up to landing attitude.   **B** Watch for signs of approaching stall.   **C** Lower nose and apply full power.

Fig. 12-5. Approach to a stall.

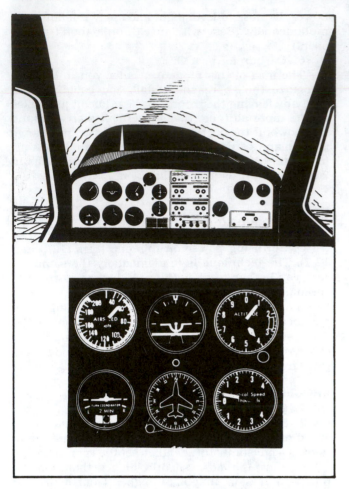

Fig. 12-6. Power-on partial (imminent) stall. The airspeed is near stall. Heading indicator, no turn. Attitude indicator, nose high, wings level. Turn coordinator, straight balanced flight. Altitude, steady or increasing slightly.

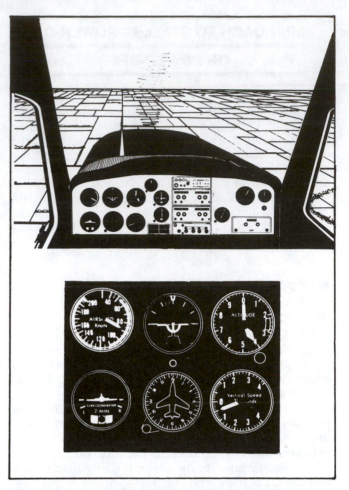

Fig. 12-7. Lower nose to level flight and open throttle at the same time. Airspeed increasing. Heading indicator, no turn. Attitude indicator, level flight. Turn coordinator, straight balanced flight. Altitude, constant or slightly decreasing.

better use full power; on the other hand, possibly I should . . . " (CRUNCH).

You'll also get practice in power-on approaches to stalls in banks of about 20°.

Open that throttle each time unless you are called upon by your instructor or the FAA check pilot to demonstrate a power-off recovery. Again, the addition of power is to reduce the altitude loss during the recovery. Getting the control wheel or stick forward is the move that recovers the airplane from the stall.

SOME STRAIGHT-AHEAD STALLS

■ Normal Stall—Power-Off

Another step in the stall sequence is the wings-level normal stall. This is the one you'll use for landing practice at altitude. The first part of the stall is quite similar to the imminent stall. In this case, however, you continue increasing back pressure until the

wheel or stick is full back or the stall breaks and the nose starts to drop. Recovery is effected by releasing back pressure and applying full power (easy with that throttle). Try to recover with a minimum altitude loss. Many students get eager to recover from the stall, release the back pressure, and immediately come back again with the wheel to minimize the altitude loss (they think). The plane stalls again (called a secondary or progressive stall) and the process of recovery must be redone. It is possible to do a series of secondary stalls (even the fifth stall is called a secondary stall) and, if the ground is close, you'll feel as nervous as a cat in a roomful of rocking chairs until the plane is flying normally again.

Recover firmly but don't get rough with the airplane. Minimize your altitude loss but don't get a secondary stall.

During your first series of stalls, the plane seems to have a mind of its own. The wings don't want to stay level. Maybe you push forward on the wheel a little harder than you mean to and your stomach comes up in your throat. Don't worry about it; there were several hundred thousand other people who felt the same way before you came along.

After several stalls you'll get the idea. (Do these stalls both clean and with landing-flap setting.)

PROCEDURE

1. Clear the area.
2. Carburetor heat ON—close the throttle to idle.
3. Ease the nose up to the landing attitude.
4. Pin it at that position with continued application of back pressure. If you are preparing for landings, practice looking out the left side, keeping the wings level and the plane at the correct nose-up attitude by checking it from the corners of your eyes.
5. When the stall breaks, recover by lowering the nose, applying full power as you do so, minimizing altitude loss. (Flaps up in increments, if used.)
6. Carburetor heat OFF.

More about this idea of keeping the wings level: If a wing drops, the plane is banked and will turn. This doesn't matter at high altitudes, but runways are only a couple of hundred feet wide at best (some of them will be 50 feet) and the plane will be off into the boondocks before you know it. Here's a good place to get the wings-level idea down pat. Then you won't have any trouble when landing practice starts.

If the wing drops slightly, use coordinated controls. Exact stall recovery techniques will vary from airplane to airplane, but the general principle is the same for all airplanes and airspeeds. *Get the air flowing smoothly over the wings with the minimum loss in altitude.*

If you are flying an older airplane, the ailerons may not be effective throughout the stall and use of rudder may be needed to stop rolling tendencies. Your instructor will demonstrate the best technique for maintaining lateral control during the stall for your particular airplane. However, the chances are that coordinated aileron and rudder usage will be the most effective means.

■ Normal Stall—Power-On

Fundamentally, this is the same as the stall just discussed. The nose position will be higher because of the power. As seen in Figure 12-8, when the plane uses power, its path is changed and the attitude will be more nose high, but the angle of attack at the stall will be roughly the same for a given flap setting period.

A plane will stall at a slightly slower airspeed with power on than it does with power off. The effect of the slipstream over the center section of the wing makes a slight difference. The more nose-high attitude also gives a vertical component to the thrust, which in effect lowers the wing loading. (Wing loading, you remember from Chapter 2, is the airplane weight or load being supported by each square foot of wing area.)

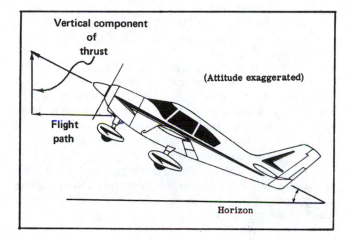

Fig. 12-8. The attitude is more nose-high than for the power-off version.

PROCEDURE

1. Clear the area. (You're probably getting tired of hearing this by now, but do it anyway.)
2. Ease the nose up to a slightly higher position than was used for the power-off normal stall; your instructor will demonstrate.
3. Keep the wings level.
4. When the stall break occurs, lower the nose and open the throttle the rest of the way. Then level the wings with coordinated aileron and rudder if you fell asleep in item 3 (Fig. 12-9).

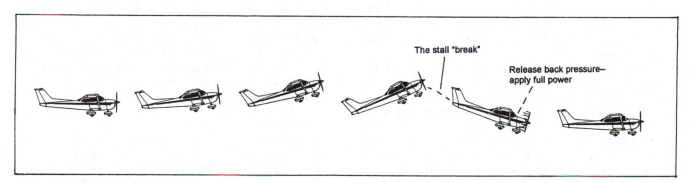

Fig. 12-9. Power-on normal stall.

■ Complete or Full Stall—Power-On and Power-Off

The stall has little application other than as an exercise in keeping the wings level during the stall, but if you're having that problem, a few of these will help you see the light.

In earlier days this was sometimes referred to as a "rudder exercise stall" because the wings were kept level or bank corrections were made by using the opposite rudder to raise a wing, as mentioned at the beginning of the chapter.

PROCEDURE—POWER-OFF

1. Clear the area.
2. Carburetor heat ON—throttle at idle.
3. Ease the nose up to an attitude of 30° above the horizon. Keep the wheel back until the nose falls to the horizon. Keep the wings level during the recovery.
4. As the nose crosses the horizon, effect a normal recovery. Release the back pressure and apply power. You will notice that the nose will be somewhat lower during the recovery from the complete stall compared to the normal stalls.
5. Carburetor heat OFF.

The power-on complete stall is fundamentally the same except for the increased difficulty in keeping the nose straight and the wings level because of torque and gyroscopic effects.

■ Characteristic Stall

This stall is a good demonstration of the built-in stability of the airplane. It disproves the nonflyer's idea that the airplane is just waiting for a chance to "fall."

This stall can be done either with power on or off, as your instructor may want to demonstrate for you. For discussion purposes assume that you are going to make a power-off stall. The idea is to start a normal power-off stall (trimmed for cruising flight), getting the break, but instead of recovering as usual take your hands and feet off the controls. The nose will drop and the plane will make its own recovery. (You can go back up and get your stomach later.) In most lightplanes you will have to ease the plane out of the dive in the power-off characteristic stall. It would come out of the dive by itself, but the airspeed might be in the region of the red line, or "never exceed" speed, at the bottom of the first oscillation.

If the plane had cruising power throughout the maneuver, the initial dive would not be nearly as steep as the power-off one and the recovery would be back to level cruising flight. Generally, it is not necessary to pull the plane out of the dive when you are using cruising power. The oscillations that result are called phugoid oscillations, which sounds very technical. Sometime during the course of an ordinary conversation, work the talk around so you can mention "phugoid oscillations"—it's quite impressive.

Put bluntly, the idea of this maneuver is to show that most of the trouble that occurs during stalls is caused by some heavy-footed, ham-handed pilot groping around in the cockpit. The characteristic stall is not required on the private practical (flight) test but is a good maneuver for building up your confidence in the airplane.

PROCEDURE

1. Clear the area.
2. Start a normal power-on or power-off stall.
3. When the stall breaks, take your hands and feet off the controls.
4. Ease the plane out of the dive and return to level flight.

COMMON ERRORS

1. Not getting the stall break or having the wheel or stick back to the stop before releasing the controls.
2. Letting the airplane pick up excessive speed in the dive.

ELEMENTARY TURNING STALLS

■ Takeoff and Departure Stalls—Climbing Turns

Many a pilot has taken off in front of friends on a bright Sunday afternoon and decided to give them a little extra show—a takeoff, followed by a steep climbing turn. The result—a little more show than bargained for, with a stall at the top of one of these colorful maneuvers.

Or suppose you are landing at a very short airport with trees at the far end. You are just about to touch down when you realize that you're almost out of runway. The trees are close. You open the throttle and try to climb and turn out of the way. What happens then? If you've practiced a few departure stalls, it will be evident just what can happen and you will have learned a method of recovery. Better to have used headwork in the first place and not gotten into such a predicament.

The takeoff and departure stall is designed to give you practice in recognizing and coping with such a situation. You will do these stalls both from straight flight and moderately banked (20°) turns with recommended takeoff power. In both types the climb will be initiated at lift-off speed and the angle of attack slowly increased until the stall occurs. Power-on, wings-level stalls have already been covered, so this section will only discuss the climbing-turn stall.

The climbing-turn stall is done by throttling back and slowing the plane as if on a landing approach. At about 5 K above the stall open the throttle to recommended takeoff power and ease the plane up into a climbing turn of approximately a 20° bank in either direction. Continue to increase the angle of attack until the stall occurs. For the majority of airplanes the higher wing will normally tend to stall first, particularly in a right-climbing turn because of torque effects, and the plane will roll in that direction. Here's why the top wing tends to stall first: As you pull up into the stalled condition, the wings start losing lift and the plane mushes and starts to slip. In the right-climbing turn, torque is working to yaw the nose against the turn and aggravates the slipping condition.

As shown by Figure 12-10, the higher wing receives turbulence when a slip occurs, while the lower wing is comparatively free; consequently the high wing will stall first. When the break occurs, the airplane rolls in the direction of the high wing. In such a case the ball of the turn and slip (needle and ball) will indicate a slip just as the break occurs. The rigging of the wings of your particular airplane can affect its reaction. Also, if the plane is skidding at the stall break, the low or inside wing will be the first to go. If your plane has a needle and ball, you can note that the roll at the stall break is "away from the ball." (If the ball is to the right, the roll will be to the left; if the ball can be kept centered, any tendency to roll is minimized, but it still might roll in either direction, so be ready for it. The only problem is that with most airplanes the rapid decay of rudder effectiveness as the stall is approached may not allow the pilot to keep the ball centered.)

When you are practicing these turning stalls, dual or solo, you might notice that the nose will usually drop at the stall break, an instant before the roll occurs. This follows the analysis shown by Figure 12-10.

Relax back pressure to recover. You already have a fair amount of power on throughout the maneuver, but you should increase it to maximum. But relaxing the back pressure is the main stall-recovery aid here, as always. In a few airplanes with good rudder effectiveness, in keeping the ball centered, the *right*-climbing turn can result in the low (right) wing dropping! This is, of course, the reason you practice these stalls—to see what *your* airplane will do.

With more practice you'll be able to recognize the start of the roll and the break and recover in level flight by catching the stall at the right point.

The angle of bank must be moderate in order to get a good break. If you bank the plane too steeply, the nose may drop, tending to make the stall harder to attain because of the increase in airspeed. In many cases the nose will be low enough that the plane will merely mush, and you'll be turning in a tight circle, losing altitude with the plane shuddering but not getting a definite stall break. This too-steep bank will give you more trouble during solo practice of these stalls. It's a matter of stall speed being a function of wing loading. Without the instructor's weight, the wing loading is lower and the plane is harder to stall from an airspeed standpoint.

In doing these stalls, note the altitude at the point of assumed takeoff and compare it with the altitude on recovery.

PROCEDURE

1. Since you will be turning, it is not necessary to make special clearing turns. Keep looking around during the approach to the stall.

2. Pull the carburetor heat (if required for your airplane), reduce the power, and slow down to lift-off speed.

3. When the airspeed drops to 5–10 K above the stall, apply recommended climb power (carburetor heat OFF) and attempt a comparatively steep climbing turn (the climb is comparatively steep; the bank is shallow) in either direction.

4. When a wing or the nose starts to drop and the break is definite, relax back pressure and apply *full* power.

5. Level the wings with coordinated controls after the recovery is started. (Don't just go by the airspeed; you should feel and hear this pickup in speed as well.)

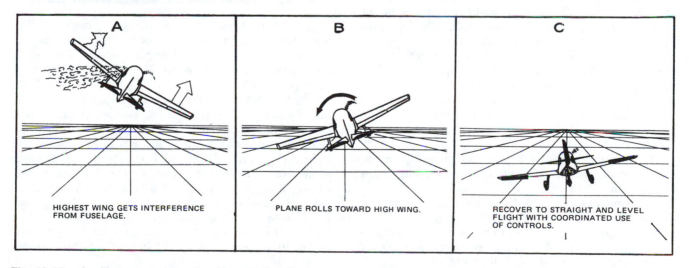

Fig. 12-10.   A rolling moment may be set up as the plane stalls.

COMMON ERRORS

1. Too steep a bank—the plane doesn't stall but mushes instead.

2. Too early a recovery—the plane does not get a chance to break before the recovery is started.

3. Too late a recovery—the plane may rotate too far before you get a recovery started.

4. Nose too low—the climb is not steep enough; the plane mushes.

■ Approach to Landing Stalls—Gliding Turn

This is the power-off version of the takeoff and departure stalls, and like them will be done from both straight glides and moderately banked (approximately 30°) gliding turns in landing configuration (full flaps and trim set). Power-off, wings-level stalls were covered earlier, so only the gliding-turn stall will be covered here (Fig. 12-11).

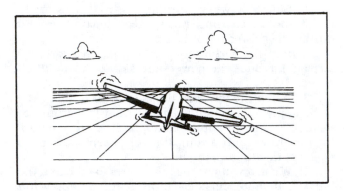

Fig. 12-11. The approach to landing stall (gliding turn).

You will notice that the rolling tendency is not nearly so great with the power off. The plane tends to mush and shudder and it is harder to get a sharp stall break, particularly if you are solo. It may require a faster rate of applying back pressure to get the break as compared with the departure stall. When you fly heavier planes, you will find that in most cases you don't have to work to make them stall. The faster planes will stall and generally give comparatively little warning. You are more interested in the technique of recovery than in the effort required to get into the stall. So don't get complacent about stalls.

PROCEDURE

1. Carburetor heat ON. Set the plane up in the landing configuration. Establish a normal gliding turn in either direction.

2. Continuously add back pressure, raise the nose to about the landing position or slightly higher.

3. When the break occurs (or you have full up-elevator), stop any rotation and recover by releasing back pressure and applying full power.

4. Carburetor heat OFF.

5. Clean up the airplane (gear and flaps, as appli-

cable) and climb to an altitude at least 300 feet above the altitude at which full control effectiveness was regained.

PROBABLE ERRORS

1. Too steeply banked—plane mushes in a turn rather than stalling.

2. Nose not high enough—again the plane mushes with no definite stall break.

3. Improper throttle handling—too fast or too timid application of power.

FLAPS AND STALLS

Figure 12-12 shows the coefficient of lift versus angle of attack (also see Fig. 2-8 and the discussion there) for a wing—clean and with landing flaps extended (20° in this case). Note that the maximum coefficient of lift is noticeably higher with flaps extended. Using, for example, the 0° angle of attack as a common starting point, you'd see that by slowly increasing the angle of attack the flapped condition would reach its critical angle of attack (stall) well before the clean wing would have problems.

The point is this: Pilots tend to think that stalls can only occur with the nose well above the horizon. You can see that some airplanes with full flaps, approaching landing in a turn, might stall with the nose well below the horizon. This is because the airplane's "normal" approach attitude is much lower, and the stall attitude will also be equivalently lower compared to the nonflapped condition (same power setting).

FLIGHT AT MINIMUM CONTROLLABLE AIRSPEED (SLOW FLIGHT)

This is designed to show you how to recognize the stall and to help you get to know your airplane. Someday you may have to fly the plane close to the stall, and knowing how the plane flies in that region may save your neck.

Such flight (sometimes called "slow flight") is usually considered to be level, sustained flight at a speed close enough to the stall (about 5 K above) so that if power was chopped, you would get immediate indications of a stall. You will be expected to demonstrate minimum-speed flying to the FAA examiner during the private practical (flight) test. You will be asked to make minimum-speed climbing, gliding, and level turns in various configurations and banks. The flight test tolerances in straight and level slow flight will be to maintain within 10° of heading and 100 feet of altitude.

Don't get the idea that the speed for slow flight (or minimum controllable airspeed) is always the same

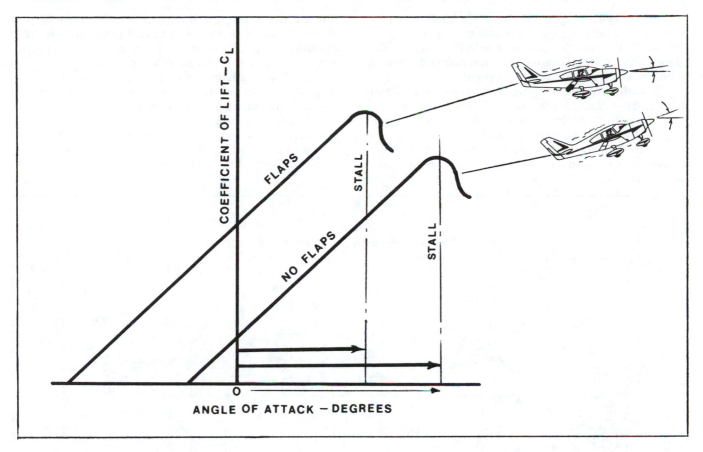

Fig. 12-12. Coefficient of lift and stall break attitudes for a fictitious airplane—clean and with landing flaps down.

for your airplane. The *angle of attack* at the stall *is* always the same (for a given flap setting), but the airspeed at the stall varies with the *weight* of the airplane. The less a particular airplane weighs, the less its airspeed when that fixed maximum angle of attack is reached. The stall airspeed is also lowered by the lowering of *flaps*. Take a look back at Figure 9-5. The stall speeds given there are at the maximum certificated weight of 1600 pounds and are power-off. Note that the stall speeds decrease with an increase in flaps for any condition of bank. For instance, looking at the effects of flaps at 0° bank (wings level), the stall speed decreases from 55 mph to 49 mph at 20° of flaps, and then down to 48 mph at the 40° flap setting. Note that the second 20° only lowered the speed by another 1 mph. Added drag is a greater factor than added coefficient of lift in this area (20°–40°) of flap setting.

Okay, take an airplane that stalls at 55 mph (no flaps) at 1600 pounds with the *power at idle*. If you wanted to fly slow flight in this condition (idle), you would probably set up a speed about 5 mph faster, *or 60 mph calibrated* airspeed.

Now use 20° of flaps. At this same airplane weight you would add 5 mph to the stall speed (49 mph) and set up a descent at about 54 mph. For 40°, 53 mph (calibrated airspeed) would be a good number.

At a *lighter weight* of 1300 pounds you might use

55, 49, or 48 mph respectively in slow flight, to have a 5 mph margin.

A little earlier in the chapter it was noted that the use of power decreased the stall speed (Fig. 12-8) so that the airspeed, as an example, could be lowered another 3 mph for each condition. Now the new speeds would be *52, 46,* and *45* mph CAS respectively for slow flight in the three flap settings, at a lighter weight, *and using power.*

While numbers have been cited here to make the point that *flaps, lower airplane weights,* and *more power* will all *lower* the airspeed given for stall (and hence the required slow-flight speed if the same margin is kept), the idea is to *fly* the airplane here by your feel and senses. Figure 9-5 and the discussion about it in this chapter use miles per hour as examples. The principle remains the same whether expressed in miles per hour or knots.

■ **Procedure**

1. Throttle back to a power setting much less than required to maintain level slow flight. After the first time you'll have an approximate idea of what's needed. Maintain altitude as the plane slows. This means that the nose must be slowly raised. As the

required speed is approached, start adding power as necessary to keep the altitude constant. You can see in Figure 12-13 that the power required to fly the airplane *increases* again after the airspeed is decreased from cruise down past a certain point.

2. Maintain your heading. You have power on and a low airspeed, and torque must be taken care of.

3. Notice that elevators and throttle are coordinated in maintaining airspeed and altitude. If you are losing altitude, add power and adjust the nose position to maintain the proper airspeed. If you are too slow

but are maintaining altitude, ease the nose slightly lower and adjust your power to assure staying on that altitude. Keep checking the altimeter and airspeed. *You'll also be watching for other planes,*

4. Make a shallow turn in each direction. Maintain altitude in the turn. This will mean increased power and a slight raising of the nose.

5. Level the wings and gradually throttle back to idle as you lower the nose to maintain a glide at the minimum control speed of 5 K above the stall. Make 20°–30° banked turns in each direction.

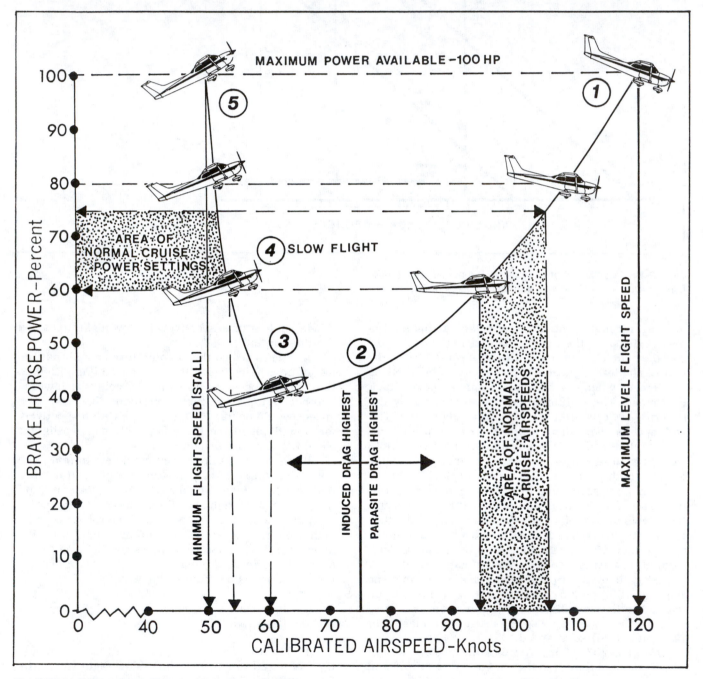

Fig. 12-13. The power required to maintain a constant altitude (sea level) for a fictitious, very efficient light trainer at a particular weight (attitudes exaggerated).

6. Return to level, slow flight by applying power and easing the nose up. Try to keep the airspeed from varying during this transition. You'll note in slow flight, as in other phases of your flying, that an increase in power tends to make the nose rise, which will make it seem as if all you have to do as power is applied is to *think* about the nose easing up.

7. Increase the power and ease the nose up to a climb at minimum speed. Make shallow turns in each direction. (Don't climb too long, since the engine will overheat at the low airspeed.)

Making the transition from level flight to glide to climb, etc., without varying the airspeed more than 5 knots takes some heavy concentration, but when you can do it you'll have an excellent feel for the aircraft.

The instructor will probably not do any more before solo than have you fly the plane at the minimum speed to get the feel of slow flight. You will not be required to maintain as close tolerances of altitude as later in your training. However, if you seem to get the hang of slow flight, you may go on with the transition to climbs and glides. You will probably review slow flight before going on the cross-country solo.

Remember that you should also be able to do slow flight by reference to the instruments. In this condition you don't rely on your senses, but on the instruments. The instructor will give you a chance to practice this both hooded and with visual references during periods of dual.

You should get slow-flight practice in as many combinations of flaps and power as possible. If your trainer has retractable gear (it's doubtful), you should fly at minimum controllable speeds in cruise and landing configurations. You might repeat steps 1 through 7 in various flap configurations. (Don't try to climb very long at the low speed with the flaps extended.)

Your instructor may cover the airspeed indicator during the visual slow-flight practice so that you can fly the airplane by pitch attitude and power and may also cover the tachometer so that you will use power as needed to maintain altitude rather than staring at the rpm numbers.

■ **Common Errors**

1. Losing or gaining altitude in the transition from cruise to slow flight.
2. Poor speed control in the transitions.
3. Stalling the airplane.
4. Poor altitude control during the level portions of slow flight.

(Slow flight is also called "maneuvering at critically slow airspeeds.")

■ **Power versus Airspeed**

You might be interested to know that in slow flight you will be operating in a condition that pilots call the "back side of the power curve." Figure 12-13 shows the attitudes and power required to maintain a *constant altitude* at various airspeeds for a typical light trainer at one particular weight at sea level. Other weights and/or altitudes would have separate curves. To simplify matters, Figure 12-13 is set for brake horsepower and indicated airspeed because these will be the two factors you'll be working with as a pilot. The engine of this plane is able to develop 100 brake horsepower at the altitude used here (sea level).

All the vehicles you've been operating before (bicycles, motorcycles, automobiles) always require *more power to go faster* at any particular road condition or grade; in your flying of the airplane in straight and *level cruising flight* this has also seemed to be the case. The kicker is that the added power in the normal cruise area doesn't directly make the airplane go faster, but *allows it to go faster without losing altitude.* This is important to remember. The airplane in the example *can* go faster than the maximum speed shown but only at a cost of altitude. Road vehicles are supported by the ground and don't have to rely on producing lift, hence their horsepower is used basically only to overcome parasite drag.

The power curve gets its characteristic U shape because the airplane is subjected to *two* types of drag. Looking back to Chapter 2, you remember that *total drag* is composed of *parasite drag*, which increases with airspeed, and *induced drag*, which becomes much stronger as the airplane slows up. You've been doing all of your straight and level flying in that area of the power curve (normal cruise) where parasite drag is the big factor. Because you're more used to talking in terms of horsepower, Figure 12-13 shows the horsepower required to overcome a combination of the two types of drag throughout the speed range of the airplane.

Note in Figure 12-13 that the area of normal cruise power settings for the airplane in the example runs from 60 horsepower (60 percent power) to 75 horsepower (75 percent power). This is the normal range of cruise power settings for most airplanes— from *60 to 75 percent* of the normal rated power, the power the engine can use continuously at sea level. To make things easier, assume that the airplane in the example can use full power (100 hp) continuously (this is not the case for all airplanes).

Point 1 shows the maximum level-flight speed of the airplane. It requires 100 horsepower to fly this airplane at a constant altitude (sea level) at this speed, and that's all that this particular engine has available. From that point, as the airplane flies at slower airspeeds, less horsepower is required to maintain a constant altitude; moving down the curve, point 2 is reached. This is the point where parasite drag and induced drag are equal. Incidentally, this is the point of maximum cruise efficiency, or maximum range, of the airplane. You'll get more miles to the gallon while maintaining altitude at that speed, which is about 150 percent of, or 1.5 times, the stall speed (for trainers). This would be the airspeed at which to fly if you got in a bind for fuel and had to stretch it to make an airport some distance away. This 1.5 ratio is calibrated airspeed.

As you move back down the curve, point 3 is the airspeed at which the minimum power is required to maintain altitude. This would be the airspeed you'd use if you were looking for maximum endurance, or longest time airborne, and is roughly about 120 percent of the stall speed (or 1.2 times the stall speed, however you like to think of it).

Decreasing the airspeed below point 3 results in the power required to fly the airplane (at a constant altitude) starting to increase sharply; induced drag is beginning to make itself known. The part of the curve from point 3 back to the stall is called the "backside of the power curve," which implies that something unnatural is occurring. It's only unnatural if you don't have the full picture of what happens to drag at various airspeeds, and the term is a rather poor one (like "elevators"). Note that this airplane requires as much power (60 horsepower) for slow flight, 5 K above the stall, (point 4) as it does to cruise at an indicated airspeed of 96.

Finally, as the stall is approached at point 5, all of the power is needed to maintain altitude; if the angle of attack is increased further, the stall occurs even though full power is being used. (The attitudes of the airplane at the extreme ends of its speed range are exaggerated in Fig. 12-13.)

Practice-flying at minimum controllable airspeeds is extremely important; it shows you what to expect when you are flying at speeds fairly close to the stall (as might be done on a power approach to a short field). If you find yourself too low and slow and close to the stall on the power approach, you would *not* pull the nose up. This would demand more horsepower and make you sink faster. You would instead add power and ease the nose lower. In less critical situations you wouldn't even have to add power but would just ease the nose down to pick up a slightly higher airspeed (where less power is required).

Notice also how steeply the curve goes up at the maximum level-flight speed, point 1. If this same airplane had another 30 horsepower, it wouldn't have a much greater top speed. In fact, an added 30 horsepower (an added 30 percent) would only mean an increase in top speed of about 10 percent. As a rule of thumb for high cruising speeds or top speeds, the increase in airspeed (percentage) is about one-third of the increase in power (percentage) for a particular airplane. Speed can be expensive.

Chapter 23 will go into more detail as to how to set up specific *cruise* power conditions for your airplane.

SPINS

While stalls are fresh in your mind, it's a good idea to discuss spins. The FAA does not require the demonstration of spins for any certificate or rating other than the flight instructor's certificate, but you'll have ground instruction for them and your instructor may demonstrate and have you practice spins in a properly certificated airplane. But sometime you may want to have a spin demonstration by an instructor; it's a good idea for anybody to know what they are and how to cope with them.

Spins are to be practiced in a properly certificated airplane, of course. Before getting a demonstration, you should have a good idea of what to expect. The spin is an aggravated stall resulting in autorotation. In short, it is a condition where one wing stalls first and the plane "falls off" in that direction. One wing has more lift left and it is chasing the other stalled wing like a dog after its tail.

You've had experience with this in the departure and approach stalls. One wing stalls and the plane starts to rotate in that direction. If you had kept the wheel back, not recovering from the stall, and held rudder in the direction of rotation, a spin would have resulted. *If you relaxed the back pressure, at least in the first turn or so, the spin would have turned into a tight diving turn and you would know how to get out of that.*

Okay, so a spin occurs when you stall one wing before the other and continue to hold back pressure, not allowing the plane to recover from the stalled condition.

The normal spin procedure is to start a normal power-off stall (after clearing the area); just before, or as, the stall occurs, apply *full rudder* in the direction you want to spin. It's a good idea to begin no lower than 4000 feet above the ground. A one-turn spin in the average lightplane takes about 1000 feet of altitude from start to finish. Have enough altitude that you will recover at least 3000 feet above the ground. *Never deliberately spin with flaps down.* A spin to the left is covered below.

■ Procedure

1. Clear the area and start a normal power-off stall. Carburetor heat ON.

2. Just before the break occurs, apply full left rudder and continue to bring the stick or wheel full aft. The left wing stalls (Fig. 12-14).

3. Hold the wheel fully back and keep that left rudder in. Sometimes it may be necessary to apply a

Fig. 12-14. Just before or *as* the stall occurs (depending on the airplane), apply rudder in the desired direction of spin.

burst of power to start the spin. The prop blast gives the rudder added effectiveness to yaw the plane. Because of the yaw created by your use of the rudder, the left wing "moves back" and has less relative lift. The right wing, being moved forward, has relatively higher lift. The result is a roll to the left, causing the left wing to increase its angle of attack (which was right at the stall, anyway), so the left wing has stalled and is now developing much less lift (and more drag) than the right wing. A spin can occur only if the airplane (or at least one wing) has passed the stall angle of attack and stays there. So the spin is created by a too-high angle of attack with yawing or rolling (the rolling can be induced by the yaw *or* other factors). For instance, look back to Figure 12-10. You're out practicing takeoff and departure stalls solo one day and are making a right-climbing turn entry as shown. The airplane rolls to the left briskly and you use full right aileron to stop the roll—before you release the back pressure. Also, maybe your feet were lazy and you didn't use rudder against the roll like you should have. The down-aileron acts as a flap and the left wing is suddenly shifted to the left-hand "flap" curve shown in Figure 12-12; in addition, the angle of attack on that down-moving wing is increasing. Things meet in the middle (at the stall) pretty fast, and the wing starts the autorotation part. You'll have to break it by closing the throttle, neutralizing the ailerons, using opposite rudder, and moving the wheel forward briskly. But let's get back to talking about a deliberate spin.

4. The nose drops, but the full up-elevator does not allow the plane to recover from the stall. The unequal lift of wings gives the plane its rotational motion (Figs. 12-15 and 12-16).

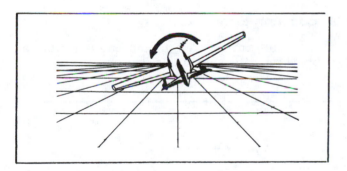

Fig. 12-15. The airplane is rolling and yawing into the spin.

The nose is not pointed straight down, though it will certainly appear this way to you during the first spin (Fig. 12-17).

Figure 12-18 shows the developed spin as seen from the cockpit.

Figure 12-19 shows the spin as it may look to *you* from the cockpit that first time. Don't stare over the nose, but keep your eyes moving, checking different objects on the ground.

The plane's rotational motion tends to keep the imbalance of lift as seen in Figure 12-20.

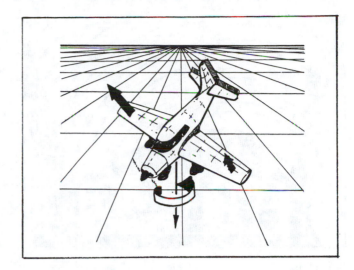

Fig. 12-16. Wings have unequal lift.

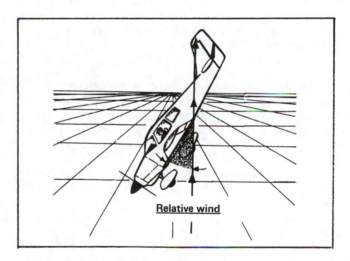

Fig. 12-17. One or both of the wings is stalled even though the airplane appears to be pointed nearly straight down. The angle of attack is the key.

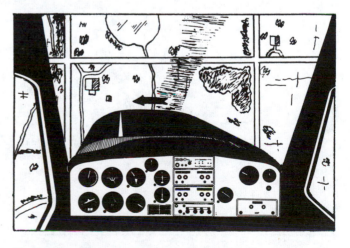

Fig. 12-18. The spin as seen from the cockpit.

Fig. 12-19. The spin as it will probably be seen from the cockpit your first time. The ball in the turn coordinator is to the left side of the instrument, you will notice, not to the right as might be assumed by theory. (You are holding full *left* rudder, which under normal flying conditions would put the ball to the right—but not in a developed spin with the instrument in the position shown.)

As you can see, the relative wind caused by the plane's path toward the ground is the same speed and direction for both wings, and the relative wind caused by rotation is the same velocity but from an opposite direction for each wing—giving a great difference in the angles of attack of the wings.

Once the spin is established, you can maintain it as long as you like with full up-elevator and full left rudder.

Relaxing the back pressure will result in some airplanes going from the spin into a spiral.

A properly executed spin is not any harder on an airplane than a stall. The airspeed remains low. (If you get a chance during the demonstration, look at the airspeed indicator. It will be showing a speed in the vicinity of the normal stalling speed or even "zero" in some trainers.)

A sloppy recovery puts far more stress on the airplane than the spin itself.

■ Recovery

At this point we'll just look at the rudder and elevator use during the spin recovery. A full recovery procedure will be discussed later.

1. Apply full opposite rudder as needed to stop rotation. Opposite rudder alone would stop rotation on older, lighter trainers. Newer planes may require full opposite rudder and then brisk forward pressure before rotation will stop.

If in an accidental spin you're confused as to the direction of rotation, use rudder opposite to the indications of the turn needle or the small airplane in the turn coordinator as shown in Figure 12-21B. These indicators are deflected to the left (a left spin), so opposite (right) rudder would be applied.

2. Relax the back pressure. Again, older lightplanes need only for you to relax the back pressure. A newer model may require the wheel or stick to be moved forward briskly and held there. But more about this shortly.

3. When the rotation stops, neutralize the rudder.

4. Ease the plane from the dive. You converted the plane's actions from a rotating stall to a straight dive. Don't wait too long to pull the plane out of the dive because the airspeed will be building up. On the other hand don't jerk it out, since you don't want to put undue stress on the airplane.

■ Common Errors

1. Failure to get a clean spin entry—getting a spiral instead. (More about this shortly.)

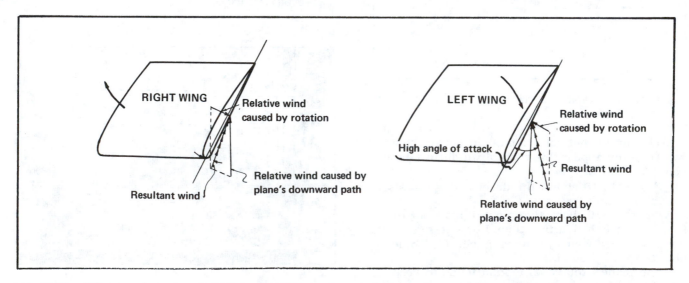

Fig. 12-20.   Differences in angles of attack in a spin.

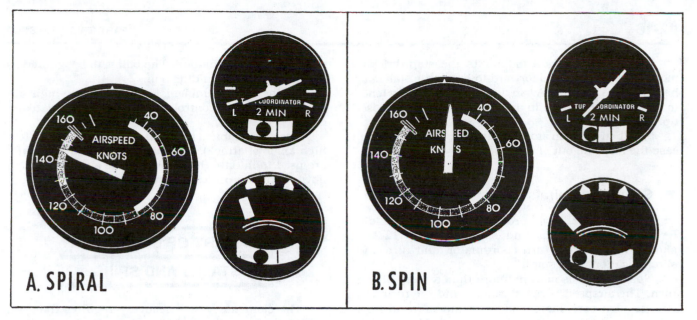

Fig. 12-21. Instrument indications in a spiral and a developed spin (to the left, in both cases). For the spin recovery, you would apply full rudder opposite to the turn coordinator airplane or the needle. (Pay no attention to the ball.) The instruments shown here are on the *left* side of the instrument panel. The ball in a slip indicator on the *right* side of the instrument panel always went to the *right* in different trainers spun by the author.

2. Overenthusiasm on the releasing of back pressure, that is, pushing forward abruptly on the wheel and causing a negative load factor to be put on the plane.

3. *Not* releasing the back pressure enough. Some students get so eager to get out of the left spin, for instance, that they stop the rotation by full right rudder and forward wheel, then pull right back with the wheel to pull out of the dive. They are astounded to notice that they restalled the airplane and that the still held right rudder caused them to enter a spin to the right. This is called a progressive spin but is really the same as any other spin. In this case it requires that you start a new recovery procedure for a spin to the right. This is embarrassing unless you can convince the instructor that it was a deliberate demonstration of your skill. Flight instructors are pretty skeptical people.

In discussing spin recovery, it was noted that most older lightplanes need only a little relaxation of rudder or back pressure to ease them out of a spin. However, improper loadings may cause even a "docile" airplane (or maybe it's a type of light airplane that doesn't know it's supposed to get out so ridiculously easy) to surprise you. One recommended type of recovery would be to apply full opposite rudder, followed immediately by a brisk forward motion of the control wheel or stick, which is held until the rotation stops. If you've tried the "relaxation of control" type of recovery and the spin is continuing, go back to pro-spin controls and institute the type of recovery just cited to get a "running start." In some cases how *fast* the wheel or stick is moved forward is almost as important as how *far* it's moved forward. You could get into

an accidental spin at altitude during a stall, and you should also know that *closing the throttle helps the recovery.*

If you do accidentally get into a spin situation someday when you are at altitude practicing stalls, your initial instinct will be (as the airplane rolls and the nose drops to a very low pitch attitude) to hold the wheel or stick all the way back to pull the nose up. *Don't.* You'll need to get the wheel or stick forward to break the stall, and sometimes this is hard to make yourself do.

One instructor experienced in spins, when asked to recommend a "spin warner" (that is, when the airplane is being stalled and yaw or roll is being induced), answered that a boxing glove should come out of the instrument panel hitting the pilot and forcing release of the wheel.

■ Summary of a Spin Recovery

If you think that during a practice stall things are going too far, as a general approach you should take the following steps:

1. Close the throttle. This makes initial recovery much easier for most airplanes.

2. Neutralize the ailerons. In some airplanes, ailerons *against* the spin make conditions worse; in others, ailerons *with* the spin make conditions worse. In the confusion you might use ailerons the wrong way and delay or preclude recovery, so keep the ailerons neutral.

3. Apply full rudder opposite to the roll or yaw.

4. Then when the rudder hits the stop, briskly move the stick or wheel forward to break the stall *and* help stop the rotation. Again, you'll automatically tend to hold back the wheel in an attempt to pull the nose up. *(Don't.)*

5. When rotation stops, neutralize the rudder and ease the airplane out of the dive.

■ Spins and Spirals

Some people have problems knowing the difference between spins and spirals. Figure 12-21A shows the airspeed, turn coordinator, and turn and slip indications in a spiral.

Spiral—This is nothing more than a steep diving turn. The airspeed is rising rapidly and the turn and slip (or turn coordinator) shows a moderate to high rate of turn in the well-developed spiral. The ball will probably be to the left in most airplanes because it takes left rudder in a dive to keep it centered; you've gotten into this situation unconsciously, so you aren't correcting for it. (Right rudder is needed at low speeds and high power settings to keep the ball centered.) The spiral is a low angle of attack/high-speed situation and the recovery procedure is to use *normal control pressures*. You'd level the wings with coordinated aileron and rudder and then ease the nose back to level. You'd also throttle back as soon as you realized the airspeed was picking up to keep the rpm under the red line (fixed-pitch prop). But the point is that the spiral is nothing more than an exaggerated combination of two of the Four Fundamentals—a turn and a descent—and the average 2-hour student should be able to recover from it.

Spin—This is a high angle of attack/low-airspeed situation; the airspeed will be hovering somewhere down in the stall area (Fig. 12-21B). The recovery requires an apparently unnatural use of the controls. You don't use aileron and rudder together to "level" the wings. You neutralize the ailerons just to keep from complicating matters, as noted earlier, and use rudder alone against the rotation. Instead of "naturally" getting the nose up by added back pressure, you'll have to try to push it even farther down to decrease the angle of attack and get out of the stalled condition. Then bring the nose up by back pressure after the airplane is out of the spin (rudder neutral) and in the dive.

Note that the ball is well to the left side of the two turn instruments in Figure 12-21B. It's assumed that this is a side-by-side trainer and those instruments are on the left side of the panel; this could well be the indications of the ball in a *left* spin. The slip indicator (ball) tends to go to the left side of the instrument in spins in either direction. Usually the rate of yaw and/or roll is much greater than in the spiral, with the indicator(s) pegged in the direction of spin. You, however, will have no comparison as shown by A and B in Figure 12-21, and the low airspeed (constant or oscillating in a low-speed range) will be the big clue for

detecting a spin situation. The ball is to be ignored as an indication of spin direction.

So the spiral problem is that of a too-high airspeed. "Normal" control *pressures* are used to recover.

The spin problem is that of a too-low airspeed (too-high angle of attack). Mechanical control *movements* are used to initiate the recovery. The airplane is then flown normally out of the dive.

SUMMARY OF ELEMENTARY STALLS AND SPINS

The elementary stall recovery requires that you get the air flowing smoothly over the wing again. Control pressure or abruptness of control movements will vary between lightplanes and heavy planes and will even vary between different types of lightplanes. Whatever the type of plane, the required result is the same.

An airplane can be stalled at any speed and attitude.

The fact that spins were covered right after the elementary stalls does not necessarily mean that this is the phase in which you will practice spins. Your instructor will likely want to wait until after you solo or may not demonstrate them at all. It is a good idea to be able to recognize and recover from the spin and move on to other maneuvers after you have done this.

All "normal" category airplanes that are capable of spinning are placarded against spins, and if your trainer is so certificated you won't be doing them. Some "utility" category airplanes are also placarded against spins.

The airplane's attitude and heading will be changing rapidly during these maneuvers, and you'd better keep those eyeballs moving. The practice area is no place to "run into an old friend" (or anybody else, for that matter) (Fig. 12-22).

A stall warning indicator may be an electrical or a reed type (sound). The electrical type is attached to a cockpit warning device such as a light or horn, depending on the make or model. A general idea is shown in Figure 12-23.

The tab is set at the correct angle for the particular airfoil being used. The indicator "warns" of any approaching stall. Generally it is set slightly ahead of the stall so that under normal conditions, such as a landing, the pilot is warned 5–8 knots before the stall. The warning may be given anytime the angle of attack becomes too great, regardless of the airspeed.

The pneumatic-type stall warning system used on some current trainers has an inlet on the leading edge of the left wing, connected by tubing to an air-operated horn within the pilot's hearing. The air pressure variations at changing, higher angles of attack draw

Fig. 12-22. Other airplanes may be using "your" practice area.

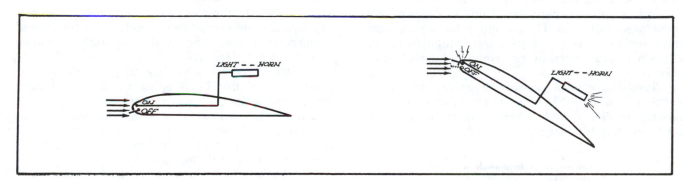

Fig. 12-23. At normal angles of attack the relative wind keeps the switch in the OFF position. As the airfoil approaches the critical angle of attack, the relative wind pushes the tab to ON. The horn or light operates as long as the critical angle of attack continues.

air through the horn, giving warning of the impending stall several knots before the stall break occurs. The tone usually gets higher in pitch as the stall approaches. (If you notice that it's starting to play a hymn, you could be in a little trouble.)

The military uses angle of attack indicators for performance information, particularly for approaches. The advantage of this is that the airplane always stalls at the same angle of attack (for a given flap setting) regardless of weight or g forces acting on the airplane. The angle of attack indicator doesn't have general acceptance for light general aviation airplanes but may be used in the future.

■ A Few Last Notes on Spins

1. The *Pilot's Operating Handbook* and an instructor current in spins in that model are the final authorities on spins for a particular airplane.

2. *In an accidental spin with the flaps down, get them up as soon as possible for easier recovery and to avoid exceeding the flaps-extended airspeed,* V_{FE},

(Fig. 3-28) in the postspin dive.

3. *If you believe you are beginning to get into an accidental spin situation and have forgotten everything else you've read, been told, or practiced,* GET THAT CONTROL WHEEL OR THE STICK FORWARD *to break the stall condition—and the sooner the better. This is not always the most efficient method of spin recovery, but it can stop an impending problem if done soon enough. However, there's absolutely no substitute for being familiar with the spin and recovery characteristics of your airplane as demonstrated by and practiced with a savvy instructor.*

FAR Part 61 requires that the applicant for a recreational or private pilot certificate has had ground instruction in stall awareness, spin entry, spins, and spin recovery techniques. If you don't actually get to do spins, your instructor will give you such ground instruction so that you'll at least have an understanding of the factors involved. There are questions about spins in the Private Pilot Knowledge Test (see Chapter 27).

13

TAKEOFFS AND

LANDINGS

After you have about five or six hours and are proficient in the Four Fundamentals, stalls, wind correction maneuvers, and elementary emergencies, the periods of shooting takeoffs and landings will begin.

Maybe you feel that takeoffs and landings are the most important parts of flying, but as far as your training is concerned they are just another maneuver. Students may place even more weight on landings than takeoffs.

Takeoffs are not important, they think, but *LANDINGS* are. So with this attitude, they find themselves having plenty of trouble when it comes to getting the bird smoothly into the air.

NORMAL TAKEOFF

The instructor will have you follow through on the takeoffs during the first flight or two, and probably by the third or fourth flight you will be making the takeoffs yourself.

Most of the trainers being used today are the tricycle-gear type. As in taxiing, the lower nose position gives better visibility during the ground roll parts of the takeoff and landing.

Because the airplane is in level flight position, the takeoff becomes a matter of opening the throttle smoothly, keeping the airplane straight and, as the controls become firm applying back pressure to ease the nosewheel up to attain the takeoff attitude and letting the airplane fly itself off. As you found in taxiing, most tricycle-gear airplanes have nosewheel steering, and during the initial part of the ground roll right rudder pedal pressure will be needed to take care of the left-turning tendency always existing in a low-speed, high-power situation. While the nosewheel is on the ground, the steering is divided between the nosewheel and the aerodynamics of the deflected right rudder; as the nosewheel is lifted, be ready to use more right rudder to compensate for the loss of nosewheel steering.

■ Tricycle-Gear Airplane Takeoff Procedure

1. See that no airplanes are landing. Taxi to the center line of the runway.

2. Line up with the center line or, if there isn't one (as on a grass strip), use a reference at the far end of the runway and keep scanning the surface between you and that reference.

3. Apply power smoothly to get rolling, then move on to full throttle. Keep your hand on the throttle and check the engine instruments.

4. As the controls become firm (particularly the elevators), use back pressure to obtain the proper angle of attack for takeoff. Be ready to add right rudder as the nosewheel lifts clear.

5. After the lift-off, relax back pressure slightly to assure that the nose maintains the proper climb attitude. Maintain that attitude and climb straight ahead, correcting for torque and keeping your hand on the throttle to make sure that it's at climb power. Ease the flaps up (if used) at a safe altitude (usually at 100–200 feet above the ground) (Fig. 13-1).

COMMON ERRORS

1. During the initial part of the roll, allowing (or "helping") the control wheel to move forward so that the elevators are slightly or well down. This could put too much weight on the nosewheel and could cause a delay in the takeoff or steering problems—called "wheelbarrowing" (to be covered later in the chapter).

2. The average student at first is usually timid in applying back pressure at the right time to ease the nose up to the takeoff attitude. This results in the airplane staying on the ground with too high an airspeed, thus wasting runway and possible getting some nosewheel "chatter." Some King Kong types give a mighty heave too early and pull the airplane abruptly into the air before it's quite ready to go. You'll find that after a couple of takeoffs you'll be able to apply the right amount of back pressure at the right time.

3. Not relaxing back pressure after the lift-off (or continuing to hold the back pressure used to bring the airplane off) so that the airplane's nose rises sharply and doesn't maintain the best rate of climb speed (V_Y).

Fig. 13-1. (A) Line up with the center line and apply power to start rolling, then smoothly go to full power for takeoff, keeping the airplane straight by smooth application of right rudder. (B) As the controls become firm, start applying more back pressure to ease the nose up to the takeoff attitude.

Figure 13-2 shows the side view of a takeoff for a tricycle-gear airplane.

■ Takeoff for Airplanes with Tailwheels

You are trying to leave the ground as smoothly and efficiently as possible. Taxi onto the runway, first making sure no other planes are on final approach, and line up with some object such as a tree at the far end (Fig. 13-3). In most planes with tailwheel-type landing gear it is difficult to see over the nose in the three-point attitude, so you must look at your object alongside the nose for the first part of the takeoff run. You may ask, "What if there's no tree?" Lacking a reference point at the end, watch the left side of the runway. The takeoff consists of three phases: (1) the initial or three-point position part of the run; (2) the roll on the front two wheels, in which the plane is

streamlined to pick up speed quickly; and (3) the lifting off, or transition to flight.

PHASE 1. Open the throttle slowly to get the plane rolling then continue to apply full power smoothly.

The torque situation is this: You are now using full power, but the plane is barely moving. The rudder itself is comparatively ineffective, and the greater part of the steering must be done with the tailwheel. As the speed picks up, the rudder begins to get more effective and less right pedal pressure is needed. Don't worry about the change in rudder pedal pressures, just keep that tree lined up correctly. Of course, you will be watching the runway as well.

The best method of handling the elevators is to have them at neutral or slightly ahead of neutral. There's no need to force the tail up abruptly. In fact, holding the wheel full forward to get the tail up too soon might cause loss of directional control. If the tail

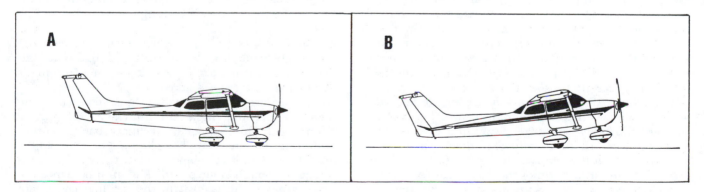

Fig. 13-2. A side view of the initial roll (A) and lift-off (B) attitudes for a tricycle-gear airplane.

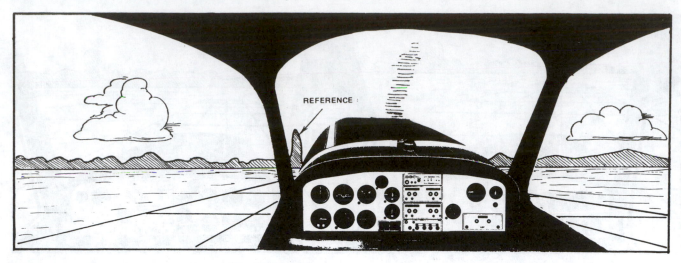

Fig. 13-3. Pick an easy to recognize object to line up on.

comes up abruptly before you have effective rudder, the plane may turn to the left before it can be stopped with rudder. If the elevator trim tab has been set at neutral (and you saw that it was during the pretakeoff check), the tail will come up by itself when the time gets ripe. You can, however, help the process by exerting slight forward pressure on the elevators as the plane picks up speed.

Don't "walk" the rudder, hoping to average out the swing of the nose—you'll never get the feel of the takeoff that way.

PROBABLE ERRORS IN PHASE 1

1. Abrupt throttle usage resulting in poor directional control at the beginning of the run.
2. Not keeping the plane lined up with the object at the end of the runway.
3. "Walking" the rudders.

PHASE 2—THE TAIL-UP PART OF THE ROLL. Many students have trouble in the transition from phase 1 to phase 2. While the plane is rolling in the three-point position, you have both tailwheel steering and, in the latter part of phase 1, fairly effective relative air moving past the rudder. When the tail comes up, the tailwheel no longer helps and a further deflection of the rudder is required for torque correction. Students sometimes have trouble correcting for torque and keeping the nose lined up. Their takeoffs are smooth rolls interrupted by a period of mad maneuvering because they use the rudder too enthusiastically. The plane weaves down the runway, putting great side stresses on the tires and landing gear. This type of rudder action slows the takeoff and, if it goes too far, may result in structural damage to the airplane.

Watch for any swinging of the nose and catch it early with the minimum of rudder pressure.

The tail slowly comes up and you are keeping the plane straight. The tree hasn't moved and now you can see the whole runway ahead (Fig. 13-4).

PROBABLE ERRORS IN PHASE 2

1. Getting eager to get the tail up and having steering problems.
2. Trying to force the airplane off too soon by excessive back pressure, resulting in slowing the airplane and delaying the takeoff.

A good demonstration of the effects of gyroscopic precession can be given by forcing the tail up suddenly on takeoff. The rotating propeller acts as a gyro wheel. When the tail is raised, the rotational plane of the propeller is changed. As the tail comes up, the effect is that of a force being exerted against the top of the propeller arc from behind. You know that a gyro reacts as if the force had been exerted at a point 90° around the wheel (measured in the direction of rotation). The airplane reacts as if the force had been exerted at the right side of the arc, and this tends to turn the plane to the left. The faster the tail is raised, the more abrupt this left-turning action. This acts only as long as the plane's attitude is being changed. Once the takeoff attitude is reached and maintained, precession is not an important factor. You will not be worried so much about what causes certain effects on takeoff as about correcting them. If the nose veers to the left, no matter what the cause, you'll have to straighten it out. This is the most important thing to remember.

PHASE 3—LIFT-OFF. As the plane picks up speed, the controls become firmer. The attitude is that of a shallow climb. This makes it possible for the plane to lift itself off as flying speed is reached.

As the airspeed approaches the takeoff value, the plane will become "light" and may start skipping or bouncing. If this happens, use more back pressure and bring the plane off. The skipping or bouncing is an indication that the plane is ready to fly but there is not quite enough angle of attack to lift it off. Once the plane breaks free, keep it in the air. It doesn't help matters to hit again. If there is a crosswind, you may be drifting across the runway the next time you hit,

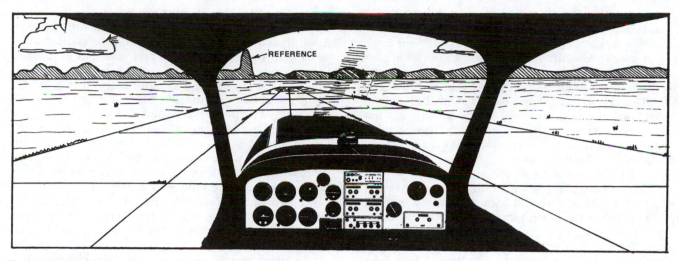

Fig. 13-4. Now all you have to do is keep it straight and wait.

thus putting a heavy side load on the landing gear.

After you are off the ground with no danger of hitting again, relax back pressure slightly so that the nose does not rise too steeply as the plane accelerates and the elevators become more effective (Fig. 13-5). After this point it's no longer a takeoff but a normal climb, and you've been able to cope with that for some time.

As soon as the wheels leave the ground, the rudder no longer steers the airplane. Use coordinated aileron and rudder to turn until you get back on the ground again.

PROCEDURE

1. See that no planes are landing. Taxi to the center line of the runway. Use all the runway available; that portion behind doesn't help you.
2. Line up with some object at the far end of the runway.
3. Apply power smoothly to get it rolling, then apply full throttle smoothly.
4. Keep the plane lined up with the object at the end of the runway.
5. As the tail comes up, be prepared for the need of added right rudder.

6. As controls become firm, bring in gentle back pressure, if necessary, to lift the plane off; when it is off, keep it off.
7. After takeoff, level off to pick up climb speed. Remember the takeoff speed is slightly above stall speed and well below proper climb speed.

If the runway is smooth and you use proper techniques, it should be hard to tell when the plane leaves the ground.

NORMAL LANDING

The normal landing is nothing more than a normal power-off stall, with the stall occurring just as the plane touches the ground. Why stall the airplane? You can put a lightplane on the ground at 100 K, but this is hard on the plane (and on you) and also uses up more runway than is practical. The best idea is to land at the slowest speed possible and still have control of the airplane.

The first thing you want to learn is that the importance of landings is vastly overrated by the public (and probably by you too at this stage). If you make a

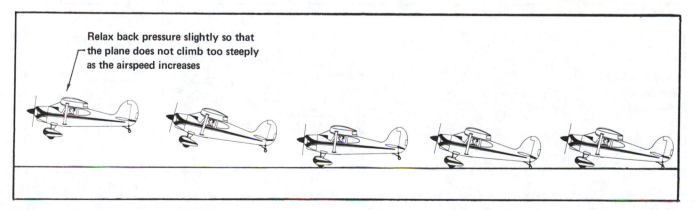

Relax back pressure slightly so that the plane does not climb too steeply as the airspeed increases

Fig. 13-5. The normal takeoff (tailwheel type).

smooth landing, you're the greatest pilot alive to your passengers. There are several reasons for this: (1) Landings are always made at the airport (you hope) where everybody can see them and know what they are; (2) this flying is still somewhat mysterious to the average person, who figures it must take a high degree of skill to be able to come back down after defying gravity and the elements; and (3) the landing is a maneuver done close to the ground where the smallest mistakes may look like near crashes. At 3000 feet you may skid a half mile, but since there are no references up there is doesn't show up.

This disease of sweating landings even strikes experienced pilots who should know better. It is interesting to note that the perfect landing is always made on a day when the airport is deserted. But bounce your way in on a sunny Sunday afternoon, with somebody special along for that first ride, and there'll be a fence full of jeerers as you try to park as inconspicuously as possible.

But speaking on a pilot's level, the idea of a landing is to get the plane on the ground without fanfare. *That's all that it is.*

■ **Procedure**

The landing is started from a normal glide at a height of about 20 feet (or about hangar height) above the ground. This starting height depends on the size of the airplane and is also affected by the rate of descent. Large planes may start the transition at 75 feet, but hangar height is about right for the lightplane. Start the landing transition or "break the glide" at hangar height by gradually easing the wheel or stick back to stall the airplane. About 19 of that 20 feet is used to change the airplane's attitude from that of the normal glide to the landing attitude.

Transition must be gradual. If you pull the wheel back abruptly, the plane will go up. This is called "ballooning." Ease the nose up to the landing attitude, disturbing the plane's flight path as little as possible. *This takes a certain amount of skill, as you will dis-*

cover. The plane is slowing up all this time as the angle of attack is being increased. By the time you are down to about a foot off the ground, the transition to landing attitude is complete and your job is to keep that attitude by using more and more back pressure. *The best way to make a landing is to try to hold the plane off as long as possible.* The faster the plane settles, the faster the rate of application of back pressure must be. The plane continues to settle and slow up. Remembering stalls, you know that the slower the plane gets, the more elevator is needed to hold its attitude, which slows it, etc. You should run out of elevator just as the plane touches the ground.

Thermals, gusts, or downdrafts affect the transition. Only in theory, or on a perfectly calm day, will the wheel be moved smoothly back at a nearly constant or predictable rate.

In other words, there may be times during the transition when you have to relax back pressure. Never *push* forward. Remember you are at a low speed and a high angle of attack. If you decrease the angle of attack suddenly, there will be a lag before the speed picks up. During this time, lift will be practically nothing and you might find that you've made a serious mistake. There may be times when the back pressure must be stopped where it is and other times when it seems that you must pull back rapidly on the wheel or stick at the last part of the landing process. Well, that's the way it is. But one thing is important—start the landing from a normal glide. Generally a poor approach means a poor landing.

Land the tricycle-gear plane so that the initial contact is made on the main wheels. Try to hold the nosewheel off so that the airplane's attitude causes aerodynamic drag and helps slow it down on the landing roll (Fig. 13-6).

Later you'll find that there is an optimum time to ease the nosewheel down to get the maximum drag on the landing roll for each plane.

The tailwheel airplane should be landed so that the tailwheel and main gear touch simultaneously (Fig. 13-7), although for some airplanes (some pilots argue) the tailwheel should touch an instant early.

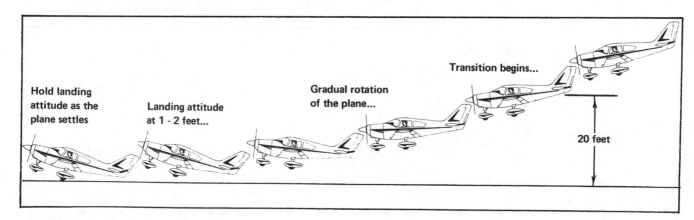

Fig. 13-6. Landing the tricycle-gear airplane.

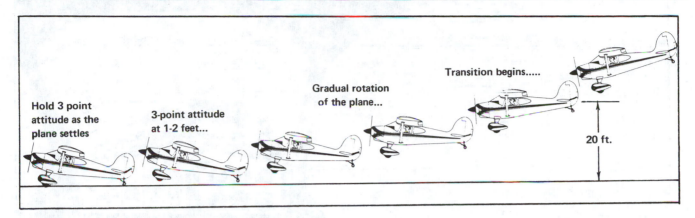

Fig. 13-7. The three-point landing (tailwheel type).

■ Where to Look

In order to know when and how much back pressure is to be applied, you must know where the plane is in relation to the ground (Fig. 13-8). The nose will be coming up higher and higher as you approach the landing attitude. It will be impossible to see the ground over the nose in some airplanes in the last stage of the landing transition, as you will see during the landing demonstrations in the early part of your training. Some trainers have "low noses," and it's possible to look directly ahead during the landing. But getting into the habit of looking along the left side of the nose works for *all* airplanes. You will be sitting on the left in a side-by-side airplane; the stick and throttle arrangement in a tandem plane also makes it easier for you to turn your head to the left. So you will look out this side after the transition begins. (When you get to be a flight instructor and sit on the right, you'll naturally have to look out the *right* side and this is a tough problem at first for new instructors.)

So the best place to look is to the left, alongside of the nose and far enough ahead so that the ground is not blurred. Don't stare at one spot. Scan the ground; your depth perception depends on a lot of eye movement. If you look only straight down out the side window, the relative movement of the ground is great and you may have a tendency to stall the airplane while it's still a fair distance off the ground. The plane will "drop in" and could be damaged. If you look too far ahead, the error in your depth perception may cause you to fly the plane into the ground.

As you reach about 20 feet, or hangar height, begin the transition (Figs. 13-9, 13-10, and 13-11).

Again, don't try to look over the nose. If you can see over the nose during the landing, in most airplanes you don't have the correct landing attitude and will fly into the ground. If you get the nose at the right position and stare blindly up at it, you won't have any idea where the ground is. You'll find out exactly where when you hear (and feel) the sickening thud.

Keep the plane lined up with the runway and have

Fig. 13-8. Maintain a constant attitude/airspeed during the latter (power-off) part of the approach.

Fig. 13-9. After starting the transition at about 20 feet, continue to ease smoothly back on the wheel (altitude shown here about 15 feet).

Fig. 13-10. Landing attitude is reached at a height of about 1 foot; continue back pressure to maintain this attitude.

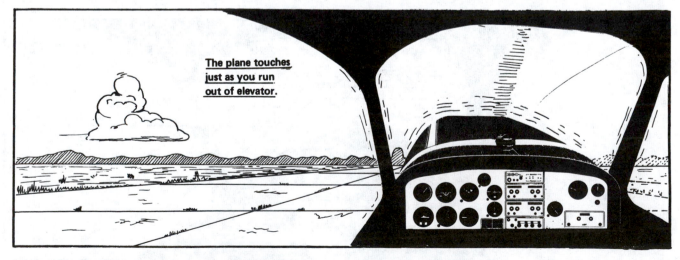

Fig. 13-11. The plane continues to settle as you keep easing back on the wheel and it touches in the correct landing attitude. Keep the wheel full back and maintain directional control with the rudder.

the wings level when you touch down. The stall warner should be making its presence known under normal conditions.

■ Ground Loops

In Figure 13-12 (tailwheel) and Figure 13-13 (tricycle-gear) you see that the momentum of the plane will act through the center of gravity, and the center of friction with the runway may be considered to act at a point midway between the wheels.

If the plane isn't pointed the way it's flying when it hits, a ground loop might result. A ground loop is an abrupt turning of the airplane on the ground. If the ground loop is violent enough, the plane may rock over and drag a wing. The best way to avoid a ground loop is to make sure the plane is lined up with the runway when the touchdown is made.

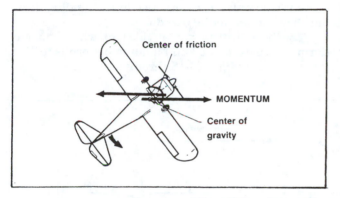

Fig. 13-12. Center of gravity, momentum, center of friction (tailwheel plane).

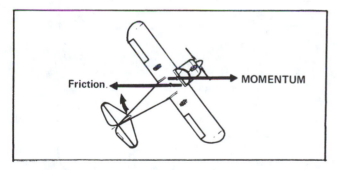

Fig. 13-13. The tricycle-gear plane tends to straighten out.

If the ground loop has started and a wing has dropped, don't get any ideas of bringing it up with rudder. This works in the air but makes matters worse on the ground. The plane rocks over because of the sharpness of the turn. Applying rudder to "speed up" that down wing in hopes of bringing it up only results in a sharper turn. Use all controls including the throttle to stop or slow down the ground loop. A burst of power coupled with prompt rudder and/or brake ac-

tion against the turn is usually effective as a means of recovery.

If a wing is down during the landing transition, the plane will turn. You'll have to concentrate on keeping those wings level, the plane straight, and making the transition all at the same time. Later you'll be shown how holding a wing down and applying opposite rudder will help in a crosswind landing.

In moving the wheel back during the landing process, students sometimes pull *up* on the left side of the control wheel as well as moving it back (the wheel is generally held on the left side by the left hand). This means that aileron is being applied, the wing is lowered, and the airplane "drifts" toward the right edge of the runway. The stick control, being used by the right hand, is often pulled to the right with the same results.

When you're flying cross-country, a slightly lowered wing for a short time may mean that the plane's path will vary several hundred feet to that side. It's not noticeable up there, but because many runways aren't even a hundred feet wide, you may end up out in the gravel pits and tall grass alongside the runway.

One common error students make during the landing roll is to relax back pressure and allow the control wheel to move forward. In the tailwheel type, this doesn't keep that tailwheel down firmly enough on the ground for good steering; in the tricycle-gear airplane it will allow too much weight on the nosewheel, causing "wheelbarrowing." Wheelbarrowing occurs when the added weight on the nosewheel causes it to become extra sensitive to rudder pressures during the ground roll. Your correction for a slight nose movement on the roll-out may be exaggerated and the airplane overreacts, abruptly turning toward the direction of that corrective rudder. In other words, relaxing of back pressure during the landing roll-out is bad for both types of airplanes.

The first hour of landings will keep both you and the instructor busy. You'll be worn out at the end of the hour but will discover that landings are quite interesting.

SHOOTING TAKEOFFS AND LANDINGS

Most students have three to four hours of takeoffs and landings before solo. Some may have two hours and others six. This will be the most enjoyable part of your presolo work. You won't have to take the instructor's word for it if you foul up. You can see and feel it for yourself.

■ The Traffic Pattern

Now that you know the theory of takeoffs and landings, it's good to have an idea of how to go about practicing them.

The airport traffic pattern is established so that

planes flying from that field will have some standard of operation for safety's sake.

A typical traffic pattern for a small airport is shown in Figure 13-14. Some airports have a right-hand pattern for some runways, as will be shown by their segmented circles, but the principle is the same.

Back in the olden days of flying, most lightplane approaches were made from a normal power-off glide so that if the engine quit during any part of the approach the airplane could still make it to the runway. Nowadays, the aircraft engine is so reliable that this is an extremely minor factor; the recommended method of making the approach is to reduce power (as recommended by your instructor) opposite to the point of intended landing and make further reductions as the approach continues. You may have all the power off just after turning final, or you may have to carry power until reaching the runway—it depends on your situation. You'll also make some approaches from a power-off glide to learn your airplane's glide characteristics, but most of the time you'll make approaches with a gradual reduction of power all the way around. Power-off approaches will be covered later in the chapter.

You'll be introduced to the takeoff and landing se-

ries gradually. You've probably made the takeoffs without any help for the last two or three flights and have been following through on the controls during the last few landings. Generally, takeoffs and landings are started with the instructor demonstrating two or three at the end of a regular period, allowing you to practice more takeoffs and landings each time until the whole period is spent on these maneuvers.

The first couple of hours of landing practice will probably be full-stop landings. That is, you will let the plane slow down; then you'll turn off the runway and taxi back to make another takeoff. This is good practice in learning to control the airplane as it slows up on the landing roll, as well as smoothing out any taxiing problems. Later you'll probably shoot "touch-and-go's"—opening the throttle, pushing the carburetor heat off (if it was used throughout the approach; your instructor will have recommendations), raising the flaps, and taking off again. Touch-and-go's allow more landings during a practice period and are good for the student who only needs practice in the landing itself, but they should not be overdone.

The best way to get an idea of what happens during a typical takeoff and landing pattern is to take it step by step (Figs. 13-15 through 13-18).

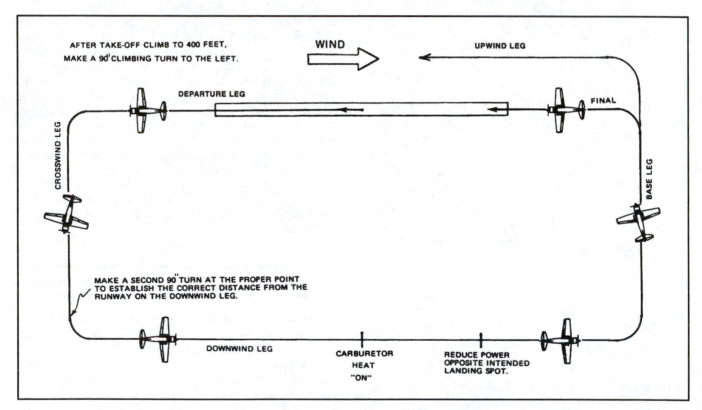

Fig. 13-14. An example traffic pattern. Shown is a "closed pattern" as used in shooting takeoffs and landings. Your instructor will demonstrate, and you will practice, departing and entering the pattern. (Substitute your own numbers.) The leg after takeoff, as you climb out tracking the runway center line, is the *departure leg*. When making a go-around or flying the pattern into the wind, the term is *upwind leg*. When making a go-around or aborted landing, fly to the *right* of the runway as you climb out, as shown here and at the top of Figure 13-39(A), so you can keep an eye out for airplanes taking off and climbing out. For a right-hand pattern go-around, it's particularly important that you watch for airplanes turning into your path.

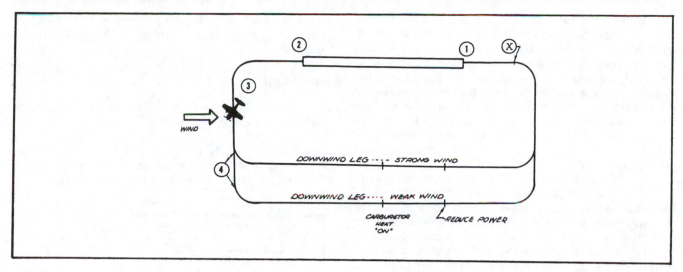

Fig. 13-15. Varying the downwind leg position for different winds.

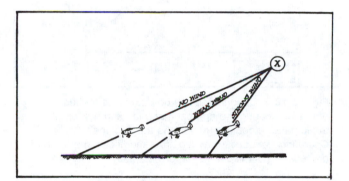

Fig. 13-16. Variation of approach angle with wind. Constant glide attitude and airspeed (angles exaggerated).

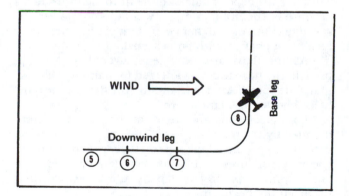

Fig. 13-17. Positions on the downwind and base leg.

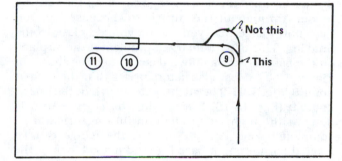

Fig. 13-18. The turn onto the final leg and the last part of the approach.

Assume that you are sitting at the takeoff end of the runway, the pretakeoff check is complete, there are no planes landing, and the wind is moderate (8–10 K) and right down the runway.

1. Taxi into position and make a normal takeoff.

2. Make a normal climb straight ahead to the altitude set by your local rules. Look back occasionally to see that you are climbing in line with the runway and not headed out cross-country.

3. When the turn altitude is reached, look behind again to make sure you aren't going to turn into a faster plane coming up from behind. Make a 90° climbing turn to the left. Actually, the turn should not be quite 90°, since you will have to compensate for the wind. Incidentally, don't make any turns inside the airport boundary. It's better to be at least 1000 feet past the end of the runway before turning.

4. Make the second 90°, left-climbing turn whenever it is necessary in order to establish the correct distance from the runway on the downwind leg. The principle of establishing a distance from the runway is this: If the wind is strong, the approach angle is steep on the final approach for a particular power setting; therefore, more altitude is needed at point X (Figs. 13-15 and 13-16). Maybe you say, "Why not stay the same distance from the runway for every wind velocity and just make that first turn sooner?" It can be demonstrated that you will be critically low at the point of the final turn if the wind is strong and also you'll have no "straightaway" or final to establish yourself for the landing.

5. Climb to 1000 feet. Keep the plane flying parallel to the runway (Fig. 13-17).

6. Pull the carburetor heat a few seconds before you reach the position directly opposite the intended landing spot. Give it a chance to clear out any ice before you start reducing power. The use of carburetor heat will depend on the airplane and conditions: (a) You may not use carburetor heat for certain airplanes if conditions preclude icing, (b) you may be told to apply carburetor heat on the downwind leg to check for icing and then push it off if no evidence of icing is found, or (c) you may use full carburetor heat throughout the approach.

7. Reduce power and set up the approach speed.

8. Make a slightly more than 90° turn to the left. Continue to reduce the power as necessary to control the approach path. One problem students have is making this turn too shallow, so the airplane is nearly over to the final leg before the first 90° of turn is completed. This would then require a steep 90° turn onto final, at a low altitude, to line up with the runway center line—a disconcerting and possibly dangerous practice. This is more likely to happen in a high-wing trainer because the down-wing may keep the runway hidden throughout the turn onto the base leg. You may not realize that you are moving so close to the final leg. This is particularly a problem if the airplane on the base leg has a tailwind component as shown in Figure 13-31. Most instructors prefer that the turn at point 8 (Fig. 13-17) be slightly steeper than the one at point 9 (Fig. 13-18) because (a) the steeper turn is done at the higher altitude, and (b) the turn will be completed soon enough so that the wings can be leveled for a straight base leg (and a good look at the runway). More about this later in the chapter.

9. Make the final turn as needed to line up with the center line of the runway. A common fault of students is overshooting the runway on the final turn. This happens when the final turn is not started soon enough and the plane must be turned back into the runway. Another error is making the final turn too soon, angling toward the runway. Shoot for at least a ½-mile final.

10. When your height above the ground is approximately 20 feet, begin the transition to the landing attitude. Keep the plane straight and the wings level. At about 1–2 feet the landing attitude should be reached. Continue the back pressure, maintaining this attitude as the plane settles. The wheel or stick should reach its full back travel just as the wheels touch.

11. Keep the plane rolling straight down the runway. As it slows to taxi speed, turn off the runway and taxi back for another go.

Don't try to turn off the runway at too high a speed—you may ground loop.

Don't try to slow the airplane too quickly by using brakes. It's not good for a tricycle-gear-equipped plane, and you may nose over in a tailwheel-type plane; it takes experience to be able to apply brakes on the landing roll without nosing over. It's hard on tires too.

Figure 20-2 gives suggested traffic patterns as in the *Aeronautical Information Manual*, and you might

want to look it over as you prepare to fly to other airports during the cross-country portion of your flight training.

■ **Flaps**

You've seen the flaps during the preflight checks, and the instructor has shown you how to use them in the approach to landing stalls as well as during the approach demonstrations. But now take a closer look at how *you* are to use them during these takeoff and landing sessions and in future pattern work (Fig. 13-19).

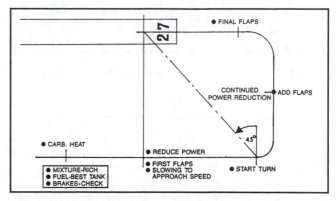

Fig. 13-19. A suggested use of a checklist and flap operations during the approach and landing portion of the traffic pattern. Your instructor and the airplane's *Pilot's Operating Handbook* may require a modification of this procedure. (*The Flight Instructor's Manual*)

By extending the flaps you increase the camber (curvature) of the wing, which results in more lift *and* more drag for a certain nose position. This means you can have a steeper angle of descent to clear obstacles because of the drag, and you will land at a lower airspeed because of a higher coefficient of lift. There may be a pitch change with flap extension.

As the flaps are extended (lowered) in the approach, the nose must be lowered to maintain the normal glide speed. After the landing, the drag of the flaps helps shorten the landing roll.

For takeoff the flaps are set at the position recommended by the manufacturer. This may vary from no deflection for normal takeoff to 10 degrees or more for special procedures such as for soft- or short-field takeoffs. If your airplane has flaps, you'll be shown their use from the beginning.

The flap operation range is marked by that white arc on the airspeed indicator. The bottom of the white arc is the flaps-down, power-off stalling speed at the maximum certificated weight of the airplane. The high value (top of the white arc) is the maximum speed at which full flaps may be used at any weight. (For safety, think of this speed as the maximum at which *any* flaps may be used.) If the flaps are lowered at an *indicated* (calibrated) airspeed higher than noted at the top of the white arc, possible structural and trim problems could occur. Since the flaps add both drag and lift for any given airspeed, you will find that your angle of descent will be steepened by their use; you will also find that the landing speed is measurably lowered. Generally it's best *not* to combine slips and flaps; some airplanes don't like it. You might check this for your airplane at a safe altitude with an instructor.

You'll get instruction on landing *without* flaps in an airplane equipped with flaps, unless the operations limitations prohibit no-flap landings. If you've been using flaps a lot lately, that first practice no-flap landing will give the impression that you're going to float all the way down the runway; you'll tend to let the airspeed get too great on final approach. For most airplanes the actual landing technique is easier without flaps; the nose can be eased up with less effort or trim. You should be familiar with how the airplane handles on landing both with and without flaps.

While flaps are a great aid to the pilot on approach and landing, they can be a problem when a sudden go-around is required. Flap drag eats up horsepower that could be used for needed climb. Someday you may have to make a go-around from a touchdown attitude with full flaps, starting at not more than a couple of feet above the runway. This would be about the most critical condition for a go-around; you're low and slow and the airplane is "dirty."

First, on any go-around *add full power before starting to clean up the airplane.* The power will stop or slow your rate of descent. Then start getting the gear and flaps up. You'll be surprised, but the change in stall speed from idle to full power almost offsets the difference in stall speed between flaps up and flaps down (it depends on the particular airplane). In other words, apply full power and figure on starting to get the flaps up right afterward. This doesn't mean that you instantly go from full flaps to no flaps, but get *some* flaps off as soon as full power is applied and continue to get them up as soon as safely possible. Many pilots have been warned so much about the danger of suddenly pulling the flaps up that they spend valuable time—and attention—just seeing how smoothly and slowly the flap cleaning-up process can be done. For some airplanes the manufacturer recommends that flaps be brought up (in increments) before the landing gear for best go-around results. Most trainers have fixed gear anyway, so this is a decision you probably won't have to make—now.

■ Slip

The slip is a very important maneuver for airplanes not equipped with flaps and is used to increase the angle of descent without increasing the airspeed. If you are high on the final approach and dive the plane (Fig. 13-20A), you will get down to the runway but will also have excess speed that will cause the plane to float down the runway. In the slip the pilot turns the plane so that an unstreamlined shape is presented to the relative wind (Fig. 13-20B). Airspeed will stay low even in a steep descent because of this drag.

There are two kinds of slips, the sideslip and the

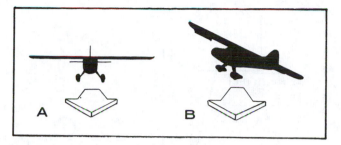

Fig. 13-20. (A) Silhouette of the drag area of a plane in a dive. (B) Silhouette of the drag area of a plane in a slip.

forward slip. The difference is a matter of the path over the ground. The idea is the same for both.

Take a slip in general. The technique is to use aileron in the direction of slip. When aileron alone is used, adverse yaw occurs, as you know from experience by now. If you give it left aileron, the plane banks to the left and the nose yaws to the right. Stop the nose at the peak of its yaw with right rudder. Continue to hold left aileron and right rudder and whatever back pressure is needed to maintain glide speed. The more aileron and opposite rudder used, the steeper the slip, which means a steeper angle of descent (Fig. 13-21).

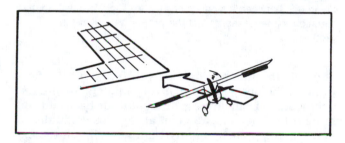

Fig. 13-21. A forward slip on final.

The steeper the slip, the more the airplane's path varies from the way the nose is pointing. Bank an airplane 15° and notice the difference between the forward slip and the sideslip. *Remember, for a 15° bank the flight path of the airplane will vary from the way the nose is pointing by a certain number of degrees.* Assume, for instance, that in the case of a 15° bank the variation is 10° (Fig. 13-22).

The forward slip is used to lose altitude while following your original flight path and is considered as a means of losing altitude faster without picking up excess speed, as would be done in a dive.

The sideslip is normally used to correct for wind drift (as you will see later in crosswind landings) *or to change the flight path of the airplane as you increase the angle of descent.*

Right now let's discuss the *forward slip.* It's a handy maneuver in a forced landing. After you are able to do slips, the forced landing becomes a matter of picking a field, making a slightly high approach,

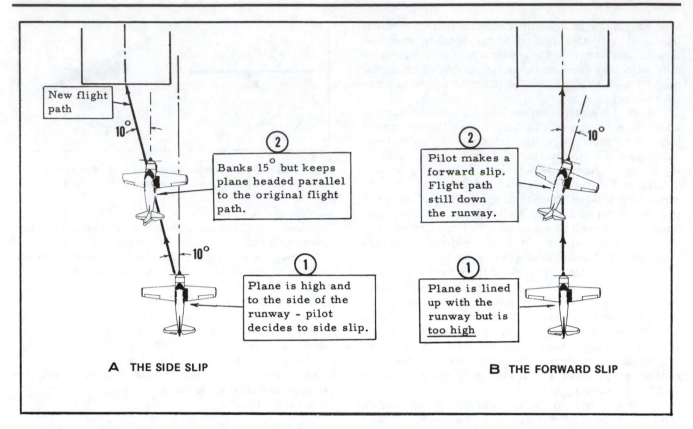

A THE SIDE SLIP

New flight path

② Banks 15° but keeps **plane** headed parallel to the original flight path.

① Plane is high and to the side of the runway - pilot decides to side slip.

B THE FORWARD SLIP

② Pilot makes a forward slip. Flight path still down the runway.

① Plane is lined up with the runway but is <u>too high</u>

Fig. 13-22. Aerodynamically, the airplane doesn't know the difference between a sideslip or forward slip. You'll set up the situation as necessary to get the proper path over the ground.

and slipping into the field. If you try to hit the field right on the nose, make an error in judgment and undershoot, you can never get that altitude back and will hit short of the field. You can always lose altitude, so make the approach slightly high and slip off the excess altitude as you need to.

If there is a slight crosswind on the runway that you are slipping into, always slip (bank) into the wind. This way you can make a slight sideslip so that your path is straight down the runway. Say, for instance, that for a 30° banked slip the flight path is 20° from the nose. Just don't turn the nose so far from the wind (Fig. 13-23).

Also, it's easier to convert the forward slip to a sideslip for wind drift correction on the landing (Fig. 13-24).

If you slipped to the right with a left crosswind, it would be necessary in recovering to roll from a right bank to a left wing-down position. You would be drifting across the runway at a good clip by the time you got the wing lowered and drift correction set up.

A left slip means using left aileron. A right slip means the use of right aileron. The aileron is considered to be the main control in the slip, with the rudder used as an auxiliary, or "heading lock."

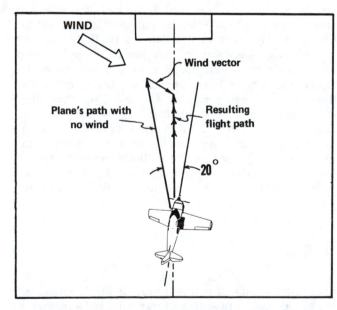

Fig. 13-23. Combining a forward slip and sideslip to correct for wind drift as you lose extra altitude.

PROCEDURE. For a forward slip to the right from a normal glide:

1. Right aileron is applied. As the nose yaws left, stop it at that position with left rudder. If added steep-

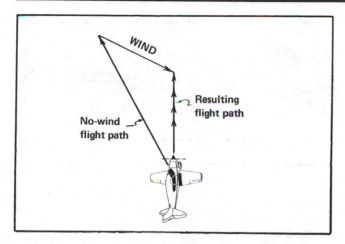

Fig. 13-24. The slideslip as used to correct for wind drift on landing. (That's quite a crosswind shown here, and in such a real situation you'd be better off going to another airport with a runway more into the wind—but the example shows up better this way.)

ness of bank is needed, add more right aileron and left rudder.

2. Maintain normal glide pitch attitude.

3. When you are ready to recover, neutralize the controls and maintain normal glide speed with the elevators.

COMMON ERRORS IN THE FORWARD SLIP

1. Yawing the plane so that the flight path is changed.

2. Letting the nose come up as the slip is entered, resulting in a low airspeed. Some airplanes require that you hold a slight forward pressure to keep the nose from coming up too high in the slip.

3. Picking up speed in the slip—nose too low.

4. Picking up speed on the recovery from the slip. *Don't slip too low, it takes altitude to recover.*

A COUPLE OF ADDED NOTES ON THE SLIP. The slip will be fairly hard at first. You've been taught for several hours that aileron and rudder go together and now have to "uncoordinate." Many instructors don't allow their students to slip the airplane if they are high on one of the first solo approaches, and some don't instruct slips until sometime after solo. You can get into trouble if the airplane is allowed to get too slow during the slip. It's always a good idea to practice at a safe altitude before using slips on a landing.

Some trainers have a static port only on the left side, as shown back in Figure 3-5. This can cause some errors of airspeed in slips. Take a good look at Figure 3-5. You can see that the static pressure in the static tube should exactly cancel the static pressure in the pitot tube (it says here), leaving the dynamic pressure to be properly indicated as airspeed. In a forward slip, or sideslip to the *left,* dynamic (or ram) pressure will be entering the static tube. The pressure in the airspeed instrument case will be higher than normal.

This will tend to keep the diaphragm from expanding properly and the airspeed will read *low.* (Dynamic pressure coming in the wrong way tends to cancel out the "proper" dynamic pressure coming in the pitot tube.) In a slip to the right, a low-pressure area tends to be formed on the left side of the airplane, *lowering that static pressure* so that it doesn't "cancel" the static pressure in the pitot tube (and diaphragm). The airspeed indicator "thinks" that the airplane is going faster than it really is and so reads *high.* The point is, if you rely on the airspeed indicator rather than the airplane's attitude, you may be too fast in the slip and lose the benefit of the maneuver, or you may be too slow and have an abnormally high sink rate. Some airplanes have static ports on both sides, which tends to cancel out static pressure problems.

Interestingly enough, some *Pilot's Operating Handbook*s advise against slipping with the flaps down (apparently meaning a *forward* slip) but say nothing against using the wing-down crosswind landing, which is a sideslip. Again, for any given banked slip the aerodynamics are the same. (Crosswind landings will be covered shortly.)

RECOVERIES FROM BAD
SITUATIONS DURING THE LANDING

As everyone knows, you will not always make perfect landings, so you will be shown the corrective procedure if you happen to foul up a little on an approach or landing.

■ Flying into the Ground

This is generally caused by your trying to look over the nose. Naturally you don't want to pull the nose up in the way when you're trying to look over it—the plane hits on the main wheels and bounces. You've probably seen planes hit on the wheels and go up 10 or 15 feet in the air and figured that the plane hit mighty hard to jump that high. Here's what happened. The plane hit the ground with quite a bit of flying speed left because the pilot hadn't gotten the nose up to slow the plane down. The wheels hit and the rebound forced the nose up suddenly, giving the wings added lift. The result was that most of the height of the bounce was caused by this lift—not by the bouncing rubber tires (tailwheel airplanes).

To recover, *ease* the nose over and if the height is 5 feet or more and you've gotten pretty slow it's a good idea to add power (Fig. 13-25). If there's not enough runway left for landing, open the throttle full, push the carburetor heat OFF, and make another pattern and landing. And *this time don't bounce.*

Many students realize that they are about to hit and make a last ditch jerk on the stick or wheel just as the plane touches. This aggravates the situation con-

Fig. 13-25. The bounce recovery.

siderably. It's also poor form to sit there doing nothing after the bounce, with the plane pointed up and the airspeed dropping.

Most bounces are minor—maybe only a foot or so. If this is the case it's generally not necessary to add power; reland, making sure the plane has the correct landing attitude when it touches again. Most instructors will forgive *one* bounce per landing but *not* two or three (or more).

Pilots have the problem of getting out of phase with the bounces, thus making matters worse. Even the tricycle-gear plane can produce a series of impressive actions when the pilot touches down on the nosewheel, yanks back on the wheel, and manages to be 180° out of phase with the airplane. . . . *Land again, making sure the airplane has the correct landing attitude when it touches.*

■ **Dropping the Plane In**

When the plane is stalled too high above the ground, the resulting back-wrenching maneuver is called "dropping it in."

Students, after flying a plane into the ground, to the accompanying anguished screams of the instructor, may go to such efforts to keep from repeating this mistake on the next landing that they "drop it in." This usually results in even louder cries from the instructor.

Dropping in is caused by not knowing where the ground is. You are looking too close to the airplane and the ground is blurred. Your depth perception is poor, and you are perfectly happy landing 5 feet off the ground. Many a flight school owner has wished for a runway jack when seeing a favorite, shiny new airplane being stalled several feet above the ground (Fig. 13-26).

It's very important that you be able to recognize the factors leading up to the drop-in.

As soon as you realize that you and the runway have not made connections and the plane is stalling, *apply power!* You won't have room to completely recover, but the power may slow the descent enough to soften the impact and keep the plane from being damaged.

Fig. 13-26. Sometimes a runway jack is needed for the pilot who keeps landing up in the air.

Another cause of dropping in is ballooning (coming back on the wheel too abruptly) during the transition and then not relaxing back pressure, thus letting the speed get too slow. Being too low on the approach while trying to stretch the glide is another cause. You unconsciously pull the nose up as you see that it's going to be a pinch making the runway; by the time you start breaking the glide, the airspeed is very low (the angle of attack is very high) and you run out of airspeed and inspiration before the airplane reaches the ground.

In some cases of ballooning or bouncing the tendency is to add power while the airplane is *pointed up*, thereby dragging it up even higher at a too-low airspeed. The power is then cut, giving the airplane a further distance to "fall" than was the situation in the first place. The addition of power *at the wrong time* can aggravate an already bad situation. *Basically, power is added to decrease or stop a rate of descent, either immediately as the descent starts and/or during the process.*

■ **Too Low in the Approach**

There'll be quite a few times during the presolo part of your flying when you'll misjudge the approach and realize the plane is going to hit short of the runway. Maybe it was caused by the downwind leg being too far out from the runway or because that first approach turn wasn't started soon enough or you chopped off all the power too soon. At any rate you're going to run out of altitude before reaching the runway. *Don't try to stretch the glide; power is needed.*

As soon as you realize that you are low, apply power. Don't wait until the plane is skimming over that cornfield a fourth of a mile from the runway before doing something about it.

Apply enough power to get to the runway and still maintain the normal approach speed. In applying the power you'll raise the nose slightly. A mistake some students make is forgetting to lower the nose again when they have the runway made and closing the throttle. With the nose up and power off, the plane loses airspeed quickly and the end result is that the student drops the plane in.

Another mistake made by some students is to apply full power and fly the plane to the runway at about cruising speed and then wonder, when the power is chopped off, why the plane won't land but floats to the far end of the runway. The sad thing is that it happens all the time.

The sooner you recognize the fact that you are low and do something about it, the more credit you'll be given for judgment.

■ **Too High in the Approach**

On a hot summer day a light airplane will be affected by updrafts and may end up too high to land; or

you may have flown the downwind leg too close to the runway. With your inexperience, the only thing to do is open the throttle and take it around. Again, the sooner you recognize that a go-around (called a *balked landing*) is required and do something about it, the more credit you'll be given for headwork.

Later, as you become proficient in slipping the plane (or using more flaps), you'll be able to land under conditions that now require a go-around.

Once you've decided to go around, do it! Use full power. Sometimes a student forgets this simple rule and staggers out at partial power, nearly causing heart failure for himself, the instructor, and any spectators present. Open the throttle all the way, get the carburetor heat OFF (if used), and get the flaps up, using the procedure recommended by your instructor. The instructor will be watching your reactions during the latter stages of presolo landing practice, wondering, "Can I trust this student to do the right thing when bouncing or coming in too high or too low on the first solo?" How you take action when action is needed will help the instructor decide when you are ready to solo.

The *Pilot's Operating Handbook* for your airplane will give the best go-around or balked landing procedure for your airplane, and you should use that and the instructor's suggestions. As mentioned in the section on flaps, the general procedure is to apply full power to stop the descent (then carburetor heat OFF, if used) and get the flaps up in increments. If the airplane has retractable landing gear, don't be too hasty about getting it up because if the airplane is inadvertently allowed to contact the runway during the go-around process, it's better that the gear is down and locked. The landing gear poses comparatively low drag at that low airspeed and may be retracted after a safe altitude is reached.

EMERGENCIES ON TAKEOFF
AND IN THE PATTERN

Right after takeoff is the worst time to have an engine failure and the most likely time for it to happen. You are climbing out after an extra good takeoff. Suddenly the engine starts kicking up and then quits cold. What now?

The worst possible move on your part would be to try to turn back to the runway. Sure, there's some pretty rough country right off the end of the runway. But you'll have a better chance for survival there than by trying to turn back to the field. Analyze what happens. The engine quits. The plane is in the climb attitude with no power. If you had time to think about it now, the situation would look like the first part of the normal power-off stall—and that's what it is. While you sit there, the airspeed is dropping fast. You realize what's happening and quickly whip the plane around to get back to the runway. A few seconds later you're sitting in a pile of junk wondering what happened. You gambled and lost. The plane was very slow, and when you turned it the speed was less than the stalling speed for that bank.

Experience has shown that an attempt to make a 180° turn to get back to the runway after engine failure on the climb-out is fraught with peril. It depends on the altitude at which the failure occurs, but statistics argue against trying to turn back. This doesn't mean that you have to just sit there when a shallow turn would let you miss that crockery factory. A rule of thumb for lightplanes is: Don't make 90° of turn at less than 400 feet and 45° of turn at less than 200 feet. Note that the rule of thumb is for light planes, or the average two-place trainer, and heavier airplanes with higher stall speeds may require much more. Approach turns in the traffic pattern were done in a normal glide with the nose down and plenty of altitude. In the takeoff emergency, you start with the nose up and sit there for a while before you realize what's happened. The 400 and 200 feet limits mentioned above take this (and your probable nervousness and roughness in an actual emergency) into consideration.

Get the nose down as soon as possible after the engine failure and make as good a landing as possible under the conditions.

In case of a partial power failure after takeoff and there's no runway left for stopping, fly straight ahead and try to gain altitude without slowing the plane critically. Make all turns shallow. Don't try to turn back to the airport too soon. You're headed into the wind and, if the engine quits completely, will land at a much slower groundspeed this way.

In any actual emergency landing, once you are definitely committed, cut off the mixture, fuel, ignition, and master switches before you land. This will lessen chances of fire after impact if things don't work out as you plan.

Although you probably won't have much time for debating in this situation, it's probably better to turn the master switch off *last* so that communications and the flaps (if electrical) could be used during the last part of the approach.

One problem with students is that once they have opened the throttle for takeoff they feel the takeoff *must* be made. If you are rolling down the runway and things don't feel right to you (perhaps the engine doesn't sound like it should or the trim doesn't feel right), throttle back immediately and keep the airplane straight, hold the wheel back, use careful braking as necessary, and taxi back to find out what the problem is. *Don't be afraid to abort a takeoff.*

Your instructor may cover the airspeed indicator during the presolo landing sessions to let you see that you can make safe approaches by checking the airplane's attitude and power combinations as you complete the pattern. It could be that sometime you may neglect to check the pitot tube during the preflight and discover after you've lifted off that it's not working—mud daubers built a home in there. Or maybe

water had gotten in and froze when the airplane was pulled out of a warm hangar to subfreezing conditions. You'll find after demonstration and practice that making an approach and landing without an airspeed indicator isn't as bad as you imagined. Remember also, once you get to the point of transition to landing you won't be looking at the airspeed indicator anyway.

The instructor also may include a pattern or two with the altimeter covered as well to allow you to estimate the various pattern altitudes *and* airspeeds.

CROSSWIND TAKEOFFS AND LANDINGS

The wind may be directly down the runway on all of your presolo flying days. But the chances are slim that this will happen, so you'll doubtless get some instruction in crosswind landings during one of the presolo landing periods. At any rate, there'll be many times during your flying career when crosswind takeoffs and landings must be made.

■ Crosswind Takeoff

As a review, look at the diagram of a tailwheel airplane sitting crosswind on the ground (Fig. 13-27). Because all the weight is centered near the wheels and because the fin and rudder present a large area and

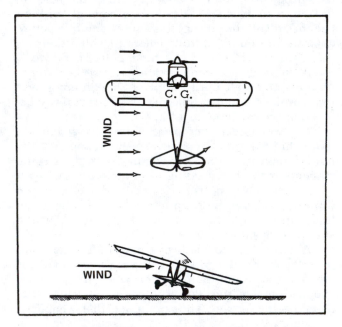

Fig. 13-27. The airplane tends not only to lean away from the wind but also to turn into it (weathercock).

the tail is light, the plane wants to weathercock, or turn into the wind. Since the points of friction (the wheels) are low, the plane also tends to "lean over." These two actions must be taken care of any time the plane is on the ground in a crosswind—the amount of compensation depends on the strength of the wind.

PROCEDURE. The tricycle-gear airplane is lined up with the center line of the runway as with a normal takeoff. Assume in this case a moderate (8 K) crosswind is from the left. As you smoothly apply power, hold full left aileron and be prepared to use enough rudder to correct for both torque and the weathercocking tendency. The controls for both types will look like Figure 13-28.

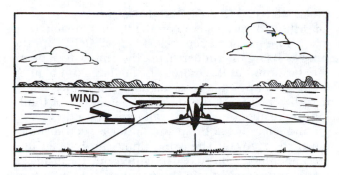

Fig. 13-28. Starting the crosswind takeoff roll.

If the wind is directly across the runway, the ailerons will be ineffective until the plane picks up speed. At low speeds, any effect you get from the aileron will be comparatively small.

As the plane picks up speed, the ailerons and rudder will become more and more effective and less control deflection will be needed.

If you ease the nosewheel off early, the tendency to turn to the left will be much greater than usual because not only have you lost the nosewheel steering and now must depend on the rudder alone but also the nosewheel friction no longer fights the weathercocking tendency.

The tailwheel airplane should have the tail lifted slightly higher than normal, and you should also be prepared to add even more right rudder than in a no-wind or no-crosswind condition. The tailwheel is free and its friction no longer opposes the wind's sidewise force. The tricycle-gear type, all other factors equal, has less tendency to weathercock during the early part of the run because of the greater weight on the (large) nosewheel. In a crosswind from the right, the two forces (torque and weathercocking) tend to counteract each other. You want to make sure of sufficient flying speed at lift-off so that the plane doesn't start skipping. As soon as the plane gets into the air, it will want to start drifting. If it hits the ground again, heavy side loads may be put on the landing gear. So, for the tailwheel type, raise the tail higher than usual and hold the plane on the ground longer.

Gradually ease off the aileron deflection as the ailerons become more effective. As soon as you are sure the plane is ready to fly, lift it off with definite, but not abrupt, back pressure. At the takeoff point there will be some aileron deflection to the left, and the plane will start to bank slightly in that direction after it leaves the ground. Apply left rudder as well and

make a balanced turn to the proper drift correction angle. Level the wings, then continue a normal climb (V_Y). Look back at the runway during the climbout to see that you have the right amount of correction angle (Fig. 13-29). Retract the flaps (and the gear, as applicable) at a safe altitude.

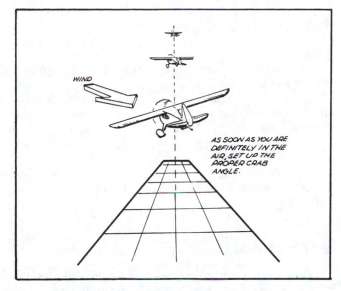

Fig. 13-29. The crosswind takeoff.

You'll hear talk of "the crosswind component" from time to time. This is the component of the wind vector that is actually trying to push you sideways. A strong wind does not necessarily mean that there will be a large crosswind component. When pilots talk of a strong crosswind, they usually are referring to the fact that there is a large crosswind component. Check Figure 13-30.

Some manufacturers furnish crosswind component charts for their airplanes, and you can find whether a particular runway wind condition would

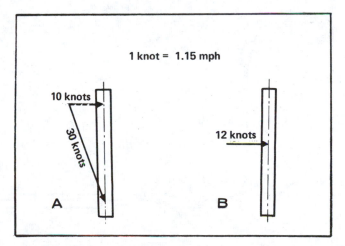

Fig. 13-30. (A) Although the wind has a velocity of 30 K, it has a crosswind component of only 10 K. (B) The wind velocity is 12 K, all of which is crosswind component.

exceed the demonstrated crosswind component for your airplane. Usually the maximum cited (at 90° to the runway) is 0.2 V$_{so}$, or 20 percent of the stall speed with landing flaps.

A crosswind slightly complicates matters for all legs of the traffic pattern. This is where the rectangular course pays off (Fig. 13-31).

PROBABLE ERRORS

1. Not enough aileron into the wind at the beginning.

2. Pulling the airplane off before it's ready to fly—letting the airplane skip and drift.

3. Holding too much aileron at the latter stages of the takeoff so that the plane banks steeply into the wind as soon as the wheels leave the ground.

4. Not making a turn to set up a crab—holding the wing down and slipping after takeoff.

5. Not looking back at the runway occasionally on climbout—resulting in not flying in line with the runway.

■ Crosswind Landing

You know that the plane must be headed straight down the runway when it lands, and in normal landings it was just a matter of keeping the nose straight and the wings level. It's a different proposition if there is a crosswind. If the nose is kept straight and wings level on final, the plane may drift completely away from the runway before you have a chance to land. Or, at best, it will land drifting across the runway and put heavy stresses on the landing gear.

The plane will tend to ground loop because not only are the forces of the impact at work but the plane wants to weathercock as well (Fig. 13-32).

CRAB METHOD OF CORRECTION. One way to correct for drift is to set up a crab on the final approach, then just before the plane touches skid it straight with the runway using the rudder. True, the crab will compensate for drift, but it requires no small amount of judgment to know when to straighten the plane. You've had the experience during normal landings of thinking you're about to touch down and then a thermal or ground effect holds the plane up for a good distance down the runway. The same thing can happen in a crab approach. You may straighten out the plane. It starts drifting again and there you sit—too little time to set another crab—so the plane lands drifting. For the next few seconds you're as busy as a one-eyed cat watching two mouse holes.

Another common error is not straightening the plane soon enough and landing while still in a crab. This doesn't do the landing gear any good either. It's best, then, that the crab method be avoided at this point of experience.

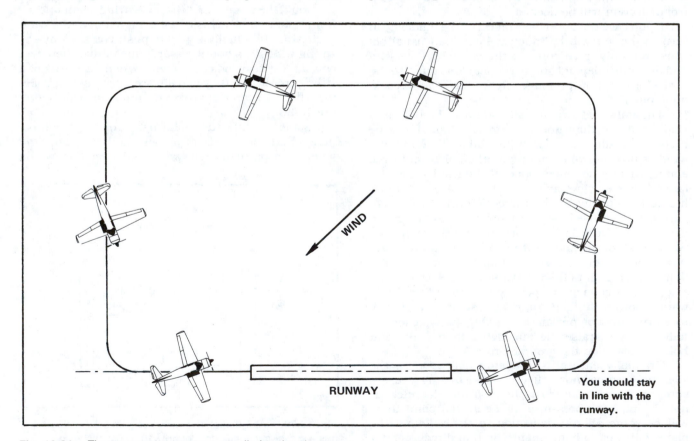

Fig. 13-31. The rectangular course as applied to the airport traffic pattern.

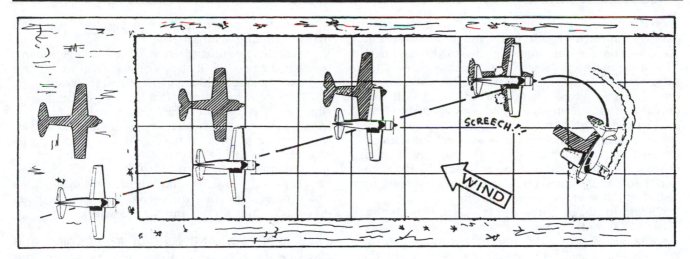

Fig. 13-32. *Hoping* that you'll touch down before the wind drifts you off the runway is no way to take care of a crosswind. A ground loop could result.

WING-DOWN METHOD. The simplest way for the student to make a crosswind landing is by the wing-down or slip correction method. This is accomplished by lowering the upwind wing and holding opposite or downwind rudder. When the plane is banked, it wants to turn. The opposite rudder stops any turn and causes the plane to slip.

The procedure is this: As you roll out on final approach, lower the upwind wing the amount needed to correct for drift and apply opposite rudder to stop any turn. Watch for any signs of over- or undercorrecting and adjust the bank and opposite rudder accordingly. *The lower the wing, the more drift correction is being applied and the more opposite rudder is needed* (Fig. 13-33).

On gusty days you may be continually varying the bank as the crosswind's velocity changes.

Continue to hold the wing down as necessary throughout the landing. There's no need to try to raise the wing at the last second. As the plane slows during the landing process, more aileron and rudder deflection is necessary because (1) the controls are less effective as the speed lowers, and (2) the actual bank should be steeper, since the ratio of wind speed to your speed is increasing (more correction is needed). As an example, if you were approaching at 100 K, a 10-K crosswind would mean a fairly minor correction. But by the time the airplane touches down at 50 K, the 10 K of wind would represent a high percentage of your speed. Of course, if your plane's normal stalling speed

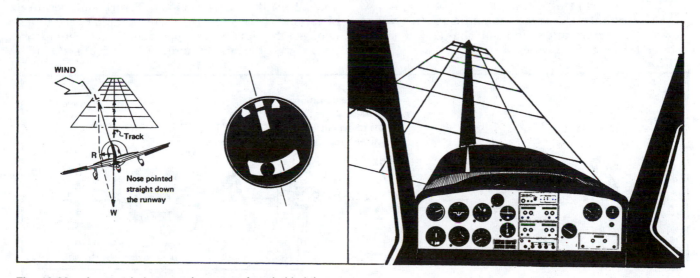

Fig. 13-33. A crosswind approach as seen from behind the airplane and from the cockpit. The slipping resultant (R) compensates for the crosswind component. If you had time to check it, the needle in the turn and slip indicator would show no turn, but the ball would show a slip.

is 50 K, you have no business approaching at such an excessive speed as 100 K. But it makes a good example. Land on the one main wheel and the tailwheel. The other main wheel will touch immediately after. After you are on the ground, the controls are used in the same way as for the crosswind takeoff—ailerons into the wind and opposite rudder as needed to keep the plane straight. Land the tricycle-gear plane on the one main wheel also. Ease the nosewheel down immediately after touchdown for effective ground steering in the crosswind. This method is so simple and effective that it is the most popular way of correcting for drift during the approach and landing.

Some fixed-gear airplanes have nosewheels that "fall free" from the steering mechanism so that when there is no weight on the wheel (anytime it's off the ground) the wheel will be free to track along the ground path when it touches. In other words, in the wing-down crosswind approach the deflection of the rudder does not have the nosewheel "cocked" at touchdown. Other airplanes have a continual direct connection from rudder to nosewheel so that in a left crosswind (left aileron, right rudder) the airplane could have a tendency to veer to the right as the cocked nosewheel touches. Your instructor will discuss the best crosswind procedure for your airplane.

As you are slipping, a slightly steeper glide results, but this is of no great significance.

The only time that slipping is used to correct for a crosswind is on the final approach and landing. After shooting crosswind landings some students climb with a wing down after takeoff. All this does is cut down the rate of climb.

PROBABLE ERRORS

1. Letting the plane turn when the wing is lowered.

2. Not recognizing drift.

3. Mechanical corrections—putting the wing down at a certain angle and not changing it, even though the crosswind varies.

4. Getting so wrapped up in the drift correction that the landing is forgotten.

If you bounce during a crosswind landing, remember that the plane will start drifting. Lower the wing and use opposite rudder as you reland the plane.

The crosswind landing technique is a form of sideslip. The sideslip, you remember, is a slip where the nose is pointed at an object (say, for instance, the runway) and the plane slips to one side—toward the lowered wing. In this case, the plane's movement to the side is counteracted by the wind. If you overcorrected for drift or if there was no wind, the sideslip idea would be evident. In strong winds the two methods may be combined (crab and wing-down).

CRAB APPROACH AND WING-DOWN LANDING. You may like this idea better than the long slipping approach, which can be uncomfortable to both pilot and passengers.

As shown in Figure 13-34, this technique is to set up the proper crab angle (it will have to change as the wind changes on the approach) and maintain a straight path over the ground by this method until the point of round-out is reached. The airplane is then lined up with the runway, and the wing is lowered into the wind (and opposite rudder is used) as necessary to correct for drift as the landing is completed. It takes time getting used to making the transition, but with practice you'll probably find this method more to your liking than the long wing-down final.

You may have a problem getting the idea of the wing-down crosswind correction because in most cases you'll have this condition set up for only a few seconds during the very last of the final approach and the landing. Some instructors, for a training condition, will set up the landing approach 5 or 6 miles out on final, controlling the power themselves, and let the student see that the sideslip *does* work and, because of varying crosswinds, the correction may need to be changed throughout the approach. This technique is used to give the student more time to get used to the

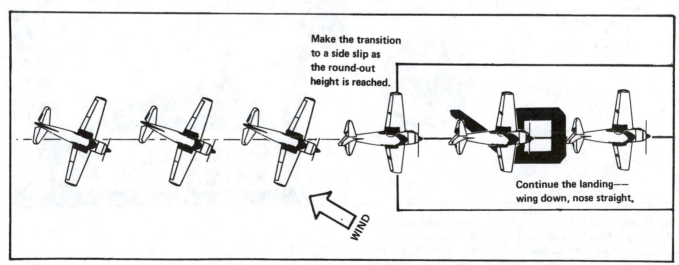

Make the transition to a side slip as the round-out height is reached.

Continue the landing—wing down, nose straight.

WIND

Fig. 13-34. A crab approach with a wing-down landing.

cross-control effect. After the idea is firmed up the instructor will usually go back to the crab approach and wing-down landing (or you may prefer the wing-down method for all of the final leg).

POWER-OFF APPROACHES

Before you solo, your instructor will demonstrate and have you practice some approaches and landings with the power reduced to idle on the downwind leg as the airplane is abeam of the point of intended landing. (This is good experience in judging the flight path for landing on a prechosen spot without use of power. This doesn't mean that you can't add power to make the runway if you misjudged and are undershooting during one of the approaches; in fact, it's again strongly suggested that you don't get the airplane slow and cocked up trying to stretch the glide.)

Figure 13-35 shows a typical traffic pattern for engine-at-idle approaches.

■ Downwind Leg

In earlier days, the type of approach shown in Figure 13-35 was called the *180° accuracy approach* and was basically the "normal" approach for most light trainers.

DOWNWIND AND BASE KEYS. Figure 13-36 shows how the power-off pattern may be played to compensate for a too-wide or too-close downwind leg. You'll also fly the pattern to cope for varying wind conditions.

If the wind is strong, the downwind leg should be closer to the runway; the wind effects on final will be greater because of the lower groundspeed and hence the airplane will spend a longer time in the headwind (or retarding) condition. In other words, a 30-second descending downwind leg (after the power is retarded to idle) is not matched by a 30-second final leg. Because of the headwind on final, depending on its strength, the final leg might be 1 minute or more (as an example), with that extra time of retarding effects resulting in not making the runway unless power is used.

With experience you will be able to judge a proper distance from the runway (called "abeam distance") for these approaches, but some instructors use aids such as masking tape on the wing (low-wing airplane) or on the strut (Fig. 13-37).

If there is a crosswind as shown in Figure 13-31, you will have to make corrections for wind on the crosswind, downwind, and final legs as well as that expected on base.

The first turn, onto base, should be (as mentioned earlier in the chapter) fairly steep (up to 30° bank), particularly in a high-wing airplane. The 90° (plus) of turn should be completed without delay so that you

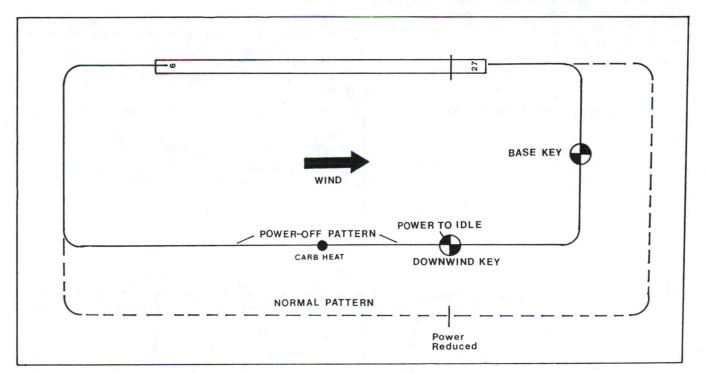

Fig. 13-35. A power-off (engine at idle) approach. The exact use of carburetor heat will depend on the *Pilot's Operating Handbook* of the airplane and/or your flight instructor's suggestions, but here full carb heat is applied midway on the downwind leg and used throughout the approach.

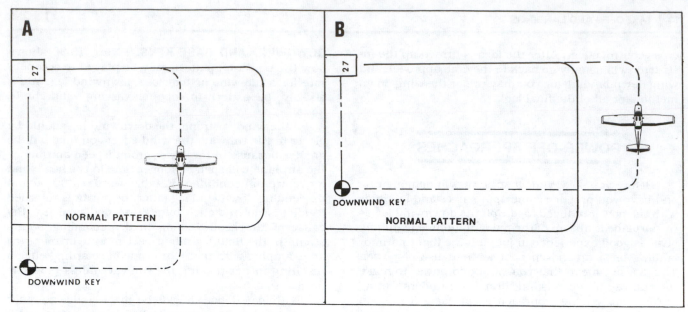

Fig. 13-36. Power-off approach (dashed lines). At the downwind key you will check to see if the airplane is too far from or too close to the runway.

(A) If the airplane is too far out from the runway, the descending part of the downwind leg could be shortened slightly to ensure making the selected touchdown point. (You should have at least one-fourth mile on final, though, because the final turn should not be a too-low altitude.)

(B) If the airplane is too close to the runway at the downwind key position, the downwind leg is extended as necessary to ensure that the airplane is not too high on final.

The temptation is to make a *quick* turn to move in or out to get over to the normal abeam distance when on the downwind leg. It's best, however, not to make radical turns in the traffic pattern because of your airplane's blind spots and the possibility of a midair.

Fig. 13-37. The instructor may put tape on the wing or strut if you're having problems finding the proper abeam distance. These references can change with various student heights (and the wings must be level each time). After some experience you'll be able to judge the abeam distance without the tape.

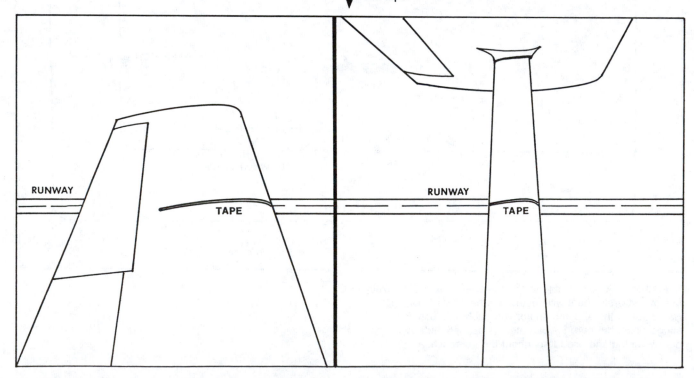

will arrive at the base key position with the wings level and a good view of the runway to check for needed corrections. On base *always look to the right* (in a left-hand pattern) *to check for airplanes that might be making a long straight-in approach.*

A problem instructors find on the base leg is that students tend to stare over the nose straight ahead, and *not* look at the runway until they *think* they are ready to turn onto final. This habit often leads to a late turn and flying past the runway center-line extension (Fig. 13-38). This could lead to the dangerous situa-

tion of "cheating" on the turn with a cross-control stall resulting (see Chapter 14).

So it's better in most cases to make the turn onto the base leg with a fairly steep bank (up to 30°) and the turn onto final with a more shallow one (20° or less). See Figure 13-39.

Fig. 13-38. Students (and other pilots) tend to stare straight over the nose while on base. Of course, you *should* check ahead here in case somebody is on an opposite base leg, but don't get locked on it.

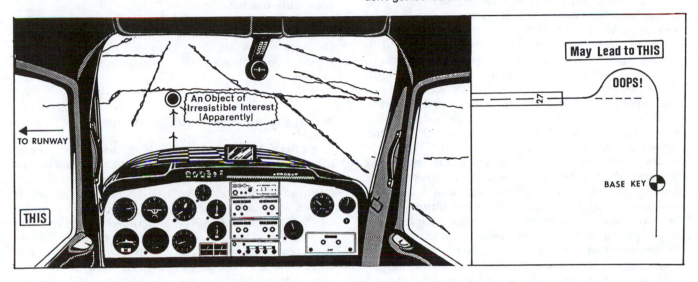

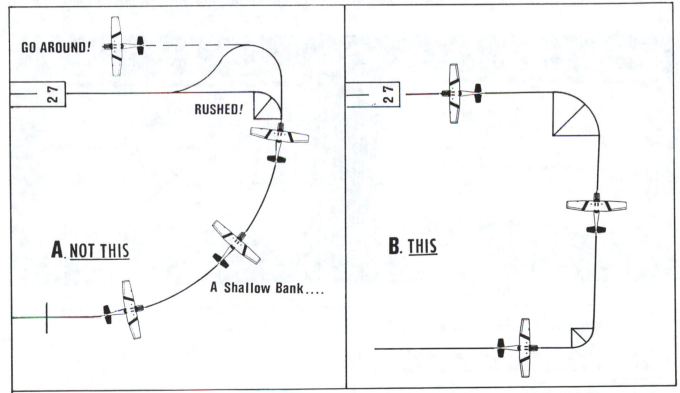

Fig. 13-39. (A) The turn onto base is extra shallow, and the wing is obscuring the runway. When the wings are leveled, the pilot finds that either a very steep (and dangerous) turn onto final is required or it is necessary to make an S-turn (also a bad idea) in order to get lined up with the runway. A go-around may be necessary.

(B) The proper way is to make a comparatively steep balanced turn of about 30° bank so that the 90° of turn is completed fairly soon, the wings leveled, and the runway checked. A shallower (and less rushed) turn is then made at the proper point to roll out on final.

Figure 13-40 shows the results of that too-shallow bank onto base, as seen from the cockpit.

THE TOO-HIGH APPROACH. If you see that you are high at the base key you have several options in an actual engine failure situation. You may (1) add flaps, if available, (2) deliberately S-turn past the center line of the landing area, or (3) slip the airplane after turning on final.

1. Assuming that you have flaps and are using them in increments (just before turn onto base, more flaps on base, with last flap setting on final), you should be able to have a setting or two to spare if you are high at the base key position, or on final.

2. Deliberately flying past the center line and S-turning back is not used on a "normal" approach, since it could put the airplane in a wrapped-up (steeply banked) turn at a low altitude. Later, as you gain experience, that procedure when used with care could be a valuable aid in a real engine-out situation, but your instructor will not want you to use it for the normal power-off approach at the airport.

3. When you're too high, slipping is *the* major aid for the airplane not equipped with flaps or if you have used up all the flaps. To repeat what was said in the section on Slips, the *Pilot's Operating Handbook*s of some airplanes warn against steep slips with flaps extended past a certain setting because of the effect of the disturbed air flow on the elevators.

A slipping turn to the final leg was used in earlier days in emergencies when the pilots saw on "base"

that they were high and the final leg would be too short to allow time to set up a slip there. Your instructor will discourage use of the slipping turn in your normal practice sessions but may discuss it for emergency use during some of your ground sessions. Again, use of it will depend on pilot experience and the airplane (Figure 13-41).

The main thing in these practice sessions is not to get the airplane too wrapped up turning final and to realize that if you're feeling pushed or rushed you have power available to make a go-around using the procedure discussed earlier.

TOO LOW. If you're too low in the power-off approach, the best and safest thing is to add power (as for the reduced-power approach discussed earlier in the chapter).

■ Gusty Air and Wind Gradients

If the velocity of the wind changes suddenly, it can affect your airplane. You know that it is possible to have different wind directions and velocities at different altitudes, so take the theoretical situation shown in Figure 13-42.

You are gliding at 70 K and the plane stalls at 50 K. Your speed in relation to the air is, of course, 70 K. But your speed in relation to the ground and to the calm wind is 70 minus 30, or 40 K. What happens as you suddenly fly into the area of calm air? The air-

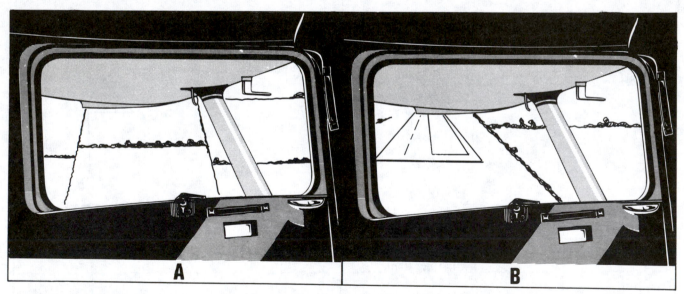

Fig. 13-40. The problem of Figure 13-39 as seen from the cockpit.

(A) You're making that nice, very shallow, banked turn onto base. The runway is out there somewhere, but the wing is obscuring it.

(B) As you are rolling out of the turn on the base leg, you look for the runway to get set up for the turn onto final (uh-oh). A tailwind component on base (Fig. 13-31) can make matters worse.

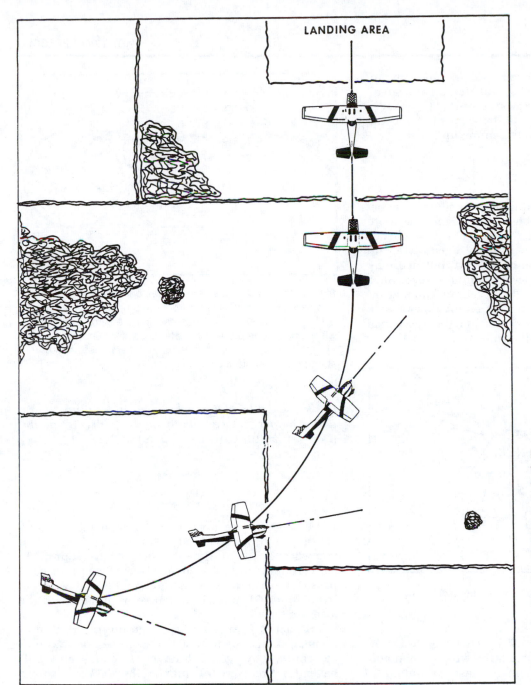

LANDING AREA

Fig. 13-41. A slipping turn to the left being used in an emergency situation (bank exaggerated). The airplane is allowed to turn toward the lowered wing by using comparatively more (left) aileron deflection and less (right) rudder deflection than would be used to maintain a straight slip. At this point in your experience the slipping turn is only *discussed* as *one* of the methods used in losing altitude on an approach, particularly in an emergency situation.

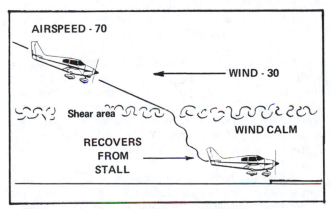

AIRSPEED - 70

WIND - 30

Shear area

WIND CALM

RECOVERS FROM STALL

Fig. 13-42. A shear area could cause problems on approach.

speed or speed relative to the calm air mass is only 40 K. The balance between lift and weight is broken (lift decreases suddenly) and the airplane accelerates downward, increasing the angle of attack to the point that the airplane stalls—which aggravates the situation.

The above illustration is exaggerated, but the principle applies on any gusty day. Many a pilot has been set for a landing on a gusty day and had the bottom drop out, then clambers from the wreckage and wonders why the airplane "fell." It's not unusual to have fluctuations of plus or minus 10 K on the airspeed indicator during an approach.

If there are gusty winds, fly your approach 5–10 K faster than normal. That way you won't be in such a bind when the wind speeds kick back and forth.

If you hit sudden turbulence on an approach, apply power first and then ask, "Was that a shear area? Was the wind dropping or picking up?" The chances are that the plane won't stall, but you may find the airspeed low with the ground coming up fast.

■ **Wake Turbulence**

Always be leery of taking off or landing close behind another plane. The bigger the plane, the more cautious you should be. Lightplanes always come out second best in a contest with an airliner's wing tip vortices. If the wind is calm, wake turbulence may hang over the runway for several minutes—even after the plane has gone out of sight. If you *have* to take off or land behind another plane, get on the upwind side if there is a crosswind (Fig. 13-43). In light crosswinds one vortex may stay over the runway.

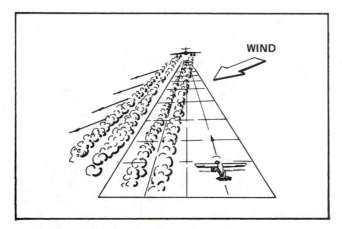

Fig. 13-43. Take off or land on the upwind side of the runway if there is a crosswind.

Better yet—don't take off for a while. Always be prepared for wake turbulence if the wind is calm and big planes have taken off in the last few minutes. Lightplane wake turbulence is bad enough, but airliners and heavy aircraft can set up a fatal crash for you.

The wing tip vortices are shown in Figures 13-44 and 2-16. But not so well shown is the fact that the vortices descend after being formed, moving downward at 400–500 feet per minute, and tend to level off about 900 feet below the aircraft. Flying behind and below a large jet when it's clean and slow could cause big problems for you.

Read Appendix A of this book, which covers information you should know about wake turbulence. Look it over carefully for some facts that could save your neck someday.

Don't fly at cruise speed directly behind and below a large aircraft or cross its flight path if you can avoid it. The turbulence may cause load factors so great as to cause structural failure of your plane.

Again, don't be proud—take it around if you

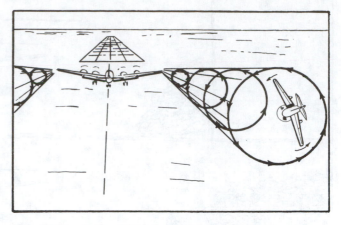

Fig. 13-44. In the olden days it was believed that "propwash" or the slipstream from the propeller(s) was the culprit, but then jets came out, causing the same problem, which couldn't be *propwash!* Later flight tests showed that the wing tips were the culprit and that the effects were much stronger than earlier considered.

think there'll be wake turbulence on landing. Wait a while longer before taking off if you're in doubt. (Don't sit on the runway; taxi out of the way.)

SOLO

When you are able to make good takeoffs and landings and use good judgment in your flying, then you will solo. By this time you should have your medical certificate and student pilot certificate and have passed a written examination on Federal Aviation Regulations 61 and 91 and on your airplane and the local traffic rules.

Students, particularly in group training programs, get anxious to solo. (Jane or Joe Doe soloed in 12 hours, how come I have 16 hours and still haven't?) Don't worry, maybe their 6-hour slump wasn't as pronounced as yours or maybe they didn't have one. Very few students solo without hitting a slump somewhere along the line. It hits a large number of students at about the 6-hour point—and oddly enough it's called a 6-hour slump.

Usually it strikes after you've had an hour or two in the pattern and had a good first period of shooting landings. You feel pretty good and are all set to get these landings down pat and solo earlier than that Horace Numbskull who thinks he's such a hot pilot. Things don't go too well, and suddenly you can't find the ground with both hands. You're right in the middle of the 6-hour slump and feel lower than a whale's stomach.

You go on shooting landings, and it seems for a couple of hours that absolutely no progress is being made. Then suddenly one day you start landing like a pro; the instructor gets out—and you solo.

The instructor will tell you that the plane will get

off quicker, climb better, and want to stay up longer with you alone—and so it will. You'll be so busy flying the plane that you won't notice that the other seat is empty. The first solo is the point where you suddenly discover that flying is pretty wonderful; when you get down and receive the congratulations of the people in the airport office, you're standing about a mile high.

The procedure for solo will generally be this: The instructor will get out and you'll shoot one landing and taxi back to where you started. The instructor may wave you on or stop you to discuss points of the solo takeoff and landing. In most cases you'll shoot three landings—for this reason: If you shoot one and then lay off for a few days you'll probably start thinking, "That one good landing could have been luck. . . ." If you shoot three you'll know it wasn't luck and won't talk yourself into a state of nervousness before the next solo flight.

Students sometimes worry about what the instructor will think if they have to take it around on that first solo approach. Traffic may require that you have to make a go-around during one or more of the first-solo approaches. So what? More than one first-solo student has had to circle (this writer did) while somebody else made an approach and landing. So don't worry about it. The instructor demonstrated,

and you've practiced this enough so that you'll know what to do. The main thing is to not talk yourself into a dither if you have to go around. The instructor will think it good judgment on your part, even if you lean to the cautious side. There will be plenty of times during your flying career when you'll have to take an airplane around, so do it now as necessary.

On a go-around, make a slight right turn and then parallel the runway so you can keep an eye on departing traffic as noted earlier in the chapter. If you think the final turn required is too steep, take it around as shown in Figure 13-39(A).

The instructor will probably ride around a couple of times the next time you fly and then turn you loose to shoot two or three periods of landings before taking you out to the area to introduce more advanced work.

Incidentally, students usually make their best landings during that first solo.

Don't get cocky and think you've learned it all now that the first solo is behind you.

Don't ask when you will solo. You've bounced and jounced your instructor while causing a lot of aspirin chewing, for the past few hours. When you are ready, the instructor will be *more than glad* to get out. Seriously, it's as much a matter of pride for the instructor as it is for you.

PART THREE

POSTSOLO MANEUVERS

14

ADVANCED STALLS

CROSS-CONTROL STALL

This stall is a perfect illustration of the dangers of abusing the controls. A typical case of a cross-control stall accident is this: A pilot sees in the turn onto final that it will require a sharper turn to line up with the runway. (Look at Figs. 13-38, 13-39, and 13-40 again.) Pilots have always been told the dangers of banking too steeply at low altitude, so to avoid this and still make the turn, they do something much worse.

A small voice inside says, "Go on, use rudder, skid it around and you won't have to bank it."

So the pilot skids it. As you know full well, when a plane is skidded, the outside wing speeds up, gets more lift than the inside one, and the plane starts to increase its bank. The pilot realizes this and unconsciously holds aileron against the turn. The down aileron on the inside of the turn helps drag that wing back more, slowing it up and decreasing the lift,

which requires more aileron to hold it up—and so the cycle goes. As the plane is banked, use of bottom rudder and opposite aileron will tend to cause the nose to drop. The pilot helpfully holds more back pressure to keep the nose up. This, then, is the perfect setup for one wing to stall before the other. Figures 14-1 through 14-4 show the maneuver as seen from outside the airplane and from the cockpit.

The practice cross-control stall (at a safe altitude) is aimed at teaching you to recognize how such a stall occurs and giving you practice in effectively recovering from such a situation. The cross-control stall, like any other, can occur at any speed, altitude, or power setting.

The bad thing about this stall is that the forces working on you as it is being approached are "normal." That is, you'll feel pushed to the outside of the turn (as is the case every time you make a turn in a car) so may not be warned of impending trouble. The slipping type of stalls (over the top) usually have peo-

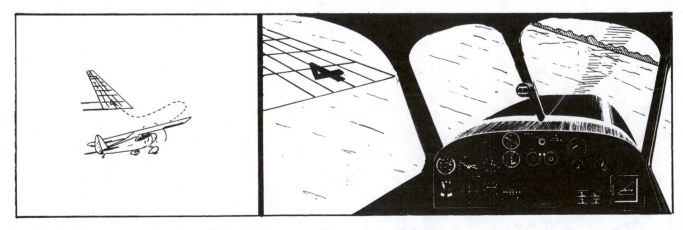

Fig. 14-1. Pilot sees that the plane will fly past the runway if the rate of turn is not increased.

14-1

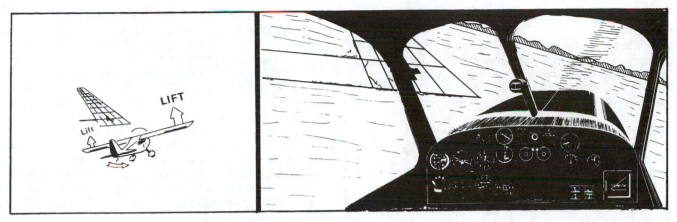

Fig. 14-2. The pilot doesn't want to make a steep bank at low altitude, knowing that the stalling speed increases with angle of bank, so applies rudder to cheat a little. Lift is greater on the outside wing because of its increased speed. The bank starts to increase because of difference in lift. The pilot doesn't want the bank any steeper, so opposite aileron is applied.

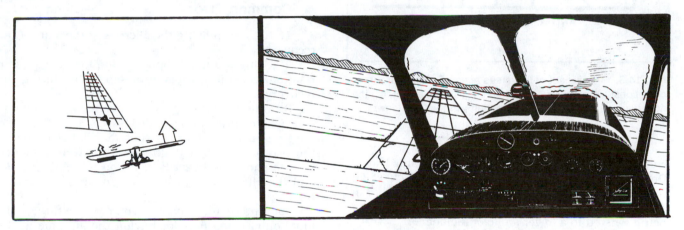

Fig. 14-3. The nose tends to drop because of rudder and aileron, so up-elevator is added. Down-aileron drags wing back causing a steeper bank. More opposite aileron is applied, etc.

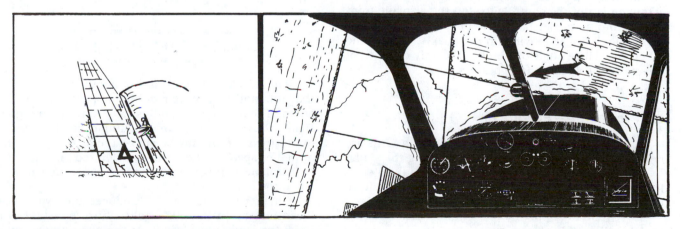

Fig. 14-4. The inside wing stalls and the plane rotates abruptly toward the low wing. The altitude is insufficient for recovery.

ple scrambling around in the cabin and putting things to rights.

Figure 14-5 shows why the inside wing (with the down aileron) tends to stall first when by "common sense" the down aileron should hold that wing up.

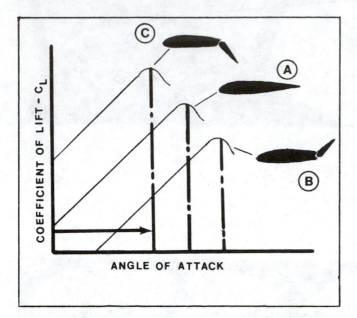

Fig. 14-5. Coefficient of lift versus angle of attack for three conditions: (A) airfoil with no aileron or flap deflection, (B) the right wing (aileron up), and (C) the left wing (aileron down).

Looking back at Figure 12-12, you can see the effect on the coefficient-of-lift curve of extended flaps. Down ailerons will give the same general effect (although they're farther out on the wing). The up aileron (B) in Figure 14-5 gives the effect of the wing being twisted to a lower angle of incidence.

As you increase the angle of attack, as indicated by the arrow in Figure 14-5, the inside wing with the down aileron (C) will reach its "break" first and will stall. The question comes up: Shouldn't there be more lift on that wing and therefore no stall? By a bizarre arrangement of the controls to stop the airplane from rolling into a steeper bank, the pilot is assuring that the actual lift of each wing (in pounds) is equal to the other. And since angle of attack is the only criterion of the stall, that inside wing goes first.

■ Procedure—Power-Off

1. This is a turning stall, so the clearing turns are not necessary. Keep looking around during the stall.
2. Pull the carburetor heat ON and make a shallow gliding turn. (The direction of turn will be left up to you—practice them in both directions.)
3. Gradually apply more and more inside rudder to "cheat" on the turn.
4. Use opposite aileron as necessary to keep the bank from increasing.

5. Keep the nose up by increasing back pressure.
6. When the break occurs, neutralize the ailerons. Stop any further rotation with opposite rudder. Relax back pressure.
7. Roll out of the bank with coordinated controls.

The rotation toward that inside wing is so fast that it is quite possible that the bank will be vertical or past vertical before it is stopped.

In many ways the recovery from this stall is close to the recovery from a spin. The lift of the wings must be made equal. This is done by neutralizing the ailerons and stopping further rotation with rudder *at the same time* as the back pressure is released. The neutralizing of the ailerons speeds the equalizing of lift.

Because of the low position of the nose when the rotation is stopped the speed will build up quickly. Recover to straight and level flight with coordinated controls as soon as possible without overstressing the plane or getting a secondary stall.

■ Common Errors

1. Not neutralizing the ailerons at the start of the recovery.
2. Using too much opposite (top) rudder to stop rotation, causing the plane to slip badly during the recovery.
3. Waiting too long to roll out with coordinated controls; too great an altitude loss.

Cross-control stalls, as such, are not required on the private practical (flight) test but were included here to show the hazards of trying to "cut corners" and the effects of neglecting coordination at critical times.

Your instructor may have you practice these stalls (dual only) using flaps or having the airplane in the clean configuration. Naturally if flaps are used, the instructor will be responsible for retracting them after the stall breaks to avoid exceeding the flaps-extended airspeed (V_{FE})—the maximum value (top) of the white arc.

ACCELERATED STALLS

The accelerated stall is a maneuver for proving that the stall is a matter of angle of attack, not airspeed. The accelerated stall is to be started no higher than 1.25 times the normal stalling speed because of extra stresses that may be put on the plane by stalling at higher speeds. The term "accelerated" means that the airplane is under forces of acceleration, or g forces, when the stall occurs.

Chapter 9 "The Turn" introduced the expression "load factor." In the 60° banked turn (constant altitude) the load factor was 2, or the airplane was subjected to 2 g's. As far as the wings were concerned the airplane weighed twice as much. By working out the math it was shown that the stall speed was increased by the square root of the load factor, which turned out to be 1.414. In the 60° banked turn the airplane stalled at a speed over 41 percent higher than normal stall speed. The same increase in stall speed effect would have occurred if instead you had loaded the plane to *twice* its normal weight and done an old-fashioned straight-ahead stall. (Don't load it that way!)

The pilot in Figure 14-6 can't understand why evidences of a stall show up when the airspeed is so high. (In a 60° banked constant-altitude turn the stall speed is increased by a factor of 1.414, from 62 to 88 K. The pilot is feeling a load factor of 2 g's.)

Fig. 14-6. The stall is a function of angle of attack, not airspeed. This pilot is learning that the load factor is 2 g's and the stall speed in a 60° banked level turn is 1.414 times that of normal 1-g flight.

You can also encounter positive acceleration forces when pulling up from dives or brisk pull-ups from straight and level; acceleration doesn't just happen in a turn, although in everyday flight this is the most likely place to meet it. Of course, you could run into strong negative acceleration forces (or negative g's) if you were to shove the wheel forward abruptly at high speeds. (Don't do *that* either!) You've been used to thinking of acceleration in terms of increasing speed. This is not the type of acceleration of interest here.

Your elevators are not only the airspeed control but are also your g control. By rapidly pulling back on the wheel or stick, positive g forces are created. (They may be too positive at speeds well above the stall, and problems could result.) Pulling back abruptly means

that excess lift is temporarily in existence. The airplane will start slowing immediately to reestablish the old lift value (*the airplane always tends to remain in equilibrium as far as lift and weight are concerned*). You may, however, exceed the airplane's limit load factor (maximum allowed g's) before the airplane has a chance to slow and equilibrium can be established. Normal category airplanes are required to have a maximum limit load factor of 3.8 positive and 1.52 negative g's. Utility category planes have limit load factors of 4.4 positive and 1.76 negative g's. Your trainer will fall into one of these two categories. If you pulled hard enough to stall the airplane at twice normal stall speed, you'd pull 4 g's (2 squared is $2 \times 2 = 4$ g's). And this would be over the limit for a normal category airplane. (A more complete explanation of the various category requirements will be covered in Chapter 23.) If you pulled hard enough to stall the airplane at *three* times the normal stall speed, 9 g's would result (3 squared = 9), and you would be flying a wingless—or tailless—wonder.

The stalls you've done up to now have been at a normal load factor of 1 g. You *gradually* increased the angle of attack until the stall occurred.

The accelerated maneuver stalls will be done best from 45° banked turns with reduced power in cruising configuration. You'll increase the angle of attack smoothly in a level or slightly climbing turn until the stall occurs just above the normal stalling speed. Don't start increasing the angle of attack at a speed of more than 25 percent above the normal stall speed.

This stall is done in the turn for two reasons: (1) This is the position in which most lightplane pilots run into accelerated stalls when they try to "tighten the turn" by rapidly pulling back on the wheel or stick, and (2) it is done to keep from getting into a whip stall or "tail slide," as might happen if the pilot pulled the plane sharply up from a wings-level position. A whip stall occurs when the plane stalls in a vertical nose-up position—the plane slides backward before recovery can be effected. This puts a severe strain on the structure, particularly the elevators, and could cause structural failure. If the plane is banked, the position of the center of gravity and the effects of the fin and rudder wouldn't allow the plane to get into such a predicament; the nose will fall to one side.

The stall break and recovery is fundamentally the same as the turning departure and approach stalls. The high wing will generally stall first unless you are skidding. Release back pressure, and after the recovery is started return the plane to level flight with coordinated controls. The flight instructor may want you to recover (1) as soon as you recognize the stall and (2) after the stall develops and the nose falls down through the horizon.

In (1) the recovery can be simple and quick. Relaxing the back pressure at the right time will cause the plane to recover in straight and level flight if the high wing stalled first and you can effect a recovery just as the wings roll level.

In (2) the plane may have rolled over into a bank

in the opposite direction and the nose will be low. Recover from the stall, *then* bring the plane to wings-level flight. Too many students try to do everything at once in this case and end up stalled again. The main thing is to get a complete and clean stall recovery with a minimum loss of altitude—the first time.

Bring the wheel back smoothly. Don't jerk the wheel back and put a lot of unnecessary stress on the plane. You will be expected to be able to recover from the accelerated stall both with and without application of power. Use power to recover unless specifically asked not to.

■ Procedure

1. At reduced power in cruise configuration make a turn with about 45° of bank. Slow the airplane to 25 percent above normal power-off stall speed by slowly increasing the back pressure.

2. When a speed of less than 25 percent above normal stall is reached, increase the back pressure rapidly but smoothly until the stall occurs.

3. Release back pressure, open the throttle, and recover to straight and level flight using coordinated controls. Also practice recoveries without use of power.

Another procedure for an accelerated stall is to set up the 45° bank and *gradually* slow up the airplane at a constant altitude or a moderate rate of climb, reducing the power until the stall occurs at a higher-than-normal airspeed.

■ Common Errors

1. Jerking the wheel back—putting too much stress on the plane.

2. The other extreme—too slow in applying back pressure so that the stall isn't an accelerated one.

3. Too brisk a forward pressure on recovery so that negative load factors are applied.

STABILATOR OR ELEVATOR

TRIM TAB STALL

When the elevator trim tab was discussed in Chapter 8, it was noted that the elevator trim tab or the stabilator trim could get you into trouble under certain conditions.

The combination of high power setting and full nose-up elevator (or stabilator) trim can result in a tendency for the airplane to assume some pretty impressive nose-up attitudes. (These spectacular reactions are most likely to occur accidentally when close to the ground.)

You saw the effects of power when you were first practicing recoveries to level flight from a glide. The nose wanted to rise as cruise power was applied, and it took a definite concentration on your part to keep it down where it belonged. You had the plane trimmed for level flight those times. Picture the situation if you had been using *full power* as would be done on a takeoff or a go-around with the trim in the *full nose-up* condition. Add to these problems outside distractions (plus the fact that you are at a low altitude), and you have a bad combination.

The instructor will demonstrate this stall at a safe altitude, as for all the other stalls. The conditions will be set up and the instructor will apply full power and probably keep hands off the wheel to show you what the plane will do. (You guessed it; the nose will claw upward and the stall break, when it comes, will most likely be to the left because of torque effects.)

You'll then get a chance to practice recovering before things get out of hand. The main idea is to keep that nose down. For most light trainers the nose can be held down without too much trouble. The hazard is in the surprise associated with an actual situation. Practicing these stalls will give you quick recognition of the problem and the best means of recovery for your particular airplane. The following will generally apply:

1. *Recognize the problem immediately* (improper trim).

2. *Get the nose down to no higher than the normal climb position with whatever forward pressure is needed.* Then get the trim back to neutral as soon as practicable. If the forward pressure required is so strong as to require both hands on the wheel, keep them there and climb to a safe altitude before trying to get on the trim. Then you can use one hand intermittently to get the trim squared away.

3. *Do not chop the power after the nose has risen.* You'll be too low to clear the problem and add power again. If the situation has progressed, the chopping of power could cause an immediate stall.

EXCESSIVE TOP-RUDDER STALL

Rough usage of the controls during a slip is a good way to discover the excessive top-rudder stall. By using excessive top rudder in a slip or turn, the high wing is stalled and the stall is quite similar to the departure and approach stalls. As you remember, in these stalls the high wing stalled first because the plane started slipping and part of that wing's airflow was disturbed or blanked out. In the excessive top-rudder stall you speed up the process by holding top rudder and causing the plane to slip. The high wing drops a little faster because of this "help," but the recovery is the same as for the departure or approach stall, or any stall for that matter. Relax back pressure as you apply coordinated control pressure to return to straight and level flight.

EXCESSIVE BOTTOM-RUDDER STALL

This one is close to the cross-control stall except that in the turn rudder alone is used to stall the low wing. You skid the plane as back pressure is increased, and the high wing has more lift so that when the plane stalls the low wing is the first to go. Recovery is standard. Relax the back pressure as you stop further rotation with the rudder and return to level flight by coordinated control usage.

GENERAL TALK ABOUT STALLS

By now you should realize that the plane can be stalled flying straight down or straight up. You can make the high wing or low wing stall in a turn. You see that one thing has remained true in all stall recoveries—*get the air flowing smoothly over the wing.* This may mean relaxing the back pressure in a light-plane or using a brisk forward movement of the stick or wheel in a heavy plane.

The problem with any stall is that at the break the nose pitches down and your instinct is to hold back on the stick or wheel to bring it back up. At low altitudes in an inadvertent stall this is even more the case. It may take some doing to move the control wheel or stick forward to recover from the stall. (You'll have to move the wheel back again to stop loss of altitude; but don't get a secondary stall.)

By the time you finish these advanced stalls you'll know a great deal more about the airplane and will be confident in your ability to fly it.

15

EMERGENCY FLYING BY

REFERENCE TO INSTRUMENTS

The idea of giving student pilots emergency instrument instruction was debated in aviation circles for several years before the ruling was made. Some argued that "a *little* knowledge is a dangerous thing" and felt that students or private pilots might get just enough confidence in their instrument flying ability to attempt to fly in marginal weather. Others said that it was up to the individual. Some might try it, but many more lives would be saved by this training than would be lost by student overconfidence.

These are emergency maneuvers like the low- and high-altitude emergencies and simulated engine failure on takeoff. You would not deliberately cut the ignition to practice those emergencies nor would you deliberately fly into conditions where flying by reference to instruments is necessary.

In requiring this training the FAA still does not authorize instrument flying by anyone other than a properly rated instrument pilot. If you accidentally fly into Class B, C, or D controlled airspace (see Chapter 21) when the weather is below visual flight minimums, the FAA will be interested in talking to you, since you would be a collision hazard for airplanes on authorized instrument flight.

Flying under actual instrument conditions is quite different from simulated instrument flying. If you foul up during the practice sessions, you can always pull the hood up and recover by looking out (if the instructor hasn't already recovered for you). "Practicing" actual instrument flying is like "practicing" actual parachuting. There's no practice about it—it must be done correctly each time. As the Navy would say, "This is no drill!"

To ensure that you have a good grasp of emergency instrument flying, you should get instruction on the following items:

1. Recovery from the start of a power-on spiral.
2. Recovery from the approach to a climbing stall.
3. Normal turns of 180° duration left and right to within 20° of a proper 180° heading.
4. Shallow climbing turns to a predetermined altitude and heading.

5. Shallow descending turns at reduced power to a predetermined altitude and heading.
6. Straight and level flight.

These will probably be introduced and practiced as definite exercises during the dual flights of the second or postsolo phase of the syllabus. A portion of each dual period will be spent on the above requirements as well as the strictly visual flight maneuvers as covered in Chapters 14, 16, 17, and 18.

One thing will be stressed at all times. *Fly the airplane in exactly the same manner as in normal flight. There are no gimmicks, shortcuts, or special techniques. The instruments "see" the plane's attitude and give you the information.* Instrument practice is to teach you to interpret the information given by the instruments and to make the plane respond in the desired manner by use of the indications.

The six maneuvers are based on the premise that you have just flown into worsening weather conditions and suddenly the ground has disappeared and there's nothing left but a gray mist. Which way is up? Or down? Even though for the first few seconds the plane may continue flying straight and level, your body may want to lie to you. The balance center in the middle ear may get a bum signal. (It's surprising how much that balance depends on sight.) You are *convinced* that the plane is in a left-climbing turn or right-diving turn or any other maneuver that may strike your fancy at the moment. The number of flying hours has no bearing on a pilot's ability to fly the plane by "feel" alone. *No pilot can do it.*

Nearly all trainers used today have a full panel of flight instruments as shown in Figure 15-1, but for training purposes some of the explanations will also assume that a heading indicator and attitude indicator are not available. *Coordinated use of the ailerons and rudder is needed, as always, and the turn indicator and ball should be thought of as working together. The term "turn indicator" as used in this book will mean either the needle in the turn and slip or the small airplane in the turn coordinator as applicable to your airplane.* The wheel and throttle will also be coordinated throughout these maneuvers.

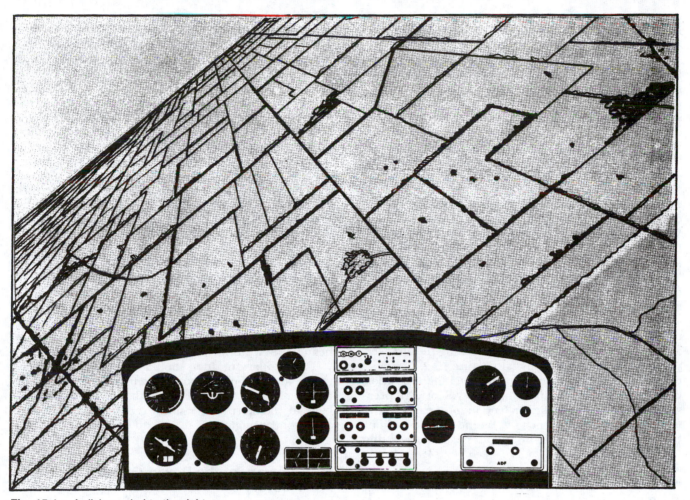

Fig. 15-1. A diving spiral to the right.

Again, if the plane you are using for training has an attitude indicator and a heading indicator, you would use them in all of the following maneuvers.

RECOVERY FROM A POWER-ON SPIRAL

The power-on spiral is probably the most common result of trying to fly the plane under instrument conditions without proper training and/or instrumentation.

Many fatalities have been caused by pilots entering this flight condition when visual references are lost. The plane picks up speed rapidly in the diving turn; the pilot sees the airspeed increasing and tries to stop the descent by increasing back pressure, which actually tightens the turn and makes the situation worse. Two things usually happen with this type of recovery attempt: (1) The plane strikes the ground while still in the spiral or (2) it comes apart in the air due to the high stresses imposed by the excessive back pressure (which, as you know, causes a high load factor). The maneuver is aptly named the "graveyard spiral."

You learned in the "720° Power Turn" in Chapter 10 that if the nose started dropping in the steep turn there was little use in trying to bring it up by back pressure alone. The situation was generally made worse instead of better. The bank had to be shallowed to stop the loss of altitude.

The bank in the power-on spiral under actual conditions may be approaching vertical. The "basic" instruments would indicate the following in the power-on spiral (Fig. 15-1):

1. Turn indicator—shows a great rate of turn; the ball may or may not be centered. It is shown as not centered in the diagram. Knowing the effects at higher speeds, the ball is more likely to be to the left because of right yaw effects. See "Torque," Chapter 2.

2. Airspeed—high and increasing.

3. Altimeter—showing a rapid loss of altitude.

■ Recovery (Altitude and Heading Indicators Inoperative)

To recover from the condition as effectively as possible, do the following:

1. Center the turn indicator and ball through co-

ordinated use of ailerons and rudder. If the needle or small airplane is to the right, apply left rudder and aileron until the indicator is centered.

This coordination is necessary because through harsh application of top rudder the turn might be stopped (the turn indicator centered) and the plane may stay in the bank, with the result that back pressure would still put stresses on the plane with little action toward recovery. A glance at the ball, however, would immediately show the imbalance.

2. Check the airspeed. As you roll out of the bank apply back pressure (easy!) to stop the increasing airspeed. A rough approximation is that in a spiral or dive recovery, at the instant the airspeed starts to decrease, the plane's nose is in the level flight attitude. When this is indicated, ease on forward pressure so that the nose is not pulled up into a steep climb or stall attitude. As the plane returns to steady, straight and level flight, the airspeed will continue to decrease until the cruise speed is attained (assuming the power is at cruise setting).

The reason for applying back pressure only until the airspeed shows the change is that in trying to go any further toward recovery by watching the airspeed you may get the result shown in Figure 15-2. The nose will want to continue up past the level flight position because of the excess speed of the dive; forward pressure is needed to stop this. In fact, in most cases as the wings are leveled, the nose will tend to rise without back pressure from you, but you'll need to check this for your airplane.

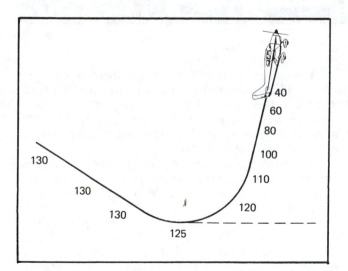

Fig. 15-2. Don't chase the airspeed.

In the 130-K dive in Figure 15-2 the plane has a lot of energy to dissipate. If you pull the nose up (remembering that there's no outside visibility) or let the nose come up until cruise speed is reached, you will find that the attitude is quite nose high and the airspeed drops rapidly to the stall, requiring forward pressure or a stall recovery.

3. The third main step in the recovery is to "stop the altimeter." This is the finer adjustment to be made immediately after getting the plane to the approximate level flight attitude through use of the airspeed indicator. Little attention will be paid to the altitude of leveling other than recovering as smoothly and quickly as possible. You will find that stopping the altimeter, with the power setting at cruise, will result in a stable cruise airspeed in a short while. Check the airspeed in your scan of the instruments, but realize that it will settle down shortly of its own accord. Don't try to rush it. After the plane is under control, altitude and directional adjustments may be made.

During the latter phases of the recovery the turn indicator must not be neglected, and a constant scan of all instruments is needed for a safe, firm recovery. The plane must be kept under control directionally while the descent (or ascent) is being controlled or stopped.

You will note that nothing was mentioned about heading in this case. It doesn't matter. Recover from the spiral, let the compass settle down, and make a timed turn back toward the good weather area. Timed turns will be discussed later in this chapter. (Reset the heading indicator as soon as feasible.)

One important thing should be mentioned. As you know, the fixed-pitch prop will windmill in a dive and the engine will overspeed. It is better if you throttle back at the start of the spiral recovery to avoid this abuse of the engine.

■ Summary of the Power-On Spiral Recovery

1. Retard power.
2. Center the turn indicator.
3. Start back pressure smoothly as needed. Check airspeed for the first indication of a change. Stop the back pressure.
4. Stop the altimeter.
5. Maintain the instrument scan throughout the recovery.
6. When the plane is under control, make adjustments in power, altitude, and direction as needed.

Figure 15-3 shows the steps used in the recovery from a power-on spiral.

You should know how to effect a recovery by use of the basic instruments shown in Figure 15-3 but will, in most cases, have the use of an attitude indicator. (Of course, you could have tumbled it with your outrageous maneuverings while getting into this situation.) Figure 15-4 shows that the attitude indicator can be used to level the wings and bring the nose up to level flight. You would readjust the power as the airspeed approaches the cruise value.

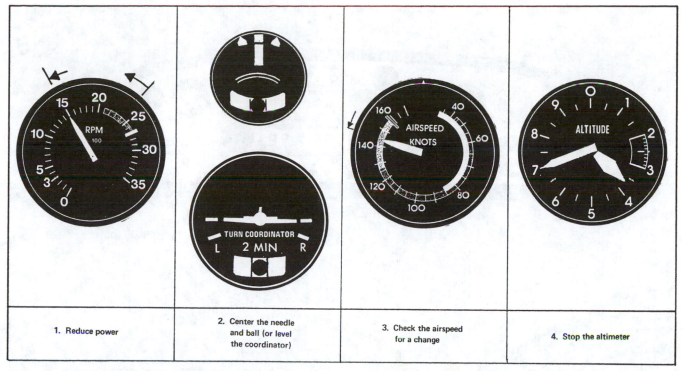

1. Reduce power	2. Center the needle and ball (or level the coordinator)	3. Check the airspeed for a change	4. Stop the altimeter

Fig. 15-3. The steps in recovery from a power-on spiral, using the basic instruments.

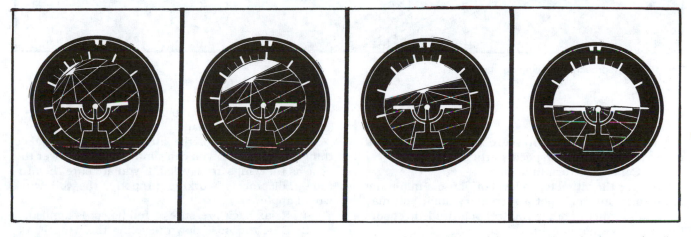

Fig. 15-4. The attitude indicator gives a better picture of the spiral situation. You would retard the power, then simultaneously roll out as you bring the nose up to the level flight attitude. One disadvantage of the *simultaneous* roll-out pull-up is that high stresses may be put on the airplane at high speeds.

RECOVERY FROM THE APPROACH
TO A CLIMBING STALL

The climbing stall sometimes occurs when visual references are lost, or it may be a result of letting the nose rise too high in the spiral recovery. In either case the recovery technique is the same. You have had a chance to practice stalls visually, as well as noting the instruments during the maneuver. Assume that the attitude and heading indicators are inoperative for the first part of the discussion.

As a review, check the indications as the stall is approached (at cruising power) (Fig. 15-5).

1. Turn indicator—may or may not be centered.
2. Airspeed—decreasing (probably rapidly).
3. Altimeter—increasing altitude, or steady in the last stage of stall approach.

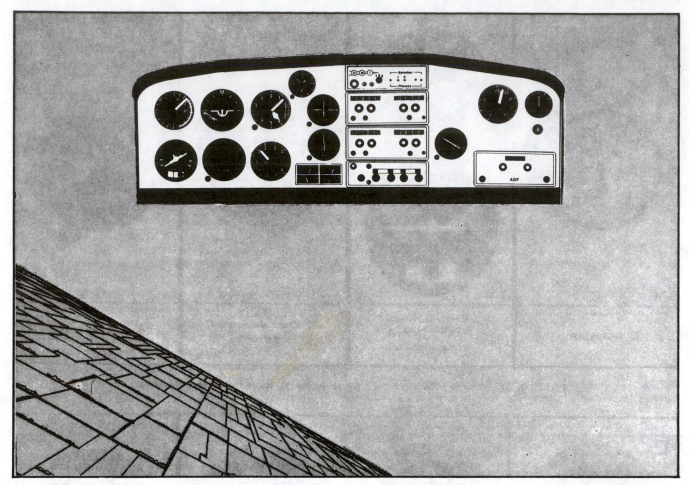

Fig. 15-5. Approach to a climbing stall.

■ Recovery

1. Relax the back pressure (or push forward if necessary) until the airspeed starts to increase.

2. Center the turn indicator.

3. Use the altimeter to level off. Lose a minimum of altitude but don't get a secondary stall; you may want to sacrifice another 100 feet of altitude to ensure that this does not happen.

4. Maintain the instrument scan throughout the recovery. Keep the turn indicator centered.

5. After recovery to cruising flight is made, make power, altitude, and directional adjustments as needed.

Use of the airspeed for recovery from an approach to a stall follows the same general idea as that in the spiral. In this case, however, the airspeed starts *increasing* when the attitude of the plane is approximately level. To be on the safe side, this point should be passed so that the plane will be in a slightly nose-down attitude to ensure a definite recovery. This is more important when the stall breaks than in an approach where the plane still has some flying speed.

You will notice that in the case of the stall recovery the primary need is to get the airspeed back into a safe range, and any deviation in direction can be corrected after the recovery is made.

Again, the use of the attitude indicator makes things much easier; you can simultaneously lower the nose as the wings are leveled. If you are sure the attitude indicator is working properly, the following would apply:

1. Relax back pressure as full power is applied.

2. Ease the nose down (leveling the wings) to a pitch attitude below level flight.

3. As the airspeed approaches cruise value, ease the nose up to level, readjust power, and make altitude and heading corrections as necessary.

The reason for the lowering of the nose below the normal level flight attitude is to assure that no new stall is encountered by trying to stop it at the level flight attitude (Fig. 15-6).

180° TURN

You had a chance to practice turns with reference to instruments during the presolo phase and will now have a chance to make turns using the turn indicator,

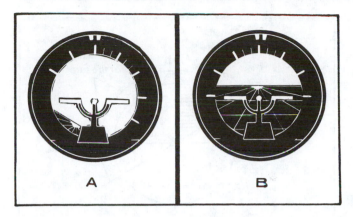

Fig. 15-6. (A) The attitude indicator shows that the airplane is in a nose-high left turn. (B) Level the wings as you ease the nose below the normal level flight attitude until the airspeed approaches the normal cruise. Then bring the nose to the level position and adjust power.

airspeed, and altimeter as well as using the full panel for these maneuvers.

A heading indicator makes 180° turns (or any turn) simple (Fig. 15-7); just turn and roll out when it reads correctly. (Your problem with the older type of instrument may be figuring out just *what* heading is 180° from the present one. See Fig. 3-22.)

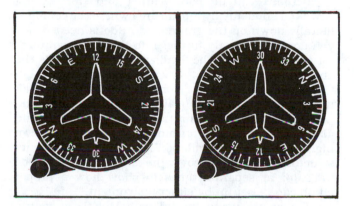

Fig. 15-7. With the most popular type of heading indicators, finding your reciprocal heading and turning to it is easy. Other methods of turning will be discussed anyway.

■ **Timed Turn—Recommended Method**

The recommended method of making a 180° turn if you don't have a good heading indicator is by timing the entire turn using the turn indicator (and, of course, your airspeed and altimeter). One advantage here is that the turns may be made in either direction without the necessity of making separate calculations for each. Another advantage is that you can merely set up a standard-rate turn and hold it for 60 seconds. A timepiece is needed.

Test programs showed that students had trouble with the timed turn in that they would forget the time the turn started. One other thing that may bother you—a minute can be a long time. Students some-

times lose track of the time and think that they have been turning a minute and 30 seconds instead of an actual 30 seconds. Or you may check the clock and continue the turn for 2 minutes instead of the 1 minute required.

In bumpy weather the needle or turn coordinator will oscillate, and you will have to fly the average position rather than trying to keep the indicator exactly at the standard-rate turn position.

Remember that the turn indicator is not the only instrument to be watched during this maneuver. Keep up your scan. If the airspeed and altitude start getting away, stop the turn, recover to straight and level flight, and after the plane is under control, continue the turn.

When you are flying cross-country, always have the reciprocal of your course in mind. *This is the heading you will want to turn to.* If the plane stays under control after entry into the weather, then a 180° turn is desired. On the other hand, if you have inadvertently turned while in the clouds or fog, a 180° turn could put you deeper into the weather. Here is an illustration.

You are on a heading of 120° and suddenly fly into a condition where outside visibility is lost. The plane enters a spiral and you recover to straight and level. After recovery the plane's heading is 280°. You don't particularly notice the heading but mechanically set up a 180° turn, which will actually carry you further into the weather, when a turn of 20° to the right would put you on the desired reciprocal heading. Even holding the 280° heading would be accurate enough to leave the weather area (Fig. 15-8).

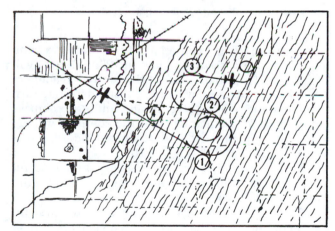

Fig. 15-8. (1) A power-on spiral is accidentally entered. (2) Recovery is effected on a heading of 280°. (3) Without thinking, the pilot mechanically makes a 180° turn. (4) The pilot would have been out of the weather in a short time.

■ **Using the Four Main Directions**

In a *shallow*-banked turn the magnetic compass is fairly accurate on the headings of East or West. It lags by about 30° as the plane passes the heading of North and leads by about 30° as the plane passes the heading of South. (The exact lag or lead will have to be checked for your situation, which will include latitude and other variables. For illustration purposes here it is assumed to be 30°.) This is called "northerly turning error" and affects the compass while the plane is turning.

Assume that you are flying on a heading of 120° and want to make a 180° turn. The desired new heading will be 300°. By turning to the right, you will have a cardinal heading (West) reasonably close to the new heading. A standard-rate turn is made to the right, but no timing is attempted. You will be watching for West on the compass, and as the plane reaches that heading you will start timing. The desired heading is 30° (300° minus 270°) past this and will require 10 seconds at the standard rate. The timing can be done by counting "one thousand and one, one thousand and two," and so forth, up to ten, at which time the turn indicator is centered (Fig. 15-9).

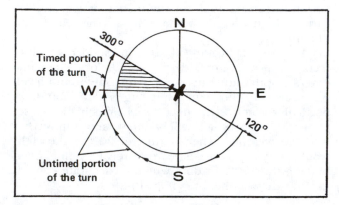

Fig. 15-9. The 180° turn to the right.

A turn to the left could have been made realizing that the nearest major compass heading (North) will be 60°, or 20 seconds short of the desired heading of 300°. In that case you will set up a standard-rate turn to the left and will not start timing until the compass indicates 030°. Remember that the compass will lag on a turn through North and will be behind the actual heading by about 30° (Fig. 15-10).

When the 030° indication is given, the 20-second timing begins, either by the sweep-second hand of the aircraft clock or your watch or by counting.

There are several disadvantages to this system, the major one being that it requires a visual picture of the airplane's present and proposed heading—and mental calculations. You may not have time or may be too excited for a mental exercise at this point. Also, in bumpy weather the compass may not give accurate readings on the four main headings.

The main advantage is that reasonably accurate turns may be made without a timepiece, as the new heading will never be more than 45°, or 15 seconds, away from a major heading. This method is presented as something to be filed away and used if necessary.

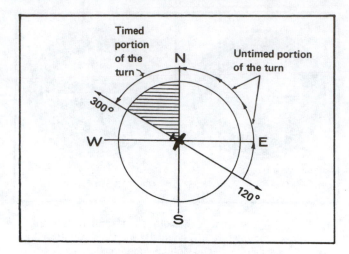

Fig. 15-10. The 180° turn to the left.

SHALLOW-CLIMBING TURNS
TO A PREDETERMINED
ALTITUDE AND HEADING

Perhaps you hit bad weather and start letting down but suddenly fly into an area where the ceiling is literally down on the ground. Altitude is needed, yet you don't want to fly further into the unknown territory as you climb. The shallow-climbing turn is the answer. You may have passed over some fairly high terrain or obstructions before letting down to sneak under and now need more altitude before flying back.

Figure 15-11 shows a typical climb attitude and airspeed indication for a trainer, as shown by the two flight instruments (climb power being used). Your instructor will probably cover the airspeed and attitude indicator at separate times to show how a proper climb can be set up by either instrument.

If you are using the full panel of flight instruments, the instructor may set up a spiral and require

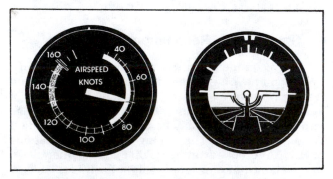

Fig. 15-11. At climb power a certain pitch attitude will give the correct climb airspeed. You should use this fact in initially setting up the climb. You may hear the expression, "Power plus attitude equals performance," among the instrument-rated pilots at the airport.

that after recovery you climb back up to a specific altitude and heading, as would be the case in a real situation. (You need to recover lost altitude; you also need to turn out of the bad weather.)

Suppose that at the end of a spiral recovery you are at 2800 feet and your heading is 035°. You need 3500 feet of altitude and a heading of 125° to get safely back to good weather. The airplane climbs at 700 fpm and you'll be back up to the safe altitude in 1 minute. A 90° turn to the right is required, so using a standard-rate turn you'll get there in 30 seconds. Fine. Set up your climb and turn. You'll be at your heading of 125° when you still have 350 feet of altitude to go. Roll out and make a straight climb until the required altitude is reached. *Keep an eye on that heading indicator during the straight climb.* Students sometimes get so interested in the altimeter that they let the airplane turn right on around—back into the imaginary, or actual, bad weather. ("Torque" sometimes is a culprit too.) So if you get to the heading first, maintain it as you climb; if you get to the altitude first, level off and continue the turn. The problem is not complete until you are at both the required altitude and heading (Fig. 15-12).

The following explanation will assume that your heading and attitude indicators are temporarily out of action. One point: While the attitude indicator may be tumbled and out of action for several minutes (if you don't have an instrument with a caging knob), you can reset the *heading indicator* immediately and use it as an aid to keep your wings level (if you are getting indications of a turn to the right, that wing is down, etc.). You'll be surprised at how much aid the heading indicator can give you. Your first resetting of it with the magnetic compass will be a rough one, but you can reset it later after the compass has settled down.

■ **Procedure (Basic Instruments)**

Right rudder will be needed to keep the ball centered, as always in a climbing turn.

If climbing power is set, the turn kept shallow, and the airspeed is held at that of climb, the rate of climb will take care of itself. The rate of climb indicator is a very useful reference here. But "chasing" the vertical speed indicator will only result in a varying flight path. *Corrections must be smooth and deliberate.*

As in the case of the shallow-climbing turn with

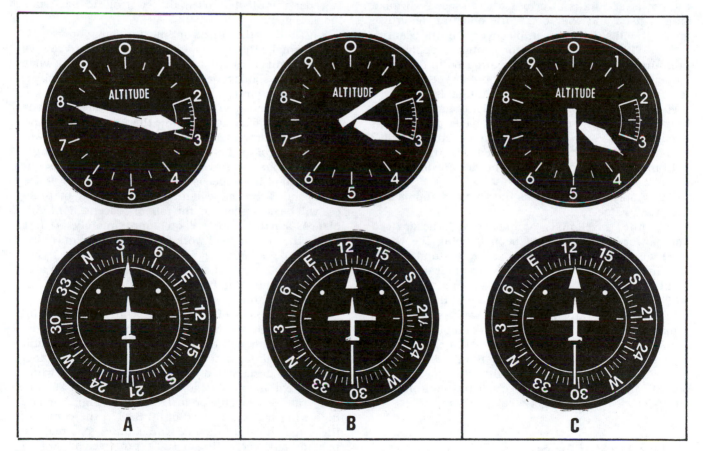

Fig. 15-12. (A) Control of the airplane is recovered on the heading (035°) and altitude (2800 MSL) shown. The desired heading is 125° and the required altitude is 3500 feet. A standard-rate turn and a climb is started. (B) The heading is reached and the turn is stopped. A straight climb is continued to the required altitude. (C) The requirements are complete when the airplane is on both the required heading and altitude.

visual references, you will ease the nose up to the required position and apply climbing power. When you could see outside, the nose position, power setting, airspeed, and altimeter showed that the plane was in a normal climb. You found that by combining the proper nose position and power setting the altimeter and airspeed took care of themselves. Here you will use the airspeed and power setting to assure that the proper climb is maintained (take care of the airspeed and power, and the nose attitude will take care of itself as long as the other instruments are not neglected). It would be advisable to establish the climb first before the turn is started.

The idea of starting the level-off about 20 feet below the desired altitude still stands (for most lightplanes). This allowance depends on the rate of climb; a fast-climbing jet may need several hundred feet to level off.

In visual flying the leveling-off was accomplished by easing the nose over to the straight and level position, checking the altimeter, and, as the airspeed approached the cruise value, throttling back to cruise rpm and trimming the airplane. You paid little attention to the airspeed during the leveling process. The same applies here. If the altitude is kept constant and the power setting is at cruise, the airspeed will take care of itself and will soon settle down at the cruise value. It will still, however, be covered in the scan.

1. With the turn indicator centered, apply back pressure slowly to decrease the airspeed to the climb value. It will take a little time for the airspeed to drop to that of the climb. The best technique is to apply a certain amount of back pressure, let the airspeed stabilize, and apply more or less back pressure as needed. After you have practiced a few climb entries, this back pressure can be estimated. The trim tab can help maintain the climb, and you are encouraged to use it.

2. Apply climb power as the airspeed approaches the climb value.

3. After the climb is established with the airspeed and power setting (you've been keeping the wings level with the turn indicator), set up a standard-rate turn in the desired direction. At your plane's probable climb speed the bank required for a standard-rate turn will be in the vicinity of 10°–15° so that the standard-rate turn is easy to maintain as well as giving the proper shallow bank required for a climbing turn.

Make all turns standard-rate. Later you may start the turn and climb at the same time. It may be desired to climb to a specific altitude and turn to a definite heading, so stop the turn but continue the climb, or vice versa, to get what you want.

■ **Leveling-Off Procedure**

1. At the proper distance below the chosen altitude start the leveling procedure. Relax back pressure to stop the climb.

2. Fly the altimeter; maintain the chosen altitude.

3. Center the turn indicator as you "stop the altimeter."

4. As the airspeed increases, throttle back to cruise power. Trim the airplane.

5. Make any heading corrections when the airplane is flying level (trimmed).

SHALLOW DESCENDING TURNS AT REDUCED POWER TO A PREDETERMINED ALTITUDE AND HEADING

A descent (or ascent) at a given airspeed can be controlled by the amount of throttle used. The simplest form of descent is one you have used in losing altitude by maintaining the cruise airspeed and throttling back until the desired rate of descent was obtained. You probably didn't measure the exact rate in feet per minute but used the controls to get what "looked good."

As in the other turns on instruments, you are advised to make these descending turns standard-rate. A steeper banked turn may get out of hand, and what started as a shallow descending turn may wind up as a graveyard spiral.

The descent could be made at any airspeed, but you wouldn't want to be slower than the normal glide or much faster than cruise because of the possibility of loss of control at the extremes of the airspeed range. The plane may be overstressed if turbulence is encountered at high speeds or a stall might occur at the lower end of the airspeed range. The leveling-off process will be dependent on the rate of descent. For lightplanes, starting the level-off approximately 50 feet above the altitude will work for an average controlled descent of 500–800 feet per minute. You will be flying the turn indicator, airspeed, and altimeter (plus attitude and heading indicators, if available) for attitude, rate of turn, and descent.

After leveling off, the power will have to be increased to that of cruise in order to maintain altitude. Your instructor will have a recommended speed and power setting for descent available for your airplane. A 500 fpm descent is a good one to shoot for. It won't be so great as to risk loss of control (this is a "standard" rate of descent for instrument flying). For one high-wing trainer, a speed of 80 K is in the center of the green arc on the airspeed indicator (a good place to be), and a power setting of 1900 rpm gives a 500 fpm descent. (You'll find that the rate of descent at your given airspeed and rpm will vary slightly with altitude and/or airplane weight, but the difference will not be enough to cause any problems.)

This exercise is complete when the airplane is on both the chosen altitude and heading.

Figure 15-13 shows the combination that might be used on descent on a partial, or emergency, panel.

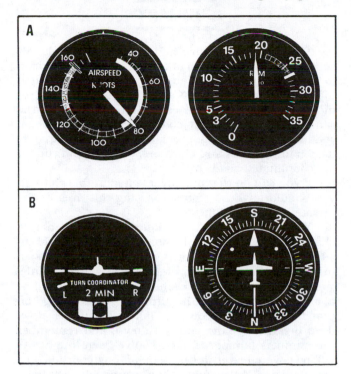

Fig. 15-13. (A) Power setting and airspeed of one trainer for a descent of approximately 500 feet per minute. (B) You would keep the wings level with the turn indicator and/or the heading indicator if you could get it back in action.

STRAIGHT AND LEVEL

Although you have been practicing straight and level flying by reference to instruments for some time, it might be well to mention a few other points.

Talking about the full flight panel, it would seem that flying straight and level using attitude and heading indicators should be very easy. But the problem is that the student sees the means to keep a precise heading (the heading indicator) and so works too hard and overcompensates. You might have this problem: You see that the airplane has turned a few degrees off heading. You want to get back on the heading right away, so you roll into a bank to accomplish this. You find that you've overshot and now have to reverse the procedure. This game can go on as long as the instructor allows it—which is not too long. This is known as "S-turns across the route." The thing to keep in mind is that you will make *minor* corrections (more about that shortly)—and this will also apply later as you work on that instrument rating—and *ease* up (or down) to an airspeed, altitude, or heading if you've slipped off a little.

You may have to fly several minutes to get out of the weather. The ability to fly a straight and level path and not wander means a quicker return to a place where visual references are available. The turn indicator is the primary indicator for keeping a fixed direction of flight if the other gyro instruments aren't available. Fly the turn indicator and use the compass for a cross-check. If the compass has settled down and shows that you are off heading, correct with a balanced turn in the proper direction. For minor variations from heading (up to 15°) a half standard-rate correction is preferred. If you are 10° off, make a one-half standard-rate turn (1½° per second) for 7 seconds. In straight and level flying the student sometimes has a tendency to stare at the turn indicator, neglecting the rest of the scan. It's important that the instruments be continually cross-checked with each other.

You may get practice in flying straight and level using one or two of the instruments to show how a reasonable heading and altitude can be maintained. The procedure might be like this: You'll be under the hood, and the instructor will set up the airplane on a heading of, say, 090° and altitude of 3500 feet. When everything has "settled down" the instructor will, for instance, cover all of the flight instruments except the attitude indicator and have you fly the airplane, keeping the proper pitch and wings-level position by this instrument only. After 2 minutes the instructor will uncover the altimeter and heading indicator, and you can see how close you are to the original requirements.

Combinations of instruments may be used to make the point that you *can* keep the airplane under control when ground references are lost. It's a good confidence-building exercise.

STRAIGHT DESCENTS USING
THE MAGNETIC COMPASS

Suppose you've lost all three gyro instruments and need to let down through a cloud layer. As a last resort and if there is a good ceiling below, the magnetic compass may be used as a combination attitude indicator and heading indicator. The letdown should be made on a heading of South because the compass will react "normally" if a wing drops and a turn is started. In other words, on that heading the compass will show an increase in heading numbers if a right turn is starting or will show a decrease for a left turn. On a North heading you recall that the compass will start moving the "wrong" way at the initiation of a turn—which could be fatally confusing. The South heading gives a quick and proper reference to keep the wings level. You might ask your instructor to demonstrate this on the way back from the practice area sometime.

SIMULATED RADAR ADVISORIES

If your airport traffic permits, your instructor on dual flights may bring you back from the practice area "on radar" several times. You'll go under the hood and the instructor will be "approach control radar" and direct you back into the traffic pattern and, perhaps, down to a few hundred feet on final. Then you will get problems that use the instrument training you've had. It will be good practice if you should later accidentally get into actual instrument weather and need radar service.

ASR—Airport surveillance radar is a nonprecision radar that allows the controller to vector you into the pattern and tell you what altitude you should be at various distances on final. (The controllers have no altitude information on you, just direction and distance.) Usually the weather minimums for ASR approaches are a 500-foot ceiling and 1 mile. Its primary purpose, however, is to expedite traffic flow in the terminal area.

PAR—Precision approach radar is used as a *landing* aid. The operator can check your approach path and advise whether the airplane is on course and/or on the glide path. This is the system used on the Late Late Show where the crippled jet has 5 minutes of fuel and the weather has just gone to zero-zero at the only airport available within several hundred miles. After much suspense (did they make it?) the jet taxies out of the fog. The minimums for the PAR approach are naturally lower than the ASR, but never as low as shown on the Late Late Show. PAR systems are usually found on military/civilian joint-use airports.

Later you will also get practice in tracking to or from a VOR station (see Chapter 21) while flying the airplane by reference to the instruments.

SUMMARY

The instruments give information on the plane's actions and attitudes and take the place of the horizon in visual flight. A certain amount of practice is needed to be able to see the plane's actions through the instruments. A plane's responses to the controls are the same under instrument conditions as under visual conditions; hence, no special control technique is used, but throttles and elevators, ailerons and rudder are coordinated as always.

If you can do the maneuvers cited in this chapter, you should have no trouble passing this area of the private practical (flight) test.

You may use the techniques discussed in this chapter to save your neck someday. If you should inadvertently find yourself in a situation of being on actual instruments, your worst enemy will be panic. You'll have practiced under the hood (dual) and will be able to keep control of the airplane *before* you go on that first solo cross-country. Take a couple of slow, deep breaths to slow yourself down, and remember the numbers (airspeeds, etc.) you've been taught. As will be covered later in this book and by your instructor, there are FAA people on the ground who can help, but *you* will have to fly the airplane and make the decisions (often the person on the ground is not a pilot).

Make a 180° turn *before* getting into the bad weather and save yourself some sweat.

16
POSTSOLO PRECISION
MANEUVERS

On every flight you'll have to cope with the wind (or lack of it), and these exercises are intended to help you compensate for drift so that *you* are in command of the airplane's path over the ground.

ANOTHER BASIC WIND DRIFT MANEUVER

■ Turns Around a Point

This maneuver is very close to the S-turn across a road because you're correcting for wind while in a turn. It is suggested that the steepest bank be approximately 45° and the altitude not be lower than 500 feet above the highest obstruction in the maneuver pat-

tern. The idea of the turns about a point is to show the check pilot how well you can fly the plane when your attention is directed outside. Your check pilot will also want to find out if you understand wind drift correction principles (Fig. 16-1).

PROCEDURE

1. Pick a tree or some other small but easily seen object and set up the radius small enough so that the steepest bank is approximately 45°.

2. Enter the pattern downwind. Practice both left and right turns around the point. Many students only practice left turns and get a rude awakening when they are asked to make a turn around a point to the right on the check flight. The best altitude for keeping the point in sight is about 700 feet above the surface.

3. Vary the turn as needed to maintain a constant

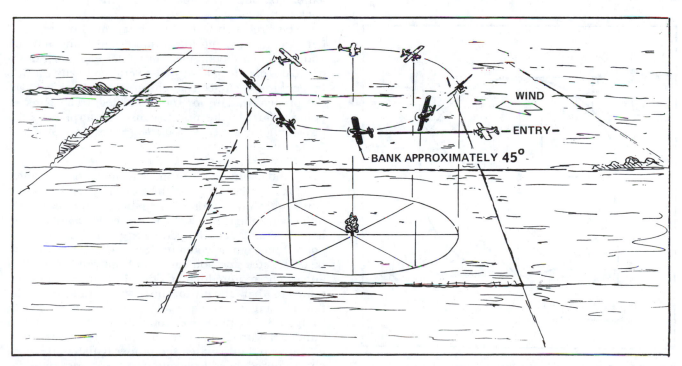

Fig. 16-1. Turns around a point. The steepest bank is required when the plane is flying directly downwind.

distance from the point—the greater the ground speed, the steeper the bank.

　4.　Maintain a constant altitude.

　5.　Keep your eyes open. This is a perfect place to have a low-altitude emergency thrown at you.

　6.　The check pilot may give you a free turn to get set up, but will keep up with the number of turns after the maneuver is started.

COMMON ERRORS

　1.　Starting the maneuver with the plane too far from the point so that the steepest bank does not reach 45°.

　2.　Poor altitude control.

　3.　Pointing the wing at the tree—trying to keep the plane's attitude the same in relation to the point rather than having the plane's path constant, as it should be.

　4.　Failure to recognize wind drift or, if recognized, not doing anything about it.

　5.　Coordination problems.

Students sometimes have trouble with this maneuver because they can't seem to convince themselves that the steepest bank is required when the airplane is headed *directly* downwind, as indicated in Figure 16-1.

Suppose an airplane is flying tangent (wings level) to the circle at the positions shown by A, B, C, and D in Figure 16-2. At A, the airplane is moving at its true airspeed *plus* the wind speed. This results in the highest relative speed to the reference point of any position around the circle. At C, its relative speed is the lowest because it is moving into a headwind. The

arrow (vector) behind each airplane gives a picture of the relative speed at the four positions. The dashed lines show the comparative distances the airplane would be "away" from the circle for any given interval of time if no turn was made.

The rate of turn, then, must be greatest at A if it is to "follow the circle." Since this is a coordinated maneuver, the bank must be greatest at that position (and the bank is most shallow at C). The banks at B and D will be comparable.

VECTOR APPROACH TO THE MANEUVER. If you are interested in the mathematics of turns around a point, Figure 16-3 shows wind triangles for the airplane at eight points around the circle. As noted, the airplane's true airspeed is 100 K and the wind is from the top of the illustration (North) at 40 K. The solid thin arrow represents *groundspeed, or the airplane's path and speed with respect to the ground.* This groundspeed vector varies in size (speed) because of the wind, but at each of the eight points shown (or at *any* point on the circle) it must be tangent to the circle or else the plane would not be following the prescribed path; so this is the first consideration.

The true airspeed is indicated by the dashed arrows and is always 100 K for this problem. (Disregard slowing the airplane in the steeper banks.) The 100 K true airspeed vector, however, must be "pointed" in such a manner that the result of the plane's heading, plus the wind, makes the airplane's path tangent to the circle at any position. The wind is a constant velocity and direction.

Maybe you haven't done any work with the navigation computer yet, but you will, and you can come back to this later.

The circle can be thought of as an "infinite number of short, straight lines," and the airplane is flying "one leg" of a rectangular course for each one. In order to do this, the plane must be crabbed to fly the line tangent to the circle. The reason for the bank is to get the proper heading for the next "leg." The bank at point 1 must be steepest because the airplane is approaching the next "leg" at the greatest rate. At point 5 the opposite is true.

Of course, practically speaking, you fly the airplane and maintain the proper distance from the point by looking at it. But given the true airspeed and wind, you could work out on a navigation computer the required headings for each of the eight points given here. For instance, the course at point 1 is 180°. At point 2 the course would be 135°. At point 3 it would be 090°, etc., and you could find the wind correction angle. The required banks at each point could be obtained mathematically for the radius of the circle to be flown, using a turn equation similar to that given in Chapter 3. The maneuver could be theoretically flown "under the hood" once the airplane is established in it. You would just match the proper banks to the various required headings you found for the eight (or more) points around the circle.

Maybe this approach to the maneuver doesn't in-

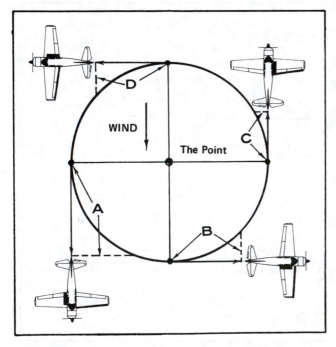

Fig. 16-2. A comparison of rates of "leaving the circle" for an airplane flying at the same true airspeed at four positions on the circle.

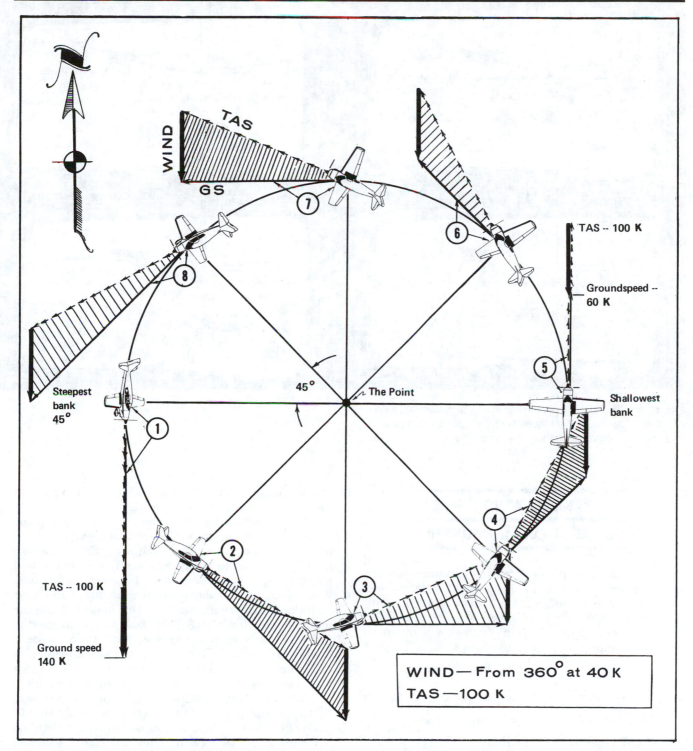

Fig. 16-3. The turns around a point maneuver when seen as a problem in vectors. You may want to come back to this one *after* you've done some work with the computer.

terest you, and you can certainly do it well if you don't know a vector from a victor. But the main point here is that the rectangular course, S-turns across a road, this maneuver, and others are designed to teach you to fly the airplane on a certain path or track, correcting for wind.

Figure 16-4 shows what you might see from a high-wing airplane at points 1, 3, 5, and 7 in Figure 16-3.

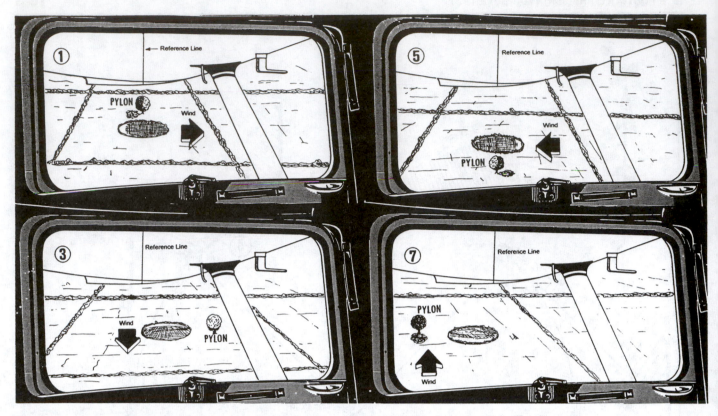

Fig. 16-4. The views from the cabin of a high-wing airplane at points 1, 3, 5, and 7 in Figure 16-3. The pond would help you know the wind direction (see Figure 18-2).

ADVANCED WIND DRIFT CORRECTION MANEUVERS

The following maneuvers aren't required on the private practical test. You'll do the rectangular course, S-turns across a road, and/or turns around a point. (See Chapter 28.) They are given here for your information in case you and your instructor want to do some more complex wind drift correction.

■ Eights Around Pylons (Around-Pylon Eights)

SHALLOW EIGHTS. This maneuver is closely akin to the turns around a point but, as the name implies, a figure eight is flown around *two* points, or pylons. The pylons are picked so that an imaginary line between them is perpendicular to the wind (Fig. 16-5).

Shown in Figure 16-5 is a shallow around-pylon eight maneuver. The steepest bank should be approximately 30°. Your instructor will note that you may use the same pylons for steep or shallow pylon eights, using a straightaway portion between the pylons as shown. While at first you will enter downwind between the pylons, you will later enter from any point

of the maneuver but will fly it so that the outer parts of the turns are made into the wind.

Note in Figure 16-5 that the entry is set up (point 1) and will be slightly wider from the pylon(s) than turns around a point (which required 45° maximum banks). You should be flying directly downwind and tracking to a position exactly between the pylons. When point 2 is reached, a turn is started in either direction (to the right, here). At this stage it's like the maneuver in Figure 16-3. (The bank will be steepest at point 2.) Note that at points 3, 4, and 5 the airplane is crabbed in the turn. Point 6 has the shallowest bank and points 7, 8, and 9 require crabbing. When point 10 is reached (or point 2 again), the airplane is rolled into a turn in the opposite direction and the second pylon is used (points 11, 12, etc.). *Not only should each circle be of constant radius, but both should be the same size.* One of the common errors students make is to have a too-wide turn on one pylon and a too-narrow turn on the other. Because of this tendency careful planning and orientation are needed to get back to the spot exactly between the two pylons. You may repeat the pattern any number of times, but usually four or five are the maximum done without a break.

The blacked-out ellipse in the center of the illustration might be called the "zone of confusion" because, even though the circles are supposed to be tangent, it's impossible for an airplane to be rolled

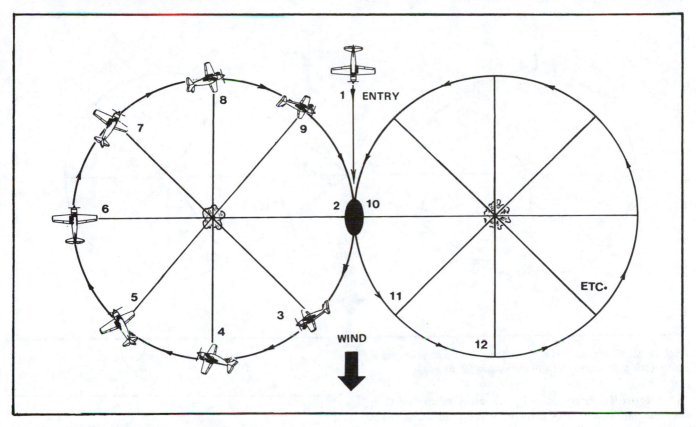

Fig. 16-5. Shallow around-pylon eights. This maneuver is a more complicated version of turns about a point.

from one turn to the other instantly. So in actual practice they can't be tangent, but you should make them as near tangent as *safely* possible.

Some common errors are:

1. Poor pylon picking (which will be covered shortly).

2. Coordination problems (slipping or skidding).

3. Altitude problems. The altitude should be the same as used for turns around a point and kept *constant*. The usual problem is a tendency to climb as the airplane is rolled from one bank to another, and then lose altitude as the new turn (steep bank) is established.

4. Orientation loss, or "losing" a pylon. You may continue well on past the point to roll in for the opposite turn. If you do this, you could make another 360° turn around the pylon, attempting to divert the attention of the instructor (or check pilot) by pointing out objects of local interest as you go around. (But you won't get away with it.)

5. Nonsymmetrical circles—the turn around one pylon steep, the other shallow. This, too, can usually be attributed to losing the other pylon as well as losing the center point of the pattern (point 2 or 10 in Fig. 16-4).

6. Having too large an eight; the airplane is too far from the pylons and very shallow banks are required.

STEEP EIGHTS. Figure 16-6 shows the steep around-pylon eight maneuver. The turn pattern is closer to each pylon, and a straight leg is used to get set up for the other pylon. This maneuver uses wind drift corrections in straight flight and a turn. You'll have to plan the roll-out from one pylon so as to have the proper drift correction set up for the leg to the other pylon.

Note the steps at the following points in Figure 16-5:

Point 1. The airplane is in a position to enter the maneuver.

Point 2. Drift correction is set up in the straight leg.

Point 3. When the proper radius is reached, a turn is set up. Points 3 and 9 will require the steepest banks in this maneuver since the groundspeeds will be greatest at these positions.

Points 4–9. The bank is varied to maintain a constant radius. As point 9 is approached, the airplane is rolled out to set up the proper crab angle, points 10 and 11, to be at the required radius at point 12. This loop of the eight uses the same principle as the first.

The four angles indicated by the two A's may be those desired by the instructor. Later, you will pick the pylons and set the maneuver up so that each of these angles will be 45°. (You can figure that the steeper the bank around a given pair of pylons, the smaller these angles will be.)

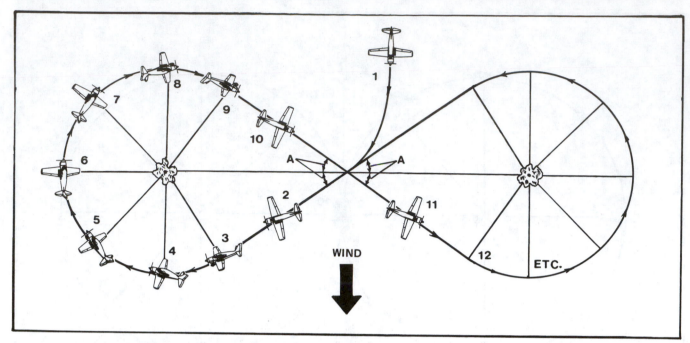

Fig. 16-6. Steep around-pylon eights. Wind drift directions in both turning and straight flight are needed in this one.

Figure on encountering the same errors as listed for the shallow eights plus a couple that would especially apply to this maneuver:

1. The turn in the entry may be too soon or too late, making the first loop too tight or too wide. You'll find that at first you may take several patterns for each eight before settling down.

2. Problems during the straight portions of the maneuver—poor wind drift correction so that the airplane is not at the proper distance from the pylon as the turn is started.

Another pylon should be selected if the "back-to-back" idea (no straightaway) is used for steep *and* shallow eights (Fig. 16-7).

The principles involved are the same as for the other wind drift correction maneuvers discussed earlier. The turning portions of the eights follow the theory shown in Figure 16-3—the airplane is crabbing in the turn. The straight flight parts use the principles you learned when you first flew parallel to a road, correcting for a crosswind.

The eights are more complicated than turns around a point because you have another pylon to consider and must be planning your transition from one to the other, even while maintaining a constant radius around one pylon. It's easy to lose that other pylon by getting too involved in the one you are working with. It makes for a sinking feeling on a check flight when you roll out and don't see anything that looks like the other pylon you selected earlier. Picking your pylons is very important; isolated trees near roads or other outstanding landmarks are good references (Fig. 16-8). Choosing one of several trees in a large field can lead to results you might imagine.

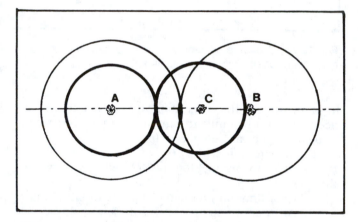

Fig. 16-7. You'll need an extra pylon if you plan to fly both steep and shallow around-pylon eights "back-to-back." (Or you could pick an entirely new pair if you want a break.) A and B are the pylons for the shallow eights; A and C would be used for the steep ones.

These maneuvers are done at the same altitude as you used for the other wind drift correction exercises.

THE PERPLEXING PROBLEM OF PROPER PYLON PICKING. Probably one of your biggest problems is finding pylons the proper distance apart and oriented perpendicular to the wind. Students have gone on pretty good cross-country flights looking for pylons that were "just right," while the instructor or check pilot grew restless in that right seat. A good method of picking pylons is to fly crosswind, looking to the downwind side. As soon as a proper pair of pylons is spotted, the airplane can be turned downwind to en-

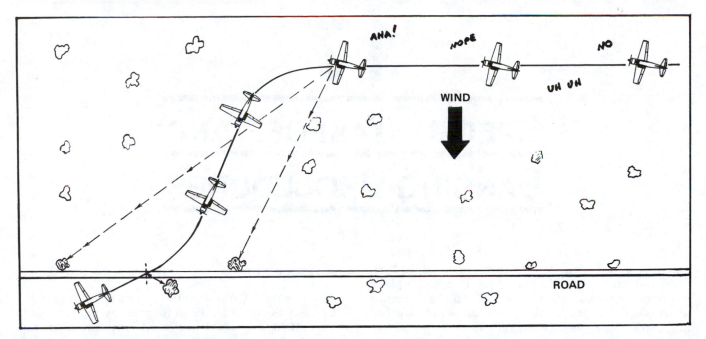

Fig. 16-8. Picking pylons. Points beside a road are best for keeping your orientation. If there is no good reference exactly between the pylons, you should find the center with relation to some object, such as the tree shown here.

ter. Picking pylons beside a road helps a great deal in orientation. (Pylons don't have to be trees, they can be other *immobile* objects; one student picked a cow for one and, needless to say, the pattern was affected as the animal walked across the pasture. An automobile that drives off is also considered a poor choice in most quarters.)

■ Spirals (Constant-Radius Descending Turns)

This is a maneuver that uses the wind drift correction skills you learned in turns around a point; in this case, however, you're descending with power at idle (Fig. 16-9). The practical aspect is that with practice you can spiral down around a particular point to set up a traffic pattern in the event of a power loss at altitude. (Fig. 18-3 shows the idea also.) You'll get some practice with this during your pre- and postsolo dual when the flight instructor gives you simulated high-altitude emergencies.

The spiral is a little more difficult than turns around a point because the airplane will be getting closer to the reference as you descend and the perspective will be changing. (The bank is steepest when flying directly downwind and shallowest when headed upwind, as discussed in turns around a point.) Hold the maximum bank to 40°.

The main problem in a spiral is that the airspeed may tend to get away from you—the usual error is to be too fast. Hold the airspeed to within ±10 K of that desired.

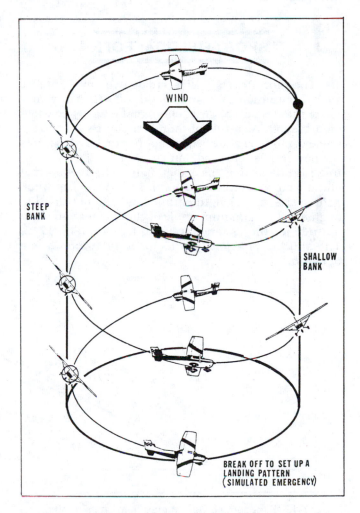

Fig. 16-9. The constant-radius spiral.

17

SPECIAL TAKEOFF AND LANDING PROCEDURES

Now you've done it! You're a victim of that old tried and true aviation saying, "You can land a plane in places where you can't get it out." The field seemed short when you landed, and now the wind has died and things look pretty grim. There are some 50-foot trees at the other end of the runway that you'll have to get over. A special type of takeoff is needed here. (*Or maybe you shouldn't take off.*)

SHORT-FIELD TAKEOFF

The short-field takeoff is a maneuver aimed at just such a situation. You must take off efficiently and make the steepest angle of climb that is safely possible after takeoff. As a rule of thumb, a plane's maximum angle of climb is at a speed of approximately 1.3 times the power-off stall speed, but check the *Pilot's Operating Handbook* for the exact figure. In the normal climb you are interested in the maximum altitude gain over a given length of time (and still not ruin the engine). The maximum *angle* of climb gives you more altitude for the distance traveled forward (Fig. 17-1). *The situation obviously calls for a maximum angle of climb.*

■ Procedure

1. Start the takeoff at the extreme end of the runway. Set the flaps as recommended by the manufacturer. Hold the wheel back and smoothly open the throttle.

2. As soon as you are certain the engine is developing full power, and the plane is accelerating properly, proceed with a tail-low takeoff. This is to let the plane get into the air sooner than would be the case if the airplane is "held on." The plane will accelerate faster when airborne because of the lesser drag of the air. Don't try to rush the process, because this will slow the takeoff as explained in "Normal Takeoff" (Chapter 13). The attitude of the plane during the takeoff will naturally vary among airplanes.

3. The plane will lift off when the minimum flying speed is reached.

Some pilots suggest that the airplane should be held on the ground until V_x (the max angle of climb speed), but this can cause two possible minor problems: (a) If the field is rough, this can be hard on the landing gear; if the surface is soft or consists of taller grass, you'll take a lot of room to get V_x. (b) For some nosewheel airplanes a shimmy could develop at high speeds on the ground, even on smooth, firm runways.

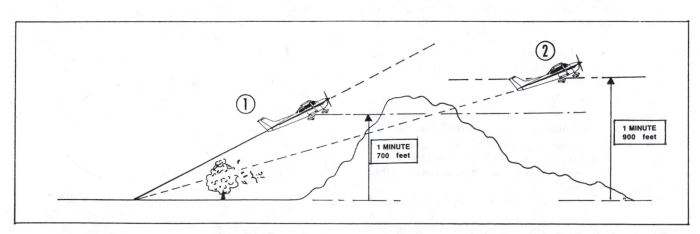

Fig. 17-1. Exaggerated view of maximum angle climb (1) versus maximum rate of climb (2).

The instructor will have suggestions for your particular airplane.

4. Attain and maintain the recommended maximum angle climb speed. Keep the throttle wide open. If conditions require this sort of takeoff, they also require all the power you can get from the engine.

5. At about 100 feet above the ground assume a normal climb (V_Y) and use normal climb power.

For check-ride purposes the obstacle to be cleared is considered to be 50 feet high and the field is considered to be firm. Use the takeoff flap setting as recommended by the Airplane Flight Manual or *Pilot's Operating Handbook*.

Some pilots argue for making a 90° rolling takeoff in preference to the straight-roll technique. The fixed-pitch propeller of the lightplane is inefficient at low speeds. This means comparatively slow acceleration as the plane starts rolling. But the added efficiency gained by the 90° rolling takeoff does not offset the chances of an inexperienced pilot losing directional control. Also, there have been instances where the sharp turn moved fuel away from a wing tank outlet (called "unporting") so that the engine suffered fuel starvation during the takeoff or initial climb.

Your instructor may demonstrate various techniques.

■ **Common Errors**

1. Poor directional control when the brakes are released.

2. Trying to hurry the plane off the ground—resulting in high drag and slowing the takeoff.

3. Letting the airspeed pick up past the best angle of climb speed after takeoff and not getting a maximum angle of climb.

SOME ADDED NOTES. The *max angle of climb airspeed* is not necessarily the same as the *best obstacle-clearance airspeed* for some airplanes. Without getting involved in the mathematics or aerodynamics, the two airspeeds would be the same in a fixed-gear airplane with no flaps. For airplanes with retractable gear and flaps or for fixed-gear airplanes with flaps, the best obstacle-clearance airspeed given in the *Pilot's Operating Handbook* may be *lower* than that cited for the best angle of climb in that same publication. The best rate and best angle of climb airspeeds are usually found, for the plane operating in the cleanest condition (flaps up and gear up, if possible), by the manufacturer through flight test and graphical representation. The manufacturer may find by flight test that the use of *some* flaps can help in the short-term requirement of clearing a 50-foot obstacle. Also, in a retractable-gear airplane, the wheels are likely to still be down at the time the obstacle is cleared. The *main thing is to set up the airspeed and configuration recommended in the POH for the short-field or obstacle-clearing takeoff because that's the problem*

at hand. For particular airplanes there may be other numbers given in the *Handbook* for the longer-term max angle climb (and there may not).

It may seem when practicing short-field takeoffs that you are using *all* of the available length, but for one popular trainer at sea level the use of *another 12 feet of runway* length will have the result of clearing the *obstacle by another 1 foot*. At a takeoff at 5000 feet density-altitude in that airplane, each extra 20 feet of runway used will give another 1 foot of clearance over the obstacle. This would be assuming constant outside factors and pilot skills.

Of course, you don't plan on cutting it this close, but maybe an *inch* or less obstacle clearance sometimes could mean the difference between a safe takeoff or an accident. There's the old aviation adage: "Runway behind you, altitude above you, and fuel back in the fuel truck are three very useless items indeed."

SHORT-FIELD LANDING

■ **More about Power Approaches**

The approach you've been making by gradually reducing the power to make the runway is a crude version of the power approach. (Sometimes you had to add power again to make the runway after you were too hasty in reducing it.) Power is a means of controlling the airplane's path on an approach and makes it possible to land on a particular spot. In the presolo work little effort was made to make a "spot landing" or landing at a chosen point on the runway, but now you'll start making precision landings.

The power approach is a general term. It doesn't say at what speed the approach is made, but only that power is used to control the glide path. There are two power approach speeds in which you will be particularly interested, however. These are the power approach at normal glide speed and the short-field approach with a speed of approximately 1.3 times that of the stall speed.

Before, you were able to tell fairly soon whether you were too high or too low to make the runway, but now you want to predict your landing place within a few feet—and land there. By controlling the airspeed, you will be able to estimate the point of landing. By controlling the airspeed *and* the rate of descent, you will be able to pick the point of landing.

On a day of steady wind in a normal power-off approach, your point of landing has already been determined. But you don't know what this point is. How do you tell? Assume that you are trying to land at a particular spot on the runway—maybe it's a big clump of grass or oil spot that is easily seen. You want to land there, so watch it. If the spot apparently moves toward the nose, you will glide over it. If it moves out away from the nose, the plane will be short of the point. So you say, "By controlling the airspeed I can set up means of telling whether I'm overshooting or under-shooting; now what?" You can slip, add flaps, or add power as needed to hit the spot (Fig. 17-2).

You keep a perfect glide speed, and the approach is such that the oil spot or clump of grass doesn't move. As soon as you start breaking the glide, the nose will start moving up and the spot will move under the nose. Let it; you've done everything possible, so forget the spot and make the landing.

Allowing for round-off, your path will look something like Figure 17-3. The landing will be slightly past the spot.

You could hit the spot exactly if the plane followed its original path all the way to the clump of grass, but who wants to glide right into the ground? It's hard on airplanes for one thing. When maneuvered properly, *the airplane will always land slightly past the spot because of the rounding-off of the landing. There's always a certain amount of float to a landing, even*

one made from a normal glide, because of ground effect.

Ground effect becomes a factor when the airplane is about at one-half wingspan distance from the ground, or about 18–20 feet for most training airplanes. The downwash of the wing is changed with the result that induced drag (look back to Figs. 2-14 and 2-16) is decreased by up to 48 percent in some cases. Since the induced drag makes up a considerable part of the total drag at the lower airspeeds at takeoff and approach/landing, there may be a noticeable decrease in total drag just above the runway and the airplane "floats." In effect, the glide (approach) angle gets shallower.

Ground effect can be a factor on takeoff as well, as pilots have found when trying to take off at a high elevation on a hot day (high density-altitude) in an overloaded airplane. The airplane may get off, but as it climbs to about one-half the wingspan above the run-

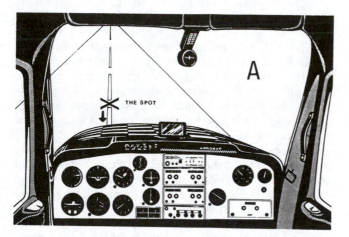
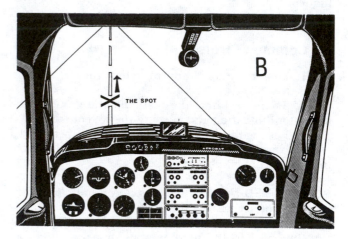

Fig. 17-2. (A) If the spot appears to be moving toward the airplane, you will be too high to land at the spot. (B) If the spot is "gaining" on the airplane, you will land short of the spot.

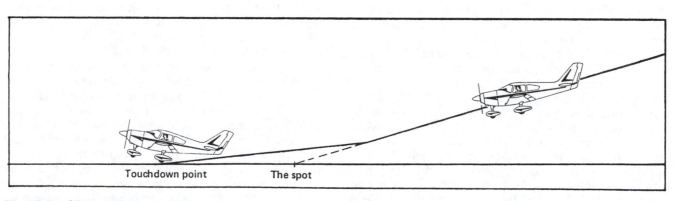

Fig. 17-3. Glide angle and touchdown point.

way, the induced drag rises sharply. There have been extreme cases of airplanes flying into obstacles after takeoff because they were unable to climb out of ground effect.

Your best bet is an approach speed that has a minimum amount of float plus a definite margin of safety. There is such a speed, and a rule of thumb gives it as about 1.3 times the power-off stalling speed (CAS). It's suggested that the airspeed on a short-field approach be *no greater* than this figure. For safety's sake don't get too close to the stall speed—remember the gradient wind and gust effects. *There is no way of predicting the point of landing from a power-off approach on a gusty day.* Reviewing the wind effects, the plane's glide angle will be different for different wind strengths. The glide angle in relation to the ground may change from approach to approach or it may change several times during one approach (Fig. 17-4). By making a power approach the pilot can compensate for wind variation and errors in judgment.

■ Short-Field Approach and Landing

The short-field approach is the most useful type of power approach. It allows the pilot to control the glide path and make a landing with a minimum of float; therefore, as the name implies, it is a good approach to make to short fields. The approach is made at a maximum speed of 1.3 times the power-off stall speed. (If your plane stalls at 50 K with landing flaps set, the maximum approach speed is 1.3 times 50, or 65 K.) At this speed the plane's glide path is particularly sensitive to power adjustments because there is little excess airspeed to cause floating when power is removed. The 50-K and 65-K airspeeds used here are *calibrated* airspeeds.

The traffic pattern will be slightly wider and longer than normal. In essence you'll fly the pattern so that using a normal glide would cause undershooting of the runway; otherwise, if power is used in a normal pattern you'll overshoot.

PROCEDURE

1. From a slightly wider downwind leg, start a power-off approach. Keep the airspeed at normal glide or slightly below.

2. After the first 90° turn, slow the plane to 1.3 times the stall speed and control the angle of descent with the throttle. Set final flaps.

3. Use power as needed to make the landing. It may be necessary to keep power on all the way to the ground if you get low and slow.

After the landing, hold the control wheel (or stick) full back as you apply braking. This helps put most of the weight on the main wheels of the tricycle-gear plane and makes for more braking efficiency. If it's not too much of a distraction, upping the flaps will kill some of the residual lift and put more weight on the braking wheels. One problem, however, is that sometimes the pilots of retractable-gear airplanes have intended to pull up the flaps but moved the landing gear handle instead. Of course, an airplane on its belly will come to a stop in a reasonable distance—it's just a more expensive practice, that's all.

COMMON ERRORS

1. Gliding too fast in the normal part of the approach so that when the nose is raised to slow to the final approach speed, the airplane balloons and is too high to make a power approach.

2. Not keeping a constant airspeed during the power approach. Letting the speed pick up and drop as the power is increased or decreased.

3. Not establishing a constant angle of descent—either using full power or idle. You are free to use as much or as little power as needed, but it looks pretty bad on a check ride if you're a throttle jockey.

4. Closing the throttle while the plane is nearly stalled and still off the ground—letting the plane drop in.

With more experience you'll be able to leave the throttle at a constant setting once the path is established. This is under constant wind and thermal conditions, of course (Fig. 17-5).

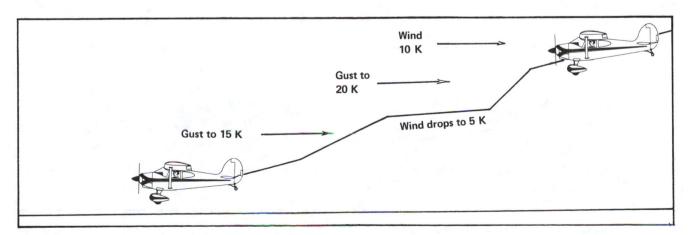

Fig. 17-4. Approach path on a gusty day.

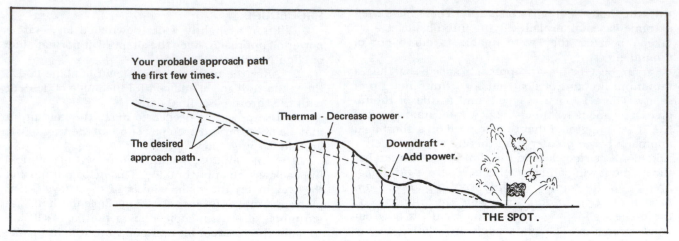

Fig. 17-5. Only in smooth air, constant-wind conditions would the approach path be constant.

SOFT-FIELD TAKEOFF

In mud, snow, or tall grass the drag on the wheels of the airplane tends to make acceleration suffer badly, and a special technique is required for taking off from a soft field.

The soft-field takeoff requires that the tail of the plane be held lower than normal for two reasons: (1) to lessen the chances of nosing over (tailwheel type) or the nosewheel digging in (tricycle-gear type) and (2) to have the angle of attack such that the weight is taken from the main wheels as soon as possible. This attitude will vary among planes, and you will have to have the correct nose position shown to you. Wheel pants are easily clogged by mud or snow and are usually removed from planes operating under these conditions.

■ Procedure

Use flaps as recommended by the *Pilot's Operating Handbook.*

1. Taxi onto the takeoff area and keep the plane

rolling—make a "rolling takeoff." On a soft field the plane will require a great deal of power to start it rolling again, and at high rpm on the ground the prop will pick up mud and gravel and be damaged.

2. Hold the wheel back to stop any tendency to nose over as you apply full power.

3. Put the nose in the proper attitude.

4. Lift the plane off as soon as possible without stalling.

5. As soon as the plane is flying, lower the nose and assume a normal climb. If obstacles are to be cleared, use V_x or the recommended obstacle-clearance airspeed.

The tailwheel airplane should be set up in a tail-low attitude as the takeoff run progresses (Fig. 17-6).

■ Common Error

Not raising the tail on the tailwheel-type airplane—trying to take off in a three-point attitude.

The airplane's attitude – soft or rough field takeoff

The attitude for a normal takeoff.

Fig. 17-6. Soft or rough field and normal takeoff attitudes (tailwheel airplane).

SOFT-FIELD LANDING

The tendency for the nose to dig in on the soft-field landing is a problem, as on the takeoff from a soft field. On the takeoff the force trying to nose the plane over is normally not a sudden thing but may build up as the plane accelerates. As the plane lands, this force may be applied suddenly as the wheels take the weight of the plane. Although this effect will always be there, you can make it as small as possible.

Land the plane as slowly as possible. Use power to decrease the landing speed and to have as much nose-up attitude as safely possible at touchdown. Keep the tail down. Hold full back pressure.

The same principle applies on the soft-field landing for the tricycle-gear equipped plane. Land as slowly as possible and try to keep the nosewheel off. The drag of the main wheels will force it down, but you can keep as much weight off the nosewheel as possible by back pressure.

GUSTY WIND LANDINGS

■ Tricycle-Gear Procedures

The glide speed should be slightly higher to take care of the variance in wind velocity and, for some airplanes, the use of full flaps should be avoided for gusty conditions.

The airplane should be landed at a lower nose attitude than for smooth air, though you shouldn't land it on all three wheels at once (or the nosewheel first). The airplane will tend to rotate forward on the nosewheel at the landing impact; this decrease in angle of attack generally results in the airplane staying firmly on the ground—a decided advantage in gusty winds.

■ Wheel Landing—Tailwheel-Type Airplane

The wheel landing (Fig. 17-7) is used when the wind is strong and/or gusty and gives you a means of having good control all the way down to the landing. The airplane is literally flown onto the ground, landing on the main wheels to keep a low angle of attack and stop any tendency for a sudden gust to pick the airplane off the ground, as might happen in a three-point landing under such conditions.

The technique at one time was to use power all through the landing, but this resulted in the pilot "juggling" the throttle, using up runway and fouling up the landing while trying to control the landing time with power (Fig. 17-8).

Use only enough power to control your rate of descent and get to the position to make your power-off landing. Some pilots advocate using 1300–1600 rpm for a wheel landing, but you have found out that the average lightplane will maintain altitude at a power setting only slightly higher.

Use power only as needed to make the landing. The less used at touchdown the better. Without power the plane will land—not float halfway down the

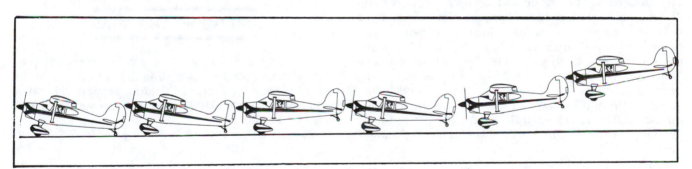

Fig. 17-7. The wheel landing.

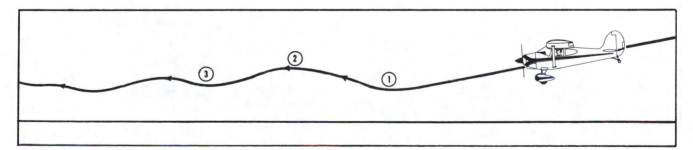

Fig. 17-8. (1) You see that you are settling fast, add more power (oops, too much). (2) Now you throttle back. (3) Plane is now settling fast—add more power again, etc.

runway. *Elevators and throttle go together. Don't be mechanical with either one.*

True, the wind will be strong and the groundspeed will be low. This will keep the plane from using as much runway as it would under lesser wind conditions, but you can still eat up a lot of runway using poor techniques.

PROCEDURE. The approach is made at an airspeed slightly faster than the normal glide, and the transition is made at a lower height for two reasons: (1) The plane must touch down at a higher speed, and (2) the attitude will be only slightly tail-low, not three-point. Power is used to control the angle and/or rate of descent.

As the plane reaches the transition height, start applying back pressure to "round off" the glide path so that the plane will contact the runway with the main wheels in a slightly tail-low attitude. After the plane has touched, apply slight forward pressure to keep the tail up. This is to keep gusts from causing the plane to leave the ground again.

As the plane slows up, more and more forward pressure will be required to keep the tail up. When the wheel is all the way forward and the tail drops, move and hold the wheel completely back so that the up elevator will now hold the tail down. The reason for holding the tail up is to keep a low angle of attack as long as possible. When the elevators become ineffective, the plane's speed is slow enough that the wings have lost all lift. If you try to put the tail on the ground too soon after landing, you may find yourself several feet in the air, with very little airspeed. One disadvantage in keeping the tail up as long as possible is that directional control may be lost before the steerable tailwheel contacts the surface. However, generally the plane with this characteristic has a higher wing loading and is not as likely to be lifted by gusts during the landing roll. A suggested technique in this case is to maintain the forward pressure applied at touchdown, allowing the tail to lower gradually as the speed diminishes. The tailwheel will contact the runway before loss of rudder control occurs and at a speed allow-

ing a smooth steady movement of the wheel to the full rear position without danger of becoming airborne again.

Too many students get impatient as the plane is skimming just above the ground and apply forward pressure to "put it on." This sets off a series of bounces that would make a kangaroo turn white. The trouble is that once you start bouncing, it usually gets worse until you take it around again or something (usually the landing gear) gives (Fig. 17-9).

Another common mistake students make is to hold the plane off too long. The airspeed drops and the plane settles fast on the two front wheels. Result—more bouncing.

The crosswind correction for a wheel landing is the same as for the three-point landing. Lower the wing, hold opposite rudder as needed, and land on one wheel. The other wheel will come down immediately. Hold aileron into the wind and apply rudder as needed to keep it straight.

COMMON ERRORS

1. Too fast an approach—the plane floats.
2. Too slow an approach—the plane settles fast and bounces.
3. Getting impatient—shoving forward on the wheel "to make it stay on" (Fig. 17-9).

If the plane starts bouncing, open the throttle and take it around and make a new landing.

Although wheel landings are best for gusty air, a good three-point landing can be made under these conditions if care is taken.

DRAGGING THE AREA

Before you go on a solo cross-country, the instructor will demonstrate how to drag the area, a procedure used in checking a strange field for landing. The idea is to pick a field and land before you run out of gas and have no choice.

Suppose that you become lost on a cross-country.

"I'LL PUT THIS #-₩Ⓡ! THING ON!"

Fig. 17-9. Some people are bound and determined to land on the first approach.

You have about 15 minutes of gas left and there are no airports or recognizable towns in sight. Or maybe you locate yourself but realize there's no airport close enough to reach with the fuel remaining. Should you fly on to get as close to an airport as possible before the engine quits? Or should you pick a good field, look it over, and land there? Obviously this is a loaded question. You should pick a field.

■ Procedure

1. Pick a likely looking field and circle it at about traffic pattern altitude. Know the wind direction. Pick a field near a road or farmhouse if possible.

2. Set yourself up on the downwind leg and start a normal approach to the portion of the field you want to land on.

3. Fly at a height of about 100 feet, just to the right of the intended landing path.

4. Look for obstructions or hazards that could not be seen from traffic pattern altitude.

5. If the field is suitable, make another traffic pattern with a short-field approach.

6. After the landing, go to the farmhouse and ask if you may use the phone.

Dragging the area is done any time you feel that for safety's sake the plane should be on the ground and you aren't close to an airport. If the engine is in imminent danger of quitting, you may not have time to drag the area but must land immediately in the first likely looking field. This will be covered more thoroughly in ''Problems and Emergencies,'' Chapter 25.

TAKEOFF AND LANDING PERFORMANCE

It's very important to know the required takeoff and landing roll distances of your airplane plus the *total* distances required to clear a 50-foot obstacle. Figure 17-10 shows typical takeoff and landing performance charts. Look at the *Pilot's Operating Handbook* for your airplane and be able to use these and other performance charts (climb, cruise, etc.).

The airplane (and engine) is only aware of the *density*-altitude (standard), and this results from a combination of pressure altitude and temperature. (Pressure altitude is that shown on the altimeter when it is set to 29.92 inches of mercury—review Chapter 3.) The sea level standard temperature is 59°F (15°C) and the normal lapse rate, or normal temperature drop, is 3½°F (2°C) per thousand feet. A higher temperature than normal would mean that the air density is less for that particular altitude. The air could be as ''thin'' as that found several thousand feet higher under standard conditions. The altimeter is only capable of measuring air pressure, not air density, so you must take variations from standard temperature into account.

In using the takeoff and landing charts (Fig. 17-

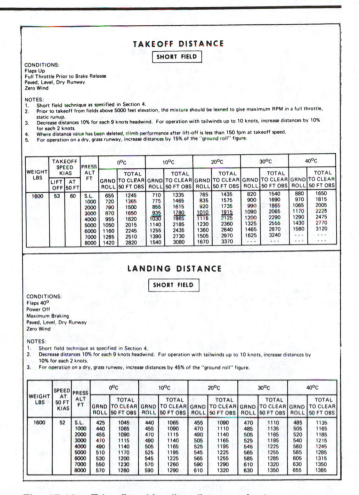

Fig. 17-10. Takeoff and landing distance charts.

TAKEOFF DISTANCE — SHORT FIELD

CONDITIONS:
Flaps Up
Full Throttle Prior to Brake Release
Paved, Level, Dry Runway
Zero Wind

NOTES:
1. Short field technique as specified in Section 4.
2. Prior to takeoff from fields above 5000 feet elevation, the mixture should be leaned to give maximum RPM in a full throttle, static runup.
3. Decrease distances 10% for each 9 knots headwind. For operation with tailwinds up to 10 knots, increase distances by 10% for each 2 knots.
4. Where distance value has been deleted, climb performance after lift-off is less than 150 fpm at takeoff speed.
5. For operation on a dry, grass runway, increase distances by 15% of the ''ground roll'' figure.

WEIGHT LBS	TAKEOFF SPEED KIAS		PRESS ALT FT	0°C		10°C		20°C		30°C		40°C	
	LIFT OFF	AT 50 FT		GRND ROLL	TOTAL TO CLEAR 50 FT OBS	GRND ROLL	TOTAL TO CLEAR 50 FT OBS	GRND ROLL	TOTAL TO CLEAR 50 FT OBS	GRND ROLL	TOTAL TO CLEAR 50 FT OBS	GRND ROLL	TOTAL TO CLEAR 50 FT OBS
1600	53	60	S.L.	655	1245	710	1335	765	1435	820	1540	880	1650
			1000	720	1365	775	1465	835	1575	900	1690	970	1815
			2000	790	1500	855	1615	920	1735	990	1865	1065	2005
			3000	870	1650	935	1780	1010	1915	1090	2065	1170	2225
			4000	955	1820	1030	1965	1116	2125	1200	2290	1290	2475
			5000	1050	2015	1140	2185	1230	2360	1325	2555	1430	2770
			6000	1160	2245	1255	2435	1360	2640	1465	2870	1580	3120
			7000	1285	2510	1390	2730	1505	2970	1625	3240	---	---
			8000	1420	2820	1540	3080	1670	3370	---	---	---	---

LANDING DISTANCE — SHORT FIELD

CONDITIONS:
Flaps 40°
Power Off
Maximum Braking
Paved, Level, Dry Runway
Zero Wind

NOTES:
1. Short field technique as specified in Section 4.
2. Decrease distances 10% for each 9 knots headwind. For operation with tailwinds up to 10 knots, increase distances by 10% for each 2 knots.
3. For operation on a dry, grass runway, increase distances by 45% of the ''ground roll'' figure.

WEIGHT LBS	SPEED AT 50 FT KIAS	PRESS ALT FT	0°C		10°C		20°C		30°C		40°C	
			GRND ROLL	TOTAL TO CLEAR 50 FT OBS	GRND ROLL	TOTAL TO CLEAR 50 FT OBS	GRND ROLL	TOTAL TO CLEAR 50 FT OBS	GRND ROLL	TOTAL TO CLEAR 50 FT OBS	GRND ROLL	TOTAL TO CLEAR 50 FT OBS
1600	52	S.L.	425	1045	440	1065	455	1090	470	1110	485	1135
		1000	440	1065	455	1090	470	1110	485	1135	505	1165
		2000	455	1090	470	1115	490	1140	505	1165	520	1185
		3000	470	1115	490	1140	505	1165	525	1195	540	1215
		4000	490	1140	505	1165	525	1195	545	1225	560	1245
		5000	510	1170	525	1195	545	1225	565	1255	580	1285
		6000	530	1200	545	1225	565	1255	585	1285	605	1315
		7000	550	1230	570	1260	590	1290	610	1320	630	1350
		8000	570	1260	590	1290	610	1320	630	1350	655	1385

10) you may have to interpolate between values given. For instance, assume that you will be taking off from a field with a pressure altitude of 3000 feet with a 14-K headwind at a temperature of 15°C: First, you'd interpolate between 10° and 20°C at that pressure altitude for the ground run and total distances to clear a 50-foot obstacle. At 10°C the values are 935 and 1780 respectively. At 20°C the distances are 1010 and 1915 feet, so you could expect them at 15°C to be 975 and 1850 feet (rounded off). As the Notes in Figure 17-10 indicate, you should subtract 10 percent for each 9 K of headwind, so at 14 K (about 1.5 times 9 K) the reduction should be 15 percent for final figures of 830 and 1575 feet (rounded off). The same idea would be used for the landing distance. *These distances are for a hard-surface runway; the takeoff run could be doubled by mud, snow, or high grass.* Also, these are ''average'' values; you can use a lot more runway with poor pilot technique.

Figure 17-11 is a climb chart for the airplane just discussed.

Figure 17-12 shows a graphical method of computing the ground roll required under different conditions. This airplane also has a chart for total distance over a 50-foot obstacle under varying conditions (not included here).

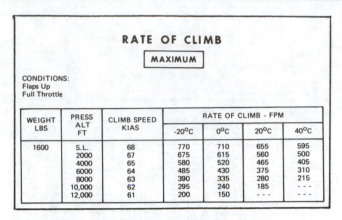

Fig. 17-11. Maximum rates of climb chart.

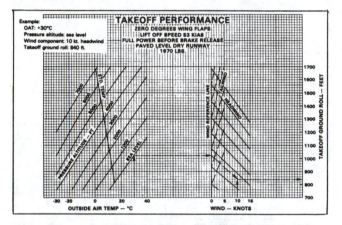

Fig. 17-12. A graphical presentation of takeoff performance.

Figure 17-13 shows the takeoff distances (ground roll and to clear a 50-foot obstacle) for a current trainer. Two propeller options are available for this airplane: (1) a climb propeller of lower (flatter) pitch, which is more efficient at lower airspeeds, and (2) a cruise propeller with a higher average pitch, which is not as efficient in lower-speed regimes but comes into its own at cruise speeds. Note that at a pressure altitude of 2000 feet and a temperature of 20°C (at a weight of 1600 pounds) the climb prop–equipped airplane uses less distance to lift off or to clear the 50-foot obstacle (arrows).

While the subjects of special takeoff and landing procedures and performance charts are being covered, it might be well to take a look at Figure 17-14.

The example shown gives an answer to 35-K headwind and 20-K crosswind components respectively. Looking at another problem, suppose an airplane has a manufacture-demonstrated crosswind component of 0.2 V_{so} (see Fig. 3-27) and its V_{so} is 60 K. Can it legally land in a 30-K wind at a 25° angle to the runway? The actual value of the demonstrated crosswind is 0.2 × 60 = 12 K. Using Figure 17-14, you'll find that the crosswind component is 12.5–13 K, which is over the demonstrated limit. (The headwind component is about 27 K.) The problems

you'll be given on the written test will require that you find the angle ("you are landing on Runway 27 with a wind from 310° at 20 K") and then use the chart. The problems apparently assume that the runway is lined up exactly with 270°, and the wind and runway are both given in the same terms (magnetic or true directions—see Fig. 6-3).

■ Simulated High-Altitude Takeoffs

Before you get to the private practical (flight) test you should make a simulated high-altitude takeoff. The instructor will limit your power for takeoff, and you can see how the airplane would do (or not do) at some high altitude. The biggest problem you'll have is the very strong tendency to rush the airplane off. (This is also the tendency the first time you fly the airplane at maximum certificated weight from a comparatively short field.) The airplane seems to be dragging its feet, and you'll want to pull it off before it's ready. After you become airborne, don't pull the nose up "to get extra climb." You might put yourself in position 4 or 5 back in Figure 12-13, and performance will suffer even more.

Talking a little more about the density-altitude idea, every 15°F (8½°C) above the standard temperature for your pressure altitude bumps the density-altitude up another 1000 feet. Suppose you are going to take off at an airport at an elevation of 2200 feet. The altimeter setting is 30.12 inches of mercury, which makes your pressure altitude right about 2000 feet (1 inch of mercury equals approximately 1000 feet of altitude). The pressure corrected to sea level is 0.20 inches of mercury higher than standard, so the pressure altitude is (0.20 × 1000) = 200 feet lower than the elevation. The pressure altitude is 2000 feet (which could also have been found by rolling 29.92 into the setting window).

The temperature is 82°F, which certainly is not unusual in the summer. The standard temperature for this pressure altitude is 52°F (59°F − [2 × 3½°F] = 52°F), since the normal lapse rate is 3½°F per thousand feet. The temperature (82°F) is 30°F above the standard. Since each added 15°F makes an additional 1000 feet of density-altitude, your real density-altitude is the pressure altitude (2000 feet) *plus* 2000 feet, or *4000 feet.* You think you're sitting at 2000 feet, but the airplane knows that it is at 4000 feet—and acts accordingly. If the temperature had been 97°F at that airport (entirely possible), another thousand feet would have been tacked on and the density-altitude would have been *5000 feet.*

At an airport in the summer it's not at all unusual for the temperature to be 15°F (or much more) above its "standard." The temperature, rather than atmospheric pressure changes, is the big factor. The not unusual temperature of 82°F in the example added 2000 feet of density-altitude. To get that effect by pressure decrease (assuming a standard temperature) would

TAKEOFF DISTANCE, CLIMB PROPELLER

ASSOCIATED CONDITIONS:
Power — Maximum
Flaps — Up
Runway — Hard surface (level & dry)
Fuel Mixture — Full throttle climb, mixture leaned above 5000 feet to smooth engine operation

NOTES:
1. Decrease distance 5% for each 5 knots headwind. For operation with tailwinds up to 10 knots increase distance by 10% for each 2.5 knots.
2. Where distance value is shaded, climb performance after lift-off, based on the engine operating at takeoff power at takeoff speed, is less than 150 feet per minute.
3. If takeoff power is set without brakes applied, then distances apply from point where full power is attained.

WEIGHT LBS	TAKEOFF SPEED KIAS (MPH) LIFT OFF	TAKEOFF SPEED KIAS (MPH) CLEAR 50 FT.	PRESS. ALT. FT.	0°C (32°F) GROUND ROLL	0°C (32°F) CLEAR 50 FT.	10°C (50°F) GROUND ROLL	10°C (50°F) CLEAR 50 FT.	20°C (68°F) GROUND ROLL	20°C (68°F) CLEAR 50 FT.	30°C (86°F) GROUND ROLL	30°C (86°F) CLEAR 50 FT.	40°C (104°F) GROUND ROLL	40°C (104°F) CLEAR 50 FT.
1600	57 (66)	66 (76)	S.L.	719	1313	798	1455	883	1607	973	1769	1069	1940
			2000	857	1554	951	1722	1051	1901	1159	2092	1273	2295
			4000	1022	1842	1135	2042	1255	2254	1383	2481	1520	2772
			6000	1224	2190	1358	2427	1502	2680	1655	2949	1819	3236
			8000	1468	2610	1629	2892	1802	3194	1986	3515	2182	3856
1500	56 (64)	64 (74)	S.L.	616	1132	684	1254	756	1385	834	1524	916	1672
			2000	734	1339	814	1483	901	1638	993	1802	1090	1977
			4000	876	1587	972	1759	1075	1942	1185	2137	1302	2345
			6000	1048	1887	1163	2091	1286	2308	1418	2541	1558	2787
			8000	1257	2248	1395	2491	1543	2751	1701	3028	1869	3322
1400	55 (63)	62 (71)	S.L.	522	965	580	1069	641	1181	706	1299	776	1425
			2000	622	1141	690	1265	763	1397	841	1537	924	1686
			4000	742	1353	824	1500	911	1656	1004	1822	1103	1999
			6000	888	1609	986	1783	1090	1968	1201	2166	1320	2376
			8000	1065	1917	1183	2124	1308	2345	1441	2581	1584	2832

TAKEOFF DISTANCE, CRUISE PROPELLER

ASSOCIATED CONDITIONS:
Power — Maximum
Flaps — Up
Runway — Hard surface (level & dry)
Fuel Mixture — Full throttle climb, mixture leaned above 5000 feet to smooth engine operation

NOTES:
1. Decrease distance 5% for each 5 knots headwind. For operation with tailwinds up to 10 knots increase distance by 10% for each 2.5 knots.
2. Where distance value is shaded, climb performance after lift-off, based on the engine operating at takeoff power at takeoff speed, is less than 150 feet per minute.
3. If takeoff power is set without brakes applied, then distances apply from point where full power is attained.

WEIGHT LBS	TAKEOFF SPEED KIAS (MPH) LIFT OFF	TAKEOFF SPEED KIAS (MPH) CLEAR 50 FT.	PRESS. ALT. FT.	0°C (32°F) GROUND ROLL	0°C (32°F) CLEAR 50 FT.	10°C (50°F) GROUND ROLL	10°C (50°F) CLEAR 50 FT.	20°C (68°F) GROUND ROLL	20°C (68°F) CLEAR 50 FT.	30°C (86°F) GROUND ROLL	30°C (86°F) CLEAR 50 FT.	40°C (104°F) GROUND ROLL	40°C (104°F) CLEAR 50 FT.
1600	58 (67)	66 (76)	S.L.	762	1365	846	1513	963	1670	1031	1838	1133	2017
			2000	908	1615	1007	1789	1114	1976	1228	2175	1349	2386
			4000	1083	1915	1202	2122	1330	2344	1466	2579	1610	2830
			6000	1296	2277	1439	2523	1591	2786	1754	3067	1927	3365
			8000	1555	2714	1726	3008	1909	3321	2104	3655	2312	4011
1500	57 (66)	64 (74)	S.L.	653	1176	725	1303	801	1439	883	1583	970	1737
			2000	777	1391	863	1541	954	1702	1052	1873	1155	2055
			4000	928	1650	1030	1828	1139	2019	1255	2221	1379	2437
			6000	1110	1961	1232	2173	1363	2400	1502	2641	1650	2898
			8000	1332	2338	1479	2590	1635	2860	1802	3148	1980	3454
1400	56 (64)	62 (71)	S.L.	553	1003	614	1111	679	1227	748	1350	822	1481
			2000	659	1186	731	1314	809	1451	891	1597	979	1752
			4000	786	1406	873	1558	965	1721	1064	1894	1169	2078
			6000	841	1672	1044	1853	1155	2046	1273	2252	1398	2470
			8000	1129	1993	1253	2208	1386	2438	1527	2684	1678	2944

Fig. 17-13. Takeoff distances for a two-place trainer for climb and cruise propellers. (*Grumman American Aviation Corp.*)

mean that the pressure would have to be *2 inches* of mercury below normal, or the sea level pressure would be 27.92 instead of 29.92 inches of mercury. This *is* unusual, and such pressure conditions could only be found in the eye of a very strong typhoon.

Talking in terms of Celsius for the earlier example problem, the standard sea level temperature is 15°C and the normal lapse rate is 2°C per thousand feet, so the following would apply: Standard temperature for the 2000 feet is 11°C; the actual temperature is 28°C, or 17°C above standard. Since each 8½°C adds another 1000 feet of density-altitude, the answer is again 4000 feet.

One thing that you may not have considered is that the *moist air is less dense than dry air* (all other factors equal), so the airplane will not perform as well in takeoffs and climbs when the air is moist. (You might figure on up to about 10 percent less performance under wet conditions.)

One last thing about density-altitude: You may not know what it is at a particular time, but the airplane *always* does.

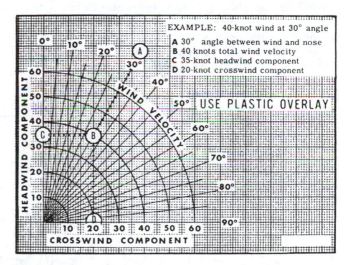

Fig. 17-14. A crosswind-headwind component chart. (*FAA*).

18

HIGH-ALTITUDE

EMERGENCIES

The high-altitude emergency gives you more time and a greater choice of fields. In earlier stages, however, this is a handicap, so practice is usually reserved for the postsolo period. The average student, when given a first high-altitude emergency, is as busy looking around as a one-eyed man watching a beauty contest. There are *too many* places to land. You may get confused and change your mind on the choice of a field several times and finally end up too low to do anything but land in the worst field within a 15-mile radius (Fig. 18-1).

The high-altitude emergency requires that you do the following:

1. Set up the max distance glide. Start easing the nose up immediately to conserve altitude as well as to get to the best glide speed. At the same time:

2. Pull the carburetor heat ON (use it before the residual engine heat is lost).

3. Pick a field and head for it.

4. Find the wind direction.

5. Try to find the cause of the problem—carburetor ice, fuel tank run dry (switch tanks), etc. Set up a glide pattern so that you will hit a "key position" at about the same point and at a slightly higher altitude above the ground than the point at which you closed the throttle to make a 180° power-off approach (the windmilling prop will hurt the glide).

This key position is chosen because by this time you've shot many power-off approaches from this position. It's literally impossible to glide for a distant field from 2000 or 3000 feet above the ground and consistently hit it. It's been found to be easier to judge the glide angle if the approach involves a turn. (Review power-off approaches in Chapter 13.)

Let's take the emergency point by point as discussed below.

TELLING WIND DIRECTION

There are many ways of telling the wind direction when you are away from the airport: smoke, waves

Fig. 18-1. The instructor will discuss your errors on the forced landing.

moving across a field of grain or waves on lakes, dust blowing, trees, etc. (Fig. 18-2).

Know the wind direction and approximate velocity at all times. Don't be like the student who didn't know the wind direction and stuck a hand out the plane window to find out (and figured that the wind was right on the nose at about 100 K).

It's better not to waste time trying to find out the wind direction *after* the emergency has occurred.

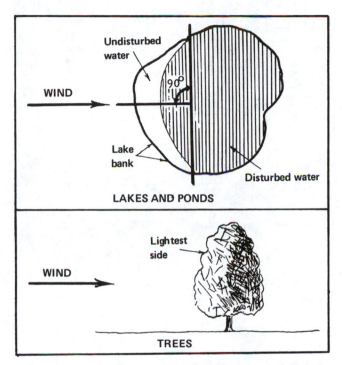

Fig. 18-2. Lakes or ponds help you tell the wind direction, even if you are too high to see the actual wave movement; the bank on the windward side keeps the water undisturbed. The wind turns tree leaves over, showing their lighter side in the direction from which it is blowing.

PICKING YOUR FIELD

Know the types of fields in your part of the country. Has it been raining a lot lately? If so, that field of young wheat may be too soft to land on.

Look for electric or telephone poles across the field because you won't see the wires themselves from your altitude.

Look for fences across the middle of pastures. These are sometimes hard to see because the fields on both sides look alike.

If two fields are equally near and good for landing, pick the one closest to the highway or houses for several reasons. If the landing is a success (and with good planning it's bound to be), you will want to call the airport. If the plane is damaged or the engine is in bad shape (maybe there was an oil leak and the engine

seized), the airport personnel may want to dismantle the plane and haul it back. If the landing was not a success and you are injured, people will see the crash and come to your aid. The main thing is to get down without injury to you or (later) your passengers.

If your airplane has flaps, they'll be a great help in hitting the field. The idea is to add flaps in increments as necessary to hit the field. The usual student error in practice emergencies is to add flaps too soon (and *full* flaps at that) and sit there helplessly while the airplane starts to sink into the woods a quarter of a mile short of the chosen field. *If you have added flaps and start losing ground* (it looks as if the field is gaining on you), *get a notch or a few degrees of flaps up and watch things start improving. Then when you get to the proper position, put 'em back down again.*

The big misconception about flaps concerns the idea that once the flaps are down, they should never be retracted because the airplane will "fall" because of a loss of "lift." If you are close to the flaps-down stall speed and suddenly raise the flaps, you might find that the airspeed is *below* the flaps-*up* stall speed, and a stall could result. *But* if you are in a typical trainer at the normal glide speed and at a reasonable altitude (say, 300 feet or more), you'll be a lot better off in getting distance by easing off a notch or so of flaps (easy), even if that initial slight sink bothers you. Putting the flaps down in increments is the best way to do it to avoid flap juggling (up and down).

PROCEDURE

1. When the engine stops, establish a normal glide. Carburetor heat ON. Pick the best field and turn to it (Fig. 18-3). (Assume for this example that you can't correct the cause of the problem.) Maintain the normal glide speed throughout the exercise. This will give you the best glide ratio and the only means of knowing your glide path. If the speed is 70 K one minute and 50 K the next, who knows what the glide path will be?

Figure 18-4 shows the maximum glide ratio of a current training airplane. As noted, from 6000 feet above the surface a distance of approximately 8 nautical miles may be covered with the conditions stated. Naturally, headwinds or tailwinds will affect these figures.

2. Get over the field. Spiral to the key position or S down. Try to find the problem.

3. Hit the key position at the traffic pattern altitude or slightly higher—better slightly high than too low at this point. Remember that a windmilling prop creates a great deal of drag, hence your glide ratio will suffer.

4. Plan your approach after the key position, thinking that a shallow slip or flaps may be required to make the field.

5. Once you have the field made, remember the landing is just like the one back at the airport (Fig.

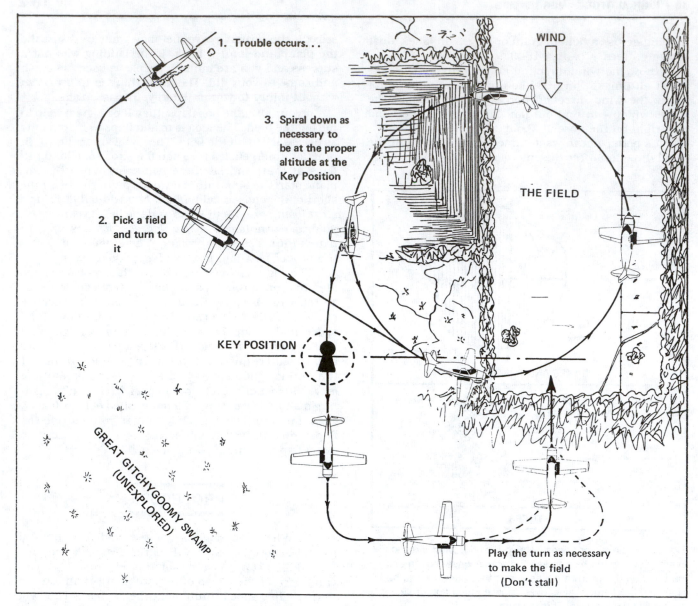

Fig. 18-3. The key position helps to turn an unusual situation into one more familiar. Keep that glide speed constant.

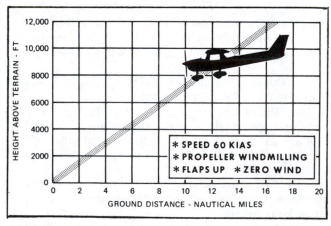

Fig. 18-4. Maximum glide performance.

18-5). A normal landing will minimize the damage if the field is rougher than it looked from 3000 feet.

6. The *Pilot's Operating Handbook* for your airplane has an emergency section in it (read it *now*—don't wait until you have a problem), and an emergency checklist is available in most current airplanes. *Know where it is;* a high-altitude emergency will probably give you time to look it over (if your hands aren't shaking so badly that the instructions are blurred). The shutdown procedure varies among airplane types and models, but you'll want to have the mixture, fuel, and ignition and master switches OFF before touchdown if you're definitely going to put it in a field. Even if the landing area looks fine, you don't want to take a chance on hitting an unseen hole or stump and having a fire result. As mentioned in Chap-

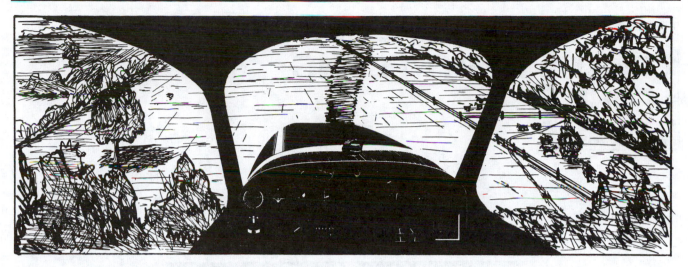

Fig. 18-5. Well, it's no Dulles International Airport, but because you used proper techniques, you'll make it without any damage to the airplane or occupants.

ter 13, the master switch (electrical) may be the last thing turned OFF because you may be wanting to talk to someone on the radio (121.5 MHz is the emergency frequency monitored by the FAA and military facilities) and you may want to "squawk" 7700, the emergency code on the transponder. (More about radio frequencies and transponder codes in Chapter 21.) If your flaps are electrically operated you may have to change the setting (increase or decrease) at the last moment. But be sure everything is OFF when you touch down.

7. After the plane is stopped and your knees have stopped shaking, see to the security of the plane. Set the parking brake if there is no tie-down kit in the plane, or better yet, chock the wheels with rocks or blocks. Get to the nearest phone and let the people at the airport know what happened.

An actual emergency is a nerve-wracking thing. You'll be scared, but if you've had practice in emergencies and gone over the procedure in your mind, you'll make out fine and it'll make good hangar talk after it's over.

■ **Common Errors in Practice Emergencies**

1. Changing fields too many times—if you see that the original field you picked has a big ditch or wires across the middle and there is another field within easy gliding distance, by all means take it. However, once you hit the key position at a field you're pretty well committed to it.

2. Not maintaining a constant glide speed—the result is usually that the field is missed and an airplane is "pranged."

3. Poor wind correction after the key position, that is, being too high or too low to make the field.

■ **Discussion**

After the field is "made," the instructor will take over and discuss the emergency with you.

Keep the engine cleared on the descent. The high-altitude emergency means an extended glide and cooling of the engine, particularly in cold weather. Use carburetor heat as recommended by the manufacturer.

You will not practice emergencies solo. But while we're on the subject of extended glides, if sometime in cold weather you've glided from 3000 feet down to 600 feet above the ground the engine may cool and not want to take throttle at first. Don't ram the throttle open—open it slowly. Give the engine a shot of primer if it's particularly hard to get going again. One quick shot will usually do the trick. In hot weather the engine tends to load up during an extended glide and will require careful clearing.

■ **Emergency Descent**

The *Pilot Operating Handbook*, or your instructor, will cover the emergency descent as it applies to your airplane. There are several situations where an immediate descent is required to save the airplane and occupants: (1) Loss of oxygen, or cabin decompression, at high altitudes (not likely a problem for the light general aviation trainer). (2) Engine fire. (3) Electrical fire. (4) Suspected carbon monoxide poisoning or the sudden critical illness of a passenger. (Don't make matters worse by a too-rapid rate of descent.) The point is that you should know and periodically review the procedure (which may require some fairly radical attitudes and airspeed) and always have a checklist available in the airplane.

PART FOUR

CROSS-COUNTRY AND NIGHT FLYING

19

THE NAVIGATION IDEA

A flight is a success or failure before you leave the ground. Preflight planning is important at all times but particularly so before a cross-country flight. Too many pilots jump into a plane and head for distant airports with little thought of weather, the condition of the destination airport, or what kind of fueling or repair service can be obtained there.

There are several types of navigation used by pilots: (1) *pilotage,* or flying by reference to landmarks; (2) *dead* (deduced) *reckoning,* which is drawing vectors of the wind and your true airspeed and computing the heading, groundspeed, and estimated time of arrival at the destination; (3) *celestial navigation,* using a sextant to measure angles to heavenly bodies; and (4) *radio navigation,* or navigation through the use of radio aids.

Pilotage is the major means of navigation for the student pilot, but radio navigation is coming into more use as trainers become better equipped.

Unfortunately, too many pilots rely on radio navigation and find themselves rusty in pilotage or use of the sectional chart if the radios go out.

MERIDIANS AND PARALLELS

The earth is laid off in imaginary lines called meridians and parallels. The meridians run north and south and divide the earth like the sections of an orange (Fig. 19-1). The prime meridian goes through Greenwich, England. All longitude (east or west measurements of position) is measured east or west from this line to the 180° meridian on the opposite side of the earth. Since the earth turns 360°, or a complete revolution, in 24 hours, each 15° of longitude means a difference of 1 hour as far as the sun is con-

cerned. You can see this roughly in the time zones in the United States.

If someone said that a ship is at 15° west longitude, it would be almost impossible to locate. It could be anywhere on that meridian between the North and South poles. It would be much like telling a friend to meet you on 54th Street in New York City on a busy day. If you said instead, "Meet me at 54th and Blank Street," it would help matters considerably. The conterminous United States extends from about 67° west longitude to about 125° west longitude.

For navigational purposes, time is standardized to that at Greenwich. This coordinated universal time (UTC) is based on the 24-hour clock. Looking at Figure 19-1, you see that the earth rotates toward the "east." The day starts at 0000UTC or 0000Z (or 0000 Zulu), and since parts of the United States are from 67° to 125° "behind" Greenwich, as the earth rotates, our time will also be behind accordingly. If you are at 90° west longitude, your standard time will be 6 hours behind that of Greenwich (each 15° longitude = 1 hour of time). If the time is 0600 (6:00 A.M.) at your position, add 6 hours to obtain the Universal Time (1200 or 12:00 noon). If you are in the eastern standard time zone add 5 hours to get UTC (or Zulu) time, add 6 in the central standard time zone, and so on. For *daylight savings time,* you'd decrease the addition by 1 hour to get Zulu time. (For central daylight time you'd add 5; for eastern daylight time you'd add 4 hours.)

The *Airport/Facility Directory* (to be discussed in the next chapter and also in Appendix B, which is the legend for the *A/FD*—see item 5 there) gives the proper conversions from Zulu (UTC) to local time or vice versa.

The earth is further divided north and south by parallels—so called because, instead of converging at

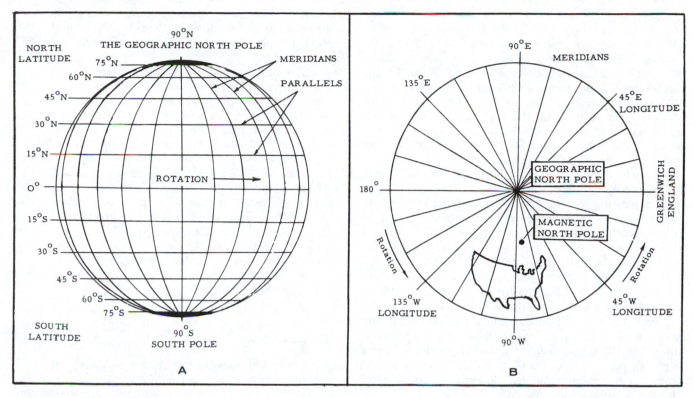

Fig. 19-1. Navigation basics.

two points as the meridians do, they are parallel to each other (Fig. 19-1). The equator is the prime parallel, and latitude is measured north and south from it. The equator is 0° latitude and the United States runs from about 25° north latitude at the Florida tip to about 49° north latitude on the Canadian border. The North Pole is at 90° north latitude and the South Pole at 90° south latitude.

Each degree of latitude and longitude is broken down into 60 minutes, and each minute of *latitude* equals a nautical mile, or 6080 feet. You can measure a distance on the map by laying the length to be measured north and south on the map and reading the minutes of latitude that it extends. This gives the distance in nautical miles; to convert these to statute or land miles, multiply the result by 1.15. In other words, a nautical mile equals 1.15 or about 1¹⁄₇ statute (land) miles. (See Chapter 3.)

Don't measure distances between meridians this way because the equator would be the only place where 1 minute of longitude equals a nautical mile. The meridians converge as they get farther north or south, so while the distance between them in degrees is always the same, the distance in miles is not. Most aeronautical charts have a scale in statute *and* nautical miles, but all have nautical mile scales.

The meridians converge on the "true" North Pole so that any angle measured from a meridian is the angle from true North. If you were at a point on a map and wanted to know the direction to another point, you would measure it as an angle clockwise from true North, using a meridian midway between the two

points. The reason for choosing a middle meridian is because of the type of map you'll be using, as will be discussed later under "Plotter."

Then directions are measured clockwise from true North using 360° (a full circle). East is 90°, South is 180°, West is 270°, and North is 0° (or 360°). Northeast is 45°, Southeast 135°, etc. This means that the pilot can plan a course to the nearest degree.

The idea is to always use three numbers in speaking or writing of directions; 90° should be 090°.

MORE ABOUT THE COMPASS

Certain errors of the compass were discussed earlier in the introduction to the instrument and emergency flying by instruments. You know of the properties such as northerly turning error, which affect the compass in a turn, or the acceleration errors that affect it with speed changes. Other factors important in aircraft navigation should be considered in using the compass in straight, unaccelerated flight.

■ Variation

The fact that the Magnetic North Pole and the True North Pole are not the same means a little more work for you in navigating. The Magnetic North Pole is in Canada, and your compass points to this magnetic pole rather than the true one.

The angle between your direction to the True North Pole and the Magnetic North Pole is called variation. If at your position the Magnetic North Pole is 6° farther "east" than the True North Pole, you say that a 6° east variation exists. The variation in the United States runs from about 22° West in Maine to about 20° East in Oregon. Isogonic lines, or lines of equal variation, are shown on the aeronautical charts and are given in 1° increments. If you are flying in an area where the two poles are "magnetically" in line, no correction for variation is needed. The isogonic line will show as 0° variation, or more properly, the 0° line is called the agonic line (Fig. 19-2).

In plotting a course, you will measure the *true course* at a middle meridian (nearest to the midpoint of your course line). Because you will be referring to the compass, the course must be considered in relation to the Magnetic North Pole. You will add or subtract variation to accomplish this. *Remember: East is least and West is best.* This means that you will look at the isogonic or variation line that is nearest to the halfway point of your course and subtract 5° from the true course if the variation is 5°E or add 5° if the variation is 5°W. *Going from true to magnetic, subtract easterly variation and add westerly variation.* How much to subtract or add will depend on the value of the midpoint isogonic line.

The British have a saying, "Variation East, Magnetic Least; Variation West, Magnetic Best," which could also help you to remember.

If you measured the true course with a protractor and found it to be 120° and the midpoint isogonic line showed 5°E, your magnetic course would be 115° (Fig. 19-3).

The terms "east" and "west" used for variation may be misleading to the newcomer. "East" and "west" might appear to have something to do with the direction the airplane is flying. *This is not the case at all.* Since the magnetic compass starts *its* measurements at the Magnetic North Pole, in Figure 19-3 it starts "counting" 5° later than your map reference (true North). This is the case whether you are flying on a course of 320°, 120°, or what have you. It might be better to speak in terms of "plus 5° variation" or "minus 5° variation," but the current method seems to be permanent. *Variation at a particular geographic position is the same for all types of airplanes, whether J-3 Cubs or B-52s.*

Figure 19-4 is part of a sectional chart showing how variation is depicted.

■ **Deviation**

The compass will have instrument error called deviation, due in part to attraction by the ferrous metal parts of the plane. This error will vary between headings, and is usually noted for every 30° on a compass correction card, which is located on the instrument panel near the compass (Fig. 19-5).

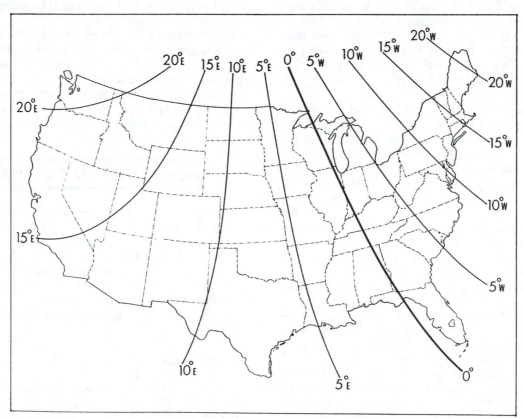

Fig. 19-2. Approximate location of isogonic lines in the United States.

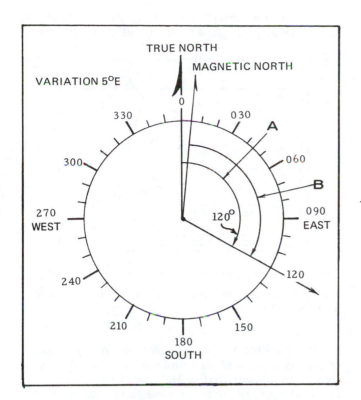

VARIATION 5°E

Fig. 19-3. True and magnetic courses and variation. You measure the course from the map reference (True North Pole) and find that the angle is 120° (A). The Magnetic North Pole, in this example, is 5° "east" of the True North Pole, and your compass measures its angle from this and starts 5° "later" (B). The angle with reference to the Magnetic North Pole to the course line is 5° less than the true course and is 115° magnetic. This is the condition for *any* course in the geographic area of 5° E variation. If the problem had been for a place where the variation is 5° W, the magnetic North Pole would have been 5° "west" of the True North Pole and the 120° *true* course would mean a *magnetic* course of 125°.

FOR (MAGNETIC)	N	30	60	E	120	150
STEER (COMPASS)	O	26	58	87	118	150
FOR (MAGNETIC)	S	210	240	W	300	330
STEER (COMPASS)	183	214	243	275	303	332

Fig. 19-5. A typical compass correction card.

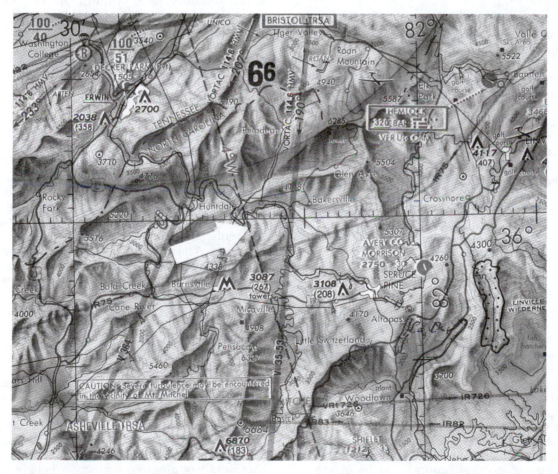

Fig. 19-4. An isogonic line (4° W) as depicted on a sectional chart.

The compass is swung or corrected on a compass rose, a large calibrated circle painted on the concrete ramp or taxiway away from metal interference such as hangars. The plane is taxied onto the rose and corrections are made with a nonmagnetic screwdriver. Attempts are made to balance out the errors as much as possible. The engine should be running and radio and electrical equipment should be on to give a true picture of the deviation as it would be experienced in flight. To be completely accurate, the plane should be in level flight attitude (no problem with tricycle gear) with the tail supported by a sawhorse or short ladder (tailwheel type).

Suppose, for example, your magnetic course is 115°. This is not shown on the correction card, so some averaging will be necessary. Practically speaking, pick the reading on the card that is 120°. You see that for 120° you must fly 118° or subtract 2°. Subtract 2° from 115° and the compass course is 113°.

Unlike variation, which is a function of your geographic position, *deviation varies between individual airplanes* (and may vary in a particular airplane, depending on what electrical or radio equipment is currently in operation).

For the steps used to go from a true course (or true heading) to the course (or heading) with respect to the *compass*, remember that "True Virgins Make Dull Company":

1. *True* course (or heading)
2. Plus or minus *Variation* gives
3. *Magnetic* course (or heading)
4. Plus or minus *Deviation* gives
5. *Compass* course (or heading).

Remember that the "east is least . . . " idea is true only when going from a true course (or heading) to a magnetic course or heading. Working from magnetic to true, you would *add* easterly variation and *subtract* westerly variation. The normal procedure is to go from true to magnetic courses or headings, but written tests have asked questions on the reverse procedure.

PLOTTER

The plotter is a small transparent plastic circle or semicircle marked from 0° to 359° (the semicircular plotter has a double row of figures at each point) to measure course angles on the map. Plotters have an attached plastic rule in scale miles for the sectional chart (1 inch equals about 7 NM) and other charts.

When you plan a cross-country, you'll draw a line between your departure airport and the destination airport and measure the true course with the plotter.

Measure the course at a midway meridian because the sectional chart is a Lambert conformal conic projection; that is, the meridians are closer together at the top of the map. By measuring from the middle meridian, a more accurate course will be obtained. However, the error is so small that little trouble will be

caused by not doing this—but, again, you might as well learn another good habit now (Fig. 19-6).

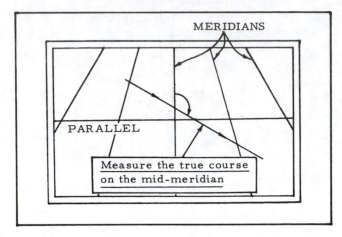

Fig. 19-6. Exaggerated drawing of meridian convergence on a sectional chart.

It's possible to get confused when laying off a course by using a plotter, and students sometimes arrive at an answer that is 180° off. *Draw the course line and figure about what the direction is* (Jonesville is east-southeast of Smithtown, so you would figure somewhere between 90° and 135° true course). *Don't mechanically measure a course but think about what you're doing.* You'll soon be able to look at a course line on a map and guess within 5° of it. You should *very carefully read the plotter when checking the course.*

You will notice azimuth circles around each VOR station on the sectional chart. These are calibrated from 0° to 360° in 5° increments and are oriented on *magnetic* north. These are an aid to the pilot who is combining pilotage and radio navigation, as will be covered later. If you can't find your plotter and one of those omni roses is near your course line on the map, you may measure the magnetic course by laying a straight edge parallel to the course line and through the center of the circle (or use a parallel rule if you have one). The *magnetic* course is then read because of the orientation with magnetic north rather than a meridian, and you've saved the step of adding or subtracting variation.

COMPUTER

A computer is a useful addition to any student pilot's equipment. It's a circular slide rule with the inside circle calibrated in time (hours and minutes) and the outside circle in velocity (miles per hour or knots) or gallons. With it you can check fuel consumption or speed. You learned in Chapter 3 that the 2 percent per thousand feet airspeed correction is only good for a standard day (59°F or 15°C at sea level) and for a normal temperature lapse rate, or drop, of 3½°F or 2°C per thousand feet. If you know the outside air temperature at your pressure altitude, the true air-speed can be readily found on the computer.

■ **Finding True Airspeed**

For navigation purposes you may want to check your true airspeed in flight. The steps are (as covered in Chapter 3) from indicated airspeed (IAS) to cali-brated airspeed (CAS) to true airspeed (TAS). Remem-ber that the IAS is not always correct. An airspeed correction table can be used to get the CAS (Fig. 19-7).

Since the *calibrated* airspeed is the "correct" in-dicated airspeed, it is used to find the true airspeed. Assume that you are flying (flaps-up) at a pressure al-titude of 4500 feet (indicating 103 K), the outside air temperature is +5°C, and you want to find the TAS. Looking at Figure 19-7 you see that the IAS and CAS can be considered the same at 103 K. That takes care of the first step.

AIRSPEED CORRECTION TABLE
(Flaps Up)

IAS	40	50	60	70	80	90	100	110	120	130	140
CAS	51	57	65	73	82	91	100	109	118	127	136

(Flaps Down)

IAS	40	50	60	70	80	90	100				
CAS	49	55	63	72	81	89	98				

Fig. 19-7. An airspeed correction table. The indicated airspeeds for normal cruise for this airplane will be in the vicinity of 100 K (depending on power setting and altitude). This is the most accurate area of the airspeed indications. (Example only.)

Figure 19-8 shows the method of getting the TAS. Practically speaking, you'll seldom get your TAS in flight. You will be more interested in finding out your groundspeed. The cruise information in the air-

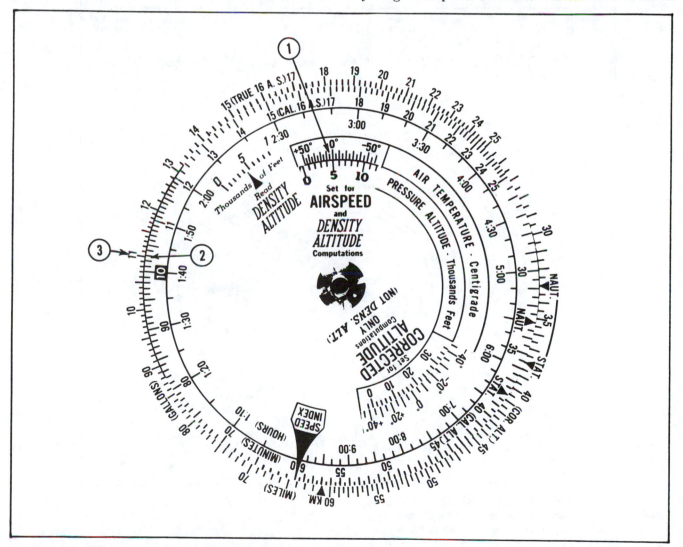

Fig. 19-8. Finding the true airspeed. (1) The temperature (+5°C) and pressure altitude (4500 feet) are set together; opposite the calibrated airspeed (2) of 103 K, the true airspeed (3) of 110 K is found.

plane's *Pilot's Operating Handbook* is given as TAS for various power settings and altitudes (see Chapter 23). However, you can check your TAS to see if you're getting the proper performance for your power setting and altitude or if the airspeed indicator has developed an error.

Some pilots use their indicated altitude instead of setting the altimeter to 29.92 for the pressure altitude to get the TAS. There is an error of about 2 percent for each thousand-foot difference between the indicated and pressure altitudes. Since this difference is generally much less than 1000 feet, this method is considered accurate enough for practical use.

■ Getting the Groundspeed

You'll find that the computer is used most often in practical situations to find groundspeeds and esti-mated times of arrival.

The Knowledge Test for the private pilot's cer-tificate contains questions that would concern a practical cross-country flight with multiple choice answers. A question might be: "You are over the cen-ter of the town of Jonesville at 1652 (4:52 p.m.) CST and over the center of Smithville at 1714 (5:14 p.m.) CST. What is your estimate of time and distance to the destination airport at Greenville?" Using the sectional chart issued to you for the examination, measure the distance between Jonesville and Smithville—it is 36 NM. So knowing that the distance from Jonesville to Smithville is 36 NM and flying time is 22 minutes, set up the problem as shown in Figure 19-9. Your ground-speed (or the speed over the ground between those towns) is 98 K. A distance of 59 NM from Smithville on to the airport would require 36 minutes. The esti-mated time of arrival (ETA) over the airport is 1750 CST. Your actual time over the airport should be close

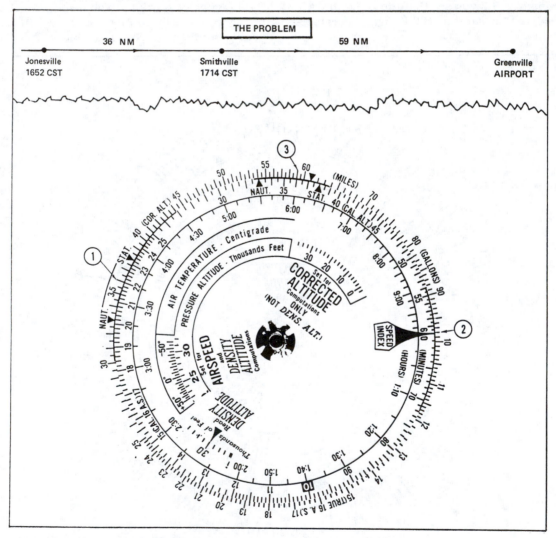

Fig. 19-9. Find the groundspeed. (1) Set up the time (22 minutes) opposite the distance traveled (36 NM) to get the groundspeed (98 K here) at (2). Once the relationship is established, the time required to fly other distances (59 NM in this case) can be found (36 minutes) as shown at (3). It is not necessary to actually know that the groundspeed is 98 K, but you can go directly to (3). The estimated time of arrival over Greenville would be 1714 + 36 minutes, or 1750 CST.

to this unless you turn back, get lost, or the wind changes radically. This method is more accurate than working a wind triangle problem before the flight, since the wind information still may not be accurate for your area.

Notice also: In the area of point 3 in Figure 19-9 is the scale for converting nautical miles (or knots) to statute miles or miles per hour. Note that 54 NM is equal to 62.5 statute miles. You could set the computer to whatever values you wanted to convert.

■ Finding Fuel Consumption

Figure 19-10 shows how to find how much fuel you'll use flying from Smithville to Greenville. As Figure 19-9 shows, after checking the groundspeed as 98 K, you'd expect to take 36 minutes to fly the 59 NM. A typical light trainer uses 4.9 gallons an hour at 65 percent normal rated power. You would set up 4.9 on the outer scale opposite 60 minutes (speed index). Opposite 36 minutes you'll note that 2.95 gallons will be used. The 4.9 gallons per hour could have been 49 or

even 490 if your airplane had this sort of fuel consumption (and the fuel consumed in 36 minutes would be 29.5 or 295 gallons respectively). The computer merely sets up ratios; *you* keep up with what the numbers really are (whether 4.9, 49, 490, or even 4900).

■ Density-Altitude

Density-altitude, which is used for calculating aircraft performance, may be found on the computer. As an example, suppose you are flying at a pressure altitude of 4500 feet and the outside air temperature gauge indicates +15°C. Figure 19-11 shows how to set up a problem.

■ Correcting Indicated or Pressure Altitude

Also shown on the computer is a *corrected altitude* function. The altimeter may read incorrectly because of a nonstandard temperature (which would af-

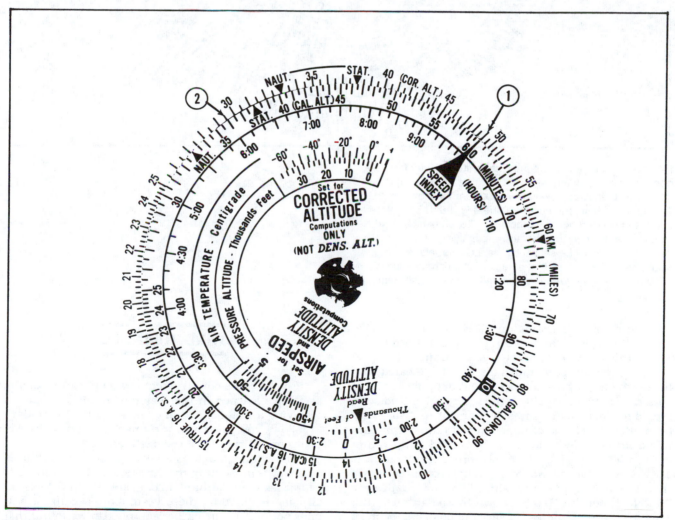

Fig. 19-10. Find fuel consumption.

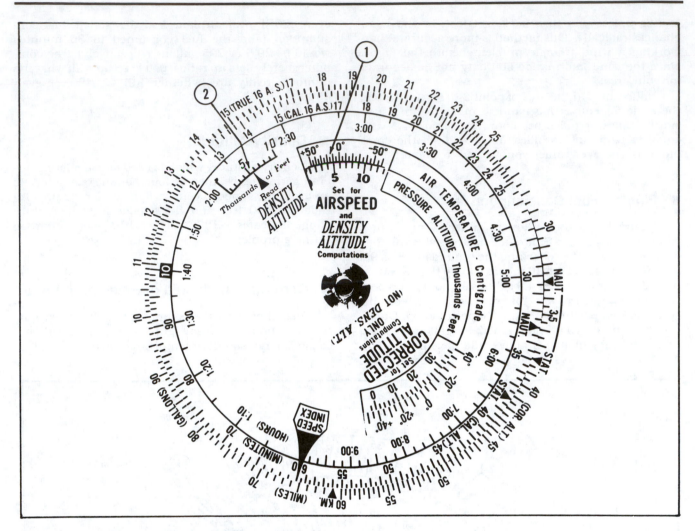

Fig. 19-11. (1) Set up the pressure altitude (4500 feet) opposite the temperature (+15°C) at that altitude. The density-altitude is approximately 5500 feet as shown at (2). The standard temperature at 4500 feet is +6°C so that the actual temperature for this problem (+15°C) is high. The warm air is "thinner" and the airplane is flying at a lower air density than would be shown by the altimeter alone and would be performing as if at an actual standard altitude of 5500 feet, since performance depends on air density (among other things).

fect the pressure working on it). Figure 19-12 shows how the altimeter can be corrected for this error.

In this case, the altimeter goes through the same steps as the airspeed indicator (from indicated altitude, which is what the instrument actually reads, to *calibrated*, which would be the indicated altitude corrected for instrument and system error to "true" altitude after the effects of nonstandard temperature are taken into consideration). The standard temperature for 10,000 feet is −5°C, so the actual temperature is 10°C lower than standard (1). The computer shows that the actual pressure altitude is approximately 9630 feet, not the 10,000 shown by the altimeter (2). This would be of interest in clearing obstacles such as mountain ranges.

WIND TRIANGLE

Flying cross-country requires application of the principles learned in the rectangular course. Instead of having a well-defined line·such as a field boundary to follow, you must use the imaginary line between your home airport and the destination.

The wind triangle is a vector system. A vector is an arrow representing the direction and magnitude of a force, as seen in Chapters 8, 9, and 16. In the case of airplane navigation, these vectors are set up for a 1-hour period. On the earlier sample trip we found that the true course was 120°. Winds aloft are always

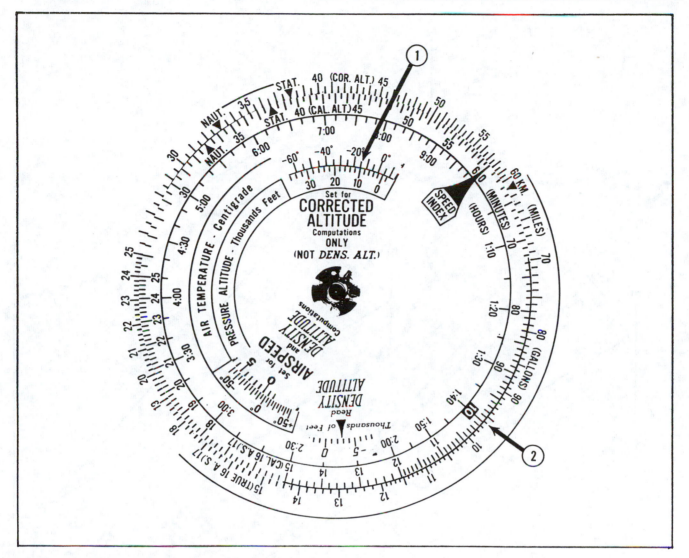

Fig. 19-12. Correcting indicated or pressure altitude for nonstandard temperature conditions.

given in true directions and in knots, and the altitudes given are above sea level. Rather than go through the steps to convert the true course to compass course, work the wind triangle with respect to true North and then convert the answer—your *true heading*—to compass heading, using corrections for variation and deviation.

Given:

1. The measured true course from Dover to Erin is 120°.

2. The Federal Aviation Regulations require that planes flying at 3000 feet or more above the surface fly at indicated (MSL) altitudes of odd thousands plus 500 if the magnetic course to be flown is from 0° to 179° (3500, 5500, 7500, etc.) and at even thousands plus 500 from 180° to 359° (4500, 6500, 8500, etc.) (VFR—cruising altitudes). You see that the variation is 5°E, so the magnetic course is 115° (which is well within the eastern semicircle). Above 3000 you must fly odd plus 500 feet. You checked the winds aloft at

the National Weather Service Office or FAA Flight Service Station and found that the wind at 5000 feet is from 220° true at 30 K. Because this gives you the best tailwind, you decide to fly at 5500 MSL, which is the closest odd plus 500 altitude.

3. You've checked the *Pilot's Operating Handbook* and found that the TAS for this altitude will be 95 K.

On a blank piece of paper draw a line representing a true North line or meridian. Measure the true course with a plotter through some point on the true North line (Fig. 19-13A).

This line is of indefinite length (1). Pick a scale and draw a wind vector in the correct direction and at the proper length from some point on the course line. The wind given for this altitude is in knots, so that length is drawn to scale and in the proper direction (2). We are assuming in this case that the wind is the same at 5500 feet as that at 5000. However, many times you can interpolate between altitudes if there is a large

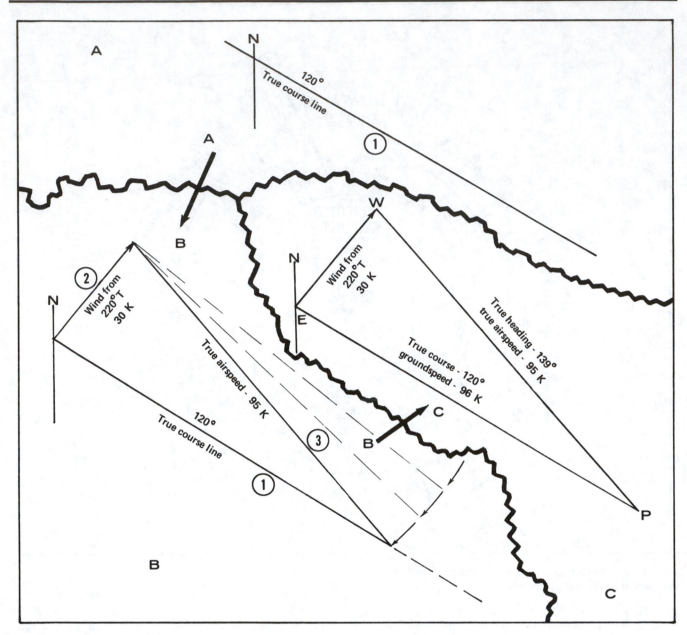

Fig. 19-13. Working a wind triangle.

variation between the wind's direction and/or velocity. For instance, if the wind is 270° and 20 K at 5000, and 310° and 30 K at 6000, a reasonable assumption would be that the wind at 5500 is halfway between these values, or 290° at 25 K.

From the wind arrow point swing a line the length of your TAS (95 K) until it hits the course line (3) (Fig. 19-13B). This gives you the vector picture (Fig. 19-13C).

The line EW (earth-wind) represents the wind with respect to the earth. The line WP represents the movement of the plane with respect to the wind, and its length (the TAS) and angle (the true heading) tell you of the plane's movement and heading within the air mass.

The line EP (earth-plane) represents the plane's

movement and speed with respect to the earth, and its direction is the true course you originally measured. The length of the line gives your groundspeed from Dover to Erin at 96 K. You could figure the return trip groundspeed and heading by extending the reciprocal course line through the point E and swinging another 95-K line from W to strike the course line, then reading the information as before.

Figure 19-14 shows another method of drawing the wind triangle for the Dover-Erin trip. For some people this makes the situation easier to see.

Figure 19-15 shows the Dover-Erin flight as applied to the airplane.

A point of interest: A plane does not make a round trip with wind in the same time as it can be done with no wind. It would seem that the plane having a

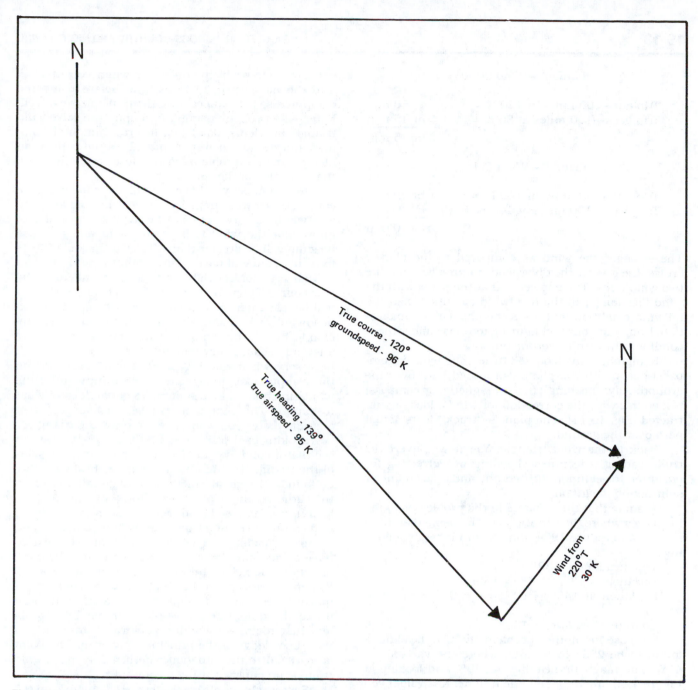

Fig. 19-14. Another version of the wind triangle solution for Figure 19-13.

In the figure: True course - 120°, groundspeed - 96 K; True heading - 139°, true airspeed - 95 K; Wind from 220°T 30 K

headwind one way and a tailwind the other would average the same as a plane making the round trip under no wind conditions—but it won't.

Assume a round trip of 100 NM each way. Both planes cruise at 100 K true airspeed. Plane A has a 20-K tailwind on the outbound leg and a 20-K headwind on the return trip. Plane B flies the same trip under no-wind conditions.

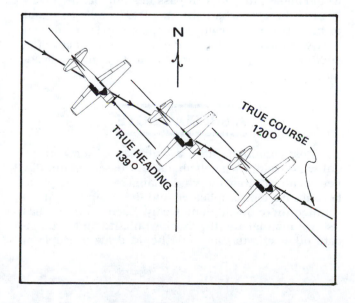

Fig. 19-15. The wind triangle solution for Figure 19-13 as applied to the airplane.

Plane A—Wind 20 K

Trip out—100 miles at 120 K =	50 min
Trip back—100 miles at 80 K =	1 hr 15 min
	2 hr 5 min

Plane B—Wind Calm

Trip out—100 miles at 100 K =	1 hr 0 min
Trip back—100 miles at 100 K =	1 hr 0 min
	2 hr 0 min

The stronger the wind as compared to the plane's cruise, the greater the divergence in time between the two conditions. The answer is that the plane with the wind flies longer in the headwind condition than the tailwind condition and loses the chance at equalization. Don't cut your fuel figuring too close under these conditions; *it won't average out.*

Remember any "course" (true, magnetic, or compass) deals with the plane's proposed path over the ground. Any "heading" (true, magnetic, or compass) is a course with the correction for wind added or subtracted (Fig. 19-14). The plane's "track" is its actual path over the ground.

Back to the wind triangle. You now convert the true heading to compass heading by correcting for *variation* to get *magnetic heading* and then adding or subtracting *deviation.*

Assume that the airplane in the problem is flying in an area where the variation is 4°E. Using that information and Figure 19-5, you would make the following steps:

True heading = 139°

Variation = 139° − 4° = 135°

Deviation at 140° (Fig. 19-5) = 0° (use nearest value)

Compass heading = 135°

Suppose in another problem the true heading is found to be 235° in an area where the variation is 4°W. The magnetic heading is 239°, and looking at Figure 19-5 at 240° magnetic (the nearest value) you would add 3° to get a compass heading of 239° + 3° = 242°. In cases where the magnetic heading is found to be between the values on the compass card, you may want to interpolate and split the difference in the addition or subtraction done to get the compass heading.

A SPECIAL NOTE

Most computers have a "wind side" for working out wind triangles (such as just discussed). Naturally, you won't be drawing wind triangles in the airplane because it's too complicated and the computer can do it much more quickly and easily. Each computer has its own manual and this book won't attempt to duplicate other efforts. *But* you should draw a couple of

wind triangle problems on paper when you first get into the navigation business, and before using the computer, so that you can see the principles involved. Later, when you are using the computer to solve wind triangle problems, don't just insert a bunch of numbers and get an answer; *it may be wrong.* Have an idea of how the problem looks before setting it up on the wind side of the computer.

For instance, you have a problem as follows: true course, 090°; wind from 135° true. The true airspeed and wind velocity need not be available in order to know that the wind is from the right and the true heading will be greater than 090° if a track of 090° is to result. (You will have a correction to the right.) You will have to correct right and "add" numbers to the true course if you are to track 090°. The *groundspeed* will be *less* than the *true airspeed* because there is a headwind component also. Too many students mechanically use the computer and come up with wrong answers because they couldn't "see" the problem. If necessary when working practice wind problems on the computer at first, make a crude drawing of the wind direction and true course to get an idea of what to expect. The computer can be used to get exact answers. Later you won't need to draw the triangle.

Looking back at Figures 19-13 and 19-14, you can see that if you have the information on any two sides of the triangle the information on the third side may be found. The usual practical problem is as given in the earlier example of known wind (direction and velocity), true airspeed, and true course, and you are to find groundspeed and true heading (and magnetic and compass headings). The computer manual will also have sample problems such as how to find the wind direction and velocity if the other factors are known. You might get a question on the written exam requiring the finding of things other than groundspeed or true heading. In other words, you may get a problem that requires "working backward" to check on your knowledge of the principles involved. The point is to visualize the situation whether drawing a wind triangle on paper or using a computer.

Several electronic computers or calculators on the market are specifically designed for use in aviation, and problems may be worked on navigation, performance, weight and balance, and other requirements of flight. As mentioned a couple of paragraphs back, computers have their own manuals and sample problems, so this book won't go into detail on how to work them. If you have that type of talent, you can also set up a program for using the "ordinary" hand-held calculator in your flying. Electronic calculators can be used during the FAA Knowledge Test, subject to certain precautions.

When you start flying, or shortly thereafter, your instructor will suggest what type of computer and plotter you will need. Listen to that advice because you might otherwise buy cheap equipment that won't stand up or items that are more expensive or more complicated than you'll need.

20

THE CHART AND
OTHER PRINTED AIDS

These printed aids will assume more and more importance as you get out flying on your own. While you won't memorize the material, you should know what is contained in each publication—and where to find it.

SECTIONAL CHART

The sectional chart, printed every six months, is the backbone of cross-country flying for the student and private pilot. It is printed by the National Oceanic and Atmospheric Administration and is extremely accurate. The scale is 1:500,000; that is, 1 inch on the chart equals 500,000 inches on the ground, or, as noted in the last chapter, the scale is about 7 NM to the inch. The chart contains such items as:

1. Aeronautical symbols used on the chart.
2. A map of the United States showing each sectional chart's coverage (named for a principal city on the chart such as Atlanta, Charlotte, etc.).
3. Topographical symbols—an explanation of the topographical or terrain symbols used on the chart.
4. Radio aids to navigation and airspace information plus data on obstruction symbols.
5. A list of the frequencies used by control towers in the area of the chart.
6. A list of prohibited, restricted, warning, and alert areas on the chart.

Figure 20-1 shows the sectional charts available for the 48 states portion of the United States.

Terminal areas of heavy air traffic concentration and complicated approach requirements (such as Chicago, Washington, Atlanta, etc.) publish VFR local area charts using the same typography and symbols as the sectional, but in more detail. These charts use a scale of 1:250,000. If you plan to go into such a congested area, you should have one of these up-to-date charts along.

Part of a sectional chart with a sample cross-country flight is reproduced in the back of this book.

AERONAUTICAL INFORMATION
MANUAL (AIM)—OFFICIAL
GUIDE TO BASIC FLIGHT INFORMATION
AND ATC PROCEDURES

The *AIM* contains instructional, educational, and training material—things that are basic and not often changed, some of which are discussed below (chapter references given are for *this* book).

Navigation aids—Backgrounds on aids to navigation such as the VHF omnirange, nondirectional beacon, and other aids to be covered in Chapter 21 of this book. There is also coverage of the Air Traffic Control Radar Beacon System, which you'll be using more and more as you gain experience.

Aeronautical lighting and airport marking aids—Includes the latest information on all types of runway lighting, airport rotating beacons, obstruction lighting, and runway markings.

Airspace—Controlled and uncontrolled airspace boundaries and altitudes are discussed, with notes on Class B, C, and D airspace (see Chapter 21). There's also coverage of special-use airspace such as military operations areas and restricted and prohibited areas. Airspace controlled by the tower (see Chapter 21) is covered in addition to information on temporary flight restrictions and other airspace requirements.

Air Traffic Control—There is a listing of services available to pilots, including Air Route Traffic Control Centers, towers, and Flight Service Stations (see Chapter 21) plus information on flying to and from controlled and uncontrolled airports. There's a section on radio communications with tips on radio techniques, the phonetic alphabet, and communications with the tower if you've lost part of your transmitting/ receiving capability.

Operations at controlled and uncontrolled airports, plus traffic patterns, visual indicators, (tetrahe-

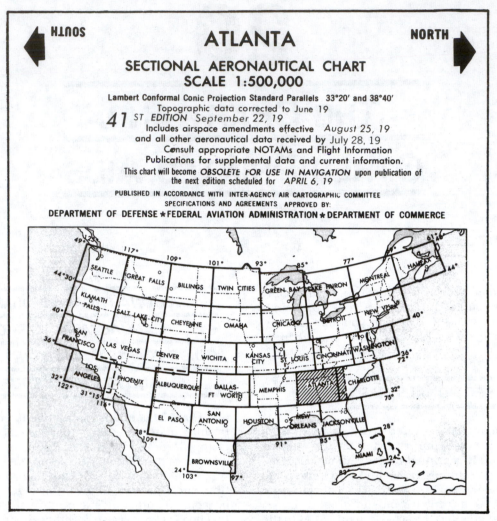

Fig. 20-1. Sectional charts available.

drons, wind socks, and segmented circles; Chapter 6), and traffic control light signals (Chapter 21).

You'll also get the word here on ATC clearances and separation required between airplanes in the airport environment.

Air traffic procedures—Preflight action, including flight plan filing and closing is included here. Departure, en route, and arrival procedures are covered step by step in great detail. The pilot/controller roles and responsibilities are covered separately within this section.

Emergency procedures—Here the coverage pertains to your responsibility and authority as a pilot in an emergency: how to operate the transponder (Chapter 21), how to get found if you are "temporarily dislocated with respect to the preplanned route" (lost), and what to do if you are flying IFR and have a radio communications failure; this won't be a problem for you for a while.

This section also includes what to do under various emergency situations, search and rescue procedures and visual emergency signals and codes, and radio communications failure procedures.

Safety of flight—Has good information on meteorology, including thunderstorms, wind shear, and microbursts (Chapter 22). There is a section on altimeter setting procedures with added information on altimeter errors.

Wake turbulence is thoroughly covered and has been inserted as Appendix A in this book.

Birds and other potential flight hazards are discussed and the admonition is given to not fly under unmanned balloons. (Not covered there, but also to be considered, is that you probably should not fly under large birds either.)

Medical facts for pilots—This chapter has very good coverage and is included as Appendix D of this book.

Aeronautical charts and related publications—Here you'll find out what charts and publications are available and how to obtain them.

En route IFR—The VOR airways systems and special control areas. (See Chapter 21.) Communications and operating procedures en route (instrument flight plan). Diagrams of cruising altitude requirements (VFR and IFR).

Arrival IFR—The use of approach control, approach procedures, radar-controlled approaches, and IFR weather minimums.

General—Airports of entry and departure, and procedures. ADIZ (Air Defense Identification Zone) procedures.

(This is only a part of the material in the *AIM*; the airport office should have a copy for you to study.)

Figure 20-2 is from the *AIM* and shows recommended traffic patterns at nontower airports.

AIRPORT/FACILITY DIRECTORY

This publication is available in seven subscriptions, each for a certain part of the United States (Fig. 20-3). The small (5⅜ × 8¼ inches) size and compactness make it handy to carry in the airplane.

The *A/F Directory* also includes VOR receiver checkpoints for the states covered in that particular section of the country and major changes in aeronautical (sectional) charts. There are Special Notices including telephone numbers for Flight Service Stations and National Weather Offices.

Appendix B of this book is the legend and information on the Sewanee (Tenn.), Columbia–Mount Pleasant (Tenn.), and Huntsville (Ala.) airports. These airports will be used in the sample cross-country discussed in Chapters 24 and 25.

NOTICES TO AIRMEN

By regulation, you are responsible for all factors that could affect the safety of your flight (FAR 91.103), and a knowledge of the Notices to Airmen (NOTAMs) that affect the route and destination of your flight is required. NOTAM information is classified into three categories:

NOTAM (D)—These are considered the ''distant'' NOTAMs and are given for all navigational facilities that are part of the National Airspace System (NAS), all IFR airports with approved instrument approach procedures, and for those VFR airports listed as such in the *Airport/Facility Directory* (see item 2 in the Appendix B legend).

NOTAM (L)—This information is distributed *locally* only and isn't put out on the hourly weather reports. A separate file of local NOTAMs is maintained at each Flight Service Station for facilities in its area only. If you were planning a longer cross-country to an airport served by another FSS, *your* Flight Service Station, at your request, would get the information from the FSS having responsibility for your destination airport. More about FSS duties in the next chapter.

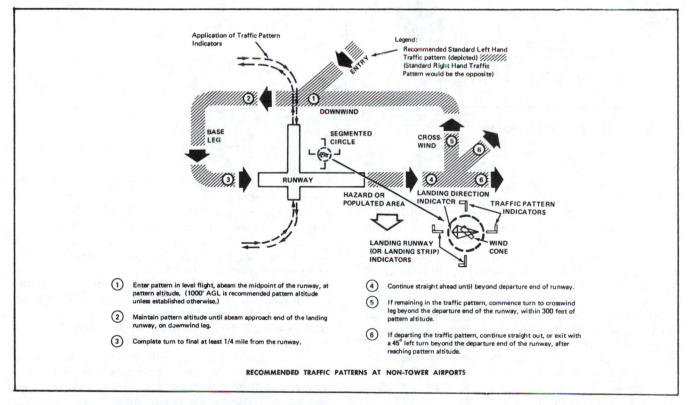

RECOMMENDED TRAFFIC PATTERNS AT NON-TOWER AIRPORTS

Fig. 20-2. Recommended traffic patterns (*Aeronautical Information Manual*)

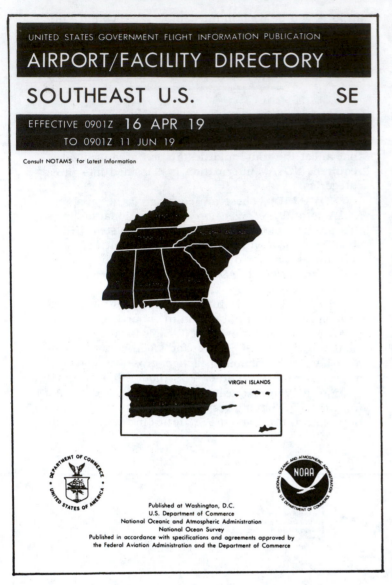

Fig. 20-3. The *Airport/Facility Directory*.

FDC NOTAMs—These can be considered as furnishing information that is regulatory in nature and contain such things as amendments to published instrument approach procedures and other current aeronautical charts. They may also be used to advise of temporary flight restrictions caused by such things as natural disasters or large-scale published events that may cause a congestion of air traffic over a specific area.

An integral part of the NOTAM system is the biweekly *Notices to Airmen* publication. Data is included in this pubilcation to reduce congestion on the telecommunications circuits. Once this is published, the information is not provided during pilot weather briefings unless specifically requested by the pilot. This publication contains two sections:

1. The first section has notices that are primarily distant (D) notices, which are expected to remain in effect for an extended period. NOTAM (L) information may be included if it will contribute to flight safety.

2. The second section contains special notices that are too long for the first section or affect a wide or unspecified geographic area.

Again, *you* as a pilot are responsible for knowing about NOTAMs that will affect your flights. You should inquire about those that will affect your flight when talking to the FSS during the preflight briefing for your cross-country flying.

Note: There are frequent changes in sectional chart symbols or items or presentations in the *AIM* and *A/FD*. Your instructor will be sure that you have the latest printings of aids available.

21

USING THE RADIO

Nowadays few airplanes are not equipped with radio communications and navigation equipment; chances are that you'll be introduced to it very soon after you start flying. The FAA has set up an extensive system of communications and navigation facilities. The personnel at those facilities are there for only one reason, to *help the pilot*, and you'll find that they can save you a lot of trouble (and perhaps help you out of a bind someday after you've messed up).

FREQUENCIES

Your AM radio dial has a range of from 550 to about 1600 kilohertz (kHz) and you dial the frequency (station) you want. The FM band uses frequencies of from 88 to about 107 megahertz (MHz). The AM is MF (medium frequency) and is more likely to get interference from nearby thunderstorms than the FM stations, which broadcast on VHF (very high frequencies). The following listing gives an idea of the frequency bands in use now:

Very low frequencies (VLF)	10–30 kHz
Low frequencies (LF)	30–300 kHz
Medium frequencies (MF)	300–3000 kHz
High frequencies (HF)	3–30 MHz
Very high frequencies (VHF)	30–300 MHz
Ultra high frequencies (UHF)	300–3000 MHz

At the HF band the term megahertz (MHz) is introduced (1 megahertz = 1000 kilohertz). This is done to keep things from getting too cumbersome as would be done by going on with the kilohertz idea. The term "hertz" is named after Heinrich Hertz, German physicist, and stands for "cycles per second." In your aviation radio work you'll be working mostly with VHF, LF, and MF.

RADIO NAVIGATION AIDS

■ Radio Beacon

Radio beacons or nondirectional radio beacons (NDB) are assigned broadcast frequencies between 190 and 535 kHz, which makes them LF/MF facilities.

Each station continually transmits its particular assigned three-letter identifier in code except during voice transmissions.

Low-powered radio beacons called compass locators are used in conjunction with instrument landing systems (ILS). The compass locators have a two-letter identification and a range of at least 15 miles. The en route or "H" (homing) facilities may have a range of up to 75 miles for the more powerful installations. The *Aeronautical Information Manual (AIM)* gives more details on designations and ranges for radio beacons and compass locators.

You will use an ADF (automatic direction finder) in the airplane to navigate with the radio beacon (Fig. 21-1). You would select REC (receive) for better audio reception and set in the proper frequency for the desired radio beacon or broadcast station. Some sets have a selection of ANT (antenna) for this purpose. After the station is coming in loud and clear, you'd switch to the ADF setting.

The needle of the ADF points to the station you have selected and gives a relative bearing to that station. Always make sure that the station has been properly identified—it would be embarrassing on the flight test if you homed in on the wrong station. It can happen!

Figure 21-2 shows some magnetic headings and relative bearings to fictitious radio beacons. (Assume no deviation for these illustrations.)

(A) Shows that the magnetic heading is 060° and the relative bearing is 060°, for a magnetic bearing to the station of 120° (60° + 60°).

(B) Indicates a magnetic heading of 290° with a relative bearing of 240°. You can find the magnetic bearing to the station in two ways: You can add the two values and subtract 360 from the answer (290° + 240° = 530°; 530° − 360° = 170° to the station); or you note that the needle is pointing 120° to the left of the nose, so you can subtract this amount from the heading (290° − 120° = 170°). If the addition results in a number greater than 360°, subtract 360° from the result.

(C) The heading is 135° with a relative bearing of 195°. You can add the two numbers (135° + 195° = 330°) to the station. If you wanted to "visualize" the situation, you could say that the reciprocal of your heading is 315° and the station is 15° "farther

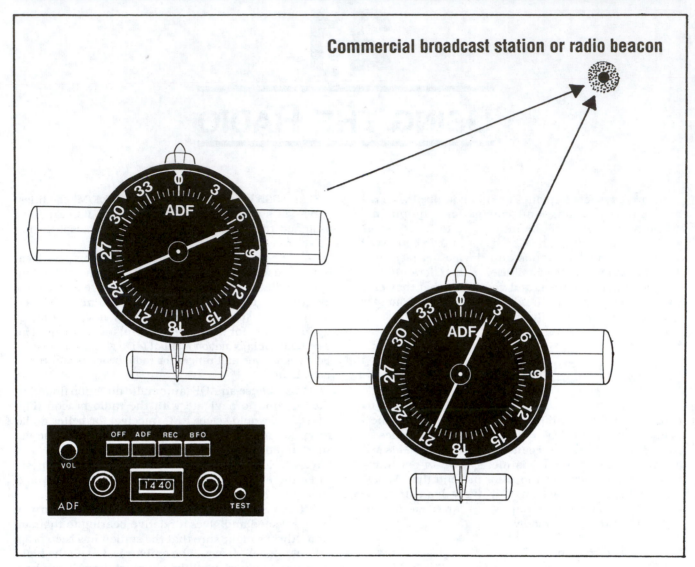

Commercial broadcast station or radio beacon

Fig. 21-1. The automatic direction finder needle points to the radio beacon or commercial broadcast station (if properly tuned).

around," so this would also give a result of 330°.

Looking at the lower part of Figure 21-2, you'll see a different presentation as found in newer ADF sets. The knob marked HDG (heading) is used to mechanically turn the card to set the airplane's heading under the indicator or lubber line. The needle then indicates the actual bearing to the station.

In (A) the airplane is heading 060°; this is set up on the card and the needle shows that the bearing to the station is 120°. This answer was found in the top illustration (A) by addition.

(B) Indicates that the bearing to the station is 170°.

(C) The bearing to the station is 330° as found before.

This presentation is an improvement over the addition and subtraction method required by the first type, but you have to be sure that the number you set

in on the card is the same as the heading indicator shows. Anytime you change the airplane's heading, the ADF card has to be mechanically reset to that new figure if you want to have an accurate relative bearing to the station.

It would be good if this could be done automatically for you, so the RMI (radio magnetic indicator) was designed. It's a combination of heading indicator and ADF and VOR pointers. The heading indicator can be slaved to a remote magnetic compass so that precession is continually being corrected. You'll probably use an RMI when you start working on that instrument rating.

You "home" into a station by turning the airplane until the ADF needle points to a relative bearing of 000° (the station is straight ahead) and by keeping the needle in this position. A crosswind will result in a curved path (Fig. 21-3). This is the simplest but not

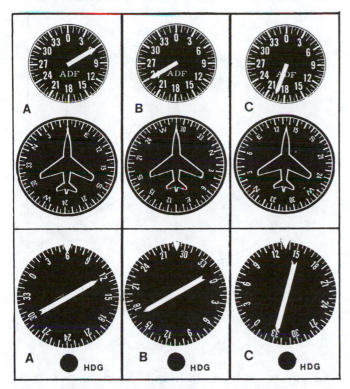

Fig. 21-2. Finding magnetic bearings to the station.

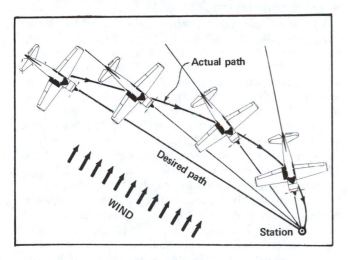

Fig. 21-3. A curved path results if there is a crosswind during "homing" (the nose is kept pointed to the station). The effect shown is exaggerated for purposes of illustration.

the most efficient way of getting to the station.

As you pass the station, the needle will swing and point to the rear (180° position on the dial). Later you will learn how to *track* into the station, setting up the proper wind drift correction that would result in a straight flight path (the same principle covered in "Wind Triangle" in Chapter 19).

Your ADF set is able to pick up stations in the frequency range of 190–1750 kHz, which means that you can also home into commercial broadcasting stations. While those stations are normally more power-

ful than the radio beacons or H facilities, they have the disadvantage of not having continuous identification. You may have to suffer through several minutes of rock music before learning that you've been tuned to the right (or wrong) station.

Because the ADF operates in the LF/MF bands, it is subject to the troubles of the regular AM home radio when thunderstorms are in the vicinity. Precipitation static and thunderstorms may make reception difficult or impossible. Precipitation static is the result of static electricity generated when the airplane flies through rain, snow, etc.

The radio beacon has one advantage over VHF navigational equipment; it is *not* "line of sight." You can pick up the signals even when you are low and the station is over the horizon (just like the home radio).

If your trainer does not have an ADF, you might ask your instructor for a ground checkout on the operation of that equipment in one of the other planes on the field.

■ VOR

The most useful of the en route radio navigation aids is the VHF omnirange, or VOR as it is sometimes called. The VOR frequency band is from 108.00 to 117.95 MHz and uses the principle of electronically measuring an angle. The VOR puts out two signals. One is all-directional (or omnidirectional) and the other is a rotating signal. The all-directional signal contracts and expands 30 times a second, and the rotating signal rotates clockwise at 30 revolutions per second. The rotating signal has a positive and a negative side.

The all-directional or reference signal is timed to transmit at the same instant the rotating beam passes *magnetic north.* These rotating beams and the reference signal result in radial measurements.

Your omni receiver picks up the all-directional signal. Some time later it picks up the maximum point of the positive rotating signal. The receiver electronically measures the time difference, and it is indicated in degrees as your *magnetic* bearing in relation to the station (Fig. 21-4). For instance, assume it took a minute instead of 1/30 of a second for the rotating signal to make one revolution. You receive the all-directional signal and 20 seconds later you receive the rotating signal. This means that your position is 20/60 or 1/3 of the way around. (One-third of 360° is 120° and you are on the 120 radial.) The VOR receiver does this in a quicker, more accurate way.

The aircraft VOR receiver presentation is composed of four main parts: (1) a dial to select the frequency of the station you want to use; (2) an azimuth or omni bearing selector (OBS) calibrated from 0 to 360; (3) a course deviation indicator (CDI), a vertical needle that moves left or right; and (4) a TO-FROM indicator. Figure 21-5 is one type of VOR receiver.

Suppose you want to fly to a certain VOR, say, 30 miles away. First you would tune the frequency and

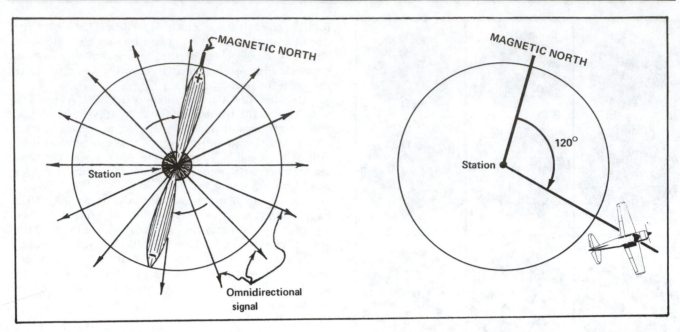

Fig. 21-4. Principle of the VOR.

Fig. 21-5. A VOR receiver. (*Narco Avionics*)

identify the station. You should have some idea where you are in relation to the station, but if not, turn the azimuth or direction selector until the deviation indicator or needle is centered and the TO-FROM indicator says TO. Read the OBS. This is your course TO the VOR. If you turn on that magnetic course and keep the needle centered, you'll fly right over the VOR. If the TO-FROM says TO and you are going to the station, fly the needle. If the needle moves to the left, the selected

bearing is to the left and you will turn the plane in that direction and fly until the needle is centered again. You will have to correct for wind to stay on the selected bearing (Fig. 21-6).

In this case your position from the VOR looks something like that shown in Figure 21-6A.

While your bearing to the station is 300°, *you are on the 120 (one-two-zero) radial.* (The radials are like spokes from the VOR.) (See Fig. 21-6A.) The radials are numbered from 0 through 359, so if the station asked where you were you'd say, "I'm inbound to the station on the 120 (one-two-zero) radial." For example, if there's a westerly wind, the plane may drift from the selected bearing as shown in Figure 21-6B, and the LEFT-RIGHT needle would look like Figure 21-7.

The angle of correction or the "cut" you'll take will depend on the amount that you've drifted from course. Usually 30° would be the maximum even at some distance from the station. It may take some time for the needle to center again if you're far out. After the needle returns to center, turn back toward the original heading, but this time include an estimated wind correction on your compass or heading indicator. Watch the needle and make further corrections as needed.

When you cross over the station, the TO-FROM indicator will oscillate, then fall to FROM. The receiver now says that you are on a bearing of 300° FROM the station. Always make sure your OBS is set close to your compass heading. This way the needle always points toward the selected radial. If you turned around and headed back to the station after passing it on a course of 300° and did not reset the bearing selector to coincide with your heading, the needle would work in reverse. *Always set your OBS to the course to be followed; then the needle senses correctly.*

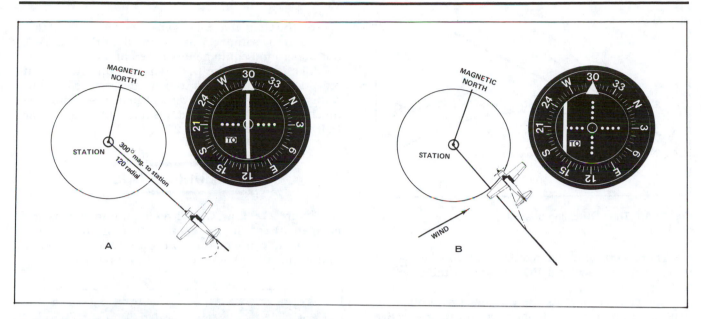

Fig. 21-6. Using a VOR receiver to track to the station.

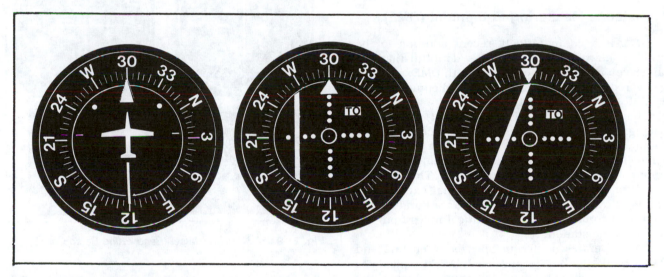

Fig. 21-7. A heading indicator and two types of VOR indicators. The selected radial is to the left.

The sensing is incorrect if you correct toward the needle, and the needle moves farther away from the center as you fly.

But, back to the station passage: The TO-FROM says FROM and your plane is on a course of 300° FROM the VOR, so the needle is correct. Continue to fly the needle as you did before the VOR was reached.

Most omni needles are set up so that a full deflection from center is 10° or more. If the needle is deflected halfway, you can figure that you are about 5° from your selected bearing.

VORs are identified by Morse code and/or by the automatic recorded voice identification ("Airville VOR").

The accuracy of the VOR ground facility is generally ± 1°, but some stations in mountainous terrain may have errors greater than this for some radials or may be unusable below certain altitudes; this is duly noted in the *Airport/Facility Directory.*

SUMMARY

Frequency Band—108.0–117.95 megahertz.

Use—Navigation and instrument approaches. (A few don't have voice facilities. See the *Aeronautical Information Manual* and *Airport/Facility Directory.*)

Identification—Continual code or code and voice.

Advantages—(1) Very high frequency; not as affected by weather as LF/MF facilities. (2) Omni or all-directional signals; pilot is not limited to four legs, as for the old LF/MF range.

Disadvantages—Line of sight; cannot be picked up if the plane is low (Fig. 21-8).

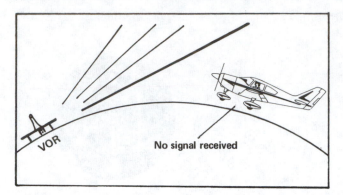

Fig. 21-8. The VOR is line of sight.

The check pilot will require that you be able to tune, identify, and fly a VOR radial on the practical test.

Appendix B contains the various classifications of VORs and other navigational aids and their effective altitudes and ranges.

■ Distance Measuring Equipment (DME)

DME is a comparatively recent addition to the family of air navigation aids. It's doubtful that the trainer you are flying is equipped with DME, so it will be covered only briefly. Special ground equipment for the VOR station is needed.

The DME set in the aircraft sends out, at a specific spacing, interrogating pulses that are received by the ground station. The ground station then transmits a paired pulse back to the aircraft at a different pulse spacing and on a different frequency. The time required for the round trip of this signal exchange is measured by the DME unit in the aircraft and is indicated on a counter or dial as nautical miles from the aircraft to the station.

DME operates on frequencies between 962 and 1213 MHz (UHF); but because each DME frequency is always paired with an associated omni frequency, the selector of the set may be marked in terms of the omni frequency band. All you have to do is to know the frequency of the associated omni and set up the selector of the DME on this number, rather than having to remember the various DME frequencies or "channels."

VORTAC stations are combinations of two components: the VOR, giving azimuth information, and the TACAN equipment, giving distance (DME) information. TACAN is the term for tactical air navigation, a military innovation being used in FAA facilities. Note in the sectional chart in the back of this book that Huntsville has a VORTAC (named *Rocket*), and you would set up 112.2 MHz on the VOR set and the same on the DME equipment and get directions, groundspeeds, and distances from that station. The "Channel 59" is for TACAN equipment, and all VORs in the United States having the frequency 112.2 MHz would have a channel 59 assigned. (See Appendix B.) Note

also on the chart that Shelbyville has a VOR/DME that is used in the same way as indicated for the VORTAC, but the ground equipment is different.

Like the VOR, the DME depends on line-of-sight reception. Because it is a transmitter (it sends signals rather than words, however), airplanes so equipped must have an Aircraft Radio Station License for the DME. (See Chapters 3 and 23.)

COMMUNICATIONS

Figure 21-9 is a COMM/NAV (communications and navigation) set. In the communications part of the equipment, you would set up the desired frequency that would allow you to transmit *and* receive.

Fig. 21-9. A NAV/COMM set (*Bendix/King Radio Corp.*)

The console (A) consists of the transceiver (this is insider talk for transmitter-receiver) set at 123.00 MHz for communications. (You'd transmit and receive with a Unicom on this setting—more about Unicom shortly.) The right half (NAV) is set at 115.95 MHz here to pick up a VOR station. The NAV receiver shown can pick up VOR/LOC (localizer) signals for navigation *and* approaches. The receiver range is from 108.00 to 117.95 MHz.

The localizer is used for instrument approaches (you can also use it in good weather if you like) and utilizes the *odd* frequencies (tenths, or hundredths, depending on the precision of your receiver) from 108.10 to 111.95 MHz (108.10, 108.15, 108.30, etc.). The localizer has a 3-letter Morse Code identifier preceded by I(··).

In Figure 21-9 (B) is a VOR/LOC indicator, and (C) is a VOR/LOC with a GS (glide slope) needle added. As a student pilot you will normally use only the vertical needle of this indicator. Note that this is COMM 1 and

NAV 1, which would usually be the uppermost set on the instrument panel. (See Fig. 3-1.) This bottom COMM/NAV set is cited as "No. 2."

Your instructor will check you out on the COMM/NAV equipment for your airplane.

■ Unicom

Unicom, or aeronautical advisory, is an aid to the pilot operating into smaller airports with no tower. The Unicom frequencies are the "private pilot's frequencies" and you can call into an airport so equipped and get the surface wind, traffic, and other information. You can also call in and get transportation if needed.

Unicom is merely an advisory service by the airport operator. It cannot give traffic clearances to take off or land as would be done by a control tower.

The primary Unicom frequencies in use are:

1. For airports *without* an operating control tower—122.7, 122.8, and 123.0 (or 122.700, 122.800, and 123.00 MHz, depending on how far you want to carry the zeroes). Other frequencies at these airports that are being brought into use are 122.725, 122.975, 123.050, and 123.075 MHz.

2. Airports with a control tower or FSS use the Unicom frequency 122.950 MHz. Make sure that you aren't on Unicom frequency asking for a cab when you should be listening and talking to approach control or the tower.

3. Multicom—This is a frequency (122.900 MHz) used to announce yourself (and your intentions) at airports without Unicom, a Flight Service Station *or* a tower. You (and other airplanes) would keep up with each other's actions using this frequency around and on the airport. It is also used for Search and Rescue.

4. *Air-to-air frequency*—122.750 MHz is to be used if you and your friend in that other airplane need to talk to each other. (This frequency is often abused.)

Incidentally, frequencies given in this book are subject to change, so you should keep up-to-date, checking the *AIM* and the *Airport/Facility Directory* for the latest frequencies assigned to the various facilities.

■ Flight Service Station (FSS)

The Flight Service Station has several functions in its job of assisting the pilot (*you*).

Flight plan service—You file all flight plans with an FSS in person, by telephone, or by radio. (If possible, avoid filing by radio; it ties up the frequencies too much.) A discussion of flight plans and how to file them will be covered in more detail in Chapter 24, which covers navigation planning.

Pilot briefing service—You may obtain preflight briefing on weather (existing and forecast) along your route and information on the operating status of navigation aids and airport conditions either in person or by phone. Incidentally, when getting flight weather information by phone from either Flight Service Stations or Weather Service Offices, always tell them that you are a pilot, give the aircraft identification, and also note the approximate departure time, route, and expected time en route to your destination and request a "standard briefing" if you haven't received a previous briefing or haven't gotten information from other FAA sources. Your briefing will normally cover the following information: (1) adverse weather; (2) current weather at the departure point, destination, and en route; (3) forecasts for the points just mentioned; and (4) advisories, pilot reports (PIREPs) (see Chapter 22), winds aloft along the route, and NOTAMs (notices to airmen) plus military training routes within 100 NM of the FSS. About NOTAMs: ask for them every time. Ask for a *weather briefing* when you talk to the facilities, rather than taking it upon yourself to just request information on individual stations along your route. The briefer may use the phrase "VFR flight is not recommended" if he or she thinks conditions are not good enough for VFR (visual flight rules) flying. The latest telephone numbers for Weather Service and Flight Service Stations can be found in the *Airport/Facility Directory*. Tell the FSS people what you need and let them help you from there.

For in-flight briefing call the FSS on the proper frequency and let them know on which frequency you'll be listening. It is suggested that you call Flight Service Stations on 122.1 MHz (which is a one-way frequency *from* aircraft to them) and listen to one of the VOR frequencies on which they can transmit. This practice cuts down interference from other stations. Don't be impatient, because the people in the FSS could be tied up on another frequency. Also let them know that you are a student pilot; they'll appreciate the information and will give you extra consideration. (This goes for contacting approach controls and towers also.)

Assume you are on a cross-country and the weather ahead doesn't look too good. You can give a call to the nearest FSS and get the latest weather and the forecast for your destination.

You: "BLANKTOWN RADIO, THIS IS CESSNA SEVEN FIVE FIVE SEVEN LIMA, STUDENT PILOT, LISTENING JONES VOR, OVER." (This is not official, but you might repeat the station name twice on the initial call; such as, "BLANKTOWN, BLANKTOWN RADIO . . ." The reason for this is that the station name would sometimes be missed by the FSS personnel and they might only hear " . . . RADIO." This would mean that the personnel in every FSS within reception distance of your transmitter think that you're calling them and time is wasted until the matter is straightened out.)

Blanktown radio: "CESSNA SEVEN FIVE FIVE SEVEN LIMA, THIS IS BLANKTOWN RADIO, OVER." (Notice that you'll make an initial contact, letting them know *who* you are and on *what* frequency you'll be listening for their reply. The FSS has identified itself to you and by saying "over" has handed the ball back. You should *always* let the FSS know on which frequency you expect a reply because the stations can transmit on several frequencies and may have to transmit simultaneously on all of them to make sure they get you. This causes interference with other stations and other aircraft. Keep in mind that some Flight Service Stations have jurisdiction over several VOR stations, so if you are listening on a VOR frequency, tell them which one.)

You: "THIS IS FIVE SEVEN LIMA, I'M ABOUT FIVE MILES EAST OF BLANKTOWN AT FOUR THOUSAND FIVE HUNDRED (4500 feet), EN ROUTE TO JONESVILLE. I'D LIKE THE LATEST JONESVILLE WEATHER AND THE FORECAST FOR THE NEXT TWO HOURS, OVER."

(Once you have established contact it is not necessary to use your full number each time.)

Blanktown: "FIVE SEVEN LIMA, WAIT, OUT." (The "out" is to let you know that a reply to this transmission is not expected.)

Blanktown: "THE LATEST JONESVILLE WEATHER IS . . ." (gives the Jonesville weather and forecast plus the Blanktown altimeter setting).

You: "FIVE SEVEN LIMA, THANK YOU. OUT."

You acknowledge receipt of transmissions or instructions by the last three digits (and/or letters) of your registration number.

Although it's not required, you should give your position and altitude when contacting an FSS, even if you're not on a flight plan because later if you don't show up at the destination your friends may start checking. The FSS will have on file a record of having talked to you at such-and-such a time, at such-and-such a position and altitude, and the search can start from there. You might as well get in the habit of giving your position and altitude in each contact.

En route services—You can report your position at any time to an FSS whether on a flight plan or not. If you are on a VFR flight plan the Flight Service Stations are the facilities to which you would normally make your position reports. In a bind you could report to a tower and they would relay the message. Position reports for IFR flight plans can be made either directly to Air Traffic Control Centers or to Flight Service Sta-

tions (for forwarding to the Centers), but IFR position reports won't be something for you to worry about yet.

Another en route service available from Flight Service Stations is that of giving aid in emergencies arising in flight. The personnel are trained to help in situations such as orienting the lost pilot (using radar, direction finding, omni receiver manipulation, or their knowledge of prominent landmarks reported by the pilot). If you have a problem, call the FSS and let *them* worry. (Of course, you are allowed to worry also if you like.) Emergency frequencies are discussed in the *AIM.*

Local Airport Advisory—This is provided by Flight Service Stations at airports not served by a control tower.

For local airport advisory service you would normally make initial contact with the FSS about 15 miles out and transmit and receive on 123.6 MHz.

Remember that it's only an advisory service. No airport control can be as complete as a control tower. Such information as wind, favored runway, field conditions, and *known* traffic will be given to you. If you have two-way radio, it's a good idea to maintain communications with the FSS within 10 statute miles of such airports. If you only have a receiver, you can maintain listening watch on the appropriate frequency when within 10 statute miles of the airport. You could also call them when taxiing out for departure. (AAS is *not* mandatory.)

Broadcast service—Continuous weather broadcasts are made over selected LF/MF navigation aids. (Chapter 22 goes into more detail.)

Pilot weather reporting service—The FSS acts as a clearinghouse between pilots for exchanging information concerning turbulence, significant weather encountered, or other pertinent information. These pilot reports, or PIREPs, will be passed on by the FSS to other pilots in the area by direct communication or through the Teletype and scheduled weather broadcasts.

Check the latest *Airport/Facility Directory* for frequencies, since the sectional chart data may be out of date. Note that not all Flight Service Stations have all of the same frequencies (Fig. 21-10).

Civil Communications Frequencies—Civil communications frequencies used in the FSS air/ground system are now operated simplex on 122.0, 122.2, 122.3, 122.4, 122.6, 123.6; emergency 121.5; plus receive-only on 122.05, 122.1, 122.15, and 123.6.
  a. 122.0 is assigned as the Enroute Flight Advisory Service channel at selected FSS's.
  b. 122.2 is assigned to most FSS's as a common enroute simplex service.
  c. 123.6 is assigned as the airport advisory channel at non-tower FSS locations, however, it is still in commission at some FSS's collocated with towers to provide part time Airport Advisory Service.
  d. 122.1 is the primary receive-only frequency at VOR's. 122.05, 122.15 and 123.6 are assigned at selected VOR's meeting certain criteria.
  e. Some FSS's are assigned 50 kHz channels for simplex operation in the 122-123 MHz band (e.g. 122.35). Pilots using the FSS A/G system should refer to this directory or appropriate charts to determine frequencies available at the FSS or remoted facility through which they wish to communicate.
Part time FSS hours of operation are shown in remarks under facility name.

Emergency frequency 121.5 is available at all Flight Service Stations, Towers, Approach Control and RADAR facilities, unless indicated as not available.
Frequencies published followed by the letter "T" or "R", indicate that the facility will only transmit or receive respectively on that frequency. All radio aids to navigation frequencies are transmit only.

Fig. 21-10. Flight Service Station frequencies. (*Airport/Facility Directory*)

The FAA is planning automated Flight Service Stations at various locations around the United States. This changeover may take some time; your instructor can keep you posted on what's happening in your area.

Common Traffic Advisory Frequency (CTAF)

The common traffic advisory frequency is a service frequency designed for the purpose of airport advisories at uncontrolled airports. The CTAF may be a Unicom, Multicom, FSS, or tower frequency and is identified in appropriate aeronautical publications.

Figure 21-11 shows how the CTAF might be presented in the *Airport/Facility Directory* airports with different facilities. Note that Muscle Shoals (Ala.) has both an FSS and a Unicom, but for traffic advisories the FSS frequency is suggested. At the Sewanee-Franklin County (Tenn.) Airport only Unicom is available, so this is the CTAF.

MUSCLE SHOALS REGIONAL (MSL) 1 E UTC–6(–5DT) N34°44.72' W87°36.61' ATLANTA
550 B S4 FUEL 100LL, JET A TPA—See Remarks ARFF Index A H-4H, L-14G
RWY 11-29: H6693X150 (ASPH-GRVD) S-59, D-98, DT-160 HIRL IAP
RWY 11: REIL. VASI(V4R)—GA 3.0°TCH 51'. Tree. RWY 29: MALSR. Tower.
RWY 18-36: H4000X100 (ASPH) S-30, D-50 MIRL
RWY 18: Tree.
AIRPORT REMARKS: Attended continuously. Flocks of birds on and in vicinity of arpt. Turtles; deer; coyotes; and other wildlife on and in vicinity of rwys and taxiways. HIRL Rwy 11–29 and MIRL Rwy 18–36 opr dusk–0500Z‡. After 0500Z‡ ctc MUSCLE SHOALS FSS—CTAF. TPA—1550 (1000) light acft 2050 (1500) turbo prop and jets. REIL Rwy 11 controlled by FSS. During calm winds preferred rwy is 29. Fuel avbl ngts call 205–383–2270 or 205–386–7597. No general aviation parking on air carrier ramp.
COMMUNICATIONS: CTAF 123.6 UNICOM 122.95
FSS (MSL) on fld. 123.6 122.4 122.2. LC 383–6541. NOTAM FILE MSL.
MEMPHIS CENTER APP/DEP CON 120.8
RADIO AIDS TO NAVIGATION: NOTAM FILE MSL. VHF/DF ctc FSS.
(L) VORTACW 116.5 MSL Chan 112 N34°42.41' W87°29.49' 290° 6.3 NM to fld. 580/01E.
ILS 109.7 I-MSL Rwy 29

SEWANEE
FRANKLIN CO (UOS) 1 E UTC–6(–5DT) N35°12.24' W85°53.92' ATLANTA
1950 B FUEL 80, 100LL L-14H
RWY 06-24: H3300X50 (ASPH) S-20 LIRL
RWY 06: SAVASI(S2L)—GA 3.5°TCH 28'. Trees. RWY 24: VASI(V2L)—GA 4.0° TCH 31'. Trees.
AIRPORT REMARKS: Attended dalgt hours. For svc after hours call 615–598–0745. Sporadic crosswinds and turbulence. Deer on and in vicinity of rwy.
COMMUNICATIONS: CTAF/UNICOM 122.8
NASHVILLE FSS (BNA) TF 1-800-WX-BRIEF. NOTAM FILE BNA.
RADIO AIDS TO NAVIGATION: NOTAM FILE BNA.
SHELBYVILLE (L) VOR/DME 109.0 SYI Chan 27 N35°33.72' W86°26.35' 130° 34.1 NM to fld. 810/01W.
SEWANEE NDB (MHW) 275 UOS N35°12.26' W85°53.75' at fld. (VFR only).

Fig. 21-11. The CTAF of two different facilities.

Control Tower

If you've never been into a controlled field, you'll face the prospect with some misgiving. It's best if you have a brief idea as to the various functions of the tower so that things won't be so confusing the first time. You can think of operations of the tower as being divided into five main services:

1. *Local control*—The function that pilots think of as "The Tower." Local control has jurisdiction over the aircraft taking off and landing and, to quote the *Aeronautical Information Manual* (see Chapter 20 again), is "established to provide for a safe, orderly, and expeditious flow of traffic on and in the vicinity of an airport. when the responsibility has been so delegated, towers also provide for the separation of Instrument Flight Rules aircraft in the terminal area."

Towers are established in Class B, C, and D airspace (Fig. 21-12).

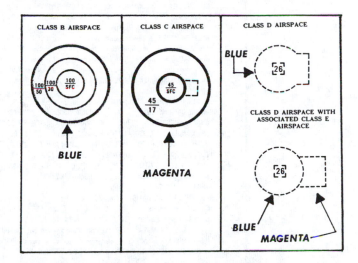

Fig. 21-12. Class B, C, and D airspace as depicted on a sectional chart (all have operating control towers). The pilot and equipment requirements are most rigid for B and C and each will be discussed later in this book.

A simpler explanation of the tower might be that the tower controls the traffic pattern entry and the traffic pattern itself, including takeoffs and landings. Local controllers are in the glassed-in top level of the tower because their control depends on visual identification of aircraft for takeoffs and landings. The local control (tower) frequencies are in the *Airport/Facility Directory* (see Fig. 21-13).

2. *Automatic Terminal Information Service* (ATIS)—ATIS is the continual broadcast of recorded noncontrol information for certain *high-activity* airports. For instance, Sewanee-Franklin County Airport (see Fig. 21-11) does *not* have ATIS.

Information such as the time of the latest weather sequence, ceiling, visibility (if the ceiling is above 5000 feet and visibility is 5 miles or more, this information is optional), obstructions to vision, temperature, dewpoint (if available), wind direction (magnetic) and velocity, altimeter setting, other pertinent remarks, instrument approach, and runways in use is continuously transmitted on the voice feature of the VOR/VORTAC located on or near the airport or on a particular, separate VHF/UHF frequency.

Pilots of arriving aircraft are expected to tune in ATIS to get the just-mentioned information plus the appropriate approach control frequencies to be used. You'll also be expected to listen to ATIS before taxiing at an ATIS airport.

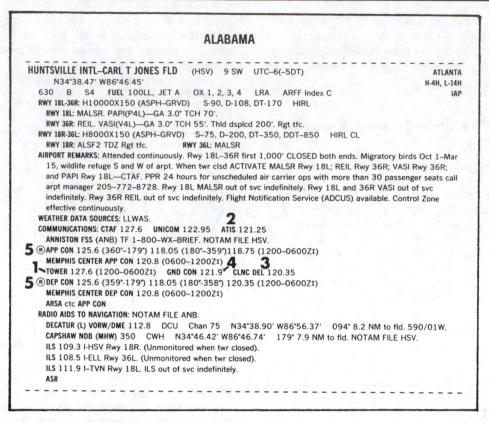

ALABAMA

HUNTSVILLE INTL–CARL T JONES FLD   (HSV)   9 SW   UTC–6(–5DT)   ATLANTA
      N34°38.47' W86°46.45'   H-4H, L-14H
  630   B   S4   FUEL 100LL, JET A   OX 1, 2, 3, 4   LRA   ARFF Index C   IAP
  RWY 18L-36R: H10000X150 (ASPH–GRVD)   S-90, D-108, DT-170   HIRL
    RWY 18L: MALSR. PAPI(P4L)—GA 3.0° TCH 70'.
    RWY 36R: REIL. VASI(V4L)—GA 3.0° TCH 55'. Thld dsplcd 200'. Rgt tfc.
  RWY 18R-36L: H8000X150 (ASPH–GRVD)   S-75, D-200, DT-350, DDT-850   HIRL CL
    RWY 18R: ALSF2 TDZ Rgt tfc.   RWY 36L: MALSR
  AIRPORT REMARKS: Attended continuously. Rwy 18L–36R first 1,000' CLOSED both ends. Migratory birds Oct 1–Mar
    15, wildlife refuge S and W of arpt. When twr clsd ACTIVATE MALSR Rwy 18L; REIL Rwy 36R; VASI Rwy 36R;
    and PAPI Rwy 18L—CTAF. PPR 24 hours for unscheduled air carrier ops with more than 30 passenger seats call
    arpt manager 205–772–8728. Rwy 18L MALSR out of svc indefinitely. Rwy 18L and 36R VASI out of svc
    indefinitely. Rwy 36R REIL out of svc indefinitely. Flight Notification Service (ADCUS) available. Control Zone
    effective continuously.
  WEATHER DATA SOURCES: LLWAS.
  COMMUNICATIONS: CTAF 127.6   UNICOM 122.95   **2** ATIS 121.25
    ANNISTON FSS (ANB) TF 1–800–WX–BRIEF. NOTAM FILE HSV.
5 ®APP CON 125.6 (360°–179°) 118.05 (180°–359°)118.75 (1200–0600Z‡)
    MEMPHIS CENTER APP CON 120.8 (0600–1200Z‡)   **4**   **3**
1 TOWER 127.6 (1200–0600Z‡)   GND CON 121.9°   CLNC DEL 120.35
5 ®DEP CON 125.6 (359°–179°) 118.05 (180°–358°) 120.35 (1200–0600Z‡)
    MEMPHIS CENTER DEP CON 120.8 (0600–1200Z‡)
    ARSA ctc APP CON
  RADIO AIDS TO NAVIGATION: NOTAM FILE ANB.
    DECATUR (L) VORW/DME 112.8   DCU   Chan 75   N34°38.90' W86°56.37'   094° 8.2 NM to fld. 590/01W.
    CAPSHAW NDB (MHW) 350   CWH   N34°46.42' W86°46.74'   179° 7.9 NM to fld. NOTAM FILE HSV.
    ILS 109.3 I-HSV Rwy 18R. (Unmonitored when twr closed).
    ILS 108.5 I-ELL Rwy 36L. (Unmonitored when twr closed).
    ILS 111.9 I–TVN Rwy 18L. ILS out of svc indefinitely.
  ASR

Fig. 21-13. Communications and other information for the Huntsville International Airport. (*Airport/Facility Directory*)

At the beginning and end of each broadcast you'll be reminded that the information given is (was), for example, "Information Delta." When the information changes (the wind shifts, etc.) the broadcast data will be given as "Information Echo."

Let Clearance Delivery, Ground Control, or Approach Control (as applicable) know that you have "Information Delta" (or the latest information) upon *initial* contact.

3. *Clearance Delivery*—This is a separate frequency used for pretaxi clearances, either VFR or IFR. At busy airports, contact Clearance Delivery after listening to ATIS. Clearance Delivery will give you frequencies and instructions necessary to depart the busy terminal area. After you have the information given by CD and read it back correctly, you'll be told to change to the ground control frequency for taxi instructions.

4. *Ground Control*—Regulates traffic moving on taxiways and the runways not being used for takeoffs and landings. (Ground Control will coordinate with the tower if you have to cross a "hot," or active, runway.) It is on a different frequency than that of the tower because you can imagine what radio clutter would result if some pilots were asking for taxi direction to Ace Flying Service while others were calling in for landing instructions.

For simplification and summary of local (tower) and Ground Control duties, note that local control has jurisdiction of aircraft in the process of landing and

taking off. This includes aircraft while in the pattern and *on* the active runway. Ground Control is used for ground traffic on the airport *other than on the active runway during the takeoff or landing process*. The ground controller will be in the tower beside the local controller. (And in some cases the same person may talk to you in both capacities, but on a different frequency, of course.)

5. *Approach/Departure Control*—The busier controlled airports have another position called "Approach/Departure Control." This may be a radar or nonradar setup. For a nonradar setup, the approach controller is in the glassed-in portion of the tower with the other two positions. The primary duty is to coordinate IFR traffic approaching the control zone, but the controller may coordinate IFR and VFR traffic during marginal VFR conditions as well as coordinating VFR traffic at busy terminals, even in CAVU (ceiling and visibility unlimited) conditions.

Approach/departure controllers using radar are usually in an IFR room that is located in the tower building but not necessarily in the glassed-in portion. Approach controllers work directly with the tower and their duties are rotated between tower and IFR room positions.

■ Class C Airspace

Figure 21-14 shows the Class C airspace at and around the Huntsville International Airport. You must

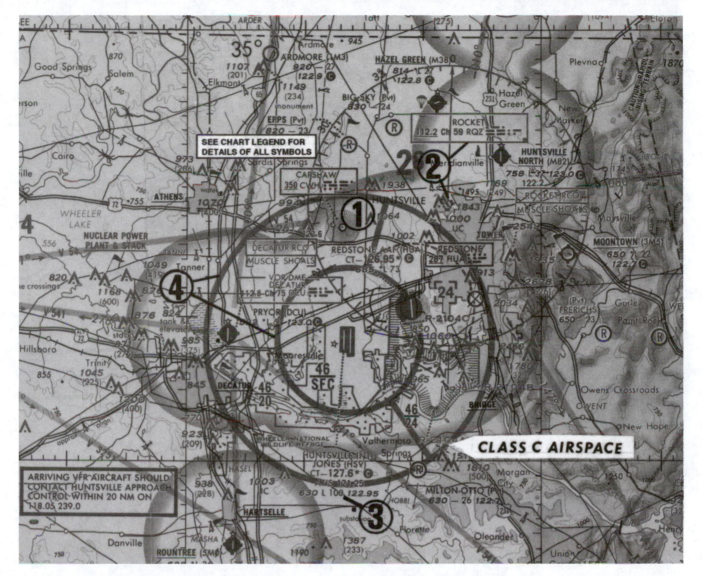

Fig. 21-14. Facilities near Huntsville. (1) The Capshaw radio beacon—identification CWH, frequency 350 kHz. The line under the "350" indicates that there is *no* voice facility on that frequency. (2) The Rocket VORTAC site and its frequency box. (3) The airport information for the Huntsville International–Carl Jones Field. The frequency for the tower is 127.6 MHz; the frequency for the Automatic Terminal Information Service (ATIS, just discussed) is 121.25 MHz. The airport elevation is 630 feet above mean sea level (MSL), it has lights, the longest runway is 10,000 feet, and there's a Unicom on the airport (122.95 MHz). (4) The boundaries of the 2000- to 4600-foot (MSL) portion of the Class C airspace (Class C airspace extends from the airport to the outer circle). Note that the smaller portion of the airspace, east of the airport, has altitude limits of 2400–4600 feet (MSL). There is a wildlife refuge south and southwest of the airport, and you are expected to remain at least 2000 feet above the surface in such an area.

establish two-way radio communications with ATC (approach control or other required facilities) before entering Class C airspace.

You may not have a chance to enter Class C airspace until after getting your private certificate, but you should have some idea of the procedure for entering.

Being in Class C airspace or being radar vectored *does not relieve you of the responsibility to avoid other traffic. Also, keep in mind that the radar controller may not know the weather conditions in your* vicinity *and could vector you into conditions where visual references would be lost.* You are responsible for maintaining visual flight conditions and must notify the controller of a marginal situation. Remember that when you are solo you are the pilot in command. The final responsibility for safety is *yours*. ATC people and FSS personnel are conscientious, but they can make mistakes. Don't hesitate to speak up or deviate from their instructions if it looks as if your safety is involved. Keep them posted on your plans, though. Figure 21-15 is a look at *Class B* airspace.

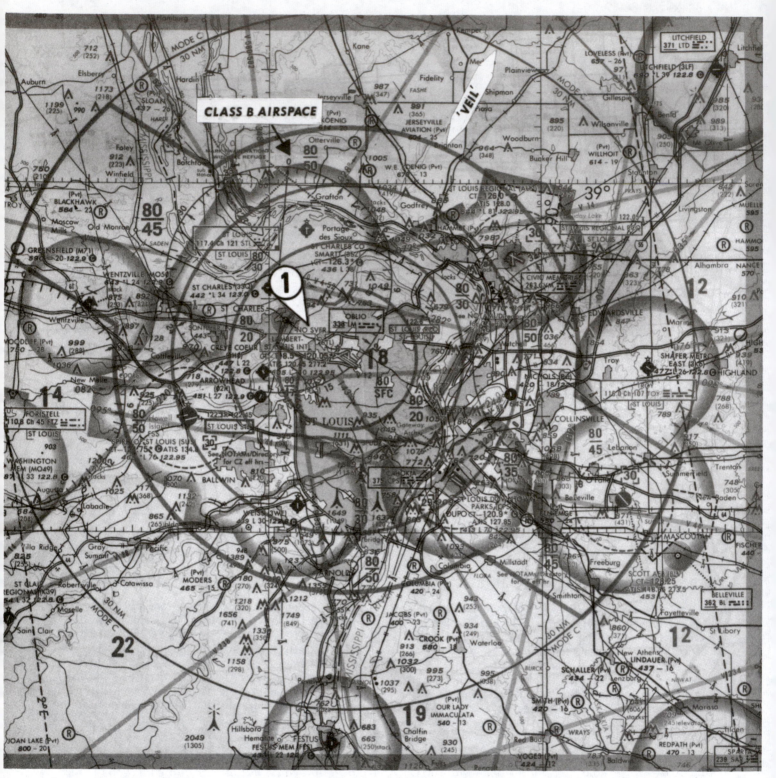

Fig. 21-15. Class B airspace at St. Louis Lambert Field. The heavier-lined circles are the boundaries of the various altitude limits. The thin-lined circle ("Mode C, 30 NM") is the outer limit of the 10 miles of "veil" around the airspace with the veil extending *generally* from a 20-NM to a 30-NM radius from the major airport. You may *not* enter Class B airspace without an air traffic control clearance but may fly in and through the veil portion without talking to anyone, as long as your airplane has an operating transponder with altitude reporting capabilities. (More about transponder operation later in the chapter.)

The Class B airspace is in the form of an inverted wedding cake, with some modifications as necessary for expediting traffic.

The "box" around the St. Louis area indicates that there is an area chart with more detail available (one inch equals 250,000 inches).

When the weather is less than that required for VFR operations in Class B, C, and D airspace (see Fig. 21-12 again), you, as a VFR pilot, may request and be given a clearance to enter, leave, or operate within it. This, of course, depends on traffic and nondelay of IFR operations (Special VFR—FAR 91.157). You'll get a clearance from a tower, or if it is Class E (without a tower) a clearance may be obtained from an FSS, the nearest tower, or an Air Traffic Control Center. The minimum visibility requirements are 1 mile flight visibility for operations in the control zone and 1 mile ground visibility for takeoffs and landings. You'll stay clear of clouds, but remember that you might end up so low that you fracture the FAR concerning clearance from ground objects. Only instrument-rated pilots and IFR-equipped airplanes can get a Special VFR clearance at night.

Some airports have too much traffic to allow Special VFR, and they have a notation on the sectional chart to this effect (NO SVFR) next to the airport name box (see item 1 in Fig. 21-15).

As a student or low-time private pilot the Special VFR clearance may allow you to get legally into an airport as the weather is deteriorating, but to use it to get out in marginal conditions on a VFR flight should be rejected. You could get into worse weather and lose control of the airplane. Wait for better weather before departing; spend that ground time studying for the instrument rating, so that you can safely use more and more of the air traffic control facilities as you get experience.

Suppose you are approaching Huntsville from the northwest and plan to land at Huntsville International Airport (Fig. 21-14):

1. Listen to *ATIS* to get the information on active runway(s), wind, and other information as covered earlier (item 2, Fig. 21-13).

2. *Contact Approach Control* (item 5, Figure 21-13) on one of the frequencies cited in the ATIS or as shown in the *Airport/Facility Directory*. Note that since you are approaching from the northwest (*from* about 315°), the frequency to be used is 118.05 MHz, since you are approaching in the Huntsville sector between 180° and 359°. Tell them your position, altitude, type of aircraft, that you are landing at Huntsville International, and have "*Information Foxtrot*" (or whatever). If you have a transponder, they will have you select a discrete code (you'll be the only one using it at this particular time), which you will dial in. (See Fig. 21-17 and accompanying text about transponder operation.)

3. The *tower* (item 1, Fig. 21-13) is to be contacted when Approach Control wants to hand you over and, as you can see from Figure 21-13, the frequency is 127.6 MHz. After landing, you will be asked to switch to *Ground Control after* clearing the active runway (item 4, Fig. 21-13). You can ask for taxi directions if you are new to the airport. The sample cross-country in Chapter 24 will go through the steps of approaching, landing, and taxiing at Huntsville International (it's one of the two stops) plus the steps

required in getting out and on your way again. But now let's take a quick look at your probable steps in approaching and leaving a busy airport.

APPROACHING
1. *ATIS*—Listen and get the pertinent information.
2. *Approach Control*—Call Approach Control at about 20 miles out after you have the numbers, giving your position, altitude, and type of aircraft (plus "Information Delta," etc.). They may have you squawk a certain transponder code.
3. *Tower*—Contact the tower at Approach Control's instructions.
4. *Ground Control*—Contact Ground Control after clearing the active runway, at the tower's instructions.

DEPARTING
1. *ATIS*—Get the information on runways in use, wind, etc.
2. *Clearance Delivery*—Contact Clearance Delivery and tell them you have "Foxtrot" or whatever. Get departure instructions, transponder "squawk," and Departure Control frequency. Write it down.
3. *Ground Control*—Tell them your position on the airport and request taxi clearance.
4. *Tower*—When you're ready for takeoff, say so and give your full aircraft number.
5. *Departure Control*—Contact Departure Control when told to by the tower, using the frequency given by Clearance Delivery.

If you are going back the way you came, note that the Departure Control frequency is the same as you used with Approach Control on the way into the airport. Note that coming *from* or departing *to* the northwest the frequency is 118.05 MHz for Huntsville (Fig. 21-13).

Sometimes it seems that the Approach Control is vectoring as shown in Figure 21-16.

Figure 21-17 is a side view of the various classes of airspace just discussed, plus Class A (above 18,000 feet MSL and not likely a factor for you as a student or private pilot just yet) and Class G (uncontrolled airspace, found only below 700 feet or 1200 feet in most of the eastern portions of the continental 48 states).

Note that Class B airspace (the top view shown in Fig. 21-15) may be multi-layered; Class C is two-layered and Class D is a cylinder with some extensions for approaches.

Figure 21-18 shows the requirements and limitations of the various classes of airspace just discussed.

Figure 21-19 shows the basic visual flight rules (VFR) minimums as shown in FAR part 91.

TRANSPONDER. Basically the transponder is a radar "transceiver" that picks up interrogations of the Air Traffic Control radar beacon systems and transmits or "replies." The two modes of operational types of equipment you'll be encountering are "A" (which is a

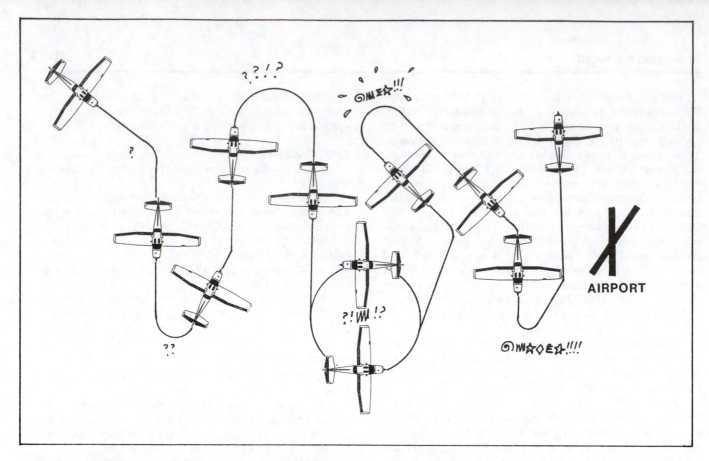

Fig. 21-16. Sometimes it appears as if approach control is trying to help you build flying time, or gain meaningful aeronautical experience. (There are other airplanes out there to be considered.)

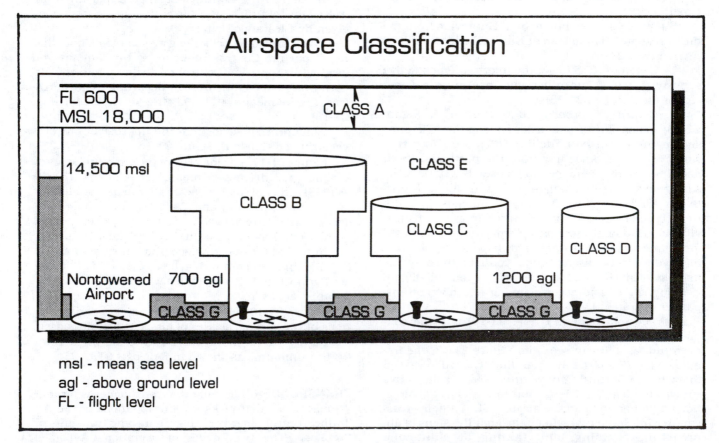

Fig. 21-17. Elevation (side) view of the Class A, B, C, D, E, and G airspace. (The United States does not have a Class F airspace as do other countries.) The Class G airspace (shaded) on the left side of the illustration, extending up to 14,500 feet MSL, is uncontrolled airspace in the western United States. (*Aeronautical Information Manual*)

Airspace Summary

Airspace Features	Class A	Class B	Class C	Class D	Class E	Class G
Operations Permitted	IFR	IFR and VFR	IFR and VFR	IFR and VFR	IFR and VFR	IFR and VFR
Entry Prerequisites	ATC clearance	ATC clearance	ATC clearance for IFR. Radio contact for all.	ATC clearance for IFR. Radio contact for all.	ATC clearance for IFR. Radio contact for all IFR.	None
Minimum Pilot Qualifications	Instrument rating	Private or student certificate	Student certificate	Student certificate	Student certificate	Student certificate
Two-way Radio Communications	Yes	Yes	Yes	Yes	Yes for IFR	No
VFR Minimum Visibility	N/A	3 statute miles	3 statute miles	3 statute miles	*3 statute miles	**1 statute mile
VFR Minimum Distance from Clouds	N/A	Clear of clouds	500' below, 1,000' above, and 2,000' horizontal	500' below, 1,000' above, and 2,000' horizontal	*500' below, 1,000' above, and 2,000' horizontal	**500' below, 1,000' above, and 2,000' horizontal
Aircraft Separation	All	All	IFR, SVFR, and runway ops	IFR, SVFR, and runway ops	IFR and SVFR	None
Conflict Resolution	N/A	N/A	Between IFR and VFR ops	No	No	No
Traffic Advisories	N/A	N/A	Yes	Workload permitting	Workload permitting	Workload permitting
Safety Advisories	Yes	Yes	Yes	Yes	Yes	Yes

* Different visibility minima and distance cloud requirements exist for operations above 10,000' MSL
**Different visibility minima and distance from cloud requirements exist for night operations, operations above 10,000' MSL, and operations below 1,200' AGL

Fig. 21-18. A summary of the airspace classes and the requirements and limitations of each. There is no Class F airspace in the United States as found in other countries. (Courtesy Phyllis Duncan, *FAA Aviation News*.)

straight reply type) and "C" (which replies, and through special encoding altimeter equipment allows the ground controller to read off your altitude on the radarscope). Figure 21-20 shows a transponder control panel.

The IDENT selection in Figure 21-20 makes an airplane stand out on the controller's radarscope and is used momentarily to identify one target among many to ATC. The reply lamp flashes when the transponder replies to each interrogation from the ground station.

Primary radar depends on the reflection and return of its impulses; the transponder in the airplane is triggered by these impulses and, in effect, boosts them back for a better target indication on the ground radar.

For VFR flying (and when not in voice contact with a radar facility) you would set up 1200 on the transponder and present a clearer target as you fly—for *their* traffic-avoidance purposes.

For instance, if you were lost but in contact with an FSS or other nonradar FAA facility, they would probably have you switch frequencies to talk to a radar-equipped facility that would ask you to "squawk"

a certain setting on your transponder. They could readily locate you among the many targets and vector you to an airport.

The emergency code is 7700; if you're in bad trouble in flight, this setting will get the attention of every radar facility in the area. (It's best to talk to the people, but if you can't, 7700 will get you help.) The lost communications code, used primarily for IFR problems, is 7600. There's even a code you'd use if someone is hijacking your Cessna 152, or whatever. Your instructor will discuss and demonstrate the use of the transponder (if the airplane has one) before you go solo cross-country.

The transponder is "line of sight" like the VOR and DME.

A Mode C transponder (a transponder that also gives altitude information to Air Traffic Control [ATC]) is required to fly in the "veil" of Class B airspace as noted earlier. They are also required when operating above the altitude of the ceilings and within the lateral boundaries of Class B or Class C airspace. So looking at Figure 21-14, if you want to fly *over* the Huntsville Class C airspace (above 4600 feet MSL) or if you

Airspace	Flight Visibility	Distance from clouds
Class A	Not Applicable	Not Applicable.
Class B	3 statute miles	Clear of Clouds.
Class C	3 statute miles	500 feet below. 1,000 feet above. 2,000 feet horizontal.
Class D	3 statute miles	500 feet below. 1,000 feet above. 2,000 feet horizontal.
Class E		
Less than 10,000 feet MSL	3 statute miles	500 feet below. 1,000 feet above. 2,000 feet horizontal.
At or above 10,000 feet MSL	5 statute miles	1,000 feet below. 1,000 feet above. 1 statute mile horizontal.
Class G:		
1,200 feet or less above the surface (regardless of MSL altitude)		
Day, except as provided in §91.155(b)	1 statute mile	Clear of clouds.
Night, except as provided in §91.155(b)	3 statute miles	500 feet below. 1,000 feet above. 2,000 feet horizontal.
More than 1,200 feet above the surface but less than 10,000 feet MSL		
Day	1 statute mile	500 feet below. 1,000 feet above. 2,000 feet horizontal.
Night	3 statute miles	500 feet below. 1,000 feet above. 2,000 feet horizontal.
More than 1,200 feet above the surface and at or above 10,000 feet MSL	5 statute miles	1,000 feet below. 1,000 feet above. 1 statute mile horizontal.

Fig. 21-19. Visibilities and cloud clearances for the various classes of airspace (FAR 91). Student pilots are further limited to 3 miles (day) and 5 miles (night), even in Class G airspace. Also see FAR 91.155.

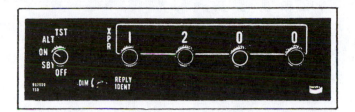

Fig. 21-20. Transponder control panel. (*Bendix/King Avionics Division*)

needed to fly *over* the St. Louis Class B airspace (above 8000 feet MSL—see Fig. 21-15), you have to be using a Mode C even though you are not actually penetrating that particular airspace. In addition, at an altitude at or above 10,000 feet MSL in the 48 continental United States, a *working* Mode C is required.

■ Light Signals

There may be a time when the airplane's radio is inoperative and you have to land at a controlled airport. You should know the light signals used by the control tower for planes not equipped with radio. The tower controller uses a portable light and is able to single out a particular aircraft for instructions. The signals are shown in Figure 21-21.

Color and Type of Signal	On the Ground	In Flight
STEADY GREEN	Cleared for take-off	Cleared to land
FLASHING GREEN	Cleared to taxi	Return for landing (to be followed by steady green at proper time)
STEADY RED	Stop	Give way to other aircraft and continue circling
FLASHING RED	Taxi clear of landing area (runway) in use	Airport unsafe-do not land
FLASHING WHITE	Return to starting point on airport	
ALTERNATING RED & GREEN	General Warning Signal—Exercise Extreme Caution	

Fig. 21-21. The light signals used by local control.

■ Visual Approach Slope Indicator (VASI)

The VASI is located at some airports to provide a visual light path within the approach zone; the two-light system is set at 3°, although this may be higher at some airports to give proper obstacle clearance.

The simplest VASI system has two-light systems

at different distances on the sides of the runway near the approach end with the glide slope reference line being between the two light units.

Each unit projects a beam of light with a white color in the upper part and a red color in the lower part. The indications would be:

Above glide slope	White	White
	White	White
On glide slope	Red	Red
	White	White
Below glide slope	Red	Red
	Red	Red

Some smaller airports have two units on the left side of the runway only.

When on the proper glide slope, the airplane in the system described here is in effect overshooting the bars at the threshold and undershooting the bars farther down. You will see the upper half (white) of the closer bars and lower half (red) of the far set.

VASI can normally be seen from 4 to 5 miles out on the final approach.

The *AIM* gives details on the more complex VASI systems in use.

■ Emergency Locator Transmitter (ELT)

The ELT is required by Federal Aviation Regulations for all aircraft *except* those training within a 50-mile radius of the airport used for that training, flight test aircraft, new aircraft being ferried, and other special uses listed in FAR 91.207. It's likely that the airplane you are using now has one, and the instructor may point out its location and antenna. (It may be so located in the airplane that the antenna is the only part you can see without removing panels.)

Basically, an ELT is a self-contained two-frequency transmitter with its own battery power. An impact resulting in 5 g's or more (or 5 times the acceleration of gravity) activates the unit, which automatically starts transmitting on 121.5 MHz (civilian international distress frequency) and 243.0 MHz (military).

The ELT, being VHF like the VOR, is line of sight and has the same advantages and disadvantages of that frequency range.

The ELT is designed to aid in locating downed airplanes, and pilots in flight often turn their receiver to 121.5 to check for an ELT signal. If one is heard, the nearest FAA facility should be advised so that a search for the transmitter can be started. ELTs may be triggered by hard landings or lightning strikes or in airplanes with a need for attention while just quietly sitting on a ramp or hangar. It sets off quite a frustrating search when a signal is heard in the midst of 100 or more tied-down aircraft (with many of the owners out of town with the keys).

The sets usually have three switch settings:

1. ON—activates the transmitter immediately. It's used for testing or if the ''g'' function is not working.

The pilot may activate it manually in an emergency such as a "soft" landing in an inaccessible spot when rescue is needed.

2. ARM—is the normal setting, which would start the signals under an impact.

3. OFF—is used for shipping and storing (and af-ter the St. Bernard with the keg gets there).

An ELT is tested *only* within the first 5 minutes past the hour and for only three audio sweeps, so that a massive Search and Rescue effort is not launched.

FAR 91.207 also covers replacement require-ments of the batteries in an ELT.

22

WEATHER INFORMATION

You're well ahead of the student pilots of pre-TV days as far as weather knowledge is concerned because you've been seeing the frontal systems and weather patterns on your screen every day. For instance, you know of the existence of high- and low-pressure areas and warm and cold fronts and the effects of circulation, but maybe a little review is in order.

WEATHER PATTERNS

■ Pressure

The standard sea level pressure is 29.92 inches of mercury (abbreviated Hg). The approximate drop in pressure is about 1 inch of mercury per thousand feet of altitude. (This is valid up to about 10,000 feet; the *decrease* in pressure is less as the altitude increases. In the 18,000- to 20,000-foot levels, for instance, the drop in pressure is only about 0.60 inch per thousand feet.)

The pressure may also be given in millibars, and at standard sea level conditions the value is 1013.2 millibars (a millibar is a unit of pressure based on the metric system). Since the standard pressure in inches of mercury is 29.92, it would follow that 1 inch of mercury pressure is equal to about 34 millibars—if you should need to convert from one to the other system. In the United States the altimeter settings are given in inches of mercury. Although the pressure is also given in millibars on the hourly sequence reports, it is *not* reported on the scheduled weather *broadcast*.

Standard sea level pressure may be given in the following ways: (1) 29.92 inches of mercury, (2) 1013.2 millibars, (3) 14.7 pounds per square inch, or (4) 2116 pounds per square foot.

■ Pressure Areas

If the pressure stayed the same everywhere, the weather wouldn't be very interesting—and people would have a hard time making conversation.

Figure 22-1 shows the circulation around high- and low-pressure systems in the Northern Hemisphere. The opposite circulation occurs in the South-

ern Hemisphere; the earth's rotation (sometimes called Coriolis force) causes this.

A high-pressure area *usually* means good weather (except for the circulation problem just covered), and low-pressure *usually* means bad weather, but you'll want to get more information from all possible sources (which is the purpose of the last part of this chapter). An elongated high-pressure area is a "ridge"; the equivalent low-pressure shape is a "trough."

■ Frontal Systems

A front is defined as a "zone of transition between two air masses of different densities."

The National Weather Service publishes charts of observed weather, and Figure 22-2 shows the symbols used to depict the various types of fronts (plus squall lines) on the charts.

You may set up your own memory aids to remember the symbols (cold is blue and sharp, etc.).

■ Cold Front

The cold front is usually characterized by a comparatively narrow weather band and by more violent weather than is associated with the warm front. The worst part of the cold-front weather is normally less than 100 miles from beginning to end. Cumulus (vertically developed clouds), heavy precipitation, and turbulence are usually associated with it. Fast-moving cold fronts may have squall lines, or a line of thunderstorms, 50 to 300 miles ahead and roughly parallel to the front. Your move, if encountering a squall line, is to get out of its area as expeditiously as possible.

After the cold-front passage there will usually be rapid clearing, with lower temperatures and strong gusty winds on the surface; this applies particularly to a fast-moving cold front. The wind will shift from the southwest or the west to northwest or north (about a 90° wind shift) as the front passes.

Figure 22-3 shows the circulation about a frontal system. Note that the "typical" system is "pivoting" around a low-pressure system.

Isobars are lines joining points of equal pressure. The closer the isobars are together the steeper the pressure gradient and the stronger the winds in that

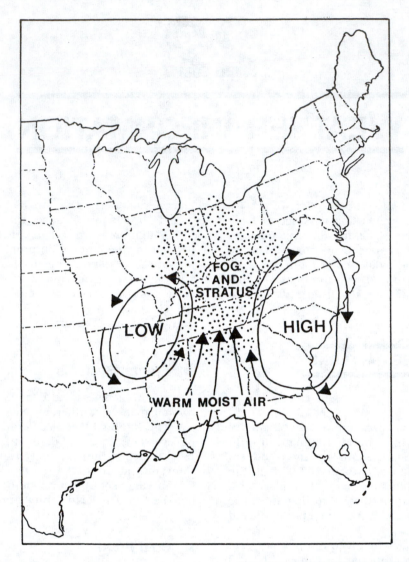

Fig. 22-1. High- and low-pressure areas and circulation. While most of our weather is considered to be caused by frontal systems, circulation can bring warm, moist air into cool areas, causing fog, low stratus, and precipitation. The Coriolis force, a result of the earth's rotation, causes deflection of the winds to the right in the Northern Hemisphere (and finally parallel to the isobars) at altitudes out of surface friction effects. The Coriolis force is directly proportional to the wind speed.

Fig. 22-2. Front symbols used on a weather chart.

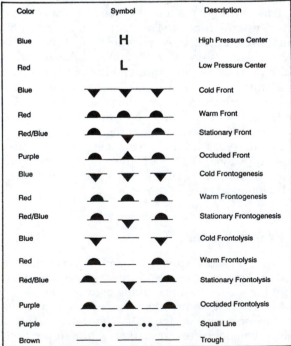

Color	Symbol	Description
Blue	H	High Pressure Center
Red	L	Low Pressure Center
Blue		Cold Front
Red		Warm Front
Red/Blue		Stationary Front
Purple		Occluded Front
Blue		Cold Frontogenesis
Red		Warm Frontogenesis
Red/Blue		Stationary Frontogenesis
Blue		Cold Frontolysis
Red		Warm Frontolysis
Red/Blue		Stationary Frontolysis
Purple		Occluded Frontolysis
Purple		Squall Line
Brown		Trough

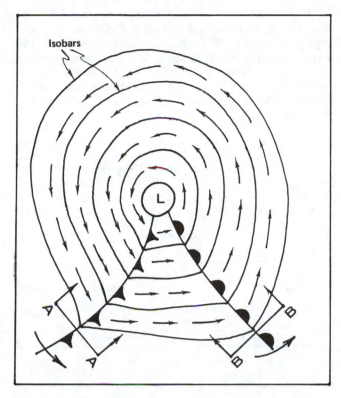

Fig. 22-3. Circulation in the vicinity of a frontal system.

area. You might keep this in mind when you look at a surface weather chart in a Weather Service Office.

The cold front moves faster than the warm front associated with the system, which can lead to a situation known as an occlusion (which will be covered later).

Figure 22-4 shows the cross section of an "average" cold front as it would appear if you sliced

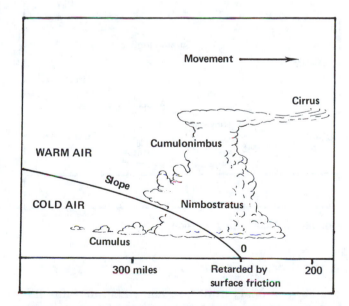

Fig. 22-4. Cross section of a typical cold front. The slope varies from 1:50 to 1:150 and averages about 1:80.

through the front as shown by A-A in Figure 22-3. (You are looking the way the arrows are pointing.)

The average cold front moves at a speed of 15–20 K but may under extreme conditions move at 50 K or more. It seems to move at a much higher speed than that if you happen to be in a 90-K trainer trying to outrun it.

■ Warm Front

The warm front usually produces a much wider band of weather consisting of stratus-type clouds, with widespread areas of low ceilings, rain, and fog. However, if the warm air is unstable, cumulus-type clouds and thunderstorms may be found in the stratus layer. Many instrument pilots have had unexpected excitement flying through what they thought was going to be stable, smooth air. This, of course, is only of academic interest to you as a student pilot since one of your primary aims is to avoid getting within *any* type of cloud.

The warm front usually moves at about one-half the speed of the cold front and this, plus the wider spread weather area, can sometimes mean a number of days when the birds are walking at your airport.

Figure 22-5 shows the cross section of an "average" warm front as indicated by the slice B-B in Figure 22-3.

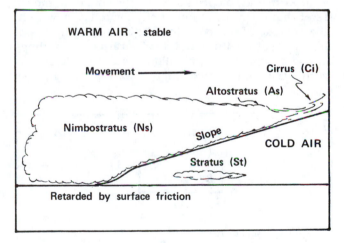

Fig. 22-5. Cross section of a typical warm front with the warm air stable. The slope of the warm front may vary from 1:50 to 1:200, the average being about 1:100.

While you won't be planning on flying within the clouds, the fall or wintertime warm front may pose problems in addition to low ceilings, as can be seen by Figure 22-6.

You can be flying well below the cloud level and suddenly encounter freezing rain. It's a very serious situation because ice can cover the windshield (and the entire airplane) in a very short time. A student pilot has no business flying if there is any possibility of freezing rain. (And this also applies for private and

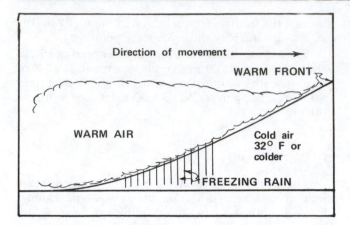

Fig. 22-6. Freezing rain associated with a warm front.

commercial pilots.) If you start running into freezing rain, get back out of the area as quickly as possible. You may find that climbing to just below the clouds will keep you in warm enough air so that the rain hasn't started to freeze. Remember, though, that you may have to let back down through the freezing levels for landing, which can be a hazardous operation.

■ Occluded Front

The occluded front generally contains weather of both warm and cold fronts. Since the cold front moves faster than the warm front, it often catches up with it. The occluded front contains all the disadvantages of the two types, with widespread stratus, low ceilings and poor visibilities with build-up, and thunderstorms within the frontal area. Figure 22-7 shows the occluded front.

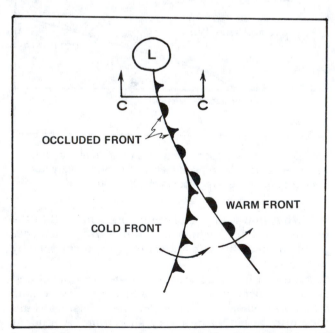

Fig. 22-7. The occluded front.

Figure 22-8 shows the cross section of an occluded front as indicated by C-C in Figure 22-7. There are two types of occlusions, and Figure 22-8 shows the warm type. If the air in front of the system is colder (and denser) than the air behind the cold front overtaking the warm front, it will move up over the cold air as shown. Most of the weather will be found ahead of the surface front.

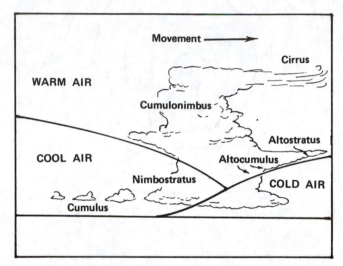

Fig. 22-8. A warm-type occlusion.

The cold-type occlusion is a situation in which the air behind the cold front is colder than that ahead of the system. It slides under the cool air as shown by Figure 22-9. Most of the weather for this type of occlusion will be found near the surface front.

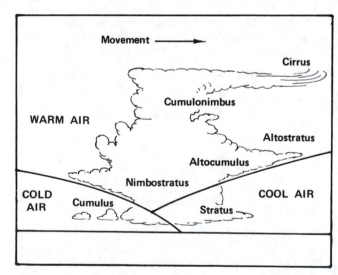

Fig. 22-9. A cold-type occlusion.

■ Stationary Front

The stationary front is so named because it is stationary, which makes pretty good sense when you stop to think about it. The weather associated with the

Cloud Types

In discussing the weather indicated on the weather map, it would be well to take a general look at cloud types you'll expect to encounter when flying.

Clouds are broken down into four families: High, middle, and low clouds and clouds with large vertical development. The clouds are further described as to their form and appearance. The puffy or billowy-type clouds are "cumulus," the layered types are "stratus."

The term "nimbo" (rain cloud) is added to clouds that would be expected to produce precipitation.

Figure 22-10 shows some representative types of clouds.

Normally, when flying near clouds of stratus-type formation, you would expect fairly smooth air. Cumulus clouds by their very nature are the product of air conditions that indicate the presence of vertical currents.

Clouds are composed of minute ice crystals or water droplets and are the result of moist air being cooled to the point of condensation. The high clouds (cirrus, cirrostratus, and cirrocumulus) are composed of extremely fine ice crystals. The biggest puzzle to the average person is that if they are composed of ice why don't they fall? For that matter, since the lower clouds are composed of water droplets, why don't they fall also? The answer is, of course, that the moisture is comparatively less dense than the ambient air. When the water droplets become a certain size, rain results (or snow or sleet depending on the conditions). Hail is a form of precipitation associated with cumulonimbus-type clouds and is the result of rain being lifted by vertical currents until it reaches an altitude where it freezes and is carried downward again to gain more moisture; the cycle may be repeated several times, giving the larger hailstones their characteristic "layers," or strata.

Clouds may be composed of supercooled moisture and the impact of your airplane on these particles causes them to immediately freeze on the airplane. (Stay out of *any* clouds until you get that instrument rating later.)

As discussed, clouds are formed by moist air being cooled to the point of condensation, and this leads to the subject of lapse rates.

For air, the dry adiabatic lapse rate is 5½°F per thousand feet. (Adiabatic describes a process during

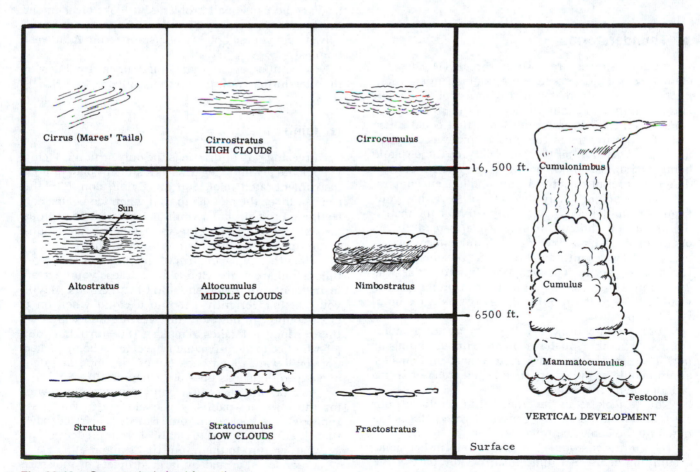

Fig. 22-10. Some typical cloud formations.

which no heat is withdrawn or added to the system or body concerned.) The normal lapse rate of "average" air is 3½°F (2°C). The moist adiabatic lapse rate is produced by convection in a saturated atmosphere such as within a cumulus cloud. At high temperatures it will be in the vicinity of 4–5°F. The dew point lapse rate is about 1°F per thousand feet.

For cumulus-type clouds that are formed by surface heating, the base of the clouds may be estimated by the rate at which the dry lapse rate "catches" the dew point. (Dry lapse rate is 5.4°F and dew point drop is 1°F per thousand, so the temperature is dropping 4.4° faster than the dew point per thousand feet.) Assume the surface temperature is 76°F and the dew point 58°F, a difference of 18°F. Dividing this number by 4.4, you find that the temperature and dew point make connections at about 4000 feet—the approximate base of the clouds. This only works for the type of cloud formed by surface heating.

HAZARDS TO FLIGHT

Freezing rain and hail were briefly covered earlier, but you'll encounter other weather problems as you gain flying experience.

■ Thunderstorms

Thunderstorms are a triple threat: (1) They are clouds, so an inexperienced or noninstrument-rated pilot could lose control because of loss of outside references, but that's usually the least of the problems because of (2) the extreme turbulence that could cause loss of control by even the most experienced instrument pilot, as well as the possibility of the airplane being overstressed and coming apart, and (3) there is danger of aircraft structural icing and hail encounters.

The turbulence may also be so great that your head is being jolted so fast and hard that the instruments become difficult or impossible to read. It was once noted that an "average" thunderstorm (whatever that is) was continually releasing the energy of 50 Hiroshima-type nuclear weapons. Whether this was ever confirmed is not known, but most people wouldn't like to fly into the energy of *one* Hiroshima-type bomb.

Your lifetime project will be to avoid thunderstorms—whether you have an instrument rating or not. They can cause airliners to crash, and smaller aircraft could be pulled apart and *then* crash—the end product being the same.

Take a look at an "average" thunderstorm from start to finish: There are generally three stages, as noted by the book *Aviation Weather*:

1. *Cumulus stage*—The thunderstorm begins as a cumulus cloud. An updraft is the key part of the thunderstorm. The cumulus may be growing at a rate of 3000 feet per minute and you won't be able to outclimb it to go over the top in a light aircraft.

The water droplets at this stage are small but are lifted past the freezing level, so icing could be a problem. As the raindrops get bigger, they fall and the rain creates a cold downdraft coexisting with the updrafts. When this occurs, the thunderstorm has entered the second stage.

2. *Mature stage*—When precipitation starts falling out of the base, the mature stage is reached. The downdraft may exceed 2500 feet per minute; it spreads outward at the surface and may be felt there as the "first gust." Meanwhile, updrafts may be moving up at 6000 feet per minute, and the shear area with its turbulence at the edge of these divergent drafts is better imagined than experienced.

All thunderstorm hazards are worse during the mature stage.

3. *Dissipating stage*—Don't get the idea that the dissipating stage means that all is fine for penetration. This stage is indicated by the anvil head (Fig. 22-10) and is now mostly downdrafts. However, since there is no defined line between the stages, the anvil may be seen while there are still updrafts, and hail may fall out of the anvil into the clear air below. A rule of thumb would be to stay at least 5 miles away from a thunderstorm.

Don't fly under a thunderstorm. Hail and turbulence could cause bad problems. It looks clear under there and you can see 15 miles, but go around anyway, giving yourself plenty of distance from it (at least 5 miles).

Remember again that squall lines are lines of thunderstorms, multiplying the problems.

■ Icing

Freezing rain or drizzle is the only structural icing you should get into "legally" as a student pilot or non-instrument-rated pilot. You are flying along "in the clear" (under the clouds) in rain when ice begins collecting on the airplane; you'll get out of it as soon as possible, perhaps climbing to just below the clouds and/or making a 180° turn as noted earlier.

The other types of structural icing you won't get unless you are in the clouds and, unless you have an instrument rating, you shouldn't be in clouds. Maybe you will inadvertently get into a cloud when on a cross-country in deteriorating weather and will be in there as long as it takes to make a 180° turn, but your job won't be to identify what type of ice is being picked up; you'll want to get back out of the cloud as safely and expeditiously as possible (Chapter 15).

As a private pilot, after getting back out and wiping those sweaty palms, you could say to your passenger(s), "We picked up some rime (or clear) ice, back there." This could tend to establish your credentials as a "weather" pro to them, but that's about all.

Clear ice—Formed by large droplets of moisture such as found in rain or cumuliform clouds. It is hard and glossy and difficult to remove. It's the type of ice you'd find in ice cubes—solid.

Rime ice—Formed by small droplets such as found in stratiform clouds or drizzle. There's air trapped in it and its appearance is close to that of the ice forming on the trays or the inside walls of a freezer. It's comparatively easy to remove. You may feel tempted later, when flying IFR, to ask that particularly obnoxious passenger (the one who keeps asking, "Where are we?") to go out and remove it—in flight.

■ **Frost**

You've left the airplane tied down in the fall or winter and find very early in the morning that it is covered with a very thin layer of frost. You figure that the weight of the frost is certainly negligible, and it's so extremely thin that there will be no problems aerodynamically, so you load up everybody and go—for a while. You find yourself (and airplane) roaring down the street of a housing development off the far end of the runway. (Housing developments are always built in that location immediately after the runway is completed.) After untangling the airplane from fences and playground equipment (fortunately nobody was out yet) and surveying the damage and the problems of getting the airplane back for repair, you wonder if maybe that little frost hasn't cost you a lot of money.

Don't take off with frost on the wings or horizontal tail surfaces; you may not get off.

■ **Fog**

While the other hazards are more readily avoided, fog can be one of the greatest dangers to any pilot because it may rapidly cover wide areas and develop while you are airborne and are concentrating on other matters. As you do more night flying, fog will become a factor to be very seriously considered under certain meteorological conditions.

Fog, according to FAA and Weather Service publications, is a surface-based cloud of either water droplets or ice crystals.

Fog may form by cooling air to its dew point or by adding moisture to air near the ground.

Radiation fog—A shallower fog; one type is called "ground fog." You'll see ground fog from the air as a gray-white sea with taller objects sticking up through it. The top half of the hangar and the wind sock with towers may be quite visible, but the runway and surrounding area are covered. This fog usually occurs at night or in early morning with a clear sky, little or no wind, and a small temperature–dew point spread. The lack of overcast allows the ground to lose heat; the cooler (or cold) ground cools the highly moist air next to it, lowering the temperature to the dew point, and visible moisture is formed.

In addition, nuclei from industrial exhausts hasten the production of fog; an airport in a river bottom or on a lake shore near industrial emissions usually has a higher number of mornings with ground conditions of zero visibility. This fog will usually burn off before noon on clear days.

Wind may disperse the fog or increase its depth, depending on the velocity.

Radiation fog is formed over land because land cools rapidly as compared to water.

If you are shooting landings and the runway lights begin to have haloes (at night) or the runway or other ground objects start becoming slightly less distinct in outline, you'd better consider terminating the flight early, unless you have a good alternate airport in mind.

Advection fog—Formed when moist air moves over colder ground or water. If you are flying in the west coast area of the United States, you are well aware of how "sea fog" can form and move inland at times. Wind may also move fog up into the United States from the Gulf of Mexico during the winter.

These fogs look alike from the air, but advection fog may be found under an overcast, unlike radiation fog. If you ever have to race advection fog to an airport, you'll see how fast it can spread.

Upslope fog—When moist air is moved up a slope, cooling it, this type of fog is formed. The Sewanee (Tenn.) Airport, the point of origin for the example cross-country in this book, is on the edge of a plateau. When warm moist air is moved up the slope under certain wind conditions, Sewanee can be fogged in solid on some winter days—the victim of upslope fog.

Ice fog—The temperature has to be very cold for this fog to form because the water vapor sublimates directly to ice. It's mostly an arctic problem but can happen in very cold spots in the United States.

Precipitation fog—You've seen this fog when it's raining, or has been, and the moisture falls into cool air. When you're flying in light rain and look down to see fog forming (particularly in the wooded areas where the trees hold moisture), you'd better plan on getting to an airport. Warm fronts are usually the causes, and widespread areas of very dense fog can be formed in a short while. Also, if you're flying over a layer of scattered clouds, and rain is falling out of an overcast above, you'd better watch for that scattered layer to very soon become broken and then overcast. There's nothing like flying along without a care in the world and looking down to discover that the undercast is nearly total—and you don't have an instrument rating!

Recommended books for you to read on weather theory and Weather Services are *Aviation Weather* and *Aviation Weather Services*, available from the U.S. Government Printing Office. (See the Bibliography in the back of this book.) As you progress in aviation, you'll be branching out into areas of weather knowledge that may not apply to you now as a student pilot. Your job will be to assure that you don't stop the learning process.

WEATHER INFORMATION SOURCES

The most important things for you to know about weather and where to get information about it and how to understand what's being said about it. You can get the latest forecasts or hourly reports from either the FAA Flight Service Stations or the National Weather Service Offices.

You aren't expected to analyze and forecast the weather. But you must be able to use the services. If you can't read hourly reports or forecasts, all of that expensive weather-watching system will be of small value to you. The people in the Flight Service Stations and Weather Services Offices are always willing to help. They will explain the symbols and will give the weather in everyday language if you request it. You're better off if you know how to read the information yourself, since you can browse through the data at your leisure and compare the various reports. You can then ask more concise and worthwhile questions.

When you contact these agencies, tell them you're a pilot and give them your time of departure, route (if you aren't going direct), destination, and expected time of arrival.

HOURLY WEATHER REPORT (METAR)

The METAR (Meteorological Aviation Routine Report) is available each hour at Flight Service Stations and Weather Service Offices. The list of abbreviations on this page are used in interpreting the surface observations. (These will be also used for forecasts later in the chapter).

Figure 22-11 shows actual METARs for Nashville, Tennessee, and Huntsville, Alabama, starting at the 1300Z weather for the 13th of the month.

Look at the Nashville information:

(1) The hourly report is for Nashville (KBNA) and is issued for the 13th at 1253Z (131253Z). The wind at 3 knots is variable in direction (VRB03KT). Visibility is 5 statute miles (5SM) in haze (HZ). The ceiling is broken clouds at 7500 feet (BKN075) and at 9000 feet

Abbreviations

AO1	Automated Observation without precipitation discriminator (rain/snow)
AO2	Automated Observation with precipitation discriminator (rain/snow)
AMD	Amended Forecast (TAF)
BECMG	Becoming (expected between 2 digit beginning hour and 2 digit ending hour)
BKN	Broken 5-7 octas (eighths) cloud coverage
CLR	Clear at or below 12,000 feet (ASOS/AWOS report)
COR	Correction to the observation
FEW	>0-2 octas (eighths) cloud coverage
FM	From (4 digit beginning time in hours and minutes)
LDG	Landing
M	In temperature field means "minus" or below zero
M	In RVR listing indicates visibility less than lowest reportable sensor value (e.g. M0600)
NO	Not available (e.g. SLPNO, RVRNO)
NSW	No Significant Weather Note: NSW only indicates obstruction to visibility or precipitation previously noted has ended. Low ceilings, wind shear, and other weather conditions may still exist.
OVC	Overcast 8 octas (eighths) cloud coverage
P	In RVR indicates visibility greater than highest reportable sensor value (e.g. P6000FT)
P6SM	Visibility greater than 6 SM (TAF only)
PK WND	Peak wind
PROB40	Probability 40 percent
R	Runway (used in RVR measurement)
RMK	Remark
RY/RWY	Runway
SCT	Scattered 3-4 octas (eighths) cloud coverage
SKC	Sky Clear
SLP	Sea Level Pressure (e.g., 1001.3 reported as 013)
SM	Statute mile(s)
SPECI	Special Report
TEMPO	Temporary changes expected (between 2 digit beginning hour and 2 digit ending hour)
TKOF	Takeoff

T01760158, 10142, 20012 and 401120084
In Remarks—examples of temperature information

V	Varies (wind direction and RVR)
VC	Vicinity
VRB	Variable wind direction when speed is less than or equal to 6 knots
VV	Vertical Visibility (Indefinite Ceiling)
WS	Wind shear (In TAFs, low level and not associated with convective activity)

Descriptors

BC	Patches
BL	Blowing
DR	Low Drifting
FZ	Supercooled/freezing
MI	Shallow
PR	Partial
SH	Showers
TS	Thunderstorm

Weather Phenomena

BR	Mist
DS	Dust Storm
DU	Widespread Dust
DZ	Drizzle
FC	Funnel Cloud
+FC	Tornado/Water Spout
FG	Fog
FU	Smoke
GR	Hail
GS	Small Hail/Snow Pellets
HZ	Haze
IC	Ice Crystals
PL	Ice Pellets
PO	Dust/Sand Whirls
PY	Spray
RA	Rain
SA	Sand
SG	Snow Grains
SN	Snow
SQ	Squall
SS	Sandstorm
UP	Unknown Precipitation (Automated Observations)
VA	Volcanic Ash

Cloud Types

CB	Cumulonimbus
TCU	Towering Cumulus

Intensity Values

-	Light
no sign	Moderate
+	Heavy

(BKN090). The lowest broken or overcast cloud level is the ceiling.

The temperature is 25°C [Celsius] and the dewpoint is 21°C (25/21). The altimeter setting is 29.96 inches of mercury ["Hg] (A2996). Remarks (RMK): The information is an automated observation (AO2)

---

(1) METAR KBNA 131253Z VRB03KT 5SM HZ BKN075 BKN090 25/21 A2996 RMK AO2 SLP137 T02510206

(2) METAR KBNA 131353Z 23005KT 6SM HZ BKN075 26/21 A2996 RMK AO2 SLP138 T02610206

(3) METAR KBNA 131453Z 23007KT 9SM FEW150 BKN250 27/21 A2997 RMK AO2 SLP140 T02710206

(4) METAR KHSV 131256Z 16003KT 3SM BR FEW028 SCT070 BKN100 22/22 A3001 RMK AO2 SLP154 T02220217

(5) METAR KHSV 131356Z 00000KT 3SM BR SCT030 BKN100 22/22 A3001 RMK AO2 SLP154 T02220217

(6) METAR KHSV 131456Z 19003KT 4SM -RA BR FEW008 24/22 A3002 RMK AO2 RAB48 SLP159 T02390217

Fig. 22-11. Actual METARs for Nashville and Huntsville issued for the 13th of the month.

equipped with a precipitation discriminator (rain/snow). [See the abbreviations.] The sea level pressure is 1013.7 hectoPascals (SLP137) [known earlier as millibars]. For the use of selected stations, the temperature and dewpoint are given to the nearest 0.1°. The temperature is 25.1°C and the dewpoint is 20.6°C (T02510206). [The "zeros" after the "T" and before the "2" are separators; you may use dashes if you wish.]

(2) This METAR for KBNA was issued at 1353Z. The wind [always in true directions] is from 230° at 5 knots (23005KT). Visibility is 6 statute miles (6SM) in haze (HZ). There is a broken layer [ceiling] of clouds at 7500 feet (BKN075). The temperature and dewpoint are 26°C and 21°C (26/21), respectively, and the altimeter setting is 29.96″ Hg (A2996). Remarks (RMK): The information is an automated observation (A02) with a precipitation discriminator, the sea level pressure is 1013.8 hectoPascals (SLP138), and the temperature and dewpoint are 26.1°C and 20.6°C (T02610206).

(3) This is a Nashville METAR issued at 1453Z. The wind is from 230° at 7 knots (23007KT). Visibility is 9 statute miles (9SM). The clouds are few [0 to 2/8 sky coverage] at 15,000 feet (FEW150) and broken at 25,000 feet (BKN250). The temperature and dewpoint are 27°C and 21°C (27/21). The altimeter setting is 29.97″ Hg (A2997). Remarks (RMK): The AO2 still indicates an automated observation, and the sea level pressure is 1014.0 hectoPascals (SLP140). The temperature is 27.1°C and the dewpoint is 20.6°C (T02710206).

Looking at Huntsville (KHSV):

(4) Issued at 1256Z on the 13th, the wind is from 160° true at 3 knots (16003KT). Visibility is 3 statute miles (3SM) in mist (BR). [BR is an abbreviation for *brouillard*, the French word for mist.] There are a few clouds [0 to 2/8 coverage] at 2800 feet (FEW028), scattered clouds at 7000 (SCT070), and a broken layer at 10,000 feet (BKN100). The temperature and dewpoint are 22°C each (22/22), and the altimeter setting is 30.01″ Hg (A3001). Remarks (RMK): AO2 is an automated observation. The sea level pressure is 1015.4 hectoPascals (SLP154), and the temperature and dewpoint are 22.2°C and 21.7°C (T02220217).

(5) At 1356Z the wind at Huntsville is calm (00000KT). Visibility is 3 statute miles (3SM) in mist (BR). There are scattered clouds [3/8 to 1/2 sky coverage] at 3000 feet (SCT030) and broken clouds [5/8 to 7/8 coverage] at 10,000 feet (BKN100). Both the temperature and dewpoint are 22°C (22/22). The altimeter setting is 30.01″ Hg (A3001). Remarks (RMK): AO2, and the sea level pressure is 1015.4 hectoPascals (SLP154). The temperature and dewpoint are 22.2°C and 21.7°C (T02220217).

(6) At 1456Z the wind at KHSV is from 190° true at 3 knots (19003KT). Visibility is 4 statute miles (4SM) in light rain (-RA) and mist (BR). There are a few clouds [0 to 2/8 coverage] at 800 feet (FEW008). The temperature and dewpoint are 24°C and 22°C (24/22). The altimeter setting is 30.02″ Hg (A3002). Remarks (RMK): AO2 [again], and the rain began at 48 minutes

past the last hour (RAB48). The sea level pressure is 1015.9 hectoPascals (SLP159), and the temperature and dewpoint are 23.9°C and 21.7°C (T02390217).

Obtaining the hourly reports should trigger a reminder to check the NOTAMs (Notices to Airmen) for that station and enroute facilities. (Review Chapter 20, Notices to Airmen.)

WEATHER FORECASTS

■ TAF

TAF stands for Terminal Airport Forecast, or a forecast for a terminal point rather than for an area. [The area forecast will be covered later.] The TAF contains a definitive forecast for specific time periods and is issued at 0000Z, 0600Z, 1200Z, and 1800Z for each 24-hour period. The following is part of an FAA/DOT explanation of the TAF. [It is paraphrased in some parts.] The descriptors and abbreviations are the same as those given for the METARs.

TAF (TAF AMD is Amended Forecast when included)
KPIT 091730Z 091818 22020KT 3SM -SHRA BKN020

    FM2030 30015G25KT 3SM SHRA OVC015 TEMPO 2022 1/2SM TSRA OVC008CB

    FM0100 27008KT 5SM -SHRA BKN020 OVC040 PROB40 0407 00000KT 1SM -RA BR

    FM1000 22010KT 5SM -SHRA OVC020 BECMG 1315 20010KT P6SM NSW SKC

TAF
KPIT 091730Z 091818 22020KT 3SM -SHRA BKN020

    <u>FM2030</u> 30015G25KT 3SM SHRA OVC015 WS015/30045KT <u>TEMPO</u> <u>2022</u> 1/2SM TSRA OVC008CB

    <u>FM2300</u> 27008KT 5SM -SHRA BKN020 OVC040 <u>PROB40</u> <u>0407</u> 00000KT 1SM -RA BR

    FM1000 22010KT 5SM -SHRA OVC020 <u>BECMG</u> <u>1315</u> 20010KT P6SM NSW SKC

TAF
KPIT 091730Z 091818 22020KT

<u>**Where**</u>

<u>KPIT</u> is the ICAO station identifier. The 3-letter identifiers are preceded by a "K" for the contiguous United States as was the case for METARs. Alaska and Hawaii will use 4-letter identifiers beginning with "PA" and "PH" respectively.

When

091730Z This is the forecast for the **9th day of the month** with an issuance time of **1730Z** or UTC. This is a 2-digit date and 4-digit time.

091818 is the valid period with the first two digits containing the day of the month (**09**).

091**8**18 The second two digits specify the hour beginning the forecast period (**1800Z**).

091**8**18 The last two digits are the hour ending the forecast period (**1800Z** on the next day, the 10th).

Wind
22020KT

See the description under METAR.

WS015/30045KT means at 1500 feet we expect wind to be **300** degrees at **45 KT**. This indicates low level wind shear, not associated with convective activity.

Time Periods, Etc.

FM2030 From 2030Z or UTC time. Indicates hours and minutes.

TEMPO 2022 Temporary changes expected between 2000Z and 2200Z.

FM2300 FROM 2300Z.

PROB40 0407 There is a **40–49 percent probability** of this condition occurring between **0400Z and 0700Z**. You will also see PROB30, or there's a 30–39% probability.

FM1000 FROM 1000Z.

BECMG 1315 Conditions **Becoming** as described between **1300Z and 1500Z**.

Once the specific time periods can be discerned, the sequence of **wind, visibility, significant weather, cloud cover and cloud height** follows and is repeated for each time block. The only exception is after qualifiers such as **PROB40, TEMPO,** and **BECMG,** some of the components may be omitted if these are not expected to change. Notice after **TEMPO 2022,** there is no wind given and after **PROB40 0407,** there is no cloud cover listed. Note: When No Significant Weather (**NSW**) appears it only indicates obstruction to visibility or precipitation previously noted has ended. (*See* Abbreviations on page 22-8.)

Figure 22-12 is an actual TAF for Nashville.

The forecast for Nashville (KBNA) was issued at 2333Z on the 12th (122333Z) and covers the 13th from 0000Z to 2400Z (130024).

Taking the rest of Figure 22-12 step by step:

At the start of the TAF, the wind is forecast to be from 200° [always a true direction] at 3 knots (20003KT), the visibility is greater than 6 statute miles (P6SM) [think of the "P" as *plus*, or greater], clouds are scattered at 4000 feet (SCT040) and broken at 25,000 feet (BKN250).

From (FM) 0600[Zulu], the forecast is for calm winds (00000KT) with visibility of 5 statute miles (5SM) in mist (BR) and broken clouds at 25,000 feet (BKN250).

From (FM) 1400[Z], the forecast is for winds from 240° at 7 knots (24007KT), visibility is greater than 6 statute miles (P6SM), and there are scattered clouds at 4000 feet (SCT040) with a broken layer at 25,000 feet (BKN250).

From (FM) 1700[Z], the forecast is for a wind from 240° at 10 knots (24010KT), visibility is 5 statute miles (5SM) in haze (HZ), with scattered clouds at 4500 feet, plus cumulonimbus (SCT045CB), and broken clouds at 25,000 feet (BKN250). There's a 30 percent probability (PROB30) between 1700[Z] and 2400[Z] (1724) of a visibility of 3 statute miles (3SM), with thunderstorms with rain (TSRA), mist (BR), and an overcast at 4500 feet with cumulonimbus clouds (OVC045CB=).

```
TAF
KBNA 122333Z 130024 20003KT P6SM SCT040 BKN250
    FM 0600 00000KT 5SM BR BKN250
    FM 1400 24007KT P6SM SCT040 BKN250
    FM 1700 24010KT 5SM HZ SCT045CB BKN250
        PROB30 1724 3SM TSRA BR OVC045CB=
```

Fig. 22-12. An actual TAF for Nashville for the 13th of the month from 0000Z to 2400Z (or 0000Z on the 14th).

Area Forecast

The Area Forecast (FA-Forecast, Area) covers several states and the coastal waters of those states, as required. Discussed here is an FA for the Salt Lake City (SLC) area, which covers the area as shown by Figure 22-13. Note that there are also five other areas with issuances cited in Miami (MIA), Dallas (DFW), San Francisco (SFO), Chicago (CHI), and Boston (BOS).

Area forecasts are normally issued 3 times daily for the six areas shown but may be at different times for different parts of the country.

Figure 22-14 is an area forecast for the Salt Lake City area as given in Aviation Weather Services.

Reviewing Figure 22-14 line by line, the heading indicates that the report is from Salt Lake City (SLC), it contains a clouds and weather forecast [the added C], and it's an area forecast (FA) issued on the 14th of the month at 1045Z (141045).

The SYNOPSIS AND clouds and weather are for VFR (VFR CLDS/WX).

The SYNOPSIS is VALID UNTIL 0500Z on the 15th (150500).

The clouds and weather (CLDS/WX) forecast is VALID UNTIL 2300Z on the 14th (142300). The outlook (OTLK) is VALID from 2300Z on the 14th to 0500Z on the 15th (142300–150500).

SYNOPSIS—The HIGH pressure (PRES) over (OVR) northern (NERN) Montana (MT) will be continuing (CONTG) to move eastward (EWD) gradually (GRDLY). LOW pressure (PRES) over (OVR) Arizona (AZ), New Mexico (NM), and western (WRN) Texas (TX) will be remaining (RMNG) generally (GENLY) stationary (STNRY). Aloft (ALF), the trough (TROF) that extends (EXTDS) from western (WRN) Montana (MT) into southern (SRN) Arizona (AZ) will be remaining (RMNG) stationary (STNRY) for the valid period.

Idaho (ID) and Montana (MT)—SEE AIRMET SIERRA FOR mountain (MTN) obscuration (OBSCN) information.

Locations FROM Medicine Hat, Alberta, Canada (YXH), TO Sheridan, Wyoming (SHR), TO 30 miles southeast (30SE) of Bozeman, Montana (BZN), TO 60 miles southeast (60SE) of Pocatello, Idaho (PIH), TO Salmon, Idaho (LKT), TO Cranbrook, British Columbia (YXC), back TO Medicine Hat (YXH) outline the area in question.

In the area outlined, conditions are forecast to be scattered to broken clouds at 7000 to 9000 feet (SCT-BKN070-090) mean sea level [MSL] with TOPS of clouds at 12,000 to 15,000 feet (120–150). There will be widely (WDLY) scattered (SCT) rain showers (SHRA). The TOPS of the showers (SHOWRS) are at the flight level of 18,000 feet (FL180). The outlook (OTLK) is for VFR weather. [Note above that the out-

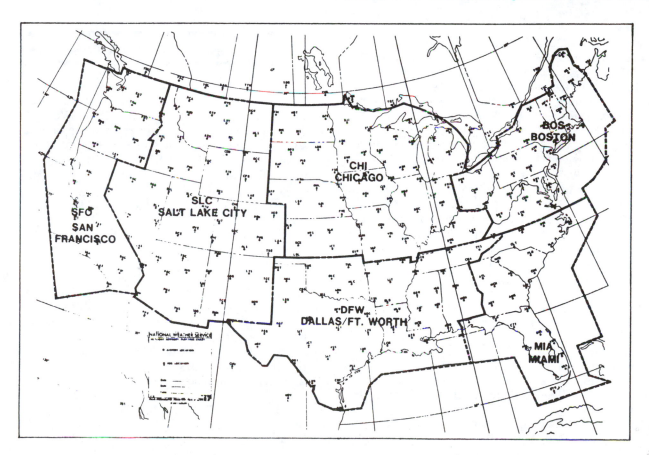

Fig. 22-13. Area forecast locations. (*Aviation Weather Services*)

```
SLCC FA 141045
SYNOPSIS AND VFR CLDS/WX
SYNOPSIS VALID UNTIL 150500
CLDS/WX VALID UNTIL 142300...OTLK VALID 142300-150500

SYNOPSIS...HIGH PRES OVR NERN MT CONTG EWD GRDLY. LOW PRES OVR
AZ NM AND WRN TX RMNG GENLY STNRY.  ALF...TROF EXTDS FROM WRN
MT INTO SRN AZ RMNG STNRY.

ID MT
SEE AIRMET SIERRA FOR MTN OBSCN.
FROM YXH TO SHR TO 30SE BZN TO 60SE PIH TO LKT TO YXC TO YXH
SCT-BKN070-090 TOPS 120-150.  WDLY SCT SHRA.  TOPS SHOWRS
FL180.  OTLK...VFR.
RMNDR AREA...SCT100-120.  ISOLD SHRA MNLY ERN PTNS AREA.
OTLK...VFR.

UT NV
NRN...SCT080. WDLY SCT SHRA.  TOPS SHWRS FL180. 15Z SKC.
18Z SCT100
OTLK...VFR.
```

Fig. 22-14. An area forecast for the Salt Lake City area.

look is from 142300Z to 150500Z.]

The remainder (RMNDR) of the AREA has forecast conditions of scattered clouds at 10,000–12,000 feet (SCT100–120). Isolated (ISOLD) rain showers (SHRA) are expected mainly (MNLY) in the eastern (ERN) portions (PTNS) of the AREA. The outlook (OTLK) is for VFR conditions.

Utah (UT) and Nevada (NV)—The northern (NRN) portion will have scattered clouds at 8000 feet (SCT080). There's a forecast for widely (WDLY) scattered (SCT) rain showers (SHRA). The TOPS of the showers (SHWRS) will be at flight level of 18,000 feet (FL180). After 1500Z (15Z), the sky will be clear (SKC).

After 1800Z (18Z), the forecast is for scattered clouds at 10,000 feet (SCT100).

The outlook (OTLK) is for VFR conditions.

The Categorical Outlooks are as follows:

VFR (Visual flight rules)—Ceiling greater than 3000 feet and visibility greater than 5 miles; includes sky clear.

MVFR (Marginal VFR)—Ceiling 1000–2000 feet and/or visibility 3–5 miles inclusive.

IFR (Instrument flight rules)—Ceiling 500 feet to less than 1000 feet and/or visibility 1 mile to less than 3 miles.

LIFR (Low IFR)—Ceiling less than 500 feet and/or visibility less than 1 mile.

IN-FLIGHT WEATHER ADVISORIES

The National Weather Service (NWS) issues in-flight weather advisories designated as AIRMETs (WA), SIGMETs (WS), Convective SIGMETs (WST), Center Weather Advisories, and Severe Weather Alerts. The advisories most useful to you as a student or low-time private pilot will be AIRMETs, SIGMETs, and Convective SIGMETs.

Weather phenomena that are potentially hazardous to the aircraft require the issuance of an AIRMET.

1. AIRMET SIERRA—This is to alert the pilot checking the area forecast (FA) that, in the state or states listed, there may be a problem with IFR conditions such as low ceilings (less than 1000 feet), low visibility (less than 3 miles over 50 percent of the area at one time), or extensive mountain obscuration. In other words, when you see this, be prepared to check further for these phenomena.

2. AIRMET TANGO—This designator is given in the flight precautions in the FA to warn the pilot that there may be a problem with turbulence or sustained winds of 30 K or more in the area cited.

3. AIRMET ZULU—This alerts the pilot to information concerning icing and freezing problems in the FA states or areas.

```
CHIS WA 300800
AIRMET SIERRA FOR IFR VALID UNTIL 301400

AIRMET IFR...IA MO
FROM FOD TO CID TO COU TO MKC TO FOD
CIG BLO 010 AND VIS BLO 3SM SN.  CONDS CONTG BYD 1400Z AND
SPRDG OVR ERN IA ERN MO AND IL.

IFR OTLK VALID 1400-2000...IA MN WI IL
CIG BLO 010 AND VIS BLO 3SM SN SPRDG INTO NERN IA SRN MN SWRN
WI AND NWRN IL BTWN 1600Z AND 2000Z CONTG BYD 2000Z.
```

Fig. 22-15. An AIRMET for Iowa and Missouri.

AIRMETs may be issued between the normal release times if conditions warrant. Check with the National Weather Service or Flight Service Stations for the current normal times of issuance. The alphabet letters S, T, and Z are reserved for use only in AIRMETs.

Figure 22-15 is an AIRMET issued on the 30th at 0800Z for the Chicago area (CHIS), and the added "S" [SIERRA] indicates that ceilings and visibilities are involved.

Looking at the figure, the AIRMET SIERRA FOR IFR is VALID UNTIL 1400Z on the 30th (301400).

AIRMET IFR for Iowa (IA) and Missouri (MO) goes as follows:

FROM Fort Dodge, Iowa (FOD), TO Cedar Rapids, Iowa (CID), TO Columbia (COU), Missouri, TO Kansas City, Missouri (MKC), back TO Fort Dodge (FOD).

Ceilings (CIG) will be below (BLO) 1000 feet (010) AND visibilities (VIS) below (BLO) 3 statute miles (3SM) in snow (SN). Conditions (CONDS) will be continuing (CONTG) beyond (BYD) 1400Z AND spreading (SPRDG) over (OVR) eastern (ERN) Iowa (IA), eastern (ERN) Missouri (MO), and Illinois (IL).

The IFR outlook (OTLK) is VALID from 1400Z to 2000Z (1400–2000) for Iowa (IA), Minnesota (MN),

Wisconsin (WI), and Illinois (IL).

Ceilings (CIG) below (BLO) 1000 feet (010) AND visibilities (VIS) below (BLO) 3 statute miles (3SM) in snow (SN) will be spreading (SPRDG) INTO northern (NERN) Iowa (IA), southern (SRN) Minnesota (MN), southwestern (SWRN) Wisconsin (WI) AND northwestern (NWRN) Illinois (IL) between (BTWN) 1600Z AND 2000Z and will be continuing (CONTG) beyond (BYD) 2000Z.

SIGMETs are also issued to cover weather phenomena considered to be of greater danger to aircraft than those situations listed above. The first issuance of a SIGMET is identified as an urgent weather SIGMET (UWS), and any that follow are given as plain WS. Items covered are

1. Severe icing not associated with thunderstorms.

2. Severe or extreme turbulence or clear air turbulence (CAT) not associated with thunderstorms.

3. Duststorms, sandstorms, or volcanic ash lowering surface or in-flight visibilities to below 3 miles.

4. Volcanic eruption.

Figure 22-16 is a SIGMET issued for the Dallas/Ft. Worth area (DFW). (See Fig. 22-13 again for the area covered.) Let's look at the SIGMET.

```
DFWP UWS 051700
SIGMET PAPA 1 VALID UNTIL 052100
AR LA MS
FROM MEM TO 30N MEI TO BTR TO MLU TO MEM
MDT TO OCNL SVR ICG ABV FRZLVL XPCD.  FRZLVL 080 E TO 120 W.
CONDS CONTG BYD 2100Z.
```

Fig. 22-16. An initial SIGMET (UWS) issued for the Dallas/Fort Worth area.

This was the first issuance from the Dallas/Ft. Worth area at 051700 (DFWP UWS 051700). It's SIGMET PAPA 1 and VALID UNTIL 2100Z on the 5th of the month. [The nonconvective SIGMETs now have alphabetic designators that run from November (N) through Yankee (Y), excluding the AIRMET designators SIERRA (S), TANGO (T), and ZULU (Z), which were covered earlier.] Continuing through the figure:

Arkansas (AR), Louisiana (LA), and Mississippi (MS)—the area covered is FROM Memphis, TN (MEM), TO 30 miles north (30N) of Meridian, MS (MEI), TO Baton Rouge, LA (BTR), TO Monroe, LA (MLU), and back TO Memphis (MEM).

There is moderate (MDT) TO occasionally (OCNL) severe (SVR) icing (ICG) above (ABV) the freezing level (FRZLVL) expected (XPCD). The freezing level (FRZLVL) is 8000 feet (080) in the eastern (E) portion TO 12000 feet (120) in the western (W) portion of the area.

The conditions (CONDS) will be continuing (CONTG) beyond (BYD) 2100Z.

The first time a SIGMET is issued for a phenomenon associated with a particular weather system, it will be given the next alphabetic designator in the series and will be numbered as the first for that designator (e.g., PAPA 1). Subsequent advisories will retain the same alphabetic designator until the phenomenon ends (PAPA 1, PAPA 2, etc.). In the conterminous United States, that phenomenon will retain that designator as it moves within an area or into one or more other areas.

Convective SIGMETs are issued in the 48 conterminous states for any of the following phenomena:
1. Severe thunderstorms due to:
   a. surface winds greater than or equal to 50 K.
   b. hail, at the surface, greater than or equal to ¾ inch in diameter.
   c. tornadoes.
2. Embedded thunderstorms.
3. A line of thunderstorms.
4. Thunderstorms greater than or equal to VIP level 4 affecting 40% or more of an area at least 3000 square miles.

There's a tie-in between the radar echo intensity and the turbulence in a thunderstorm (Called VIP levels).

Level 1—weak.
Level 2—moderate.
Level 3—strong.
Level 4—very strong.
Level 5—intense.
Level 6—extreme.

Your job will be to give *all* thunderstorms a wide berth of several miles.

Figure 22-17 is a sample Convective SIGMET for the *central* U.S. region.

Taking the figure line by line, Kansas City has issued a CONVECTIVE SIGMET. Its designator is 20C.

It is VALID UNTIL 2055Z.

North and South Dakota (ND SD)—FROM 90 miles west of Minot, ND (90W MOT), to Grand Forks, ND (GFK), to Aberdeen, SD (ABR), to Rapid City, ND (RAP), to 90 miles west of Minot (90W MOT):

There is an intensifying (INTNSFYG) AREA of severe (SVR) thunderstorms (TSTMS) moving (MVG) FROM 250° [true] at 30 knots (25030KT). The TOPS are above (ABV) the flight level of 45,000 feet (FL450).

A TORNADO was reported (RPRTD) at 1820Z at 50 miles southwest (50SW) of Dupree, South Dakota (KDPR). There is HAIL up TO ½ inch diameter (1/2IN). WIND GUSTS TO 45 knots (45KT) have been reported (RPRTD). TORNADOES, HAIL TO 1 inch (1IN), and WIND GUSTS up TO 65 knots (65KT) are possible (PSBL).

When you first start reading area forecasts and AIRMETs/SIGMETs, the sentence structure seems that of an ex-Luftwaffe pilot with punctuation problems, but with practice *yu cn rd thm without mch trbl.*

■ Transcribed Weather Broadcasts (TWEBs)

At selected locations, higher-powered radio beacons used as outer market (OM) compass locators for the ILS broadcast TWEB information in the 190–535 kHz frequency range. The information broadcast includes weather reports, NOTAMs, special notices, and forecasts. Especially helpful are the route forecasts given between selected terminals. Figure 22-18 has TWEBs for two different routes.

(1) The route is 383 TWEB and is valid from 1200Z

```
CONVECTIVE SIGMET 20C
VALID UNTIL 2055Z
ND SD
FROM 90W MOT-GFK-ABR-RAP-90W MOT
INTSFYG AREA SVR TSTMS MOVG FROM 25030KT.  TOPS ABV FL450.
TORNADO RPRTD 1820Z 50SW KDPR. HAIL TO 1/2IN...WIND GUSTS TO 45KT
RPRTD.  TORNADOES...HAIL TO 1 IN...WIND GUSTS TO 65KT PSBL.
```

Fig. 22-17. A Convective SIGMET issued for North and South Dakota.

```
    383 TWEB 171203 KSLC-KENV-KEKO.  ALL HGTS MSL XCP CIGS.  KSLC-KENV
1   BKN-SCT100 TOPS AOB 150...WDLY SCT SHSN WITH MTNS LCL
    OBSCD.  RMDR OF RTE NOT AVBL.


    239 TWEB 181001 KMKC-KDSM-KMCW.  ALL HGTS MSL XCP CIGS.  TIL 17Z
2   SCT040-050...AREAS OVC030 VIS 1SM TSRA.  AFT 17Z BECMG
    SCT-BKN040-050.  KMKC VCNTY-70NM NW KMKC AFT 18Z AREAS
    OVC015-025 VIS 1/2-1SM TSRA.
```

Fig. 22-18. TWEBs issued for (1) the western United States and (2) the Midwest. See the body of the text for the translation.

on the 17th of the month to 0300Z (171203) on the 18th. The route is from Salt Lake City (KSLC) to Wendover, UT (KENV), to Elka, NV (KEKO). ALL heights (HGTS) cited are mean sea level (MSL) except ceilings (XCP CIGS). The route from Salt Lake City to Wendover (KSLC-KENV) will have broken to scattered clouds at 10,000 feet (BKN-SCT100), TOPS at or below (AOB) 15,000 feet (150). There will be widely (WDLY) scattered (SCT) snow showers (SHSN) with mountains (MTNS) locally (LCL) obscured (OBSCD). The remainder (RMDR) OF the route (RTE) is NOT available (AVBL).

(2) Route 239 TWEB is for the 18th of the month from 1000Z to 0100Z on the 19th (181001). The route is from Kansas City, MO (KMKC), to Des Moines, IA (KDSM), to Mason City, IA (KMCW). Again, ALL heights (HGTS) are mean sea level (MSL) except ceilings (XCP CIGS). Until (TIL) 1700Z (17Z) there is a forecast for scattered clouds at 4000–5000 feet (SCT040-050). There will be AREAS of overcast at 3000 feet (OVC030) and VISIBILITIES of 1 statute mile (1SM) in thunderstorms and rain (TSRA). After 1700Z (AFT 17Z) the conditions are forecast to become (BECMG) scattered to broken at 4000 to 5000 feet (SCT-BKN040-050). In the Kansas City (KMKC) vicinity to 70 nautical miles (VCNTY-70NM) northwest of Kansas City (NW KMKC), after 1800Z (AFT 18Z) there will be AREAS of overcast at 1500–2500 feet (OVC015-025). Visibility (VIS) from ½ to 1 statute mile (1/2-1SM) in thunderstorms and rain (TSRA).

WIND INFORMATION

Figure 22-19 is part of an actual winds aloft forecast. The altitudes given are above *mean sea level;* some of the western stations may not issue the winds at 3000 or even 6000 feet because of the elevation (see DEN—Denver in Fig. 22-19). Winds aloft information

```
FD KWBC 151640
Based on 151200Z DATA
VALID 151800Z FOR USE 1700-2100Z. TEMPS NEG ABV 24000
```

FT	3000	6000	9000	12000	18000	24000	30000	34000	39000
BHM	1928	2231+07	2437+05	2539-02	2541-14	2756-26	267140	277749	288159
BNA	2043	2246+05	2346+03	2446+02	2448-15	2559-28	267842	268451	278561
RDU	9900	2207+05	2310+04	2612-01	2931-18	3047-31	316644	317352	317260
RIC	9900	2109+04	2314+02	2513-03	2926-19	3039-32	316145	316653	316360
ROA	1917	2220+03	2226+03	2328-03	2738-18	2850-30	296644	297453	297562
DBQ	2139	2341-02	2343+00	2353-04	2370-19	2382-32	740448	740756	750161
DEN			3126-13	3234-21	3136-31	3157-37	318244	309051	298955

Fig. 22-19. Part of an actual winds aloft forecast for several selected stations, based on 0000Z data on the 12th of the month valid the 12th at 1200Z for use 0900Z to 1800Z. As noted, all temperatures (which are Celsius) are negative above 24,000 feet to save the space of putting a minus (—) in each case.

is given to the nearest 10° *true* and in *knots*. Temperature are *Celsius*.

Looking at the forecast winds at Raleigh-Durham (RDU) and Richmond, Va. (RIC), at 3000 feet, the winds are forecast to be light and variable (5 K or less) as given by the digits 9900. (No temperatures are given for any station at 3000 feet as you can see.)

AT 6000 feet MSL at Nashville (BNA) the wind is forecast to be from 220° true at 46 K and the temperature is +5°C.

At Dubuque (DBQ) the forecast wind at 30,000 feet is from 740° at 4 K and . . . *hold it!* The compass only includes 360°, but this is no misprint. When the wind is over 100 K, 50 is added to the direction (this way they can keep the number of digits down to six for wind and temperature), and 100 is subtracted from the wind velocity, given as 4 (04) K here. So you would subtract 50 from the 74, which gives a direction of 24 (240° true), and add 100 to the force to get 104 K. The temperature is a cool –48°C at that altitude.

■ Winds Aloft Chart

The National Meteorological Center plots and transmits wind data obtained from observations and is given as 12-hour prognostics twice daily.

The winds aloft chart is a map of the continental United States for each particular altitude (or pressure level) of interest wih an "arrow, barb, and pennant" presentation at selected stations to show the wind direction and velocity and the temperature (in degrees Celsisus) for that altitude.

Figure 22-20 shows the "arrow, barb, and pennant" idea. Each pennant is 50 K, a full barb is 10 K, a half barb is 5 K. Remember that winds aloft are given in *true* directions and knots.

The numbers associated with the arrows are the middle digit of the direction from which the wind is coming. The "6" in the left-hand arrow system in Figure 22-20 confirms that the wind is from 260° true. (You could guess by the arrow that it is west-southwest but now know it's 260°.) The right-hand arrow in

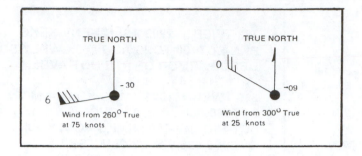

Fig. 22-20. Barb and pennant presentation for the winds aloft chart. The wind is blowing the way the arrow is going. The temperatures are –30°C and –9°C. The levels covered for the winds aloft charts are (thousands of feet MSL) 6, 9, 12, 18, 24, 30, 34, and 39.

the figure has a "0" noting that the wind is from 300° true.

PILOT REPORTS

One of the most valuable sources of information is pilot reports, or PIREPs. These are on-the-spot reports *in the air* and are not forecasts or information based on observations from the ground. Figure 22-21 is an actual PIREP. One way to remember the symbol for pilot reports is that it is an *Up* in the *Air* (UA) report. UUA is an *urgent* pilot report.

OTHER SERVICES

There are some other weather services available to you:

1. *PATWAS*—Pilots Automatic Telephone Weather Answering Service. The telephone numbers are listed in the *Airport/Facility Directory*, and by dialing you can get a transcription of current and forecast weath-

---

ICT UA/OV SLN /TM 0450 /FL060 /TP C172 /TA +10 /TB MDT AT 040 LGT BLO 030 /RM at 30 TEMP +2

Fig. 22-21. A pilot report (UA) released by Wichita (ICT). The aircraft was over Salina (SLN). Time (TM) is 0450, flight level (FL060), TP (type) Cessna 172 (C172), temperature of the air (TA) +10˚C. There is no report of sky cover (SK), flight visibility and weather (WX), or wind velocity (WV). The pilot does report turbulence (TB) as being moderate at 4000 feet (MDT AT 040) and light below 3000 feet (LGT BLO 030). RM stands for remarks. The temperature at 3000 feet is +2°C (at 30 TEMP +2). An *Urgent* pilot report is UUA.

er. If you require more specific information, a forecaster is available.

2. *Transcribed weather broadcasts*—To repeat, certain LF/MF radio facilities throughout the country broadcast current weather, winds, forecasts, and PIREPs continuously plus forecasts updated every few hours. The printed version was discussed earlier.

3. *FLIGHT WATCH* (En Route Flight Advisory Service—EFAS)—This service will provide timely weather information *in flight* and is available throughout the conterminous United States along prominent and heavily traveled flyways. You should be able to get the service at 5000 feet above the ground level within 80 miles from a FLIGHT WATCH outlet. Selected Flight Service Stations will provide the service, using remote communications facilities, as necessary. (FLIGHT WATCH is not intended for flight plan filing or position reporting, but for weather information only.)

FLIGHT WATCH will use 122.0 MHz and you would call the particular FSS controlling the FLIGHT WATCH in the area. ("CHATTANOOGA FLIGHT WATCH. THIS IS ZEPHYR SIX FIVE FOUR FOUR TANGO," etc.) If you don't know which FSS controls the service, just call FLIGHT WATCH and give your position relative to the nearest VOR; they will contact you.

Figure 22-22 shows the EFAS facilities in the southeastern United States.

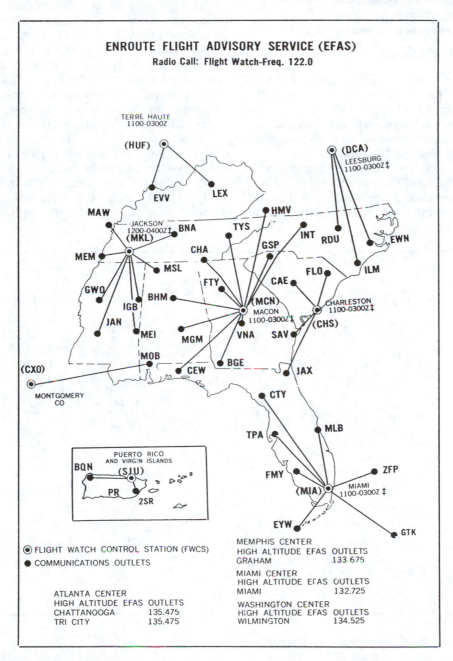

Fig. 22-22. En Route Flight Advisory Service stations in the southeastern United States. *(Airport/Facility Directory)*

4. *Hazardous In-Flight Weather Advisory Service (HIWAS)*. This is a continuous broadcast of in-flight weather advisories including AWW (Severe Weather Watch Bulletins and Alerts), SIGMETs, CWAs (Center Weather Advisories), Convective SIGMETs, AIRMETs, and urgent PIREPs. HIWAS is *not* a replacement for preflight or in-flight briefings or real time updates from Flight Watch (EFAS).

These other services are covered briefly so that you will have an idea of what is available. As you gain experience you'll make more and more use of these aids.

WEATHER CHARTS

■ Surface Analysis Chart

The surface analysis chart shows weather at various stations, and a date-time group (ZULU) indicates the time of the observation (valid time).

The isobars, pressure systems, and fronts are indicated on the chart as mentioned earlier in the chapter and as shown in Figures 22-2 and 22-3.

This chart is a good first reference when you go to a Weather Service Office of FSS for a check of the general layout of the weather. Since weather in the United States moves generally from west to east, you can check fronts or pressure systems to the west of your route and get a general idea of what might be a factor later.

After you've seen the chart, take a look at the hourly sequence reports along the route and compare them with what had been forecast. The presence of a cold or warm front nearly always means some kind of ceiling and visibility restriction, but occasionally "dry" fronts will move through an area. By looking at the chart only you might decide to "cancel because of the front."

Remember that the chart may be 2 or 3 hours out of date before *you* get to it, and you'll have to take the weather movement into consideration.

Figure 22-23 is redrawn from an actual surface analysis chart. The solid line isobars are spaced at 4-millibar intervals, but if the pressure gradient is weak, dashed isobars are put in at 2-mb intervals to better define the system. A "24" is 1024.0 mb and "92" stands for 992 mb. The Highs and Lows have a two-digit underlined number ("22" means that the pressure at the center is 1022 mb, etc.). Look back at Figure 22-2 for color presentations of fronts and squall lines.

A three-digit number is placed by a frontal system on the chart to show type, intensities, and character of the front. For instance, "453" by a front means that it is a cold front at the surface (4), moderate, little, or no change in intensity (5), with frontal activity in-

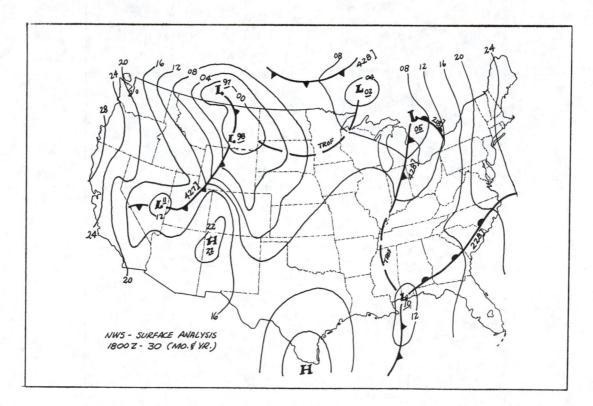

Fig. 22-23. The major value of the surface analysis chart is to give an overall view of fronts and pressure systems. It's issued every 3 hours. *(Instrument Flight Manual)*

creasing (3). These codes are in the Weather Service Offices so don't try to memorize them.

On the chart are "station models" (not included here) showing sky coverage, temperature, dewpoint, and other information. You'd be better off to get this information on the latest hourly report for a particular station, since the map may be a couple of hours old when you see it.

■ Weather Depiction Chart

This chart is useful for getting a good look at restrictions to visibility and low ceilings that would not actually be shown on the surface analysis chart.

Cross-hatched or shaded areas indicate IFR conditions (ceiling less than 1000 feet and/or visibility less than 3 statute miles). MVFR areas are enclosed within smooth contoured lines and indicate weather conditions of ceilings between 1000 and 3000 feet and visibility of 3–5 statute miles. No contours indicate VFR areas with ceilings greater than 3000 feet and visibility greater than 5 statute miles.

Figure 22-24 is a sample weather depiction chart,

showing areas of IFR and MVFR weather. Only a few of the reporting stations are shown for example purposes.

Station (1) is overcast as shown by the solid black circle and the visibility is 5 (statute) miles in smoke (5 K). The ceiling is 1400 feet (MVFR).

Station (2) is overcast at 1300 feet with a visibility of 6 miles in smoke (MVFR).

Station (3) has a ceiling of 600 overcast, visibility of 1 mile in smoke, which puts it properly in the IFR boundary.

Station (4) has a broken layer at 10,000 feet and the visibility is *more* than 6 miles, since it is not reported at the station. (Visibility is reported if it is 6 miles or less.)

Station (5) is clear with a visibility over 6 miles.

Station (6) has scattered clouds at 25,000 feet, visibility more than 6 miles.

■ Radar Summary Chart

Figure 22-25 is part of an actual radar summary chart. The shading shows radar echo areas and the

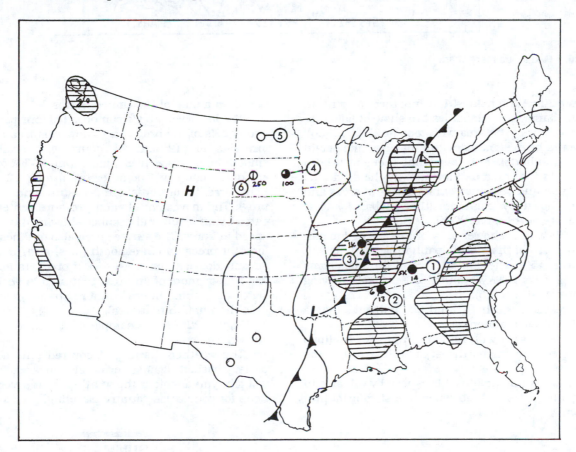

Fig. 22-24. Weather depiction chart. (A) Shaded areas—IFR with ceilings less than 1000 feet and/or visibility less than 3 statute miles. (B) Contoured without shading—VFR areas with ceilings greater than 3000 feet and visibility greater than 5 miles.

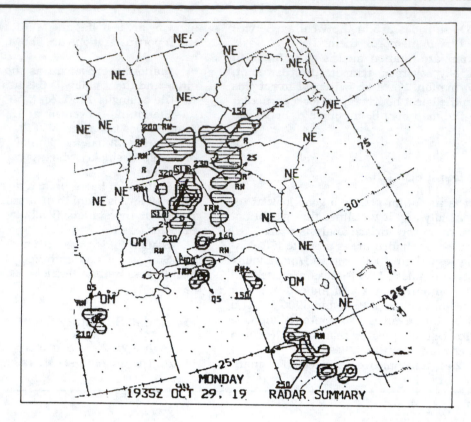

Fig. 22-25. Radar summary chart.

contours outline intensities. For instance, over north central Alabama, the intensities are show building toward the center of the area with tops at 32,000 (320). Bases would be designated with a bar *over* the height value (in hundreds). Note also that there may be a squall line in that area as indicated by the solid line. The direction and velocities of movement of areas and lines are shown by pennants (that Alabama line is moving east-southeast at 20 K).

There is a rain shower area of increasing intensity (RW +) south of the Florida panhandle, with tops at 15,000 feet (150). There is an area of thunderstorms and rain showers (TRW) south of Pensacola moving south (arrow) at 5 K (05).

NE means "no echoes" and OM indicates that a station is out for maintenance.

There is an area with rain and rain showers (tops at 23,000 feet) moving northeast at 25 K (arrow) in the Atlanta area.

The radar summary chart is especially valuable in showing intensities and movements of significant precipitation.

■ Weather Prognostic Chart

The low-level prognostic chart would be of the most value to you as a student or new private pilot. Like the other charts, the weather areas are superim-

posed on a map of the United States.

The low-level prog is composed of four panels. The two panels of the prog chart show *signficiant weather* forecasts for 12 and 24 hours with scalloped or smooth lines, respectively, indicating MVFR or IFR areas of the country. The other two are for 12- and 24-hour forecasts of fronts and pressure centers (surface prog). The movement of each pressure center is indicated by an arrow and number showing direction and speed in knots. The surface prog also outlines areas of forecast precipitation (smooth lines) and/or thunderstorms (dot-dash lines ·--·--·). If precipitation will affect half or more of an area, that area is shaded. Absence of shading indicates more sparse precipitation (less than half in that area).

Figure 22-28 is a sample of the charts just discussed.

The weather charts just covered plus actual reports, checked against terminal and area forecasts, will give you a look at the weather from several directions for your cross-country planning.

SUMMARY

Your probable first reaction to this chapter on weather is that it is impossible to know everything about what is available to you as far as weather ser-

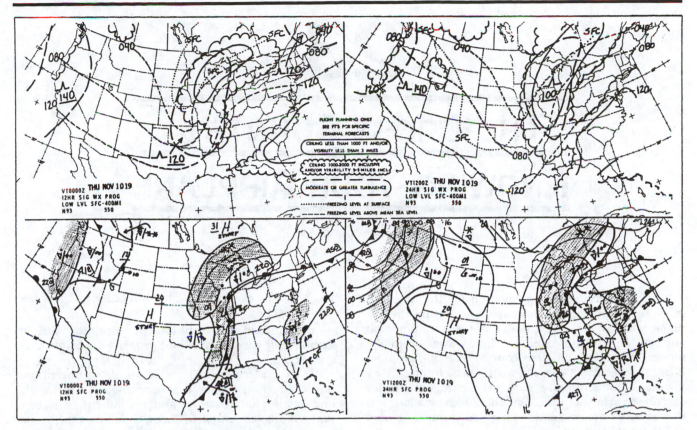

Fig. 22-26. U.S. low-level significant weather prognostic chart (surface—400 millibars). *(Aviation Weather Services)*

vices are concerned. You don't have to remember that the terminal forecast is available at such and such a time. You do need to have some idea that sequence reports, winds aloft forecasts, area forecasts, and terminal forecasts are *available* to you through either a National Weather Service Office or FSS. Tell them that you're a (student) pilot and you're planning to fly from A to B, leaving A at such and such a time and expecting to get to B at about so and so. Will they give you the word? The specialists will help every way they can, so don't expect to understand weather facilities like those who spend 8 hours a day working at it.

This chapter is aimed for practical use by you as a student or noninstrument-qualified private pilot. To include more weather theory and/or high-altitude information would be like shooting flies with a shotgun.

The weather information sources and NOTAM services are in a continuing state of flux. Your flight instructor will be the major source for the latest changes.

If you are in doubt about the weather before takeoff—don't go. If you are in doubt about the weather ahead when you're flying—make a 180° turn before you are also in doubt about the weather behind you.

23

THE CROSS-COUNTRY—
KNOWING YOUR AIRPLANE

Before planning the navigation for the trip, you must be introduced to facts about operating the airplane, which will be of particular importance now that you will be making extended flights away from the home airport. On the cross-country you will be more likely to encounter adverse weather, including flying in areas of moderate to severe turbulence, perhaps for extended periods of time. You should know about setting exact power and conserving fuel, a minor or non-existent problem for local flying.

When you are off by yourself at a strange airport, you'll have added responsibility and will have to know more about your airplane systems than is needed at home base where your instructor is available to answer questions.

The following information is both for your present cross-country planning and for future reference. For instance, it's unlikely that you'll have problems with weight and balance of the airplane on your solo cross-country. But later, after you get that private certificate and decide to fly on vacation with your family or friends, and their baggage, knowledge of this area of flying will be of vital interest to you.

AIRPLANE PAPERS

In Chapter 3 you noted that the airplane must have certain documents on board at all times. Make sure that these are in the aircraft before you leave the home airport.

1. The *Certificate of Aircraft Registration* must be current.

2. The *Airworthiness Certificate* must also be *displayed* so that it can be easily read. Your airplane will have a *Standard Airworthiness Certificate*, because it will meet either "normal" or "utility" airworthiness requirements. (Airplanes can meet both requirements, but this will be covered a little later.) "Aerobatic" and "transport" category airplanes also have a Standard Airworthiness Certificate, but it's un-

likely that as a student pilot you'll be flying these last two categories.

A Standard Airworthiness Certificate is used in aircraft that carry people and property for compensation or hire.

3. The *Airplane Flight Manual*, or *operations limitations form*, must be aboard; the airplane weight and balance information (and *equipment list*) are a part of these documents. Check to see that the *major repair and alteration forms* (if any) are with the weight and balance information. These forms show major repairs. Their effects on empty weight and center of gravity of the airplane are indicated in the aircraft logbook.

Since you will be flying away from the home base, you'd better have the *aircraft and engine logbooks* along. (There's a separate book for the airframe and each engine.) *Make sure that the logbooks are up-to-date.* Your instructor, however, may prefer to keep them at the home base, with the understanding that the books can still be available to FAA or NTSB representatives within a reasonable time.

AIRPLANE CATEGORIES
AND LOAD FACTORS

The airplane you will be flying will be of the normal or utility category, as mentioned in the last section. The primary difference between the two categories is that the normal category airplane normally has "limit load factors" of 3.8 positive g's and 1.52 negative g's, and the utility category airplane has limit load factors of 4.4 positive and 1.76 negative g's. Intentional spins are prohibited for normal category airplanes but may be done in *some* airplanes in the utility category. (Some airplanes are licensed as being "characteristically incapable of spinning.")

The term "g" is a unit of acceleration based on that of gravity, as discussed in Chapter 9 in "The

Turn." The airplane in straight and level flight and un-accelerated glides and climbs has 1 g acting upon it (and so do you). In the utility category airplane you can create positive acceleration forces up to 4.4 times its weight without exceeding safe limits (if the airplane hasn't been previously damaged). If you should encounter an upward gust during your 4.4-g maneuver, the lift is increased even more sharply, and instead of 4.4 g's you may suddenly have a load factor of 6 or 7. Airplanes get bent this way.

Some airplanes are certificated in both categories. At the maximum certificated weight the airplane is in the normal category and must be held to a load factor of 3.8 positive and 1.52 negative g's, or less. At a specified lighter weight it can be flown as a utility category airplane and can safely cope with a load factor of up to 4.4 positive g's (and 1.76 negative g's). This can be seen by an example: One airplane having a maximum certificated weight of 2300 pounds is classed as normal category at that weight; when it is flown at a reduced weight of 2000 pounds, it can be operated as a utility category airplane. The maximum force acting on the wings would be approximately the same in both cases. At a weight of 2300 pounds a load factor of 3.8 g's (normal category limit) would be 3.8 × 2300, or 8740 pounds. At 2000 pounds a load factor of 4.4 g's (utility category limit) would result in a force of 8800 pounds. Very nearly the same force would be supported by the wings, and 8800 pounds would be the maximum they would be expected to carry without showing signs of permanent deformation. Not all airplanes would work out this close in total force but would have been checked out to sustain the *maximum* force expected from either of the two categories. The wings are not the only structures subject to failure; the horizontal tail or other components might break first if you exceed the limit load factor.

The load factors just mentioned are the *total* g's imposed. You start at 1 g, so only 3.4 g's more are needed to reach the positive limit load factor of 4.4. As far as the negative load factors are concerned, the normal positive 1 g is to the good; you'll have to go farther to get to the negative limit load factor. The airplane can go up to 2.76 g's on the negative side (from 1+g) without permanent deformation.

Knowing that vertical gusts (up or down) can impose g loads on the airplane, you should be aware of several points. Take a look at your airspeed indicator markings and review the discussion of the airspeed indicator in Chapter 3. The green arc is the *normal operating range,* and the lower limit is the flaps-up, power-off stall speed at maximum certificated weight. The upper limit (where it meets the yellow) is called the *maximum structural cruising speed.* The yellow arc is the caution range and, of course, the red line speaks for itself. Even if the air is only mildly turbulent, keep the airspeed in the green arc. Although your airplane at normal cruise will be indicating in the green, on letdowns you will have a tendency to let it slip up into the yellow, and the sudden encountering of turbulence at those speeds could result in excessive load factors. Figure 23-1 shows what happens when an up-gust is encountered.

In (A) in Figure 23-1 the airplane is flying in normal cruising flight at a certain required angle of attack. (You wouldn't know the angle but would have the proper cruise power setting and would be maintaining a constant altitude.) In (B) a sharp up-gust of 20 K is encountered. The vector diagram in (B) shows that the angle of attack has been sharply increased and lift is at a much higher value than normal. *A high positive load factor is imposed.* A measure of the load factor is the ratio of lift to weight. You remember that in the steep turn at 60° of bank you had to double the lift in order to maintain a constant altitude, which resulted in a load factor of 2.

The airplane would quickly return to the original lift-to-weight ratio of 1, since drag would also rise sharply, but in the meantime the load factor would have already been at work.

Figure 23-2 shows what would happen if the airplane was flying at a much lower airspeed when it encountered the same 20-K upward gust.

You'll note in (A) that angle of attack is higher to maintain this lower airspeed of 70 K. (The lift-to-weight ratio is still 1, or lift equals weight, as before.) At (B) the 20-K up-gust results in the stalling angle of attack being reached; the lift starts to increase radically, but the airplane stalls before the limit load factor is reached. The gust in both cases would have the

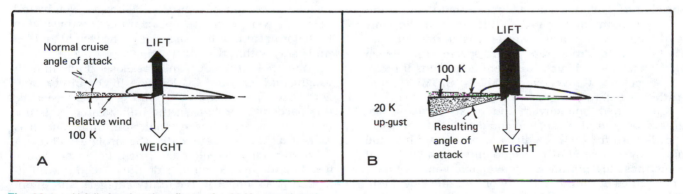

Fig. 23-1. Airfoil showing the effects of a 20-K upward gust on an airplane cruising at 100 K.

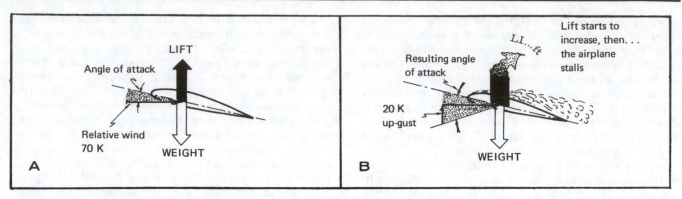

Fig. 23-2. Effects of a 20-K upward gust on the airplane cruising at 70 K.

same effect—as if you had suddenly increased back pressure. The stall resulting from the situation in Figure 23-2 would be momentary, and the airplane would soon return to equilibrium. (The stall and recovery would be so fast that it would be unlikely that you would have time to even start your own recovery procedures.) Such a stall situation *would* be very bad on a low, slow approach on a gusty, turbulent day. You would make the approach several knots faster to lower the chances of an inadvertent stall at a critical point. This was also covered in "Gusty Air and Wind Gradients" in Chapter 13. Because of the comparatively low speeds used in the approach, the stall, rather than high load factors, is the problem encountered. At cruise and higher speeds the possibility of overstress is the major factor; as everyone will agree, it's better to have a temporary stalled condition than to overstress the airplane, so in very turbulent air you would slow the airplane to assure that it would stall before this could occur. The speed that divides the stall area from the area of possible overstress is called the "maneuvering speed" and has the designation V_A. The maneuvering speed varies from approximately twice the stall speed, as given by the gross weight calibrated stall speed (at maximum certificated weight), to about 1.6 times that figure at minimum flyable weight for most lightplanes. As an all-around figure when encountering moderate to severe turbulence, slow the airplane to indicate 1.6 times the calibrated stall speed *at all weights*. This will mean that the airplane will stall momentarily rather than exceed the limit load factor; it will also mean an easier ride for you. Pilots' ideas of "moderate" and "severe" turbulence vary widely. Avoid such turbulent conditions if possible—check for Convective SIGMETs, SIGMETs, and AIRMETs before you go.

If you had "maneuver" and "gust" envelope diagrams for your airplane, you could pick out a range of speeds within which the airplane would neither stall nor be overstressed by expected maximum gusts, but since these are not normally available you may have to use rules of thumb and stay on the safe side. Vertical gusts don't just affect the airplane in cruise; they work during climbs, glides, or any other maneuvers.

■ **Summary**

1. In light turbulence or if you expect that you may encounter it, keep the airspeed in the green arc.

2. In moderate to severe turbulence slow the airplane to 1.6 times the stall speed as given in the *Pilot's Operating Handbook* (calibrated airspeed). If you have more detailed information from the manufacturer on gust penetration speeds, use that (calibrated airspeed). Remember that the 1.6 figure is aimed only at assuring that the airplane will stall rather than be overstressed. Stalls at low altitudes are dangerous also.

3. Check Convective SIGMETs, SIGMETs and AIRMETs, and avoid areas of turbulence if possible.

WEIGHT AND BALANCE

It's very important that you have an understanding of airplane loading. Some pilots have only a hazy conception of what can happen when the maximum allowable weight and/or center of gravity limits are exceeded.

The "center of gravity" (CG) is the point at which the airplane's entire weight is assumed to be concentrated. The manufacturer sets limits on weight *and* the CG location so that the airplane will not be in a dangerous condition. Figure 23-3 shows a weight and balance envelope for a two-place trainer certificated in both normal and utility categories. The weight and CG must stay within the envelope.

You see that the airplane has closer limits in both weight *and* rearward CG location in the utility category. The less rearward CG limit is present because this particular airplane in the utility category is *not* restricted against intentional spinning. The less rearward CG limit assures that the airplane will retain good spin recovery characteristics. If the airplane's weight and balance fall outside the dashed line envelope, it is no longer in the utility category and intentional spins are prohibited. Figure 23-3 is taken from the airplane's *weight and balance form*. (The arrow

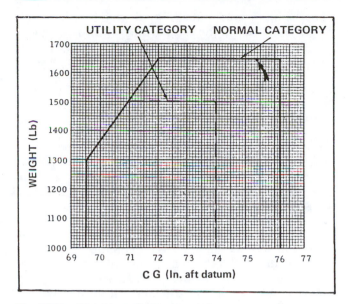

Fig. 23-3. Weight and balance envelope.

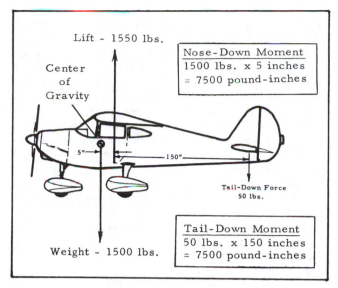

Fig. 23-5. Lift and weight forces and moments acting on an airplane in flight at a particular weight and CG position.

shows a point to be discussed later.)

The term "moment" will become important when you start working with airplane loading. A moment is a measure of a force (or weight) times an arm and is normally expressed in pound-inches or pound-feet. For airplane weight and balance computations moments are discussed in terms of pound-inches. Figure 23-4 shows how two equal moments acting in opposite directions can cancel each other so that equilibrium exists.

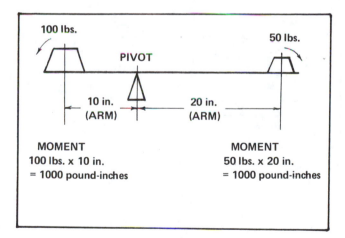

Fig. 23-4. A system of two equal moments.

You've applied the principle of moments by using a long pole in levering. (Automobile bumper jacks use this principle.) Figure 23-5 shows the lift and weight moments acting on an airplane in straight and level cruising flight.

The forces and moments must be in balance for equilibrium to exist. A tail-down force normally exists in flight. To simplify matters, rather than establish the

moments from the center of gravity (which is the usual case), for this problem they will be measured fore and aft from the center of lift. Assume for now that the lift force is a string holding the airplane "up"; its value will be found later.

The airplane of Figure 23-5 weighs 1500 pounds and the CG is 5 inches ahead of the center of lift. This results in a *nose-down* moment of 7500 pound-inches (5 × 1500). To balance this, a moment acting in the opposite direction is required and will be furnished by a tail-down force. The center of "lift" (down-acting) of the horizontal tail is 150 inches behind the wing center of lift, so in order to have an equal *nose-up* moment, a tail-down force of 50 pounds is necessary (50 × 150 = 7500 pound-inches). The airplane nose does not tend to pitch either up or down; the weight and tail-down moments are balanced. This is a simplified approach; there are other moments in existence (the wing has a moment of its own, which was neglected here). The tail-down force varies with CG position and/or airspeed.

Earlier in this book it was stated that in straight and level flight lift equals weight. This was done to keep from complicating the issue in the introduction to flight and the Four Fundamentals. The *vertical* forces working on the airplane must be equal. Looking at Figure 23-5, you'll note that the down forces are weight, 1500 pounds, and the tail-down force, 50 pounds. In order to fly straight and level, the total up force (lift) must equal these down forces. Hence lift must be 1550 pounds, not 1500 pounds as would be expected from the weight of the airplane. But in flight you don't worry about this. You make the airplane do its job by varying the tail force as necessary to obtain the desired angle of attack—and airspeed. You use trim or wheel pressure as necessary (plus power) to attain the end performance. There are thrust and drag moments also existing in flight, but these are comparatively minor and were neglected for this problem. So as far as flying the airplane and the g-force discussion are concerned, the tail-down force could be considered as weight and is important to the stability of the plane.

The position of the CG has an effect on the stall speed for a given weight because the higher the tail-down force (as would be required at a forward CG position), the greater the total load carried or the more the airplane "weighs." Look at Figure 23-5 to see that if the CG is moved far enough forward so that the tail-down force increases to, say, 150 pounds total, the wings would have to support 1650 pounds, not 1550. The stall speed increases as the square root of the "weight" increases (actual or aerodynamic down load, as is the case for the example here). The stall speed marked on the airspeed indicator is based on maximum certificated weight at the most forward CG that would give the highest stall speed, a conservative approach.

If you place a lot of weight too far aft, a condition known as "longitudinal instability" could arise. First, look at the correctly loaded airplane; you know that the airplane, as you have flown it to date, has been inherently stable. That is, you can trim it at airspeeds from just above the stall to just below the red line and it tends to stay at that speed. You also know that if you trim it for cruise, for instance, slowing it down requires more and more back pressure as the speed decreases. Conversely, to increase the speed above the trimmed speed requires more and more forward pressure. This is the "weigh" it should be.

If you load it so that the CG is too far aft, the required tail-down force decreases. If you move the CG to a certain point, the requirement for a tail-down force may disappear completely. In effect, the airplane's nose may be displaced up or down with little or no effort on your part. In extreme cases, if offset by bumpy air, the nose could pitch up (or down) at an increasing rate, and an extremely dangerous situation would be in effect. In short, the airplane would *not* tend to return to its trim speed and would not want to maintain any particular airspeed.

If you observe the rearward CG limits as shown on the weight and balance envelope of your airplane, you'll avoid this condition.

Notice in Figure 23-3 that there is a forward CG limit on the envelope; this is to be observed also. However, just from the general characteristics of most light trainers, this one is sometimes hard to exceed, simply because there's just too little space up front to put baggage and other excess weight. (However, there are always those who will manage to put an anvil or two under the pilot's feet or pull some other unlikely stunt.) The main reason for the forward CG limit is that of control rather than stability. The CG may be so far forward that you don't have enough up-elevator to get the nose up to the proper position on landing—and end up bending something.

You'll notice that the airplane's baggage compartment is placarded for maximum weight. This is done for two reasons: (1) to prevent the center of gravity

from being moved outside approved limits, which could be the case if the compartment was overloaded, and (2) structural considerations. The baggage compartment floor will be able to take its placarded weight up to the positive limit load factor, but suppose instead of putting in 100 pounds as placarded, you put in *200 pounds*. This would mean that at 3.8 g's *760* pounds would be exerted instead of the 380 pounds it was designed to take. You just might have a new observation window where the baggage compartment floor used to be. *Observe all placards in the airplane.*

■ Running a Weight and Balance

The next step is to assure yourself that your airplane will stay within approved CG limits.

The "datum" is the point from which all measurements are taken for weight and balance computations. It may be a point on the airplane or it may be well in front of the airplane. Figure 23-6 shows a high-wing trainer with the datum located 60 inches in front of the wing leading edge. It is standard practice to establish the datum at a fixed distance ahead of a well-defined part of the airplane, or in some cases the front side of the firewall may be used. For trainers so equipped, the reference point is the straight leading edge.

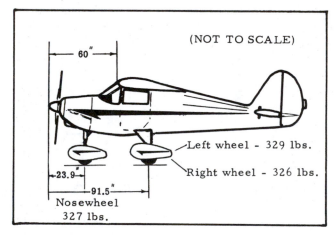

Fig. 23-6. The datum, with weighing point locations and weights. The arms have been rounded off here.

Suppose that the empty weight CG is required. This is determined initially at the factory but may change if equipment is added. In such an event it would be duly noted in the logbook by the mechanic who makes the change.

The airplane is weighed in a level attitude with a scale under each wheel. The illustrated airplane's empty weight is a total of the three scale weights, or 982 pounds. The empty weight CG is found by using the datum and summing up moments.

The nosewheel center line is 23.9 inches aft of the datum, so its moment is $23.9 \times 327 = 7815$ pound-inches. The two main wheel center lines are 91.5 inches aft of the datum, so 91.5 × (329 + 326) = 91.5 inches × 655 pounds. The moment of the two main wheels is 59,932 pound-inches. The total moment is 59,932 + 7815, or 67,747 pound-inches. To find the position of the CG, the total moment is divided by the total weight, which gives the position where the *center of the total empty weight is acting to create the total moment.* This works out to be 69 inches (rounded off) (67,747 ÷ 982). (Actually, if you worked it out on a calculator, the answer would be 68.988798 inches, but you don't measure points that closely.) Hence, the empty center of gravity is 69 inches aft of the datum, or 9 inches behind the leading edge of the wing. The basic empty weight of an airplane includes fixed ballast, unusable fuel, full oil, full engine coolant (naturally, not applicable to air-cooled engines), and hydraulic fluid. The unusable fuel for this particular airplane was negligible and did not add any weight to the empty weight "as weighed." If the airplane had not yet been painted at the time of the weighing, the weight of the paint (this may be from 5 to 20 pounds, depending on the size of the airplane) plus unusable fuel (if applicable) would have to be added to get the basic *empty weight* and CG. This is a favorite subject for questions on the Private Pilot Knowledge Test, and the student usually forgets that *unusable fuel and full oil are included in the basic empty weight.* Item 1 in Figure 23-7 gives the basic empty weight. Included in that item is 3.5 gallons of unusable fuel (the total tank capacity is 26.0 gallons with 22.5 usable).

Now you could compute the effects of adding fuel, pilot, passengers, and baggage. The following would be a computation as given in the example airplane's weight and balance form. The arms here are rounded off to the nearest inch and would be given on the form.

Item	Weight		Arm (aft of datum)		Moment (pound-inches)
Basic empty weight (includes full oil)	982	×	69	=	67,747
Usable Fuel (36 gal)	216	×	84	=	18,144
Pilot	180	×	81	=	14,580
Passenger	160	×	81	=	12,960
Baggage	100	×	101	=	10,100
	1638				123,531

(You may find a slight error in the total pound-inches because the empty CG was rounded off.)

Again, dividing the *total moment* (123,531 pound-inches) by the *total weight* (1638 pounds), it is found that the CG is located at 75.4 inches aft of the datum. Looking back to Figure 23-3, the weight and balance envelope for this airplane, you see by the dot that the CG is within limits for the *normal* category (you have 12 pounds to spare in weight and about 0.8 inch in CG range). The airplane as loaded in this problem is outside the utility category envelope, so intentional spins would be prohibited.

You could work out combinations of pilot, fuel, baggage, and passengers to allow the airplane to be operated in the utility category.

The method used earlier to find the empty weight CG is not the shortest way to arrive at the proper answer; the weight and balance form will give simple preset equations to arrive at the answer. The method used here shows the principle. (Also, some of the figures were rounded off for simplicity.)

The FAA considers fuel to weigh 6 pounds per gallon and oil 7½ pounds per gallon. In practice you'd use actual passengers' weights to be accurate. (Natu-

SAMPLE LOADING PROBLEM	SAMPLE AIRPLANE		YOUR AIRPLANE	
	Weight (lbs.)	Moment (lb.-ins. /1000)	Weight (lbs.)	Moment (lb.-ins. /1000)
1. Basic Empty Weight (Use the data pertaining to your airplane as it is presently equipped. Includes unusable fuel and full oil)	1125	36.6		
2. Usable Fuel (At 6 Lbs./Gal.) Standard Tanks (22.5 Gal. Maximum)	135	5.7		
Long Range Tanks (35 Gal. Maximum)				
Reduced Fuel (As limited by maximum weight)				
3. Pilot and Passenger (Station 33 to 41)	340	13.3		
4. Baggage - Area 1 (Or passenger on child's seat) (Station 50 to 76, 120 Lbs. Max.)				
5. Baggage - Area 2 (Station 76 to 94, 40 Lbs. Max.)				
6. TOTAL WEIGHT AND MOMENT	1600	55.6		
7. Locate this point (1600 at 55.6) on the Center of Gravity Moment Envelope, and since this point falls within the envelope, the loading is acceptable.				

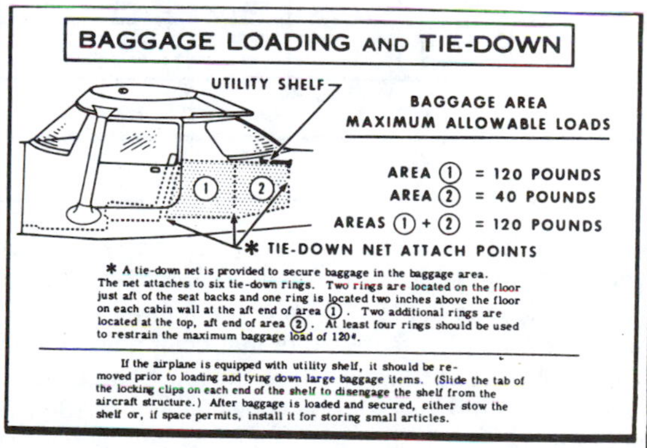

Fig. 23-7. A sample loading problem. If the airplane had been weighed at the factory without full oil or unusable fuel, this would have been "added" mathematically to get the basic empty weight of 1125 pounds.

rally, you'd better not be flying nonrated persons until you get that private certificate.)

Other airplanes use a slightly different approach to the weight and balance problem. Figures 23-7, 23-8, and 23-9 show the steps as would be applied to a Cessna 150.

The sample problem in Figure 23-7 shows that the loaded airplane weighs 1600 pounds and has a moment of 55,600 pound-inches. This would place it in the envelope, as shown in Figure 23-9.

Note in Figure 23-7 that the total baggage capacity of the airplane is 120 pounds. You can either put it all in area 1, or 80 pounds in area 1 and 40 pounds (*its* maximum allowable) in area 2.

Figure 23-8 shows how the various moments were obtained. The shallower the slope of the line, the more effect each pound has in increasing the moment. As an example, 100 pounds of baggage in area 1 furnishes the same moment (6500 pound-inches) as a 167-pound pilot or front-seat passenger.

Don't try to beat the system by overloading the airplane (and having it in a critical aft CG condition)

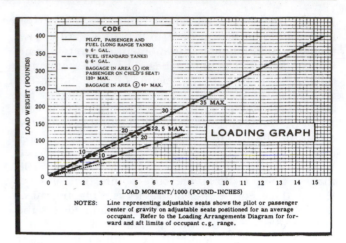

Fig. 23-8. A loading graph.

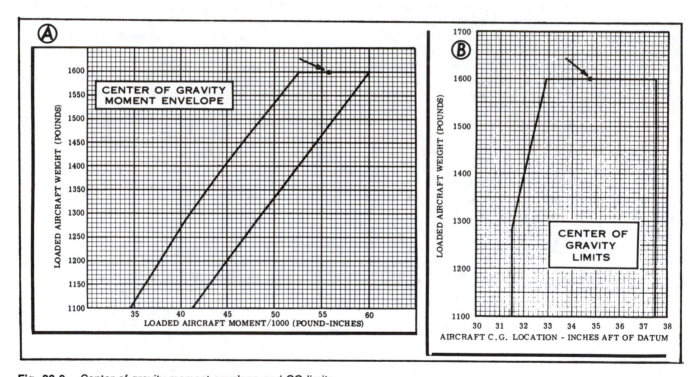

Fig. 23-9. Center of gravity moment envelope and CG limits envelope for the sample airplane in Figure 23-7. The datum for that airplane is the front face of the firewall. In this arrangement weights rearward of that point have "positive" arms; weights forward would have "negative" arms and this is basically the same idea as that shown in Figure 23-4. The total of the "negative" moments would be subtracted from the total of the "positive" moments to get some answer with the CG as (A), a positive moment acting "behind" the datum, or (B), the weight and CG in inches aft of the datum (firewall) as shown by the arrows in the two envelopes. To find the location of the CG, divide the total moment (55,600) by the total weight (1600 pounds) as indicated by the arrow (B) to get a position of 34.75 inches aft of the datum.

and thinking that you'll "just get it a foot or so off the ground and if it doesn't feel right, I'll put it back on." An interesting phenomenon is that an airplane for a given weight and CG position is more stable in ground effect; that is, it doesn't want to nose up abruptly as it might up at altitude at a rearward CG. Also, as discussed in Chapter 17, ground effect starts to make itself known on approach and landing at a height of about one-half wingspan above the ground and becomes stronger as the airplane descends. The "downwash" from the wing is altered in ground effect; the direction of the flow at the horizontal tail is changed and the effect is slightly more of a tendency for the nose to stay lower, or the airplane feels more stable. This could fool a pilot who is *taking off;* the airplane may be marginal as far as stability is concerned in the ground effect; but as it moves upward after lift-off, the stabilizing influence is lost, the nose rises higher and higher, and a stall—and accident— occurs. This problem often occurs when people are leaving on vacation and "need" all that gear in the back seat and baggage compartment.

Also, you'll find that overloading affects takeoff performance radically. (A 10 percent increase in weight means a 21 percent increase in takeoff run distance.) The overloaded airplane manages to get off with the aid of ground effect, which reduces the induced drag, but is unable to climb more than a few feet. Obstructions at the far end of the runway could mean a hasty end to the flight. With the aid of ground effect the airplane's performance is marginal, and when it starts to move upward the increase in induced drag stops any further climb. Too many accidents are caused by the pilot trying to beat aerodynamic laws.

FUEL AND OIL INFORMATION

When you refuel at a strange airport, it will be your responsibility to see that the proper grade of fuel and oil are used.

■ Fuel

Never use fuel rated below the minimum grade recommended for your airplane. You may go above as a temporary arrangement, but even then, stay as close to the recommended grade as you can.

The numbers are the antiknock quality of the fuel—the higher the number, the better the antiknock qualities. For instance, 80/87: The first number (80) is the minimum antiknock quality of the fuel in the lean mixture; the last (87) is the minimum antiknock quality in the rich mixture.

Grade 80 is not available at some airports in the United States. Some operators are stocking only one grade, 100 Low Lead, for use by airplanes requiring as a minimum either 80/87 or the 100/130 grade higher-leaded fuel.

You'll hear aviation fuel referred to as, for instance, grade 80 or grade 100, without the qualifying second figure (80/87). This term use was recommended by the American Society for Testing and Materials (ASTM) for simplification purposes, but the rich and lean antiknock ratings still stand.

The various octanes and performance-numbered fuels contain dyes of different colors to aid in identification.

The colors are valid in the United States but may not be the same in foreign countries.

Grade	Fuel Color
80/87	Red
100LL	Blue
100/130	Green
115/145	Purple

In planning the cross-country, check in the *Airport/Facility Directory* to be sure that at least the minimum rated fuel for your engine is available at any place you intend to refuel.

■ Oil

The viscosity of oil will be indicated in one of two ways. You are probably most familiar with the SAE (Society of Automotive Engineers) numbers.

SAE Number	Commercial Aviation Number
40	80
50	100
60	120
70	140

Notice that in every case the commercial aviation number is exactly twice the SAE "weight." To get the SAE equivalent, divide the aviation number by two.

Know what type of oil your airplane has been using. You'll hear the terms "detergent" and "nondetergent," but the oil primarily used in airplane engines today is one of two types: (1) *mineral oil* (called nondetergent), which is plain oil with no additives, and (2) *ashless dispersant (AD) oil* (often mistakenly called "detergent"), which is mineral oil with additives that give better wear qualities and multiviscosity and also pick up small particles (contamination), keeping them suspended in the oil so that sludge and carbon deposits aren't formed. The old detergent oils could wash out carbon if added to an engine that had been using straight mineral oil and cause sludge problems in filters.

There are many multiviscosity oils available and your airplane may use one of these. Check before using cross-brands of this type of oil too.

Know and add *only* the brand and type of oil the

airplane has been using. Some brands or types may not mix, but some oil manufacturers say that adding straight mineral oil to AD oils is normally all right if you feel you really need the oil to get home safely.

CRUISE CONTROL

Probably the term "cruise control" brings to your mind a vision of a vast array of knobs and dials requiring continual monitoring on your part. Cruise control actually means that in cruising flight you know such things as the amount of power being carried and the fuel consumption. This section is based on the premise that your trainer uses a fixed-pitch prop.

Up to now you've probably paid little attention to what power has been carried or the specific fuel consumption, but you have used your instructor's recommendations for tachometer settings for various maneuvers; and, of course, you always checked to make sure you had more than enough fuel to make those local flights. (And if asked for the fuel consumption, you probably could answer vaguely "five or six gallons an hour, or somewhere in there.")

On extended flights it is vital that you *know* how much fuel you are using and how to plan use and conserve fuel if necessary.

Back in Chapter 5 the mixture control was introduced, but it could be that for local flying you haven't been checked out on mixture leaning techniques and have flown the airplane in the full-rich setting.

Figure 23-10 is a time, fuel, and distance to climb tabulation for a particular airplane. Pressure altitudes and standard temperatures are given. Suppose the elevation of your airport is 1000 feet (and the pressure altitude and temperature are standard). You want to depart on a cross-country to cruise at 5500 MSL. Your problem might be worked out like this: You would interpolate between 5000 and 6000 feet and get totals from *sea level* of 10 minutes, 1.45 gallons of fuel (call it 1.5 gallons), and a distance of 11.5 NM (call it 12 NM). However, you're starting from 1000 feet so would subtract those values (2 minutes, 0.2 gallons, and 2 NM) from the 5500 figure and get an answer of *8 minutes, 1.3 gallons, and 10 NM to climb from your airport at 1000 feet elevation to the cruise level of 5500 MSL.*

Note that for each 8°C above standard for the altitudes you would add 10 percent to the three values of time, fuel used, and distance.

The distances shown are based on zero wind conditions. Wind would only affect distance (assuming that turbulence isn't hurting your climb performance), since the airplane would require the same amount of time and fuel to get to a certain altitude. Headwind or tailwind components would affect the number of miles covered in that time. (More about that in Chapter 24.)

TIME, FUEL, AND DISTANCE TO CLIMB

MAXIMUM RATE OF CLIMB

CONDITIONS:
Flaps Up
Full Throttle
Standard Temperature

NOTES:
1.   Add 0.8 of a gallon of fuel for engine start, taxi and takeoff allowance.
2.   Increase time, fuel and distance by 10% for each 8°C above standard temperature.
3.   Distances shown are based on zero wind.

WEIGHT LBS	PRESSURE ALTITUDE FT	TEMP °C	CLIMB SPEED KIAS	RATE OF CLIMB FPM	FROM SEA LEVEL		
					TIME MIN	FUEL USED GALLONS	DISTANCE NM
1600	S.L.	15	68	670	0	0	0
	1000	13	68	630	2	0.2	2
	2000	11	67	590	3	0.5	4
	3000	9	66	550	5	0.7	6
	4000	7	65	510	7	1.0	8
	5000	5	65	470	9	1.3	10
	6000	3	64	425	11	1.6	13
	7000	1	64	385	14	1.9	16
	8000	-1	63	345	17	2.3	19
	9000	-3	63	305	20	2.7	23
	10,000	-5	62	265	23	3.2	27
	11,000	-7	62	220	27	3.7	32
	12,000	-9	61	180	33	4.3	38

Fig. 23-10.   Time, fuel, and distance to climb chart.

SETTING POWER

In Chapter 12 you had a brief encounter with cruise control when it was mentioned that airplanes cruised at 75 percent of the rated power or below (65 and 75 percent are the most popular settings and 65 percent is a lot easier on the engine, for only a very small loss in airspeed).

To review some theory of airplane performance: Figure 23-11 shows a performance chart for a light trainer with a fixed-pitch prop.

Point (1) is the top speed of the airplane. For airplanes with unsupercharged engines the *top speed is always found at sea level* because this is the place where the engine develops its maximum power. As altitude increases, the air becomes less dense and power is lost so that the maximum possible airspeed decreases. (You lose power faster than TAS is gained.)

At point (2) you'll note that the power has dropped off so that only 75 percent of the sea level power is being developed. Because 75 percent is the maximum legitimate cruising power setting, you'd get the most true airspeed per horsepower for extended operations at about 7000 feet (standard, or density-altitude). If your airplane has a fuel consumption of, say, 7 gallons

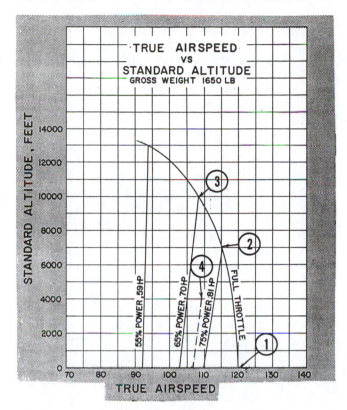

Fig. 23-11. True airspeed versus standard (density) altitude for a light trainer.

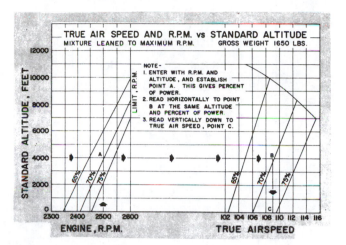

Fig. 23-12. True airspeed and RPM versus standard (density) altitude.

per hour at 75 percent power, you'll do better at that power by cruising at 7000 feet (115) than at sea level (110). This is, of course, taking winds, ceiling, and other outside factors into consideration.

If you are a believer in the use of 65 percent as a cruise power setting, you can see that point (3) is best for this setting as far as getting the maximum TAS is concerned. An altitude of 10,000 feet is the highest at which you will be able to develop 65 percent of the sea level power.

Notice that TAS is gained with altitude at 65 and 75 percent—until you run out of throttle. You'll have to open the throttle as you increase altitude to maintain the same percentage of power. (You'll need a higher rpm to get 75 percent at 7000 feet than at sea level.) Obviously, the first thought is that if you are carrying more throttle (rpm) you must be burning more fuel at the higher altitudes. That's where cruise control comes in. Because of the decrease in air density the mixture can be progressively leaned with altitude increase so that 75 percent power at the higher altitudes results in fuel consumption no higher than that same power setting at sea level.

Next comes the question as to how to obtain the various power settings; the *Pilot's Operating Handbook* will help you out here. Figure 23-12 shows a graphical method of setting up power for various altitudes.

If you wanted to carry 70 percent power at 4000 feet, it would require 2500 rpm and the airplane

would have a TAS as shown in Figures 23-11 and 23-12.

A careful look at Figure 23-12 shows that a power setting of 2410 rpm is required for 70 percent at sea level, and 2500 rpm at a standard (density) altitude of 4000 feet. This means an increase of 90 rpm, which works out to be approximately 1 percent increase in rpm per thousand feet. For airplanes having rpm settings for cruise at sea level of 2300–2500 rpm you could, as a rule of thumb, add 25 rpm per thousand feet to maintain the sea level percentage of power. As an example, if 70 percent power at sea level requires 2410 rpm, you'd be very close to the *Pilot's Operating Handbook* figure for *4000 feet* by adding 100 rpm (4 × 25) to the sea level figure. This comes out to be 2510 rpm, as compared to the 2500 rpm in Figure 23-12. (It's extremely difficult to read a difference of 10 rpm on the tachometer, anyway.) Obviously, it is always better to use the manufacturer's figures, but if you know the sea level rpm for the desired power setting you can get within the ball park for your altitude if you don't have a power setting chart along. You will be limited by the maximum allowable rpm as you reach a certain altitude. (Don't exceed the red line on the tachometer at any time.)

■ Leaning the Mixture

How do you know your fuel consumption? Figure 23-13 is a fuel consumption graph for the airplane just discussed. The mixture control can make quite a difference in fuel consumption. At 2500 rpm there is a difference of about 1.1 gallons; in this case you'd burn about 16 percent more fuel in the full-rich setting than in best lean. (Nonturbocharged, direct drive engines can normally be leaned at *any* altitude in cruise at 75 percent power and below.)

There are several techniques, some of which include the use of cylinder-head temperature gauges or even special gauges designed specifically for use in

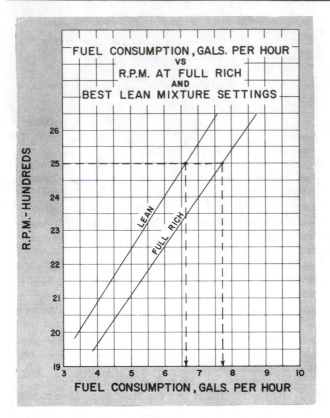

Fig. 23-13. Fuel consumption versus rpm.

leaning. For your present situation, however, the following technique, while not as accurate as some, is simple to use.

After you have reached the desired altitude, let the airplane establish the cruise airspeed and set the power. Then slowly move the mixture control aft until the engine begins to roughen slightly. Move the mixture control forward *just enough* to smooth out the engine again.

As the airplane climbs, the air becomes less dense, and less fuel is needed for a proper mixture as you increase altitude. If the engine starts running rough as altitude is gained (it will normally require at least 5000 feet density-altitude for this problem to make itself known), lean the mixture until smooth operation occurs. Use a full-rich mixture for climbs unless your instructor or the *Pilot's Operating Handbook* gives specific procedures for leaning in the climb.

Because of the loss of power involved with an overly rich mixture, at airports at higher elevations you may need to lean to obtain the best power for takeoff. Do this at full throttle at static run-up. Lean the mixture until maximum rpm occurs. Richen the mixture slightly from this maximum rpm setting to ensure proper cooling. (Don't do this over sand or gravel.)

One of the problems you may encounter on one of the solo cross-countries is forgetting to richen the mixture when descending from cruising altitude. You'll

probably be so engrossed in getting into the traffic pattern and watching for other airplanes (plus maybe having a little stage fright about going into a controlled field) that the need for richening the mixture on the descent doesn't occur to you. The engine may start running rough—or even quit momentarily—to remind you of this oversight. *Use a checklist.*

Figure 23-14 is a cruise performance chart for a current trainer. You can interpolate between altitudes, horsepower, and temperatures to get the conditions you're flying in. For instance, you're flying at a pressure altitude of 5500 feet and the temperature is 5°C above normal for that altitude. You want to carry 69 percent power and so would go through the following process: You'd start with the standard temperature numbers. At 4000 feet use 69 percent to get a TAS of 100 K and fuel consumption of 5.2 gallons per hour. At 6000 feet (standard temperature) the values are 102 K and 5.2 gallons per hour (again interpolated between 73 and 64 percent brake horsepower).

So the values at 5500 feet would be three-fourths of the difference between 4000 and 6000, or 101.5 K and 5.2 gallons per hour. (Practically speaking, you

CRUISE PERFORMANCE

CONDITIONS:
1600 Pounds
Recommended Lean Mixture

PRESSURE ALTITUDE	RPM	20°C BELOW STANDARD TEMP			STANDARD TEMPERATURE			20°C ABOVE STANDARD TEMP		
		% BHP	KTAS	GPH	% BHP	KTAS	GPH	% BHP	KTAS	GPH
2000	2650	- - -	- - -	- - -	78	103	5.9	72	102	5.4
	2600	80	102	6.0	73	101	5.5	68	100	5.1
	2500	70	97	5.3	65	96	4.9	60	95	4.6
	2400	62	92	4.7	57	91	4.3	53	91	4.1
	2300	54	87	4.1	50	87	3.9	47	86	3.7
	2200	47	83	3.7	44	82	3.5	42	81	3.3
4000	2700	- - -	- - -	- - -	78	105	5.8	72	104	5.4
	2600	75	101	5.6	69	100	5.2	64	99	4.8
	2500	66	96	5.0	61	95	4.6	57	95	4.3
	2400	58	91	4.4	54	91	4.1	50	90	3.9
	2300	51	87	3.9	48	86	3.7	45	85	3.5
	2200	45	82	3.5	42	81	3.3	40	80	3.2
6000	2750	- - -	- - -	- - -	77	107	5.8	71	105	5.3
	2700	79	105	5.9	73	104	5.4	67	103	5.1
	2600	70	100	5.2	64	99	4.8	60	98	4.5
	2500	62	95	4.7	57	95	4.3	53	94	4.1
	2400	54	91	4.2	51	90	3.9	48	89	3.7
	2300	48	86	3.7	45	85	3.5	42	84	3.4
8000	2700	74	104	5.5	68	103	5.1	63	102	4.8
	2600	65	99	4.9	60	99	4.6	57	98	4.3
	2500	58	95	4.4	54	94	4.1	51	93	3.9
	2400	52	90	4.0	48	89	3.7	45	88	3.5
	2300	46	85	3.6	43	84	3.4	40	82	3.2
10000	2700	69	103	5.2	64	102	4.8	59	102	4.5
	2600	61	99	4.6	57	98	4.3	53	97	4.1
	2500	55	94	4.2	51	93	3.9	48	92	3.7
	2400	49	89	3.8	45	88	3.6	43	87	3.4
12000	2650	61	100	4.6	57	99	4.3	53	98	4.1
	2600	58	98	4.4	54	97	4.1	50	96	3.9
	2500	52	93	4.0	48	92	3.7	45	91	3.5
	2400	46	89	3.6	43	87	3.4	41	84	3.3

Fig. 23-14. Cruise performance chart. Note, as mentioned earlier, that the power required for a specific horsepower must be increased by 25 rpm per 1000 feet of altitude. The underlined figures for 54 percent power for 4000, 8000, and 12,000 feet show this (2400, 2500, and 2600 rpm respectively).

could get the information for 6000 feet and be very close.) Further checking by interpolation of the effect of the addition of about 5°C above standard would show a small addition of about 1 K.

One point should be brought out: The airplane will use basically the same amount of fuel for, say, 70 percent power at any altitude, assuming that the mixture has been leaned properly in each case, but the advantage of altitude (once you get there) is that you are getting more miles to the gallon because of increased TAS. Take a look under the *standard temperature column* at 65 percent power at 2000 feet and at 8000 feet. At 2000 feet pressure altitude the TAS is 96 K at a fuel consumption of 4.9 gallons per hour. Interpolating to get 65 percent at 8000 feet, the same fuel consumption (4.9 gph) is found but with a TAS of 101+ K. Note that the chart is given to the nearest knot and 0.1 gallon.

Figure 23-15 includes range and endurance profile charts. Note (1) in each chart indicates that the chart allows for the fuel used for engine start, taxi, takeoff and climb distance (range profile), and climb time (endurance profile) as would be taken from the time, fuel, and distance to climb chart (Fig. 23-10).

You could use Figures 23-10 (for climb) and Figure 23-14 (for level cruise) to make your own graphs.

Looking at Figure 23-15 you can see some interesting points.

Range is not so much affected by altitude (at lower altitudes) as endurance is. The range stays constant with altitude until certain altitudes are reached and then decreases with altitude. This is because the time and fuel required to climb to higher altitudes cut down on the total range available. (The miles covered during the climb are considerably less than those covered in straight and level flight—and at higher fuel consumption.)

Note that endurance is greatest at sea level and starts dropping immediately with altitude increase. Maximum endurance is found at lower altitudes (and at power settings of 40–45 percent power for this airplane (look at Fig. 23-15).

As an example, using the range profile, you would expect to have a range of 400 NM (including climb) at 7000 feet density-altitude (standard pressure and temperature at that altitude) at 55 percent power (94 K TAS) in zero wind. At sea level at 45 percent power you could expect a range of 440 NM at 81 K (TAS).

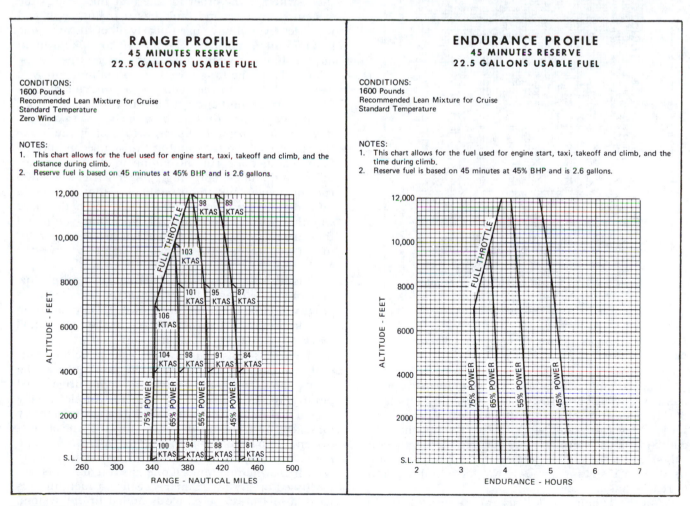

Fig. 23-15. Range and endurance profile charts for the airplane of Figures 23-10 and 23-14.

A point you might notice by checking Figure 23-15 is that the maximum range (miles) for the various altitudes is found at a power setting of 45 percent of the rated brake horsepower. Looking back at Figure 12-13, you note that this is the area indicated by point (2) for maximum range conditions. The thumb rule cited for maximum range was a calibrated airspeed of 1.5 times the power-off, flaps-up stall speed. The stall speed for this example airplane (flaps-up, etc.) is 47 K. Converting the true airspeeds for the various altitudes, you would find that the thumb rule is close to optimum.

Figure 23-16 shows another method of computing fuel, time, and distance to climb.

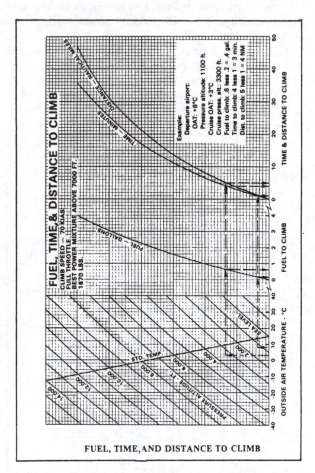

FUEL, TIME, AND DISTANCE TO CLIMB

Fig. 23-16. Graphical presentation of fuel, time, and distance to climb.

The example shows how you would work a problem from an airport of 1100 feet pressure altitude at 8°C temperature for a climb of 3300 feet. A quick way to see how the graph works is to assume that the climb is from sea level to 8000 feet (standard temperatures). You'd see that approximately 2.2 + gallons are used, with 17 minutes and 21 NM (no wind) required to reach that altitude.

Another point: Look back at Figure 23-10. To climb from sea level to 9000 feet would require 20 minutes and 2.7 gallons of fuel. This works out to be

8.1 gallons per hour used in the climb, or more than a 65 percent higher rate of consumption than is found at 65 percent power at cruise and leaned (4.9 gph). The point is that if you have to do a lot of climbing, your fuel consumption (and hence fuel remaining) may fool you if *POH* figures aren't available.

■ Fuel Management

If your plane has selectors for two tanks, cross-country flying will likely be your introduction to switching tanks in flight.

If your airplane has two separate fuel tanks, one may be listed as the "main" tank and the other the "auxiliary." The manufacturer in this case will recommend use of the main tank (both may have the same fuel capacity) for takeoff and landing and use of the auxiliary only in level flight. A recommended procedure in this case would be to use the main tank until cruising flight is established, then switch to the other one. This will give you a check as to whether fuel can be gotten from that second tank. There's no law that says you couldn't use most of the fuel in one tank and *then* switch to the other for the first time, *but* if the system of that second tank has a malfunction and won't feed you'll have used nearly all of the available fuel and might not have enough left to make it to an airport. If you switch tanks right after cruise is established and the tank does have a malfunction, you can switch back to the good one and have sufficient fuel to turn around and get back to the airport.

It's important that the run-up before takeoff be done on the tank to be used for takeoff. If you make the run-up on one tank and at the last second decide to use another one, run the engine on the new tank for at least a minute at fairly high power settings (at 1500 rpm or above). You might switch to that new tank, immediately start your takeoff, and learn the hard way that it won't feed. (Roughly speaking, you'll find that there'll be just enough fuel left in the carburetor to get you to exactly the worst point to have the engine quit.)

Figure 23-17 shows the schematic of a simple fuel system for a current high-wing trainer. The engine is gravity fed (the tanks are in the wings). The fuel from both tanks simultaneously moves to the fuel shutoff valve, which has two positions, OFF and ON.

Figure 23-18 is a fuel system schematic for a current low-wing trainer. Because the fuel tanks in the wings don't give gravity feed, an engine-driven fuel pump must be a part of the system. An electrically driven fuel pump also is installed as a safety standby for takeoff and landing, or at any time that the engine-driven pump should fail. Some *Pilot Operating Handbooks* recommend turning the electrical fuel pump ON when switching tanks if one has run dry. This was discussed briefly in Chapter 7. Since a fuel pump is used, a fuel pressure gauge is added to the system. This particular airplane uses fuel from either the left or right tank (you don't have a setting to use both

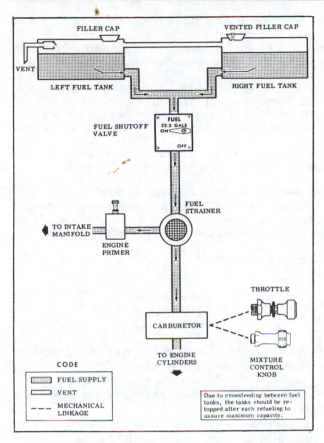

Fig. 23-17. Fuel system schematic for a high-wing trainer.

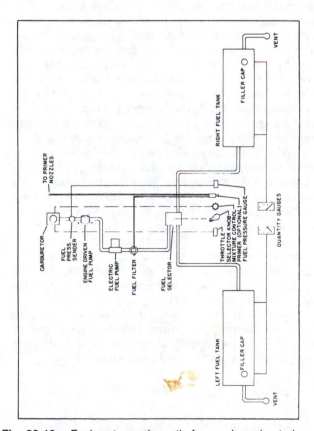

Fig. 23-18. Fuel system schematic for one low-wing trainer.

tanks simultaneously). There is also an OFF position on the fuel selector valve.

An important factor is the amount of *usable* fuel in the tank. If a tank has a capacity of 20 gallons yet only 18 gallons are usable, you'll have to do your planning on the basis of 18 gallons, no matter what it says on the tank caps.

Have an idea ahead of time about how long you can expect to run on each tank at the planned power setting. Fuel gauges may not always be truthful and, also, will go out if that system fails if they are a part of the electrical system.

If at all possible, when the airplane is being fueled at *any* time, you should personally oversee it. This is particularly important when at a strange airport. There have been cases of jet fuel being put in the tanks of an airplane with a reciprocating engine. If undetected, this will almost certainly cause a crash. While your small airplane is not likely to be mistaken for a jet or turboprop, fuel in service trucks has been inadvertently switched. In checking the fuel during the preflight of the airplane, *smell* it. You might be able to pick up the kerosenelike smell of jet fuel in the tanks.

There have been instances of fuel being put in the wrong tanks of an airplane; the pilot tells the line service people to "fill up the mains," and they fill up the auxiliary tanks with the result that the airplane may run out of fuel in the main tanks on takeoff or another critical point. (*But* the pilot should have checked the fuel level in the mains during the preflight.)

There have also been situations in which the "proper" fuel was put in *another airplane!* The pilot, not checking the fuel level before the flight, took off, figuring that "the gauges are wrong," and the flight was shorter than anticipated.

The problems just mentioned can be avoided if you oversee the fueling and check the fuel level visually (with a dipstick) during a thorough preflight check.

NOTES ON CONSTANT-SPEED
PROPELLER OPERATION
(NONTURBOCHARGED ENGINES)

A study has shown that the vast majority of flight schools use only smaller trainers with fixed-gear and fixed-pitch propellers for their students and that most pilots are introduced to constant-speed propellers well after receiving the private certificate. However, a brief discussion should be given here so that if you get a chance to fly an airplane so equipped you will have some background.

An analogy might be that the propeller pitch control is the "gear shift" and the throttle is the "foot pedal" of an automobile.

The prop control is moved forward for low pitch (high rpm) for takeoff and other high-power requirements and would be the equivalent of first or low gear. The engine would be revving up, developing higher horsepower. When the control is moved aft, the rpm decreases (the pitch or angle of attack *increases*) and less horsepower is being developed; also, the propeller is more efficient at cruise airspeeds at lower rpm. The constant-speed propeller acts as if you were automatically changing fixed-pitch props as needed.

Since, within the constant-speed operating range, the propeller automatically maintains the rpm set by the pilot, checking the mags in that range would mask any drop. (The propeller would automatically flatten pitch to maintain constant rpm, even with the slight power loss of running on one mag.) In the pretakeoff check the mags are normally checked below 2000 rpm so that the prop reacts to the rpm drop like the fixed-pitch trainer you've been flying.

Taking a typical flight, after checking the mags, you would "exercise" the prop by moving the control rearward and then forward a couple of times. This is done to ensure that it's responding properly and to recirculate the oil in the prop dome in systems using engine oil as a method of changing pitch.

Before the takeoff you'd make sure that the prop control was in the full-forward (high rpm) position and the mixture was full rich (or leaned as necessary for density altitude effects—see Chapter 5, *SPFM*), then, after checking that the area was clear, you would open the throttle fully.

Sometime during the climbout you would, as the manufacturer requires, throttle back to a specific manifold pressure setting, *then* ease the prop control back to get recommended climb rpm (usually leaving the mixture rich until a particular density altitude is reached).

After reaching the chosen altitude, climb power would be left on until cruise speed is reached, then the procedure for reducing power would be to (1) ease the throttle back to the cruise setting, (2) move the prop control back to reduce the rpm to the cruise setting, and (3) lean the mixture.

Note that this is the safe way to do it. The throttle is moved back first because a lot of fuel/air mixture pressure (manifold pressure) and low rpm could cause a rise in pressure in the cylinders to such a point that damage could occur. (Figure what would happen if you were in fourth gear in a very powerful car and jammed on the gas to try to burn rubber.) The mixture is leaned *last* as a safety procedure when power is reduced.

If you suddenly had to climb, you would reverse the procedure just discussed to avoid detonation and damage. To *increase power*: (1) mixture—rich, (2) prop rpm—increase, and (3) throttle—open. To *decrease power*: (1) throttle—back, (2) prop control—back, and (3) mixture—lean (as desired).

For further reading on the subject you might check *Advanced Pilot's Flight Manual*, Chapter 12, and *Flight Instructor's Manual*, Chapter 17.

Now look at a couple of questions.

01. When adding full power in an airplane with a constant-speed propeller, the order of control usage would be

1—throttle forward, mixture rich, propeller control forward.
2—mixture rich, throttle forward, propeller control forward.
3—throttle forward, propeller control forward, mixture rich.
4—mixture rich, propeller control forward, throttle forward.

02. When reducing power with a constant-speed propeller, which of the following procedures should be used?

1—Throttle back, reduce rpm, and set (lean) the mixture.
2—Reduce rpm, throttle back, and set (lean) the mixture.
3—Lean the mixture, reduce rpm, and throttle back.
4—Richen the mixture, reduce rpm, and throttle back.

(Answers: 01, 4; 02, 1.)

SUMMARY

You should be familiar with the following airplane systems before going solo cross-country:

Fuel system—Total capacity, usable fuel, minimum grade of fuel, recommended order of tank usage (if pertinent).

Oil—Proper brand and grade for the engine and season (ashless dispersant or straight mineral oil).

Electrical—Location of the circuit breakers and other systems that would be affected by electrical power loss; battery capacity and location.

Many airplanes are equipped with outlets for use with *external power* sources so that the engine can be started if the battery is dead. As noted in Chapter 3, some electrical systems (alternator) need a certain amount of electrical power to start the process of battery charging again. If the battery is completely dead and the engine is hand propped (*not* using an external power supply to "energize" the system), the alternator *won't* charge the battery. While the engine will run very well, electrical components (radios, lights, etc.) will still not be available. Check on this concerning your airplane. Don't prop the airplane unless you have had thorough instruction in the art, and a competent operator is in the airplane. Someone can fly to get you if necessary. Better that than an accident.

Radios—Ability to transmit and receive on all of the equipment; a knowledge of required frequencies, *plus* light signals used by the tower; how to use the radio to obtain aid in an emergency.

Power plant and cruise control—Engine operating limitations, setting up proper rpm for the power required, fuel consumption at various power settings, use of mixture control, use of carburetor heat.

24

NAVIGATION PLANNING

Most student cross-countries are in the form of a triangle, with landings at two strange airports. By the time you are ready for the private practical (flight) test, you'll have at least 5 hours of solo cross-country. Each flight will have included a landing at a place more than 50 nautical miles from the original point of departure. One flight must be of at least 150 NM with landings at a minimum of three points, one of which is at least 50 NM from the original departure point. You'll also have done three solo takeoffs and landings to a full stop at an airport with an operating control tower (FAR 61.109).

FAR 61.93 gives complete details, but basically before you will be allowed to fly solo out of the local area, you must be familiar with flight planning (plotting courses, estimating times en route and fuel required, and obtaining and evaluating weather reports). You'll have had flight instruction in crosswind and short- and soft-field takeoffs and landings as well as cross-country navigation by reference to aeronautical charts and to the magnetic compass. You'll also be checked out on emergency situations such as marginal visibility (and estimating visibility), deteriorating weather, getting lost, flying by reference to the instruments, and using radio aids and radar directives. Make sure you know about collision avoidance, wake turbulence precautions, and windshear avoidance. You'll have the word on the proper use of two-way radio communications and VFR navigational procedures and techniques plus traffic-pattern procedures.

Check back to Chapter 21 and be familiar with the various air space and clearance-from-clouds requirements.

MAP WORK

Your first cross-country flight will be with your instructor. After you've proved on the dual cross-country that you can do the job, your instructor will endorse your student pilot certificate for solo cross-country flights. This doesn't mean that you're now free to go anywhere, anytime the whim hits you. Your instructor will supervise all of your solo cross-country planning *and* flying because you aren't a pro just yet.

Most instructors will go dual with you on a specific route as discussed here, then have you fly the same route solo, so that you won't have the added pressure of flying by yourself in "unknown" territory. Later you'll make solo flights (with proper briefing) to airports that you haven't flown into dual. Speaking of planning, there are a few things you should make sure to cover.

The chart in the back of this book is set up for a flight from Franklin County Airport (Sewanee, Tenn.), to Maury County Airport (Columbia–Mount Pleasant, Tenn.), to the Huntsville International (Carl T. Jones) Airport (Ala.), and back to Sewanee. It's intended as an example to cover points that might arise on *your* cross-countries.

Each 10 NM of the route has been marked off to aid in checking distances and en route times. This probably won't be needed after the first few trips.

1. *Check the Airport/Facility Directory* for the facilities of the airports at which you plan on landing. This material is always more up-to-date than the charts in this matter.

When you check the weather and file the flight plan with the Flight Service Station (personally or by phone), get the latest NOTAMs for the facilities and airports that you'll be using or that will affect some phase of the flight.

Check Appendix B for this information on the airports and facilities to be used.

2. *Check the weather*, present and forecast, for the routes you'll be flying. This means also to check the forecast weather for the area of the home airport. You may get off and not be able to get back. (Pilots with friends of the opposite sex in other towns have been known to accidentally overlook this last point, and almost invariably get "weathered in" at one of the towns in question.)

3. *On the chart draw course lines between the airports.* Convert the true courses to magnetic and write this and the distance by each course line on the map. *The chart in the back of the book is a reduced presentation of this cross-country as drawn on a sectional chart. The scale at the top of the sample chart is reduced also so that you can use it to lay off distances with the edge of a sheet of paper. Fold out and refer to that chart as you read this chapter.*

If the courses lie close to omni "roses" on the chart (the calibrated circles around each VOR station), you can save the step of converting the true course to a magnetic course by laying off a straight edge through the omni rose parallel to the course line and find the magnetic course directly (VOR stations are oriented with magnetic north, you remember).

Always, when planning a cross-country of this type, look over the entire route generally to become familiar with landmarks on the map and to check on possible hazards.

Taking each leg of the example here, you might check the following.

Some of the checkpoints of the legs discussed next have been photographed from an airplane flying the route. Carefully compare the numbered checkpoint on the chart with its accompanying photograph. (Point 1 on the chart matches Fig. 24-1; point 2 is shown in Fig. 24-2, etc.) The points on the numbers on the chart are indicating the direction you would be looking at the checkpoint from the airplane. Sometimes railroads or streams are hard to see from the airplane, and you should compare different topographical features before jumping to conclusions about the checkpoint. For instance, maybe the town you see down there has a railroad running through but there's no fair-sized river west of town, so it's not Jonesville, after all. You may want to remove the chart from the back of the book so that you can rotate it to get a better comparison of photos and chart.

Fig. 24-2. Approximately 24 miles from Sewanee: The town of Lynchburg is 3 miles down the road to the left (south). The junction of the highways to Shelbyville and Tullahoma is just to the left of course. Note the stream to the west side of the road to Lynchburg and that it crosses just north of town (chart and photo). The secondary road between the highway and the stream is not shown on the sectional chart. Small streams usually can be picked out by the trees along the banks; this one is "open" at the northern portion (at bottom right in photo).

Fig. 24-1. A "close-up" view of Estill Springs. The road and railroad run northwest to Tullahoma.

■ Leg 1—Sewanee to Maury County Airport (4500 Feet MSL)

All of the distances noted on this flight are nautical miles and are rounded off to the nearest mile.

8 miles—You will cross a highway and power line and note that the town of Decherd is about 2 miles to your left and that Woods Reservoir is about 4 miles off to your right.

11 miles—The town of Estill Springs (and bridges) is below and slightly to the right as you cross a part of

Tims Ford Reservoir (Fig. 24-1).

19 miles—There are two inlets of Tims Ford Reservoir to the left, and the town of Tullahoma and Tullahoma Regional Airport are approximately 4 miles to the right.

24 miles—The course crosses a major highway just north of a road junction and the town of Lynchburg with its distillery and storage warehouses is 3 miles to the left (Fig. 24-2).

30 miles—There is an antenna or communications tower right on your course line about 2 miles before you cross a highway (Fig. 24-3).

32 miles—You will cross a major highway 1 mile south of where it is crossed by a transmission line.

41 miles—The route goes directly over the village of Belfast, intersecting a highway and railroad combination, and goes on from there to Lewisburg.

44 miles—The town of Lewisburg is below and to the right. Outstanding landmarks are a quarry just north and east of town (note the crossed-picks symbol on the chart) and the perimeter highway and airport north of town (arrows). Note the stream running generally north-south through town (Fig. 24-4).

Fig. 24-3. You'll fly over (or very close to) an antenna at 30 NM from Sewanee.

Fig. 24-4. Lewisburg, with the quarry on the northeast edge and the airport seen off to the north of town (arrow). The railroad runs through the western edge of the town (photo arrow and chart).

50 miles—You will cross Interstate I-65 approximately 1 mile north of an intersection. On the chart this intersection is almost obscured by the Verona (Ellington Airport nondirectional beacon) information box. You would see the intersection on your dual flight and would use it as an important landmark on your solo on this leg to Maury County Airport. A power line crosses the interstate about 2 miles north (Fig. 24-5).

Fig. 24-5. An intersection to I-65 and a secondary road 1 mile to the left (south) of the course. Note the stream in the right side of the photo (the secondary road crosses it). The stream (arrows) is not shown on the chart, but, again, you'd see it on the dual flight and make a mental note for your solo trip.

56 miles—You will cross a railroad and highway (the highway is merged with the railroad on the chart) about a mile north of Culleoka.

59 miles—A major highway is crossed (Fig. 24-6). Maury County Airport is in sight 6 miles ahead. The airport is centered in an area of mining. The *Airport/Facility Directory* (Appendix B) indicates that Unicom is available and you should give a call at this point for wind, runway, and traffic information. Note that the main runway is 5-23, which is hard surfaced and 5000 feet long; there is a shorter grass runway. Figure 24-7 shows the airport as seen from about 3 miles out. Note the mining areas.

Check on the layout of the airports you'll be flying into both as a student pilot and in your future flying as well (Fig. 24-8).

When looking at each leg, check the airports that can be used if you need to divert because of weather or other problems. On this route are Winchester, Tullahoma, Shelbyville, and Lewisburg airports. Arnold Air Force Base could be used in an emergency (it's normally restricted to military traffic except by prior per-

Fig. 24-6. Crossing a highway due south of the city of Columbia, which can be seen off to the right (north) at about 5 miles. The curves in roads and highways are not shown in great detail on sectional charts but are general representations. You'll note that the curve shown in the photograph appears to be slightly sharper than that on the chart. Part of this is due to perspective, but the chart cannot follow the sharper, smaller curves exactly.

mission), but you could get to either Tullahoma or Winchester more easily during that portion of the trip. For instance, if visibility was sharply restricted when you were south of Tullahoma, you would turn north and fly until you hit either a highway and railroad (in which case you'd turn left to follow them to the city of Tullahoma and the airport), or a highway alone running east-west (and you'd turn right to find the airport). Roads and railroads can be "brackets" to funnel you into a town. In other words, don't cross over these references, but use them to bring you to a prechosen point if visibility is down. You can look at the chart before going to see what would be a sure way to get to each of the airports, but more about that in the next chapter.

Be sure to have a good idea beforehand of the relative position of the airport to the town and other prominent landmarks. This is very important for airports having only grass landing areas; they are sometimes very hard to find.

■ Leg 2—Maury County Airport to Huntsville International Airport(5500 Feet MSL)

3 miles on course—A lake and dam should be about a mile off to your right.

9 miles—The route crosses a secondary road and power line, with a settlement about ½ mile to the left.

Fig. 24-7. Approaching Maury County Airport. The town of Mount Pleasant is in the upper left corner of the picture.

A secondary road will be to the right, more or less paralleling the course line. A larger highway will be converging with the course from behind and the left; it's about 4 miles left at this point. Look for the antenna just north of a settlement there.

18 miles—You'll cross two streams and a railroad and then intercept a major highway. There's an electric substation about a mile to the right. The city of Pulaski will be in sight 4 miles ahead.

22 miles—The route crosses the eastern edge of Pulaski and will "parallel" the railroad for a few miles. Abernathy Airport will be a little over 2 miles to the right as you pass it (Fig. 24-9). Confirm that this is Pulaski by the road and railroad patterns out of the city. In very poor visibility if you get off course, it's possible that you might, for instance, confuse Lawrenceburg with Pulaski at first glance. One railroad runs directly through Lawrenceburg, but another skirts the west edge of Pulaski after it passes the airport. There is an outdoor movie at Pulaski west of town. Lawrenceburg has no outdoor movie.

30 miles—Rivers and stream bends make good checks; look at the system in the vicinity of the 30- to 35-mile distances. There's a railroad with a tunnel off to the right, but it may be hard to see. The railroad is more or less "paralleling" the course and is off to the right (Fig. 24-10).

40 miles—You'll cross Interstate I-65 at a curve just north of where the railroad running from Athens to Ardmore goes under the highway. Figure 24-11 looks down the railroad to the southwest shortly after the interstate is crossed.

45 miles—The city of Athens is about 5 miles off to the right of your course, and you should listen to ATIS (121.25 MHz) and then contact Huntsville ap-

proach control on 118.05 MHz with your position information as indicated by the *Airport/Facility Directory* or the Airport Radar Service Area information of the sectional chart.

You will be asked to "squawk" a certain transponder code and probably to "ident." If your airplane doesn't have a transponder, you'd say so and may be given a turn or turns for identification. You'd then be vectored as necessary to avoid other traffic. You may be cleared straight in if wind and traffic permit (Fig. 24-12).

Let approach control know if you are changing altitude, since this would affect other traffic at your new level. You'll be switched to the tower frequency about 5 miles out and after clearing the active runway will contact ground control for taxiing.

You may be shy about getting taxi directions the first time or two that you go into a large airport. An "I'm a stranger here; could you direct me to _____," will always get you the right help.

■ Leg 3—Huntsville to Sewanee (5500 Feet MSL)

There's a lack of checkpoints on this leg, and you can figure before you even start that about halfway along a leg like this in poor visibility you might decide that you are "lost" and maybe ought to circle to try to find a recognizable checkpoint. *DON'T.* Look at the route; you'll see that the railroad and road(s) on the left (west) of the course can form one side of a bracket; the railroad and road coming up from the south through Sherwood will be the other side and will be

Fig. 24-8. After talking to Unicom and entering traffic at Maury County Airport, you are circling and are on the crosswind leg (heading 140°); a 90° turn will be made shortly to get on the downwind leg for Runway 23. (You are looking to the east-northeast here.) Check the layout of an unfamiliar airport as you circle. The wind is across the runway, but Unicom has noted that traffic is using Runway 23. Note: This is still the crosswind leg of the pattern, even if the tetrahedron in the photo (just this side of the center short taxiway) indicates you're flying "upwind" at this point.

Fig. 24-9. Looking southwest (to the right) at the Pulaski Airport (Abernathy Field) from over the city. Compare the sectional chart with the view here. Note on the chart that there is a road fork about a mile north of the airport (arrow). There is a quarry just off the approach end of Runway 33; it can be seen in the upper left of the photo. The quarry is not shown on the sectional chart. The airport remarks for Abernathy Field in the *Airport/Facility Directory* indicate, "Advance notice given to airport when explosives are scheduled at rock quarry approx. 1000′ from RWY 33." This is not a factor for you on this trip but shows that having an up-to-date copy of the *A/FD* on hand is a good idea. Note the railroad and well-defined stream just east of the airport.

Fig. 24-10. At 33 miles from Maury County Airport a railroad crosses the Elk River 3 miles off to the right (west) of the course. Compare the distinctive bends in the river (photo and chart).

Fig. 24-11. Over the railroad after crossing the interstate. Athens (Alabama) is 7 miles south-southwest.

enough to get back to familiar territory. (In other words, in this particular case, if you are "temporarily disoriented with respect to the preplanned route" and off to either side of the course, flying in a northerly direction up a railroad will take you either to Cowan or Winchester and you can locate yourself if you haven't already done so on the way.)

Always in planning a flight by pilotage, look for such brackets. Spend a lot of time looking over sectional charts; you'll be surprised at the amount of information available.

Use the whole sectional chart. (One student cut out the piece of the sectional chart that pertained to his flight but discovered that strong winds had caused him to drift off of that little patch of chart. He had to finally pick a field and land when fuel ran low.) Along these same lines, if one of your routes is close to the edge of the sectional chart, better take along the adjacent chart as well.

Huntsville International Airport has Automatic Terminal Information Service (ATIS), and you'd listen on 121.25 MHz before contacting clearance delivery on 120.35 MHz (Fig. 21-13). You'd make contact with clearance delivery and then tell them your aircraft number and type, indicating that you have "Information Delta" (or whatever—see ATIS in Chapter 21). You'd also give your on-course heading (approximately) and cruising altitude (exact).

Clearance delivery would give a clearance such as "MAINTAIN RUNWAY HEADING" (after takeoff), "DEPARTURE CONTROL" (frequency) "on 125.6" (MHz), and "SQUAWK 0134" (if you have a transponder). You'd read it back; and if it was correct, you'd hear:

Clearance delivery—"READBACK CORRECT. CONTACT GROUND CONTROL ON 121.9" (MHz). Incidentally, if you have a transponder you'd squawk as late as feasible prior to the takeoff.

You would then contact ground control (121.9 MHz, Appendix B) and get taxi instructions to the takeoff runway. If you need to, ask for specific taxi directions to the runway.

After taxiing to the run-up area (at Runway 18 Left for this example) and completing the run-up,

Fig. 24-12. Getting set up for a straight-in approach to Runway 18 Left at Huntsville–Madison County Jetport. It would be best to move over to the left from here in order to have about a 1-mile straightaway to check for wind drift and not angle all the way to the touchdown point.

switch to tower frequency (127.6 MHz) and request takeoff clearance. Before you're cleared for takeoff, you first may be told to hold or to taxi into position and hold. After you're airborne and about ½ mile past the end of the runway, the tower will tell you to switch to departure control. (No frequency will be given; you got that earlier, although you may ask if you forgot.) You'll get a radar vector ("Heading 020") to avoid the Redstone Restricted Area. After clearing the area, you'll be told to resume normal navigation and can track to the VOR and thence fly to Sewanee.

The VOR is on top of a knoll 12 NM from the airport, and you may have flown slightly more than this distance, depending on the radar vectors.

16–17 miles—You'll cross Highway 231 and then pass about 1 mile south of the North Huntsville Airport (Fig. 24-13).

24 miles—The town of New Market, Ala., is directly below. Note that the highway crosses a railroad just south of town and then recrosses it on the north edge.

31 miles—The town of Elora is 3 miles to the left (note the railroad junction). After passing New Market, the terrain in this area to the right of your course is rugged and rises rapidly from the flatlands, as evidenced by the warning (Fig. 24-14).

36 miles—The course line crosses a secondary road, and the town of Huntland is 3 miles to the left. Visibility permitting, you should see that you are "paralleling" a highway (U.S. 64) and a railroad about 3–4 miles to the left (Fig. 24-15).

42 miles—The village of Belvidere is 4 miles to the left. The railroad track is south of the road and swaps sides going through the town. (Historical note: Davy Crockett's first wife, Polly, is buried just southwest of Belvidere.) You'll likely be seeing Tims Ford Lake and Winchester and Decherd by now.

48 miles—Decherd and Winchester are about 5 miles to the left and Winchester Airport is 3 miles to the left (Fig. 24-16).

50 miles—Cowan is 1 mile to the left and the town of Sewanee and Franklin County Airport will be in sight (Fig. 24-17). At 5500 feet MSL you can figure on starting the descent at 500 fpm before getting to Cowan, since the traffic pattern at Sewanee is 2800 MSL (elevation 1950), a descent of 2700 feet, or about 5½ minutes, is required. Don't be letting down in the pattern, so start a little sooner.

To continue the general checklist:

4. *Using the expected true airspeed* of your plane, compute the approximate flight time involved in making the trip. (The *Pilot's Operating Handbook* gives the range for your airplane as well as covering the fuel consumption in gallons per hour.) Always allow yourself at least 45 minutes of fuel reserve—an hour is even better. Your instructor will have some suggestions; the *Pilot's Operating Handbook* also gives the amount of fuel used to climb to particular

Fig. 24-13. Just after crossing Highway 231 after passing the VOR en route to Sewanee (looking south). You'll cross Highway 231 about a mile north of where it bends slightly toward the west. The secondary road that forms a V with Highway 231 is not shown on the sectional. Looking to your left, you'll fly about ½ mile south of North Huntsville Airport.

Fig. 24-14. After you pass New Market, the terrain ahead and to the right becomes more rugged (looking south-southeast).

Fig. 24-15. The town of Huntland, 3 miles to the left of course. Not much detail here, but if you've been keeping up with checkpoints you'll know it's Huntland.

Fig. 24-17. Cowan is 1 mile to the left of course. Check the railroad running to the northwest through town. Note the quarry south of town as seen here. Quarries and mines are indicated by crossed picks on the chart.

Fig. 24-16. Winchester Airport, with the city of Winchester and Tims Ford in the upper left of the photo.

altitudes (along with recommended airspeeds and re-sulting climb rates at the various altitudes), as shown in Figures 23-10 and 23-16.

COMPUTING TIME AND
FUEL FOR THE LEGS

The trip, as shown on the sample chart, covers a total distance of 66 + 58 + 56 = 180 NM. An estimate of *flying* time at 96 K shows that 1 hour and 52 minutes would be *required just to fly the legs themselves;* another 10 minutes must be allowed at each stop for entering and leaving traffic and takeoff and landing, for a total of 2 hours and 12 minutes. The fuel to be used, based on this quick and dirty look at things, would be 10.8 gallons. It's a good idea to add another 30 percent, giving a final figure of about 14 gallons.

Plan on a *minimum* of 15 minutes on the ground at Maury County Airport and 30 minutes at a controlled field like Huntsville International, or figure that the example trip would take a minimum of 2 hours and 57 minutes (call it 3 hours).

Okay, for practice, plan the trip from Sewanee to Columbia–Mount Pleasant to Huntsville and back to Sewanee. You're using the numbers as given as an example in Figure 24-18.

The weather at Nashville, Huntsville, and surrounding areas is clear with 15 miles visibility and is forecast to be good VFR for the next 12 hours.

The winds-aloft forecast for the area is
3-1914 (at 3000 MSL, from 190° true at 14 K)
6-2312
9-2614
Assume standard temperatures.

■ Sewanee to Columbia–Mount Pleasant

You'll fly to Columbia–Mount Pleasant at 4500 MSL and to Huntsville and Sewanee at 5500 MSL using 65 percent power. Stops will be made at Columbia–Mount Pleasant and Huntsville.

Work out on a computer the groundspeed and true heading for the first leg (true course 288°). It's to be flown at 4500 feet, so you'd split the wind directions and velocities between 3000 and 6000 feet and get an "average" at 4500 feet of 210° true at 13 K.

The wind side of your computer shows that for a TAS of 98 K (65 percent power at 4500 feet at standard temperature—see Fig. 24-18) a groundspeed of 94 K and a true heading of 281° results. Note on the sectional chart that the isogonic (1° W variation) line crosses the course, so the magnetic heading will be 282°. *Use 1° W variation for all three legs.* Looking at Figure 24-18 as the compass card for this trip and interpolating, you'd see that for this heading you'd add 4°, so the compass heading would be 286°. Note that this is for the cruise part of the flight. If you worked

out the climb portion also (Fig. 24-18), you'd find that during the climb at 66 K (average TAS 69K), the groundspeed would be 70 K, the true heading 276°, and the compass heading 281°. (Use the 3000-foot wind as "average.") You'd set this up on the compass and have no worries on the trip. *Nonsense.* You'll find in actual flying, for instance, that (1) the wind was not forecast accurately, (2) the wind has changed, (3) the compass hasn't been swung or the card corrected since Montgolfier's balloon flight, or (4) you added variation or deviation when you should have subtracted—or vice versa. *You'll watch those checkpoints and make heading corrections as necessary.*

The time, fuel, and distance required to climb from Sewanee (assume a 2000-foot elevation) to 4500 feet will be *5 minutes, 0.6 gallons, and 5 NM* (Fig. 24-18—no wind).

Your groundspeed (assuming you take off and immediately head on course) for the first 5 minutes during the climb will be 70 K and you'll cover 6 miles. The cruise part of the trip, or the remaining 60 miles, will be at an estimated groundspeed of 94 K, so the time required to get there will be (5 + 38) = 43 minutes.

The letdown to Maury County Airport may be made at the same indicated airspeed as you had at cruise, and you would reduce the power to get a 500 fpm descent. You'll be letting down from 4500 to 1500 feet (elevation here is 676, and an 800-foot traffic pattern would be used for an altitude of 1476 feet—call it 1500). A 3000-foot descent at 500 fpm would take 6 minutes. Six minutes from your estimated, or recomputed, time of arrival you'll set up the descent. Since the last 6 minutes en route will be at reduced power, the fuel consumption would be slightly less, but to be on the conservative side, assume it to be the same as at cruise.

Fuel to be used then would be the start, taxi, and takeoff allowance (0.8 gallon) and climb (0.6 gallon) plus the 38 minutes at 4.9 gph (3.1 gallons), which gives a total for the leg of 4.5 gallons. Admittedly, this is carrying calculations pretty far, but later you may want to try out for the U.S. Proficiency Flight Team, or if you're in a college that is a member of the National Intercollegiate Flying Association such computations could help you to be top pilot in various air meets.

■ Columbia–Mount Pleasant to Huntsville

The climb portion will be from 676 feet to 5500 feet for this leg (you might want to use 3500 feet as a cruising altitude in an actual situation for a short trip like this). A quick estimate would be to assume a climb from 500 feet to 5500 feet. Interpolating in Figure 24-18 between sea level and 1000 feet, you'd need 1 minute, 0.1 gallon of fuel, and 1 NM. You'd then subtract this from the result of the interpolation between 5000 and 6000 feet, which is 10 minutes, 1.5 gallons (rounded off), and 12 miles. The difference would be 9 minutes, 1.4 gallons of fuel, and 11 miles (no wind).

TIME, FUEL, AND DISTANCE TO CLIMB

MAXIMUM RATE OF CLIMB

CONDITIONS:
Flaps Up
Full Throttle
Standard Temperature

NOTES:
1. Add 0.8 of a gallon of fuel for engine start, taxi and takeoff allowance.
2. Increase time, fuel and distance by 10% for each 8°C above standard temperature.
3. Distances shown are based on zero wind.

WEIGHT LBS	PRESSURE ALTITUDE FT	TEMP °C	CLIMB SPEED KIAS	RATE OF CLIMB FPM	FROM SEA LEVEL		
					TIME MIN	FUEL USED GALLONS	DISTANCE NM
1600	S.L.	15	68	670	0	0	0
	1000	13	68	630	2	0.2	2
	2000	11	67	590	3	0.5	4
	3000	9	66	550	5	0.7	6
	4000	7	65	510	7	1.0	8
	5000	5	65	470	9	1.3	10
	6000	3	64	425	11	1.6	13
	7000	1	64	385	14	1.9	16
	8000	-1	63	345	17	2.3	19
	9000	-3	63	305	20	2.7	23
	10,000	-5	62	265	23	3.2	27
	11,000	-7	62	220	27	3.7	32
	12,000	-9	61	180	33	4.3	38

CRUISE PERFORMANCE

CONDITIONS:
1600 Pounds
Recommended Lean Mixture

PRESSURE ALTITUDE	RPM	20°C BELOW STANDARD TEMP			STANDARD TEMPERATURE			20°C ABOVE STANDARD TEMP		
		% BHP	KTAS	GPH	% BHP	KTAS	GPH	% BHP	KTAS	GPH
2000	2650	---	---	---	78	103	5.9	72	102	5.4
	2600	80	102	6.0	73	101	5.5	68	100	5.1
	2500	70	97	5.3	65	96	4.9	60	95	4.6
	2400	62	92	4.7	57	91	4.3	53	91	4.1
	2300	54	87	4.1	50	87	3.9	47	86	3.7
	2200	47	83	3.7	44	82	3.5	42	81	3.3
4000	2700	---	---	---	78	105	5.8	72	104	5.4
	2600	75	101	5.6	69	100	5.2	64	99	4.8
	2500	66	96	5.0	61	95	4.6	57	95	4.3
	2400	58	91	4.4	54	91	4.1	50	90	3.9
	2300	51	87	3.9	48	86	3.7	45	85	3.5
	2200	45	82	3.5	42	81	3.3	40	80	3.2
6000	2750	---	---	---	77	107	5.8	71	105	5.3
	2700	79	105	5.9	73	104	5.4	67	103	5.1
	2600	70	100	5.2	64	99	4.8	60	98	4.5
	2500	62	95	4.7	57	95	4.3	53	94	4.1
	2400	54	91	4.2	51	90	3.9	48	89	3.7
	2300	48	86	3.7	45	85	3.5	42	84	3.4
8000	2700	74	104	5.5	68	103	5.1	63	102	4.8
	2600	65	99	4.9	60	99	4.6	57	98	4.3
	2500	58	95	4.4	54	94	4.1	51	93	3.9
	2400	52	90	4.0	48	89	3.7	45	88	3.5
	2300	46	85	3.6	43	84	3.4	40	82	3.2
10000	2700	69	103	5.2	64	102	4.8	59	102	4.5
	2600	61	99	4.6	57	98	4.3	53	97	4.1
	2500	55	94	4.2	51	93	3.9	48	92	3.7
	2400	49	89	3.8	45	88	3.6	43	87	3.4
12000	2650	61	100	4.6	57	99	4.3	53	98	4.1
	2600	58	98	4.4	54	97	4.1	50	96	3.9
	2500	52	93	4.0	48	92	3.7	45	91	3.5
	2400	46	89	3.6	43	87	3.4	41	84	3.3

FOR (MAGNETIC)	N	30	60	E	120	150
STEER (COMPASS)	0	26	58	87	118	150
FOR (MAGNETIC)	S	210	240	W	300	330
STEER (COMPASS)	183	214	243	275	303	332

Fig. 24-18. A composite of Figures 19-4, 23-10, and 23-14 for use in planning the sample trip. (Do not use for actual flight planning.) Assume IAS = CAS in the climb and that the magnetic variation is 0°.

The wind at 5500 could be interpolated (and rounded off) to be from 225° at 12 K; you would use that for your computer numbers. Using the 6000-foot wind would be accurate enough, since there is only a 500-foot difference. Again, the time required to climb and the fuel used would not be affected by wind (except for turbulence, which might cut down performance), but the distance would be, so working it out on the computer:

Wind—190° true at 14 K (average wind in the climb)

True course—160°

TAS (average in climb—see Fig. 24-18) for an IAS of 66 K—69 K

Groundspeed in climb—57 K

True heading—166°

Magnetic heading—167°

Compass heading—168° (Fig. 24-18)

The time required to climb has already been found to be 9 minutes, which at 57 K would be found on the computer to cover 8.6 NM (call it 9 miles).

The level and descent portion of the trip would cover 49 NM, so by looking at the cruise performance chart (Fig. 24-18) you'd interpolate to find that at 5500 feet (standard temperature) at 65 percent power the TAS would be 99 K (rounded off).

Working the last 49 miles at 99 K TAS (and not worrying about computing the average TAS and groundspeed in the descent), the wind is interpolated to be 225° true at 12 K, the computer indicates that the true heading is 166° (compass heading is 167°), and the groundspeed is 93 K.

The time required to fly the 49 NM leg is 32 minutes and the fuel used will be 2.6 gallons (again rounded off). The total fuel and time required for this leg will be:

Start, taxi, takeoff—0.8 gallon
Climb—1.4 gallons and 9 minutes
Level—2.6 gallons and 32 minutes
Total—4.8 gallons and 41 minutes

■ Huntsville to Sewanee

Altitude to be flown—5500 feet MSL

Winds aloft—225° true at 12 K at 5500 feet (the winds may have changed since the first part of the trip, but assume that this is a still correct interpolation).

Average wind in the climb—190° true at 14 K

Climb—no-wind climb information in Figure 24-18 from 500 feet MSL to 5500 feet is the same as for the last leg.

Count on 9 minutes in the climb at an average TAS of 69 K and a fuel consumption of 1.4 gallons. The groundspeed and true heading, after using the computer, turn out to be 81 K and 041° respectively. The magnetic heading will be 042° and the compass heading 039° (Fig. 24-18). The distance covered in the 9 minutes of climb at 81K is 12 NM.

TAS at 5500—99 K at 65 percent power
True course (climb) to VOR—036°
VOR to Sewanee—057°
Compass heading for cruise—056°
Groundspeed—111 K

Time and fuel required to fly 44 miles will be 23.8 minutes (24 minutes) and 2.0 gallons of fuel.

This leg will require 0.8 gallon for start, taxi, and takeoff; 1.4 gallons for climb; and 2.0 gallons for cruise, or a total of 4.2 gallons and a time of 33 + minutes. The flight log shows a total of 34 minutes, the result of rounding off the time between each checkpoint to the nearest minute.

The total flying time and fuel required for the three legs: 1 hour and 58 minutes and 13.5 gallons.

The climb takes more fuel than generally considered. For instance, looking at the climb chart portion of Figure 24-18, you'd see that to climb from sea level to 6000 feet would require 11 minutes and 1.6 gallons of fuel. To fly at 6000 feet at 65 percent power for 11 minutes (at 4.9 gph) would require 0.9 gallon. The climb fuel consumption is, for that example, nearly twice as much as for cruise consumption at 65 percent. The relative amount of fuel used in the climb part of a trip depends on a ratio of climb to cruise for that trip. The example here had short legs at comparatively high altitudes (maybe *too* high for a practical application in smooth air, but it gave more time for example purposes).

As noted earlier, you normally won't go into such detail on a short pilotage flight but, with experience in a particular airplane, will be able to estimate the fuel and time required to climb to various altitudes and the fuel consumption at, say, 65 and 75 percent power.

As a rule of thumb, for cruise (leaned) fuel consumption for airplanes with fixed-pitch props multiply

the actual brake horsepower being used (at 65 percent power in the 100-hp engine used in the example you'd be using 65 hp) by 0.075 to get gallons of fuel per hour being consumed. Multiplying 0.075 by 65 would give a number of 4.875 gph, which is "pretty close" to the 4.9-gph cruise consumption found on the chart (Fig. 24-18). Of course you will have some different horsepower engines to deal with (now or later), for example, one rated at 150 hp. If you wanted to find the gallons per hour consumed at 65 percent power (leaned) in this engine, you'd multiply 150 by 0.65 to find the number of horsepower (which works out to be 97.5—call it 98 hp) and multiply this by the constant *0.075* to get 7.35 gph at cruise (call it 7.4 gph). You'll find that this works out close to the *Pilot's Operating Handbook*s for most light aircraft reciprocating engines.

FLIGHT LOG

In making up the flight log and planning the flight, remember to take into consideration the rule that on *magnetic courses* of 0°–179° inclusive, the airplane is to fly at an odd altitude plus 500 feet (3500, 5500, etc.) and on *magnetic courses* of 180°–359° inclusive, even altitudes plus 500 feet (4500, 6500, etc.). *Although this must be followed if the airplane is flying 3000 feet or more above the surface, it may be used at lower altitudes.*

If you have made wind triangles or used the wind side of your computer, the *original* time estimates will be based on estimated groundspeed; otherwise use your cruise TAS. During the flight the actual groundspeed as checked on the computer will generally vary slightly because of changing wind velocities as well as the fact that you will probably round off the time (minutes) between checkpoints.

Figure 24-19 is the flight log for the trip. The distances are between checkpoints and also the cumulation for each leg. ETE is the estimated time en route and ATE is the actual time en route. You'll work out the estimated groundspeed (EST GS) before leaving. AGS is actual groundspeed. The columns Time Over, ATE, and AGS are done in flight. Assume a magnetic variation of 0°.

▶

Fig. 24-19. A flight log for the three legs of the sample cross-country. Minutes have been rounded off and time for climb has been added at the first part of each leg. Later you won't use so many checkpoints on the flight log (but will still keep up with all of the checkpoints). Sometimes it's pretty discouraging to work out a complex and lengthy flight log only to discover that the wind information was all wet and your estimated groundspeeds and estimated headings have no resemblance to what you have during the flight. The big idea is to follow that line on the map. Anyway, making out a flight log is a good way to ensure that you've checked the route and is an aid in getting ready for the Private Pilot (Airplane) Knowledge Test. Note that the checkpoints chosen are not necessarily those in the photos shown earlier.

AIRCRAFT N 7557L	WEATHER	Reported KHSV 18010KT 15SM CLR	BNA WINDS
DATE June 29		Forecast 18010KT P6SM FEW	3-1914
	Time	Reported KBNA 19008KT 15SM SKC	6-2312
PILOT S.G. Dickson	0815 CDT	Forecast 22015KT SCT 050	9-2614

CHECK POINT	TC	MC	TAS	TH	MH	CH	DIST / Rmdr	EST. GS	ETE	EST. FUEL	TIME OVER	ATE	AGS	Notes, T.O.Time, Alt., Actual Fuel Used, etc.
ROAD & POWERLINE	288°	289°	69/98	276°/281°	277°/282°	281°/286°	8 / 58	70/94	6	1.5				FUEL INCLUDES TAXI, T.O. CLIMB, SHORT CRUISE (0.1)
ESTILL SPRINGS			98k	281°	282°	286°	4 / 54	94	3	↑				T.O. TIME _____
ROAD/ LYNCHBURG							12 / 42		8					
BELFAST							18 / 24		11					
CROSS HWY							18 / 6		11					
MAURY CO. AIRPORT	↓	↓	↓	↓	↓	↓	6 / 0	↓	4	3.0				TOTAL CRUISE FUEL — 3.1 TAXI, T.O. & CLIMB — 1.4
									43	4.5				TOTAL MINUTES AND GALLONS — LEG 1
ROAD & POWERLINE	160°	161°	69k	166°	167°	168°	9 / 49	57k	9	2.2				FUEL INCLUDES TAXI, T.O. & CLIMB — 2.2 GALLONS
CROSS RR STREAMS			99k				9 / 40	95	6	↑				T.O. TIME _____
CROSS INTERSTATE							22 / 18	↑	14					
CROSS 4 LANE HWY							10 / 8		7					
HUNTSVILLE INTL AP	↓	↓	↓	↓	↓	↓	8 / 0	↓	5	2.6				TOTAL CRUISE FUEL 2.6 GALLONS
									41	4.8				TOTAL MINUTES AND GALLONS — LEG 2
HUNTSVILLE (ROCKET) VOR	036°	7°	69k	041°	042°	039°	12 / 44	81k	9	2.2				FUEL INCLUDES TAXI, T.O. AND CLIMB — 2.2 GALS
CROSS HWY N. HSV AIRPORT	056°	057°	99k	057°	058°	056°	5 / 39	111k	3	↑				T.O. TIME _____
NEW MARKET							7 / 32		4					
ELORA							7 / 25		4					
HUNTLAND							5 / 20		3					
COWAN							14 / 6		8					
SEWANEE FRANKLIN CO. AIRPORT	↓	↓	↓	↓	↓	↓	6 / 0	↓	3	2.0				CRUISE FUEL 2.0 GALLONS
									34	4.2	←			TOTAL MINUTES AND GALLONS — LEG 3
									1:58	13.5	←			TOTAL TIME AND GALLONS — ALL 3 LEGS

NOTAMS/Airspace Restrictions

HUNTSVILLE

R 2104 A/C - TO 30,000

R 2104 B - TO 2400

COMM/NAV Frequencies

MAURY CO. 122.8 MHz, NDB 365 kHz - PBC

HSV - APPROACH (FROM MAURY CO.) - 118.05

TWR - 127.6, GROUND 121.9, DEPT 125.6

VOR - RQZ - 112.2, ATIS 121.25

FILING A FLIGHT PLAN

The following are the types of flight plans available to the pilot:

VFR—A VFR flight plan, like the other types of flight plans, is filed with the nearest Flight Service Station in person or by phone. (Filing a flight plan by radio is frowned upon because it ties up needed frequencies, but it can be done.) The VFR flight plan is the simplest; basically, you follow these steps when filing.

You file the flight plan with a Flight Service Station using the format shown in Figure 24-20. It will be held by the FSS for 2 hours after the proposed departure time.

The "T" following the aircraft type (C-150/T) indicates that your airplane has a transponder *without* altitude-reporting capabilities.

If the airplane was equipped with Mode C (altitude-reporting capability) the "U" would be the letter to use there. The *Aeronautical Information Manual* has a listing of letters used to show special equipment (or lack thereof) on board when filing a flight plan.

The VFR flight plan does not require such information as altitude changes en route or alternate airports. You can use the remarks section of the flight plan for pertinent information such as "will close by phone after landing at Sewanee" or other data; you may use this space to list the names of your passengers after you get the private certificate. Closing a flight plan by radio is perfectly acceptable because it requires only a short transmission. (More about this in Chapter 25.) If you have no vital information for the remarks section, you may want to use this space to write some stirring line such as "Don't give up the airplane."

If your flight is terminated before reaching the destination, be sure to contact the nearest Flight Service Station and have them pass this information to the FSS responsible for the destination airport. If a flight plan is not closed within ½ hour of the estimated time of arrival, queries are sent to determine the location of the aircraft. If this and further checking down the line fail to locate the aircraft, Search and Rescue people are alerted 1½ hours after your ETA. (And if you want to get into a real jam, just be the guy who gets Search and Rescue alerted by forgetting to close a flight plan!)

If it looks like headwinds or other factors are going to result in a delay in arriving at your destination,

U.S. DEPARTMENT OF TRANSPORTATION
FEDERAL AVIATION ADMINISTRATION

FLIGHT PLAN

(FAA USE ONLY) ☐ PILOT BRIEFING ☐ VNR ☐ STOPOVER | TIME STARTED | SPECIALIST INITIALS

1. TYPE	2. AIRCRAFT IDENTIFICATION	3. AIRCRAFT TYPE/ SPECIAL EQUIPMENT	4. TRUE AIRSPEED	5. DEPARTURE POINT	6. DEPARTURE TIME PROPOSED (Z)	6. DEPARTURE TIME ACTUAL (Z)	7. CRUISING ALTITUDE
☒ VFR ☐ IFR ☐ DVFR	N7557L	C150/T	99 KTS	HUNTSVILLE INTERNATIONAL	1645		5500

8. ROUTE OF FLIGHT: DIRECT ROCKET VOR, DIRECT UOS

9. DESTINATION (Name of airport and city)	10. EST. TIME ENROUTE HOURS	10. EST. TIME ENROUTE MINUTES	11. REMARKS
SEWANEE/FRANKLIN CO. (UOS) SEWANEE, TN	0	34	—

12. FUEL ON BOARD HOURS	12. FUEL ON BOARD MINUTES	13. ALTERNATE AIRPORT(S)	14. PILOT'S NAME, ADDRESS & TELEPHONE NUMBER & AIRCRAFT HOME BASE	15. NUMBER ABOARD
2	40	—	S. G. DICKSON, SEWANEE, TN (615) 598-5723	1
			17. DESTINATION CONTACT/TELEPHONE (OPTIONAL): TIGER AVIATION (615) 598-1910	

16. COLOR OF AIRCRAFT: WHITE, RED & BLUE

CIVIL AIRCRAFT PILOTS. FAR Part 91 requires you file an IFR flight plan to operate under instrument flight rules in controlled airspace. Failure to file could result in a civil penalty not to exceed $1,000 for each violation (Section 901 of the Federal Aviation Act of 1958, as amended). Filing of a VFR flight plan is recommended as a good operating practice. See also Part 99 for requirements concerning DVFR flight plans.

CLOSE VFR FLIGHT PLAN WITH NASHVILLE FSS ON ARRIVAL

Fig. 24-20. VFR flight plan from Huntsville, Ala. to Sewanee, Tenn.

you'd better get on the radio and tell the nearest FSS to pass this information along.

DVFR—The DefenseVFR flight plan is necessary if you are going to fly in an ADIZ (Air Defense Identification Zone). It is a little different from the VFR, in that filing in the air is not just frowned upon, but doing so is to invite possible interception for positive identification. (This is an interesting experience, to say the least.) If you look suspicious, the interceptors could go further, but it's mighty hard to look suspicious in a trainer that cruises at 90 K or so, although perhaps looking at the interceptor pilots with a dark scowl or wearing a false red beard would help. If you are operating solely over the continental United States, this type of flight plan is not going to be a problem for you. The *AIM* gives details on procedures for filing and flying a DVFR flight plan.

IFR—The IFR flight plan is not for you yet and won't be covered here. Later as you work on the instrument rating you will be checked out on the procedures of filing and flying an IFR flight plan.

SUMMARY

Be sure before you go that you have the proper charts, flight log, simple computer, a protractor or plotter (preferably with a straight-edge ruler as a part of it), a pencil, note paper, and a watch (if the airplane doesn't have a reliable clock).

In your flight planning remember that FAR 91.151 requires that there be enough fuel to fly to the first point of intended landing, assuming normal cruising speed—during the day, to fly after that for at least 30 minutes; or at night, after reaching the first point of intended landing, to fly at normal cruising speed for at least 45 minutes. It would be well for you as a student pilot to plan on having at least twice this amount of time.

It's suggested that you fly *at least* one local area dual night session before starting on the cross-country phase of your flying. (See Chapter 26 of this book.) You'll feel a lot more comfortable if, on one of your solo cross-countries, a delay causes you to land at home (or elsewhere) after sunset. You and your instructor should plan those flights so that there is plenty of fuel, *and* daylight, reserve. It can be pretty uncomfortable when you realize that darkness is coming on and you're still in strange territory—with no night flying experience.

25

FLYING THE
CROSS-COUNTRY

You've planned the trip and checked the weather and the plane. You know how to operate the radios, know the light signals, and have checked the destination airports, so there's nothing left to do but go.

DEPARTURE

There are two ways to get established on course:
1. You may leave traffic and turn on course.
2. You may climb to altitude and fly back over the airport on course.

The first method is most commonly used, but you must realize that you'll be slightly off course in the beginning of the trip. Students generally have the most trouble at the first checkpoint. They have not had time to set up a heading. Use the compass (and heading indicator) as a reference to get you headed in the right direction. Set the heading indicator with the magnetic compass *at least once* during each leg (Fig. 25-1). There may not be winds-aloft information available, so fly the first few miles of the course with an estimated drift correction held. As soon as the first couple of checkpoints are passed, you will have some idea of the amount of correction needed.

EN ROUTE

After takeoff if you've filed a flight plan and have a radio, let the FSS know of your takeoff time (and write it on your log). Once you have leveled off and established the cruise power setting (making sure that you are at the proper altitude for your course if above 3000 feet above the surface) switch tanks, if applicable, and lean the mixture.

When you are established on course and the first checkpoint is where it is supposed to be, note the compass heading.

Hold the map so the course line on the map is parallel to the course you're flying in the airplane (Fig. 25-2). This way the towns will lie in their proper

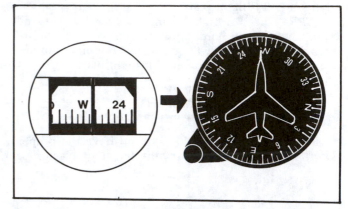

Fig. 25-1. Set your heading indicator with the magnetic compass *at least once* on each leg of the cross-country (every 15 minutes would be a good interval). Students have set the heading indicator before leaving the home airport and *not* compared it with the magnetic compass further during the flight. After a couple of landings and engine shutdowns at other airports (the heading indicator may be well off because of this), the homeward-bound leg is often very interesting and gives the student experience in using orientation, or lost aircraft, procedures.

places. Sure, you have to read the names of towns upside down if you're headed south, but this is better than getting a checkpoint's relative position fouled up when you try to turn the map around in your mind.

As you come to each checkpoint, look on the map to see how far it is from the course line. If you're more than a mile off course, better make a correction because you'll probably be even farther off course by the time you get to the next checkpoint. *Keep up with the checkpoints.* Run a groundspeed check when you get to the prechosen checkpoint.

If there are scattered or broken clouds near your altitude, you can get a rough idea of the wind direction and velocity by observing the movement of the cloud shadows on the ground.

Watch your altitude and heading. (Airplanes sometimes ease off heading while you're checking the map.)

Fig. 25-2. Hold the map so that the course line on it is parallel to the course you are flying.

Don't bury your head in the cockpit; keep an eye (or two) out for other airplanes.

Be sure to check the engine instruments at least every 5 minutes (checking more often is a lot better). See each time if the fuel gauge is showing about the right reading for the time you've been flying.

If you have a radio, listen to the transcribed weather broadcasts for your area of operation. Reset your altimeter to that of the FSS nearest to you.

If you decide to make a position report en route, use the PTA (*Position, Time,* and *Altitude*) system. Tell them what kind of flight plan you're on and from *where* to *where.* For instance you might call an FSS as follows:

You: "JONESVILLE RADIO, THIS IS CESSNA NINE EIGHT EIGHT ZERO JULIET, STUDENT PILOT, LISTENING JONESVILLE VOR, OVER."

After Jonesville Radio has acknowledged your call you might say the following:

You: "CESSNA EIGHT ZERO JULIET, (POSITION) GEORGETOWN AT (TIME) THREE TWO, (ALTITUDE) FOUR THOUSAND FIVE HUNDRED (4500 feet), VFR FLIGHT PLAN FROM PALMYRA TO CUMBERLAND CITY, OVER."

Jonesville Radio will acknowledge and give you the altimeter setting, as well as any other information they feel would be of assistance.

You'll want to contact the nearest FSS if you're on a flight plan and decide to change your destination or return to the point of departure.

DESTINATION AIRPORT

To students on their first cross-country, the new airports are strange indeed. It seems that the sky is filled with planes, and at noncontrolled fields the airport personnel seem to have hidden the wind indica-

tor. The wind indicator will be in a hard place to spot—not put in a clear place like your airport's indicators are. (Of course, the students going into your airport have the same thought.)

Start the letdown some distance out so you won't be diving into the traffic pattern. If you have a mixture control, don't forget to richen the mixture because, as noted in Chapter 23, when you descend into the denser air the mixture, as leaned at altitude, will become too lean and the engine may run rough and/or quit. (If this happens, move the mixture control to rich and give yourself about 5 minutes for your pulse rate to normalize.) Use your prelanding checklist. If it's a controlled airport, give Approach Control a call about 20 miles out or as indicated in the *Airport/Facility Directory* (Chapter 20). If the field has a Flight Service Station, the personnel can only give information such as surface wind and altimeter setting and known traffic. You should contact the FSS for Airport Advisory Service (AAS) 10 miles out. (Unicom can be used also at a non-FSS airport.)

Airport advisories also provide NOTAMs, airport taxi route, and traffic patterns. When you're ready to taxi at an airport having AAS, you should call the Flight Service Station and let them know your intentions. (Give your aircraft number, type, location on the airport, and type of flight plan, if any.) They'll give you the recommended runway and other necessary information. AAS is not mandatory and no traffic control is exercised, but you should take advantage of this service whenever possible. (The frequency 123.6 MHz is first choice for AAS, but you may use any of the other FSS frequency combinations if you choose.)

Check the field elevation on the chart. Add the traffic pattern altitude at which you want to fly to this and use the result as your indicated altitude around the airport.

Figure 25-3 shows some types of runways you may see on your trips into larger airports. When you first encounter some of the markings shown, you may

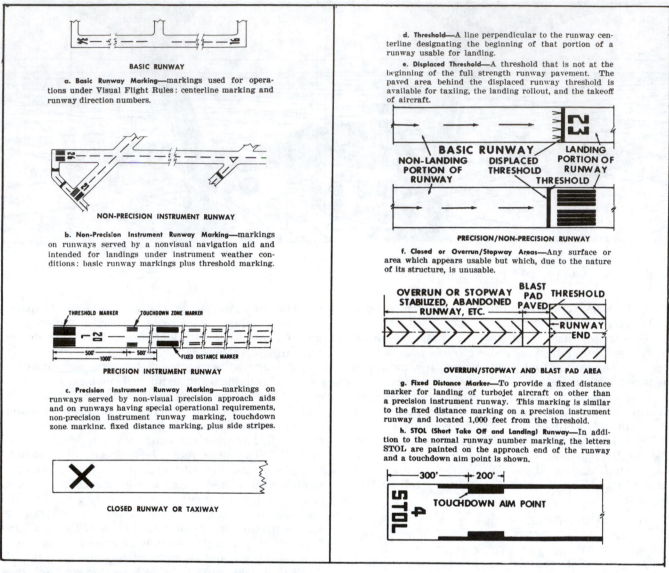

BASIC RUNWAY

a. Basic Runway Marking—markings used for operations under Visual Flight Rules: centerline marking and runway direction numbers.

NON-PRECISION INSTRUMENT RUNWAY

b. Non-Precision Instrument Runway Marking—markings on runways served by a nonvisual navigation aid and intended for landings under instrument weather conditions: basic runway markings plus threshold marking.

PRECISION INSTRUMENT RUNWAY

c. Precision Instrument Runway Marking—markings on runways served by non-visual precision approach aids and on runways having special operational requirements, non-precision instrument runway marking, touchdown zone marking, fixed distance marking, plus side stripes.

CLOSED RUNWAY OR TAXIWAY

d. Threshold—A line perpendicular to the runway centerline designating the beginning of that portion of a runway usable for landing.

e. Displaced Threshold—A threshold that is not at the beginning of the full strength runway pavement. The paved area behind the displaced runway threshold is available for taxiing, the landing rollout, and the takeoff of aircraft.

PRECISION/NON-PRECISION RUNWAY

f. Closed or Overrun/Stopway Areas—Any surface or area which appears usable but which, due to the nature of its structure, is unusable.

OVERRUN/STOPWAY AND BLAST PAD AREA

g. Fixed Distance Marker—To provide a fixed distance marker for landing of turbojet aircraft on other than a precision instrument runway. This marking is similar to the fixed distance marking on a precision instrument runway and located 1,000 feet from the threshold.

h. STOL (Short Take Off and Landing) Runway—In addition to the normal runway number marking, the letters STOL are painted on the approach end of the runway and a touchdown aim point is shown.

Fig. 25.3. Runway markings as published in the *Aeronautical Information Manual.* This illustration is not so much an item for you to memorize, but is for reference if an *AIM* isn't available. While the fixed distance marker (g) is of little interest to you now, the symbol for a *closed* runway or taxiway could be. The precision instrument runway is one of parallel runways (20 Left).

wonder whether a student pilot is allowed to land a small airplane on such areas.

It's an interesting point that pilots tend to be too high at a strange airport. Keep this in mind and make a slightly wider pattern than at home base.

Look the airport over as you circle. See how the taxiways are laid out. It's mighty easy to get lost on the ground at big airports.

After landing, as you taxi in to the ramp, locate the parking areas. If there is the slightest doubt in your mind about fuel—*gas up.*

Since you probably won't know where to park, line service may direct you to a parking place. You

should be familiar with the taxi signals before going on the cross-country (Fig. 6-8). The taxi director does *not* have to stand directly in front of the airplane to direct it, but may stand well to one side, pointing to the wheel that needs braking, and put you in the proper spot. Follow directions; the director has a reason for putting you in a parking place, even if you don't see the point as you taxi in. Park there and discuss it after the engine shutdown if you want another spot. Most student pilots don't give taxi directors trouble. It's the more experienced pilots who tend to argue—while the engine (and prop) is running in a congested area. Some operations ask on Unicom, while you are still in the air, if you will need service so that they'll know where to park you.

Tie the plane down. Don't go off and leave it untied, for several reasons: (1) Even if the wind is calm, it could pick up suddenly. (2) In the summer dust devils, or whirligigs as they are sometimes called, quite often suddenly move across the airport ramp. Some are strong enough to pick up a lightplane. (3) Propwash from other planes, particularly from big planes, can move your plane into other planes or into the side of the hangar. Later, when you have that private certificate and fly to an airport to spend the night or several days in the area, *do not set the parking brake, lock the airplane*, and leave for the motel or other parts unknown. The folks at the airport may need to move the airplane and this can't be done in that case. Make sure the airplane is well-chocked and tied-down with the *parking brake* OFF.

Take your logbook on solo cross-countries and have somebody sign it, attesting that you were at that airport.

Don't forget to close your flight plan. Sometimes you'll find it more convenient to contact the FSS and close it just after making the initial call to the tower. Or you may prefer to wait until you are on the ground and close it by phone.

It's an outside chance, but it has happened that pilots have closed their flight plans well before reaching their destination and had engine problems before contacting the tower (or landing at an uncontrolled airport), and Search and Rescue was delayed. You'll want to give yourself and your passengers the best protection, so you could cancel with the nearest FSS *after* initially contacting approach control or the tower, getting their permission to leave the tower frequency for a minute or so. If you had problems, approach control or the tower would wonder why you didn't call back or show up and could get things moving faster to find you. The tower will not automatically close your flight plan.

If you are on a solo cross-country and something happens that will delay you at an airport for a length of time (say, 30 minutes or more), also call the home airport and let them know what's happening. They have figured approximately when you should get home, allowing for time on the ground and average flight time, so don't make them have to call to find out what happened.

PROBLEMS AND EMERGENCIES

■ Getting Lost

If you encounter a problem, it will be most likely that of getting off course and/or losing checkpoints. The expression "and/or" is used because students (and experienced pilots) sometimes lose checkpoints even while *on course*. This is particularly true on parts of the trip where checkpoints are sparse or visi-

bility is somewhat restricted. Being at a low altitude aggravates the problem.

When this happens, don't start circling or heading off into what you think might be the "new" right direction. Maintain the compass heading that you were doing okay with earlier and try to identify other checkpoints. Pilots have missed a checkpoint and figured they were lost, but by continuing on their original heading found they were right on course, and checkpoints began to show up again.

If you are off course *and know it*, take a cut that will get you back to the course line at the next checkpoint (watch for it). Later when you have more experience, you won't be so hasty about getting back on course, but for now it's best that you get back to the preplanned route fairly soon (or else you might continue to drift farther off).

For the sake of argument let's say that you've been flying for a long time and nothing looks as it should. You should have reached the destination airport several minutes ago, and because of low visibility you believe (and your groundspeed estimates tend to confirm this) that you have missed it. It wouldn't be wise to continue blindly on, but you would set up definite means of locating yourself. First you'd climb, if the ceiling permits, to be able to see farther and to get better radio reception.

If the destination airport has an associated VOR and you have a working VOR receiver in the airplane, your problem is a small one. You'd look at the sectional chart (better yet, the *Airport/Facility Directory*) and get the VOR frequency. You'd turn on the set, *identify the station*, and rotate the omni bearing selector (OBS) until the needle was centered and the TO-FROM indicator showed TO. Then you'd track into the station using the OBS indicator as the base course.

If the destination airport doesn't have an associated VOR, your omni receiver can still help locate you. Look on the sectional chart and pick a couple of VOR stations in the *area* where you are. Pick some fairly outstanding landmark such as a river bend, road fork, or railroad junction and stay near it. Tune in one of the VOR stations, *identify it*, and turn the OBS until the needle centers and the TO-FROM indicator says FROM. Draw a line on the map from the chosen VOR along the bearing indicated on the OBS. Then tune in the other VOR and repeat the procedure.

You are in the *area* where the lines intersect. The term *area*, rather than *point*, is used because both your VOR receiver and the VORs may be slightly inaccurate. Find the outstanding checkpoint that you are circling by looking in the area of the intersecting lines on the map. This method of locating yourself by cross bearings is particularly interesting in turbulent air, and you may rest assured that at some time during the process you will drop (1) the pencil, (2) the straight-edge, (3) the map, or (4) all three at once.

Suppose you are on the Sewanee to Columbia–Mount Pleasant leg, the visibility has deteriorated, and you don't know where you are. You've made a couple of corrections and suspect that you may be south of course but aren't sure at all anymore. There's a very small town with a railroad track running north and south through it just ahead and you circle it as you work out the cross bearings.

1. You set up the Shelbyville VOR (109.0-SYI) on your receiver and *identify it*. Turn the OBS until the needle is centered and the TO-FROM indicator says FROM. You see that you are on the 250 radial of SYI.

Continuing to circle the small town, you note that there is a two-lane highway running north and south, just west of town, but continue your cross-bearing check.

2. Set up Graham VOR (111.6-GHM) and *identify* it. With the needle centered and the TO-FROM indicator reading FROM, you see that you are on the 140 radial of Graham. The *precise* line intersection of the two bearings shows that you are over the two Tennessee towns of Waco and Lynnville (and have an extraordinarily accurate VOR in this example).

Note on the chart that the two towns are quite close and Lynnville has a railroad running north-south through it.

Now that you have located yourself, you note that

the visibility is worse now to the south and east and your best move is to go on to Maury County Airport rather than try to find the Lawrenceburg or Pulaski airports.

Okay, the visibility is poor and you don't have an ADF. How would be the best way to get there? If you fly directly for it, there's a possibility that you might pass to the left or right of the airport and miss it. (And after deciding you had missed it—were you to the right or the left?)

Shown in Figure 25-4 are two ways of getting to Maury County Airport. Regardless of the great accuracy of dead reckoning and other types of navigation as indicated by novels and movies, roads and railroads are still very good ways to get to a point in a pinch. Notice that using route A you are going to intercept the highway well south of the airport—and will *know* which way to turn. You'll turn right, staying on the right side of the highway. It is an unwritten rule that airplanes following a highway always stay to the *right*. This makes it easier to see the highway from the left seat, and if everybody stays on the right there will be less chance of midair collisions. You would follow

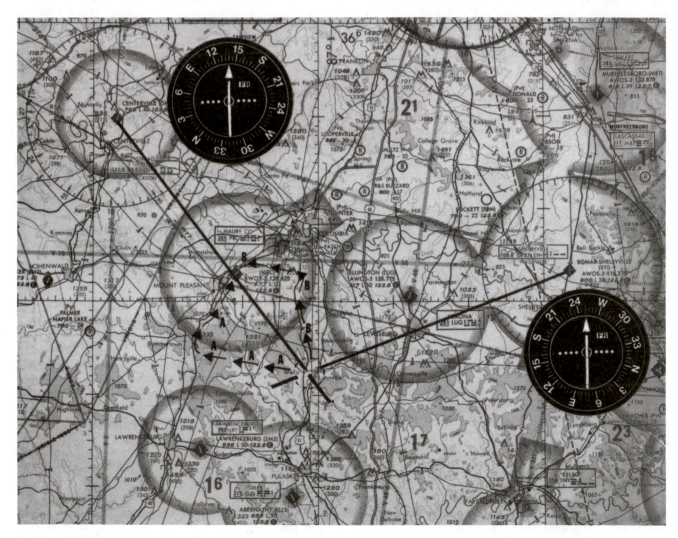

Fig. 25-4. Using VOR cross bearings to help locate yourself.

the highway north and call Unicom about 5 miles out.

Route B is another possibility. In this one you are staying over the original highway, following it north and turning to intercept the "airport highway" north of the airport. You would then follow that highway south to the airport, contacting Unicom. Following a highway if you are caught in bad weather is one way of getting to a safe place. Note the relationship of railroads and highways to each other if they are running side by side. Make sure that, for instance, the railroad is west (or east, etc.) of the highway as it should be. (You may be following the wrong highway.) Look for good checkpoints along the highway such as bridges, power line crossings, and other references. Watch for television or radio towers as shown on the chart.

Best of all, when you check the weather before the flight, if it looks like it will be deteriorating— don't go. It's better to be sitting in an airport office wondering why you *didn't* go than to be flying with a tense stomach in rapidly decreasing visibility—wishing you were back in the airport office.

Assume for this example that the radio beacon is on the airport, so (getting back to your imaginary problem with Maury County Airport) if you have an ADF on board you may tune in the radio beacon, *identify it,* and *home* into the airport (Fig. 25-5). (You'll probably have a curving path because of wind, as indicated by Fig. 21-3, but that's all right.) You could also use the ADF to help you in the highway-following process as discussed earlier. On route B you could follow the highway north from Waco–Lynnville until the ADF needle points well behind the left wing and then make your turn west to intercept the "airport's highway."

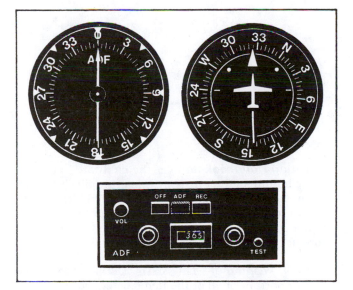

Fig. 25-5. Using the ADF to home into Maury County Airport after locating yourself over the town of Bufords. You could home into the station at the airport by changing the heading as necessary to keep the ADF needle on the 0° position in the indicator.

You may have to divert to an alternate airport because of weather or engine problems (such engine problems will be discussed shortly) and should have a good idea of the procedure before going on a solo cross-country. Your instructor, or the check pilot, may require that you leave the preplanned route for another airport on a dual flight or the check ride.

You should pinpoint your present position, if possible, estimate the heading (check the time), and then further work out heading and distances with a plotter—en route.

Figure 25-6 shows that a nearby VOR can be an aid in finding an alternate airport in poor visibility. You could find the radial that the airport is on, fly over to that radial, and fly—from TO or FROM the station (as applicable)—to get to the airport. Remember to look on both sides as well as straight ahead when looking for the airport. You may not have set up the exact radial and your VOR receiver also could be a couple of degrees off.

Suppose nothing has worked so far and you still have no idea of your location: Better make up your mind to get help before things get too far out of hand and you run out of fuel.

The FAA suggests the use of the "FOUR Cs" in such a situation:

Confess—Admit to yourself that you have a problem and *confess* it to the nearest ground station.

Climb—Altitude enables you to see farther and makes for better communications.

Communicate—You have a complex system of emergency aid no farther away than your microphone. Use 121.5 MHz if the other frequencies don't get any results.

Radar can also give you a hand. After you've called for help, the nearest radar facility may pick you up. The FSS or tower may have you switch to another frequency for radar help and you'll be requested to make various turns for radar identification or to set up a particular transponder code if you have the equipment. They will then vector you to an airport and will probably keep an eye on you to keep you on the straight and narrow.

Comply—Do what you're told by the aiding facilities.

Conserve—This a fifth "C" that is also very important. *Conserve fuel!* You'll be needing it. Don't grind around aimlessly at 75 percent power and rich mixtures when you could be using half the gas. For a situation where maximum endurance is needed (you are in communication and are being located and want to stay aloft as long as possible), a thumb rule would be to set up a speed of about 1.2 times the flaps-up, power-off stall speed as shown at the bottom of the green arc on the airspeed indicator, then adjust your power to maintain a constant altitude and lean the mixture. If your airplane has a calibrated stall speed of 50 K (flaps up) the maximum endurance speed would be at about 60 K (CAS). This would roughly give the most *time* per *gallon.*

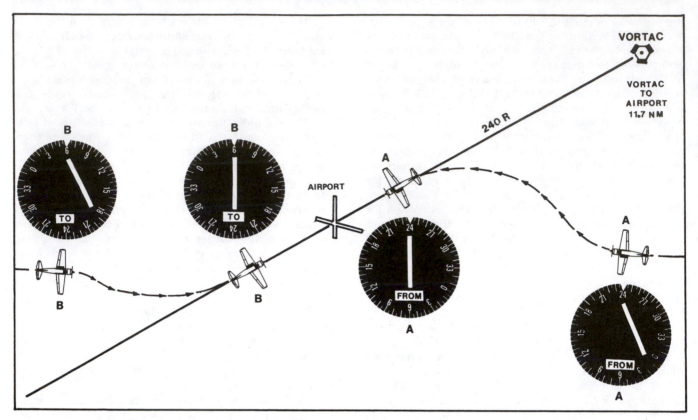

Fig. 25-6. Using a VOR to help find an airport, which for the example is on the 240 radial of the VORTAC. The pilot of airplane A would set up 240 FROM on the VOR receiver and turn toward the radial and fly until the needle is centered— then would set up a heading to keep the needle centered and "fly the radial" to the airport. The pilot of airplane B would set 060 TO on the VOR receiver and fly over to the radial, keeping the needle centered while looking for the airport. Both pilots would keep their eyes looking out of the cockpit, with only an occasional glance at the VOR receiver and heading indicator. DME equipment in the airplanes would make it easier too.

For maximum range (*miles* per *gallon*) another thumb rule is to set up an airspeed of 1.5 times the calibrated stall speed (clean). (The plane stalls at 50; 1.5 × 50 = 75 K for maximum distance per gallon.) This would be important over bad terrain or water when it is questionable whether or not you have the fuel left to get back to decent terrain. (Lean the mixture.)

Airplanes manufactured in 1976 and after have the airspeed indicators marked in knots and *indicated* airspeeds. The older airspeed indicators were marked in *calibrated* (or corrected indicated) airspeeds, and that's what these rules of thumb are based on. For the newer airplanes you would check Part 5 of the *Pilot's*

Operating Handbook and get the calibrated airspeed for the bottom of the green arc and multiply *that* number by the 1.5, or 1.2, as just given. You then might have to reconvert from the calibrated number you got by the multiplication back to indicated airspeed for more accurate flying.

Cool, keeping of—This sixth "C" is probably the most important of all of the items listed. *Take it easy.* If you have to, you can follow a road or railroad to a town and try to recognize it from the chart by landmarks such as racetracks, factories, and road and railroad complexes. The name of the town may be on a water tank, and some towns have their names painted on the roofs of large buildings. One student with a great opinion of the accuracy of VOR cross bearings, argued with the instructor: "That couldn't be Sewanee, the VOR bearings show us to be 5 miles north of there" (while looking at Sewanee prominently displayed in yellow on the roof of one of the buildings). This goes back to the idea that a VOR cross bearing is a good check of the *area* you are in. *Do not blindly depend on your navigation equipment when supposedly flying a cross-country by pilotage.*

All right, so you haven't been able to contact anybody, the ceiling is too low to pick up any help on the radio, or you have no radio and fuel is about gone. You'll have to make the decision to land the airplane before the engine quits from fuel starvation and you have no choice of fields. This is where dragging the area is a useful maneuver (Chapter 17). You'll pick the most likely looking field in the area and look it over. You've had practice in short- and soft-field landings, so do what is necessary—then get to a phone.

Probably it's unnecessary to note that once you've made an off-airport landing, you *do not* take off again, but will have your instructor or somebody come to the scene and fly the airplane out (or dismantle it for shipping home). There have been cases of pilots making off-airport landings with low fuel and/or weather problems with no damage to the airplane or themselves, and then when the weather improved, attempting to take off with a resulting—sometimes fatal—accident. Short-field *landings* are useful to a student or low-time pilot, but that would be no time to practice short-field *takeoffs*.

■ Weather

The best maneuver when in doubt about the weather ahead is a 180° turn while you still have visual references.

In the warmer months you may encounter isolated thunderstorms or fast-moving lines of thunderstorms called "squall lines."

In the case of the single thunderstorm (cumulonimbus cloud) you can circumnavigate, or fly around it. The vertical currents within a thunderstorm are able to tear an airplane apart. While you will not fly into the cell, it should be noted that violent vertical currents are also found in the areas close to the storm. Stay plenty wide of the cell, remembering these currents and the fact that hail sometimes falls out of the anvil-shaped overhang of the cloud.

If you are blocked by a squall line that is too long to fly around, there's nothing to do but go back.

Occasionally a squall line will have a break, through which the sun can be seen shining brightly on the other side. This is one version of what is known as a "sucker hole." If, in trying to go through the gap, the turbulence doesn't get you, the closing of the hole will probably do the trick.

In winter, snow and snow showers will be a problem. Visibility goes to nothing in snow, and snow areas can spread rapidly.

When the visibility is well down and you are flying a strange area, don't bore along at full cruise, but slow the airplane down. This isn't as critical now in the comparatively slow trainer you're flying, but later when you are flying that fast retractable it could make quite a difference. The radius of turn of an airplane for a given bank angle is a function of the *square* of the velocity, so if you are traveling twice as fast as you need to be when you see that ridge suddenly appear, your radius of turn will be four times as great. This doesn't mean you'd put the airplane right on the edge of a stall, but you would reduce airspeed to a value that would still give you safe control but would not maintain full-out cruise. (Trim the airplane.) Trying to remain VFR while the ceiling and visibility are getting below VFR minimums is an exotic form of Russian roulette.

If you are approaching for a landing and there's a thunderstorm in the near area, be prepared for very strong and shifting winds. Sometimes the thunderstorm may beat you to the airport. Usually they move on in a few minutes, but you will have to make the decision to wait it out or go on to another airport.

■ Engine Problems

If you have a partial power failure and can maintain altitude without overheating the engine, head for an alternate airport, but look for fields on the way. If possible, fly around poor landing areas. Usually if you get a partial power failure, a complete power failure is not far away.

The problem of dropping (or no) oil pressure was discussed in Chapter 3; as noted there, the instrument can be wrong, but you'd turn toward the nearest good landing area (preferably an airport) and watch the oil temperature gauge for a rise. If it starts to go up, reduce power to get more time from the engine to make that field.

If the engine starts overheating, head for an alternate airport. If the power loss is severe, remember that the least power is required at a speed about 1.2 times the stall speed. Conserve altitude. With such emergencies as power loss or overheating, the engine could quit at any time. This means that you'd be in a bad spot if it quit during a power approach or when dragging the area. When you reach the alternate, circle it within gliding distance. If it's a controlled field and you have a transmitter and receiver, before you get there tell them in plain language what's happening; the tower will clear traffic out of your way. If the engine is in danger of quitting, an emergency situation exists and you will have the right of way. For non-controlled fields keep your eyes open and land. (Contact Unicom if available.)

Carburetor ice (discussed in Chapter 4) may ease in while you're looking for checkpoints or have your attention diverted from engine operations. If the rpm (on a fixed-pitch prop) starts creeping back you'd better check on this. If you're using a constant-speed prop, the best indication of carburetor ice is a decrease in manifold pressure—if you are at a constant altitude and haven't moved the throttle. Follow your instructor's recommendations on use of carburetor heat (full or partial heat).

You may inadvertently fly into such heavy rain that, in addition to the expected visibility problems, it may partially block the carburetor intake screen and the engine will lose much of its power. Use carburetor heat or alternate air. Talk to your instructor and check the *Pilot's Operating Handbook* for recommendations for *your* airplane.

Engine fires in flight are very rare, and while you'll follow the instructions of your particular *Pilot's Operating Handbook* and/or instructor, generally you'll want to shut off fuel controls (mixture and fuel selector) and sources of sparks (ignition and master switches). This naturally would mean a forced landing, but that is not *immediately* the big problem. Sometimes the fire can be put out by diving. After the fire is out, you can keep your mind occupied with picking a field (Chapters 11 and 18).

You may be so absorbed in navigation that you run a tank dry en route. The sudden silence will certainly grab your attention and for a couple of seconds your mind will be blank. Throttle back and switch to the fuller tank and then reset power after the engine is running properly again. (It will take about 5 minutes after operations are normal for *you* to settle down again.) If you have been so remiss as to have flown out *all* of your fuel and have no other tank to switch to, you can get practice in picking a field (Chapters 11 and 18). Check the *POH now* for specific procedures when a tank runs dry. This may require use of electric fuel pump, richening the mixture, etc., as you switch.

■ Electrical Problems

Your instructor will go over possible electrical problems on the cross-country (or local flights, for that matter). If you smell burning insulation, better turn off the electrical master switch. If you don't need radios, etc., leave the switch off for the rest of the flight. If things are tight and you need radios and/or other electrical equipment, you might turn off the master switch to stop the immediate problem (kill the electri-

cal system) and then turn off all the radios and other electrical equipment. After the fire's out, open cabin ventilators to get rid of residual smells or smoke. One method of isolating the troublemaker is to turn the master switch back ON, and turn ON the individual systems one by one until you smell the culprit. Of course, what you will do in such a situation will actually depend on your airplane's electrical system, the *POH*, and your instructor's recommendations, based on your hours of experience. Your instructor may prefer to have you turn the master switch OFF and leave it that way and land at an airport as soon as practicable in case of "funny smells." (Remember that the electrical system has nothing to do with *keeping* the engine running; it only furnishes the power for starting.)

There have been cases of a popped circuit breaker or of inadvertently turning the master switch OFF ("somebody's" knee hit it), and the pilot looks up to see the fuel gauges on *empty*. This usually happens over water or the worst terrain in a 700-mile radius. Pilots have landed or ditched airplanes, thinking that there must have been a sudden and complete fuel leak because they should have had 2 (or 3, etc.) hours of fuel left at the expected consumption. This is not to say that a fuel leak should not be considered, but you might first check the master switch and circuit breaker(s) or fuses before landing in a strange place. If the master switch and circuit breaker(s) are okay (and the fuel gauges still indicate empty), a turn toward better terrain and airports should be in order.

Take a look at Figure 3-26 in this book, the electrical system for a light trainer. Note that the oil *temperature* gauge for that airplane has a circuit breaker. People generally don't get as excited about no oil *temperature* as they do about an indication of *no* fuel (understandably).

Probably after reading this section you're wondering if the cross-country flight will be nothing less than a hazardous operation. That isn't the case at all; it's highly unlikely that you will ever have a real problem on the cross-country. You may have to avoid bad weather or may find yourself off course temporarily, but you can minimize the chances of trouble by *careful preflight planning*. The flight is usually a success or failure before leaving the ground. The cross-country is just like the rest of flying—*a matter of headwork*.

26

INTRODUCTION TO NIGHT FLYING

You will probably face your first session of night flying with some trepidation but will find that, once you get started, it can be even more enjoyable than daytime operations. It will increase your confidence in your daytime flying, too.

Usually the air is smoother, and you can see cities and towns at greater distances because of their lights. You'll also be able to see more airborne traffic for this same reason.

There's one factor that you will likely encounter, and that is *Zilch's Law.* Zilch's Law basically states that as you get out of gliding range of the airport at night the engine starts running rough. Zilch's Law is also applicable in the daytime when flying over mountainous terrain or well out over the water (in a land plane). "The roughness of the engine is directly proportional to the *square* of the roughness of the terrain and the *cube* of the pilot's imagination." This phenomenon is also known as *automatic rough* (Fig. 26-1).

Fig. 26-1. Zilch's Law makes the engine run rough and "sound funny" over bad terrain or when out of gliding distance of the airport at night. This phenomenon starts to make itself known when darkness approaches and you are still some distance from the airport.

You'll get 3 hours of dual instruction including at least 10 takeoffs and landings to a full stop. You'll fly a dual night cross-country of over 100 NM total distance. The instructor will probably also have you fly in the local area at a good altitude before having you come back into the pattern for the 10 dual takeoffs and landings.

CONES AND RODS

Owls and cats (and other night predators) do better than people at night because their eyes are designed for it.

The retina, or layer of cellular structure at the back of the eyeball, electrochemically converts the light of the optical image into nervous impulses flowing to the brain. This conversion takes place in light-sensitive cells (millions of them), which are of two distinct types: (1) The rods (highly sensitive, activated by a minimum of light intensity) enable only black and white (but no color) perception and are mostly useful for night vision. (2) The cones, considerably less light sensitive than the rods, require higher light intensities to get activated, enable color perception as well as black and white, and are mostly useful for daylight vision.

The human retina contains both types of cells, the animal retina contains only rods. A bull gets excited by the movements of the bullfighter's muleta, not by its red color, which he is unable to perceive.

Your top visual acuity occurs by focusing your eyes on an object, in which case its optical image is projected on a tiny portion of the retina called the *fovea.* Foveas are very useful for the details needed in say, "girl watching."

You will never make out visual details at night as well as by daylight because the human fovea contains exclusively the less sensitive color-distinguishing cones and in the dark becomes much less sensitive than the rest of the retina, which contains only rods. The resulting differences in the visual perception is, in principle, illustrated in Figure 26-2, which shows an identical scene visualized by the human eye by day-

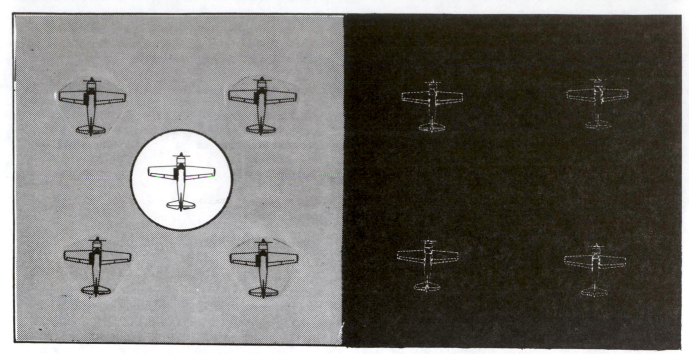

Fig. 26-2. Different views of a flight of five airplanes as seen by day and night. The vision is focused in the center of the formation in both cases. At night a blind spot exists in the center; in order to see that airplane, you would focus on one of the outer airplanes. *Scan* slowly at night.

light and at night. Owls, cats, etc. don't have this kind of visual problem (they have other visual problems). Their retinas, containing only rods, are uniformly relatively highly sensitive in the dark.

Figure 26-3 gives an idea of the physical makeup of the human eye. The light reflected from the objects in your visual field enters the eyeball; the amount of the light admitted is constantly being adjusted by the changes in diameter of the iris (which is comparable to the diaphragm of a camera). The light rays go through the optical media (lens, etc.), form an optical image of those objects, and project it on the retina (which is comparable to the film in a camera).

The optical nerves of both eyes are interconnected in a crossover and transmit the light-generated impulses to the optical centers in the brain.

The final visual perception results from interactions between the transferred image from the retina and the activity of the optical brain centers and is therefore different from the real optical image. In other words, the eyes can lie to you about what's really going on. You could have an eye disease characterized by a blind sector in the retina and not be aware of it. The brain simply fills in what it feels should be there. A blind spot exists in the normal eye where the optical nerve leaves the retina; you can find it in your own eye by the following experiment: Figure 26-4 is a section from this book with a cross and large dot marked in it. Hold the figure approximately 6 inches from your eyes—eyes perpendicular to the page—and focus with your right eye on the cross. When you

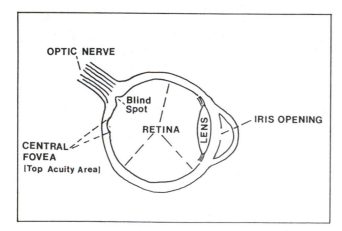

Fig. 26-3. A look at the human eye.

There's one factor that you will likely encounter, and that is *Zilch's Law*. Zilch's Law basically states that as you get out of gliding range of the airport at night, the engine starts running rough. Zilch's Law is also applicable in the daytime when flying over mountainous terrain or well out over the water (in a land plane). "The roughness of the engine is directly proportional to the *square* of the roughness of the terrain, and the *cube* of the pilot's imagination." This phenomenon is also known as

Fig. 26-4. A test for checking the blind spot in each eye.

cover or close your left eye, the dot will disappear because its image will fall on that eye's blind spot. You may have to adjust the distance from your eye to the figure. The experiment can be repeated with the left eye (right eye closed) focusing on the dot. You'll get the impression that you can see the printing that's covered by the dot or cross because the brain "fills it in." Think about that when you stare fixedly from the airplane (day or night). Another airplane might be in your blind spot and the brain fills in background details (horizon or blue sky) because it can't "see" that area and assumes that it's landscape like the rest. Of course, using two eyes tends to cover possible blind spots and the idea of two-eye *scanning* is even better—day or night.

You'll never be able to see as well at night as in the day, and Figure 26-2 is merely given to show the comparative areas of coverage.

The shift from cones to rods is gradual; as the light decreases, more and more of the rods take over. The rods require a certain period of adaptation to the dark. That's why you sat on that person's lap in the movie theater—your eyes weren't dark-adapted.

It takes about 30 minutes to become night-adapted. The rods are not as sensitive to red lights, so night fighter pilots are briefed in a room with red lights or wear red-lensed goggles before a flight.

Adaptation to the dark actually occurs in two steps: (1) A first rise in sensitivity, due to the cones, is completed in 4–5 minutes. (2) The second rise is due to rods and is completed in 40 minutes, making a *total* time of about 45 minutes. But, as noted, you'll be about as dark-adapted as practical after a total time of 30 minutes. That last 15 of the 45 minutes raises adaptation only a little.

While a theoretical look at the eyes' reactions to darkness might warrant the program of red lighting just mentioned (as well as having all cockpit and instrument lights red), for practical purposes in civil aviation white lighting is used in many of the newer airplanes.

If you are night-adapted, a bright flash of white light will ruin your careful work, and the whole adaptation procedure has to begin again.

It's impractical and unnecessary for you to wear red goggles for 30 minutes before the flight, but you should have an idea of what to do for night vision:

1. Avoid *brilliant* white lighting before flying if possible.

2. Under normal flying conditions keep the instrument and cockpit lighting as low as practicable.

3. If you are over a brightly lighted city or other bright outside lights, turn the instrument lights up in intensity.

4. When looking for objects at night, scan, don't stare. Remember that center blind spot.

5. *Always* carry a flashlight when night flying. It can be very difficult to fly while holding a cigarette lighter or matches to check the instruments. Older instruments had luminous paint on the numbers and hands that could be "excited" by shining a flashlight

on each instrument for a few seconds. The flashlight could be turned off and the hands and numbers would continue to glow for several minutes. This was helpful when cockpit lighting was lost. On the downwind leg the instruments could be "energized." You wouldn't need the flashlight any more and would have both hands free for the rest of the approach and landing. You might order a Cyalume, a "cold" lightstick that can be stored in the map compartment until needed. It can be taken out of its wrapper and energized by bending it and shaking it, then it gives several hours of constant light for chart or instrument reading. Since the light is "cold," it can be put in a pocket or back in the compartment when light is not needed.

6. Don't stare at the instruments; it tires your eyes quickly.

Oxygen helps night vision but it's doubtful that you would have it or want to use it. Alcohol and tobacco hurt night vision. If you were a night fighter, you could set up a daily routine (including a diet to ensure that there was sufficient vitamin A in your system) to help improve your night vision, and all of your cockpit lighting would be red, including your flashlight lens. (If you need to convert your flashlight to a red light type, a round piece of red cellophane placed inside the lens is an inexpensive way.)

AIRPLANE LIGHTING

Following are requirements for night lighting and equipment, in addition to the required equipment discussed in Chapter 3 (FAR 91.205):

1. Approved position lights.

2. An approved aviation red or aviation white anticollision light system.

3. If the aircraft is operated for hire, one electric landing light.

4. An adequate source of electrical energy for all installed electrical and radio equipment.

5. One spare set of fuses, or three spare fuses of each kind required.

The airplane's navigation (or position) lights, like a lot of other aviation equipment and terminology, are arranged similarly to those of ships. The lighting arrangement (red light on the left wing, green on the right, and white light on the tail) can tell you approximately what that other airplane is doing.

Suppose that you see the red and green wing lights of an airplane dead ahead and on your altitude. Is it coming or going? One clue that it is moving toward you is the fact that the white taillight is very dim or not showing. The final touch would be if the red light is on your right and the green is on your left—it is headed your way!

Remember the term, "Red, right, returning," meaning that when you see the red and green lights of another airplane and the red light is on the right, the other airplane is headed toward you—not necessarily on a collision course but at least in your general direction.

The airplane's position lights must be on *any time the airplane moves after sunset*—taxiing or flying. Also, the lights must be on if the airplane is parked or moved to within dangerous proximity of the portion of the airport used for or available to night operation if the area is not well lighted.

The red rotating beacon (or a white anticollision light, as applicable) is required for all airplanes flying at night. The anticollision light can be seen many miles away in good atmospheric conditions. Many pilots also turn on this light in the daytime when visibility is down.

The landing light is a great aid but not as necessarily vital to night operation on a lighted field as you might think. If you find yourself at night with the landing light on the blink, you'll still have no trouble at all making a good landing by watching the runway lights and using a little power in the landing.

Don't fly without adequate cockpit lighting. If you fly into hazy conditions, it's possible that you may lose all visual references and have to go on instruments to get out of that area. You may also fly into clouds or fog before realizing it. You'd be in pretty sad shape if you couldn't see the instruments.

If you inadvertently get on solid instruments, turn off the anticollision light. The "flashing" effect as it rotates can lead to disorientation. After you've completed the 180° turn and are back out in VFR conditions, turn it on as you go back home.

AIRPORT LIGHTING

One of the toughest parts of night flying when going into or out of a strange, large airport is the taxiing. This is especially rough if you haven't been there in the daytime. Even a familiar airport is strange the first time you try to taxi at night. The only thing to do is to get help from ground control if it's available. Otherwise just take it slow and easy until you find your destination on the field. One private pilot, on his first night taxi, was too proud to ask directions and managed to taxi out the airport gate and onto the freeway. He was last seen taxiing at 60 miles an hour, trying to keep from being run over by traffic (auto type). He didn't know that taxiway lights are blue. Figure 26-5 shows taxiway and runway lighting. Some bigger airports have small lights embedded in the taxiway for guidance.

The runway edge lights are white lights of controlled brightness and are usually at the lowest brightness compatible with atmospheric conditions. The runway lights in some airports may be nearly flush with the runway; others may be on stands a foot or two high. The lights marking the ends of the runway are green (Fig. 26-5). Some runway end lights are red on the runway side to warn pilots on the roll-out that The End Is Near.

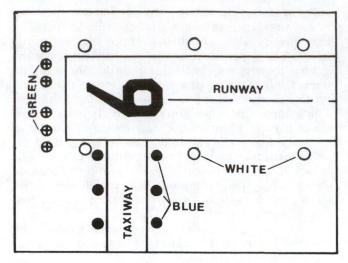

Fig. 26-5. Taxiway and runway lighting.

Airport beacons flash alternating green and white. This is because they are rotating and have the different-colored lights on opposite sides of the beacon head. The following should be noted:

1. Civilian airport beacons—green-white-green, etc.

2. Military airport beacons—green-white-white-green, etc.

The military airport beacon has a dual-peaked white flash between the green flashes to differentiate it from the civilian. Remember this because the beacons can be seen from a great distance and are used to help the pilot locate the airport at night. It would cause plenty of excitement if you were talking to a civilian tower and landed at a nearby SAC base by mistake.

If an airport beacon is on during the day in Class B, C, D, or E surface areas, it often means that the weather conditions at the airport are less than 3 miles visibility and/or 1000 feet of ceiling. It's no longer VFR.

All dangerous obstructions near the airport are marked by red lights. High-intensity flashing white lights identify tall structures (chimneys, etc.) and overhead structures of transmission lines going across rivers, gorges, etc.

TIPS FOR NIGHT FLYING

You might consider the following when you are about to start night flying.

■ Starting

Be doubly careful in starting at night. The darkness could result in your starting the engine when someone was in the propeller arc. Turn on the position (navigation) lights before starting. Granted, this won't help the battery, but it will serve to warn people on the ramp that you are going to start. If other airplanes' engines are running-up close by, no one can hear your "CLEAR." Keep all other equipment off while starting, particularly the taxi and landing lights and radios. If you can possibly work it, have someone standing outside as you start. If it's all clear, that person will give you a signal that consists of moving one of the wands or flashlights in a rotary motion and pointing to the engine to be started with the other wand (in your case, at this stage, pointing to *the only* engine).

■ **Taxiing**

After the engine (or engines) is started, turn on other needed equipment but leave the taxi or landing lights off until you are released by the taxi director. The taxi signals shown in Figure 6-8 still apply at night. The director will hold a lighted wand or flashlight in each hand.

A signal normally used at night but not shown in Figure 6-8 is crossed wands or flashlights, meaning "stop" or "hold brakes."

Taxi slowly at all times and very slowly in congested areas at night. You'll find that judgment of distances is harder when you first start night operations. Your visibility is limited to the area covered by the taxi light and you don't want to have to suddenly apply brakes while moving at a high speed.

■ **Run-Up**

Once you have reached the run-up position and have stopped, turn the taxi or landing lights off to conserve electrical power. Turn off the strobes to avoid distracting other pilots. Use a flashlight to check controls and switches not readily seen by normal instrument or cockpit lighting. Make a pretakeoff check carefully, as you should always do—day or night.

When you are ready to go, call the tower; or if you have no transmitter, turn on the landing light and turn toward the tower. (You would arrange by phone before flying in such a rare situation.) If you are at a nontower airport, clear the area carefully before taxiing out onto the runway. Make sure the cockpit lighting is at the proper level and that the navigation lights and strobes or rotating beacon are on.

TAKEOFF

You will normally use the landing light(s) during the takeoff roll, but the instructor will probably have you make several takeoffs and landings without it during one or more of the dual sessions. You'll find that the runway edge lights and the landing light will be more than sufficient to help you keep the airplane

straight. Don't just stare into the area illuminated by the landing lights. Scan ahead, using the runway edge lights as well. Your first feeling will be that of a great deal of speed because of your tendency to pick references somewhat closer than during daytime takeoffs. Because of this, make sure that you have reached a safe takeoff speed before lifting off; don't try any short- or soft-field takeoff techniques for the first few times. Actually, only in an emergency would you attempt taking off from a very short or soft field at night.

Retract the gear (if applicable) when you are definitely airborne and can no longer land on the runway ahead. Establish a normal climb and turn off the landing light(s). Turn off the boost pump (if applicable) at a safe altitude.

AIRWORK

You'll find that you are referring more to the flight instruments at night than in the daytime because of the occasional lack of a definite horizon in sparsely populated areas.

On a hazy night with few scattered lights, the ground lights may appear as a continuation of the stars and you could talk yourself into a good case of disorientation. If you're in doubt, rely on the instruments.

You may have a tendency to fumble for controls in the cockpit at first. It's wise to spend extra time sitting on the ground in the cockpit until you can find any control blindfolded. At night you may suddenly encounter IFR conditions. Make a 180° turn as discussed in Chapter 15.

Figure 26-6 shows some night lighting combinations for various relative airplane positions and headings. If you see the *Super Cub* (the bottom airplane) at this position and relative size, you could be in a little trouble.

APPROACH AND LANDING

The landing checklist should be carefully followed, taking even more time than in daylight, to ensure that you make no mistakes.

Make the downwind leg farther from the runway than in the daytime. Plan on a wider, more leisurely pattern. Since judging distances also is more difficult, plan on a pattern that requires the use of power *all the way around*. Figure 26-7 shows three views of the lighted runway after turning final.

Turn the landing lights on after turning onto final. You will notice that if the night is hazy, your visibility will be less during the early part of the final approach with the landing lights on but will improve as you get to the round-off height. When you are using landing lights, the landing technique should be very close to

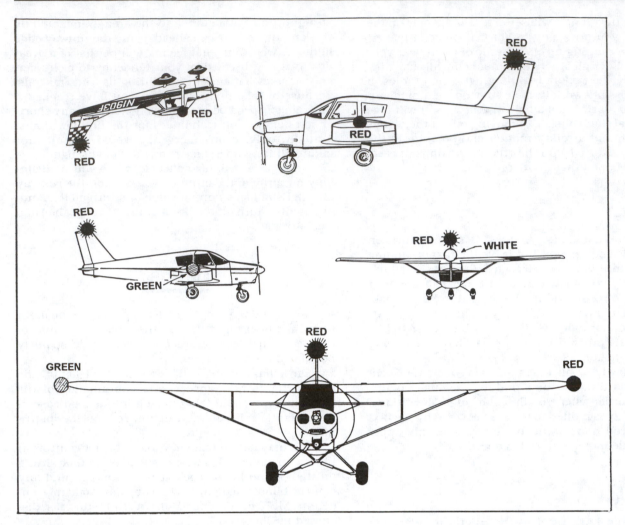

Fig. 26-6. Some of the light combinations of other airplanes you might see on a night flight.

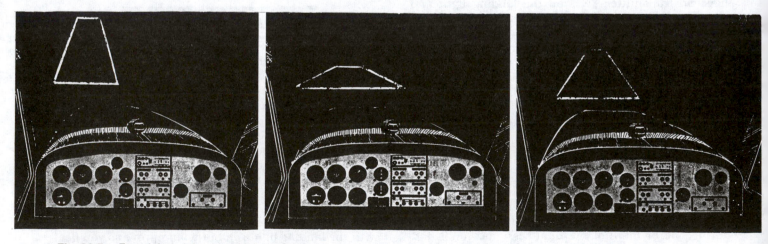

Fig. 26-7. From left to right: a too-high, too-low, and proper final approach as seen from the cockpit.

that used in daytime landings (power-off). If the landing lights are not used, make the touchdown still carrying some power. Those first few landings will seem to be a speeded-up process. Figure 26-8 shows how that first night landing might seem to you.

The common tendency is to carry too much airspeed on night landings; watch for it.

The landing light is important, but the amount of light reflected varies with the atmospheric conditions for a particular runway and will vary from airport to airport. The runway lights may be the essential reference.

Your instructor will probably have you shoot takeoffs and landings without the landing lights and may also have you shoot one or two without cockpit lighting so that you can see the difficulty of handling a flashlight and flying the airplane. The instructor will also review emergency procedures such as partial power loss, or total engine failure, and electrical problems with special attention to your particular airplane and local terrain.

SUMMARY

You'll find that after the slight confusion of the first flight has passed, you'll be sold on night flying and Zilch's Law won't be such a big factor. If you get caught away from the airport after dark, it won't be a crisis situation after you've had instruction. Night flying is like everything else: familiarity and experience ease your worry.

Fig. 26-8. On your first night landing, things will seem to be moving much more rapidly than in the daytime.

27

THE KNOWLEDGE TEST

The purpose of your flight and ground instruction is to make you an efficient and safe pilot, *not* to specifically prepare you for passing tests. However, in order to use all of that good training and to fly passengers to those vacation spots you have in mind, you'll have to pass the knowledge and practical tests. This chapter is aimed at the private pilot applicant.

Before going up for the practical test, you'll have to have passed the FAA knowledge test with a score of 70 percent or better. Some people sweat the knowledge test more than the practical test, particularly if they've been out of school and haven't taken any examinations for more years than they'd like to remember. As often happens, however, they have better self-discipline than recent high school or college graduates (applying their studying time better) and they do better on the knowledge test than expected. This doesn't ease that initial nervousness when sitting down face to face with a blank answer sheet.

HOME STUDY

You may not have an opportunity to attend a formal ground school or you may prefer to study on your own; either is a legitimate way to get ready for the knowledge test. However, you'll want to get some advice on areas to study from your instructor, who can then review your work and make a written statement that you have successfully completed the requirements so that you can take the test. Your instructor will suggest sections of the Federal Aviation Regulations (Parts 61, 91, and NTSB 830) to study.

■ Review

Following is a list of chapters with suggestions for home study that apply especially to the knowledge test:

Chapter 2—Know the four forces and how they work. Remember that lift doesn't make the airplane climb but that the axis of reference is the airplane's flight path and the forces are measured parallel and perpendicular to the flight path. Make sure that you know how the trim tabs move to get the required control (elevator, etc.) reaction.

Chapter 3—Review the actions of the various flight instruments, noting what could cause erroneous indications. (Know the various types of airspeeds and altitudes.) Be sure that you know what airplane document must be *displayed* and whether the logbooks must be kept in the airplane at all times. Know the requirements for 100-hour and annual inspections.

Chapter 4—Know *why* you check the various items in the preflight check. It will make it much easier to remember. In fact, in all of your ground instruction study, try to tie all of the information with your experiences in the airplane. You should be getting ready for the knowledge test as you fly; don't wait and cram just before the test. It's best to take the test after you've had at least one solo cross-country and had a chance to pin down in your mind that area of the training, because many of the questions are based on a typical cross-country trip.

Chapter 5—Review the mixture control, general starting and shutdown, and procedures for cold or hot engines or different outside air temperatures. Remember that the airplane should not be hand propped by anyone who has not had proper instruction in the art; there must be a competent operator inside the airplane during this effort.

Chapter 6—You may run into a question on the test about how the controls should be held when taxiing in winds from various directions relative to the airplane. Confirm that you know the various airport wind indicators and can read them properly (and know what indicates a right-hand traffic pattern).

Chapter 7—Here in the pretakeoff or cockpit check, know *why* items are checked. (What are you looking for in the mag check or carb heat check, etc.?)

Chapter 8—Know the three axes of the airplane and how the flight controls (and trim) work.

Chapter 9—Know the Four Fundamentals, and the effects of bank on the stall speeds. Understand torque and the forces and moments that operate on the airplane at lower airspeeds and higher power settings.

Chapter 12—The stall is a function of angle of attack, *not* airspeed, although the calibrated airspeed at the stall depends directly on the square root of the weight of the airplane. Know the effects of flaps and power on the stall speed (CAS) at a given weight. For light general aviation trainers altitude has no effect on the calibrated airspeed (CAS) at the stall if flaps, weight, power used, and bank angle are the same. Check Figure 12-13 for a review of power required in relation to CAS.

Chapter 13—Check with your instructor about the demonstrated maximum crosswind component for your airplane and also get the word on avoiding wake turbulence. If the *Pilot's Operating Handbook* has a crosswind chart, you should review its use.

Chapter 14—Reaffirm that the stall is the result of a too-high angle of attack, not airspeed. Review accelerated stalls.

Chapter 15—Be able to look at diagrams of flight instruments and describe what actions the airplane is taking (Figs. 15-1, 15-3, 15-4, 15-5, and 15-12).

Chapter 16—Note that in turns around a point and eights around pylons, the angle of bank, as in S-turns across the road, is proportional to the ground speed. Be able to look at any diagram of these maneuvers (with wind direction shown) and pick the point of steepest or shallowest bank.

Chapter 17—Look over Figures 17-10, 17-11, 17-12, 17-13, and 17-14 to make sure you can use the different types of takeoff and landing performance charts. Review the atmospheric factors that affect the performance (e.g., moist air, if factors of temperatures and pressure are equal, gives *poorer* performance than dry air). Temperature changes affect the density-altitude more than pressure changes.

Chapter 18—Know the significance of charts such as Figure 18-4.

Chapter 19—Review this chapter thoroughly and be able to work from true to magnetic to compass directions and back, using isogonic lines and a compass correction card (Fig. 19-5). When working from CAS to IAS on an airspeed correction table, you may have a

tendency to add when you need to subtract, and vice versa (check Fig. 19-7). You may want to have your instructor give you some sample problems in "different approaches." For example, you may be given the pressure altitude and density-altitude and required to find the outside air temperature. In other words, pin down your knowledge and use of the computer.

When measuring courses with the plotter, double check the figures you get. You might measure a number of courses between points on your sectional chart and ask the instructor to confirm their accuracy.

Be able to locate geographic points when given latitude and longitude.

Chapter 20—Sit down one evening for a few minutes with your sectional chart and look it over for symbols you might not have encountered in your cross-country planning.

You'll be expected to be able to use the *Aeronautical Information Manual* and the *Airport/Facility Directory* on the written *and* the practical test, so you should review all parts (available at the airport). (Maybe you could borrow them overnight for some home study.) You won't be expected to memorize any of the *AIM*, but a fairly brief review of what information is available and how to use the *Airport/Facility Directory* could save you some fumbling around during the knowledge test. (Such information is furnished but it's better if you've looked it over just before taking the test.)

Chapter 21—Review the uses of the VOR and ADF and their advantages and limitations. One question sometimes used on the knowledge test includes a diagram showing several airplanes at various positions around a VOR station, and you are required to match the proper VOR received indication with the airplane at a particular position. (One hint is to "turn" the airplane in your mind to the OBS given and see whether the needle is pointed toward the selected radial.) Your instructor will want to review this with you.

Know the basics of ATIS and VASI.

Chapter 22—Confirm the circulation around low- and high-pressure areas, the symbols, and weather associated with warm, cold, occluded, and stationary fronts. Be able to read hourly reports, forecasts, and wind information. Brush up on converting local time to UTC (Zulu time) and back. Know what weather information services are available to the pilot.

Chapter 23—Know the significance of maneuvering speeds and the effects of turbulence on the airplane. Brush up on the steps of running a weight and balance calculation and remember that the *basic empty weight* includes full operating fluids and unusable fuel and full oil. Be able to use both types of cruise charts as shown by Figures 23-12 and 23-14.

Chapter 24—Go over in your mind the steps involved in navigation planning, including filling out a flight log and filing a VFR flight plan. Ask your instructor to review with you the procedure for reporting an accident or an overdue aircraft.

Chapter 25—Fly one of your own cross-countries again in your mind and think of the items you had to consider on the trip, including (as applicable) taxi director signals and runway markings, as you review this chapter.

Take another look at procedures used in helping to locate yourself by use of VOR cross bearings and what FAA ground facilities are available to help.

Chapter 26—Hit this chapter for aircraft and airport lighting and the physiological aspects of night vision.

Check Appendix A for "Wake Turbulence" and Appendix C for "Medical Facts for Pilots" taken from the *AIM*. There will be questions on the knowledge test about these subjects; as a pilot you should be aware of the factors discussed.

The questions, answers, and explanations in this book are the latest at the time of printing. The questions, whether those here or those revised by the FAA, are aimed at the same areas of knowledge. The purpose of a textbook is to provide general knowledge in order to help you fly safely as a private pilot and beyond.

PRIVATE PILOT KNOWLEDGE TEST

FEDERAL AVIATION ADMINISTRATION

CONTENTS

PREFACE

The Flight Standards Service of the Federal Aviation Administration (FAA) has developed this guide to help applicants meet the knowledge requirements for private pilot certification.

This guide contains information about eligibility requirements, test descriptions, testing and retesting procedures, and sample test questions representative of those used in the official tests. Sample test questions and choices of answers are based on regulations, principles, and practices valid at the time the guide was printed. In addition, Appendix 1 of *The Student Pilot's Flight Manual*, starting on page 27-67, provides a list of reference materials and subject matter knowledge codes. Computer testing designees appear on page 27-6. The list of subject matter knowledge codes should be referred to when reviewing areas of deficiency on the airman test report. Changes to the subject matter knowledge code list will be published as a separate advisory circular.

The private pilot test question bank and subject matter knowledge code list for all airmen certificates and ratings, with changes, may be obtained by computer modem from FedWorld at (703) 321-8020. This bulletin board service is provided by the U.S. Department of Commerce 24 hours a day, 7 days per week. For technical assistance regarding computer software and modem requirements for this service, contact the FedWorld help desk at (703) 487-4608 from 7:30 a.m. to 5 p.m. EST, Monday through Friday.

This publication may be purchased from the Superintendent of Documents, U.S. Government Printing Office, Washington, DC 20402-9325 or from U.S. Government Printing Office bookstores located in major cities throughout the United States.

Comments regarding this guide should be sent to:

Federal Aviation Administration
Operations Support Branch, AFS-630
ATTN: Private Pilot Certification Area Manager
P.O. Box 25082
Oklahoma City, OK 73125

PRIVATE PILOT KNOWLEDGE TEST GUIDE

INTRODUCTION

The FAA has available hundreds of computer testing centers nationwide. These testing centers offer the full range of airman knowledge tests including military competence, instrument foreign pilot, and pilot examiner predesignated tests. Refer to the list of computer testing designees on page 27-6 of *The Student Pilot's Flight Manual*.

The knowledge test guide in this book was developed to be used by applicants preparing to take a knowledge test for the Private Pilot—Airplane, Single-engine, Land.

This is not offered as a quick and easy way to obtain the necessary information for passing the knowledge tests. There is no quick and easy way to obtain this knowledge in addition to the skills needed to transform a student into a pilot capable of operating safely in our complex national airspace system. Rather, the intent of this guide is to define and narrow the field of study, while directing the applicant to the required knowledge for obtaining a private pilot certificate.

ELIGIBILITY REQUIREMENTS

An applicant for a private pilot certificate should review FAR Section 61.83, Eligibility requirements: Student pilots, for detailed information pertaining to eligibility.

An applicant for a private pilot certificate should review FAR Section 61.103, Eligibility requirements: General, for additional detailed information pertaining to eligibility.

KNOWLEDGE AREAS ON THE TESTS

The tests are comprehensive as they must test an applicant's knowledge in many subject areas.

An applicant for a private pilot certificate or added

rating should review FAR Section 61.105, Aeronautical knowledge, for the knowledge areas on the tests.

DESCRIPTION OF THE TESTS

All test questions are the objective, multiple-choice type, with three choices of answers. Each question can be answered by the selection of a single response. Each test question is independent of other questions, that is, a correct response to one does not depend upon, or influence, the correct response to another.

The maximum time allowed for taking the test is based on previous experience and educational statistics. This amount of time is considered adequate for applicants with proper preparation and instruction.

The Private Pilot—Airplane Test contains 60 questions and 2.5 hours is allowed to take each test.

Communication between individuals through the use of words is a complicated process. In addition to being an exercise in the application and use of aeronautical knowledge, a knowledge test is also an exercise in communication since it involves the use of the written language. Since the tests involve written rather than spoken words, communication between the test writer and the person being tested may become a difficult matter if care is not exercised by both parties. Consequently, considerable effort is expended to write each question in a clear, precise manner. Make sure you carefully read the instructions given with each test, as well as the statements in each test item.

When taking a test, keep the following points in mind:

1. Answer each question in accordance with the latest regulations and procedures.

2. Read each question carefully before looking at the possible answers. You should clearly understand the problem before attempting to solve it.

3. After formulating an answer, determine which choice most nearly corresponds with that answer. The answer chosen should completely resolve the problem.

4. From the answer given, it may appear that there is more than one possible answer. However, there is only one answer that is correct and complete. The other answers are either incomplete, erroneous, or represent common misconceptions.

5. If a certain question is difficult for you, it is best to mark it for RECALL and proceed to the next question. After you answer the less difficult questions, return to those which you marked for recall and answer them. The recall marking procedure will be explained to you prior to starting the test. Although the computer should alert you to unanswered questions, make sure every question has an answer recorded. This procedure will enable you to use the available time to the maximum advantage.

6. When solving a calculation problem, select the answer nearest your solution. The problem has been checked with various types of calculators; therefore, if you have solved it correctly, your answer will be closer to the correct answer than any of the other choices.

TAKING A KNOWLEDGE TEST BY COMPUTER

You must determine what authorization requirements are necessary before contacting or going to the computer testing center. Testing center personnel cannot begin the test until you provide them with the proper authorization, if one is required. A limited number of tests require no authorization. In the case of retesting, you must present either a passed, expired passed (24 months old) or failed test report for that particular test. This policy is covered in FAA Order 8080.6, Conduct of Airmen Knowledge Tests via the Computer Medium. However, you should always check with your instructor or local Flight Standards District Office if you are unsure of what kind of authorization to bring to the testing facility.

The next step is the actual registration process. Most computer testing centers require that all applicants contact a central 1-800 phone number. At this time you should select a testing site of your choice, schedule a test date, and make financial arrangements for test payment.

You may register for tests several weeks in advance of the proposed testing date. You may also cancel your appointment up to 2 business days before test time, without financial penalty. After that time, you may be subject to a cancellation fee as determined by the testing center.

You are now ready to take the test. Remember, you always have an opportunity to take a sample test before the actual test begins. Your actual test is under a time limit, but if you know your material, there should be sufficient time to complete and review your test. Within moments of completing the test, you will receive an airman test report, which contains your score. It also lists those subject matter knowledge areas where questions were answered incorrectly. *The total number of subject matter knowledge codes shown on the test report is not necessarily an indication of the total number of questions answered incorrectly.* These codes refer to a list of knowledge areas that can be found in Appendix 1 of *The Student Pilot's Flight Manual,* starting on page 27-67. You can study these knowledge areas to improve your understanding of the subject matter.

Your instructor is required to review each of the knowledge areas listed on your airman test report with you, and complete an endorsement that remedial study was conducted in these deficient areas. The examiner may also quiz you on these areas of deficiency during the practical test.

The airman test report, which must show the computer testing company's embossed seal, is an important document. DO NOT LOSE THE AIRMAN TEST REPORT as you will need to present it to the examiner prior to taking the practical test. Loss of this report means that you will have to request a duplicate copy from the FAA in Oklahoma City. This will be costly and time consuming.

CHEATING OR OTHER
UNAUTHORIZED CONDUCT

Computer testing centers follow rigid testing procedures established by the FAA. This includes test security. When entering the test area, you are permitted to take only scratch paper furnished by the test administrator and an authorized aviation computer, plotter, etc., approved for use in accordance with FAA Order 8080.6, Conduct of Airmen Knowledge Testing via the Computer Medium, and AC 60-11, Aids Authorized for Use by Airman Written Test Applicants. The FAA has directed testing centers to stop a test any time a test administrator suspects a cheating incident has occurred. An FAA investigation will then follow. If the investigation determines that cheating or other unauthorized conduct has occurred, any airman certificate that you hold may be revoked, and you may not be allowed to take a test for 1 year.

RETESTING PROCEDURES

If the score on the airman test report is 70 or above, in most cases the report is valid for 24 calendar months. You may elect to retake the test, in anticipation of a better score, after 30 days from the date your last test was taken. Prior to retesting, you must give your current airman test report to the computer testing administrator. Remember, the score of the *latest* test you take will become the official test score. The FAA will not consider allowing anyone with a passing score to retake a test before the 30-day remedial study period.

A person who fails a knowledge test may apply for retesting before 30 days of the last test providing that person presents the failed test report and an endorsement from an authorized instructor certifying that additional instruction has been given, and the instructor finds the person competent to pass the test. A person may retake a failed test after 30 days without an endorsement from an authorized instructor.

COMPUTER TESTING DESIGNEES

The following is a list of the computer testing designees authorized to give FAA knowledge tests. This list should be helpful in choosing where to register for a test or for requesting additional information.

Aviation Business Services
1-800-947-4228
outside U.S. (415) 259-8550

Drake Prometric
1-800-359-3278
outside U.S. (612) 896-7702

Sylvan Learning Systems, Inc.
1-800-967-1100
outside U.S. (410) 880-0880, Ext. 8890

The latest listing of computer testing center locations may be obtained through FedWorld, (703) 321-8020, in the FAA library file named TST SITE. For technical assistance, contact the FedWorld help desk at (703) 487-4608.

QUESTIONS

[Questions not pertaining to private pilot (land) airplanes have been deleted (indicated by asterisks).]

The charts for the navigation questions in this book were pasted up by this writer and have letters (A, B, C, etc.) imposed to help the reader locate airports and other landmarks for use in the problems ("refer to figure 21, area C"). Other sources may use numbers for the same purpose.

As will be repeated at the beginning of the Answers and Explanations section, the FAA periodically changes not only the questions but also the order of the answers. (Answer A may have been the correct choice earlier, but that same answer may be listed as Choice C on the actual Knowledge Test.) That's fine. You're not supposed to memorize the letters for the correct choices but should know the information and understand the principles, regardless of the letter designation in a test question.

It is suggested that you obtain a copy of FAR/AIM from a commercial source (your instructor will help you obtain one) so you can check the FAR references in the Explanations. WKK

3001. With respect to the certification of airmen, which is a category of aircraft?

A— Gyroplane, helicopter, airship, free balloon.
B— Airplane, rotorcraft, glider, lighter-than-air.
C— Single-engine land and sea, multiengine land and sea.

3002. With respect to the certification of airmen, which is a class of aircraft?

A— Airplane, rotorcraft, glider, lighter-than-air.
B— Single-engine land and sea, multiengine land and sea.
C— Lighter-than-air, airship, hot air balloon, gas balloon.

3003. With respect to the certification of aircraft, which is a category of aircraft?

A— Normal, utility, acrobatic.
B— Airplane, rotorcraft, glider.
C— Landplane, seaplane.

3004. With respect to the certification of aircraft, which is a class of aircraft?

A— Airplane, helicopter, glider, hot air balloon.
B— Normal, utility, acrobatic, limited.
C— Transport, restricted, provisional.

3005. The definition of nighttime is

A— sunset to sunrise.
B— 1 hour after sunset to 1 hour before sunrise.
C— the time between the end of evening civil twilight and the beginning of morning civil twilight.

3006. Which V-speed represents the maneuvering speed?

A— V_A.
B— V_{LO}.
C— V_{NE}.

3007. Which V-speed represents maximum flap extended speed?

A— V_{FE}.
B— V_{LOF}.
C— V_{FC}.

3008. Which V-speed represents maximum landing gear extended speed?

A— V_{LE}.
B— V_{LO}.
C— V_{FE}.

3009. V_{NO} is defined as the

A— normal operating range.
B— never-exceed speed.
C— maximum structural cruising speed.

3010. V_{SO} is defined as the

A— stalling speed or minimum steady flight speed in the landing configuration.
B— stalling speed or minimum steady flight speed in a specified configuration.
C— stalling speed or minimum takeoff safety speed.

3011. Which would provide the greatest gain in altitude in the shortest distance during climb after takeoff?

A— V_Y.
B— V_A.
C— V_X.

3012. After takeoff, which airspeed would the pilot use to gain the most altitude in a given period of time?

A— V_Y.
B— V_X.
C— V_A.

3013. Preventive maintenance has been performed on an aircraft. What paperwork is required?

A— A full, detailed description of the work done must be entered in the airframe logbook.
B— The date the work was completed, and the name of the person who did the work must be entered in the airframe and engine logbook.
C— The signature, certificate number, and kind of certificate held by the person approving the work and a description of the work must be entered in the aircraft maintenance records.

3014. Which operation would be described as preventive maintenance?

A— Servicing landing gear wheel bearings.
B— Alteration of main seat support brackets.
C— Engine adjustments to allow automotive gas to be used.

3015. Which operation would be described as preventive maintenance?

A— Repair of landing gear brace struts.
B— Replenishing hydraulic fluid.
C— Repair of portions of skin sheets by making additional seams.

3016. What document(s) must be in your personal possession while operating as pilot in command of an aircraft?

A— Certificates showing accomplishment of a checkout in the aircraft and a current biennial flight review.
B— A pilot certificate with an endorsement showing accomplishment of an annual flight review and a pilot logbook showing recency of experience.
C— An appropriate pilot certificate and an appropriate current medical certificate.

3017. When must a current pilot certificate be in the pilot's personal possession or readily accessible in the aircraft?

A— When acting as a crew chief during launch and recovery.
B— Only when passengers are carried.
C— Anytime when acting as pilot in command or as a required crewmember.

3018. Private pilots acting as pilot in command, or in any other capacity as a required pilot flight crewmember, must have in his or her personal possession while aboard the aircraft a current

A— logbook endorsement to show that a flight review has been satisfactorily accomplished.
B— medical certificate and an appropriate pilot certificate.
C— endorsement on the pilot certificate to show that a flight review has been satisfactorily accomplished.

3019. Each person who holds a pilot certificate or a medical certificate shall present it for inspection upon the request of the Administrator, the National Transportation Safety Board, or any

A— authorized representative of the Department of Transportation.
B— person in a position of authority.
C— Federal, state, or local law enforcement officer.

3020. A Third-Class Medical Certificate is issued to a 36-year-old pilot on August 10, this year. To exercise the privileges of a Private Pilot Certificate, the medical certificate will be valid until midnight on

A— August 10, 2 years later.
B— August 31, 3 years later.
C— August 31, 2 years later.

3021. A Third-Class Medical Certificate is issued to a 51-year-old pilot on May 3, this year. To exercise the privileges of a Private Pilot Certificate, the medical certificate will be valid until midnight on

A— May 3, 1 year later.
B— May 31, 1 year later.
C— May 31, 2 years later.

3022. For private pilot operations, a Second-Class Medical Certificate issued to a 42-year-old pilot on July 15, this year, will expire at midnight on

A— July 31, 2 years later.
B— July 15, 2 years later.
C— July 31, 1 year later.

3023. For private pilot operations, a First-Class Medical Certificate issued to a 23-year-old pilot on October 21, this year, will expire at midnight on

A— October 31, 3 years later.
B— October 31, 1 year later.
C— October 21, 2 years later.

3024. The pilot in command is required to hold a type rating in which aircraft?

A— Aircraft operated under an authorization issued by the Administrator.
B— Aircraft having a gross weight of more than 12,500 pounds.
C— Aircraft involved in ferry flights, training flights, or test flights.

3025. What is the definition of a high-performance airplane?

A— An airplane with 180 horsepower, retractable landing gear, flaps, and a fixed-pitch propeller.
B— An airplane with more than 200 horsepower.
C— An airplane with a normal cruise speed in excess of 200 knots, flaps, and a controllable propeller.

3026. Before a person holding a Private Pilot Certificate may act as pilot in command of a high-performance airplane, that person must have

A— passed a flight test in that airplane from an FAA inspector.
B— an endorsement in that person's logbook that he/she is competent to act as pilot in command.
C— received flight instruction from an authorized flight instructor who then endorses that person's logbook.

3027. In order to act as pilot in command of a high-performance airplane, a pilot must have

A— made and logged three solo takeoffs and landings in a high-performance airplane.
B— received and logged ground and flight instruction in an airplane that has more than 200 horsepower.
C— passed a flight test in a high-performance airplane.

3028. To act as pilot in command of an aircraft carrying passengers, a pilot must show by logbook endorsement the satisfactory completion of a flight review or completion of a pilot proficiency check within the preceding

A— 6 calendar months.
B— 12 calendar months.
C— 24 calendar months.

3029. If recency of experience requirements for night flight are not met and official sunset is 1830, the latest time passengers may be carried is

A— 1829.
B— 1859.
C— 1929.

3030. To act as pilot in command of an aircraft carrying passengers, the pilot must have made at least three takeoffs and three landings in an aircraft of the same category, class, and if a type rating is required, of the same type, within the preceding

A— 90 days.
B— 12 calendar months.
C— 24 calendar months.

3031. To act as pilot in command of an aircraft carrying passengers, the pilot must have made three takeoffs and three landings within the preceding 90 days in an aircraft of the same

A— make and model.
B— category and class, but not type.
C— category, class, and type, if a type rating is required.

3032. The takeoffs and landings required to meet the recency of experience requirements for carrying passengers in a tailwheel airplane

A— may be touch and go or full stop.
B— must be touch and go.
C— must be to a full stop.

3033. The three takeoffs and landings that are required to act as pilot in command at night must be done during the time period from

A— sunset to sunrise.
B— 1 hour after sunset to 1 hour before sunrise.
C— the end of evening civil twilight to the beginning of morning civil twilight.

3034. To meet the recency of experience requirements to act as pilot in command carrying passengers at night, a pilot must have made at least three takeoffs and three landings to a full stop within the preceding 90 days in

A— the same category and class of aircraft to be used.
B— the same type of aircraft to be used.
C— any aircraft.

3035. If a certificated pilot changes permanent mailing address and fails to notify the FAA Airmen Certification Branch of the new address, the pilot is entitled to exercise the privileges of the pilot certificate for a period of only

A— 30 days after the date of the move.
B— 60 days after the date of the move.
C— 90 days after the date of the move.

3036. A certificated private pilot may not act as pilot in command of an aircraft towing a glider unless there is entered in the pilot's logbook a minimum of

A— 100 hours of pilot flight time in any aircraft.
B— 100 hours of pilot flight time in powered aircraft.
C— 200 hours of pilot flight time in powered aircraft.

3037. To act as pilot in command of an aircraft towing a glider, a person is required to have made within the preceding 12 months

A— at least three flights as observer in a glider being towed by an aircraft.
B— at least three flights in a powered glider.
C— at least three actual or simulated glider tows while accompanied by a qualified pilot.

* * *

3064. In regard to general privileges and limitations, a private pilot may

A— act as pilot in command of an aircraft carrying a passenger for compensation if the flight is in connection with a business or employment.
B— not pay less than the pro rata share of the operating expenses of a flight with passengers, provided the expenses involved only fuel, oil, airport expenditures, or rental fees.
C— not be paid in any manner for the operating expenses of a flight.

3065. According to regulations pertaining to general privileges and limitations, a private pilot may

A— not pay less than the pro rata share of the operating expenses of a flight with passengers, provided the expenses involved only fuel, oil, airport expenditures, or rental fees.
B— not be paid in any manner for the operating expenses of a flight.
C— be paid for the operating expenses of a flight if at least three takeoffs and three landings were made by the pilot within the preceding 90 days.

3066. What exception, if any, permits a private pilot to act as pilot in command of an aircraft carrying passengers who pay for the flight?

A— If the passengers pay all the operating expenses.
B— If a donation is made to a charitable organization for the flight.
C— There is no exception.

3067. The width of a Federal Airway from either side of the centerline is

A— 4 nautical miles.
B— 6 nautical miles.
C— 8 nautical miles.

3068. Unless otherwise specified, Federal Airways include that Class E airspace extending upward from

A— 700 feet above the surface up to and including 17,999 feet.
B— 1,200 feet above the surface up to and including 17,999 feet.
C— the surface up to and including 18,000 feet MSL.

3069. Normal VFR operations in Class D airspace with an operating control tower require the ceiling and visibility to be at least

A— 1,000 feet and one mile.
B— 1,000 feet and three miles.
C— 2,500 feet and three miles.

3070. The final authority as to the operation of an aircraft is the

A— Federal Aviation Administration.
B— pilot in command.
C— aircraft manufacturer.

* * *

3072. If an in-flight emergency requires immediate action, the pilot in command may

A— deviate from the FAR's to the extent required to meet the emergency, but must submit a written report to the Administrator within 24 hours.
B— deviate from the FAR's to the extent required to meet that emergency.
C— not deviate from the FAR's unless prior to the deviation approval is granted by the Administrator.

3073. When must a pilot who deviates from a regulation during an emergency send a written report of that deviation to the Administrator?

A— Within 7 days.
B— Within 10 days.
C— Upon request.

3074. Who is responsible for determining if an aircraft is in condition for safe flight?

A— A certificated aircraft mechanic.
B— The pilot in command.
C— The owner or operator.

3075. Where may an aircraft's operating limitations be found?

A— On the Airworthiness Certificate.
B— In the current, FAA-approved flight manual, approved manual material, markings, and placards, or any combination thereof.
C— In the aircraft airframe and engine logbooks.

3076. Under what conditions may objects be dropped from an aircraft?

A— Only in an emergency.
B— If precautions are taken to avoid injury or damage to persons or property on the surface.
C— If prior permission is received from the Federal Aviation Administration.

3077. A person may not act as a crewmember of a civil aircraft if alcoholic beverages have been consumed by that person within the preceding

A— 8 hours.
B— 12 hours.
C— 24 hours.

3078. Under what condition, if any, may a pilot allow a person who is obviously under the influence of drugs to be carried aboard an aircraft?

A— In an emergency or if the person is a medical patient under proper care.
B— Only if the person does not have access to the cockpit or pilot's compartment.
C— Under no condition.

3079. No person may attempt to act as a crewmember of a civil aircraft with

A— .008 percent by weight or more alcohol in the blood.
B— .004 percent by weight or more alcohol in the blood.
C— .04 percent by weight or more alcohol in the blood.

3080. Which preflight action is specifically required of the pilot prior to each flight?

A— Check the aircraft logbooks for appropriate entries.
B— Become familiar with all available information concerning the flight.
C— Review wake turbulence avoidance procedures.

3081. Preflight action, as required for all flights away from the vicinity of an airport, shall include

A— the designation of an alternate airport.
B— a study of arrival procedures at airports/heliports of intended use.
C— an alternate course of action if the flight cannot be completed as planned.

3082. In addition to other preflight actions for a VFR flight away from the vicinity of the departure airport, regulations specifically require the pilot in command to

A— review traffic control light signal procedures.
B— check the accuracy of the navigation equipment and the emergency locator transmitter (ELT).
C— determine runway lengths at airports of intended use and the aircraft's takeoff and landing distance data.

3083. Flight crewmembers are required to keep their safety belts and shoulder harnesses fastened during

A— takeoffs and landings.
B— all flight conditions.
C— flight in turbulent air.

3084. Which best describes the flight conditions under which flight crewmembers are specifically required to keep their seatbelts and shoulder harnesses fastened?

A— Safety belts during takeoff and landing; shoulder harnesses during takeoff and landing.
B— Safety belts during takeoff and landing; shoulder harnesses during takeoff and landing and while en route.
C— Safety belts during takeoff and landing and while en route; shoulder harnesses during takeoff and landing.

3085. With respect to passengers, what obligation, if any, does a pilot in command have concerning the use of safety belts?

A— The pilot in command must instruct the passengers to keep safety belts fastened for the entire flight.
B— The pilot in command must brief the passengers on the use of safety belts and notify them to fasten the safety belts during taxi, takeoff, and landing.
C— The pilot in command has no obligation in regard to passengers' use of safety belts.

3086. With certain exceptions, safety belts are required to be secured about passengers during

A— taxi, takeoffs, and landings.
B— all flight conditions.
C— flight in turbulent air.

3087. Safety belts are required to be properly secured about which persons in an aircraft and when?

A— Pilots only, during takeoffs and landings.
B— Passengers, during takeoffs and landings only.
C— Each person on board the aircraft during the entire flight.

3088. No person may operate an aircraft in formation flight

A— over a densely populated area.
B— in Class D airspace under special VFR.
C— except by prior arrangement with the pilot in command of each aircraft.

3089. Which aircraft has the right-of-way over all other air traffic?

A— A balloon.
B— An aircraft in distress.
C— An aircraft on final approach to land.

3090. What action is required when two aircraft of the same category converge, but not head-on?

A— The faster aircraft shall give way.
B— The aircraft on the left shall give way.
C— Each aircraft shall give way to the right.

3091. Which aircraft has the right-of-way over the other aircraft listed?

A— Glider.
B— Airship.
C— Aircraft refueling other aircraft.

3092. An airplane and an airship are converging. If the airship is left of the airplane's position, which aircraft has the right-of-way?

A— The airship.
B— The airplane.
C— Each pilot should alter course to the right.

3093. Which aircraft has the right-of-way over the other aircraft listed?

A— Airship.
B— Aircraft towing other aircraft.
C— Gyroplane.

3094. What action should the pilots of a glider and an airplane take if on a head-on collision course?

A— The airplane pilot should give way to the left.
B— The glider pilot should give way to the right.
C— Both pilots should give way to the right.

3095. When two or more aircraft are approaching an airport for the purpose of landing, the right-of-way belongs to the aircraft

A— that has the other to its right.
B— that is the least maneuverable.
C— at the lower altitude, but it shall not take advantage of this rule to cut in front of or to overtake another.

3096. A seaplane and a motorboat are on crossing courses. If the motorboat is to the left of the seaplane, which has the right-of-way?

A— The motorboat.
B— The seaplane.
C— Both should alter course to the right.

3097. Unless otherwise authorized, what is the maximum indicated airspeed at which a person may operate an aircraft below 10,000 feet MSL?

A— 200 knots.
B— 250 knots.
C— 288 knots.

3098. Unless otherwise authorized, the maximum indicated airspeed at which aircraft may be flown when at or below 2,500 feet AGL and within four nautical miles of the primary airport of Class B airspace is

A— 200 knots.
B— 230 knots.
C— 250 knots.

3099. When flying in the airspace underlying Class B airspace, the maximum speed authorized is

A— 200 knots.
B— 230 knots.
C— 250 knots.

3100. When flying in a VFR corridor designated through Class B airspace, the maximum speed authorized is

A— 180 knots.
B— 200 knots.
C— 250 knots.

3101. Except when necessary for takeoff or landing, what is the minimum safe altitude for a pilot to operate an aircraft anywhere?

A— An altitude allowing, if a power unit fails, an emergency landing without undue hazard to persons or property on the surface.
B— An altitude of 500 feet above the surface and no closer than 500 feet to any person, vessel, vehicle, or structure.
C— An altitude of 500 feet above the highest obstacle within a horizontal radius of 1,000 feet.

3102. Except when necessary for takeoff or landing, what is the minimum safe altitude required for a pilot to operate an aircraft over congested areas?

A— An altitude of 1,000 feet above any person, vessel, vehicle, or structure.
B— An altitude of 500 feet above the highest obstacle within a horizontal radius of 1,000 feet.
C— An altitude of 1,000 feet above the highest obstacle within a horizontal radius of 2,000 feet.

3103. Except when necessary for takeoff or landing, what is the minimum safe altitude for a pilot to operate an aircraft over other than a congested area?

A— An altitude allowing, if a power unit fails, an emergency landing without undue hazard to persons or property on the surface.
B— An altitude of 500 feet AGL, except over open water or a sparsely populated area, which requires 500 feet from any person, vessel, vehicle, or structure.
C— An altitude of 500 feet above the highest obstacle within a horizontal radius of 1,000 feet.

3104. Except when necessary for takeoff or landing, an aircraft may not be operated closer than what distance from any person, vessel, vehicle, or structure?

A— 500 feet.
B— 700 feet.
C— 1,000 feet.

3105. If an altimeter setting is not available before flight, to which altitude should the pilot adjust the altimeter?

A— The elevation of the nearest airport corrected to mean sea level.
B— The elevation of the departure area.
C— Pressure altitude corrected for nonstandard temperature.

3106. Prior to takeoff, the altimeter should be set to which altitude or altimeter setting?

A— The current local altimeter setting, if available, or the departure airport elevation.
B— The corrected density altitude of the departure airport.
C— The corrected pressure altitude for the departure airport.

3107. At what altitude shall the altimeter be set to 29.92, when climbing to cruising flight level?

A— 14,500 feet MSL.
B— 18,000 feet MSL.
C— 24,000 feet MSL.

3108. When an ATC clearance has been obtained, no pilot in command may deviate from that clearance, unless that pilot obtains an amended clearance. The one exception to this regulation is

A— when the clearance states "at pilot's discretion."
B— an emergency.
C— if the clearance contains a restriction.

3109. When would a pilot be required to submit a detailed report of an emergency which caused the pilot to deviate from an ATC clearance?

A— When requested by ATC.
B— Immediately.
C— Within 7 days.

3110. What action, if any, is appropriate if the pilot deviates from an ATC instruction during an emergency and is given priority?

A— Take no special action since you are pilot in command.
B— File a detailed report within 48 hours to the chief of the appropriate ATC facility, if requested.
C— File a report to the FAA Administrator, as soon as possible.

3111. A steady green light signal directed from the control tower to an aircraft in flight is a signal that the pilot

A— is cleared to land.
B— should give way to other aircraft and continue circling.
C— should return for landing.

3112. Which light signal from the control tower clears a pilot to taxi?

A— Flashing green.
B— Steady green.
C— Flashing white.

3113. If the control tower uses a light signal to direct a pilot to give way to other aircraft and continue circling, the light will be

A— flashing red.
B— steady red.
C— alternating red and green.

3114. A flashing white light signal from the control tower to a taxiing aircraft is an indication to

A— taxi at a faster speed.
B— taxi only on taxiways and not cross runways.
C— return to the starting point on the airport.

3115. An alternating red and green light signal directed from the control tower to an aircraft in flight is a signal to

A— hold position.
B— exercise extreme caution.
C— not land; the airport is unsafe.

3116. While on final approach for landing, an alternating green and red light followed by a flashing red light is received from the control tower. Under these circumstances, the pilot should

A— discontinue the approach, fly the same traffic pattern and approach again, and land.
B— exercise extreme caution and abandon the approach, realizing the airport is unsafe for landing.
C— abandon the approach, circle the airport to the right, and expect a flashing white light when the airport is safe for landing.

3117. A blue segmented circle on a Sectional Chart depicts which class airspace?

A— Class B
B— Class C
C— Class D

3118. Airspace at an airport with a part-time control tower is classified as Class D airspace only

A— When the weather minimums are below basic VFR.
B— When the associated control tower is in operation.
C— When the associated Flight Service Station is in operation.

3119. Unless otherwise authorized, two-way radio communications with Air Traffic Control are required for landings or takeoffs

A— at all tower controlled airports regardless of weather conditions.
B— at all tower controlled airports only when weather conditions are less than VFR.
C— at all tower controlled airports within Class D airspace only when weather conditions are less than VFR.

3120. Each pilot of an aircraft approaching to land on a runway served by a visual approach slope indicator (VASI) shall

A— maintain a 3° glide to the runway.
B— maintain an altitude at or above the glide slope.
C— stay high until the runway can be reached in a power-off landing.

3121. When approaching to land on a runway served by a visual approach slope indicator (VASI), the pilot shall

A— maintain an altitude that captures the glide slope at least 2 miles downwind from the runway threshold.
B— maintain an altitude at or above the glide slope.
C— remain on the glide slope and land between the two-light bar.

3122. Which is appropriate for a helicopter approaching an airport for landing?

A— Remain below the airplane traffic pattern altitude.
B— Avoid the flow of fixed-wing traffic.
C— Fly right-hand traffic.

3123. Which is the correct traffic pattern departure procedure to use at a noncontrolled airport?

A— Depart in any direction consistent with safety, after crossing the airport boundary.
B— Make all turns to the left.
C— Comply with any FAA traffic pattern established for the airport.

3124. Two-way radio communication must be established with the Air Traffic Control facility having jurisdiction over the area prior to entering

A— Class C.
B— Class E.
C— Class G.

3125. What minimum radio equipment is required for operation within Class C airspace?

A— Two-way radio communications equipment and a 4096-code transponder.
B— Two-way radio communications equipment, a 4096-code transponder, and DME.
C— Two-way radio communications equipment, a 4096-code transponder, and an encoding altimeter.

3126. What minimum pilot certification is required for operation within Class B airspace?

A— Recreational Pilot Certificate.
B— Private Pilot Certificate or Student Pilot Certificate with appropriate logbook endorsements.
C— Private Pilot Certificate with an instrument rating.

3127. What minimum pilot certification is required for operation within Class B airspace?

A— Private Pilot Certificate or Student Pilot Certificate with appropriate logbook endorsements.
B— Commercial Pilot Certificate.
C— Private Pilot Certificate with an instrument rating.

3128. What minimum radio equipment is required for VFR operation within Class B airspace?

A— Two-way radio communications equipment and a 4096-code transponder.
B— Two-way radio communications equipment, a 4096-code transponder, and an encoding altimeter.
C— Two-way radio communications equipment, a 4096-code transponder, an encoding altimeter, and a VOR or TACAN receiver.

3129. An operable 4096-code transponder and Mode C encoding altimeter are required in

A— Class B airspace within 30 miles of the primary airport.
B— Class D airspace.
C— Class E airspace below 10,000 feet.

3130. In which type of airspace are VFR flights prohibited?

A— Class A airspace.
B— Class B airspace.
C— Class C airspace.

3131. What is the specific fuel requirement for flight under VFR during daylight hours in an airplane?

A— Enough to complete the flight at normal cruising speed with adverse wind conditions.
B— Enough to fly to the first point of intended landing and to fly after that for 30 minutes at normal cruising speed.
C— Enough to fly to the first point of intended landing and to fly after that for 45 minutes at normal cruising speed.

3132. What is the specific fuel requirement for flight under VFR at night in an airplane?

A— Enough to complete the flight at normal cruising speed with adverse wind conditions.
B— Enough to fly to the first point of intended landing and to fly after that for 30 minutes at normal cruising speed.
C— Enough to fly to the first point of intended landing and to fly after that for 45 minutes at normal cruising speed.

* * *

3136. During operations within controlled airspace at altitudes of less than 1,200 feet AGL, the minimum horizontal distance from clouds requirement for VFR flight is

A— 1,000 feet.
B— 1,500 feet.
C— 2,000 feet.

3137. What minimum visibility and clearance from clouds are required for VFR operations in uncontrolled airspace at 700 feet AGL or below during daylight hours?

A— 1 mile visibility and clear of clouds.
B— 1 mile visibility, 500 feet below, 1,000 feet above, and 2,000 feet horizontal clearance from clouds.
C— 3 miles visibility and clear of clouds.

3138. What minimum flight visibility is required for VFR flight operations on an airway below 10,000 feet MSL?

A— 1 mile.
B— 3 miles.
C— 4 miles.

3139. The minimum distance from clouds required for VFR operations on an airway below 10,000 feet MSL is

A— remain clear of clouds.
B— 500 feet below, 1,000 feet above, and 2,000 feet horizontally.
C— 500 feet above, 1,000 feet below, and 2,000 feet horizontally.

3140. During operations within controlled airspace at altitudes of more than 1,200 feet AGL, but less than 10,000 feet MSL, the minimum distance above clouds requirement for VFR flight is

A— 500 feet.
B— 1,000 feet.
C— 1,500 feet.

3141. VFR flight in controlled airspace above 1,200 feet AGL and below 10,000 feet MSL requires a minimum visibility and vertical cloud clearance of

A— 3 miles, and 500 feet below or 1,000 feet above the clouds in controlled airspace.
B— 5 miles, and 1,000 feet below or 1,000 feet above the clouds at all altitudes.
C— 5 miles, and 1,000 feet below or 1,000 feet above the clouds only in Class A airspace.

3142. During operations outside controlled airspace at altitudes of more than 1,200 feet AGL, but less than 10,000 feet MSL, the minimum flight visibility for VFR flight at night is

A— 1 mile.
B— 3 miles.
C— 5 miles.

3143. Outside controlled airspace, the minimum flight visibility requirement for VFR flight above 1,200 feet AGL and below 10,000 feet MSL during daylight hours is

A— 1 mile.
B— 3 miles.
C— 5 miles.

3144. During operations outside controlled airspace at altitudes of more than 1,200 feet AGL, but less than 10,000 feet MSL, the minimum distance below clouds requirement for VFR flight at night is

A— 500 feet.
B— 1,000 feet.
C— 1,500 feet.

3145. The minimum flight visibility required for VFR flights above 10,000 feet MSL and more than 1,200 feet AGL in controlled airspace is

A— 1 mile.
B— 3 miles.
C— 5 miles.

3146. For VFR flight operations above 10,000 feet MSL and more than 1,200 feet AGL, the minimum horizontal distance from clouds required is

A— 1,000 feet.
B— 2,000 feet.
C— 1 mile.

3147. During operations at altitudes of more than 1,200 feet AGL and at or above 10,000 feet MSL, the minimum distance above clouds requirement for VFR flight is

A— 500 feet.
B— 1,000 feet.
C— 1,500 feet.

3148. No person may take off or land an aircraft at an airport that lies within Class D airspace unless the

A— flight visibility at that airport is at least 1 mile.
B— ground visibility at that airport is at least 1 mile.
C— ground visibility at that airport is at least 3 miles.

3149. The basic VFR weather minimums for operating an aircraft within Class D airspace are

A— 500-foot ceiling and 1 mile visibility.
B— 1,000-foot ceiling and 3 miles visibility.
C— clear of clouds and 2 miles visibility.

3150. A special VFR clearance authorizes the pilot of an aircraft to operate VFR while within Class D airspace when the visibility

A— is less than 1 mile and the ceiling is less than 1,000 feet.
B— is at least 1 mile and the aircraft can remain clear of clouds.
C— is at least 3 miles and the aircraft can remain clear of clouds.

3151. What is the minimum weather condition required for airplanes operating under special VFR in Class D airspace?

A— 1 mile flight visibility.
B— 1 mile flight visibility and 1,000-foot ceiling.
C— 3 miles flight visibility and 1,000-foot ceiling.

* * *

3153. What are the minimum requirements for airplane operations under special VFR in Class D airspace at night?

A— The airplane must be under radar surveillance at all times while in Class D airspace.
B— The airplane must be equipped for IFR and with an altitude reporting transponder.
C— The pilot must be instrument rated, and the airplane must be IFR equipped.

3154. No person may operate an airplane within Class D airspace at night under special VFR unless the

A— flight can be conducted 500 feet below the clouds.
B— airplane is equipped for instrument flight.
C— flight visibility is at least 3 miles.

3155. Which cruising altitude is appropriate for a VFR flight on a magnetic course of 135°?

A— Even thousandths.
B— Even thousandths plus 500 feet.
C— Odd thousandths plus 500 feet.

3156. Which VFR cruising altitude is acceptable for a flight on a Victor Airway with a magnetic course of 175°? The terrain is less than 1,000 feet.

A— 4,500 feet.
B— 5,000 feet.
C— 5,500 feet.

3157. Which VFR cruising altitude is appropriate when flying above 3,000 feet AGL on a magnetic course of 185°?

A— 4,000 feet.
B— 4,500 feet.
C— 5,000 feet.

3158. Each person operating an aircraft at a VFR cruising altitude shall maintain an odd-thousand plus 500-foot altitude while on a

A— magnetic heading of 0° through 179°.
B— magnetic course of 0° through 179°.
C— true course of 0° through 179°.

3159. In addition to a valid Airworthiness Certificate, what documents or records must be aboard an aircraft during flight?

A— Aircraft engine and airframe logbooks, and owner's manual.
B— Radio operator's permit, and repair and alteration forms.
C— Operating limitations and Registration Certificate.

3160. When must batteries in an emergency locator transmitter (ELT) be replaced or recharged, if rechargeable?

A— After any inadvertent activation of the ELT.
B— When the ELT has been in use for more than 1 cumulative hour.
C— When the ELT can no longer be heard over the airplane's communication radio receiver.

3161. When are non-rechargeable batteries of an emergency locator transmitter (ELT) required to be replaced?

A— Every 24 months.
B— When 50 percent of their useful life expires.
C— At the time of each 100-hour or annual inspection.

3162. Except in Alaska, during what time period should lighted position lights be displayed on an aircraft?

A— End of evening civil twilight to the beginning of morning civil twilight.
B— 1 hour after sunset to 1 hour before sunrise.
C— Sunset to sunrise.

3163. When operating an aircraft at cabin pressure altitudes above 12,500 feet MSL up to and including 14,000 feet MSL, supplemental oxygen shall be used during

A— the entire flight time at those altitudes.
B— that flight time in excess of 10 minutes at those altitudes.
C— that flight time in excess of 30 minutes at those altitudes.

3164. Unless each occupant is provided with supplemental oxygen, no person may operate a civil aircraft of U.S. registry above a maximum cabin pressure altitude of

A— 12,500 feet MSL.
B— 14,000 feet MSL.
C— 15,000 feet MSL.

3165. An operable 4096-code transponder with an encoding altimeter is required in which airspace?

A— Class A, Class B (within 30 miles of the primary airport), and Class C.
B— Class D and Class E (below 10,000 feet MSL).
C— Class D and Class G (below 10,000 feet MSL).

3166. With certain exceptions, all aircraft within 30 miles of a Class B primary airport from the surface upward to 10,000 feet MSL must be equipped with

A— an operable VOR or TACAN receiver and an ADF receiver.
B— instruments and equipment required for IFR operations.
C— an operable transponder having either Mode S or 4096-code capability with Mode C automatic altitude reporting capability.

3167. No person may operate an aircraft in acrobatic flight when

A— flight visibility is less than 5 miles.
B— over any congested area of a city, town, or settlement.
C— less than 2,500 feet AGL.

3168. In which controlled airspace is acrobatic flight prohibited?

A— Class D airspace, Class E airspace designated for Federal Airways.
B— All Class E airspace above 1200 feet AGL.
C— All Class G airspace.

3169. What is the lowest altitude permitted for acrobatic flight?

A— 1,000 feet AGL.
B— 1,500 feet AGL.
C— 2,000 feet AGL.

3170. No person may operate an aircraft in acrobatic flight when the flight visibility is less than

A— 3 miles.
B— 5 miles.
C— 7 miles.

3171. A chair-type parachute must have been packed by a certificated and appropriately rated parachute rigger within the preceding

A— 60 days.
B— 90 days.
C— 120 days.

3172. An approved chair-type parachute may be carried in an aircraft for emergency use if it has been packed by an appropriately rated parachute rigger within the preceding

A— 120 days.
B— 180 days.
C— 365 days.

3173. With certain exceptions, when must each occupant of an aircraft wear an approved parachute?

A— When a door is removed from the aircraft to facilitate parachute jumpers.
B— When intentionally pitching the nose of the aircraft up or down 30° or more.
C— When intentionally banking in excess of 30°.

* * *

3178. Which is normally prohibited when operating a restricted category civil aircraft?

A— Flight under instrument flight rules.
B— Flight over a densely populated area.
C— Flight within Class D airspace.

3179. Unless otherwise specifically authorized, no person may operate an aircraft that has an experimental certificate

A— beneath the floor of Class B airspace.
B— over a densely populated area or in a congested airway.
C— from the primary airport within Class D airspace.

3180. The responsibility for ensuring that an aircraft is maintained in an airworthy condition is primarily that of the

A— pilot in command.
B— owner or operator.
C— mechanic who performs the work.

3181. The responsibility for ensuring that maintenance personnel make the appropriate entries in the aircraft maintenance records indicating the aircraft has been approved for return to service lies with the

A— owner or operator.
B— pilot in command.
C— mechanic who performed the work.

3182. Completion of an annual inspection and the return of the aircraft to service should always be indicated by

A— the relicensing date on the Registration Certificate.
B— an appropriate notation in the aircraft maintenance records.
C— an inspection sticker placed on the instrument panel that lists the annual inspection completion date.

3183. If an alteration or repair substantially affects an aircraft's operation in flight, that aircraft must be test flown by an appropriately-rated pilot and approved for return to service prior to being operated

A— by any private pilot.
B— with passengers aboard.
C— for compensation or hire.

3184. Before passengers can be carried in an aircraft that has been altered in a manner that may have appreciably changed its flight characteristics, it must be flight tested by an appropriately-rated pilot who holds at least a

A— Commercial Pilot Certificate with an instrument rating.
B— Private Pilot Certificate.
C— Commercial Pilot Certificate and a mechanic's certificate.

3185. An aircraft's annual inspection was performed on July 12, this year. The next annual inspection will be due no later than

A— July 1, next year.
B— July 13, next year.
C— July 31, next year.

3186. To determine the expiration date of the last annual aircraft inspection, a person should refer to the

A— Airworthiness Certificate.
B— Registration Certificate.
C— aircraft maintenance records.

3187. How long does the Airworthiness Certificate of an aircraft remain valid?

A— As long as the aircraft has a current Registration Certificate.
B— Indefinitely, unless the aircraft suffers major damage.
C— As long as the aircraft is maintained and operated as required by Federal Aviation Regulations.

3188. What aircraft inspections are required for rental aircraft that are also used for flight instruction?

A— Annual and 100-hour inspections.
B— Biannual and 100-hour inspections.
C— Annual and 50-hour inspections.

3189. An aircraft had a 100-hour inspection when the tachometer read 1259.6. When is the next 100-hour inspection due?

A— 1349.6 hours.
B— 1359.6 hours.
C— 1369.6 hours.

3190. A 100-hour inspection was due at 3302.5 hours on the tachometer. The 100-hour inspection was actually done at 3309.5 hours. When is the next 100-hour inspection due?

A— 3312.5 hours.
B— 3402.5 hours.
C— 3409.5 hours.

3191. No person may use an ATC transponder unless it has been tested and inspected within at least the preceding

A— 6 calendar months.
B— 12 calendar months.
C— 24 calendar months.

3192. Maintenance records show the last transponder inspection was performed on September 1, 1997. The next inspection will be due no later than

A— September 30, 1998.
B— September 1, 1999.
C— September 30, 1999.

3193. Which records or documents shall the owner or operator of an aircraft keep to show compliance with an applicable Airworthiness Directive?

A— Aircraft maintenance records.
B— Airworthiness Certificate and Pilot's Operating Handbook.
C— Airworthiness and Registration Certificates.

3194. If an aircraft is involved in an accident which results in substantial damage to the aircraft, the nearest NTSB field office should be notified

A— immediately.
B— within 48 hours.
C— within 7 days.

3195. Which incident requires an immediate notification to the nearest NTSB field office?

A— A forced landing due to engine failure.
B— Landing gear damage, due to a hard landing.
C— Flight control system malfunction or failure.

3196. Which incident would necessitate an immediate notification to the nearest NTSB field office?

A— An in-flight generator/alternator failure.
B— An in-flight fire.
C— An in-flight loss of VOR receiver capability.

3197. Which incident requires an immediate notification be made to the nearest NTSB field office?

A— An overdue aircraft that is believed to be involved in an accident.
B— An in-flight radio communications failure.
C— An in-flight generator or alternator failure.

3198. May aircraft wreckage be moved prior to the time the NTSB takes custody?

A— Yes, but only if moved by a federal, state, or local law enforcement officer.
B— Yes, but only to protect the wreckage from further damage.
C— No, it may not be moved under any circumstances.

3199. The operator of an aircraft that has been involved in an accident is required to file an accident report within how many days?

A— 5.
B— 7.
C— 10.

3200. The operator of an aircraft that has been involved in an incident is required to submit a report to the nearest field office of the NTSB

A— within 7 days.
B— within 10 days.
C— when requested.

3201. The four forces acting on an airplane in flight are

A— lift, weight, thrust, and drag.
B— lift, weight, gravity, and thrust.
C— lift, gravity, power, and friction.

3202. When are the four forces that act on an airplane in equilibrium?

A— During unaccelerated flight.
B— When the aircraft is accelerating.
C— When the aircraft is at rest on the ground.

3203. (Refer to figure 1.) The acute angle A is the angle of

A— incidence.
B— attack.
C— dihedral.

3204. The term "angle of attack" is defined as the angle

A— between the wing chord line and the relative wind.
B— between the airplane's climb angle and the horizon.
C— formed by the longitudinal axis of the airplane and the chord line of the wing.

3205. What is the relationship of lift, drag, thrust, and weight when the airplane is in straight-and-level flight?

A— Lift equals weight and thrust equals drag.
B— Lift, drag, and weight equal thrust.
C— Lift and weight equal thrust and drag.

3206. How will frost on the wings of an airplane affect takeoff performance?

A— Frost will disrupt the smooth flow of air over the wing, adversely affecting its lifting capability.
B— Frost will change the camber of the wing, increasing its lifting capability.
C— Frost will cause the airplane to become airborne with a higher angle of attack, decreasing the stall speed.

3207. In what flight condition is torque effect the greatest in a single-engine airplane?

A— Low airspeed, high power, high angle of attack.
B— Low airspeed, low power, low angle of attack.
C— High airspeed, high power, high angle of attack.

3208. The left turning tendency of an airplane caused by P-factor is the result of the

A— clockwise rotation of the engine and the propeller turning the airplane counter-clockwise.
B— propeller blade descending on the right, producing more thrust than the ascending blade on the left.
C— gyroscopic forces applied to the rotating propeller blades acting 90° in advance of the point the force was applied.

3209. When does P-factor cause the airplane to yaw to the left?

A— When at low angles of attack.
B— When at high angles of attack.
C— When at high airspeeds.

3210. An airplane said to be inherently stable will

A— be difficult to stall.
B— require less effort to control.
C— not spin.

3211. What determines the longitudinal stability of an airplane?

A— The location of the CG with respect to the center of lift.
B— The effectiveness of the horizontal stabilizer, rudder, and rudder trim tab.
C— The relationship of thrust and lift to weight and drag.

3212. What causes an airplane (except a T-tail) to pitch nosedown when power is reduced and controls are not adjusted?

A— The CG shifts forward when thrust and drag are reduced.
B— The downwash on the elevators from the propeller slipstream is reduced and elevator effectiveness is reduced.
C— When thrust is reduced to less than weight, lift is also reduced and the wings can no longer support the weight.

3213. What is the purpose of the rudder on an airplane?

A— To control yaw.
B— To control overbanking tendency.
C— To control roll.

3214. (Refer to figure 2.) If an airplane weighs 2,300 pounds, what approximate weight would the airplane structure be required to support during a 60° banked turn while maintaining altitude?

A— 2,300 pounds.
B— 3,250 pounds.
C— 4,600 pounds.

3215. (Refer to figure 2.) If an airplane weighs 3,300 pounds, what approximate weight would the airplane structure be required to support during a 30° banked turn while maintaining altitude?

A— 1,200 pounds.
B— 3,100 pounds.
C— 3,800 pounds.

3216. (Refer to figure 2.) If an airplane weighs 4,500 pounds, what approximate weight would the airplane structure be required to support during a 45° banked turn while maintaining altitude?

A— 4,500 pounds.
B— 6,400 pounds.
C— 7,200 pounds.

3217. The amount of excess load that can be imposed on the wing of an airplane depends upon the

A— position of the CG.
B— speed of the airplane.
C— abruptness at which the load is applied.

3218. Which basic flight maneuver increases the load factor on an airplane as compared to straight-and-level flight?

A— Climbs.
B— Turns.
C— Stalls.

3219. One of the main functions of flaps during approach and landing is to

A— decrease the angle of descent without increasing the airspeed.
B— permit a touchdown at a higher indicated airspeed.
C— increase the angle of descent without increasing the airspeed.

3220. What is one purpose of wing flaps?

A— To enable the pilot to make steeper approaches to a landing without increasing the airspeed.
B— To relieve the pilot of maintaining continuous pressure on the controls.
C— To decrease wing area to vary the lift.

3221. Excessively high engine temperatures will

A— cause damage to heat-conducting hoses and warping of the cylinder cooling fins.
B— cause loss of power, excessive oil consumption, and possible permanent internal engine damage.
C— not appreciably affect an aircraft engine.

3222. If the engine oil temperature and cylinder head temperature gauges have exceeded their normal operating range, the pilot may have been operating with

A— the mixture set too rich.
B— higher-than-normal oil pressure.
C— too much power and with the mixture set too lean.

3223. One purpose of the dual ignition system on an aircraft engine is to provide for

A— improved engine performance.
B— uniform heat distribution.
C— balanced cylinder head pressure.

3224. On aircraft equipped with fuel pumps, the practice of running a fuel tank dry before switching tanks is considered unwise because

A— the engine-driven fuel pump or electric fuel boost pump may draw air into the fuel system and cause vapor lock.
B— the engine-driven fuel pump is lubricated by fuel and operating on a dry tank may cause pump failure.
C— any foreign matter in the tank will be pumped into the fuel system.

3225. The operating principle of float-type carburetors is based on the

A— automatic metering of air at the venturi as the aircraft gains altitude.
B— difference in air pressure at the venturi throat and the air inlet.
C— increase in air velocity in the throat of a venturi causing an increase in air pressure.

3226. The basic purpose of adjusting the fuel/air mixture at altitude is to

A— decrease the amount of fuel in the mixture in order to compensate for increased air density.
B— decrease the fuel flow in order to compensate for decreased air density.
C— increase the amount of fuel in the mixture to compensate for the decrease in pressure and density of the air.

3227. During the run-up at a high-elevation airport, a pilot notes a slight engine roughness that is not affected by the magneto check but grows worse during the carburetor heat check. Under these circumstances, what would be the most logical initial action?

A— Check the results obtained with a leaner setting of the mixture.
B— Taxi back to the flight line for a maintenance check.
C— Reduce manifold pressure to control detonation.

3228. While cruising at 9,500 feet MSL, the fuel/air mixture is properly adjusted. What will occur if a descent to 4,500 feet MSL is made without readjusting the mixture?

A— The fuel/air mixture may become excessively lean.

B— There will be more fuel in the cylinders than is needed for normal combustion, and the excess fuel will absorb heat and cool the engine.

C— The excessively rich mixture will create higher cylinder head temperatures and may cause detonation.

3229. Which condition is most favorable to the development of carburetor icing?

A— Any temperature below freezing and a relative humidity of less than 50 percent.

B— Temperature between 32 and 50 °F and low humidity.

C— Temperature between 20 and 70 °F and high humidity.

3230. The possibility of carburetor icing exists even when the ambient air temperature is as

A— high as 70 °F and the relative humidity is high.

B— high as 95 °F and there is visible moisture.

C— low as 0 °F and the relative humidity is high.

3231. If an aircraft is equipped with a fixed-pitch propeller and a float-type carburetor, the first indication of carburetor ice would most likely be

A— a drop in oil temperature and cylinder head temperature.

B— engine roughness.

C— loss of RPM.

3232. Applying carburetor heat will

A— result in more air going through the carburetor.

B— enrich the fuel/air mixture.

C— not affect the fuel/air mixture.

3233. What change occurs in the fuel/air mixture when carburetor heat is applied?

A— A decrease in RPM results from the lean mixture.

B— The fuel/air mixture becomes richer.

C— The fuel/air mixture becomes leaner.

3234. Generally speaking, the use of carburetor heat tends to

A— decrease engine performance.

B— increase engine performance.

C— have no effect on engine performance.

3235. The presence of carburetor ice in an aircraft equipped with a fixed-pitch propeller can be verified by applying carburetor heat and noting

A— an increase in RPM and then a gradual decrease in RPM.

B— a decrease in RPM and then a constant RPM indication.

C— a decrease in RPM and then a gradual increase in RPM.

3236. With regard to carburetor ice, float-type carburetor systems in comparison to fuel injection systems are generally considered to be

A— more susceptible to icing.

B— equally susceptible to icing.

C— susceptible to icing only when visible moisture is present.

3237. If the grade of fuel used in an aircraft engine is lower than specified for the engine, it will most likely cause

A— a mixture of fuel and air that is not uniform in all cylinders.

B— lower cylinder head temperatures.

C— detonation.

3238. Detonation occurs in a reciprocating aircraft engine when

A— the spark plugs are fouled or shorted out or the wiring is defective.

B— hot spots in the combustion chamber ignite the fuel/air mixture in advance of normal ignition.

C— the unburned charge in the cylinders explodes instead of burning normally.

3239. If a pilot suspects that the engine (with a fixed-pitch propeller) is detonating during climb-out after takeoff, the initial corrective action to take would be to

A— lean the mixture.

B— lower the nose slightly to increase airspeed.

C— apply carburetor heat.

3240. The uncontrolled firing of the fuel/air charge in advance of normal spark ignition is known as

A— combustion.
B— pre-ignition.
C— detonation.

3241. Which would most likely cause the cylinder head temperature and engine oil temperature gauges to exceed their normal operating ranges?

A— Using fuel that has a lower-than-specified fuel rating.
B— Using fuel that has a higher-than-specified fuel rating.
C— Operating with higher-than-normal oil pressure.

3242. What type fuel can be substituted for an aircraft if the recommended octane is not available?

A— The next higher octane aviation gas.
B— The next lower octane aviation gas.
C— Unleaded automotive gas of the same octane rating.

3243. Filling the fuel tanks after the last flight of the day is considered a good operating procedure because this will

A— force any existing water to the top of the tank away from the fuel lines to the engine.
B— prevent expansion of the fuel by eliminating airspace in the tanks.
C— prevent moisture condensation by eliminating airspace in the tanks.

3244. For internal cooling, reciprocating aircraft engines are especially dependent on

A— a properly functioning thermostat.
B— air flowing over the exhaust manifold.
C— the circulation of lubricating oil.

3245. An abnormally high engine oil temperature indication may be caused by

A— the oil level being too low.
B— operating with a too high viscosity oil.
C— operating with an excessively rich mixture.

3246. What effect does high density altitude, as compared to low density altitude, have on propeller efficiency and why?

A— Efficiency is increased due to less friction on the propeller blades.
B— Efficiency is reduced because the propeller exerts less force at high density altitudes than at low density altitudes.
C— Efficiency is reduced due to the increased force of the propeller in the thinner air.

3247. If the pitot tube and outside static vents become clogged, which instruments would be affected?

A— The altimeter, airspeed indicator, and turn-and-slip indicator.
B— The altimeter, airspeed indicator, and vertical speed indicator.
C— The altimeter, attitude indicator, and turn-and-slip indicator.

3248. Which instrument will become inoperative if the pitot tube becomes clogged?

A— Altimeter.
B— Vertical speed.
C— Airspeed.

3249. Which instrument(s) will become inoperative if the static vents become clogged?

A— Airspeed only.
B— Altimeter only.
C— Airspeed, altimeter, and vertical speed.

3250. (Refer to figure 3.) Altimeter 1 indicates

A— 500 feet.
B— 1,500 feet.
C— 10,500 feet.

3251. (Refer to figure 3.) Altimeter 2 indicates

A— 1,500 feet.
B— 4,500 feet.
C— 14,500 feet.

3252. (Refer to figure 3.) Altimeter 3 indicates

A— 9,500 feet.
B— 10,950 feet.
C— 15,940 feet.

3253. (Refer to figure 3.) Which altimeter(s) indicate(s) more than 10,000 feet?

A— 1, 2, and 3.
B— 1 and 2 only.
C— 1 only.

3254. Altimeter setting is the value to which the barometric pressure scale of the altimeter is set so the altimeter indicates

A— calibrated altitude at field elevation.
B— absolute altitude at field elevation.
C— true altitude at field elevation.

3255. How do variations in temperature affect the altimeter?

A— Pressure levels are raised on warm days and the indicated altitude is lower than true altitude.
B— Higher temperatures expand the pressure levels and the indicated altitude is higher than true altitude.
C— Lower temperatures lower the pressure levels and the indicated altitude is lower than true altitude.

3256. What is true altitude?

A— The vertical distance of the aircraft above sea level.
B— The vertical distance of the aircraft above the surface.
C— The height above the standard datum plane.

3257. What is absolute altitude?

A— The altitude read directly from the altimeter.
B— The vertical distance of the aircraft above the surface.
C— The height above the standard datum plane.

3258. What is density altitude?

A— The height above the standard datum plane.
B— The pressure altitude corrected for nonstandard temperature.
C— The altitude read directly from the altimeter.

3259. What is pressure altitude?

A— The indicated altitude corrected for position and installation error.
B— The altitude indicated when the barometric pressure scale is set to 29.92.
C— The indicated altitude corrected for nonstandard temperature and pressure.

3260. Under what condition is indicated altitude the same as true altitude?

A— If the altimeter has no mechanical error.
B— When at sea level under standard conditions.
C— When at 18,000 feet MSL with the altimeter set at 29.92.

3261. If it is necessary to set the altimeter from 29.15 to 29.85, what change occurs?

A— 70-foot increase in indicated altitude.
B— 70-foot increase in density altitude.
C— 700-foot increase in indicated altitude.

3262. The pitot system provides impact pressure for which instrument?

A— Altimeter.
B— Vertical-speed indicator.
C— Airspeed indicator.

3263. As altitude increases, the indicated airspeed at which a given airplane stalls in a particular configuration will

A— decrease as the true airspeed decreases.
B— decrease as the true airspeed increases.
C— remain the same regardless of altitude.

3264. What does the red line on an airspeed indicator represent?

A— Maneuvering speed.
B— Turbulent or rough-air speed.
C— Never-exceed speed.

3265. (Refer to figure 4.) What is the full flap operating range for the airplane?

A— 60 to 100 MPH.
B— 60 to 208 MPH.
C— 65 to 165 MPH.

3266. (Refer to figure 4.) What is the caution range of the airplane?

A— 0 to 60 MPH.
B— 100 to 165 MPH.
C— 165 to 208 MPH.

3267. (Refer to figure 4.) The maximum speed at which the airplane can be operated in smooth air is

A— 100 MPH.
B— 165 MPH.
C— 208 MPH.

3268. (Refer to figure 4.) Which color identifies the never-exceed speed?

A— Lower limit of the yellow arc.
B— Upper limit of the white arc.
C— The red radial line.

3269. (Refer to figure 4.) Which color identifies the power-off stalling speed in a specified configuration?

A— Upper limit of the green arc.
B— Upper limit of the white arc.
C— Lower limit of the green arc.

3270. (Refer to figure 4.) What is the maximum flaps-extended speed?

A— 65 MPH.
B— 100 MPH.
C— 165 MPH.

3271. (Refer to figure 4.) Which color identifies the normal flap operating range?

A— The lower limit of the white arc to the upper limit of the green arc.
B— The green arc.
C— The white arc.

3272. (Refer to figure 4.) Which color identifies the power-off stalling speed with wing flaps and landing gear in the landing configuration?

A— Upper limit of the green arc.
B— Upper limit of the white arc.
C— Lower limit of the white arc.

3273. (Refer to figure 4.) What is the maximum structural cruising speed?

A— 100 MPH.
B— 165 MPH.
C— 208 MPH.

3274. What is an important airspeed limitation that is not color coded on airspeed indicators?

A— Never-exceed speed.
B— Maximum structural cruising speed.
C— Maneuvering speed.

3275. (Refer to figure 5.) A turn coordinator provides an indication of the

A— movement of the aircraft about the yaw and roll axes.
B— angle of bank up to but not exceeding 30°.
C— attitude of the aircraft with reference to the longitudinal axis.

3276. (Refer to figure 6.) To receive accurate indications during flight from a heading indicator, the instrument must be

A— set prior to flight on a known heading.
B— calibrated on a compass rose at regular intervals.
C— periodically realigned with the magnetic compass as the gyro precesses.

3277. (Refer to figure 7.) The proper adjustment to make on the attitude indicator during level flight is to align the

A— horizon bar to the level-flight indication.
B— horizon bar to the miniature airplane.
C— miniature airplane to the horizon bar.

3278. (Refer to figure 7.) How should a pilot determine the direction of bank from an attitude indicator such as the one illustrated?

A— By the direction of deflection of the banking scale (A).
B— By the direction of deflection of the horizon bar (B).
C— By the relationship of the miniature airplane (C) to the deflected horizon bar (B).

3279. Deviation in a magnetic compass is caused by the

A— presence of flaws in the permanent magnets of the compass.
B— difference in the location between true north and magnetic north.
C— magnetic fields within the aircraft distorting the lines of magnetic force.

3280. In the Northern Hemisphere, a magnetic compass will normally indicate initially a turn toward the west if

A— a left turn is entered from a north heading.
B— a right turn is entered from a north heading.
C— an aircraft is accelerated while on a north heading.

3281. In the Northern Hemisphere, a magnetic compass will normally indicate initially a turn toward the east if

A— an aircraft is decelerated while on a south heading.
B— an aircraft is accelerated while on a north heading.
C— a left turn is entered from a north heading.

3282. In the Northern Hemisphere, a magnetic compass will normally indicate a turn toward the north if

A— a right turn is entered from an east heading.
B— a left turn is entered from a west heading.
C— an aircraft is accelerated while on an east or west heading.

3283. In the Northern Hemisphere, the magnetic compass will normally indicate a turn toward the south when

A— a left turn is entered from an east heading.
B— a right turn is entered from a west heading.
C— the aircraft is decelerated while on a west heading.

3284. In the Northern Hemisphere, if an aircraft is accelerated or decelerated, the magnetic compass will normally indicate

A— a turn momentarily.
B— correctly when on a north or south heading.
C— a turn toward the south.

3285. In the Northern Hemisphere, if a glider is accelerated or decelerated, the magnetic compass will normally indicate

A— a turn toward north while decelerating on an east heading.
B— correctly only when on a north or south heading.
C— a turn toward south while accelerating on a west heading.

3286. During flight, when are the indications of a magnetic compass accurate?

A— Only in straight-and-level unaccelerated flight.
B— As long as the airspeed is constant.
C— During turns if the bank does not exceed 18°.

3287. An airplane has been loaded in such a manner that the CG is located aft of the aft CG limit. One undesirable flight characteristic a pilot might experience with this airplane would be

A— a longer takeoff run.
B— difficulty in recovering from a stalled condition.
C— stalling at higher-than-normal airspeed.

3288. Loading an airplane to the most aft CG will cause the airplane to be

A— less stable at all speeds.
B— less stable at slow speeds, but more stable at high speeds.
C— less stable at high speeds, but more stable at low speeds.

3289. If the outside air temperature (OAT) at a given altitude is warmer than standard, the density altitude is

A— equal to pressure altitude.
B— lower than pressure altitude.
C— higher than pressure altitude.

3290. Which combination of atmospheric conditions will reduce aircraft takeoff and climb performance?

A— Low temperature, low relative humidity, and low density altitude.
B— High temperature, low relative humidity, and low density altitude.
C— High temperature, high relative humidity, and high density altitude.

3291. What effect does high density altitude have on aircraft performance?

A— It increases engine performance.
B— It reduces climb performance.
C— It increases takeoff performance.

3292. (Refer to figure 8.) What is the effect of a temperature increase from 25 to 50 °F on the density altitude if the pressure altitude remains at 5,000 feet?

A— 1,200-foot increase.
B— 1,400-foot increase.
C— 1,650-foot increase.

3293. (Refer to figure 8.) Determine the pressure altitude with an indicated altitude of 1,380 feet MSL with an altimeter setting of 28.22 at standard temperature.

A— 1,250 feet MSL.
B— 1,373 feet MSL.
C— 3,010 feet MSL.

3294. (Refer to figure 8.) Determine the density altitude for these conditions:

Altimeter setting 29.25
Runway temperature . . . . +81 °F
Airport elevation . . . . 5,250 ft MSL

A— 4,600 feet MSL.
B— 5,877 feet MSL.
C— 8,500 feet MSL.

3295. (Refer to figure 8.) Determine the pressure altitude at an airport that is 3,563 feet MSL with an altimeter setting of 29.96.

A— 3,527 feet MSL.
B— 3,556 feet MSL.
C— 3,639 feet MSL.

3296. (Refer to figure 8.) What is the effect of a temperature increase from 30 to 50 °F on the density altitude if the pressure altitude remains at 3,000 feet MSL?

A— 900-foot increase.
B— 1,100-foot decrease.
C— 1,300-foot increase.

3297. (Refer to figure 8.) Determine the pressure altitude at an airport that is 1,386 feet MSL with an altimeter setting of 29.97.

A— 1,341 feet MSL.
B— 1,451 feet MSL.
C— 1,562 feet MSL.

3298. (Refer to figure 8.) Determine the density altitude for these conditions:

Altimeter setting 30.35
Runway temperature . . . . +25 °F
Airport elevation . . . . 3,894 ft MSL

A— 2,000 feet MSL.
B— 2,900 feet MSL.
C— 3,500 feet MSL.

3299. (Refer to figure 8.) What is the effect of a temperature decrease and a pressure altitude increase on the density altitude from 90 °F and 1,250 feet pressure altitude to 60 °F and 1,750 feet pressure altitude?

A— 500-foot increase.
B— 1,300-foot decrease.
C— 1,300-foot increase.

3300. What effect, if any, does high humidity have on aircraft performance?

A— It increases performance.
B— It decreases performance.
C— It has no effect on performance.

3301. What force makes an airplane turn?

A— The horizontal component of lift.
B— The vertical component of lift.
C— Centrifugal force.

3302. When taxiing with strong quartering tailwinds, which aileron positions should be used?

A— Aileron down on the downwind side.
B— Ailerons neutral.
C— Aileron down on the side from which the wind is blowing.

3303. Which aileron positions should a pilot generally use when taxiing in strong quartering headwinds?

A— Aileron up on the side from which the wind is blowing.
B— Aileron down on the side from which the wind is blowing.
C— Ailerons neutral.

3304. Which wind condition would be most critical when taxiing a nosewheel equipped high-wing airplane?

A— Quartering tailwind.
B— Direct crosswind.
C— Quartering headwind.

3305. (Refer to figure 9, area A.) How should the flight controls be held while taxiing a tricycle-gear equipped airplane into a left quartering headwind?

A— Left aileron up, elevator neutral.
B— Left aileron down, elevator neutral.
C— Left aileron up, elevator down.

3306. (Refer to figure 9, area B.) How should the flight controls be held while taxiing a tailwheel airplane into a right quartering headwind?

A— Right aileron up, elevator up.
B— Right aileron down, elevator neutral.
C— Right aileron up, elevator down.

3307. (Refer to figure 9, area C.) How should the flight controls be held while taxiing a tailwheel airplane with a left quartering tailwind?

A— Left aileron up, elevator neutral.
B— Left aileron down, elevator neutral.
C— Left aileron down, elevator down.

3308. (Refer to figure 9, area C.) How should the flight controls be held while taxiing a tricycle-gear equipped airplane with a left quartering tailwind?

A— Left aileron up, elevator neutral.
B— Left aileron down, elevator down.
C— Left aileron up, elevator down.

3309. In what flight condition must an aircraft be placed in order to spin?

A— Partially stalled with one wing low.
B— In a steep diving spiral.
C— Stalled.

3310. During a spin to the left, which wing(s) is/are stalled?

A— Both wings are stalled.
B— Neither wing is stalled.
C— Only the left wing is stalled.

3311. The angle of attack at which an airplane wing stalls will

A— increase if the CG is moved forward.
B— change with an increase in gross weight.
C— remain the same regardless of gross weight.

3312. What is ground effect?

A— The result of the interference of the surface of the Earth with the airflow patterns about an airplane.
B— The result of an alteration in airflow patterns increasing induced drag about the wings of an airplane.
C— The result of the disruption of the airflow patterns about the wings of an airplane to the point where the wings will no longer support the airplane in flight.

3313. Floating caused by the phenomenon of ground effect will be most realized during an approach to land when at

A— less than the length of the wingspan above the surface.
B— twice the length of the wingspan above the surface.
C— a higher-than-normal angle of attack.

3314. What must a pilot be aware of as a result of ground effect?

A— Wingtip vortices increase creating wake turbulence problems for arriving and departing aircraft.
B— Induced drag decreases; therefore, any excess speed at the point of flare may cause considerable floating.
C— A full stall landing will require less up elevator deflection than would a full stall when done free of ground effect.

3315. Ground effect is most likely to result in which problem?

A— Settling to the surface abruptly during landing.
B— Becoming airborne before reaching recommended takeoff speed.
C— Inability to get airborne even though airspeed is sufficient for normal takeoff needs.

3316. During an approach to a stall, an increased load factor will cause the airplane to

A— stall at a higher airspeed.
B— have a tendency to spin.
C— be more difficult to control.

3317. Angle of attack is defined as the angle between the chord line of an airfoil and the

A— direction of the relative wind.
B— pitch angle of an airfoil.
C— rotor plane of rotation.

* * *

3381. Every physical process of weather is accompanied by, or is the result of, a

A— movement of air.
B— pressure differential.
C— heat exchange.

3382. What causes variations in altimeter settings between weather reporting points?

A— Unequal heating of the Earth's surface.
B— Variation of terrain elevation.
C— Coriolis force.

3383. A temperature inversion would most likely result in which weather condition?

A— Clouds with extensive vertical development above an inversion aloft.
B— Good visibility in the lower levels of the atmosphere and poor visibility above an inversion aloft.
C— An increase in temperature as altitude is increased.

3384. The most frequent type of ground or surface-based temperature inversion is that which is produced by

A— terrestrial radiation on a clear, relatively still night.
B— warm air being lifted rapidly aloft in the vicinity of mountainous terrain.
C— the movement of colder air under warm air, or the movement of warm air over cold air.

3385. Which weather conditions should be expected beneath a low-level temperature inversion layer when the relative humidity is high?

A— Smooth air, poor visibility, fog, haze, or low clouds.
B— Light wind shear, poor visibility, haze, and light rain.
C— Turbulent air, poor visibility, fog, low stratus type clouds, and showery precipitation.

3386. What are the standard temperature and pressure values for sea level?

A— 15 °C and 29.92" Hg.
B— 59 °C and 1013.2 millibars (or hectoPascals).
C— 59 °F and 29.92 millibars (or hectoPascals).

3387. If a pilot changes the altimeter setting from 30.11 to 29.96, what is the approximate change in indication?

A— Altimeter will indicate .15" Hg higher.
B— Altimeter will indicate 150 feet higher.
C— Altimeter will indicate 150 feet lower.

3388. Under which condition will pressure altitude be equal to true altitude?

A— When the atmospheric pressure is 29.92" Hg.
B— When standard atmospheric conditions exist.
C— When indicated altitude is equal to the pressure altitude.

3389. Under what condition is pressure altitude and density altitude the same value?

A— At sea level, when the temperature is 0 °F.
B— When the altimeter has no installation error.
C— At standard temperature.

3390. If a flight is made from an area of low pressure into an area of high pressure without the altimeter setting being adjusted, the altimeter will indicate

A— the actual altitude above sea level.
B— higher than the actual altitude above sea level.
C— lower than the actual altitude above sea level.

3391. If a flight is made from an area of high pressure into an area of lower pressure without the altimeter setting being adjusted, the altimeter will indicate

A— lower than the actual altitude above sea level.
B— higher than the actual altitude above sea level.
C— the actual altitude above sea level.

3392. Under what condition will true altitude be lower than indicated altitude?

A— In colder than standard air temperature.
B— In warmer than standard air temperature.
C— When density altitude is higher than indicated altitude.

3393. Which condition would cause the altimeter to indicate a lower altitude than true altitude?

A— Air temperature lower than standard.
B— Atmospheric pressure lower than standard.
C— Air temperature warmer than standard.

3394. Which factor would tend to increase the density altitude at a given airport?

A— An increase in barometric pressure.
B— An increase in ambient temperature.
C— A decrease in relative humidity.

3395. The wind at 5,000 feet AGL is southwesterly while the surface wind is southerly. This difference in direction is primarily due to

A— stronger pressure gradient at higher altitudes.
B— friction between the wind and the surface.
C— stronger Coriolis force at the surface.

* * *

3397. What is meant by the term "dewpoint"?

A— The temperature at which condensation and evaporation are equal.
B— The temperature at which dew will always form.
C— The temperature to which air must be cooled to become saturated.

3398. The amount of water vapor which air can hold depends on the

A— dewpoint.
B— air temperature.
C— stability of the air.

3399. Clouds, fog, or dew will always form when

A— water vapor condenses.
B— water vapor is present.
C— relative humidity reaches 100 percent.

3400. What are the processes by which moisture is added to unsaturated air?

A— Evaporation and sublimation.
B— Heating and condensation.
C— Supersaturation and evaporation.

3401. Which conditions result in the formation of frost?

A— The temperature of the collecting surface is at or below freezing when small droplets of moisture fall on the surface.
B— The temperature of the collecting surface is at or below the dewpoint of the adjacent air and the dewpoint is below freezing.
C— The temperature of the surrounding air is at or below freezing when small drops of moisture fall on the collecting surface.

3402. The presence of ice pellets at the surface is evidence that there

A— are thunderstorms in the area.
B— has been cold frontal passage.
C— is a temperature inversion with freezing rain at a higher altitude.

3403. What measurement can be used to determine the stability of the atmosphere?

A— Atmospheric pressure.
B— Actual lapse rate.
C— Surface temperature.

3404. What would decrease the stability of an air mass?

A— Warming from below.
B— Cooling from below.
C— Decrease in water vapor.

3405. What is a characteristic of stable air?

A— Stratiform clouds.
B— Unlimited visibility.
C— Cumulus clouds.

3406. Moist, stable air flowing upslope can be expected to

A— produce stratus type clouds.
B— cause showers and thunderstorms.
C— develop convective turbulence.

3407. If an unstable air mass is forced upward, what type clouds can be expected?

A— Stratus clouds with little vertical development.
B— Stratus clouds with considerable associated turbulence.
C— Clouds with considerable vertical development and associated turbulence.

3408. What feature is associated with a temperature inversion?

A— A stable layer of air.
B— An unstable layer of air.
C— Chinook winds on mountain slopes.

3409. What is the approximate base of the cumulus clouds if the surface air temperature is 70 °F and the dewpoint is 48 °F?

A— 4,000 feet MSL.
B— 5,000 feet MSL.
C— 6,000 feet MSL.

3410. At approximately what altitude above the surface would the pilot expect the base of cumuliform clouds if the surface air temperature is 82 °F and the dewpoint is 38 °F?

A— 9,000 feet AGL.
B— 10,000 feet AGL.
C— 11,000 feet AGL.

* * *

3412. What are characteristics of a moist, unstable air mass?

A— Cumuliform clouds and showery precipitation.
B— Poor visibility and smooth air.
C— Stratiform clouds and showery precipitation.

3413. What are characteristics of unstable air?

A— Turbulence and good surface visibility.
B— Turbulence and poor surface visibility.
C— Nimbostratus clouds and good surface visibility.

3414. A stable air mass is most likely to have which characteristic?

A— Showery precipitation.
B— Turbulent air.
C— Smooth air.

3415. The suffix "nimbus," used in naming clouds, means

A— a cloud with extensive vertical development.
B— a rain cloud.
C— a middle cloud containing ice pellets.

3416. Clouds are divided into four families according to their

A— outward shape.
B— height range.
C— composition.

3417. An almond or lens-shaped cloud which appears stationary, but which may contain winds of 50 knots or more, is referred to as

A— an inactive frontal cloud.
B— a funnel cloud.
C— a lenticular cloud.

3418. Crests of standing mountain waves may be marked by stationary, lens-shaped clouds known as

A— mammatocumulus clouds.
B— standing lenticular clouds.
C— roll clouds.

3419. What clouds have the greatest turbulence?

A— Towering cumulus.
B— Cumulonimbus.
C— Nimbostratus.

3420. What cloud types would indicate convective turbulence?

A— Cirrus clouds.
B— Nimbostratus clouds.
C— Towering cumulus clouds.

3421. The boundary between two different air masses is referred to as a

A— frontolysis.
B— frontogenesis.
C— front.

3422. One of the most easily recognized discontinuities across a front is

A— a change in temperature.
B— an increase in cloud coverage.
C— an increase in relative humidity.

3423. One weather phenomenon which will always occur when flying across a front is a change in the

A— wind direction.
B— type of precipitation.
C— stability of the air mass.

3424. Steady precipitation preceding a front is an indication of

A— stratiform clouds with moderate turbulence.
B— cumuliform clouds with little or no turbulence.
C— stratiform clouds with little or no turbulence.

3425. Possible mountain wave turbulence could be anticipated when winds of 40 knots or greater blow

A— across a mountain ridge, and the air is stable.
B— down a mountain valley, and the air is unstable.
C— parallel to a mountain peak, and the air is stable.

3426. Where does wind shear occur?

A— Only at higher altitudes.
B— Only at lower altitudes.
C— At all altitudes, in all directions.

3427. When may hazardous wind shear be expected?

A— When stable air crosses a mountain barrier where it tends to flow in layers forming lenticular clouds.
B— In areas of low-level temperature inversion, frontal zones, and clear air turbulence.
C— Following frontal passage when stratocumulus clouds form indicating mechanical mixing.

3428. A pilot can expect a wind-shear zone in a temperature inversion whenever the windspeed at 2,000 to 4,000 feet above the surface is at least

A— 10 knots.
B— 15 knots.
C— 25 knots.

3429. One in-flight condition necessary for structural icing to form is

A— small temperature/dewpoint spread.
B— stratiform clouds.
C— visible moisture.

3430. In which environment is aircraft structural ice most likely to have the highest accumulation rate?

A— Cumulus clouds with below freezing temperatures.
B— Freezing drizzle.
C— Freezing rain.

3431. Why is frost considered hazardous to flight?

A— Frost changes the basic aerodynamic shape of the airfoils, thereby decreasing lift.
B— Frost slows the airflow over the airfoils, thereby increasing control effectiveness.
C— Frost spoils the smooth flow of air over the wings, thereby decreasing lifting capability.

3432. How does frost affect the lifting surfaces of an airplane on takeoff?

A— Frost may prevent the airplane from becoming airborne at normal takeoff speed.
B— Frost will change the camber of the wing, increasing lift during takeoff.
C— Frost may cause the airplane to become airborne with a lower angle of attack at a lower indicated airspeed.

3433. The conditions necessary for the formation of cumulonimbus clouds are a lifting action and

A— unstable air containing an excess of condensation nuclei.
B— unstable, moist air.
C— either stable or unstable air.

3434. What feature is normally associated with the cumulus stage of a thunderstorm?

A— Roll cloud.
B— Continuous updraft.
C— Frequent lightning.

3435. Which weather phenomenon signals the beginning of the mature stage of a thunderstorm?

A— The appearance of an anvil top.
B— Precipitation beginning to fall.
C— Maximum growth rate of the clouds.

3436. What conditions are necessary for the formation of thunderstorms?

A— High humidity, lifting force, and unstable conditions.
B— High humidity, high temperature, and cumulus clouds.
C— Lifting force, moist air, and extensive cloud cover.

3437. During the life cycle of a thunderstorm, which stage is characterized predominately by downdrafts?

A— Cumulus.
B— Dissipating.
C— Mature.

3438. Thunderstorms reach their greatest intensity during the

A— mature stage.
B— downdraft stage.
C— cumulus stage.

3439. Thunderstorms which generally produce the most intense hazard to aircraft are

A— squall line thunderstorms.
B— steady-state thunderstorms.
C— warm front thunderstorms.

3440. A nonfrontal, narrow band of active thunderstorms that often develop ahead of a cold front is a known as a

A— prefrontal system.
B— squall line.
C— dry line.

3441. If there is thunderstorm activity in the vicinity of an airport at which you plan to land, which hazardous atmospheric phenomenon might be expected on the landing approach?

A— Precipitation static.
B— Wind-shear turbulence.
C— Steady rain.

3442. Upon encountering severe turbulence, which flight condition should the pilot attempt to maintain?

A— Constant altitude and airspeed.
B— Constant angle of attack.
C— Level flight attitude.

3443. What situation is most conducive to the formation of radiation fog?

A— Warm, moist air over low, flatland areas on clear, calm nights.
B— Moist, tropical air moving over cold, offshore water.
C— The movement of cold air over much warmer water.

3444. If the temperature/dewpoint spread is small and decreasing, and the temperature is 62 °F, what type weather is most likely to develop?

A— Freezing precipitation.
B— Thunderstorms.
C— Fog or low clouds.

3445. In which situation is advection fog most likely to form?

A— A warm, moist air mass on the windward side of mountains.
B— An air mass moving inland from the coast in winter.
C— A light breeze blowing colder air out to sea.

3446. What types of fog depend upon wind in order to exist?

A— Radiation fog and ice fog.
B— Steam fog and ground fog.
C— Advection fog and upslope fog.

3447. Low-level turbulence can occur and icing can become hazardous in which type of fog?

A— Rain-induced fog.
B— Upslope fog.
C— Steam fog.

3448. The development of thermals depends upon

A— a counterclockwise circulation of air.
B— temperature inversions.
C— solar heating.

3449. Which is considered to be the most hazardous condition when soaring in the vicinity of thunderstorms?

A— Static electricity.
B— Lightning.
C— Wind shear and turbulence.

3450. Convective circulation patterns associated with sea breezes are caused by

A— warm, dense air moving inland from over the water.
B— water absorbing and radiating heat faster than the land.
C— cool, dense air moving inland from over the water.

* * *

3452. Which weather phenomenon is always associated with a thunderstorm?

A— Lightning.
B— Heavy rain.
C— Hail.

3453. Individual forecasts for specific routes of flight can be obtained from which weather source?

A— Transcribed Weather Broadcasts (TWEB's).
B— Terminal Forecasts.
C— Area Forecasts.

3454. Transcribed Weather Broadcasts (TWEB's) may be monitored by tuning the appropriate radio receiver to certain

A— airport advisory frequencies.
B— VOR and NDB frequencies.
C— ATIS frequencies.

3455. When telephoning a weather briefing facility for preflight weather information, pilots should state

A— the aircraft identification or the pilot's name.
B— true airspeed.
C— fuel on board.

3456. To get a complete weather briefing for the planned flight, the pilot should request

A— a general briefing.
B— an abbreviated briefing.
C— a standard briefing.

3457. Which type weather briefing should a pilot request, when departing within the hour, if no preliminary weather information has been received?

A— Outlook briefing.
B— Abbreviated briefing.
C— Standard briefing.

3458. Which type of weather briefing should a pilot request to supplement mass disseminated data?

A— An outlook briefing.
B— A supplemental briefing.
C— An abbreviated briefing.

3459. To update a previous weather briefing, a pilot should request

A— an abbreviated briefing.
B— a standard briefing.
C— an outlook briefing.

3460. A weather briefing that is provided when the information requested is 6 or more hours in advance of the proposed departure time is

A— an outlook briefing.
B— a forecast briefing.
C— a prognostic briefing.

3461. When requesting weather information for the following morning, a pilot should request

A— an outlook briefing.
B— a standard briefing.
C— an abbreviated briefing.

3462. (Refer to Figure 12.) Which of the reporting stations have VFR weather?

A— All.
B— KINK, KBOI, and KJFK.
C— KINK, KBOI, and KLAX.

3463. For aviation purposes, ceiling is defined as the height above the Earth's surface of

A— lowest reported obscuration and the highest layer of clouds reported as overcast.
B— lowest broken or overcast layer or vertical visibility into an obscuration.
C— lowest layer of clouds reported as scattered, broken, or thin.

3464. (Refer to Figure 12.) The wind direction and velocity at KJFK is from

A— 180° true at 4 knots.
B— 180° magnetic at 4 knots.
C— 040° true at 18 knots.

3465. (Refer to Figure 12.) What are the wind conditions at Wink, Texas (KINK)?

A— Calm.
B— 110° at 12 knots, gusts 18 knots.
C— 111° at 2 knots, gusts 18 knots.

3466. (Refer to Figure 12.) The remarks section for KMDW has RAB35 listed. This entry means

A— blowing mist has reduced the visibility to 1½ SM.
B— rain began at 1835Z.
C— the barometer has risen .35" Hg.

3467. (Refer to Figure 12.) What are the current conditions depicted for Chicago Midway Airport (KMDW)?

A— Sky 700 feet overcast, visibility 1½ SM, rain.
B— Sky 7000 feet overcast, visibility 1½ SM, heavy rain.
C— Sky 700 feet overcast, visibility 11, occasionally 2 SM, with rain.

* * *

3472. (Refer to figure 14.) The base and tops of the overcast layer reported by a pilot are

A— 1,800 feet MSL and 5,500 feet MSL.
B— 5,500 feet AGL and 7,200 feet MSL.
C— 7,200 feet MSL and 8,900 feet MSL.

3473. (Refer to figure 14.) The wind and temperature at 12,000 feet MSL as reported by a pilot are

A— 009° at 121 MPH and 90 °F.
B— 090° at 21 knots and −9 °F.
C— 090° at 21 knots and −9 °C.

3474. (Refer to figure 14.) If the terrain elevation is 1,295 feet MSL, what is the height above ground level of the base of the ceiling?

A— 505 feet AGL.
B— 1,295 feet AGL.
C— 6,586 feet AGL.

3475. (Refer to figure 14.) The intensity of the turbulence reported at a specific altitude is

A— moderate at 5,500 feet and at 7,200 feet.
B— moderate from 5,500 feet to 7,200 feet.
C— light to moderate from 7,200 feet to 8,900 feet.

3476. (Refer to figure 14.) The intensity and type of icing reported by a pilot is

A— light to moderate.
B— light to moderate clear.
C— moderate rime.

* * *

3478. From which primary source should information be obtained regarding expected weather at the estimated time of arrival if your destination has no Terminal Forecast?

A— Low-Level Prognostic Chart.
B— Weather Depiction Chart.
C— Area Forecast.

3479. (Refer to Figure 15.) What is the valid period for the TAF for KMEM?

A— 1200Z to 1200Z.
B— 1200Z to 1800Z.
C— 1800Z to 1800Z.

3480. (Refer to Figure 15.) In the TAF for KMEM, what does "SHRA" stand for?

A— Rain showers.
B— A shift in wind direction is expected.
C— A significant change in precipitation is possible.

3481. (Refer to Figure 15.) Between 1000Z and 1200Z the visibility at KMEM is forecast to be

A— ½ statute mile.
B— 3 statute miles.
C— 6 statute miles.

3482. (Refer to Figure 15.) What is the forecast wind for KMEM from 1600Z until the end of the forecast?

A— No significant wind.
B— 020° at 8 knots.
C— Variable in direction at 4 knots.

3483. (Refer to Figure 15.) In the TAF from KOKC, the "FM (FROM) Group" is

A— forecast for the hours from 1600Z to 2200Z with the wind from 160° at 10 knots.
B— forecast for the hours from 1600Z to 2200Z with the wind from 160° at 10 knots, becoming 220° at 13 knots with gusts to 20 knots.
C— forecast for the hours from 1600Z to 2200Z with the wind from 160° at 10 knots, becoming 210° at 15 knots.

3484. (Refer to Figure 15.) In the TAF from KOKC, the clear sky becomes

A— overcast at 2,000 feet during the forecast period between 2200Z and 2400Z.
B— overcast at 200 feet with a 40% probability of becoming overcast at 600 feet during the forecast period between 2200Z and 2400Z.
C— overcast at 200 feet with the probability of becoming overcast at 400 feet during the forecast period between 2200Z and 2400Z.

3485. (Refer to Figure 15.) During the time period of 0600Z to 0800Z, what significant weather is forecast for KOKC?

A— Wind—210° at 15 knots.
B— Visibility—possibly 6 statute miles with scattered clouds at 4,000 feet.
C— No significant weather is forecast for this time period.

3486. (Refer to Figure 15.) The only cloud type forecast in TAF reports is

A— Nimbostratus.
B— Cumulonimbus.
C— Scattered cumulus.

3487. To best determine general forecast weather conditions over several states, the pilot should refer to

A— Area Forecasts.
B— Weather Depiction Charts.
C— Satellite Maps.

3488. (Refer to figure 16.) What is the forecast ceiling and visibility for Tennessee from 2300Z through 0500Z?

A— 500 feet to less than 1,000 feet, and 1 mile to less than 3 miles.
B— 1,000 to 3,000 feet, and 3 to 5 miles.
C— 3,000 feet or greater, and 5 miles or greater.

3489. To determine the freezing level and areas of probable icing aloft, the pilot should refer to the

A— Radar Summary Chart.
B— Weather Depiction Chart.
C— Area Forecast.

3490. The section of the Area Forecast entitled "SIG CLDS AND WX" contains a summary of

A— cloudiness and weather significant to flight operations broken down by states or other geographical areas.
B— forecast sky cover, cloud tops, visibility, and obstructions to vision along specific routes.
C— weather advisories still in effect at the time of issue.

3491. (Refer to figure 16.) What hazards are forecast in the Area Forecast for TN, AL, and the coastal waters?

A— Thunderstorms with severe or greater turbulence, severe icing, and low-level wind shear.
B— Moderate rime icing above the freezing level to 10,000 feet.
C— Moderate turbulence from 25,000 to 38,000 feet due to the jetstream.

3492. (Refer to figure 16.) What type obstructions to vision, if any, are forecast for the entire area from 2300Z until 0500Z the next day?

A— None of any significance, VFR is forecast.
B— Visibility 3 to 5 miles in fog.
C— Visibility below 3 miles in fog over south-central Texas.

3493. (Refer to figure 16.) What sky condition and type obstructions to vision are forecast for all the area except TN from 1040Z until 2300Z?

A— Ceilings 3,000 to 5,000 feet broken, visibility 3 to 5 miles in fog.
B— 8,000 feet scattered to clear except visibility below 3 miles in fog until 1500Z over south-central Texas.
C— Generally ceilings 3,000 to 8,000 feet to clear with visibility sometimes below 3 miles in fog.

3494. To obtain a continuous transcribed weather briefing, including winds aloft and route forecasts for a cross-country flight, a pilot should monitor a

A— Transcribed Weather Broadcast (TWEB) on an ADF radio receiver.
B— VHF radio receiver tuned to an Automatic Terminal Information Service (ATIS) frequency.
C— regularly scheduled weather broadcast on a VOR frequency.

3495. What is indicated when a current CONVECTIVE SIGMET forecasts thunderstorms?

A— Moderate thunderstorms covering 30 percent of the area.
B— Moderate or severe turbulence.
C— Thunderstorms obscured by massive cloud layers.

3496. What information is contained in a CONVECTIVE SIGMET?

A— Tornadoes, embedded thunderstorms, and hail 3/4 inch or greater in diameter.
B— Severe icing, severe turbulence, or widespread dust storms lowering visibility to less than 3 miles.
C— Surface winds greater than 40 knots or thunderstorms equal to or greater than video integrator processor (VIP) level 4.

3497. SIGMET's are issued as a warning of weather conditions hazardous to which aircraft?

A— Small aircraft only.
B— Large aircraft only.
C— All aircraft.

3498. Which in-flight advisory would contain information on severe icing?

A— Convective SIGMET.
B— SIGMET.
C— AIRMET.

3499. AIRMET's are issued as a warning of weather conditions particularly hazardous to which aircraft?

A— Small single-engine aircraft.
B— Large multiengine aircraft.
C— All aircraft.

3500. (Refer to figure 17.) What wind is forecast for STL at 6,000 feet?

A— 210° magnetic at 13 knots.
B— 230° true at 25 knots.
C— 232° true at 5 knots.

3501. (Refer to figure 17.) What wind is forecast for STL at 18,000 feet?

A— 230° true at 56 knots.
B— 235° true at 06 knots.
C— 235° magnetic at 06, peak gusts to 16 knots.

3502. (Refer to figure 17.) Determine the wind and temperature aloft forecast for DEN at 30,000 feet.

A— 023° magnetic at 53 knots, temperature 47 °C.
B— 230° true at 53 knots, temperature –47 °C.
C— 235° true at 34 knots, temperature –7 °C.

3503. (Refer to figure 17.) Determine the wind and temperature aloft forecast for 3,000 feet at MKC.

A— 050° true at 7 knots, temperature missing.
B— 360° magnetic at 5 knots, temperature –7 °C.
C— 360° true at 50 knots, temperature +7 °C.

3504. (Refer to figure 17.) What wind is forecast for STL at 34,000 feet?

A— 007° magnetic at 30 knots.
B— 073° true at 6 knots.
C— 230° true at 106 knots.

3505. What values are used for Winds Aloft Forecasts?

A— Magnetic direction and knots.
B— Magnetic direction and miles per hour.
C— True direction and knots.

3506. When the term "light and variable" is used in reference to a Winds Aloft Forecast, the coded group and windspeed is

A— 0000 and less than 7 knots.
B— 9900 and less than 5 knots.
C— 9999 and less than 10 knots.

3507. (Refer to figure 18.) What is the status of the front that extends from New Mexico to Indiana?

A— Stationary.
B— Occluded.
C— Retreating.

3508. (Refer to figure 18.) The IFR weather in eastern Texas is due to

A— intermittent rain.
B— fog.
C— dust devils.

3509. (Refer to figure 18.) Of what value is the Weather Depiction Chart to the pilot?

A— For determining general weather conditions on which to base flight planning.
B— For a forecast of cloud coverage, visibilities, and frontal activity.
C— For determining frontal trends and air mass characteristics.

3510. (Refer to figure 18.) The marginal weather in southeast New Mexico is due to

A— reported thunderstorms.
B— 600-foot overcast ceilings.
C— low visibility.

3511. (Refer to figure 18.) What weather phenomenon is causing IFR conditions along the coast of Oregon and California?

A— Squall line activity.
B— Low ceilings.
C— Heavy rain showers.

3512. (Refer to figure 18.) According to the Weather Depiction Chart, the weather for a flight from central Arkansas to southeast Alabama is

A— broken clouds at 2,500 feet.
B— visibility from 3 to 5 miles.
C— broken to scattered clouds at 25,000 feet.

3513. Radar weather reports are of special interest to pilots because they indicate

A— large areas of low ceilings and fog.
B— location of precipitation along with type, intensity, and trend.
C— location of broken to overcast clouds.

3514. What information is provided by the Radar Summary Chart that is not shown on other weather charts?

A— Lines and cells of hazardous thunderstorms.
B— Ceilings and precipitation between reporting stations.
C— Types of clouds between reporting stations.

3515. (Refer to figure 19, area A.) What is the direction and speed of movement of the radar return?

A— 020° at 20 knots.
B— East at 15 knots.
C— Northeast at 22 knots.

3516. (Refer to figure 19, area C.) What type of weather is occurring in the radar return?

A— Continuous rain.
B— Heavy rain showers.
C— Rain showers increasing in intensity.

3517. (Refer to figure 19, area D.) What is the direction and speed of movement of the radar return?

A— Southeast at 30 knots.
B— Northeast at 20 knots.
C— West at 30 knots.

3518. (Refer to figure 19, area D.) The top of the precipitation is

A— 2,000 feet.
B— 20,000 feet.
C— 30,000 feet.

3519. (Refer to figure 19, area B.) What does the dashed line enclose?

A— Areas of heavy rain.
B— Severe weather watch area.
C— Areas of hail 1/4 inch in diameter.

3520. (Refer to figure 20.) How are Significant Weather Prognostic Charts best used by a pilot?

A— For overall planning at all altitudes.
B— For determining areas to avoid (freezing levels and turbulence).
C— For analyzing current frontal activity and cloud coverage.

3521. (Refer to figure 20.) Interpret the weather symbol depicted in southern California on the 12-hour Significant Weather Prognostic Chart.

A— Moderate turbulence, surface to 18,000 feet.
B— Thunderstorm tops at 18,000 feet.
C— Base of clear air turbulence, 18,000 feet.

3522. (Refer to figure 20.) What weather is forecast for the Gulf Coast area just ahead of the cold front during the first 12 hours?

A— Ceiling 1,000 to 3,000 feet and/or visibility 3 to 5 miles with intermittent thundershowers and rain showers.
B— IFR with moderate or greater turbulence over the coastal areas.
C— Rain and thunderstorms moving northeastward ahead of the front.

3523. (Refer to figure 20.) The low pressure associated with the cold front in the western states is forecast to move

A— east at 30 knots.
B— northeast at 12 knots.
C— southeast at 30 knots.

3524. (Refer to figure 20.) At what altitude is the freezing level over northeastern Oklahoma on the 24-hour Significant Weather Prognostic Chart?

A— 4,000 feet.
B— 8,000 feet.
C— 10,000 feet.

* * *

3526. When telephoning a weather briefing facility for preflight weather information, pilots should

A— identify themselves as pilots.
B— tell the number of hours they have flown within the preceding 90 days.
C— state the number of occupants on board and the color of the aircraft.

3527. When telephoning a weather briefing facility for preflight weather information, pilots should state

A— the full name and address of the pilot in command.
B— the intended route, destination, and type of aircraft.
C— the radio frequencies to be used.

3528. When telephoning a weather briefing facility for preflight weather information, pilots should state

A— the full name and address of the formation commander.
B— that they possess a current pilot certificate.
C— whether they intend to fly VFR only.

3529. (Refer to figure 21.) En route to First Flight Airport (area F), your flight passes over Hampton Roads Airport (area B) at 1456 and then over Chesapeake Municipal at 1501. At what time should your flight arrive at First Flight?

A— 1516.
B— 1521.
C— 1526.

3530. (Refer to figure 21, area C.) Determine the approximate latitude and longitude of Currituck County Airport.

A— 36°24'N – 76°01'W.
B— 36°48'N – 76°01'W.
C— 47°24'N – 75°58'W.

3531. (Refer to figure 21.) Determine the magnetic course from First Flight Airport (area F) to Hampton Roads Airport (area B).

A— 312°.
B— 321°.
C— 330°.

3532. (Refer to figure 21.) What is your approximate position on low altitude airway Victor 1, southwest of Norfolk (area A), if the VOR receiver indicates you are on the 340° radial of Elizabeth City VOR (area C)?

A— 15 nautical miles from Norfolk VORTAC.
B— 18 nautical miles from Norfolk VORTAC.
C— 23 nautical miles from Norfolk VORTAC.

3533. (Refer to figure 21, area C; and figure 29.) The VOR is tuned to Elizabeth City VOR, and the aircraft is positioned over Shawboro. Which VOR indication is correct?

A— 5.
B— 6.
C— 8.

3534. (Refer to figure 22.) What is the estimated time en route from Mercer County Regional Airport (area C) to Minot International (area A)? The wind is from 330° at 25 knots and the true airspeed is 100 knots. Add 3-1/2 minutes for departure and climb-out.

A— 44 minutes.
B— 48 minutes.
C— 52 minutes.

3535. (Refer to figure 22, area B.) Which airport is located at approximately 47°39'30"N latitude and 100°53'00"W longitude?

A— Linrud.
B— Crooked Lake.
C— Johnson.

3536. (Refer to figure 22, area C.) Which airport is located at approximately 47°21'N latitude and 101°01'W longitude?

A— Underwood.
B— Evenson.
C— Washburn.

3537. (Refer to figure 22.) An airship crosses over Minot VORTAC (area A) at 1056 and over the creek 8 nautical miles south-southeast on Victor 15 at 1108. What should be the approximate position on Victor 15 at 1211?

A— Over Lake Nettie National Wildlife Refuge.
B— Crossing the road east of Underwood.
C— Over the powerlines east of Washburn Airport.

3538. (Refer to figure 22.) Determine the magnetic heading for a flight from Mercer County Regional Airport (area C) to Minot International (area A). The wind is from 330° at 25 knots, the true airspeed is 100 knots, and the magnetic variation is 10° east.

A—002°
B—012°
C—352°

3539. (Refer to figure 22.) What course should be selected on the omnibearing selector (OBS) to make a direct flight from Mercer County Regional Airport (area C) to the Minot VORTAC (area A) with a TO indication?

A— 001°.
B— 012°.
C— 181°.

3540. (Refer to figure 23.) What is the estimated time en route from Sandpoint (Wall) Airport (area A) to St. Maries Airport (area D)? The wind is from 215° at 25 knots and the true airspeed is 125 knots.

A— 27 minutes.
B— 30 minutes.
C— 34 minutes.

3541. (Refer to figure 23.) Determine the estimated time en route for a flight from Priest River Airport (area A) to Shoshone County Airport (area C). The wind is from 030 at 12 knots and the true airspeed is 95 knots. Add 2 minutes for climb-out.

A— 23 minutes.
B— 27 minutes.
C— 31 minutes.

3542. (Refer to figure 23.) What is the estimated time en route for a flight from St. Maries Airport (area D) to Priest River Airport (area A)? The wind is from 300° at 14 knots and the true airspeed is 90 knots. Add 3 minutes for climb-out.

A— 38 minutes.
B— 43 minutes.
C— 48 minutes.

3543. (Refer to figure 23, area C.) Determine the approximate latitude and longitude of Shoshone County airport. (A hint from this book—*SPFM*—start at Area E.)

A— 47°02'N – 116°11'W.
B— 47°32'N – 116°11'W.
C— 47°32'N – 116°41'W.

3544. (Refer to figure 23, area B.) If a balloon is launched at Ranch Aero (Pvt) Airport with a reported wind from 220° at 5 knots, what should be its approximate position after 2 hours of flight?

A— Near Hackney (Pvt) Airport.
B— Crossing the railroad southwest of Granite Airport.
C— 3-1/2 miles southwest of Rathdrum.

3545. (Refer to figure 23.) Determine the magnetic heading for a flight from Sandpoint Airport (area A) to St. Maries Airport (area D). The wind is from 215° at 25 knots and the true airspeed is 125 knots.

A— 161°.
B— 167°.
C— 181°.

3546. (Refer to figure 23.) What is the magnetic heading for a flight from Priest River Airport (area A) to Shoshone County Airport (area C)? The wind is from 030° at 12 knots and the true airspeed is 95 knots.

A— 116°.
B— 123°.
C— 130°.

3547. (Refer to figure 23.) Determine the magnetic heading for a flight from St. Maries Airport (area D) to Priest River Airport (area A). The wind is from 300° at 14 knots and the true airspeed is 90 knots.

A— 319°.
B— 325°.
C— 331°.

3548. (Refer to figure 24.) What is the estimated time en route for a flight from Allendale County Airport (area A) to Claxton-Evans County Airport (area B)? The wind is from 090° at 16 knots and the true airspeed is 90 knots. Add 2 minutes for climb-out.

A— 33 minutes.
B— 37 minutes.
C— 41 minutes.

3549. (Refer to figure 24.) What is the estimated time en route for a flight from Claxton-Evans County Airport (area B) to Hampton Varnville Airport (area A)? The wind is from 290° at 18 knots and the true airspeed is 85 knots. Add 2 minutes for climb-out.

A— 35 minutes.
B— 39 minutes.
C— 44 minutes.

3550. (Refer to figure 24.) Determine the compass heading for a flight from Allendale County Airport (area A) to Claxton-Evans County Airport (area B). The wind is from 090° at 16 knots and the true airspeed is 90 knots.

A— 200°.
B— 205°.
C— 211°.

3551. (Refer to figure 24.) Determine the compass heading for a flight from Claxton-Evans County Airport (area B) to Hampton Varnville Airport (area A). The wind is from 290° at 18 knots and the true airspeed is 85 knots.

A— 034°.
B— 038°.
C— 042°.

3552. (Refer to figure 24.) What is the approximate position of the aircraft if the VOR receivers indicate the 310° radial of Savannah VORTAC (area C) and the 190° radial of Allendale VOR (area A)?

A— Town of Guyton.
B— Town of Springfield.
C— 3 miles east of Marlow.

3553. (Refer to figure 24.) On what course should the VOR receiver (OBS) be set to navigate direct from Hampton Varnville Airport (area A) to Savannah VORTAC (area C)?

A—005°
B—183°
C—200°

3554. (Refer to figure 24.) While en route on Victor 185, a flight crosses the 248° radial of Allendale VOR at 0951 and then crosses the 216° radial of Allendale VOR at 1000. What is the estimated time of arrival at Savannah VORTAC?

A— 1023.
B— 1028.
C— 1036.

3555. (Refer to figure 25.) Estimate the time en route from Majors Airport (area A) to Winnsboro Airport (area B—arrow). The wind is form 340° at 12 knots and the true airspeed is 36 knots.

A— 55 minutes.
B— 59 minutes.
C— 63 minutes.

3556. (Refer to figure 25.) Determine the magnetic course form Airpark East Airport (area A) to Winnsboro Airport (area B—arrow). Magnetic variation is 6°30'E.

A— 075°.
B— 082°.
C— 091°.

* * *

3558. (Refer to figure 25.) Determine the magnetic heading for a flight from Majors Airport (area A) to Winnsboro Airport (area B—arrow). The wind is form 340° at 12 knots, the true airspeed is 36 knots, and the magnetic variation is 6°30'E.

A— 078°.
B— 091°.
C— 101°.

3559. (Refer to figure 25.) What is the approximate position of the aircraft if the VOR receivers indicate the 245 radial of Sulphur Springs VORTAC (right center) and the 130° radial of Blue Ridge VORTAC (upper left)?

A— Caddo Mills Airport.
B— Meadowview Airport.
C— 3 miles southeast of Caddo Mills Airport.

3560. (Refer to figure 25.) On what course should the VOR receiver (OBS) be set in order to navigate direct from Majors Airport (area A) to Quitman VORTAC (area B)?

A— 101°.
B— 108°.
C— 281°.

3561. (Refer to figure 25, area A; and figure 29.) The VOR is tuned to Blue Ridge VORTAC, and the aircraft is positioned over the town of Lone Oak, southeast of Majors Airport. Which VOR indication is correct?

A— 1.
B— 4.
C— 7.

3562. (Refer to figure 26.) What is the estimated time en route for a flight from Denton Muni (area A) to Addison (area B)? The wind is from 200° at 20 knots, the true airspeed is 110 knots, and the magnetic variation is 7° east.

A— 13 minutes.
B— 16 minutes.
C— 19 minutes.

3563. (Refer to figure 26.) Estimate the time en route from Addison (area B) to Redbird (area C). The wind is from 300° at 15 knots, the true airspeed is 120 knots, and the magnetic variation is 7° east.

A— 8 minutes.
B— 11 minutes.
C— 14 minutes.

3564. (Refer to figure 26.) Determine the magnetic heading for a flight from Redbird (area C) to Fort Worth Meacham (area D). The wind is from 030° at 10 knots, the true airspeed is 35 knots, and the magnetic variation is 7° east.

A— 266°.
B— 298°.
C— 312°.

3565. (Refer to figure 26.) Determine the magnetic heading for a flight from Fort Worth Meacham (area D) to Denton Muni (area A). The wind is from 330° at 25 knots, the true airspeed is 110 knots, and the magnetic variation is 7° east.

A— 003°.
B— 017°.
C— 023°.

3566. (Refer to figure 26, area E.) The VOR is tuned to the Dallas/Fort Worth VORTAC. The omnibearing selector (OBS) is set on the 253° radial, with a TO indication, and a right course deviation indicator (CDI) deflection. What is the aircraft's position from the VORTAC?

A— East-northeast.
B— North-northeast.
C— West-southwest.

3567. (Refer to figure 27, area B.) What is the approximate latitude and longitude of Cooperstown Airport?

A— 47°25'N – 98°06'W.
B— 47°25'N – 99°54'W.
C— 47°55'N – 98°06'W.

3568. (Refer to figure 27.) Determine the magnetic course from Breckheimer (Pvt) Airport (area A) to Jamestown Airport (area D).

A— 013°.
B— 021°.
C— 181°.

3569. (Refer to figure 27, area E.) A balloon drifts over the town of Eckelson on a magnetic course of 282° at 10 MPH. If wind conditions remain constant, where will the balloon be after 2 hours 30 minutes?

A— 3 miles south-southwest of Buchanan.
B— Over Buchanan.
C— Over the tower southwest of Fried.

3570. (Refer to figure 27, areas D and C; and figure 29.) The VOR is tuned to Jamestown VOR, and the aircraft is positioned over the town of Wimbledon. Which VOR indication is correct?

A— 1.
B— 4.
C— 6.

3571. (Refer to figure 28.) An aircraft departs an airport in the eastern daylight time zone at 0945 EDT for a 2-hour flight to an airport located in the central daylight time zone. The landing should be at what coordinated universal time?

A— 1345Z.
B— 1445Z.
C— 1545Z.

3572. (Refer to figure 28.) An aircraft departs an airport in the central standard time zone at 0930 CST for a 2-hour flight to an airport located in the mountain standard time zone. The landing should be at what time?

A— 0930 MST.
B— 1030 MST.
C— 1130 MST.

3573. (Refer to figure 28.) An aircraft departs an airport in the central standard time zone at 0845 CST for a 2-hour flight to an airport located in the mountain standard time zone. The landing should be at what coordinated universal time?

A— 1345Z.
B— 1445Z.
C— 1645Z.

3574. (Refer to figure 28.) An aircraft departs an airport in the mountain standard time zone at 1615 MST for a 2-hour 15-minute flight to an airport located in the Pacific standard time zone. The estimated time of arrival at the destination airport should be

A— 1630 PST.
B— 1730 PST.
C— 1830 PST.

3575. (Refer to figure 28.) An aircraft departs an airport in the Pacific standard time zone at 1030 PST for a 4-hour flight to an airport located in the central standard time zone. The landing should be at what coordinated universal time?

A— 2030Z.
B— 2130Z.
C— 2230Z.

3576. (Refer to figure 28.) An aircraft departs an airport in the mountain standard time zone at 1515 MST for a 2-hour 30-minute flight to an airport located in the Pacific standard time zone. What is the estimated time of arrival at the destination airport?

A— 1645 PST.
B— 1745 PST.
C— 1845 PST.

3577. (Refer to figure 29, illustration 1.) The VOR receiver has the indications shown. What is the aircraft's position relative to the station?

A— North.
B— East.
C— South.

3578. (Refer to figure 29, illustration 3.) The VOR receiver has the indications shown. What is the aircraft's position relative to the station?

A— East.
B— Southeast.
C— West.

3579. (Refer to figure 29, illustration 8.) The VOR receiver has the indications shown. What radial is the aircraft crossing?

A— 030°.
B— 210°.
C— 300°.

3580. (Refer to figure 30, illustration 1.) Determine the magnetic bearing TO the station.

A— 030°.
B— 180°.
C— 210°.

3581. (Refer to figure 30, illustration 2.) What magnetic bearing should the pilot use to fly TO the station?

A— 010°.
B— 145°.
C— 190°.

3582. (Refer to figure 30, illustration 2.) Determine the approximate heading to intercept the 180° bearing TO the station.

A— 040°.
B— 160°.
C— 220°.

3583. (Refer to figure 30, illustration 3.) What is the magnetic bearing FROM the station?

A— 025°.
B— 115°.
C— 295°.

3584. (Refer to figure 30.) Which ADF indication represents the aircraft tracking TO the station with a right crosswind?

A— 1.
B— 2.
C— 4.

3585. (Refer to figure 30, illustration 1.) What outbound bearing is the aircraft crossing?

A— 030°.
B— 150°.
C— 180°.

3586. (Refer to figure 30, illustration 1.) What is the relative bearing TO the station?

A— 030°.
B— 210°.
C— 240°.

3587. (Refer to figure 30, illustration 2.) What is the relative bearing TO the station?

A— 190°.
B— 235°.
C— 315°.

3588. (Refer to figure 30, illustration 4.) What is the relative bearing TO the station?

A— 020°.
B— 060°.
C— 340°.

3589. (Refer to figure 31, illustration 1.) The relative bearing TO the station is

A— 045°.
B— 180°.
C— 315°.

3590. (Refer to figure 31, illustration 2). The relative bearing TO the station is

A— 090°.
B— 180°.
C— 270°.

3591. (Refer to figure 31, illustration 3.) The relative bearing TO the station is

A— 090°.
B— 180°.
C— 270°.

3592. (Refer to figure 31, illustration 4.) On a magnetic heading of 320°, the magnetic bearing TO the station is

A— 005°.
B— 185°.
C— 225°.

3593. (Refer to figure 31, illustration 5). On a magnetic heading of 035°, the magnetic bearing TO the station is

A— 035°.
B— 180°.
C— 215°.

3594. (Refer to figure 31, illustration 6.) On a magnetic heading of 120°, the magnetic bearing TO the station is

A— 045°.
B— 165°.
C— 270°.

3595. (Refer to figure 31, illustration 6.) If the magnetic bearing TO the station is 240°, the magnetic heading is

A— 045°.
B— 105°.
C— 195°.

3596. (Refer to figure 31, illustration 7.) If the magnetic bearing TO the station is 030°, the magnetic heading is

A— 060°.
B— 120°.
C— 270°.

3597. (Refer to figure 31, illustration 8.) If the magnetic bearing TO the station is 135°, the magnetic heading is

A— 135°.
B— 270°.
C— 360°.

3598. When the course deviation indicator (CDI) needle is centered during an omnireceiver check using a VOR test signal (VOT), the omnibearing selector (OBS) and the TO/FROM indicator should read

A— 180° FROM, only if the pilot is due north of the VOT.
B— 0° TO or 180° FROM, regardless of the pilot's position from the VOT.
C— 0° FROM or 180° TO, regardless of the pilot's position from the VOT.

3599. (Refer to figure 26, area D.) The floor of Class B airspace at Forth Worth Meacham Field is

A— at the surface.
B— 3,200 feet MSL.
C— 4,000 feet MSL.

3600. (Refer to figure 26, area B.) The floor of Class B airspace at Addison Airport is

A— at the surface.
B— 3,000 feet MSL.
C— 3,100 feet MSL.

3601. (Refer to figure 21.) What hazards to aircraft may exist in warning areas such as Warning W-50B?

A— Unusual, often invisible, hazards such as aerial gunnery or guided missiles over international waters.

B— High volume of pilot training or unusual type of aerial activity.

C— Heavy military aircraft traffic in the approach and departure area of the North Atlantic Control Area.

3602. (Refer to figure 27.) What hazards to aircraft may exist in areas such as Devils Lake East MOA?

A— Unusual, often invisible, hazards to aircraft such as artillery firing.

B— High density military training activities.

C— Parachute jump operations.

3603. (Refer to figure 22.) What type military flight operations should a pilot expect along IR 644?

A— IFR training flights above 1,500 feet AGL at speeds in excess of 250 knots.

B— VFR training flights above 1,500 feet AGL at speeds less than 250 knots.

C— Instrument training flights below 1,500 feet AGL at speeds in excess of 150 knots.

3604. (Refer to figure 21, area C.) What is the recommended communications procedure for a landing at Currituck County Airport?

A— Transmit intentions on 122.9 MHz when 10 miles out and give position reports in the traffic pattern.

B— Contact Elizabeth City FSS for airport advisory service.

C— Contact New Bern FSS for area traffic information.

3605. (Refer to figure 22, area B.) The CTAF/MULTICOM frequency for Garrison Municipal is

A— 122.8 MHz.
B— 122.9 MHz.
C— 123.0 MHz.

3606. (Refer to figure 23, area B; and figure 32.) If Coeur D'Alene Tower is not in operation, which frequency should be used as a Common Traffic Advisory Frequency (CTAF) to self-announce position and intentions?

A—122.05 MHz.
B—122.1/108.8 MHz.
C—122.8 MHz.

3607. (Refer to figure 23, area B; and figure 32.) At Coeur D'Alene, which frequency should be used as a Common Traffic Advisory Frequency (CTAF) to monitor airport traffic?

A—122.05 MHz.
B—122.1/108.8 MHz.
C—122.8 MHz.

3608. (Refer to figure 23, area B; and figure 32.) What is the correct UNICOM frequency to be used at Coeur D'Alene to request fuel?

A— 119.1 MHz.
B— 122.1/108.8 MHz.
C— 122.8 MHz.

3609. (Refer to figure 26, area C.) If Redbird Tower is not in operation, which frequency should be used as a Common Traffic Advisory Frequency (CTAF) to monitor airport traffic?

A— 120.3 MHz.
B— 122.95 MHz.
C— 126.35 MHz.

3610. (Refer to figure 27, area B.) What is the recommended communication procedure when inbound to land at Cooperstown Airport?

A— Broadcast intentions when 10 miles out on the CTAF/MULTICOM frequency, 122.9 MHz.

B— Contact UNICOM when 10 miles out on 122.8 MHz.

C— Circle the airport in a left turn prior to entering traffic.

3611. (Refer to figure 27, area D.) The CTAF/UNICOM frequency at Jamestown Airport is

A— 122.0 MHz.
B— 123.0 MHz.
C— 123.6 MHz.

3612. (Refer to figure 27, area F.) What is the CTAF/UNICOM frequency at Barnes County Airport?

A— 122.0 MHz.
B— 122.8 MHz.
C— 123.6 MHz.

3613. When flying HAWK N666CB, the proper phraseology for initial contact with McAlester AFSS is

A— "MC ALESTER RADIO, HAWK SIX SIX SIX CHARLIE BRAVO, RECEIVING ARDMORE VORTAC, OVER."
B— "MC ALESTER STATION, HAWK SIX SIX SIX CEE BEE, RECEIVING ARDMORE VORTAC, OVER."
C— "MC ALESTER FLIGHT SERVICE STATION, HAWK NOVEMBER SIX CHARLIE BRAVO, RECEIVING ARDMORE VORTAC, OVER."

3614. The correct method of stating 4,500 feet MSL to ATC is

A— "FOUR THOUSAND FIVE HUNDRED."
B— "FOUR POINT FIVE."
C— "FORTY-FIVE HUNDRED FEET MSL."

3615. The correct method of stating 10,500 feet MSL to ATC is

A— "TEN THOUSAND, FIVE HUNDRED FEET."
B— "TEN POINT FIVE."
C— "ONE ZERO THOUSAND, FIVE HUNDRED."

3616. How should contact be established with an En Route Flight Advisory Service (EFAS) station, and what service would be expected?

A— Call EFAS on 122.2 for routine weather, current reports on hazardous weather, and altimeter settings.
B— Call flight assistance on 122.5 for advisory service pertaining to severe weather.
C— Call Flight Watch on 122.0 for information regarding actual weather and thunderstorm activity along proposed route.

3617. What service should a pilot normally expect from an En Route Flight Advisory Service (EFAS) station?

A— Actual weather information and thunderstorm activity along the route.
B— Preferential routing and radar vectoring to circumnavigate severe weather.
C— Severe weather information, changes to flight plans, and receipt of routine position reports.

3618. (Refer to figure 27, area C.) When flying over Arrowwood National Wildlife Refuge, a pilot should fly no lower than

A— 2,000 feet AGL.
B— 2,500 feet AGL.
C— 3,000 feet AGL.

3619. (Refer to figure 23, area B and legend 1.) For information about the parachute jumping and glider operations at Silverwood Airport, refer to

A— notes on the border of the chart.
B— the Airport/Facility Directory.
C— the Notices to Airmen (NOTAM) publication.

3620. (Refer to figure 23, area A.) The visibility and cloud clearance requirements to operate VFR during daylight hours over Wall Airport at less than 1,200 feet AGL are

A— 1 mile and clear of clouds.
B— 1 mile and 1,000 feet above, 500 feet below, and 2,000 feet horizontally from each cloud.
C— 3 miles and 1,000 feet above, 500 feet below, and 2,000 feet horizontally from each cloud.

3621. (Refer to figure 27, area B.) The visibility and cloud clearance requirements to operate VFR during daylight hours over the Cooperstown Airport between 1,200 feet AGL and 10,000 feet MSL are

A— 1 mile and clear of clouds.
B— 1 mile and 1,000 feet above, 500 feet below, and 2,000 feet horizontally from clouds.
C— 3 miles and 1,000 feet above, 500 feet below, and 2,000 feet horizontally from clouds.

3622. (Refer to figure 27, area A.) Identify the airspace over Lowe Airport that exists from the surface to 14,500 feet MSL.

A— Class G airspace–surface to 14,500 feet MSL.
B— Class G airspace–surface to 3,500 feet MSL; Class E airspace–3,500 feet MSL to 14,500 feet MSL.
C— Class G airspace–surface to 3,500 feet MSL; Class E airspace–3,500 feet MSL to 10,000 feet MSL; Class G airspace–10,000 feet MSL to 14,500 feet MSL.

3623. (Refer to figure 27, area F.) The airspace overlying and within 5 miles of Barnes County Airport is

A— Class D airspace from the surface to the floor of the overlying Class E airspace.
B— Class E airspace from the surface to 1,200 feet MSL.
C— Class G airspace from the surface to 700 feet AGL.

3624. (Refer to figure 26, area G.) The airspace overlying McKinney Muni is uncontrolled from the surface to

A— 700 feet AGL.
B— 1,700 feet MSL.
C— 4,000 feet AGL.

3625. (Refer to figure 26, area D.) The airspace directly overlying Fort Worth Meacham is

A— Class B airspace to 10,000 feet MSL.
B— Class C airspace to 5,000 feet MSL.
C— Class D airspace to 3,200 feet MSL.

3626. (Refer to figure 24, area C.) What is the floor of the Savannah Class C airspace at the outer circle?

A— 1,200 feet AGL.
B— 1,300 feet MSL.
C— 1,700 feet MSL.

3627. (Refer to figure 21, area A.) What minimum radio equipment is required to land and take off at Norfolk International?

A— Mode C transponder and omnireceiver.
B— Mode C transponder, VHF transmitter and receiver.
C— Mode C transponder, omnireceiver, and DME.

3628. (Refer to figure 26.) At which airports is fixed-wing Special VFR not recommended?

A— Fort Worth Meacham and Fort Worth Spinks.
B— Dallas-Fort Worth International and Dallas Love Field.
C— Addison and Redbird.

3629. (Refer to figure 23, area C.) The vertical limits of that portion of Class E airspace designated as a Federal Airway over Magee Airport are

A— 1,200 feet AGL to 10,000 feet MSL.
B— 7,500 feet MSL to 12,500 feet MSL.
C— 7,500 feet MSL to 17,999 feet MSL.

3630. (Refer to figure 22.) On what frequency can a pilot receive Hazardous In-Flight Weather Advisory Service (HIWAS) in the vicinity of area A?

A— 117.1 MHz.
B— 118.0 MHz.
C— 122.0 MHz.

3631. (Refer to figure 21, area E.) The CAUTION box denotes what hazard to aircraft?

A— Guy wires extending from radio or TV towers.
B— Tall bridge over the inlet to the body of water.
C— Cable extending from radar-outfitted blimps.

3632. (Refer to figure 21, area B.) The flag symbol at Lake Drummond represents

A— a compulsory reporting point for the Norfolk Class C airspace.
B— a compulsory reporting point for Hampton Roads Airport.
C— a visual checkpoint used to identify position for initial callup to Norfolk Approach Control.

3633. (Refer to figure 21, area B.) The elevation of the Chesapeake Municipal Airport is

A— 20 feet.
B— 36 feet.
C— 360 feet.

3634. (Refer to figure 22.) The terrain elevation of the light tan area between Minot (area A) and Audubon Lake (area B) varies from

A— sea level to 2,000 feet MSL.
B— 2,000 feet to 2,500 feet MSL.
C— 2,000 feet to 2,700 feet MSL.

3635. (Refer to figure 22.) Which public use airports depicted are indicated as having fuel?

A— Minot and Mercer County Regional.
B— Minot and Garrison.
C— Mercer County Regional and Garrison.

3636. (Refer to figure 24.) The flag symbols at Statesboro Airport and Claxton-Evans County Airport are

A— outer boundaries of the Savannah Class C airspace.
B— airports with special traffic patterns.
C— visual checkpoints to identify position for initial callup prior to entering the Savannah Class C airspace.

3637. (Refer to figure 24, area C.) What is the height of the lighted obstacle approximately 7 nautical miles southwest of Savannah International?

A— 1,500 feet AGL.
B— 1,532 feet AGL.
C— 1,549 feet AGL.

3638. (Refer to figure 24, area C.) The top of the lighted stack approximately 12 nautical miles from the Savannah VORTAC on the 350° radial is

A— 305 feet AGL.
B— 400 feet AGL.
C— 430 feet AGL.

3639. (Refer to figure 25, area A.) What minimum altitude is necessary to vertically clear the obstacle on the northeast side of Airpark East Airport by 500 feet?

A— 1,010 feet MSL.
B— 1,273 feet MSL.
C— 1,283 feet MSL.

3640. (Refer to figure 25, area B.) What minimum altitude is necessary to vertically clear the obstacle on the southeast side of Winnsboro Airport by 500 feet?

A— 823 feet MSL.
B— 1,013 feet MSL.
C— 1,403 feet MSL.

3641. (Refer to figure 26, area B.) What is the tower frequency for Addison Airport?

A— 122.95 MHz.
B— 126.0 MHz.
C— 133.4 MHz.

3642. (Refer to figure 26, area H.) What minimum altitude is required to fly over the Cedar Hill TV towers in the congested area south of NAS Dallas?

A— 2,533 feet MSL.
B— 2,849 feet MSL.
C— 3,349 feet MSL.

3643. (Refer to figure 26.) (Author's note: Look about 1 inch to the left of the letter "F.") The navigation facility at Dallas/Fort Worth International airport (DFW) is a

A— VOR.
B— VORTAC.
C— VOR/DME.

* * *

3651. What action can a pilot take to aid in cooling an engine that is overheating during a climb?

A— Reduce rate of climb and increase airspeed.
B— Reduce climb speed and increase RPM.
C— Increase climb speed and increase RPM.

3652. What is one procedure to aid in cooling an engine that is overheating?

A— Enrichen the fuel mixture.
B— Increase the RPM.
C— Reduce the airspeed.

3653. How is engine operation controlled on an engine equipped with a constant-speed propeller?

A— The throttle controls power output as registered on the manifold pressure gauge and the propeller control regulates engine RPM.
B— The throttle controls power output as registered on the manifold pressure gauge and the propeller control regulates a constant blade angle.
C— The throttle controls engine RPM as registered on the tachometer and the mixture control regulates the power output.

3654. What is an advantage of a constant-speed propeller?

A— Permits the pilot to select and maintain a desired cruising speed.
B— Permits the pilot to select the blade angle for the most efficient performance.
C— Provides a smoother operation with stable RPM and eliminates vibrations.

3655. A precaution for the operation of an engine equipped with a constant-speed propeller is to

A— avoid high RPM settings with high manifold pressure.
B— avoid high manifold pressure settings with low RPM.
C— always use a rich mixture with high RPM settings.

3656. What should be the first action after starting an aircraft engine?

A— Adjust for proper RPM and check for desired indications on the engine gauges.
B— Place the magneto or ignition switch momentarily in the OFF position to check for proper grounding.
C— Test each brake and the parking brake.

3657. Should it become necessary to handprop an airplane engine, it is extremely important that a competent pilot

A— call "contact" before touching the propeller.
B— be at the controls in the cockpit.
C— be in the cockpit and call out all commands.

3658. In regard to preflighting an aircraft, what is the minimum expected of a pilot prior to every flight?

A— Drain fuel from each quick drain.
B— Perform a walk-around inspection of the aircraft.
C— Check the required documents aboard the aircraft.

3659. Why is the use of a written checklist recommended for preflight inspection and engine start?

A— To ensure that all necessary items are checked in a logical sequence.
B— For memorizing the procedures in an orderly sequence.
C— To instill confidence in the passengers.

3660. What special check should be made on an aircraft during preflight after it has been stored an extended period of time?

A— ELT batteries and operation.
B— Condensation in the fuel tanks.
C— Damage or obstructions caused by animals, birds, or insects.

3661. Which items are included in the empty weight of an aircraft?

A— Unusable fuel and undrainable oil.
B— Only the airframe, powerplant, and optional equipment.
C— Full fuel tanks and engine oil to capacity.

3662. An aircraft is loaded 110 pounds over maximum certificated gross weight. If fuel (gasoline) is drained to bring the aircraft weight within limits, how much fuel should be drained?

A— 15.7 gallons.
B— 16.2 gallons.
C— 18.4 gallons.

3663. If an aircraft is loaded 90 pounds over maximum certificated gross weight and fuel (gasoline) is drained to bring the aircraft weight within limits, how much fuel should be drained?

A— 10 gallons.
B— 12 gallons.
C— 15 gallons.

3664. GIVEN:

	WEIGHT (LB)	ARM (IN)	MOMENT (LB-IN)
Empty weight	1,495.0	101.4	151,593.0
Pilot and passengers	380.0	64.0	----
Fuel (30 gal usable no reserve)	----	96.0	----

The CG is located how far aft of datum?

A— CG 92.44.
B— CG 94.01.
C— CG 119.8.

3665. (Refer to figures 33 and 34.) Determine if the airplane weight and balance is within limits.

Front seat occupants . . . . 340 lb
Rear seat occupants . . . . 295 lb
Fuel (main wing tanks) . . . . 44 gal
Baggage . . . . 56 lb

A— 20 pounds overweight, CG aft of aft limits.
B— 20 pounds overweight, CG within limits.
C— 20 pounds overweight, CG forward of forward limits.

3666. (Refer to figures 33 and 34.) What is the maximum amount of baggage that can be carried when the airplane is loaded as follows?

Front seat occupants . . . . 387 lb
Rear seat occupants . . . . 293 lb
Fuel . . . . 35 gal

A— 45 pounds.
B— 63 pounds.
C— 220 pounds.

3667. (Refer to figures 33 and 34.) Calculate the weight and balance and determine if the CG and the weight of the airplane are within limits.

Front seat occupants . . . . 350 lb
Rear seat occupants . . . . 325 lb
Baggage . . . . 27 lb
Fuel . . . . 35 gal

A— CG 81.7, out of limits forward.
B— CG 83.4, within limits.
C— CG 84.1, within limits.

3668. (Refer to figures 33 and 34.) Determine if the airplane weight and balance is within limits.

Front seat occupants . . . . 415 lb
Rear seat occupants . . . . 110 lb
Fuel, main tanks . . . . 44 gal
Fuel, aux. tanks . . . . 19 gal
Baggage. . . . . 32 lb

A— 19 pounds overweight, CG within limits.
B— 19 pounds overweight, CG out of limits forward.
C— Weight within limits, CG out of limits.

3669. (Refer to figure 35.) What is the maximum amount of baggage that may be loaded aboard the airplane for the CG to remain within the moment envelope?

	WEIGHT (LB)	MOM/1000
Empty weight	1,350	51.5
Pilot and front passenger	250	---
Rear passengers	400	---
Baggage	---	---
Fuel, 30 gal	---	---
Oil, 8 qt	---	−0.2

A— 105 pounds.
B— 110 pounds.
C— 120 pounds.

3670. (Refer to figure 35.) Calculate the moment of the airplane and determine which category is applicable.

	WEIGHT (LB)	MOM/1000
Empty weight	1,350	51.5
Pilot and front passenger	310	---
Rear passengers	96	---
Fuel, 38 gal	---	---
Oil, 8 qt	---	−0.2

A— 79.2, utility category.
B— 80.8, utility category.
C— 81.2, normal category.

3671. (Refer to figure 35.) What is the maximum amount of fuel that may be aboard the airplane on takeoff if loaded as follows?

	WEIGHT (LB)	MOM/1000
Empty weight	1,350	51.5
Pilot and front passenger	340	---
Rear passengers	310	---
Baggage	45	---
Oil, 8 qt	---	---

A— 24 gallons.
B— 32 gallons.
C— 40 gallons.

3672. (Refer to figure 35.) Determine the moment with the following data:

	WEIGHT (LB)	MOM/1000
Empty weight	1,350	51.5
Pilot and front passenger	340	---
Fuel (std tanks)	Capacity	---
Oil, 8 qt	---	---

A— 69.9 pound-inches.
B— 74.9 pound-inches.
C— 77.6 pound-inches.

3673. (Refer to figure 35.) Determine the aircraft loaded moment and the aircraft category.

	WEIGHT (LB)	MOM/1000
Empty weight	1,350	51.5
Pilot and front passenger	380	---
Fuel, 48 gal	288	---
Oil, 8 qt	---	---

A— 78.2, normal category.
B— 79.2, normal category.
C— 80.4, utility category.

3674. (Refer to figures 33 and 34.) Upon landing, the front passenger (180 pounds) departs the airplane. A rear passenger (204 pounds) moves to the front passenger position. What effect does this have on the CG if the airplane weighed 2,690 pounds and the MOM/100 was 2,260 just prior to the passenger transfer?

A— The CG moves forward approximately 3 inches.
B— The weight changes, but the CG is not affected.
C— The CG moves forward approximately 0.1 inch.

3675. (Refer to figures 33 and 34.) Which action can adjust the airplane's weight to maximum gross weight and the CG within limits for takeoff?

Front seat occupants . . . . 425 lb
Rear seat occupants . . . . 300 lb
Fuel, main tanks . . . . 44 gal

A— Drain 12 gallons of fuel.
B— Drain 9 gallons of fuel.
C— Transfer 12 gallons of fuel from the main tanks to the auxiliary tanks.

3676. (Refer to figures 33 and 34.) What effect does a 35-gallon fuel burn (main tanks) have on the weight and balance if the airplane weighed 2,890 pounds and the MOM/100 was 2,452 at takeoff?

A— Weight is reduced by 210 pounds and the CG is aft of limits.
B— Weight is reduced by 210 pounds and the CG is unaffected.
C— Weight is reduced to 2,680 pounds and the CG moves forward.

3677. (Refer to figures 33 and 34.) With the airplane loaded as follows, what action can be taken to balance the airplane?

Front seat occupants . . . . 411 lb
Rear seat occupants . . . . 100 lb
Main wing tanks . . . . 44 gal

A— Fill the auxiliary wing tanks.
B— Add a 100-pound weight to the baggage compartment.
C— Transfer 10 gallons of fuel from the main tanks to the auxiliary tanks.

3678. (Refer to figure 36.) Approximately what true airspeed should a pilot expect with 65 percent maximum continuous power at 9,500 feet with a temperature of 36 °F below standard?

A— 178 MPH.
B— 181 MPH.
C— 183 MPH.

3679. (Refer to figure 36.) What is the expected fuel consumption for a 1,000-nautical mile flight under the following conditions?

Pressure altitude .... 8,000 ft
Temperature .... 22 °C
Manifold pressure .... 20.8" Hg
Wind .... Calm

A— 60.2 gallons.
B— 70.1 gallons.
C— 73.2 gallons.

3680. (Refer to figure 36.) What is the expected fuel consumption for a 500-nautical mile flight under the following conditions?

Pressure altitude .... 4,000 ft
Temperature .... +29 °C
Manifold pressure .... 21.3" Hg
Wind .... Calm

A— 31.4 gallons.
B— 36.1 gallons.
C— 40.1 gallons.

3681. (Refer to figure 36.) What fuel flow should a pilot expect at 11,000 feet on a standard day with 65 percent maximum continuous power?

A— 10.6 gallons per hour.
B— 11.2 gallons per hour.
C— 11.8 gallons per hour.

3682. (Refer to figure 36.) Determine the approximate manifold pressure setting with 2,450 RPM to achieve 65 percent maximum continuous power at 6,500 feet with a temperature of 36 °F higher than standard.

A— 19.8" Hg.
B— 20.8" Hg.
C— 21.0" Hg.

3683. (Refer to figure 37.) What is the headwind component for a landing on Runway 18 if the tower reports the wind as 220° at 30 knots?

A— 19 knots.
B— 23 knots.
C— 26 knots.

3684. (Refer to figure 37.) Determine the maximum wind velocity for a 45° crosswind if the maximum crosswind component for the airplane is 25 knots.

A— 25 knots.
B— 29 knots.
C— 35 knots.

3685. (Refer to figure 37.) What is the maximum wind velocity for a 30° crosswind if the maximum crosswind component for the airplane is 12 knots?

A— 16 knots.
B— 20 knots.
C— 24 knots.

3686. (Refer to figure 37.) With a reported wind of north at 20 knots, which runway (6, 29, or 32) is acceptable for use for an airplane with a 13-knot maximum crosswind component?

A— Runway 6.
B— Runway 29.
C— Runway 32.

3687. (Refer to figure 37.) With a reported wind of south at 20 knots, which runway (10, 14, or 24) is appropriate for an airplane with a 13-knot maximum crosswind component?

A— Runway 10.
B— Runway 14.
C— Runway 24.

3688. (Refer to figure 37.) What is the crosswind component for a landing on Runway 18 if the tower reports the wind as 220° at 30 knots?

A— 19 knots.
B— 23 knots.
C— 30 knots.

3689. (Refer to figure 38.) Determine the total distance required to land.

OAT .... 32 °F
Pressure altitude .... 8,000 ft
Weight .... 2,600 lb
Headwind component .... 20 kts
Obstacle .... 50 ft

A— 850 feet.
B— 1,400 feet.
C— 1,750 feet.

3690. (Refer to figure 38.) Determine the total distance required to land.

OAT . . . . Std
Pressure altitude . . . . 2,000 ft
Weight . . . . 2,300 lb
Wind component . . . . Calm
Obstacle . . . . None

A— 850 feet.
B— 1,250 feet.
C— 1,450 feet.

3691. (Refer to figure 38.) Determine the total distance required to land.

OAT . . . . 90 °F
Pressure altitude . . . . 3,000 ft
Weight . . . . 2,900 lb
Headwind component . . . . 10 kts
Obstacle . . . . 50 ft

A— 1,450 feet.
B— 1,550 feet.
C— 1,725 feet.

3692. (Refer to figure 38.) Determine the approximate ground roll distance after landing.

OAT . . . . 90 °F
Pressure altitude . . . . 4,000 ft
Weight . . . . 2,800 lb
Tailwind component . . . . 10 kts

A— 1,575 feet.
B— 1,725 feet.
C— 1,950 feet.

3693. (Refer to figure 39.) Determine the approximate landing ground roll distance.

Pressure altitude . . . . Sea level
Headwind . . . . 4 kts
Temperature . . . . Std

A— 356 feet.
B— 401 feet.
C— 490 feet.

3694. (Refer to figure 39.) Determine the total distance required to land over a 50-foot obstacle.

Pressure altitude . . . . 7,500 ft
Headwind . . . . 8 kts
Temperature . . . . Std
Runway . . . . Dry grass

A— 1,004 feet.
B— 1,205 feet.
C— 1,506 feet.

3695. (Refer to figure 39.) Determine the total distance required to land over a 50-foot obstacle.

Pressure altitude . . . . 5,000 ft
Headwind . . . . 8 kts
Temperature . . . . 41 °F
Runway . . . . Hard surface

A— 837 feet.
B— 956 feet.
C— 1,076 feet.

3696. (Refer to figure 39.) Determine the total distance required to land over a 50-foot obstacle.

Pressure altitude . . . . 5,000 ft
Headwind . . . . Calm
Temperature . . . . 101 °F

A— 1,076 feet.
B— 1,291 feet.
C— 1,314 feet.

3697. (Refer to figure 39.) Determine the approximate landing ground roll distance.

Pressure altitude . . . . 3,750 ft
Headwind . . . . 12 kts
Temperature . . . . Std

A— 338 feet.
B— 425 feet.
C— 483 feet.

3698. (Refer to figure 39.) Determine the approximate landing ground roll distance.

Pressure altitude . . . . 1,250 ft
Headwind . . . . 8 kts
Temperature . . . . Std

A— 275 feet.
B— 366 feet.
C— 470 feet.

* * *

3705. (Refer to figure 41.) Determine the total distance required for takeoff to clear a 50-foot obstacle.

OAT .... Std
Pressure altitude .... 4,000 ft
Takeoff weight .... 2,800 lb
Headwind component .... Calm

A— 1,500 feet.
B— 1,750 feet.
C— 2,000 feet.

3706. (Refer to figure 41.) Determine the total distance required for takeoff to clear a 50-foot obstacle.

OAT .... Std
Pressure altitude .... Sea level
Takeoff weight .... 2,700 lb
Headwind component .... Calm

A— 1,000 feet.
B— 1,400 feet.
C— 1,700 feet.

3707. (Refer to figure 41.) Determine the approximate ground roll distance required for takeoff.

OAT .... 100 °F
Pressure altitude .... 2,000 ft
Takeoff weight .... 2,750 lb
Headwind component .... Calm

A— 1,150 feet.
B— 1,300 feet.
C— 1,800 feet.

3708. (Refer to figure 41.) Determine the approximate ground roll distance required for takeoff.

OAT .... 90 °F
Pressure altitude .... 2,000 ft
Takeoff weight .... 2,500 lb
Headwind component .... 20 kts

A— 650 feet.
B— 800 feet.
C— 1,000 feet.

3709. FAA advisory circulars (some free, others at cost) are available to all pilots and are obtained by

A— distribution from the nearest FAA district office.
B— ordering those desired from the Government Printing Office.
C— subscribing to the Federal Register.

3710. Prior to starting each maneuver, pilots should

A— check altitude, airspeed, and heading indications.
B— visually scan the entire area for collision avoidance.
C— announce their intentions on the nearest CTAF.

3711. The most important rule to remember in the event of a power failure after becoming airborne is to

A— immediately establish the proper gliding attitude and airspeed.
B— quickly check the fuel supply for possible fuel exhaustion.
C— determine the wind direction to plan for the forced landing.

3712. What is the most effective way to use the eyes during night flight?

A— Look only at far away, dim lights.
B— Scan slowly to permit offcenter viewing.
C— Concentrate directly on each object for a few seconds.

3713. The best method to use when looking for other traffic at night is to

A— look to the side of the object and scan slowly.
B— scan the visual field very rapidly.
C— look to the side of the object and scan rapidly.

3714. The most effective method of scanning for other aircraft for collision avoidance during nighttime hours is to use

A— regularly spaced concentration on the 3-, 9-, and 12-o'clock positions.
B— a series of short, regularly spaced eye movements to search each 30-degree sector.
C— peripheral vision by scanning small sectors and utilizing offcenter viewing.

3715. During a night flight, you observe a steady red light and a flashing red light ahead and at the same altitude. What is the general direction of movement of the other aircraft?

A— The other aircraft is crossing to the left.
B— The other aircraft is crossing to the right.
C— The other aircraft is approaching head-on.

3716. During a night flight, you observe a steady white light and a flashing red light ahead and at the same altitude. What is the general direction of movement of the other aircraft?

A— The other aircraft is flying away from you.
B— The other aircraft is crossing to the left.
C— The other aircraft is crossing to the right.

3717. During a night flight, you observe steady red and green lights ahead and at the same altitude. What is the general direction of movement of the other aircraft?

A— The other aircraft is crossing to the left.
B— The other aircraft is flying away from you.
C— The other aircraft is approaching head-on.

3718. Airport taxiway edge lights are identified at night by

A— white directional lights.
B— blue omnidirectional lights.
C— alternate red and green lights.

3719. VFR approaches to land at night should be accomplished

A— at a higher airspeed.
B— with a steeper descent.
C— the same as during daytime.

* * *

3759. To use VHF/DF facilities for assistance in locating an aircraft's position, the aircraft must have a

A— VHF transmitter and receiver.
B— 4096-code transponder.
C— VOR receiver and DME.

3760. A slightly high glide slope indication from a precision approach path indicator is

A— four white lights.
B— three white lights and one red light.
C— two white lights and two red lights.

3761. A below glide slope indication from a tri-color VASI is a

A— red light signal.
B— pink light signal.
C— green light signal.

3762. An above glide slope indication from a tri-color VASI is

A— a white light signal.
B— a green light signal.
C— an amber light signal.

3763. An on glide slope indication from a tri-color VASI is

A— a white light signal.
B— a green light signal.
C— an amber light signal.

3764. A below glide slope indication from a pulsating approach slope indicator is a

A— pulsating white light.
B— steady white light.
C— pulsating red light.

3765. (Refer to figure 48.) Illustration A indicates that the aircraft is

A— below the glide slope.
B— on the glide slope.
C— above the glide slope.

3766. (Refer to figure 48.) VASI lights as shown by illustration C indicate that the airplane is

A— off course to the left.
B— above the glide slope.
C— below the glide slope.

3767. (Refer to figure 48.) While on final approach to a runway equipped with a standard 2-bar VASI, the lights appear as shown by illustration D. This means that the aircraft is

A— above the glide slope.
B— below the glide slope.
C— on the glide slope.

3768. To set the high intensity runway lights on medium intensity, the pilot should click the microphone seven times, then click it

A— one time.
B— three times.
C— five times.

3769. An airport's rotating beacon operated during daylight hours indicates

A— there are obstructions on the airport.
B— that weather in the Class D airspace is below basic VFR weather minimums.
C— the Air Traffic Control Tower is not in operation.

* * *

3771. A military air station can be identified by a rotating beacon that emits

A— white and green alternating flashes.
B— two quick, white flashes between green flashes.
C— green, yellow, and white flashes.

3772. How can a military airport be identified at night?

A— Alternate white and green light flashes.
B— Dual peaked (two quick) white flashes between green flashes.
C— White flashing lights with steady green at the same location.

3773. (Refer to figure 49.) That portion of the runway identified by the letter A may be used for

A— landing.
B— taxiing and takeoff.
C— taxiing and landing.

3774. (Refer to figure 49.) According to the airport diagram, which statement is true?

A— Runway 30 is equipped at position E with emergency arresting gear to provide a means of stopping military aircraft.
B— Takeoffs may be started at position A on Runway 12, and the landing portion of this runway begins at position B.
C— The takeoff and landing portion of Runway 12 begins at position B.

3775. (Refer to figure 49.) What is the difference between area A and area E on the airport depicted?

A— "A" may be used for taxi and takeoff; "E" may be used only as an overrun.
B— "A" may be used for all operations except heavy aircraft landings; "E" may be used only as an overrun.
C— "A" may be used only for taxiing; "E" may be used for all operations except landings.

3776. (Refer to figure 49.) Area C on the airport depicted is classified as a

A— stabilized area.
B— multiple heliport.
C— closed runway.

3777. (Refer to figure 50.) The arrows that appear on the end of the north/south runway indicate that the area

A— may be used only for taxiing.
B— is usable for taxiing, takeoff, and landing.
C— cannot be used for landing, but may be used for taxiing and takeoff.

3778. The numbers 9 and 27 on a runway indicate that the runway is oriented approximately

A— 009° and 027° true.
B— 090° and 270° true.
C— 090° and 270° magnetic.

3779. The vertical limit of Class C airspace normally above the primary airport is

A— 1,200 feet AGL.
B— 3,000 feet AGL.
C— 4,000 feet AGL.

3780. The normal radius of the outer area of Class C airspace is

A— 5 nautical miles.
B— 15 nautical miles.
C— 20 nautical miles.

3781. All operations within Class C airspace must be in

A— accordance with instrument flight rules.
B— compliance with ATC clearances and instructions.
C— an aircraft equipped with a 4096-code transponder with Mode C encoding capability.

3782. Under what condition may an aircraft operate from a satellite airport within Class C airspace?

A— The pilot must file a flight plan prior to departure.
B— The pilot must monitor ATC until clear of the Class C airspace.
C— The pilot must contact ATC as soon as practicable after takeoffs.

3783. Under what condition, if any, may pilots fly through a restricted area?

A— When flying on airways with an ATC clearance.
B— With the controlling agency's authorization.
C— Regulations do not allow this.

* * *

3785. What action should a pilot take when operating under VFR in a Military Operations Area (MOA)?

A— Obtain a clearance from the controlling agency prior to entering the MOA.
B— Operate only on the airways that transverse the MOA.
C— Exercise extreme caution when military activity is being conducted.

3786. Responsibility for collision avoidance in an alert area rests with

A— the controlling agency.
B— all pilots.
C— Air Traffic Control.

3787. The lateral dimensions of Class D airspace are based on

A— the number of airports that lie within the Class D airspace.
B— five statute miles from the geographical center of the primary airport.
C— the instrument procedures for which the controlled airspace is established.

3788. A non-tower satellite airport with the same Class D airspace as that designated for the primary airport requires radio communications be established and maintained with the

A— satellite airport's unicom.
B— associated Flight Service Station.
C— primary airport's control tower.

3789. Prior to entering an Airport Advisory Area, a pilot should

A— monitor ATIS for weather and traffic advisories.
B— contact approach control for vectors to the traffic pattern.
C— contact the local FSS for airport and traffic advisories.

* * *

3791. Automatic Terminal Information Service (ATIS) is the continuous broadcast of recorded information concerning

A— pilots of radar-identified aircraft whose aircraft is in dangerous proximity to terrain or to an obstruction.
B— nonessential information to reduce frequency congestion.
C— noncontrol information in selected high-activity terminal areas.

3792. An ATC radar facility issues the following advisory to a pilot flying on a heading of 090°:

"TRAFFIC 3 O'CLOCK, 2 MILES, WESTBOUND..."

Where should the pilot look for this traffic?

A— East.
B— South.
C— West.

3793. An ATC radar facility issues the following advisory to a pilot flying on a heading of 360°:

"TRAFFIC 10 O'CLOCK, 2 MILES, SOUTHBOUND..."

Where should the pilot look for this traffic?

A— Northwest.
B— Northeast.
C— Southwest.

3794. An ATC radar facility issues the following advisory to a pilot during a local flight:

"TRAFFIC 2 O'CLOCK, 5 MILES, NORTHBOUND..."

Where should the pilot look for this traffic?

A— Between directly ahead and 90° to the left.
B— Between directly behind and 90° to the right.
C— Between directly ahead and 90° to the right.

3795. An ATC radar facility issues the following advisory to a pilot flying north in a calm wind:

"TRAFFIC 9 O'CLOCK, 2 MILES, SOUTHBOUND..."

Where should the pilot look for this traffic?

A— South.
B— North.
C— West.

3796. Basic radar service in the terminal radar program is best described as

A— traffic advisories and limited vectoring to VFR aircraft.
B— mandatory radar service provided by the Automated Radar Terminal System (ARTS) program.
C— wind-shear warning at participating airports.

3797. From whom should a departing VFR aircraft request Stage II Terminal Radar Advisory Service during ground operations?

A— Clearance delivery.
B— Tower, just before takeoff.
C— Ground control, on initial contact.

3798. Class B or TRSA Service (earlier known as Stage III Service) in the terminal radar program provides

A— IFR separation (1,000 feet vertical and 3 miles lateral) between all aircraft.
B— warning to pilots when their aircraft are in unsafe proximity to terrain, obstructions, or other aircraft.
C— sequencing and separation for participating VFR aircraft.

3799. If radar traffic information is desired, which action should a pilot take prior to entering Class C airspace?

A— Contact approach control on the appropriate frequency.
B— Contact the tower and request permission to enter.
C— Contact the FSS for traffic advisories.

3800. When making routine transponder code changes, pilots should avoid inadvertent selection of which codes?

A— 0700, 1700, 7000.
B— 1200, 1500, 7000.
C— 7500, 7600, 7700.

3801. When operating under VFR below 18,000 feet MSL, unless otherwise authorized, what transponder code should be selected?

A— 1200.
B— 7600.
C— 7700.

3802. Unless otherwise authorized, if flying a transponder equipped aircraft, a recreational pilot should squawk which VFR code?

A— 1200.
B— 7600.
C— 7700.

3803. If Air Traffic Control advises that radar service is being terminated when the pilot is departing Class C airspace, the transponder should be set to code

A— 0000.
B— 1200.
C— 4096.

3804. If the aircraft's radio fails, what is the recommended procedure when landing at a controlled airport?

A— Observe the traffic flow, enter the pattern, and look for a light signal from the tower.
B— Enter a crosswind leg and rock the wings.
C— Flash the landing lights and cycle the landing gear while circling the airport.

3805. (Refer to figure 50.) Select the proper traffic pattern and runway for landing.

A— Left-hand traffic and Runway 18.
B— Right-hand traffic and Runway 18.
C— Left-hand traffic and Runway 22.

3806. (Refer to figure 50.) If the wind is as shown by the landing direction indicator, the pilot should land on

A— Runway 18 and expect a crosswind from the right.
B— Runway 22 directly into the wind.
C— Runway 36 and expect a crosswind from the right.

3807. (Refer to figure 51.) The segmented circle indicates that the airport traffic is

A— left-hand for Runway 35 and right-hand for Runway 17.
B— left-hand for Runway 17 and right-hand for Runway 35.
C— right-hand for Runway 9 and left-hand for Runway 27.

3808. (Refer to figure 51.) The traffic patterns indicated in the segmented circle have been arranged to avoid flights over an area to the

A— south of the airport.
B— north of the airport.
C— southeast of the airport.

3809. (Refer to figure 51.) The segmented circle indicates that a landing on Runway 26 will be with a

A— right-quartering headwind.
B— left-quartering headwind.
C— right-quartering tailwind.

3810. (Refer to figure 51.) Which runway and traffic pattern should be used as indicated by the wind cone in the segmented circle?

A— Right-hand traffic on Runway 8.
B— Right-hand traffic on Runway 17.
C— Left-hand traffic on Runway 35.

3811. After landing at a tower-controlled airport, when should the pilot contact ground control?

A— When advised by the tower to do so.
B— Prior to turning off the runway.
C— After reaching a taxiway that leads directly to the parking area.

3812. If instructed by ground control to taxi to Runway 9, the pilot may proceed

A— via taxiways and across runways to, but not onto, Runway 9.
B— to the next intersecting runway where further clearance is required.
C— via taxiways and across runways to Runway 9, where an immediate takeoff may be made.

3813. What ATC facility should the pilot contact to receive a special VFR departure clearance in Class D airspace?

A— Automated Flight Service Station.
B— Air Traffic Control Tower.
C— Air Route Traffic Control Center.

3814. What procedure is recommended when climbing or descending VFR on an airway?

A— Execute gentle banks, left and right for continuous visual scanning of the airspace.
B— Advise the nearest FSS of the altitude changes.
C— Fly away from the centerline of the airway before changing altitude.

3815. (Refer to figure 52.) If more than one cruising altitude is intended, which should be entered in block 7 of the flight plan?

A— Initial cruising altitude.
B— Highest cruising altitude.
C— Lowest cruising altitude.

3816. (Refer to figure 52.) What information should be entered in block 9 for a VFR day flight?

A— The name of the airport of first intended landing.
B— The name of destination airport if no stopover for more than 1 hour is anticipated.
C— The name of the airport where the aircraft is based.

3817. (Refer to figure 52.) What information should be entered in block 12 for a VFR day flight?

A— The estimated time en route plus 30 minutes.
B— The estimated time en route plus 45 minutes.
C— The amount of usable fuel on board expressed in time.

3818. How should a VFR flight plan be closed at the completion of the flight at a controlled airport?

A— The tower will automatically close the flight plan when the aircraft turns off the runway.
B— The pilot must close the flight plan with the nearest FSS or other FAA facility upon landing.
C— The tower will relay the instructions to the nearest FSS when the aircraft contacts the tower for landing.

3819. When activated, an emergency locator transmitter (ELT) transmits on

A— 118.0 and 118.8 MHz.
B— 121.5 and 243.0 MHz.
C— 123.0 and 119.0 MHz.

3820. When must the battery in an emergency locator transmitter (ELT) be replaced (or recharged if the battery is rechargeable)?

A— After one-half the battery's useful life.
B— During each annual and 100-hour inspection.
C— Every 24 calendar months.

3821. When may an emergency locator transmitter (ELT) be tested?

A— Anytime.
B— At 15 and 45 minutes past the hour.
C— During the first 5 minutes after the hour.

3822. Which procedure is recommended to ensure that the emergency locator transmitter (ELT) has not been activated?

A— Turn off the aircraft ELT after landing.
B— Ask the airport tower if they are receiving an ELT signal.
C— Monitor 121.5 before engine shutdown.

3823. Below FL180, en route weather advisories should be obtained from an FSS on

A— 122.0 MHz.
B— 122.1 MHz.
C— 123.6 MHz.

3824. Wingtip vortices are created only when an aircraft is

A— operating at high airspeeds.
B— heavily loaded.
C— developing lift.

3825. The greatest vortex strength occurs when the generating aircraft is

A— light, dirty, and fast.
B— heavy, dirty, and fast.
C— heavy, clean, and slow.

3826. Wingtip vortices created by large aircraft tend to

A— sink below the aircraft generating turbulence.
B— rise into the traffic pattern.
C— rise into the takeoff or landing path of a crossing runway.

3827. When taking off or landing at an airport where heavy aircraft are operating, one should be particularly alert to the hazards of wingtip vortices because this turbulence tends to

A— rise from a crossing runway into the takeoff or landing path.
B— rise into the traffic pattern area surrounding the airport.
C— sink into the flightpath of aircraft operating below the aircraft generating the turbulence.

3828. The wind condition that requires maximum caution when avoiding wake turbulence on landing is a

A— light, quartering headwind.
B— light, quartering tailwind.
C— strong headwind.

3829. When landing behind a large aircraft, the pilot should avoid wake turbulence by staying

A— above the large aircraft's final approach path and landing beyond the large aircraft's touchdown point.
B— below the large aircraft's final approach path and landing before the large aircraft's touchdown point.
C— above the large aircraft's final approach path and landing before the large aircraft's touchdown point.

3830. When departing behind a heavy aircraft, the pilot should avoid wake turbulence by maneuvering the aircraft

A— below and downwind from the heavy aircraft.
B— above and upwind from the heavy aircraft.
C— below and upwind from the heavy aircraft.

3831. Pilots flying over a national wildlife refuge are requested to fly no lower than

A— 1,000 feet AGL.
B— 2,000 feet AGL.
C— 3,000 feet AGL.

3832. Large accumulations of carbon monoxide in the human body result in

A— tightness across the forehead.
B— loss of muscular power.
C— an increased sense of well-being.

3833. What effect does haze have on the ability to see traffic or terrain features during flight?

A— Haze causes the eyes to focus at infinity.
B— The eyes tend to overwork in haze and do not detect relative movement easily.
C— All traffic or terrain features appear to be farther away than their actual distance.

3834. The most effective method of scanning for other aircraft for collision avoidance during daylight hours is to use

A— regularly spaced concentration on the 3-, 9-, and 12-o'clock positions.
B— a series of short, regularly spaced eye movements to search each 10-degree sector.
C— peripheral vision by scanning small sectors and utilizing offcenter viewing.

3835. Which technique should a pilot use to scan for traffic to the right and left during straight-and-level flight?

A— Systematically focus on different segments of the sky for short intervals.
B— Concentrate on relative movement detected in the peripheral vision area.
C— Continuous sweeping of the windshield from right to left.

3836. How can you determine if another aircraft is on a collision course with your aircraft?

A— The other aircraft will always appear to get larger and closer at a rapid rate.
B— The nose of each aircraft is pointed at the same point in space.
C— There will be no apparent relative motion between your aircraft and the other aircraft.

3837. An ATC clearance provides

A— priority over all other traffic.
B— adequate separation from all traffic.
C— authorization to proceed under specified traffic conditions in controlled airspace.

3838. (Refer to figure 53.) When approaching Lincoln Municipal from the west at noon for the purpose of landing, initial communications should be with

A— Lincoln Approach Control on 124.0 MHz.
B— Minneapolis Center on 128.75 MHz.
C— Lincoln Tower on 118.5 MHz.

3839. (Refer to figure 53.) Which type radar service is provided to VFR aircraft at Lincoln Municipal?

A— Sequencing to the primary Class C airport and standard separation.
B— Sequencing to the primary Class C airport and conflict resolution so that radar targets do not touch, or 1,000 feet vertical separation.
C— Sequencing to the primary Class C airport, traffic advisories, conflict resolution, and safety alerts.

3840. (Refer to figure 53.) What is the recommended communications procedure for landing at Lincoln Municipal during the hours when the tower is not in operation?

A— Monitor airport traffic and announce your position and intentions on 118.5 MHz.
B— Contact UNICOM on 122.95 MHz for traffic advisories.
C— Monitor ATIS for airport conditions, then announce your position on 122.95 MHz.

3841. (Refer to figure 53.) Where is Loup City Municipal located with relation to the city?

A— Northeast approximately 3 miles.
B— Northwest approximately 1 mile.
C— East approximately 10 miles.

3842. (Refer to figure 53.) Traffic patterns in effect at Lincoln Municipal are

A— to the right on Runway 17L and Runway 35L; to the left on Runway 17R and Runway 35R.
B— to the left on Runway 17L and Runway 35L; to the right on Runway 17R and Runway 35R.
C— to the right on Runways 14 – 32.

3843. The letters VHF/DF appearing in the Airport/Facility Directory for a certain airport indicate that

A— this airport is designated as an airport of entry.
B— the Flight Service Station has equipment with which to determine your direction from the station.
C— this airport has a direct-line phone to the Flight Service Station.

3844. Which statement best defines hypoxia?

A— A state of oxygen deficiency in the body.
B— An abnormal increase in the volume of air breathed.
C— A condition of gas bubble formation around the joints or muscles.

3845. Rapid or extra deep breathing while using oxygen can cause a condition known as

A— hyperventilation.
B— aerosinusitis.
C— aerotitis.

3846. Which would most likely result in hyperventilation?

A— Emotional tension, anxiety, or fear.
B— The excessive consumption of alcohol.
C— An extremely slow rate of breathing and insufficient oxygen.

3847. A pilot should be able to overcome the symptoms or avoid future occurrences of hyperventilation by

A— closely monitoring the flight instruments to control the airplane.
B— slowing the breathing rate, breathing into a bag, or talking aloud.
C— increasing the breathing rate in order to increase lung ventilation.

3848. Susceptibility to carbon monoxide poisoning increases as

A— altitude increases.
B— altitude decreases.
C— air pressure increases.

3849. What preparation should a pilot make to adapt the eyes for night flying?

A— Wear sunglasses after sunset until ready for flight.
B— Avoid red lights at least 30 minutes before the flight.
C— Avoid bright white lights at least 30 minutes before the flight.

3850. The danger of spatial disorientation during flight in poor visual conditions may be reduced by

A— shifting the eyes quickly between the exterior visual field and the instrument panel.
B— having faith in the instruments rather than taking a chance on the sensory organs.
C— leaning the body in the opposite direction of the motion of the aircraft.

3851. A state of temporary confusion resulting from misleading information being sent to the brain by various sensory organs is defined as

A— spatial disorientation.
B— hyperventilation.
C— hypoxia.

3852. Pilots are more subject to spatial disorientation if

A— they ignore the sensations of muscles and inner ear.
B— body signals are used to interpret flight attitude.
C— eyes are moved often in the process of cross-checking the flight instruments.

3853. If a pilot experiences spatial disorientation during flight in a restricted visibility condition, the best way to overcome the effect is to

A— rely upon the aircraft instrument indications.
B— concentrate on yaw, pitch, and roll sensations.
C— consciously slow the breathing rate until symptoms clear and then resume normal breathing rate.

3854. FAA advisory circulars containing subject matter specifically related to Airmen are issued under which subject number?

A— 60.
B— 70.
C— 90.

3855. FAA advisory circulars containing subject matter specifically related to Airspace are issued under which subject number?

A— 60.
B— 70.
C— 90.

3856. FAA advisory circulars containing subject matter specifically related to Air Traffic Control and General Operations are issued under which subject number?

A— 60.
B— 70.
C— 90.

* * *

APPENDIX 1
DEPARTMENT OF TRANSPORTATION
FEDERAL AVIATION ADMINISTRATION

SUBJECT MATTER KNOWLEDGE CODES

FAR 1   **Definitions and Abbreviations**

   A01   General Definitions
   A02   Abbreviations and Symbols

FAR 23   **Airworthiness Standards: Normal, Utility, and Acrobatic Category Aircraft**

   A10   General

FAR 25   **Airworthiness Standards: Transport Category Airplanes**

   A03   General
   A04   Flight
   A05   Structure
   A06   Design and Construction
   A07   Powerplant
   A08   Equipment
   A09   Operating Limitations and Information

FAR 43   **Maintenance, Preventive Maintenance, Rebuilding, and Alteration**

   A15   General
   A16   Appendixes

FAR 61   **Certification: Pilots and Flight Instructors**

   A20   General
   A21   Aircraft Ratings and Special Certificates
   A22   Student Pilots
   A23   Private Pilots

   A24   Commercial Pilots
   A25   Airline Transport Pilots
   A26   Flight Instructors
   A27   Appendix A: Practical Test Requirements for Airline Transport Pilot Certificates and Associated Class and Type Ratings
   A28   Appendix B: Practical Test Requirements for Rotorcraft Airline Transport Pilot Certificates with a Helicopter Class Rating and Associated Type Ratings
   A29   Recreational Pilot

FAR 63   **Certification: Flight Crewmembers Other Than Pilots**

   A30   General
   A31   Flight Engineers
   A32   Flight Navigators

FAR 65   **Certification: Airmen Other Than Flight Crewmembers**

   A40   General
   A41   Aircraft Dispatchers

FAR 71   **Designation of Federal Airways, Area Low Routes, Controlled Airspace, and Reporting Points**

   A60   General
   A61   Airport Radar Service Areas
   A64   Control Areas and Extensions

FAR 91	**General Operating Rules**		D08	Aircraft Requirements
			D09	Airplane Performance Operating Limitations
B07	General		D10	Special Airworthiness Requirements
B08	Flight Rules - General		D11	Instrument and Equipment Requirements
B09	Visual Flight Rules		D12	Maintenance, Preventive Maintenance, and Alterations
B10	Instrument Flight Rules			
B11	Equipment, Instrument, and Certification Requirements		D13	Airman and Crewmember Requirements
			D14	Training Program
B12	Special Flight Operations		D15	Crewmember Qualifications
B13	Maintenance, Preventive Maintenance, and Alterations		D16	Aircraft Dispatcher Qualifications and Duty Time Limitations: Domestic and Flag Air Carriers
B14	Large and Turbine-powered Multiengine Airplanes			
			D17	Flight Time Limitations and Rest Requirements: Domestic Air Carriers
B15	Additional Equipment and Operating Requirements for Large and Transport Category Aircraft		D18	Flight Time Limitations: Flag Air Carriers
			D19	Flight Time Limitations: Supplemental Air Carriers and Commercial Operators
B16	Appendix A - Category II Operations: Manual, Instruments, Equipment, and Maintenance		D20	Flight Operations
			D21	Dispatching and Flight Release Rules
			D22	Records and Reports
FAR 97	**Standard Instrument Approach Procedures**		D23	Crewmember Certificate: International
			D24	Special Federal Aviation Regulation SFAR No. 14
B97	General			
			FAR 125	**Certification and Operations: Airplanes Having a Seating Capacity of 20 or More Passengers or a Maximum Payload Capacity of 6,000 Pounds or More**
FAR 108	**Airplane Operator Security**			
C10	General			
			D30	General
FAR 121	**Certification and Operations: Domestic, Flag and Supplemental Air Carriers and Commercial Operators of Large Aircraft**		D31	Certification Rules and Miscellaneous Requirements
			D32	Manual Requirements
			D33	Airplane Requirements
D01	General		D34	Special Airworthiness Requirements
D02	Certification Rules for Domestic and Flag Air Carriers		D35	Instrument and Equipment Requirements
			D36	Maintenance
D03	Certification Rules for Supplemental Air Carriers and Commercial Operators		D37	Airman and Crewmember Requirements
			D38	Flight Crewmember Requirements
D04	Rules Governing all Certificate Holders Under This Part		D39	Flight Operations
			D40	Flight Release Rules
D05	Approval of Routes: Domestic and Flag Air Carriers		D41	Records and Reports
D06	Approval of Areas and Routes for Supplemental Air Carriers and Commercial Operators			
D07	Manual Requirements			

FAR 135 **Air Taxi Operators and Commercial Operators**

E01 General
E02 Flight Operations
E03 Aircraft and Equipment
E04 VFR/IFR Operating Limitations and Weather Requirements
E05 Flight Crewmember Requirements
E06 Flight Crewmember Flight Time Limitations and Rest Requirements
E07 Crewmember Testing Requirements
E08 Training
E09 Airplane Performance Operating Limitations
E10 Maintenance, Preventive Maintenance, and Alterations
E11 Appendix A: Additional Airworthiness Standards for 10 or More Passenger Airplanes
E12 Special Federal Aviation Regulations SFAR No. 36
E13 Special Federal Aviation Regulations SFAR No. 38

US HMR 172 Hazardous Materials Table

F02 General

US HMR 175 Materials Transportation Bureau Hazardous Materials Regulations (HMR)

G01 General Information and Regulations
G02 Loading, Unloading, and Handling
G03 Specific Regulation Applicable According to Classification of Material

NTSB 830 Rules Pertaining to the Notification and Reporting of Aircraft Accidents or Incidents and Overdue Aircraft, and Preservation of Aircraft Wreckage, Mail, Cargo, and Records

G10 General
G11 Initial Notification of Aircraft Accidents, Incidents, and Overdue Aircraft
G12 Preservation of Aircraft Wreckage, Mail, Cargo, and Records
G13 Reporting of Aircraft Accidents, Incidents, and Overdue Aircraft

AC 61-23 **Pilot's Handbook of Aeronautical Knowledge**

H01 Principles of Flight
H02 Airplanes and Engines
H03 Flight Instruments
H04 Airplane Performance
H05 Weather
H06 Basic Calculations Using Navigational Computers or Electronic Calculators
H07 Navigation
H09 Appendix 1: Obtaining FAA Publications

AC 91-23 **Pilot's Weight and Balance Handbook**

H10 Weight and Balance Control
H11 Terms and Definitions
H12 Empty Weight Center of Gravity
H13 Index and Graphic Limits
H14 Change of Weight
H15 Control of Loading — General Aviation
H16 Control of Loading — Large Aircraft

AC 60-14 **Aviation Instructor's Handbook**

H20 The Learning Process
H21 Human Behavior
H22 Effective Communication
H23 The Teaching Process
H24 Teaching Methods
H25 The Instructor as a Critic
H26 Evaluation
H27 Instructional Aids
H30 Flight Instructor Characteristics and Responsibilities
H31 Techniques of Flight Instruction
H32 Planning Instructional Activity

AC 61-21   **Flight Training Handbook**

H50   Introduction to Flight Training
H51   Introduction to Airplanes and Engines
H52   Introduction to the Basics of Flight
H53   The Effect and Use of Controls
H54   Ground Operations
H55   Basic Flight Maneuvers
H56   Airport Traffic Patterns and Operations
H57   Takeoffs and Departure Climbs
H58   Landing Approaches and Landings
H59   Faulty Approaches and Landings
H60   Proficiency Flight Maneuvers
H61   Cross-Country Flying
H62   Emergency Flight by Reference to Instruments
H63   Night Flying
H64   Seaplane Operations
H65   Transition to Other Airplanes
H66   Principles of Flight and Performance Characteristics

AC 61-13   **Basic Helicopter Handbook**

H70   General Aerodynamics
H71   Aerodynamics of Flight
H72   Loads and Load Factors
H73   Function of the Controls
H74   Other Helicopter Components and Their Functions
H75   Introduction to the Helicopter Flight Manual
H76   Weight and Balance
H77   Helicopter Performance
H78   Some Hazards of Helicopter Flight
H79   Precautionary Measures and Critical Conditions
H80   Helicopter Flight Maneuvers
H81   Confined Area, Pinnacle, and Ridgeline Operations
H82   Glossary

Gyroplane Flight Training Manual — McCulloch

H90   Gyroplane Systems
H91   Gyroplane Terms
H92   Use of Flight Controls (Gyroplane)
H93   Fundamental Maneuvers of Flight (Gyroplane)
H94   Basic Flight Maneuvers (Gyroplane)

AC 61-27   **Instrument Flying Handbook**

I01   Training Considerations
I02   Instrument Flying: Coping with Illusions in Flight
I03   Aerodynamic Factors Related to Instrument Flying
I04   Basic Flight Instruments
I05   Attitude Instrument Flying — Airplanes
I06   Attitude Instrument Flying — Helicopters
I07   Electronic Aids to Instrument Flying
I08   Using the Navigation Instruments
I09   Radio Communications Facilities and Equipment
I10   The Federal Airways System and Controlled Airspace
I11   Air Traffic Control
I12   ATC Operations and Procedures
I13   Flight Planning
I14   Appendix: Instrument Instructor Lesson Guide — Airplanes
I15   Segment of En Route Low Altitude Chart

AC 00-6   **Aviation Weather**

I20   The Earth's Atmosphere
I21   Temperature
I22   Atmospheric Pressure and Altimetry
I23   Wind
I24   Moisture, Cloud Formation, and Precipitation
I25   Stable and Unstable Air
I26   Clouds
I27   Air Masses and Fronts
I28   Turbulence
I29   Icing
I30   Thunderstorms
I31   Common IFR Producers
I32   High Altitude Weather
I33   Arctic Weather

I34	Tropical Weather	J21	Emergency Procedures — General
I35	Soaring Weather	J22	Emergency Services Available to Pilots
I36	Glossary of Weather Terms	J23	Distress and Urgency Procedures
		J24	Two-Way Radio Communications Failure
AC 00-45	**Aviation Weather Services**	J25	Meteorology
		J26	Altimeter Setting Procedures
I40	The Aviation Weather Service Program	J27	Wake Turbulence
I41	Surface Aviation Weather Reports	J28	Bird Hazards, and Flight Over National Refuges, Parks, and Forests
I42	Pilot and Radar Reports and Satellite Pictures	J29	Potential Flight Hazards
I43	Aviation Weather Forecasts	J30	Safety, Accident, and Hazard Reports
I44	Surface Analysis Chart	J31	Fitness for Flight
I45	Weather Depiction Chart	J32	Type of Charts Available
I46	Radar Summary Chart	J33	Pilot Controller Glossary
I47	Significant Weather Prognostics	J34	Airport/Facility Directory
I48	Winds and Temperatures Aloft	J35	En Route Low Altitude Chart
I49	Composite Moisture Stability Chart	J36	En Route High Altitude Chart
I50	Severe Weather Outlook Chart	J37	Sectional Chart
I51	Constant Pressure Charts	J40	Standard Instrument Departure (SID) Chart
I52	Tropopause Data Chart	J41	Standard Terminal Arrival (STAR) Chart
I53	Tables and Conversion Graphs	J42	Instrument Approach Procedures
		J43	Helicopter Route Chart
AIM	**Aeronautical Information Manual**		
		AC 67-2	**Medical Handbook for Pilots**
J01	Air Navigation Radio Aids		
J02	Radar Services and Procedures	J52	Hypoxia
J03	Airport Lighting Aids	J53	Hyperventilation
J04	Air Navigation and Obstruction Lighting	J55	The Ears
J05	Airport Marking Aids	J56	Alcohol
J06	Airspace — General	J57	Drugs and Flying
J07	Uncontrolled Airspace	J58	Carbon Monoxide
J08	Controlled Airspace	J59	Vision
J09	Special Use Airspace	J60	Night Flight
J10	Other Airspace Areas	J61	Cockpit Lighting
J11	Service Available to Pilots	J62	Disorientation (Vertigo)
J12	Radio Communications Phraseology and Techniques	J63	Motion Sickness
J13	Airport Operations	J64	Fatigue
J14	ATC Clearance/Separations	J65	Noise
J15	Preflight	J66	Age
J16	Departure Procedures	J67	Some Psychological Aspects of Flying
J17	En Route Procedures	J68	The Flying Passenger
J18	Arrival Procedures		
J19	Pilot/Controller Roles and Responsibilities		
J20	National Security and Interception Procedures		

ADDITIONAL ADVISORY CIRCULARS

K01   AC 00-24, Thunderstorms

K02   AC 00-30, Rules of Thumb for Avoiding or Minimizing Encounters with Clear Air Turbulence

K03   AC 00-34, Aircraft Ground Handling and Servicing

K04   AC 00-54, Pilot Wind Shear Guide

K11   AC 20-34D, Prevention of Retractable Landing Gear Failure

K12   AC 20-32, Carbon Monoxide (CO) Contamination in Aircraft — Detection and Prevention

K13   AC 20-43, Aircraft Fuel Control

K20   AC 20-103, Aircraft Engine Crankshaft Failure

L10   AC 61-92, Use of Distractions During Pilot Certification Flight Tests

L34   AC 90-48, Pilots' Role in Collision Avoidance

L42   AC 90-87, Helicopter Dynamic Rollover

L50   AC 91-6, Water, Slush, and Snow on the Runway

L51   AC 61-107, Operations of Aircraft at Altitudes Above 25,000 feet MSL and/or Mach Numbers (Mmo) Greater Than .75

L52   AC 91-13, Cold Weather Operation of Aircraft

L53   AC 91-14, Altimeter Setting Sources

L57   AC 91-43, Unreliable Airspeed Indications

L59   AC 91-46, Gyroscopic Instruments — Good Operating Practices

L61   AC 91-50, Importance of Transponder Operation and Altitude Reporting

L62   AC 91-51, Airplane Deice and Anti-Ice Systems

L80   AC 103-4, Hazard Associated with Sublimation of Solid Carbon Dioxide (Dry Ice) Aboard Aircraft (This AC is not listed as a current document

M01   AC 120-12, Private Carriage Versus Common Carriage of Persons or Property

M02   AC 120-27, Aircraft Weight and Balance Control

M13   AC 121-195, Alternate Operational Landing Distances for Wet Runways; Turbojet Powered Transport Category Airplanes

M51   AC 20-117, Hazards Following Ground Deicing and Ground Operations in Conditions Conducive to Aircraft Icing

M52   AC 00-2, Advisory Circular Checklist

American Soaring Handbook — Soaring Society of America

N01   A History of American Soaring
N02   Training
N03   Ground Launch
N04   Airplane Tow
N05   Meteorology
N06   Cross-Country and Wave Soaring
N07   Instruments and Oxygen
N08   Radio, Rope, and Wire
N09   Aerodynamics
N10   Maintenance and Repair

Soaring Flight Manual — Jeppesen-Sanderson, Inc.

N20   Sailplane Aerodynamics
N21   Performance Considerations
N22   Flight Instruments
N23   Weather for Soaring
N24   Medical Factors
N25   Flight Publications and Airspace
N26   Aeronautical Charts and Navigation
N27   Computations for Soaring
N28   Personal Equipment
N29   Preflight and Ground Operations
N30   Aerotow Launch Procedures
N31   Ground Launch Procedures
N32   Basic Flight Maneuvers and Traffic
N33   Soaring Techniques
N34   Cross-Country Soaring

Taming The Gentle Giant — Taylor Publishing

O01   Design and Construction of Balloons
O02   Fuel Source and Supply
O03   Weight and Temperature

O04	Flight Instruments
O05	Balloon Flight Tips
O06	Glossary

Balloon Federation Of America — Flight Instructor Manual

O10	Flight Instruction Aids
O11	Human Behavior and Pilot Proficiency
O12	The Flight Check and the Designated Examiner

Balloon Federation of America — Propane Systems

O20	Propane Glossary
O21	Chemical and Physical Systems
O22	Cylinders
O23	Lines and Fittings
O24	Valves
O25	Regulators
O26	Burners
O27	Propane Systems — Schematics
O28	Propane References

Balloon Federation of America — Powerline Excerpts

O30	Excerpts

Goodyear Airship Operations Manual

P01	Buoyancy
P02	Aerodynamics
P03	Free Ballooning
P04	Aerostatics
P05	Envelope
P06	Car
P07	Powerplant
P08	Airship Ground Handling
P09	Operating Instructions
P10	History

Aerodynamics For Naval Aviators, NAVWEPS 00-80T-80

R01	Wing and Airfoil Forces
R02	Planform Effects and Airplane Drag

R10	Required Thrust and Power
R11	Available Thrust and Power
R12	Items of Airplane Performance
R21	General Concepts and Supersonic Flow Patterns
R22	Configuration Effects
R31	Definitions
R32	Longitudinal Stability and Control
R33	Directional Stability and Control
R34	Lateral Stability and Control
R35	Miscellaneous Stability Problems
R40	General Definitions and Structural Requirements
R41	Aircraft Loads and Operating Limitations
R50	Application of Aerodynamics to Specific Problems of Flying

AC 65-9A Airframe and Powerplant Mechanics General Handbook

S01	Mathematics
S02	Aircraft Drawings
S03	Aircraft Weight and Balance
S04	Fuels and Fuel Systems
S05	Fluid Lines and Fittings
S06	Aircraft Hardware, Materials, and Processes
S07	Physics
S08	Basic Electricity
S09	Aircraft Generators and Motors
S10	Inspection Fundamentals
S11	Ground Handling, Safety, and Support Equipment

AC 65-12A Airframe and Powerplant Mechanics Powerplant Handbook

S12	Theory and Construction of Aircraft Engines
S13	Induction and Exhaust Systems
S14	Engine Fuel and Metering Systems
S15	Engine Ignition and Electrical Systems
S16	Engine Starting Systems
S17	Lubrication and Cooling Systems
S18	Propellers
S19	Engine Fire Protection Systems
S20	Engine Maintenance and Operation

AC 65-15A   **Airframe and Powerplant Mechanics Airframe Handbook**

S21   Aircraft Structures
S22   Assembly and Rigging
S23   Aircraft Structural Repairs
S24   Ice and Rain Protection
S25   Hydraulic and Pneumatic Power Systems
S26   Landing Gear Systems
S27   Fire Protection Systems
S28   Aircraft Electrical Systems
S29   Aircraft Instrument Systems
S30   Communications and Navigation Systems
S31   Cabin Atmosphere Control Systems

EA-ITP-GB   **Aviation Technician Integrated Training Program General Section Textbook — International Aviation Publishers (IAP), Inc.**

S32   Mathematics and Physics
S33   Basic Electricity - DC
S34   Basic Electricity - AC
S35   Aircraft Drawings
S36   Weight and Balance
S37   Fluid Lines and Fittings
S38   Aircraft Hardware
S39   Aircraft Structure Materials
S40   Corrosion and Its Control
S41   Ground Handling and Servicing
S42   Maintenance Publications
S43   Federal Aviation Regulations
S44   Maintenance Forms and Records

EA-ITP-P   **Aviation Technician Integrated Training Program Powerplant Section Textbook — IAP, Inc.**

S45   Reciprocating Engine Theory
S46   Reciprocating Engine Maintenance and Operation
S47   Turbine Engine Theory
S48   Turbine Engine Maintenance and Operation
S49   Engine Ignition Systems

S50   Powerplant Electrical Systems
S51   Powerplant Electrical Installation
S52   Powerplant Instrument Systems
S53   Fire Protection Systems
S54   Aircraft Fuel Metering Systems
S55   Engine Induction Systems
S56   Engine Cooling Systems
S57   Engine Exhaust Systems
S58   Engine Starting Systems
S59   Engine Lubrication Systems
S60   Propellers

EA-ITP-AB   **Aviation Technician Integrated Training Program Airframe Section Textbook — IAP, Inc.**

S61   Aircraft Structures
S62   Sheet Metal Structural Repair
S63   Aircraft Fabric Covering
S64   Aircraft Assembly and Rigging
S65   Aircraft Electrical Systems
S66   Aircraft Hydraulic and Pneumatic Power Systems
S67   Aircraft Cabin Atmosphere Control Systems
S68   Aircraft Instrument Systems
S69   Aircraft Fuel Systems

EA-TEP-2   **Aircraft Gas Turbine Powerplants — IAP, Inc.**

S70   History of Turbine Engine Development
S71   Jet Propulsion Theory
S72   Turbine Engine Design and Construction
S73   Engine Familiarization
S74   Inspection and Maintenance
S75   Lubrication Systems
S76   Fuel Systems
S77   Compressor Anti-Stall Systems
S78   Anti-Icing Systems
S79   Starter Systems
S80   Ignition Systems
S81   Engine Instrument Systems
S82   Fire/Overheat Detection and Extinguishing Systems for Turbine Engines
S83   Engine Operation

EA-APC   Aircraft Propellers and Controls — IAP, Inc.

S84   Introduction to Propellers
S85   FAR's and Propellers
S86   Propeller Theory
S87   Fixed-Pitch Propellers and Propeller Blades
S88   Propeller Installations
S89   Reversing Propeller Systems
S90   Propeller Auxiliary Systems

EA-BAL   Aircraft Weight and Balance — IAP, Inc.

S91   Theory of Weight and Balance
S92   Data Investigation
S93   Weighing the Aircraft
S94   Aircraft Loading

The Aircraft Gas Turbine Engine and Its Operation — United Technologies Corporation, Pratt Whitney

T01   Gas Turbine Engine Fundamentals
T02   Gas Turbine Engine Terms
T03   Gas Turbine Engine Components
T04   Gas Turbine Engine Operation
T05   Operational Characteristics of Jet Engines
T06   Gas Turbine Engine Performance

Aircraft Powerplants — McGraw-Hill

T07   Aircraft Powerplant Classification and Progress
T08   Reciprocating-Engine Construction and Nomenclature
T09   Internal-Combustion Engine Theory and Performance
T10   Lubricants and Lubricating Systems
T11   Induction Systems, Superchargers, Turbochargers, and Exhaust Systems
T12   Basic Fuel Systems and Carburetors
T13   Fuel Injection Systems
T14   Reciprocating-Engine Ignition and Starting Systems
T15   Operation, Inspection, Maintenance, and Troubleshooting of Reciprocating Engines
T16   Reciprocating-Engine Overhaul Practices
T17   Gas Turbine Engine: Theory, Construction, and Nomenclature
T18   Gas Turbine Engine: Fuels and Fuel Systems
T19   Turbine-Engine Lubricants and Lubricating Systems
T20   Ignition and Starting Systems of Gas-Turbine Engines
T21   Turbofan Engines
T22   Turboprop Engines
T23   Turboshaft Engines
T24   Gas-Turbine Operation, Inspection, Troubleshooting, Maintenance, and Overhaul
T25   Propeller Theory, Nomenclature, and Operation
T26   Turbopropellers and Control Systems
T27   Propeller Installation, Inspection, and Maintenance
T28   Engine Control System
T29   Engine Indicating and Warning Systems

EA-ATD-2   Aircraft Technical Dictionary — IAP, Inc.

T30   Definitions

Aircraft Basic Science — McGraw-Hill

T31   Fundamentals of Mathematics
T32   Science Fundamentals
T33   Basic Aerodynamics
T34   Airfoils and their Applications
T35   Aircraft in Flight
T36   Aircraft Drawings
T37   Weight and Balance
T38   Aircraft Materials
T39   Fabrication Techniques and Processes
T40   Aircraft Hardware
T41   Aircraft Fluid Lines and their Fittings
T42   Federal Aviation Regulations and Publications
T43   Ground Handling and Safety
T44   Aircraft Inspection and Servicing

Aircraft Maintenance and Repair — McGraw-Hill

T45	Aircraft Systems
T46	Aircraft Hydraulic and Pneumatic Systems
T47	Aircraft Landing Gear Systems
T48	Aircraft Fuel Systems
T49	Environmental Systems
T50	Aircraft Instruments and Instrument Systems
T51	Auxiliary Systems
T52	Assembly and Rigging

EA-363   **Transport Category Aircraft Systems — IAP, Inc.**

T53	Types, Design Features and Configurations of Transport Aircraft
T54	Auxiliary Power Units, Pneumatic, and Environmental Control Systems
T55	Anti-Icing Systems and Rain Protection
T56	Electrical Power Systems
T57	Flight Control Systems
T58	Fuel Systems
T59	Hydraulic Systems
T60	Oxygen Systems
T61	Warning and Fire Protection Systems
T62	Communications, Instruments, and Navigational Systems
T63	Miscellaneous Aircraft Systems and Maintenance Information

Aircraft Electricity and Electronics — McGraw-Hill

T64	Fundamentals of Electricity
T65	Magnetism and Electromagnetism
T66	Capacitors and Inductors
T67	Alternating Current
T68	Electric Measuring Instruments
T69	Batteries
T70	Generator Theory
T71	DC Generators, Controls, and Systems
T72	AC Generators, Controls, and Systems
T73	Electric Motors
T74	Installation and Maintenance of Electrical Systems
T75	Principles of Electronics

T76	Radio Transmitters and Receivers
T77	Electric Instruments
T78	Electronic Control Systems
T79	Weather Warning Systems
T80	Communication and Navigation Systems
T81	Automatic Pilots and Landing Systems

FAA Accident Prevention Program Bulletins

V01	FAA-P-8740-2, Density Altitude
V02	FAA-P-8740-5, Weight and Balance
V03	FAA-P-8740-12, Thunderstorms
V04	FAA-P-8740-19, Flying Light Twins Safely
V05	FAA-P-8740-23, Planning your Takeoff
V06	FAA-P-8740-24, Tips on Winter Flying
V07	FAA-P-8740-25, Always Leave Yourself an Out
V08	FAA-P-8740-30, How to Obtain a Good Weather Briefing
V09	FAA-P-8740-40, Wind Shear
V10	FAA-P-8740-41, Medical Facts for Pilots
V11	FAA-P-8740-44, Impossible Turns
V12	FAA-P-8740-48, On Landings, Part I
V13	FAA-P-8740-49, On Landings, Part II
V14	FAA-P-8740-50, On Landings, Part III
V15	FAA-P-8740-51, How to Avoid a Midair Collision
V16	FAA-P-8740-52, The Silent Emergency

EA-338   **Flight Theory for Pilots — IAP, Inc.**

W01	Introduction
W02	Air Flow and Airspeed Measurement
W03	Aerodynamic Forces on Airfoils
W04	Lift and Stall
W05	Drag
W06	Jet Aircraft Basic Performance
W07	Jet Aircraft Applied Performance
W08	Prop Aircraft Basic Performance
W09	Prop Aircraft Applied Performance
W10	Helicopter Aerodynamics
W11	Hazards of Low Speed Flight
W12	Takeoff Performance
W13	Landing Performance
W14	Maneuvering Performance
W15	Longitudinal Stability and Control

| W16 | Directional and Lateral Stability and Control |
| W17 | High Speed Flight |

Fly the Wing — Iowa State University Press/Ames

X01	Basic Aerodynamics
X02	High-Speed Aerodynamics
X03	High-Altitude Machs
X04	Approach Speed Control and Target Landings
X05	Preparation for Flight Training
X06	Basic Instrument Scan
X07	Takeoffs
X08	Rejected Takeoffs
X09	Climb, Cruise, and Descent
X10	Steep Turns
X11	Stalls
X12	Unusual Attitudes
X14	Maneuvers At Minimum Speed
X15	Landings: Approach Technique and Performance
X16	ILS Approaches
X17	Missed Approaches and Rejected Landings
X18	Category II and III Approaches
X19	Nonprecision and Circling Approaches
X20	Weight and Balance
X21	Flight Planning
X22	Icing
X23	Use of Anti-ice and Deice
X24	Winter Operation
X25	Thunderstorm Flight

NOTE: AC 00-2, Advisory Circular Checklist, transmits the status of all FAA advisory circulars (AC's), as well as FAA internal publications and miscellaneous flight information such as AIM, Airport/Facility Directory, written test question books, practical test standards, and other material directly related to a certificate or rating. To obtain a free copy of the AC 00-2, send your request to:

U.S. Department of Transportation
Utilization and Storage Section, M-443.2
Washington, DC 20590

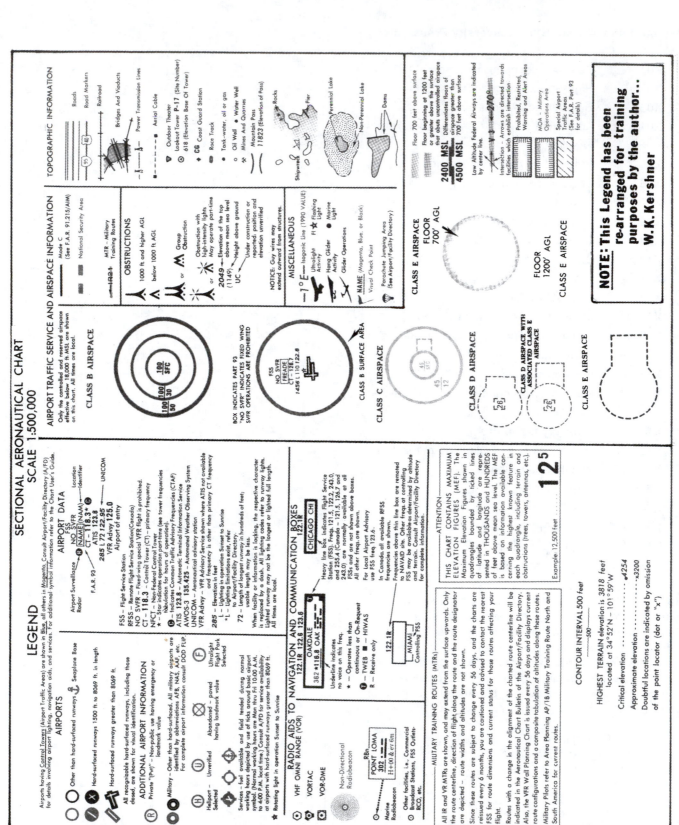

LEGEND 1.—Sectional Aeronautical Chart. The color presentation of this legend follows page 27-94.

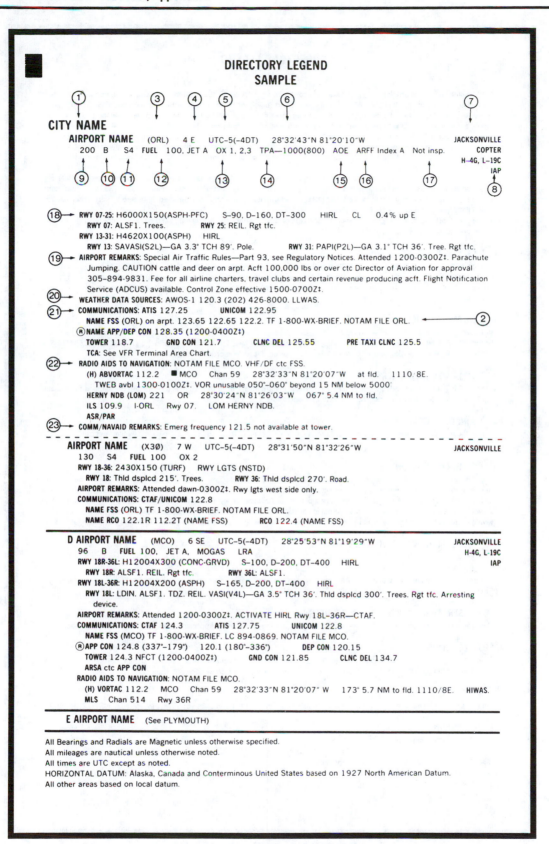

LEGEND 2.—Airport/Facility Directory.

DIRECTORY LEGEND

LEGEND

This Directory is an alphabetical listing of data on record with the FAA on all airports that are open to the public, associated terminal control facilities, air route traffic control centers and radio aids to navigation within the conterminous United States, Puerto Rico and the Virgin Islands. Airports are listed alphabetically by associated city name and cross referenced by airport name. Facilities associated with an airport, but with a different name, are listed individually under their own name, as well as under the airport with which they are associated.

The listing of an airport in this directory merely indicates the airport operator's willingness to accommodate transient aircraft, and does not represent that the facility conforms with any Federal or local standards, or that it has been approved for use on the part of the general public.

The information on obstructions is taken from reports submitted to the FAA. It has not been verified in all cases. Pilots are cautioned that objects not indicated in this tabulation (or on charts) may exist which can create a hazard to flight operation. Detailed specifics concerning services and facilities tabulated within this directory are contained in Airman's Information Manual, Basic Flight Information and ATC Procedures.

The legend items that follow explain in detail the contents of this Directory and are keyed to the circled numbers on the sample on the preceding page.

① CITY/AIRPORT NAME

Airports and facilities in this directory are listed alphabetically by associated city and state. Where the city name is different from the airport name the city name will appear on the line above the airport name. Airports with the same associated city name will be listed alphabetically by airport name and will be separated by a dashed rule line. All others will be separated by a solid rule line.

② NOTAM SERVICE

All public use landing areas are provided NOTAM "D" (distant dissemination) and NOTAM "L" (local dissemination) service. Airport NOTAM file identifier is shown following the associated FSS data for individual airports, e.g. "NOTAM FILE IAD". See AIM, Basic Flight Information and ATC Procedures for detailed description of NOTAM's.

③ LOCATION IDENTIFIER

A three or four character code assigned to airports. These identifiers are used by ATC in lieu of the airport name in flight plans, flight strips and other written records and computer operations.

④ AIRPORT LOCATION

Airport location is expressed as distance and direction from the center of the associated city in nautical miles and cardinal points, i.e., 4 NE.

⑤ TIME CONVERSION

Hours of operation of all facilities are expressed in Coordinated Universal Time (UTC) and shown as "Z" time. The directory indicates the number of hours to be subtracted from UTC to obtain local standard time and local daylight saving time UTC–5(–4DT). The symbol ‡ indicates that during periods of Daylight Saving Time effective hours will be one hour earlier than shown. In those areas where daylight saving time is not observed that (–4DT) and ‡ will not be shown. All states observe daylight savings time except Arizona and that portion of Indiana in the Eastern Time Zone and Puerto Rico and the Virgin Islands.

⑥ GEOGRAPHIC POSITION OF AIRPORT

⑦ CHARTS

The Sectional Chart and Low and High Altitude Enroute Chart and panel on which the airport or facility is located. Helicopter Chart locations will be indicated as, i.e., COPTER.

⑧ INSTRUMENT APPROACH PROCEDURES

IAP indicates an airport for which a prescribed (Public Use) FAA Instrument Approach Procedure has been published.

⑨ ELEVATION

Elevation is given in feet above mean sea level and is the highest point on the landing surface. When elevation is sea level it will be indicated as (00). When elevation is below sea level a minus (–) sign will precede the figure.

⑩ ROTATING LIGHT BEACON

B indicates rotating beacon is available. Rotating beacons operate dusk to dawn unless otherwise indicated in AIRPORT REMARKS.

⑪ SERVICING

S1: Minor airframe repairs.
S2: Minor airframe and minor powerplant repairs.
S3: Major airframe and minor powerplant repairs.
S4: Major airframe and major powerplant repairs.

LEGEND 3.—Airport/Facility Directory.

DIRECTORY LEGEND

⑫ FUEL

CODE	FUEL
80	Grade 80 gasoline (Red)
100	Grade 100 gasoline (Green)
100LL	100LL gasoline (low lead) (Blue)
115	Grade 115 gasoline
A	Jet A—Kerosene freeze point–40° C.
A1	Jet A-1—Kerosene freeze point–50°C.
A1+	Jet A-1—Kerosene with icing inhibitor, freeze point–50° C.
MOGAS	Automobile gasoline which is to be used as aircraft fuel.

CODE	FUEL
B	Jet B—Wide-cut turbine fuel, freeze point–50° C.
B+	Jet B—Wide-cut turbine fuel with icing inhibitor, freeze point–50° C.

NOTE: Automobile Gasoline. Certain automobile gasoline may be used in specific aircraft engines if a FAA supplemental type cetificate has been obtained. Automobile gasoline which is to be used in aircraft engines will be identified as "MOGAS", however, the grade/type and other octane rating will not be published.

Data shown on fuel availability represents the most recent informtion the publisher has been able to acquire. Because of a variety of factors, the fuel listed may not always be obtainable by transient civil pilots. Confirmation of availability of fuel should be made directly with fuel dispensers at locations where refueling is planned.

⑬ OXYGEN

OX 1 High Pressure
OX 2 Low Pressure
OX 3 High Pressure—Replacement Bottles
OX 4 Low Pressure—Replacement Bottles

⑭ TRAFFIC PATTERN ALTITUDE

Traffic Pattern Altitude (TPA)—The first figure shown is TPA above mean sea level. The second figure in parentheses is TPA above airport elevation.

⑮ AIRPORT OF ENTRY AND LANDING RIGHTS AIRPORTS

AOE—Airport of Entry—A customs Airport of Entry where permission from U.S. Customs is not required, however, at least one hour advance notice of arrival must be furnished.
LRA—Landing Rights Airport—Application for permission to land must be submitted in advance to U.S. Customs. At least one hour advance notice of arrival must be furnished.
NOTE: Advance notice of arrival at both an AOE and LRA airport may be included in the flight plan when filed in Canada or Mexico, where Flight Notification Service (ADCUS) is available the airport remark will indicate this service. This notice will also be treated as an application for permission to land in the case of an LRA. Although advance notice of arrival may be relayed to Customs through Mexico, Canadian, and U.S. Communications facilities by flight plan, the aircraft operator is solely responsible for insuring that Customs receives the notification. (See Customs, Immigration and Naturalization, Public Health and Agriculture Department requirements in the International Flight Information Manual for further details.)

⑯ CERTIFICATED AIRPORT (FAR 139)

Airports serving Department of Transportation certified carriers and certified under FAR, Part 139, are indicated by the ARFF index; i.e., ARFF Index A, which relates to the availability of crash, fire, rescue equipment.

FAR–PART 139 CERTIFICATED AIRPORTS
INDICES AND AIRCRAFT RESCUE AND FIRE FIGHTING EQUIPMENT REQUIREMENTS

Airport Index	Required No. Vehicles	Aircraft Length	Scheduled Departures	Agent + Water for Foam
A	1	< 90'	≥ 1	500 #DC or HALON 1211 or 450 #DC + 100 gal H₂O
B	1 or 2	≥ 90', < 126'	≥ 5	Index A + 1500 gal H₂O
		≥ 126', < 159'	< 5	
C	2 or 3	≥ 126', < 159'	≥ 5	Index A + 3000 gal H₂O
		≥ 159', < 200'	< 5	
D	3	≥ 159', < 200'	≥ 5	Index A + 4000 gal H₂O
		> 200'	< 5	
E	3	≥ 200'	≥ 5	Index A + 6000 gal H₂O

> Greater Than; < Less Than; ≥ Equal or Greater Than; ≤ Equal or Less Than; H₂O–Water;
DC–Dry Chemical.

LEGEND 4.—Airport/Facility Directory.

DIRECTORY LEGEND

NOTE: The listing of ARFF index does not necessarily assure coverage for non-air carrier operations or at other than prescribed times for air carrier. ARFF Index Ltd.—indicates ARFF coverage may or may not be available, for information contact airport manager prior to flight.

⑰ FAA INSPECTION

All airports not inspected by FAA will be identified by the note: Not insp. This indicates that the airport information has been provided by the owner or operator of the field.

⑱ RUNWAY DATA

Runway information is shown on two lines. That information common to the entire runway is shown on the first line while information concerning the runway ends are shown on the second or following line. Lengthy information will be placed in the Airport Remarks.

Runway direction, surface, length, width, weight bearing capacity, lighting, gradient and appropriate remarks are shown for each runway. Direction, length, width, lighting and remarks are shown for sealanes. The full dimensions of helipads are shown, i.e., 50X150.

RUNWAY SURFACE AND LENGTH

Runway lengths prefixed by the letter "H" indicate that the runways are hard surfaced (concrete, asphalt). If the runway length is not prefixed, the surface is sod, clay, etc. The runway surface composition is indicated in parentheses after runway length as follows:

(AFSC)—Aggregate friction seal coat	(GRVD)—Grooved	(TURF)—Turf
(ASPH)—Asphalt	(GRVL)—Gravel, or cinders	(TRTD)—Treated
(CONC)—Concrete	(PFC)—Porous friction courses	(WC)—Wire combed
(DIRT)—Dirt	(RFSC)—Rubberized friction seal coat	

RUNWAY WEIGHT BEARING CAPACITY

Runway strength data shown in this publication is derived from available information and is a realistic estimate of capability at an average level of activity. It is not intended as a maximum allowable weight or as an operating limitation. Many airport pavements are capable of supporting limited operations with gross weights of 25-50% in excess of the published figures. Permissible operating weights, insofar as runway strengths are concerned, are a matter of agreement between the owner and user. When desiring to operate into any airport at weights in excess of those published in the publication, users should contact the airport management for permission. Add 000 to figure following S, D, DT, DDT and MAX for gross weight capacity:

 S—Runway weight bearing capacity for aircraft with single-wheel type landing gear, (DC-3), etc.
 D—Runway weight bearing capacity for aircraft with dual-wheel type landing gear, (DC-6), etc.
 DT—Runway weight bearing capacity for aircraft with dual-tandem type landing gear, (707), etc.
 DDT—Runway weight bearing capacity for aircraft with double dual-tandem type landing gear, (747), etc.

Quadricycle and dual-tandem are considered virtually equal for runway weight bearing consideration, as are single-tandem and dual-wheel.

Omission of weight bearing capacity indicates information unknown.

RUNWAY LIGHTING

Lights are in operation sunset to sunrise. Lighting available by prior arrangement only or operating part of the night only and/or pilot controlled and with specific operating hours are indicated under airport remarks. Since obstructions are usually lighted, obstruction lighting is not included in this code. Unlighted obstructions on or surrounding an airport will be noted in airport remarks. Runway lights nonstandard (NSTD) are systems for which the light fixtures are not FAA approved L-800 series: color, intensity, or spacing does not meet FAA standards. Nonstandard runway lights, VASI, or any other system not listed below will be shown in airport remarks.

Temporary, emergency or limited runway edge lighting such as flares, smudge pots, lanterns or portable runway lights will also be shown in airport remarks.

Types of lighting are shown with the runway or runway end they serve.

NSTD—Light system fails to meet FAA standards.	SALS—Short Approach Lighting System.
LIRL—Low Intensity Runway Lights	SALSF—Short Approach Lighting System with Sequenced Flashing Lights.
MIRL—Medium Intensity Runway Lights	SSALS—Simplified Short Approach Lighting System.
HIRL—High Intensity Runway Lights	SSALF—Simplified Short Approach Lighting System with Sequenced Flashing Lights.
RAIL—Runway Alignment Indicator Lights	
REIL—Runway End Identifier Lights	SSALR—Simplified Short Approach Lighting System with Runway Alignment Indicator Lights.
CL—Centerline Lights	
TDZ—Touchdown Zone Lights	ALSAF—High Intensity Approach Lighting System with Sequenced Flashing Lights
ODALS—Omni Directional Approach Lighting System.	
AF OVRN—Air Force Overrun 1000' Standard Approach Lighting System.	ALSF1—High Intensity Approach Lighting System with Sequenced Flashing Lights, Category I, Configuration.
LDIN—Lead-In Lighting System.	ALSF2—High Intensity Approach Lighting System with Sequenced Flashing Lights, Category II, Configuration.
MALS—Medium Intensity Approach Lighting System.	
MALSF—Medium Intensity Approach Lighting System with Sequenced Flashing Lights.	VASI—Visual Approach Slope Indicator System.
MALSR—Medium Intensity Approach Lighting System with Runway Alignment Indicator Lights.	

NOTE: Civil ALSF-2 may be operated as SSALR during favorable weather conditions.

LEGEND 5.—Airport/Facility Directory.

DIRECTORY LEGEND

VISUAL GLIDESLOPE INDICATORS

APAP—A system of panels, which may or may not be lighted, used for alignment of approach path.

PNIL	APAP on left side of runway
PNIR	APAP on right side of runway

PAPI—Precision Approach Path Indicator

P2L	2-identical light units placed on left side of runway
P2R	2-identical light units placed on right side of runway
P4L	4-identical light units placed on left side of runway
P4R	4-identical light units placed on right side of runway

PVASI—Pulsating/steady burning visual approach slope indicator, normally a single light unit projecting two colors.

PSIL–	PVASI on left side of runway
PSIR–	PVASI on right side of runway

SAVASI—Simplified Abbreviated Visual Approach Slope Indicator

S2L	2-box SAVASI on left side of runway
S2R	2-box SAVASI on right side of runway

TRCV—Tri-color visual approach slope indicator, normally a single light unit projecting three colors.

TRIL	TRCV on left side of runway
TRIR	TRCV on right side of runway

VASI—Visual Approach Slope Indicator

V2L	2-box VASI on left side of runway
V2R	2-box VASI on right side of runway
V4L	4-box VASI on left side of runway
V4R	4-box VASI on right side of runway
V6L	6-box VASI on left side of runway
V6R	6-box VASI on right side of runway
V12	12-box VASI on both sides of runway
V16	16-box VASI on both sides of runway

NOTE: Approach slope angle and threshold crossing height will be shown when available; i.e., –GA3.5° TCH37'

PILOT CONTROL OF AIRPORT LIGHTING

Key Mike	Function
7 times within 5 seconds	Highest intensity available
5 times within 5 seconds	Medium or lower intensity (Lower REIL or REIL-Off)
3 times within 5 seconds	Lowest intensity available (Lower REIL or REIL-Off)

Available systems will be indicated in the Airport Remarks, as follows:

ACTIVATE MALSR Rwy 7, HIRL Rwy 7–25–122.8 (or CTAF).
or
ACTIVATE MIRL Rwy 18–36–122.8 (or CTAF).
or
ACTIVATE VASI and REIL, Rwy 7–122.8 (or CTAF).

Where the airport is not served by an instrument approach procedure and/or has an independent type system of different specification installed by the airport sponsor, descriptions of the type lights, method of control, and operating frequency will be explained in clear text. See AIM, "Basic Flight Information and ATC Procedures," for detailed description of pilot control of airport lighting.

RUNWAY GRADIENT

Runway gradient will be shown only when it is 0.3 percent or more. When available the direction of slope upward will be indicated, i.e., 0.5% up NW.

RUNWAY END DATA

Lighting systems such as VASI, MALSR, REIL; obstructions; displaced thresholds will be shown on the specific runway end. "Rgt tfc"—Right traffic indicates right turns should be made on landing and takeoff for specified runway end.

⑲ AIRPORT REMARKS

Landing Fee indicates landing charges for private or non-revenue producing aircraft. in addition, fees may be charged for planes that remain over a couple of hours and buy no services, or at major airline terminals for all aircraft.
Remarks—Data is confined to operational items affecting the status and usability of the airport.
Parachute Jumping.—See "PARACHUTE" tabulation for details.

LEGEND 6.—Airport/Facility Directory.

DIRECTORY LEGEND

⟨20⟩ WEATHER DATA SOURCES

AWOS—Automated Weather Observing System

AWOS-A—reports altimeter setting.
AWOS-1—reports altimeter setting, wind data and usually temperature, dewpoint and density altitude.
AWOS-2—reports the same as AWOS-1 plus visibility.
AWOS-3—reports the same as AWOS-1 plus visibility and cloud/ceiling data.
See AIM, Basic Flight Information and ATC Procedures for detailed description of AWOS.

SAWRS—identifies airports that have a Supplemental Aviation Weather Reporting Station available to pilots for current weather information.
LAWRS—Limited Aviation Weather Reporting Station where observers report cloud height, weather, obstructions to vision, temperature and dewpoint (in most cases), surface wind, altimeter and pertinent remarks.
LLWAS—indicates a Low Level Wind Shear Alert System consisting of a center field and several field perimeter anemometers.
HIWAS—See RADIO AIDS TO NAVIGATION

⟨21⟩ COMMUNICATIONS

Communications will be listed in sequence in the order shown below:

Common Traffic Advisory Frequency (CTAF), Automatic Terminal Information Service (ATIS) and Aeronautical Advisory Stations (UNICOM) along with their frequency is shown, where available, on the line following the heading "COMMUNICATIONS." When the CTAF and UNICOM is the same frequency, the frequency will be shown as CTAF/UNICOM freq.

Flight Service Station (FSS) information. The associated FSS will be shown followed by the identifier and information concerning availability of telephone service, e.g., Direct Line (DL), Local Call (LC-384-2341), Toll free call, dial (TF 800–852–7036 or TF 1-800-227-7160), Long Distance (LD 202-426-8800 or LD 1-202-555-1212) etc. The airport NOTAM file identifier will be shown as "NOTAM FILE IAD." Where the FSS is located on the field it will be indicated as "on arpt" following the identifier. Frequencies available will follow. The FSS telephone number will follow along with any significant operational information. FSS's whose name is not the same as the airport on which located will also be listed in the normal alphabetical name listing for the state in which located. Remote Communications Outlet (RCO) providing service to the airport followed by the frequency and name of the Controlling FSS.

FSS's provide information on airport conditions, radio aids and other facilities, and process flight plans. Airport Advisory Service is provided on the CTAF by FSS's located at non-tower airports or airports where the tower is not in operation.

(See AIM, Par. 157/158 Traffic Advisory Practices at airports where a tower is not in operation or AC 90 - 42C.)

Aviation weather briefing service is provided by FSS specialists. Flight and weather briefing services are also available by calling the telephone numbers listed.

Remote Communications Outlet (RCO)—An unmanned air/ground communications facility, remotely controlled and providing UHF or VHF communications capability to extend the service range of an FSS.

Civil Communications Frequencies—Civil communications frequencies used in the FSS air/ground system are now operated simplex on 122.0, 122.2, 122.3, 122.4, 122.6, 123.6; emergency 121.5; plus receive-only on 122.05, 122.1, 122.15, and 123.6.

    a. 122.0 is assigned as the Enroute Flight Advisory Service channel at selected FSS's.
    b. 122.2 is assigned to most FSS's as a common enroute simplex service.
    c. 123.6 is assigned as the airport advisory channel at non-tower FSS locations, however, it is still in commission at some FSS's collocated with towers to provide part time Airport Advisory Service.
    d. 122.1 is the primary receive-only frequency at VOR's. 122.05, 122.15 and 123.6 are assigned at selected VOR's meeting certain criteria.
    e. Some FSS's are assigned 50 kHz channels for simplex operation in the 122-123 MHz band (e.g. 122.35). Pilots using the FSS A/G system should refer to this directory or appropriate charts to determine frequencies available at the FSS or remoted facility through which they wish to communicate.

Part time FSS hours of operation are shown in remarks under facility name.

  Emergency frequency 121.5 is available at all Flight Service Stations, Towers, Approach Control and RADAR facilities, unless indicated as not available.

Frequencies published followed by the letter "T" or "R", indicate that the facility will only transmit or receive respectively on that frequency. All radio aids to navigation frequencies are transmit only.

TERMINAL SERVICES

CTAF—A program designed to get all vehicles and aircraft at uncontrolled airports on a common frequency.
ATIS—A continuous broadcast of recorded non-control information in selected areas of high activity.
UNICOM—A non-government air/ground radio communications facility utilized to provide general airport advisory service.
APP CON —Approach Control. The symbol ® indicates radar approach control.
TOWER—Control tower
GND CON—Ground Control
DEP CON—Departure Control. The symbol ® indicates radar departure control.
CLNC DEL—Clearance Delivery.
PRE TAXI CLNC—Pre taxi clearance
VFR ADVSY SVC—VFR Advisory Service. Service provided by Non-Radar Approach Control.
  Advisory Service for VFR aircraft (upon a workload basis) ctc APP CON.
STAGE II SVC—Radar Advisory and Sequencing Service for VFR aircraft
STAGE III SVC—Radar Sequencing and Separation Service for participating VFR Aircraft within a Terminal Radar Service Area (TRSA)

LEGEND 7.—Airport/Facility Directory.

DIRECTORY LEGEND

ARSA—Airport Radar Service Area
TCA—Radar Sequencing and Separation Service for all aircraft in a Terminal Control Area (TCA)
TOWER, APP CON and DEP CON RADIO CALL will be the same as the airport name unless indicated otherwise.

㉒ RADIO AIDS TO NAVIGATION

The Airport Facility Directory lists by facility name all Radio Aids to Navigation, except Military TACANS, that appear on National Ocean Service Visual or IFR Aeronautical Charts and those upon which the FAA has approved an Instrument Approach Procedure. All VOR. VORTAC ILS and MLS equipment in the National Airspace System has an automatic monitoring and shutdown feature in the event of malfunction. Unmonitored, as used in this publication for any navigational aid. means that FSS or tower personnel cannot observe the malfunction or shutdown signal. The NAVAID NOTAM file identifier will be shown as ''NOTAM FILE IAD'' and will be listed on the Radio Aids to Navigation line. When two or more NAVAIDS are listed and the NOTAM file identifier is different than shown on the Radio Aids to Navigation line, then it will be shown with the NAVAID listing. Hazardous Inflight Weather Advisory Service (HIWAS) will be shown where this service is broadcast over selected VOR's.
NAVAID information is tabulated as indicated in the following sample:

```
                  TWEB        TACAN/DME Channel      Geographical Position        Site Elevation

NAME (L) ABVORTAC 117.55  ▪ ABE  Chan 122(Y)  40°43'36"N 75°27'18"W   180°  4.1 NM to fld.  1110/8E.  HIWAS.

     ↑         ↑           ↑                    Bearing and distance   Magnetic Variation   Hazardous Inflight
   Class   Frequency   Identifier              facility to center of airport              Weather Advisory Service
```

VOR unusable 020°-060° beyond 26 NM below 3500'

Restriction within the normal altitude/range of the navigational aid (See primary alphabetical listing for restrictions on VORTAC and VOR/DME).

Note: Those DME channel numbers with a (Y) suffix require TACAN to be placed in the ''Y'' mode to receive distance information.

HIWAS—Hazardous Inflight Weather Advisory Service is a continuous broadcast of inflight weather advisories including summarized SIGMETs, convective SIGMETs, AIRMETs and urgent PIREPs. HIWAS is presently broadcast over selected VOR's and will be implemented throughout the conterminous U.S.

ASR/PAR—Indicates that Surveillance (ASR) or Precision (PAR) radar instrument approach minimums are published in U.S. Government Instrument Approach Procedures.

RADIO CLASS DESIGNATIONS

VOR/DME/TACAN Standard Service Volume (SSV) Classifications

SSV Class	Altitudes	Distance (NM)
(T) Terminal	1000' to 12,000'	25
(L) Low Altitude	1000' to 18,000'	40
(H) High Altitude	1000' to 14,500'	40
	14,500' to 18,000'	100
	18,000' to 45,000'	130
	45,000' to 60,000'	100

NOTE: Additionally, (H) facilities provide (L) and (T) service volume and (L) facilities provide (T) service. Altitudes are with respect to the station's site elevation. Coverage is not available in a cone of airspace directly above the facility.

The term VOR is, operationally, a general term covering the VHF omnidirectional bearing type of facility without regard to the fact that the power, the frequency protected service volume, the equipment configuration, and operational requirements may vary between facilities at different locations.

AB ———————	Automatic Weather Broadcast (also shown with ▪ following frequency.)
DF ———————	Direction Finding Service.
DME ———————	UHF standard (TACAN compatible) distance measuring equipment.
DME(Y)———————	UHF standard (TACAN compatible) distance measuring equipment that require TACAN to be placed in the ''Y'' mode to receive DME.
H ———————	Non-directional radio beacon (homing). power 50 watts to less than 2,000 watts (50 NM at all altitudes).
HH———————	Non-directional radio beacon (homing). power 2,000 watts or more (75 NM at all altitudes).
H-SAB———————	Non-directional radio beacons providing automatic transcribed weather service.

LEGEND 8.—Airport/Facility Directory.

DIRECTORY LEGEND

ILS	Instrument Landing System (voice, where available, on localizer channel).
ISMLS	Interim Standard Microwave Landing System.
LDA	Localizer Directional Aid.
LMM	Compass locator station when installed at middle marker site (15 NM at all altitudes).
LOM	Compass locator station when installed at outer marker site (15 NM at all altitudes).
MH	Non-directional radio beacon (homing) power less than 50 watts (25 NM at all altitudes).
MLS	Microwave Landing System
S	Simultaneous range homing signal and/or voice.
SABH	Non-directional radio beacon not authorized for IFR or ATC. Provides automatic weather broadcasts.
SDF	Simplified Direction Facility.
TACAN	UHF navigational facility-omnidirectional course and distance information.
VOR	VHF navigational facility-omnidirectional course only.
VOR/DME	Collocated VOR navigational facility and UHF standard distance measuring equipment.
VORTAC	Collocated VOR and TACAN navigational facilities.
W	Without voice on radio facility frequency.
Z	VHF station location marker at a LF radio facility.

ABBREVIATIONS

The following abbreviations are those commonly used within this Directory. Other abbreviations may be found in the Legend and are not duplicated below.

AAS	airport advisory service	ldg	landing
acft	aircraft	med	medium
apch	approach	NFCT	non-federal control tower
arpt	airport	ngt	night
avbl	available	NSTD	nonstandard
bcn	beacon	ntc	notice
blo	below	opr	operate
byd	beyond	ops	operates operation
clsd	closed	ovrn	overrun
ctc	contact	p-line	power line
dalgt	daylight	PPR	prior permission required
dsplc	displace	req	request
dsplcd	displaced	rqr	requires
durn	duration	rgt tfc	right traffic
emerg	emergency	rwy	runway
extd	extend, extended	svc	service
fld	field	tmpry	temporary, temporarily
FSS	Flight Service Station	tkf	take off
ints	intensity	tfc	traffic
lgtd	lighted	thld	threshold
lgts	lights	twr	tower

LEGEND 9.—Airport/Facility Directory.

APPENDIX 3

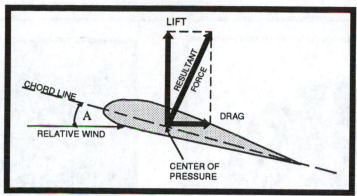

FIGURE 1.—Lift Vector.

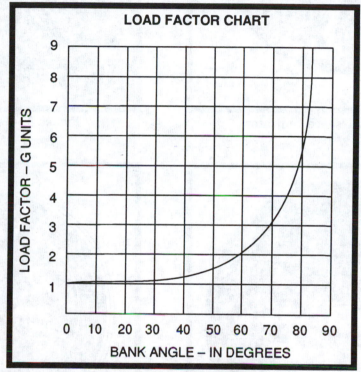

FIGURE 2.—Load Factor Chart.

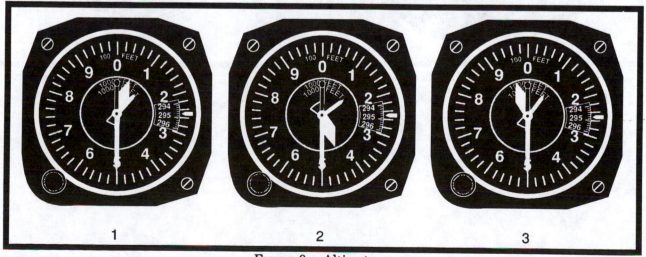

FIGURE 3.—Altimeter.

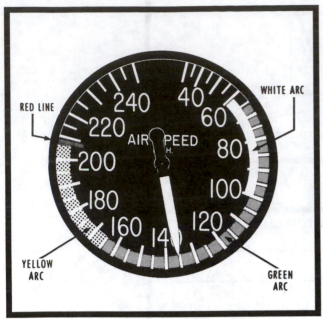

FIGURE 4.—Airspeed Indicator.

FIGURE 6.—Heading Indicator.

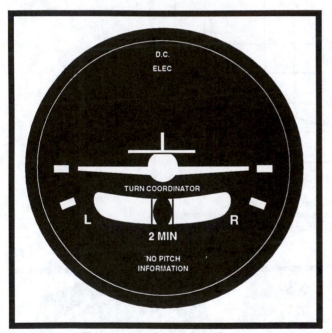

FIGURE 5.—Turn Coordinator.

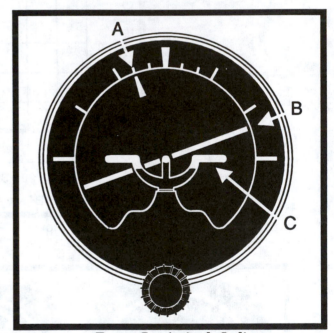

FIGURE 7.—Attitude Indicator.

DENSITY ALTITUDE CHART

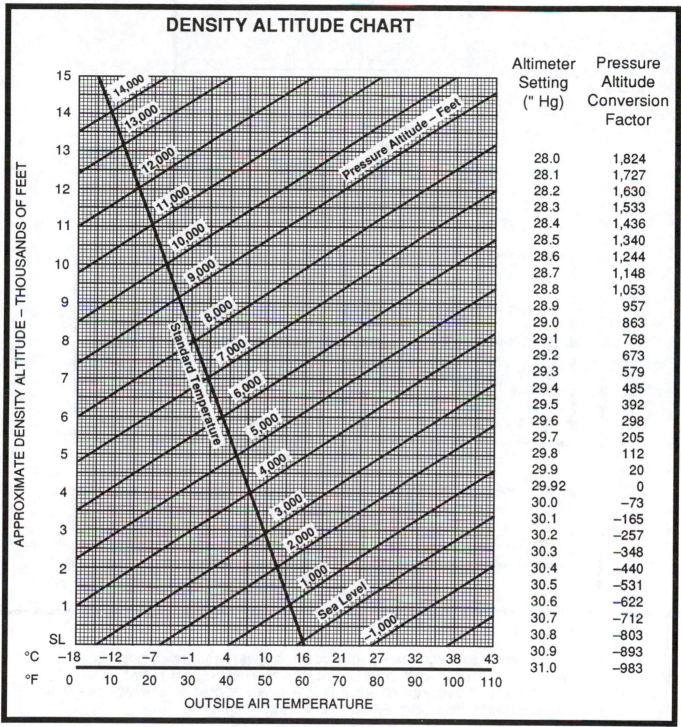

Altimeter Setting (" Hg)	Pressure Altitude Conversion Factor
28.0	1,824
28.1	1,727
28.2	1,630
28.3	1,533
28.4	1,436
28.5	1,340
28.6	1,244
28.7	1,148
28.8	1,053
28.9	957
29.0	863
29.1	768
29.2	673
29.3	579
29.4	485
29.5	392
29.6	298
29.7	205
29.8	112
29.9	20
29.92	0
30.0	−73
30.1	−165
30.2	−257
30.3	−348
30.4	−440
30.5	−531
30.6	−622
30.7	−712
30.8	−803
30.9	−893
31.0	−983

FIGURE 8.—Density Altitude Chart.

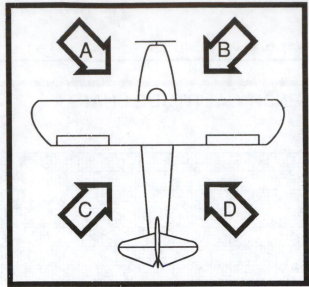

FIGURE 9.—Control Position for Taxi.

METAR KINK 121845Z 11012G18KT 15SM SKC 25/17 A3000

METAR KBOI 121854Z 13004KT 30SM SCT150 17/6 A3015

METAR KLAX 121852Z 25004KT 6SM BR SCT007 SCT250 16/15 A2991

SPECI KMDW 121856Z 32005KT 1 1/2SM RA OVC007 17/16 A2980 RMK RAB35

SPECI KJFK 121853Z 18004KT 1/2SM FG R04/2200 OVC005 20/18 A3006

FIGURE 12.—Surface Aviation Weather Report (METAR).

UA /OV OKC–TUL /TM 1800 /FL 120 /TP BE90 /SK 018 BKN 055 / /072 OVC 089 /CLR ABV /TA –9/WV 0921/TB MDT 055–072 /IC LGT–MDT CLR 072–089.

FIGURE 14.—Pilot Weather Report.

TAF

KMEM   121720Z 121818 20012KT 5SM HZ BKN030 PROB40 2022 1SM TSRA OVC008CB
         FM2200 33015G20KT P6SM BKN015 OVC025 PROB40 2202 3SM SHRA
         FM0200 35012KT OVC008 PROB40 0205 2SM -RASN BECMG 0608 02008KT NSW BKN012
          BECMG 1012 00000KT 3SM BR SKC TEMPO 1214 1/2SM FG
         FM1600 VRB04KT P6SM NSW SKC=

KOKC   051130Z 051212 14008KT 5SM BR BKN030 TEMPO 1316 1 1/2SM BR
         FM1600 16010KT P6SM NSW SKC BECMG 2224 20013G20KT 4SM SHRA OVC020
          PROB40 0006 2SM TSRA OVC008CB BECMG 0608 21015KT P6SM NSW SCT040 =

FIGURE 15.—Terminal Forecast (TAF).

```
DFWH FA Ø41Ø4Ø
HAZARDS VALID UNTIL Ø423ØØ
OK TX AR LA TN MS AL AND CSTL WTRS
FLT PRCTNS...TURBC...TN AL AND CSTL WTRS
            ...ICG...TN
            ...IFR...TX
TSTMS IMPLY PSBL SVR OR GTR TURBC SVR ICG AND LLWS
NON MSL HGTS NOTED BY AGL OR CIG
THIS FA ISSUANCE INCORPORATES THE FOLLOWING AIRMETS STILL IN
EFFECT... NONE.


DFWS FA Ø41Ø4Ø
SYNOPSIS VALID UNTIL Ø5Ø5ØØ
AT 11Z RDG OF HI PRES ERN TX NWWD TO CNTRL CO WITH HI CNTR
OVR ERN TX.  BY Ø5Z HI CNTR MOVS TO CNTRL LA.


DFWI FA Ø41Ø4Ø
ICING AND FRZLVL VALID UNTIL Ø423ØØ
TN
FROM SLK TO HAT TO MEM TO ORD TO SLK
OCNL MDT RIME ICGIC ABV FRZLVL TO 1ØØ. CONDS ENDING BY 17Z.
FRZLVL 8Ø CHA SGF LINE SLPG TO 12Ø S OF A IAH MAF LINE.


DFWT FA Ø41Ø4Ø
TURBC VALID UNTIL Ø423ØØ
TN AL AND CSTL WTRS
FROM SLK TO FLO TO 9ØS MOB TO MEI TO BUF TO SLK
OCNL MDT TURBC 25Ø-38Ø DUE TO JTSTR. CONDS MOVG SLOLY EWD
AND CONTG BYD 23Z.


DFWC FA Ø41Ø4Ø
SGFNT CLOUD AND WX VALID UNTIL Ø423ØØ... OTLK Ø423ØØ-Ø5Ø5ØØ
IFR...TX
FROM SAT TO FSX TO BRO TO MOV TO SAT
VSBY BLO 3F TIL 15Z.
OK AR TX LA MS AL AND CSTL WTRS
8Ø SCT TO CLR EXCP VSBY BLO 3F TIL 15Z OVR PTNS S CNTRL TX.
OTLK...VFR.
TN
CIGS 3Ø-5Ø BKN 1ØØ VSBYS OCNLY 3-5F BCMG AGL 4Ø-5Ø SCT TO
CLR BY 19Z. OTLK...VFR.
```

FIGURE 16.—Area Forecast.

```
FD WBC 151745
BASED ON 151200Z DATA
VALID 1600Z FOR USE 1800-0300Z.  TEMPS NEG ABV 24000
```

FT	3000	6000	9000	12000	18000	24000	30000	34000	39000
ALS			2420	2635–08	2535–18	2444–30	245945	246755	246862
AMA		2714	2725+00	2625–04	2531–15	2542–27	265842	256352	256762
DEN			2321–04	2532–08	2434–19	2441–31	235347	236056	236262
HLC		1707–01	2113–03	2219–07	2330–17	2435–30	244145	244854	245561
MKC	0507	2006+03	2215–01	2322–06	2338–17	2348–29	236143	237252	238160
STL	2113	2325+07	2332+02	2339–04	2356–16	2373–27	239440	730649	731960

FIGURE 17.—Winds and Temperatures Aloft Forecast.

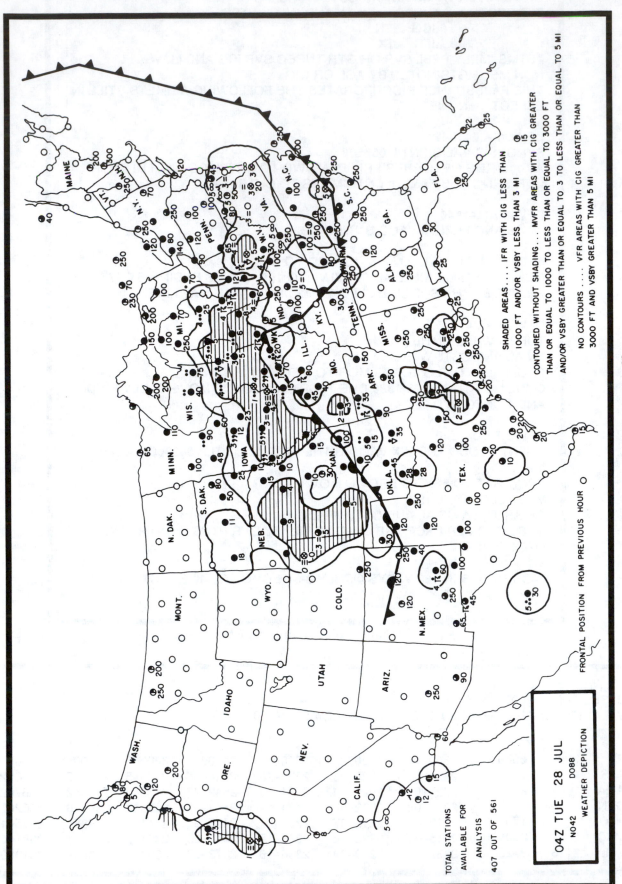

FIGURE 18.—Weather Depiction Chart.

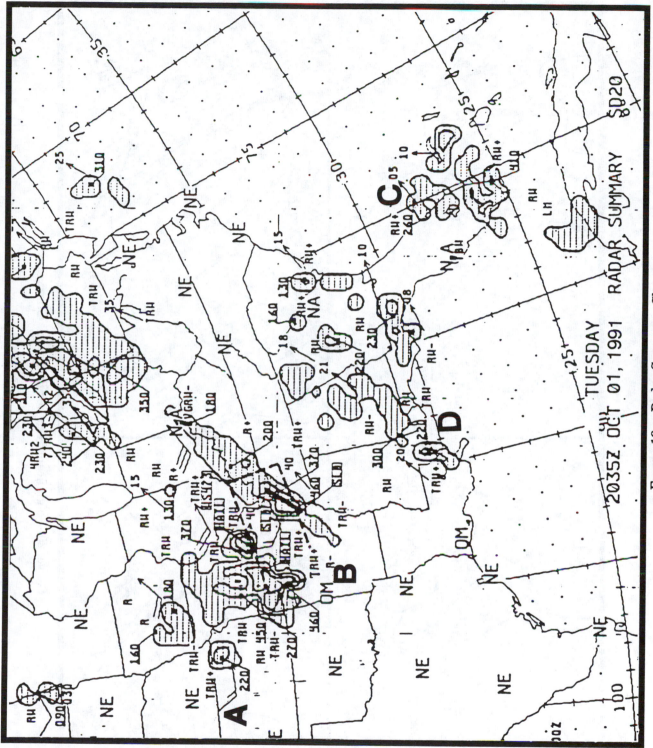

FIGURE 19.—Radar Summary Chart.

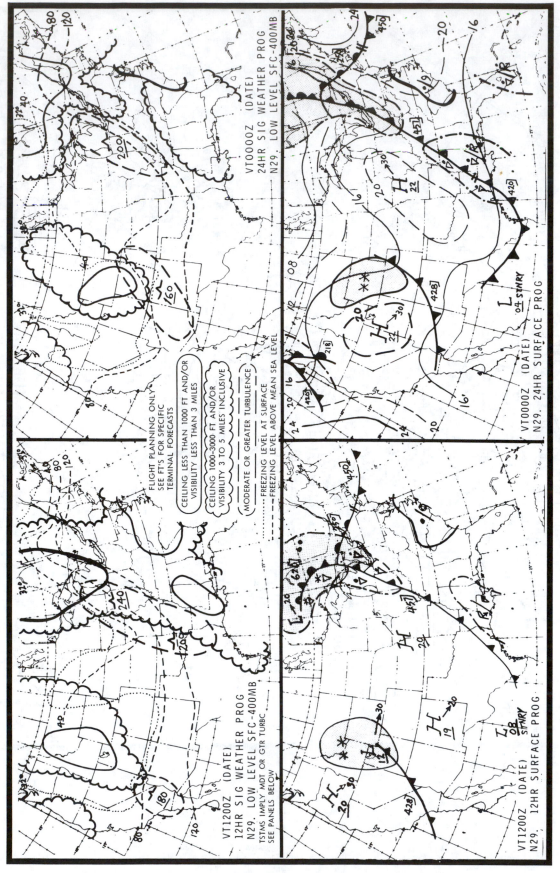

FIGURE 20.—Significant Weather Prognostic Chart.

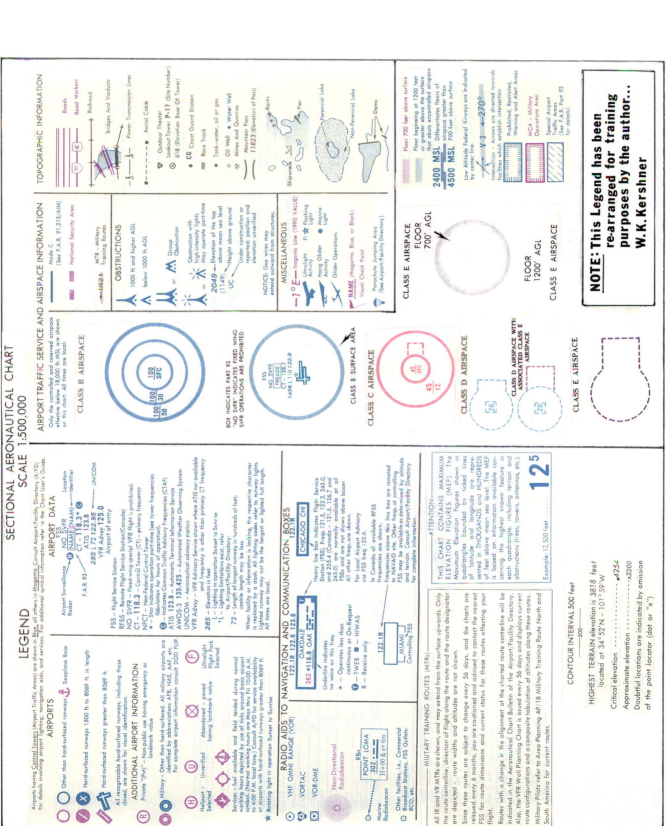

LEGEND 1.—Sectional Aeronautical Chart.

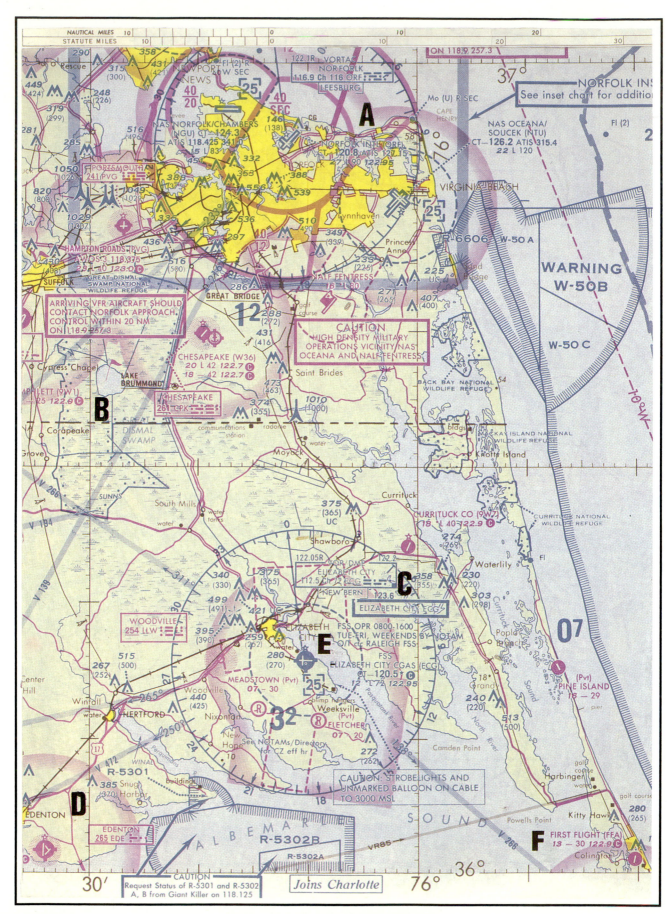

FIGURE 21.—Sectional Chart Excerpt.

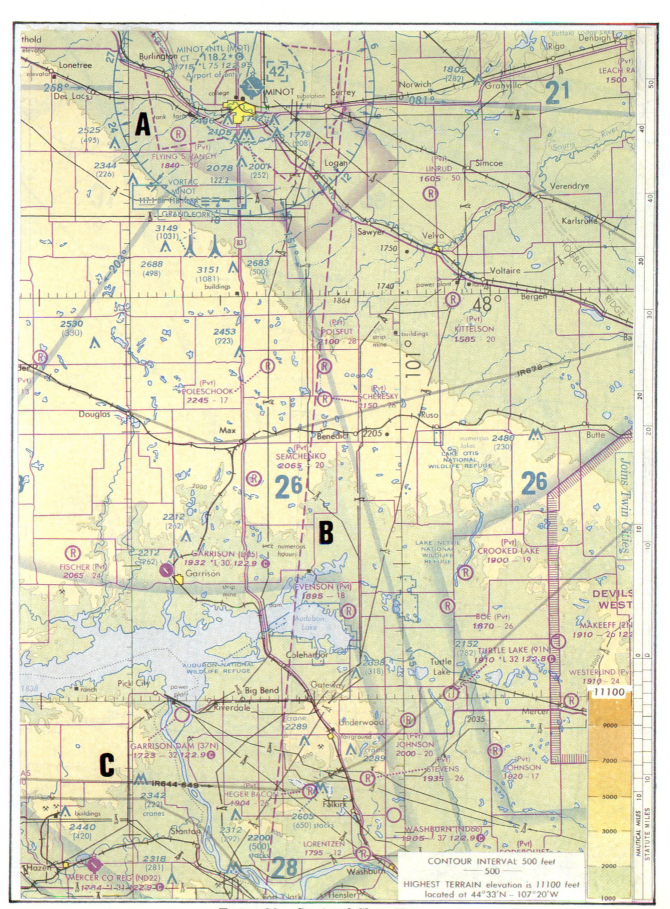

FIGURE 22.—Sectional Chart Excerpt.

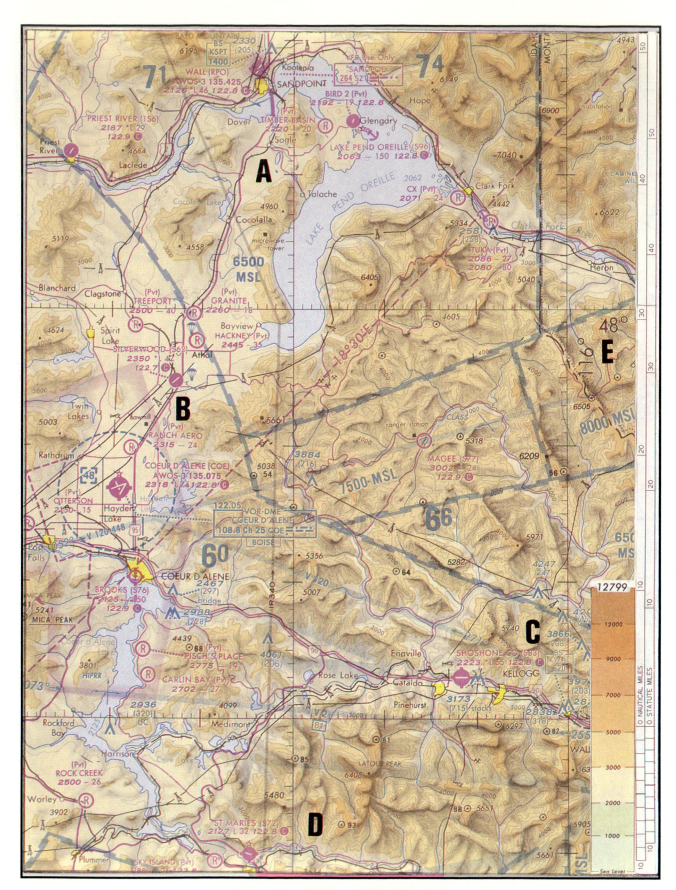

FIGURE 23.—Sectional Chart Excerpt.

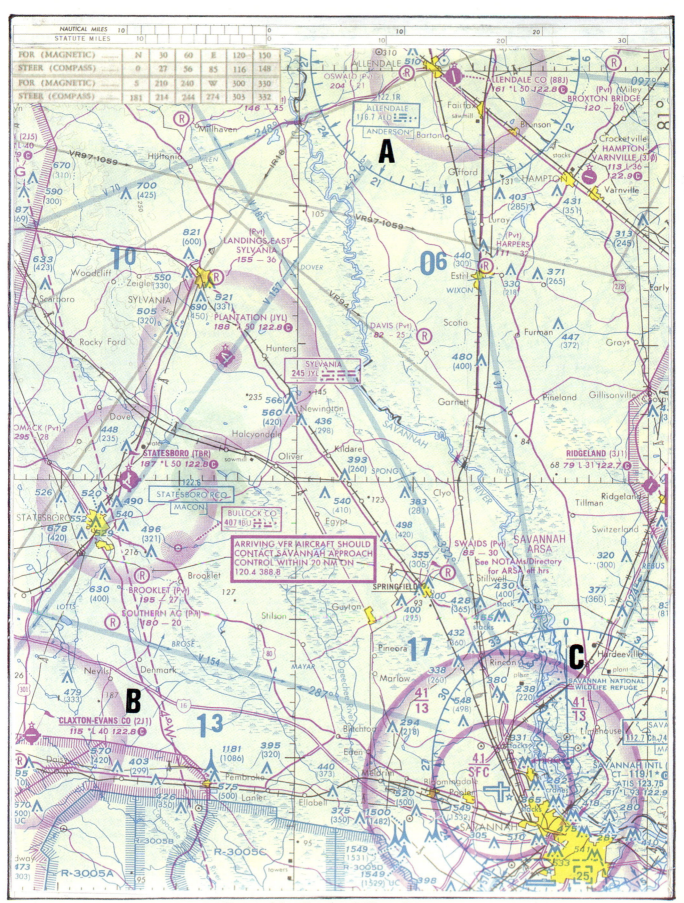

FIGURE 24.—Sectional Chart Excerpt.

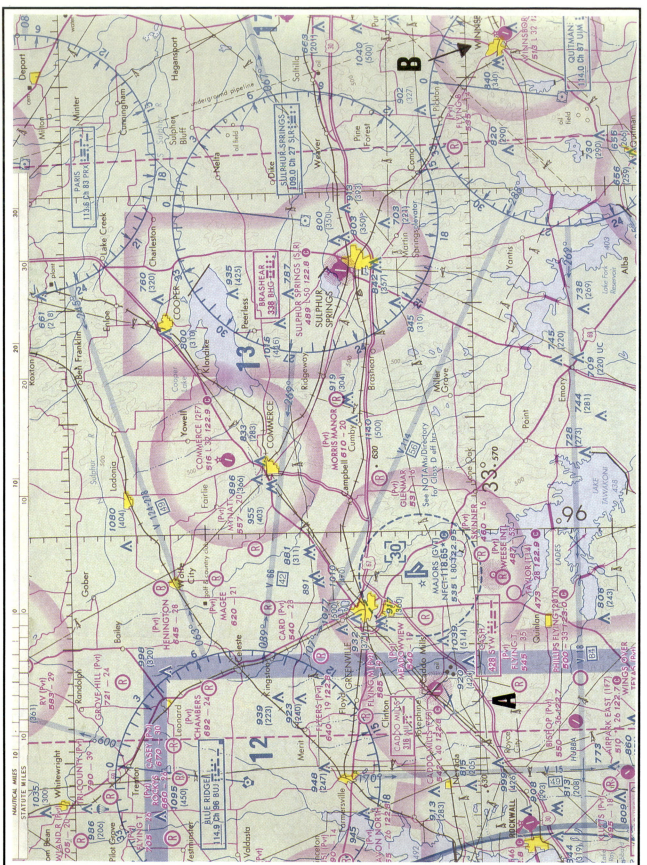

FIGURE 25.—Sectional Chart Excerpt.

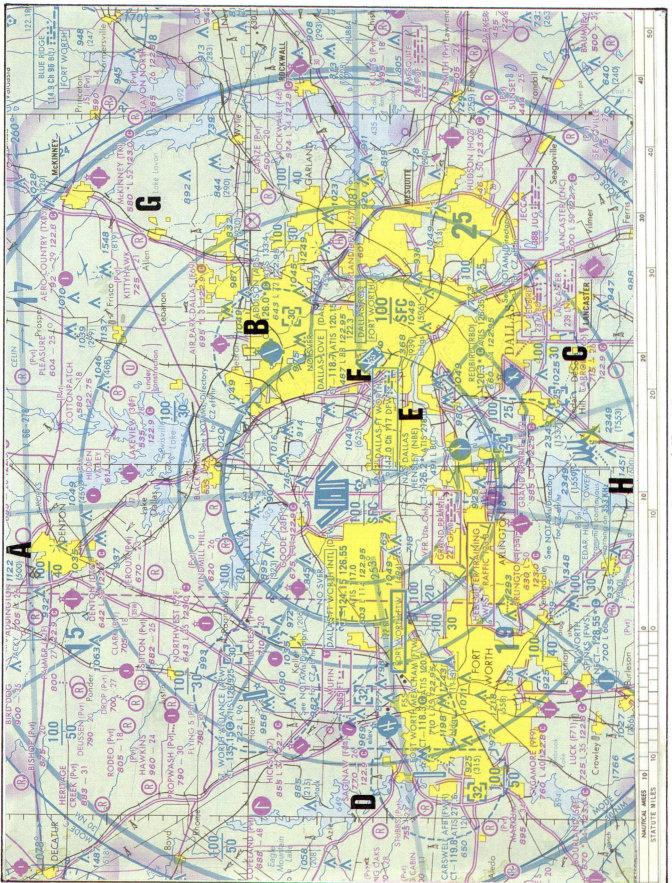

FIGURE 26.—Sectional Chart Excerpt.

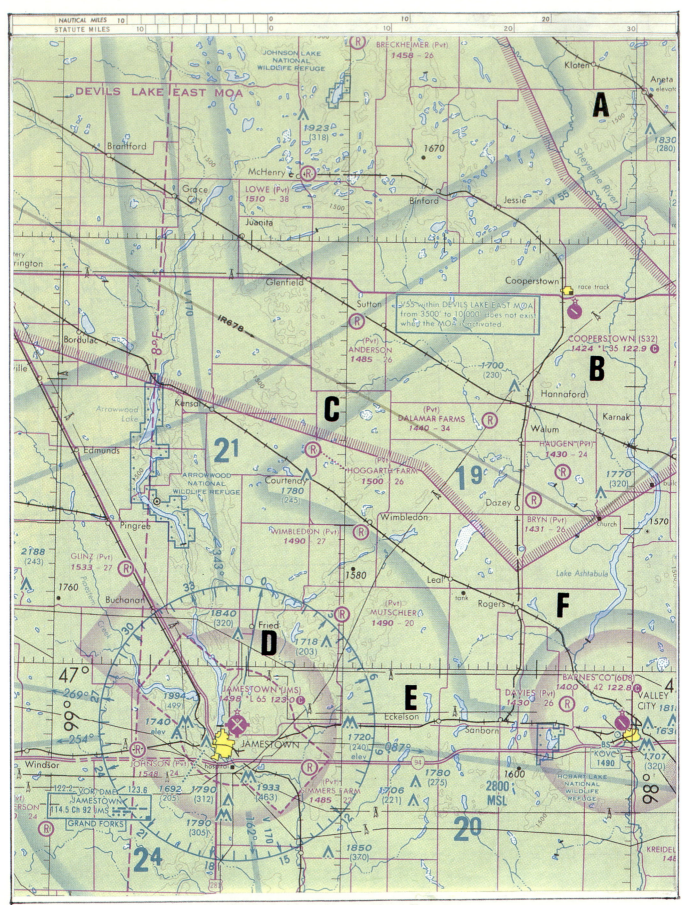

FIGURE 27.—Sectional Chart Excerpt.

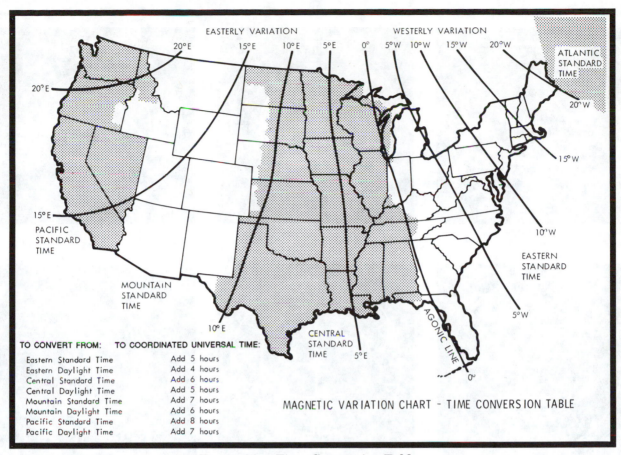

FIGURE 28.—Time Conversion Table.

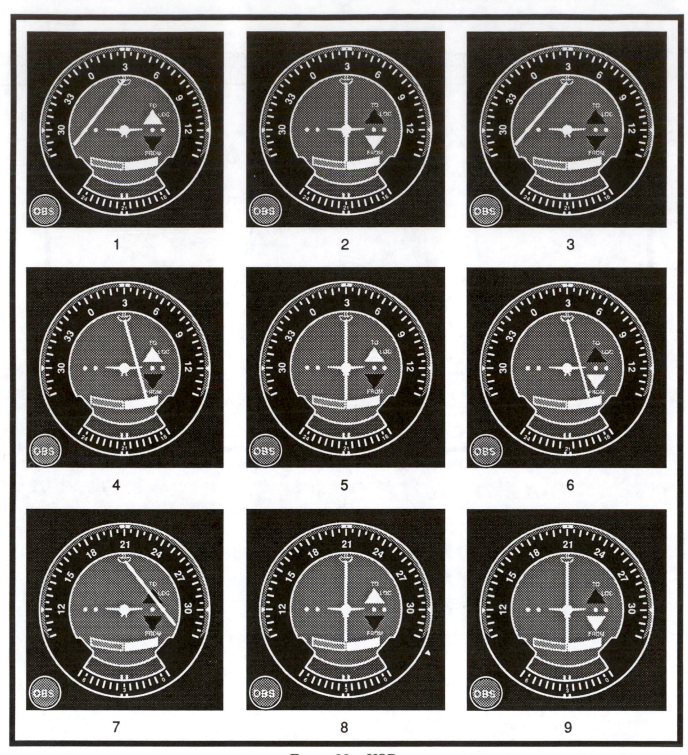

FIGURE 29.—VOR.

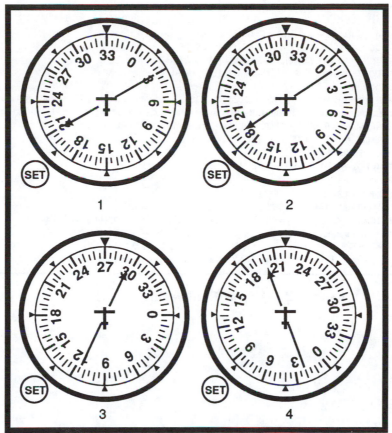

FIGURE 30.—ADF (Movable Card).

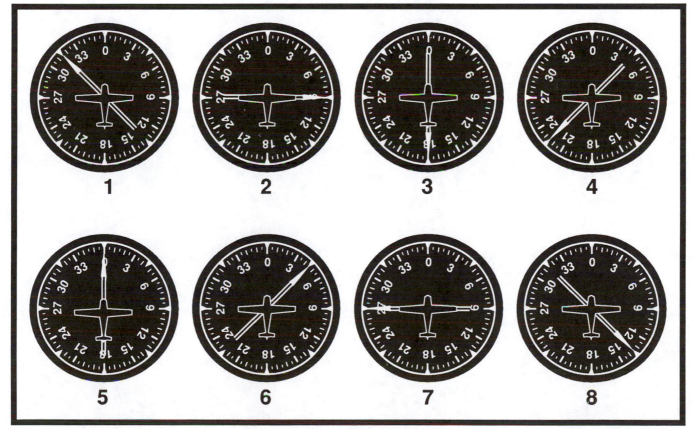

FIGURE 31.—ADF (Fixed Card).

COEUR D'ALENE AIR TERMINAL   (COE)   9 NW   UTC−8(−7DT)   N47°46.46′ W116°49.17′   **GREAT FALLS**
2318   B   S4   **FUEL** 80, 100, JET A   OX 1, 2   H−1B, L−9A
RWY 05−23: H7400X140 (ASPH−GRVD)   S−57, D−95, DT−165   HIRL   0.7%up NE   IAP
   RWY 05: MALSR.   **RWY 23:** REIL. VASI(V4L)—GA 3.0° TCH 39′.
RWY 01−19: H5400X75 (ASPH)   S−50, D−83, DT−150   MIRL
   RWY 01: REIL. Rgt tfc.
AIRPORT REMARKS: Attended Mon−Fri 1500−0100Z‡. Rwy 05−23 potential standing water and/or ice on center 3000′
   of rwy. Arpt conditions avbl on UNICOM. Rwy 19 is designated calm wind rwy. ACTIVATE MIRL Rwy 01−19, HIRL
   Rwy 05−23 and MALSR Rwy 05—CTAF. REIL Rwy 23 opr only when HIRL on high ints.
WEATHER DATA SOURCES: AWOS−3 135.075 (208) 772−8215.
COMMUNICATIONS: CTAF/UNICOM 122.8
   BOISE FSS (BOI) TF 1−800−WX−BRIEF. NOTAM FILE COE.
   RCO 122.05 (BOISE FSS)
Ⓡ **SPOKANE APP/DEP CON** 132.1
RADIO AIDS TO NAVIGATION: NOTAM FILE COE.
   (T) **VORW/DME** 108.8   COE   Chan 25   N47°46.42′ W116°49.24′   at fld. 2290/19E.
      DME portion unusable 280°−350° byd 15 NM blo 11000′ 220°−240° byd 15 NM.
   LEENY NDB (LOM) 347   CO   N47°44.57′ W116°57.66′   053° 6.0 NM to fld.
   ILS 110.7   I−COE   Rwy 05   LOM LEENY NDB. ILS localizer/glide slope unmonitored.

FIGURE 32.—Airport/Facility Directory Excerpt.

USEFUL LOAD WEIGHTS AND MOMENTS

OCCUPANTS

FRONT SEATS ARM 85		REAR SEATS ARM 121	
Weight	Moment 100	Weight	Moment 100
120	102	120	145
130	110	130	157
140	119	140	169
150	128	150	182
160	136	160	194
170	144	170	206
180	153	180	218
190	162	190	230
200	170	200	242

BAGGAGE OR 5TH SEAT OCCUPANT ARM 140

Weight	Moment 100
10	14
20	28
30	42
40	56
50	70
60	84
70	98
80	112
90	126
100	140
110	154
120	168
130	182
140	196
150	210
160	224
170	238
180	252
190	266
200	280
210	294
220	308
230	322
240	336
250	350
260	364
270	378

USABLE FUEL

MAIN WING TANKS ARM 75

Gallons	Weight	Moment 100
5	30	22
10	60	45
15	90	68
20	120	90
25	150	112
30	180	135
35	210	158
40	240	180
44	264	198

AUXILIARY WING TANKS ARM 94

Gallons	Weight	Moment 100
5	30	28
10	60	56
15	90	85
19	114	107

*OIL

Quarts	Weight	Moment 100
10	19	5

*Included in basic Empty Weight

Empty Weight ~ 2015

MOM / 100 ~ 1554

MOMENT LIMITS vs WEIGHT

Moment limits are based on the following weight and center of gravity limit data (landing gear down).

WEIGHT CONDITION	FORWARD CG LIMIT	AFT CG LIMIT
2950 lb (takeoff or landing)	82.1	84.7
2525 lb	77.5	85.7
2475 lb or less	77.0	85.7

FIGURE 33.—Airplane Weight and Balance Tables.

MOMENT LIMITS vs WEIGHT (Continued)

Weight	Minimum Moment 100	Maximum Moment 100	Weight	Minimum Moment 100	Maximum Moment 100
2100	1617	1800	2600	2037	2224
2110	1625	1808	2610	2048	2232
2120	1632	1817	2620	2058	2239
2130	1640	1825	2630	2069	2247
2140	1648	1834	2640	2080	2255
2150	1656	1843	2650	2090	2263
2160	1663	1851	2660	2101	2271
2170	1671	1860	2670	2112	2279
2180	1679	1868	2680	2123	2287
2190	1686	1877	2690	2133	2295
2200	1694	1885	2700	2144	2303
2210	1702	1894	2710	2155	2311
2220	1709	1903	2720	2166	2319
2230	1717	1911	2730	2177	2326
2240	1725	1920	2740	2188	2334
2250	1733	1928	2750	2199	2342
2260	1740	1937	2760	2210	2350
2270	1748	1945	2770	2221	2358
2280	1756	1954	2780	2232	2366
2290	1763	1963	2790	2243	2374
2300	1771	1971			
2310	1779	1980	2800	2254	2381
2320	1786	1988	2810	2265	2389
2330	1794	1997	2820	2276	2397
2340	1802	2005	2830	2287	2405
2350	1810	2014	2840	2298	2413
2360	1817	2023	2850	2309	2421
2370	1825	2031	2860	2320	2428
2380	1833	2040	2870	2332	2436
2390	1840	2048	2880	2343	2444
			2890	2354	2452
2400	1848	2057	2900	2365	2460
2410	1856	2065	2910	2377	2468
2420	1863	2074	2920	2388	2475
2430	1871	2083	2930	2399	2483
2440	1879	2091	2940	2411	2491
2450	1887	2100	2950	2422	2499
2460	1894	2108			
2470	1902	2117			
2480	1911	2125			
2490	1921	2134			
2500	1932	2143			
2510	1942	2151			
2520	1953	2160			
2530	1963	2168			
2540	1974	2176			
2550	1984	2184			
2560	1995	2192			
2570	2005	2200			
2580	2016	2208			
2590	2026	2216			

FIGURE 34.—Airplane Weight and Balance Tables.

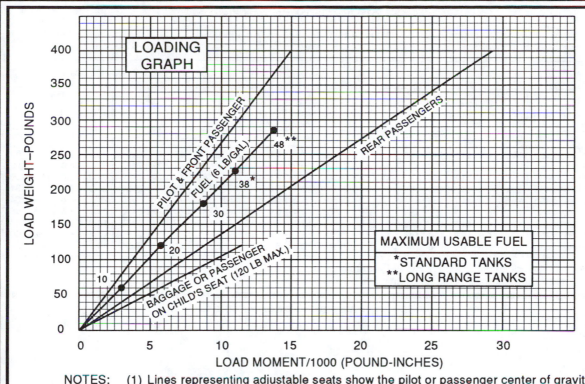

LOADING GRAPH

LOAD WEIGHT—POUNDS

LOAD MOMENT/1000 (POUND-INCHES)

PILOT & FRONT PASSENGER

FUEL (6 LB/GAL)

REAR PASSENGERS

BAGGAGE OR PASSENGER ON CHILD'S SEAT (120 LB MAX.)

MAXIMUM USABLE FUEL
*STANDARD TANKS
**LONG RANGE TANKS

NOTES: (1) Lines representing adjustable seats show the pilot or passenger center of gravity on adjustable seats positioned for an average occupant. Refer to the Loading Arrangements diagram for forward and aft limits of occupant CG range.
(2) Engine Oil: 8 Qt. = 15 Lb at −0.2 Moment/1000.

NOTE: The empty weight of this airplane does not include the weight of the oil.

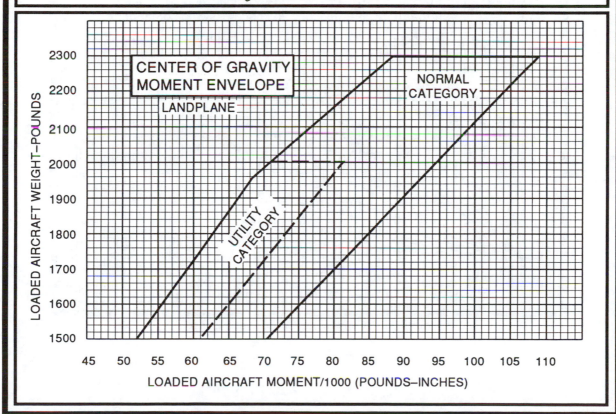

CENTER OF GRAVITY MOMENT ENVELOPE

LANDPLANE

NORMAL CATEGORY

UTILITY CATEGORY

LOADED AIRCRAFT WEIGHT—POUNDS

LOADED AIRCRAFT MOMENT/1000 (POUNDS−INCHES)

FIGURE 35.—Airplane Weight and Balance Graphs.

CRUISE POWER SETTINGS

65% MAXIMUM CONTINUOUS POWER (OR FULL THROTTLE)
2800 POUNDS

PRESS ALT.	ISA −20 °C (−36 °F)								STANDARD DAY (ISA)								ISA +20 °C (+36 °F)							
	IOAT		ENGINE SPEED	MAN. PRESS	FUEL FLOW PER ENGINE		TAS		IOAT		ENGINE SPEED	MAN. PRESS	FUEL FLOW PER ENGINE		TAS		IOAT		ENGINE SPEED	MAN. PRESS	FUEL FLOW PER ENGINE		TAS	
FEET	°F	°C	RPM	IN HG	PSI	GPH	KTS	MPH	°F	°C	RPM	IN HG	PSI	GPH	KTS	MPH	°F	°C	RPM	IN HG	PSI	GPH	KTS	MPH
SL	27	-3	2450	20.7	6.6	11.5	147	169	63	17	2450	21.2	6.6	11.5	150	173	99	37	2450	21.8	6.6	11.5	153	176
2000	19	-7	2450	20.4	6.6	11.5	149	171	55	13	2450	21.0	6.6	11.5	153	176	91	33	2450	21.5	6.6	11.5	156	180
4000	12	-11	2450	20.1	6.6	11.5	152	175	48	9	2450	20.7	6.6	11.5	156	180	84	29	2450	21.3	6.6	11.5	159	183
6000	5	-15	2450	19.8	6.6	11.5	155	178	41	5	2450	20.4	6.6	11.5	158	182	79	26	2450	21.0	6.6	11.5	161	185
8000	-2	-19	2450	19.5	6.6	11.5	157	181	36	2	2450	20.2	6.6	11.5	161	185	72	22	2450	20.8	6.6	11.5	164	189
10000	-8	-22	2450	19.2	6.6	11.5	160	184	28	-2	2450	19.9	6.6	11.5	163	188	64	18	2450	20.3	6.5	11.4	166	191
12000	-15	-26	2450	18.8	6.4	11.3	162	186	21	-6	2450	18.8	6.1	10.9	163	188	57	14	2450	18.8	5.9	10.6	163	188
14000	-22	-30	2450	17.4	5.8	10.5	159	183	14	-10	2450	17.4	5.6	10.1	160	184	50	10	2450	17.4	5.4	9.8	160	184
16000	-29	-34	2450	16.1	5.3	9.7	156	180	7	-14	2450	16.1	5.1	9.4	156	180	43	6	2450	16.1	4.9	9.1	155	178

NOTES: 1. Full throttle manifold pressure settings are approximate.
        2. Shaded area represents operation with full throttle.

FIGURE 36.—Airplane Power Setting Table.

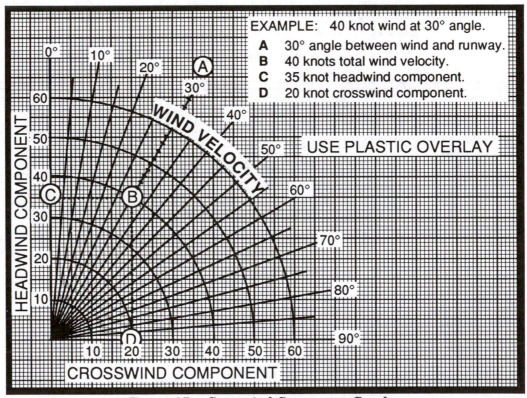

FIGURE 37.—Crosswind Component Graph.

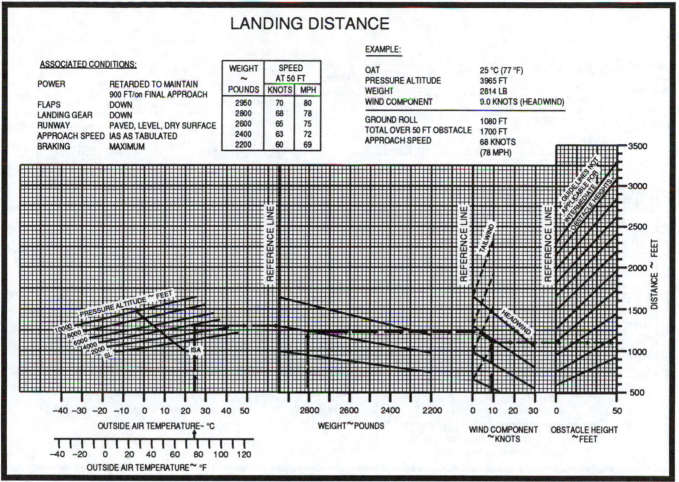

LANDING DISTANCE

ASSOCIATED CONDITIONS:

POWER	RETARDED TO MAINTAIN 900 FT/on FINAL APPROACH
FLAPS	DOWN
LANDING GEAR	DOWN
RUNWAY	PAVED, LEVEL, DRY SURFACE
APPROACH SPEED	IAS AS TABULATED
BRAKING	MAXIMUM

WEIGHT ~ POUNDS	SPEED AT 50 FT	
	KNOTS	MPH
2950	70	80
2800	68	78
2600	65	75
2400	63	72
2200	60	69

EXAMPLE:

OAT	25 °C (77 °F)
PRESSURE ALTITUDE	3965 FT
WEIGHT	2814 LB
WIND COMPONENT	9.0 KNOTS (HEADWIND)
GROUND ROLL	1080 FT
TOTAL OVER 50 FT OBSTACLE	1700 FT
APPROACH SPEED	68 KNOTS (78 MPH)

FIGURE 38.—Airplane Landing Distance Graph.

LANDING DISTANCE

FLAPS LOWERED TO 40° - POWER OFF
HARD SURFACE RUNWAY - ZERO WIND

GROSS WEIGHT LB	APPROACH SPEED, IAS, MPH	AT SEA LEVEL & 59 °F		AT 2500 FT & 50 °F		AT 5000 FT & 41 °F		AT 7500 FT & 32 °F	
		GROUND ROLL	TOTAL TO CLEAR 50 FT OBS	GROUND ROLL	TOTAL TO CLEAR 50 FT OBS	GROUND ROLL	TOTAL TO CLEAR 50 FT OBS	GROUND ROLL	TOTAL TO CLEAR 50 FT OBS
1600	60	445	1075	470	1135	495	1195	520	1255

NOTES: 1. Decrease the distances shown by 10% for each 4 knots of headwind.
2. Increase the distance by 10% for each 60 °F temperature increase above standard.
3. For operation on a dry, grass runway, increase distances (both "ground roll" and "total to clear 50 ft obstacle") by 20% of the "total to clear 50 ft obstacle" figure.

FIGURE 39.—Airplane Landing Distance Table.

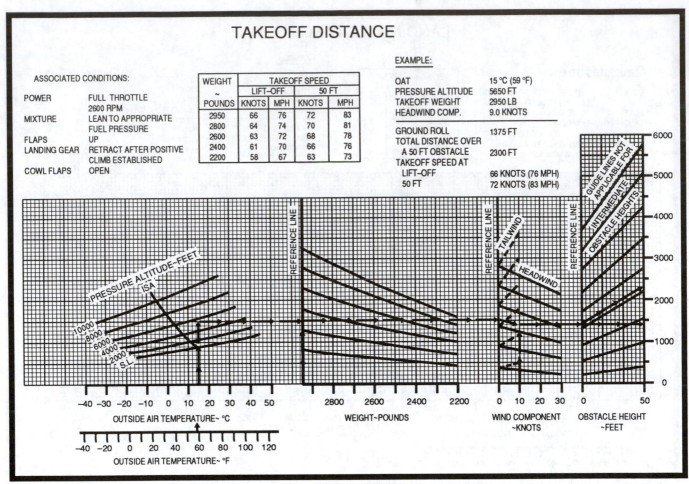

FIGURE 41.—Airplane Takeoff Distance Graph.

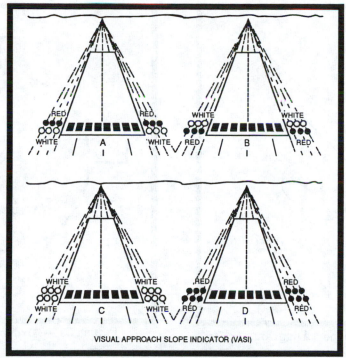

FIGURE 48.—VASI Illustrations.

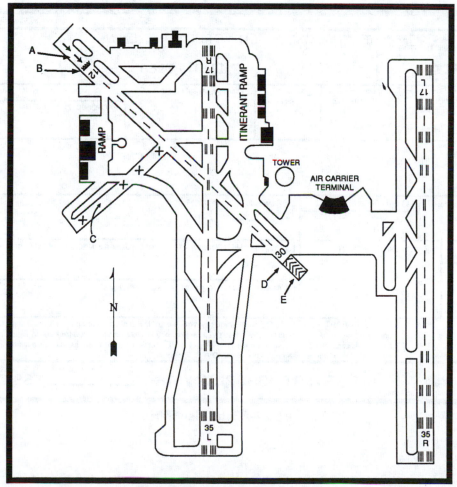

FIGURE 49.—Airport Diagram.

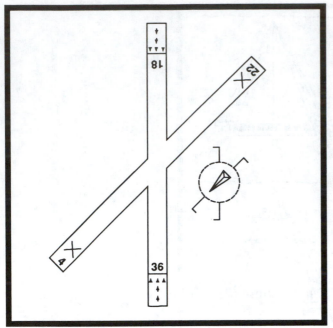

FIGURE 50.—Airport Diagram.

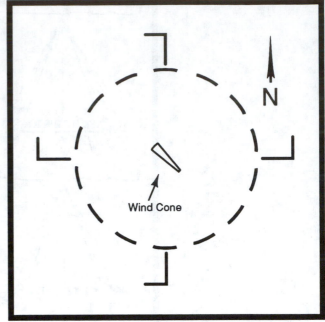

FIGURE 51.—Airport Landing Indicator.

						Form Approved: OMB No. 2120-0026
U.S. DEPARTMENT OF TRANSPORTATION FEDERAL AVIATION ADMINISTRATION	(FAA USE ONLY) ☐ PILOT BRIEFING ☐ VNR				TIME STARTED	SPECIALIST INITIALS
FLIGHT PLAN	☐ STOPOVER					

1 TYPE	2 AIRCRAFT IDENTIFICATION	3 AIRCRAFT TYPE/ SPECIAL EQUIPMENT	4 TRUE AIRSPEED	5 DEPARTURE POINT	6 DEPARTURE TIME		7 CRUISING ALTITUDE
VFR					PROPOSED (Z)	ACTUAL (Z)	
IFR							
DVFR			KTS				

8 ROUTE OF FLIGHT

9 DESTINATION (Name of airport and city)	10 EST. TIME ENROUTE		11 REMARKS
	HOURS	MINUTES	

12 FUEL ON BOARD		13 ALTERNATE AIRPORT(S)	14 PILOT'S NAME, ADDRESS & TELEPHONE NUMBER & AIRCRAFT HOME BASE	15 NUMBER ABOARD
HOURS	MINUTES			
			17 DESTINATION CONTACT/TELEPHONE (OPTIONAL)	

16 COLOR OF AIRCRAFT	CIVIL AIRCRAFT PILOTS. FAR Part 91 requires you file an IFR flight plan to operate under instrument flight rules in controlled airspace. Failure to file could result in a civil penalty not to exceed $1,000 for each violation (Section 901 of the Federal Aviation Act of 1958, as amended). Filing of a VFR flight plan is recommended as a good operating practice. See also Part 99 for requirements concerning DVFR flight plans.

FAA Form 7233-1 (8-82)   CLOSE VFR FLIGHT PLAN WITH _____ FSS ON ARRIVAL

FIGURE 52.—Flight Plan Form.

NEBRASKA

LINCOLN MUNI   (LNK)   4 NW   UTC–6(–5DT)   40°51'03"N 96°45'32"W   **OMAHA**
   1214   B   S4   FUEL 100LL, JET A   TPA—2214 (1000)   ARFF Index C   H-1E, 3F, 4F, L-11B
   RWY 17R-35L: H12901X200 (ASPH-CONC-AFSC)   S-100, D-200, DT-400   HIRL   IAP
     RWY 17R: MALSR. VASI(V4L)—GA 3.0° TCH 55'. Rgt tfc. Arresting device.
     RWY 35L: MALSR. VASI(V4L)—GA 3.0° TCH 55'. Arresting device.
   RWY 14-32: H8620X150 (ASPH-CONC-GRVD)   S-80, D-170, DT-280   MIRL
     RWY 14: REIL. VASI(4VL)—GA 3.0° TCH 48'.
     RWY 32: VASI(4VL)—GA 3.0° TCH 53'. Thld dsplcd 431'. Pole.
   RWY 17L-35R: H5500X100 (ASPH-CONC-AFSC)   S-49, D-60   HIRL   0.8% up N
     RWY 17L: VASI(V4L)—GA 3.0° TCH 33'.   RWY 35R: VASI(V4L)—GA 3.0° TCH 35'. Light standard. Rgt tfc.
   AIRPORT REMARKS: Attended continuously. Arresting barrier located 2200' in from thld 17R and 1500' in from thld
     35L. Arresting barrier in place departure end Rwy 17R-35L during military operations and approach end during
     emergencies. Airport manager advise 43000 lbs GWT single wheel Rwy 17L-35R. For MALSR Rwy 17R and 35L
     ctc Twr.; When Twr clsd MALSR Rwy 17R and 35L preset to Med intst. NOTE: See SPECIAL
     NOTICE–Simultaneous Operations on Intersecting runways.
   WEATHER DATA SOURCES: LLWAS
   COMMUNICATIONS: CTAF 118.5   ATIS 118.05   UNICOM 122.95
     COLUMBUS FSS (OLU) TF 1–800–WX–BRIEF. NOTAM FILE LNK.
     RCO 122.65 (COLUMBUS FSS)
   ®APP/DEP CON 124.0 (170°-349°) 124.8 (350°-169°) (1200-0600Z‡)
   ®MINNEAPOLIS CENTER APP/DEP CON 128.75 (0600-1200Z‡)
     TOWER 118.5 125.7 (1200-0600Z‡)   GND CON 121.9   CLNC DEL 120.7

   RADIO AIDS TO NAVIGATION: NOTAM FILE LNK. VHF/DF ctc COLUMBUS FSS
     (H) VORTACW 116.1   LNK   Chan 108   40°55'26"N 96°44'30"W   181° 4.5 NM to fld. 1370/9E
     POTTS NDB (MHW/LOM) 385   LN   40°44'50"N 96°45'44"W   355° 6.2 NM to fld. Unmonitored when twr
     clsd.
     ILS 111.1 I-OCZ Rwy 17R. MM and OM unmonitored.
     ILS 109.9 I-LNK Rwy 35L LOM POTTS NDB. MM unmonitored. LOM unmonitored when twr clsd.
   COMM/NAVAID REMARKS: Emerg frequency 121.5 not available at tower.

LOUP CITY MUNI   (NE03)   1 NW   UTC–6(–5DT)   41°17'25"N 98°59'25"W   **OMAHA**
   2070   B   FUEL 100LL   L-11B
   RWY 15-33: H3200X50 (ASPH)   S-8   LIRL
     RWY 33: Trees.
   RWY 04-22: 2100X100 (TURF)
     RWY 04: Tree.   RWY 22: Road.
   AIRPORT REMARKS: Unattended. For svc call 308-745-0328.
   COMMUNICATIONS: CTAF 122.9
     COLUMBUS FSS (OLU) TF 1–800–WX–BRIEF. NOTAM FILE OLU.
   RADIO AIDS TO NAVIGATION: NOTAM FILE OLU.
     WOLBACH (H) VORTAC 114.8   OBH   Chan 95   41°22'33"N 98°21'12"W   250° 29.3 NM to fld.
     2010/10E.

MARTIN FLD   (See SO SIOUX CITY)

McCOOK MUNI   (MCK)   2 E   UTC–6(–5DT)   40°12'23"N 100°35'29"W   **OMAHA**
   2579   B   S4   FUEL 100LL, JET A   ARFF Index Ltd.   H-2D, L-11A
   RWY 12-30: H5998X100 (CONC)   S-30, D-38   MIRL   0.6% up NW   IAP
     RWY 12: MALS. VASI(V4L)—GA 3.0° TCH 33'. Tree.   RWY 30: REIL. VASI(V4L)—GA 3.0° TCH 42'.
   RWY 03-21: H3999X75 (CONC)   S-30, D-38   MIRL
     RWY 03: VASI(V2L)—GA 3.0° TCH 26'. Rgt tfc.   RWY 21: VASI(V2L)—GA 3.0° TCH 26'.
   RWY 17-35: 1350X200 (TURF)
   AIRPORT REMARKS: Attended daylight hours. Deer on and in vicinity of arpt. Bird activity on and in vicinity of arpt.
     Arpt closed to air carrier operations with more than 30 passengers except 24 hour PPR, call arpt manager
     308–345–2022. Rwy 17–35 turf rwy rough. Avoid McCook State (abandoned) arpt 7 miles NW on the MCK
     VOR/DME 313° radial at 8.3 DME. ACTIVATE VASI Rwys 12 and 30 and MALS Rwy 12—CTAF. Control Zone
     effective 1100-0500Z‡, except holidays.
   COMMUNICATIONS: CTAF/UNICOM 122.8
     COLUMBUS FSS (OLU) TF 1–800–WX–BRIEF. NOTAM FILE MCK.
     RCO 122.6 (COLUMBUS FSS)
   ®DENVER CENTER APP/DEP CON 132.7
   RADIO AIDS TO NAVIGATION: NOTAM FILE MCK.
     (L) VORW/DME 116.5   MCK   Chan 112   40°12'14"N 100°35'38"W   at fld. 2550/11E.

FIGURE 53.—Airport/Facility Directory Excerpt.
Lincoln Municipal is in Class C airspace.

ANSWERS AND EXPLANATIONS
FOR FAA QUESTIONS

■ Introduction to Answers and Explanations—Read This!

The following answers and explanations refer to illustrations and/or descriptions used in this book, *The Student Pilot's Flight Manual* (*SPFM*). Examples:

3209. B. *SPFM*, Chapter 2, Thrust.
3212. B. *SPFM*, Figs. 9-9 and 9-21.

Other books referred to will be spelled out the first time (Aeronautical Information Manual) and initials (AIM) used thereafter.
Example:

3142. B. AIM. (No specific paragraphs for the references since they are subject to change. Check the AIM Index for the subject matter.)

Other abbreviations:

The Advanced Pilot's Flight Manual—APFM
Aviation Weather—AW
Aviation Weather Services—AWS
Federal Aviation Regulation—FAR 91.303 (for example)
Flight Training Handbook—FTH
National Transportation Safety Board Part 830—NTSB 830
Pilot's Handbook of Aeronautical Knowledge—PHAK
Airport/Facility Directory—A/FD

If there could be confusion in the reader's mind about the wrong answers for a question, the reasons *why* they are wrong will be cited. When the steps leading to the correct answer are spelled out, in many cases it will not be necessary to explain why the other two choices were wrong.

The Student Pilot's Flight Manual was written for the pilot working on the private pilot certificate for *airplanes*. Only questions on airplanes from the FAA test are answered and explained here. Questions and figures for other aircraft categories (hot air balloons, rotorcraft, blimps, and gliders) have been eliminated with a few exceptions. For instance questions 3318 through 3380 are not included here because they are not about *airplanes*. It's handled this way:

3317. Angle of attack is defined as the angle between the chord line of an airfoil and the

A.—direction of the relative wind. (The correct answer)
B.—pitch angle of an airfoil.
C.—rotor plane of rotation.

* * *

The asterisks (*'s) indicate that one or more questions not applying to airplanes and private pilots have been deleted.

Reference figures for the knowledge test have been deleted if they don't apply to airplanes (figures 10, 11, 40, 42-47, 54-58).

There are 60 questions (airplane) on the FAA test and those 60 questions will be taken from the questions here.

Too many people treat the knowledge test as a chore that must be "gotten out of the way" so that training can proceed. You should use the review of these several hundred questions as a way to improve your knowledge of flying an airplane safely.

In earlier years, the questions used on any of the CAA/FAA tests were *not* published in advance. Applicants knew only the *areas* of knowledge they would encounter when they sat down to take the test. It's believed that now, given time enough and good coaching, a person who had never even *been* to an airport could make a passing (or better) grade by studying the question books and explanations available. It is hoped that the old method of nonpublication of test questions will be reinstated.

As was noted earlier, from time to time the FAA may change the order of the three choices for the established questions. They may also change the questions themselves. The point is for you to learn the correct material and understand the principles, not to memorize the letters for the correct answers, which may now be choice C instead of the earlier A.

■ Answers and Explanations

Questions 3001–3200 concern the Federal Aviation Regulations, most of which are in the back of this book. In such cases the reference will be noted: FAR 61.3 etc.

3001. B. FAR Part 1, Definitions.

3002. B. Part 1.

3003. A. Part 1.

3004. A. Part 1.

3005. C. Part 1. You might be thrown off by answer B, which is a requirement for recent experience as noted in FAR 61.57.

3006. A. Part 1. *SPFM*, Chapter 23, Airplane Categories and Load Factors. V_A is the maneuvering speed, which decreases as the square root of the weight decrease of the airplane.

3007. A. Part 1. This is marked as the top of the white arc.

3008. A. Part 1. This V-speed is not marked on the airspeed indicator.

3009. C. This can be remembered as the "top of the Normal Operations (green arc)."

3010. A. Part 1.

3011. C. Part 1. V_x is the airspeed for *best angle* of climb (or best *gain* for a given *distance*). V_Y is the *best rate*. Remember best angle by the fact that an X has *more angles* than a Y.

3012. A. Part 1. Best rate (max altitude *gain* with *time*) is V_Y.

3013. C. FAR 91.417.

3014. A. Part 1.

3015. B. Part 1.

3016. C. FAR 61.3.

3017. C. FAR 61.3.

3018. B. FAR 61.3.

3019. C. FAR 61.3.

3020. B. FAR 61.23.

3021. C. FAR 61.23.

3022. A. FAR 61.23.

3023. A. FAR 61.23.

3024. B. FAR 61.31. Part 1: "Large aircraft means aircraft of more than 12,500 pounds maximum certificated takeoff weight."

3025. B. FAR 61.31. Note that 61.31 states that the airplane has more than 200 horsepower.

3026. C. FAR 61.31.

3027. B. FAR 61.31.

3028. C. FAR 61.56.

3029. C. FAR 61.57.

3030. A. FAR 61.57.

3031. C. FAR 61.57.

3032. C. FAR 61.57.

3033. B. FAR 61.57.

3034. A. FAR 61.57.

3035. A. FAR 61.60.

3036. B. FAR 61.69. The pilot needs 100 hours of flight time in powered aircraft.

3037. C. FAR 61.69.

* * *

3064. B. FAR 61.113.

3065. B. FAR 61.113.

3066. B. FAR 61.113.

3067. A. FAR 71.75. A Federal Airway is 8 NM wide or extends 4 NM each side of the center line.

3068. B. FAR 71.75. That Regulation states, "Each Federal airway includes that airspace extending upward from 1,200 feet above the surface of the earth to, but not including, 18,000 feet MSL, except that Federal airways for Hawaii have no upper limits." (B is the correct answer because it is the rule rather than the exception for Hawaii.)

3069. B. FAR 91.55. The visibility minimum for Class D airspace is 3 statute miles and the minimum ceiling is 1000 feet. These minimums also apply for *normal* (no SVFR) operations in Class B, C, and E airspace.

3070. B. FAR 91.3. That's what it says, although it seems that more and more A is becoming the answer.

* * *

3072. B. FAR 91.3.

3073. C. FAR 91.3

3074. B. FAR 91.3.

3075. B. FAR 91.9.

3076. B. FAR 91.15. When you get the private certificate, you may want to push that obnoxious passenger who asks embarrassing (and sometimes unanswerable) questions such as "Where are we?" out of the airplane. *Don't.* You might injure people or damage property on the ground—and violate FAR 91.15.

3077. A. FAR 91.17. Don't be like the late Horatio Botts Finsterhoffer who thought that it was "twenty minutes from bottle to throttle."

3078. A. FAR 91.17.

3079. C. FAR 91.17.

3080. B. FAR 91.103.

3081. C. FAR 91.103.

3082. C. FAR 91.103.

3083. A. FAR 91.105.

3084. C. FAR 91.105.

3085. B. FAR 91.107.

3086. A. FAR 91.107.

3087. B. FAR 91.107. Okay, okay! Enough already about safety belts. If you make your passengers (and yourself) secure their safety belts and shoulder harnesses before you start the engine and keep them that way throughout the flight and until the airplane is stopped and shutdown is complete, you won't have to worry about all this.

3088. C. FAR 91.111.

3089. B. FAR 91.113. An aircraft in distress has the right of way over all other air traffic. This is the FARs, but what about that guy or gal in the hot air balloon who happens to be sitting there right on final when you need to get onto the runway with a sick or failed engine? (Of course, balloons would be *soft* to hit.)

3090. B. FAR 91.113. The airplane on the other's left shall give way to the right, or the aircraft to the other's right shall have the right of way.

3091. A. FAR 91.113.

3092. A. FAR 91.113.

3093. B. FAR 91.113. The other two choices are engine-driven, so the aircraft towing other aircraft has the right of way.

3094. C. FAR 91.113. Both aircraft shall give way to the right.

3095. C. FAR 91.113.

3096. B. FAR 91.115(b). Right-of-way rules: *Water operations.* "Crossing: When aircraft, or an aircraft and a vessel, are on crossing courses, the aircraft or vessel to the other's right has the right of way."

3097. B. FAR 91.117.

3098. C. FAR 91.117.

3099. A. FAR 91.117.

3100. B. FAR 91.117.

3101. A. FAR 91.119.

3102. C. FAR 91.119.

3103. B. FAR 91.119.

3104. A. FAR 91.119.

3105. B. FAR 91.121.

3106. A. FAR 91.121.

3107. B. FAR 91.121.

3108. B. FAR 91.123.

3109. A. FAR 91.123.

3110. B. FAR 91.123.

3111. A. *SPFM*, Chapter 21, Light Signals, also FAR 91.125.

3112. A. *SPFM*, Chapter 21, Light Signals, also FAR 91.125.

3113. B. *SPFM*, Chapter 21, Light Signals, also FAR 91.125.

3114. C. *SPFM*, Chapter 21, Light Signals, also FAR 91.125.

3115. B. *SPFM*, Chapter 21, Light Signals, also FAR 91.125.

3116. B. *SPFM*, Chapter 21, Light Signals, also FAR 91.125.

3117. C. *SPFM*. Check the chart legend (Legend 1).

3118. B. FAR, Part 1—Definitions. If the control tower is not in operation, Class D airspace does not exist.

3119. A. FAR 91.129.

3120. B. FAR 91.129.

3121. B. FAR 91.129.

3122. B. FAR 91.129.

3123. C. FAR 91.127.

3124. A. FAR 91.130.

3125. C. FARs 91.130 and 91.215.

3126. B. FARs 91.13 (b)(i) and (ii) and 61.95(a)(2).

3127. A. FARs 91.131 and 61.95.

3128. B. FAR 91.131.

3129. A. FAR 91.215.

3130. A. Class A airspace is that airspace at or above 18,000 feet MSL, and no VFR is allowed. It's doubtful that at this stage of your flying you will be operating up there.

3131. B. FAR 91.151.

3132. C. FAR 91.151.

* * *

3136. C. FAR 91.155.

3137. A. FAR 91.155. Again, Class G airspace below 700 feet or 1200 feet AGL, as applicable, is uncontrolled.

3138. B. FAR 91.155.

3139. B. FAR 91.155.

3140. B. FAR 91.155.

3141. A. FAR 91.155.

3142. B. FAR 91.155. Student pilots cannot fly with less than 5 statute miles visibility at night (FAR 61.89).

3143. A. FAR 91.155.

3144. A. FAR 91.155.

3145. C. FAR 91.155.

3146. C. FAR 91.155.

3147. B. FAR 91.155.

3148. C. FAR 91.155.

3149. B. FAR 91.155.

3150. B. FAR 91.157.

3151. A. FAR 91.157.

* * *

3153. C. FAR 91.157.

3154. B. FAR 91.157.

3155. C. FAR 91.159.

3156. C. FAR 91.159.

3157. B. FAR 91.159.

3158. B. FAR 91.159. Remember, it's the magnetic course, not magnetic heading, that 91.159 cites.

3159. C. FAR 91.203.

3160. B. FAR 91.207.

3161. B. FAR 91.207.

3162. C. FAR 91.209.

3163. C. FAR 91.211.

3164. C. FAR 91.211.

3165. A. FAR 91.215.

3166. C. FAR 91.215.

3167. B. FAR 91.303.

3168. A. FAR 91.303.

3169. B. FAR 91.303.

3170. A. FAR 91.303.

3171. C. FAR 91.307.

3172. A. FAR 91.307.

3173. B. FAR 91.307.

* * *

3178. B. FAR 91.313.

3179. B. FAR 91.319.

3180. B. FAR 91.403.

3181. A. FAR 91.405.

3182. B. FAR 91.407.

3183. B. FAR 91.407.

3184. B. FAR 91.407.

3185. C. FAR 91.409. Remember *any* FAA requirements citing *calendar months* is valid to the end of that month cited.

3186. C. FAR 91.407 and 91.409. The Airworthiness Certificate is valid as long as the aircraft is maintained in accordance with FARs. See question 3187.

3187. C. *SPFM*, Fig. 3-27, the Standard Airworthiness Certificate. It's hard to read but item 6, Terms and Conditions, states, "Unless sooner surrendered, suspended, revoked or a termination date is otherwise established by the Administrator, this airworthiness certificate is effective as long as the maintenance, preventive maintenance and alterations are performed in accordance with Parts 21, 43 and 91 of the Federal Aviation Regulations as appropriate and the aircraft is registered in the United States."

3188. A. FAR 91.409. *Every* airplane must have a valid annual inspection date. Those used for hire or instruction must have a 100-hour inspection. *SPFM*, Chapter 3, Logbooks.

3189. B. FAR 91.409. It's assumed that the 10-hour grace period cited in there is not a factor in this question.

3190. B. FAR 91.409. Sorry, but the extra 7 hours has no effect on the next required 100-hour period (3302.5–3402.5 hours).

3191. C. FAR 91.413.

3192. C. FAR 91.413.

3193. A. FAR 91.417.

Questions 3194–3200 concern NTSB reports.

3194. A. NTSB 830.

3195. C. NTSB 830.

3196. B. NTSB 830.

3197. A. NTSB 830.

3198. B. NTSB 830.

3199. C. NTSB 830.

3200. C. NTSB 830.

Questions 3201–3317 concern the principles of flight.

3201. A. Read *SPFM*, Chap. 2, with particular attention to Fig. 2-1.

3202. A. When the airplane is in equilibrium the forces acting parallel to the flight path (thrust and drag) must be balanced, and the forces acting perpendicular to the flight path (lift and the component of weight) must cancel each other out. There can be some pretty complicated analyses of the four forces at different phases of flight, but answer A says it simply and best.
 B. *No.* The answer can't be this because if the aircraft is accelerating, things aren't in equilibrium.
 C. *No.* This isn't the answer because the only force acting there is weight.

3203. B. The angle of attack is that angle between the relative wind and the airfoil chord line. See angle of attack. *SPFM*, Chapter 2.
 A. Wrong because the angle of incidence is a *fixed* angle between the fuselage reference line and the wing chord line (Fig. 2-5).
 C. Dihedral is the fixed angle of the wings, as shown in Fig. 27-1, so C isn't the answer.

3204. A. *SPFM*, Chapter 2 and Fig. 2-7.
 B. Not right because Fig. 2-7 and the accompanying text describe the climb angle. The description of answer C is that of the angle of incidence (Fig. 2-5).

3205. A. *SPFM*, Fig. 2-1.
 B. *No.* This answer would mean a state of strange flying, for sure!
 C. *No.* Lift and/or weight will have values 3–10 times those of thrust and/or drag for the average light airplane, depending on the make and model and airspeed regime.

3206. A. *SPFM*, Chapter 22, Frost. Frost won't change the camber of the wing. Answers B and C are the reverse of the truth of aerodynamics.

3207. A. *SPFM*, Chapter 2, Torque. Look at Fig. 2-12. Also see Chapter 9, The Normal Climb.
 B. Wrong because of the "low power and low angle of attack."
 C. *Nope.* This answer is shot down by mentioning "high airspeed."

3208. B. *SPFM*, Chapter 2. This question is for U.S. powered airplanes (prop turning clockwise as seen from the cockpit).
 A. *No.* That's "equal and opposite reaction," as covered in SPFM, Chapter 2.
 C. *No.* That's not P-factor. See *SPFM*, Chapter 2.

3209. B. Asymmetric loading (P-factor) acts at slow speeds (high angles of attack), as described in *SPFM*, Chapter 2.

3210. A. An inherently stable airplane will want to stay where it is trimmed and will require an effort to stall.
 B. *No.* The inherently stable airplane will require *more* effort to control (to *maneuver*). See *APFM*, Chap. 10. This choice is ambiguous. Is "to control" a term for ability "to maneuver"?
 C. *No.* An airplane can be *forced* to spin even if it is inherently stable.

3211. A. A forward CG means a more longitudinally stable airplane; as the CG is moved aft the airplane becomes less stable, so its location determines the stability. *SPFM*, Chapter 23, Weight and Balance.
 B. *No.* The rudder and rudder trim tab aren't part of the longitudinal system.
 C. *No.* That would be more of a *performance* function.

3212. B. *SPFM*, Figs. 9-9 and 9-21.
 A. *No.* The CG can only be affected by an actual weight shift.
 C. *No.* The airplane rotates or pitches down when power is reduced. Under the conditions cited here, the airplane would be accelerating downward. Besides, for general aviation airplanes, thrust is *always* less than weight.

3213. A. *SPFM*, Fig. 8-4.
 B. *No.* The rudder isn't used for that.
 C. *No.* Sometimes rudder is used in some older airplanes to control wing drop or roll but that's not the answer here. See *SPFM*, Chapter 12.

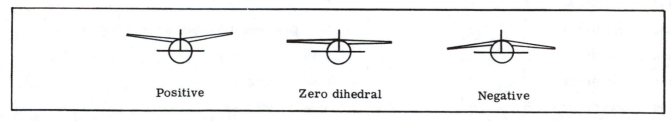

Positive　　　　Zero dihedral　　　　Negative

Fig. 27-1. Dihedral. (From *The Advanced Pilot's Flight Manual*)

3214. C. Looking at figure 2, the Load Factor Chart, in and moving up to intercept the load factor line and then moving across to the g-unit value, you'll see that 2 g's are occurring. To find the stall speed multiplier at various load factors, use Fig. 9-4. The stall speed goes up as the *square root* of the load factor.

   A. *No.* That would be the value for 1-g flight and this can't be done while making a balanced turn at a constant altitude.

   B. *No.* Look at *SPFM*, Fig. 9-4, which shows that there will be an *increase* in stall speed by a factor of about 1.4 and *wrongly* using this multiplier, it would work out to 3250 pounds of support required.

3215. C. Referring to figure 2 here, it looks as if at a 30° bank the load factor is increased to about 1.15 that of level flight, so that $1.15 \times 3300 = 3795$ pounds. Working it out with a computer for the 30° bank and using trigonometry, the actual multiplier is 1.1547 and the answer is 3810 pounds. The answer of 3800 is the closest to this.

   A. *No.* The airplane in a constant altitude balanced turn will always "weigh" more than in level flight, so 3100 is too low.

   B. *No.* The same reasoning as A. This answer of 1200 pounds is too low.

3216. B. The multiplier as shown by figure 2 is approximately 1.4 (mathematically solving for the bank angle, the answer is 1.414), and this multiplied by the weight of 4500 pounds would be 6363 pounds.

   A. *No.* As noted for the last question, the airplane couldn't make the 45° banked turn with only 4500 pounds (its weight) imposed on it.

   C. *Nope.* 7200 pounds is too high.

3217. B. The faster the airplane, the more load can be put on it by a zealous and strong pilot. See *SPFM*, Chapter 23, Airplane Categories and Load Factors.

   A. *No.* An aft position of the CG can cause the stick or wheel forces to be so light that the pilot could pull more g's than he intended to, but damage would only occur if the airspeed is above the maneuvering speed V_A *for its particular weight.*

   C. *No.* It doesn't matter how abruptly the controls are moved if the airplane is below the maneuvering speed V_A *for its particular weight.*

3218. B. Turns. *SPFM*, Fig. 9-3. Load Factors in the Turn. *Climbs and stalls* are imposing approximately 1g on the pilot (except if, in getting the nose down to recover from a stall, the pilot makes an enthusiastic recovery moving to *less* than 1 g and is getting light in the seat.

3219. C. The angle of descent is steeper for a given airspeed with flaps. Another function is that they also lower the stall speed. See *SPFM*, Fig. 9-5 and Chapter 13, Flaps.

   A. *No.* The angle of descent is *increased* as just mentioned.

   B. *No.* The stall or touchdown is at a *lower* indicated airspeed.

3220. A. This is correct and is about the same answer as C in 3219.

   B. *No.* Trim does that.

   C. *No.* The wing area is *increased* for some types (Fowler flaps).

3221. B. *PHAK*, 1980 edition, indicates that this is indeed the case.

   A. This can be a long-term result of long-term lack of proper cooling and in addition some of the ignition leads on rear cylinders may be damaged, but B is the proper answer here.

   C. *No.* Excessively high oil and cylinder head temperatures definitely affect an aircraft engine.

3222. C. A too-high power setting and a too-lean mixture will cause high oil and cylinder head temps and the pilot should richen the mixture and get more air moving across the cylinders and oil cooler radiator by leveling off if in a climb. See *SPFM*, App. D, for some background.

   A. *No.* A too-rich mixture might cause rough running under some conditions but is not normally a contributor to high oil and cylinder head temps.

   B. *No.* A higher-than-normal oil pressure could cause other problems, but not the ones cited.

3223. A. *SPFM*, Chapters 4 and 7.

   B. *No.* Two magnetos help in uniform burning in the cylinder(s) but not for engine heat distribution.

   C. *No.* While the more even fuel burning distribution in the cylinder because of two magnetos *might* make a more balanced brake mean effective pressure in the cylinder(s), this is not the best answer.

3224. A. This is probably the "right" answer, but reading most *Pilot Operating Handbook* procedures, you'll see that if a tank is run dry, the selector is switched to a "good tank" and *then* the boost pump turned on to facilitate the new fuel flow.

   B. This *could* happen, but only after a long period of running for *some* systems.

   C. This *could* happen too, particularly with submerged pumps and much junk on the bottom of the tank, but like B, it is highly unlikely. Remember, the *Pilot's Operating Handbook* procedures for a particular airplane have precedence over the general suggestions as given in textbooks or these questions.

3225. B. *SPFM*, Chapter 4, and App. D, Carburetion. As the air moves from the air inlet through the venturi, the pressure drops, controlling the metering of the fuel/air mixture.

   A. *No.* A mixture control has to be used to "meter the air" at the proper ratio of fuel/air mixture. See question 3226.

   C. *No.* The increase in velocity *decreases* the sidewise air pressure in the throat of the venturi.

3226. B. *SPFM*, Chapter 5, Mixture Control.
   A. *No.* You *don't* decrease the amount of fuel to compensate for *increased* air density, but instead for a decreased air density.
   C. *No.* You don't want to *increase* the amount of fuel in the mixture at altitude. This would be rich indeed.

3227. A. The combination of rich mixture *and* application of carburetor heat (which richens the mixture further) at a high-altitude airport can mean a rich mixture indeed. See *SPFM*, Chapter 7 and Chapter 23, Leaning the Mixture.
   B. *No.* Try leaning the mixture and running the checks before taxiing back for maintenance. If the leaning doesn't help, taxi back for maintenance. Your instructor will demonstrate the procedure.
   C. *No.* If the mixture is overly rich, detonation is not likely under these circumstances. Besides, maybe your airplane doesn't have a manifold pressure gauge.

3228. A. *SPFM*, Chapter 5, Mixture Control, and Chapter 23, Learning the Mixture.
   B. *No.* The opposite is the situation. The mixture is too lean. (See the reference for A.)
   C. No. Again, the mixture will be excessively lean.

3229. C. *SPFM*, Chapter 4 and Chapter 25, Engine Problems, last paragraph. It ain't necessarily the heat (or lack of it), it's the humidity. The float carburetor acts as a refrigerator and the moist air coming in gets a good temperature drop. The conditions cited in this answer choice are very good (or bad) for icing conditions, but a range of 50–70°F is more realistic for carb ice.
   A. *No.* A low temperature and low relative humidity are not the problem. This condition is usually best for *not* getting carb ice.
   B. *No.* The lower temperature and low humidity are less apt to cause carb ice than choice C.

3230. A. Same reason as in C in question 3229. Note the use of the word "even."
   B. *No.* This temperature is a little high. The carburetor is less likely to get ice than in A.
   C. *No.* The humidity is *not* going to be high at 0°F.

3231. C. *SPFM*, Chapter 4.
   A. *No.* This is *not* an indication of carb ice.
   B. *No.* Usually, at least in the early stages, the decrease in rpm is smooth and subtle. See the reference above.

3232. B. *SPFM*, Chapter 4.
   A. *No.* Not more air, but *unfiltered* air. *SPFM*, Chapter 7.
   C. *No.* Just the opposite.

3233. B. *SPFM*, Chapters 4 and 7.
   The fuel/air mixture becomes richer.
   A. *No.* The mixture is *richened* when carb heat is applied.
   C. *No.* Same reason as A.

3234. A. *SPFM*, Chapters 4 and 7. There is a loss of power with carb heat on, and you know that on a go-around, after full throttle is applied, the carb heat must be pushed OFF to get the max power available.

3235. C. *SPFM*, Chapter 7. Sometimes the rpm increase is dramatic.
   A. *No.* Just the opposite.
   B. *No.* The first part of the answer is okay but the last half isn't.

3236. A. *SPFM*, Chapter 7, and App. D, Fuel Injection.
   B. *No.* As discussed in the references.
   C. *No.* It's that *invisible* moisture that can also cause carb ice problems.

3237. C. *SPFM*, Chapter 23, Fuel and Oil Information, and APP. D, see *detonation* under Spark Plugs. Don't use a grade of fuel below that recommended. It could cause detonation.
   A. *No.* The fuel grade doesn't have any effect on fuel distribution.
   B. *No.* Just the opposite.

3238. C. *SPFM*, App. D, see *detonation* under Spark Plugs.
   A. *No.* Like App. D says in detonation, the burning is more like a hammer blow than a smooth push.
   B. *No.* Dagnabit, that's *preignition*. (See *SPFM*, App. D, *preignition* under Spark Plugs.)

3239. B. *SPFM*, App. D, *detonation* under Spark Plugs. This would be one corrective action. Richening the mixture is another.
   A. *No.* For Pete's sake, *don't* lean the mixture. That's one cause of detonation.
   C. *No.* Don't do *that* either. Don't climb with carb heat ON. Engine temps could get too high.

3240. B. *SPFM*, App. D, *preignition* under Spark Plugs. That's *preignition* (makes sense).
   A. *No.* Well, there's combustion all right, but it's too soon. As the young woman said to her boyfriend in church, "This ain't the time or the place."
   C. *No. Detonation* was covered in the explanations for questions 3237–3239.

3241. A. *SPFM*, Chapter 23. Fuel and Oil Information. A *too-low* fuel rating can cause detonation *and* the problem cited in this question.
   B. *No.* This could cause spark plug fouling if the engine is operated at low rpm at length, but not the symptoms described.
   C. *No.* This should have no effect on the temperatures listed.

3242. A. *SPFM*, Chapter 23, Fuel and Oil Information. The next higher octane is the *only* substitute if the recommended octane is not available, but it should be a temporary arrangement.
- B. *No.* This could invite detonation and/or the problems in question 3241.
- C. *No.* If your airplane does not have a Supplemental Type Certificate (STC) for automotive fuel, stay away from it. Your airplane engine might get indigestion.

3243. C. *SPFM*, Chapter 4.
- A. *No.* Water will go to the *bottom* of the tank and to the bottom of the fuel checker.
- B. *No.* A liquid such as fuel has little expansion properties, but as the day heats up, the fuel would dump overboard if the tanks were full.

3244. C. *SPFM*, App. D, Oil System, last paragraph. Approximately 5–10 percent of the heat is removed by the oil. This is the best answer of the 3 choices.
- A. *No.* What kind of thermostat? Oil? Water?
- B. *No.* By the time the burning (and burned) mixture gets to the exhaust manifold, the engine has already had the hots.

3245. A. If oil is important in engine cooling, a too-low-level oil volume would hold heat and contribute to overtemp. In hot weather flying, be sure that the oil level is kept up high enough to help carry off the heat.
- B. *No.* Though this might contribute somewhat to higher operating temperatures.
- C. *No.* A rich mixture helps cool the oil and cylinder head temps.

3246. B. *SPFM*, Chapter 23, Setting Power. Propeller efficiency is reduced because of the less dense air. To get the same horsepower for an engine with a fixed prop, the power must be increased by approximately 25 rpm per 1000 feet of density-altitude. The "thinner" air requires more revs to get the same amount of thrust (horsepower). See the accompanying cruise control chart (Fig. 27-2) that shows this effect.
- A. *No.* Efficiency is not increased.
- C. *No.* Efficiency is reduced, but not for the reasons cited.

3247. B. *SPFM*, Fig. 3-5.
- A. *No. SPFM.* Turn and Slip Indicator, and Fig. 3-17. The turn and slip indicator is not part of the pitot-static system. It may be electrical or a vacuum/pressure pump system.
- C. *No.* The attitude indicator and turn and slip may be electrically or vacuum-driven and are not part of the pitot-static system, which is "self-contained."

3248. C. *SPFM*, Fig. 3-5. The airspeed indicator is the only instrument that depends on the pitot tube.

3249. C. *SPFM*, Fig. 3-5.
- A. *No.* The airspeed and vertical speed also depend on the static system (Fig. 3-5).
- B. *No. SPFM*, Fig. 3-5.

3250. C. Looking at altimeter 1, the small (10,000 foot hand) indicates that the altitude is above that number. The 1000 foot hand shows approximately another 500 feet and the 100 foot hand confirms it.

3251. C. Altimeter 2 has the small (10,000 foot hand) above that number, with the 1000 foot hand indicating between 4000 and 5000 feet and the 100 foot hand confirming the 500 feet value, or summed up, 10,000 + 4000 + 500 = 14,500 feet.
- A. *No.* If you thought the small 10,000 foot hand was the 1000 foot hand, and ignored the real 1000 foot hand, this might seem like a logical choice (1500 feet).
- B. *No.* You would have to ignore the 10,000 foot hand to get this answer.

CRUISE PERFORMANCE

CONDITIONS:
1600 Pounds
Recommended Lean Mixture

PRESSURE ALTITUDE	RPM	20°C BELOW STANDARD TEMP			STANDARD TEMPERATURE			20°C ABOVE STANDARD TEMP		
		% BHP	KTAS	GPH	% BHP	KTAS	GPH	% BHP	KTAS	GPH
2000	2650	- - -	- - -	- - -	78	103	5.9	72	102	5.4
	2600	80	102	6.0	73	101	5.5	68	100	5.1
	2500	70	97	5.3	65	96	4.9	60	95	4.6
	2400	62	92	4.7	57	91	4.3	53	91	4.1
	2300	54	87	4.1	50	87	3.9	47	86	3.7
	2200	47	83	3.7	44	82	3.5	42	81	3.3
4000	2700	- - -	- - -	- - -	78	105	5.8	72	104	5.4
	2600	75	101	5.6	69	100	5.2	64	99	4.8
	2500	66	96	5.0	61	95	4.6	57	95	4.3
	2400	58	91	4.4	54	91	4.1	50	90	3.9
	2300	51	87	3.9	48	86	3.7	45	85	3.5
	2200	45	82	3.5	42	81	3.3	40	80	3.2
6000	2750	- - -	- - -	- - -	77	107	5.8	71	105	5.3
	2700	79	105	5.9	73	104	5.4	67	103	5.1
	2600	70	100	5.2	64	99	4.8	60	98	4.5
	2500	62	95	4.7	57	95	4.3	53	94	4.1
	2400	54	91	4.2	51	90	3.9	48	89	3.7
	2300	48	86	3.7	45	85	3.5	42	84	3.4
8000	2700	74	104	5.5	68	103	5.1	63	102	4.8
	2600	65	99	4.9	60	99	4.6	57	98	4.3
	2500	58	95	4.4	54	94	4.1	51	93	3.9
	2400	52	90	4.0	48	89	3.7	45	88	3.5
	2300	46	85	3.6	43	84	3.4	40	82	3.2
10000	2700	69	103	5.2	64	102	4.8	59	102	4.5
	2600	61	99	4.6	57	98	4.3	53	97	4.1
	2500	55	94	4.2	51	93	3.9	48	92	3.7
	2400	49	89	3.8	45	88	3.6	43	87	3.4
12000	2650	61	100	4.6	57	99	4.3	53	98	4.1
	2600	58	98	4.4	54	97	4.1	50	96	3.9
	2500	52	93	4.0	48	92	3.7	45	91	3.5
	2400	46	89	3.6	43	87	3.4	41	84	3.3

Fig. 27-2. Cruise Performance Chart. (From *The Student Pilot's Ground School Manual*)

3252. A. Altimeter 3 indicates 9500 feet. Note that the (small) 10,000 foot hand is close to the "1" (10,000 feet), the 1000 foot hand is between 9000 and 10,000 feet, and the 100 foot hand is at 500 feet; therefore 9500 feet.

3253. B. Altimeter 1 indicates 10,500 feet and altimeter 2 indicates 14,500 feet.
   A. *No.* Altimeter 3 indicates 9500 feet.
   C. *No.* See B above.

3254. C. *True* altitude is the height above sea level and *if* the setting given is correct and *if* the altimeter is right on the money, the field elevation (true altitude) should be indicated.
   A. *No.* Assume that the altimeter is correctly calibrated, as indicated in C.
   B. *No.* The *absolute* altitude is the height *above* the ground. (Maybe 6 feet here? It depends on the size of the airplane, I reckon.)

3255. A. One way to remember the temperature effects is the following. HALT (*High Altimeter* because of *Low Temperature*)—HALT, you may run into something. *SPFM*, Chapter 3, Altimeter. As the illustration shows, terrain clearance might be a problem in IMC conditions. The altimeter would read *low* in higher-than-normal temperatures. See Figs. 19-12 and 27-3.

3256. A. *SPFM*, Fig. 3-3. *True altitude is the vertical distance above mean sea level.*
   B. *No.* See p. 15; this choice is *absolute* altitude.
   C. *No. SPFM*, Fig. 3-3.

3257. B. *SPFM*, Fig. 3-3. *Absolute altitude is the vertical distance of the airplane above the surface.*
   A. *No. SPFM*, Chapter 3, Altimeter. That's indicated altitude.
   C. *No. SPFM*, Fig. 3-3.

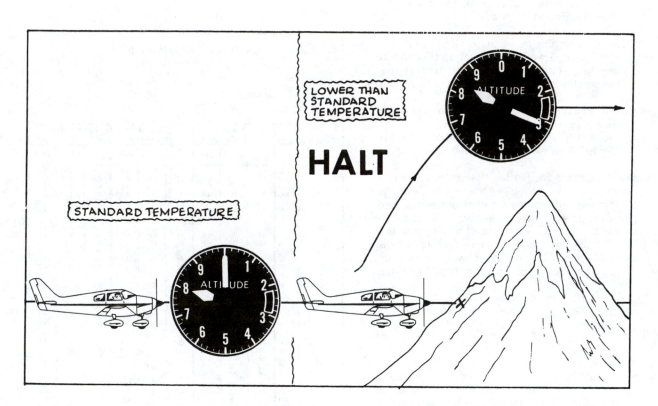

Fig. 27-3. HALT. There is a *High Altimeter* because of *Low Temperature*. The altimeter shows that the airplane could clear that mountain when actually it wouldn't. This will be of even more importance when you start working on that instrument rating and are concerned about terrain clearance when on solid instruments.

3258. B. *SPFM,* Chapter 3, Altimeter, and Fig. 19-11.
   A. *No.* Check *SPFM,* Fig. 3-3, again.
   C. *No.* Dagnabit! That's *indicated* altitude. *SPFM,* Chapter 3, Altimeter.

3259. B. *SPFM,* Chapter 3, Altimeter.
   A. *No.* See reference above.
   C. *No.* This sounds like an attempt to get the density-altitude, but you use *pressure* (not indicated) *altitude* for that.

3260. B. This is the best choice, assuming that the altimeter *setting* obtained is correct.
   A. *No.* Non-standard conditions could be existing even if the altimeter is mechanically correct.
   C. *No.* This is a required altimeter setting for operations at or above 18,000 feet MSL. AIM.

3261. C. For each inch of barometric pressure change, the altimeter is affected by approximately 1000 feet at lower altitudes. If you reset the altimeter from 29.15 to 29.85, you would *increase* the setting by 0.70 (seventy-hundredths) of an inch, or the altimeter reading would *increase* by 700 feet. (*Increasing* the window setting *increases* the altitude reading and vice versa.) If the altimeter was not corrected when flying from the low to the higher pressure area just cited, the altimeter would indicate 700 feet *lower* than the actual altitude. (See Fig. 27-4 for a high-pressure to low-pressure situation.)

3262. C. *SPFM,* Fig. 3-5. Note that the altimeter and vertical speed indicator use only the *static* system.

3263. C. *SPFM,* Chapter 3, Airspeed Indicator.

3264. C. *SPFM,* Figs. 3-28 and 3-29.

3265. A. SPFM, Figs. 3-28 and 3-29.
   B. *No.* Good grief! At 208 mph, the flaps will have long departed the airplane.
   C. *No.* This doesn't compute, Captain Kirk.

3266. C. *SPFM,* Figs. 3-28 and 3-29.
   A. *No.* That airplane can't fly in the range of 0-60 mph.
   B. *No.* That's part of the green arc (normal operations).

3267. C. *SPFM,* Figs. 3-28 and 3-29 and Chapter 23, Airplane Categories and Load Factors.
   A. *No.* That's the maximum flaps-extended speed. *SPFM,* Figs. 3-28 and 3-29.
   B. *No.* That airspeed (165 mph) is the maximum structural cruising speed, V$_{NO}$.

3268. C. The red line is V$_{NE}$ *SPFM,* Figs. 3-28 and 3-29.
   A. *No.* That's V$_{NO}$.
   B. *No.* That's V$_{FE}$, the max flap-extended speed.

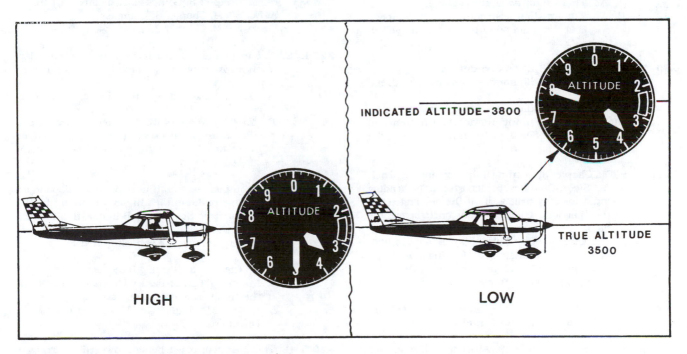

Fig. 27-4. Flying from a high-pressure area to a low-pressure area without resetting the altimeter results in the altimeter indicating erroneously high.

3269. C. The FARs refer to that as V_{S1} and it's usually the power-off flaps-up clean condition, but not always. (That's why a "specified condition" is included in FAR Part 1, Abbreviations.)
  A. *No.* That's V_{NO} where the green and yellow arcs meet.
  B. *No.* That's V_{FE}.

3270. B. The top of the white arc (100 mph here) is V_{FE}.
  A. *No.* That's V_{S1}. (See question 3269.)
  C. *No.* That's V_{NO}.

3271. C. *SPFM*, Figs. 3-28 and 3-29.
  A. *No.* V_{FE} is 100 mph, so if the flaps were extended above that, structural damage could occur.
  B. *No.* The green arc is the normal (clean) operating range, V_{S1} to V_{NO}.

3272. C. *SPFM*, Figs. 3-28 and 3-29. That's V_{SO}.
  A. *No.* That's V_{NO}.
  B. *No.* That's V_{FE}.

3273. B. *SPFM*, Figs. 3-28 and 3-29. The top of the green arc is V_{NO}, the maximum structural cruising speed (Normal Operations).
  A. *No.* That's V_{FE}.
  C. *No.* That's V_{NE}.

3274. C. The maneuvering speed V_A varies with the square root of the weight change. *SPFM*, Chapter 23, Airplane Categories and Load Factors.
  A. *No.* That's *marked* as the red line.
  B. *No.* The maximum structural cruising speed V_{NO} is at the junction of the green and yellow arcs.

3275. A. *SPFM*, Chapter 3, Turn Coordinator.
  B. *No.* The turn coordinator does not directly indicate bank.
  C. *No.* The turn coordinator gives no information on pitch (or attitude of the aircraft with reference to the lateral axis). See the note at the bottom of the instrument in figure 5.

3276. C. *SPFM*, Chapter 3, Heading Indicator, and Fig. 25-1.
  A. *No. SPFM*, Chapter 25, Departure. The heading indicator may drift well off the heading set to the magnetic compass only once *before* the flight.
  B. *No.* That wouldn't help because, as was indicated in *SPFM*, Chapter 3, Heading Indicator, it "has no brain" and wouldn't remember" any realignment on a compass rose.

3277. C. When the airplane is flying straight and level, the miniature airplane should be aligned with the horizon bar.
  A. *No. SPFM*, Chapter 3, Attitude Indicator. You can't mechanically align the horizon bar for most modern instruments. Older instruments could be "caged" to set up the proper alignment (straight and level) and then be uncaged.
  B. *No.* Again, you don't adjust the horizon bar to the airplane, but the other way around.

3278. C. The little airplane is *your* airplane and its relationship to the horizon bar shows you the *direction* of bank, just as in the case with visual reference to the horizon.
  A. *No.* The banking scale works backward here and should be used to check the *angle* of bank.
  B. *No.* You'd have to think backward to use this concept.

3279. C. *SPFM*, Chapter 19, Deviation.
  A. *No.* This is not the case. See answer C.
  B. *No.* That's Variation that can distort the earth's lines of magnetic force. *SPFM*, p.163, Variation.

3280. B. *SPFM*, Chapter 3, Compass.
  A. *No.* A left turn with the compass on a heading of north would indicate an easterly heading. (It goes backward initially under the conditions cited.)
  C. *No. SPFM*, p. 19, col. 1, para. 4. There is no noticeable effect on the compass in accelerations/decelerations on north or south headings.

3281. C. *SPFM*, Chapter 3, Compass. See answer A in question 3280.

3282. C. *SPFM*, Chapter 3, Compass—remember ANDS there.

3283. C. *SPFM*, Chapter 3, Compass. Remember ANDS.
  A. *No. SPFM*, Chapter 3, Compass.
  B. *No. SPFM*, Chapter 3, Compass.

3284. B. The magnetic compass has no noticeable acceleration error when the airplane is headed north or south.
  A. *No.* The heading isn't mentioned (see answer B).
  C. *No.* The compass would show a turn toward the south when *decelerating* on an *east* or *west* heading, but the heading is not mentioned.

3285. B. The compass accelerating errors don't depend on what type of aircraft it's in, so this would be the proper answer. See answer B in question 3284.
  A. *No.* Just the opposite—remember ANDS.
  C. *No.* Again, just the opposite—ANDS.

3286. A. *SPFM*, Chapter 3, Heading Indicator.
  B. *No.* In a climb or descent (constant airspeed), the compass is not exactly accurate.
  C. *No. SPFM*, Chapter 15, Using the Four Main Directions.

3287. B. The nose will resist being "lowered" to recover from the stall (and if the stall is carried far enough, a flat spin could result).
  A. *No.* If anything, an airplane with an aft CG could "leap off" the ground before *you* are ready.
  C. *No. SPFM*, Chapter 23, Weight and Balance.

3288. A. *SPFM*, Chapter 23, Weight and Balance.
  B. *No.* An airplane is generally less stable in slow speeds than at high speeds, but an aft CG makes it less stable at all speeds, everything else being equal.
  C. *No.* See the explanation in B.

3289. C. *SPFM*, Chapter 3, Altimeter.
  A. *No. SPFM*, Chapter 17, Takeoff and Landing Performance and Simulated High-Altitude Takeoffs.
  B. *No.* Same as the explanations for A.

3290. C. Remember the 4 H's: *High* and/or *Hot* and/or *Humid Hurts.*
  A. *No.* This makes for good performance.
  B. *No.* This combination gives better performance than answer C, but it's not as good as A.

3291. B. A high density-altitude (the hyphen makes things clearer) means that the air is "thin," so performance of both airplane and engine suffers.

3292. C. Move down and to the left on the 5000 foot pressure altitude line, directly above the 25°F value (A). The density-altitude is approximately 3850 feet (B). Move to the right along the pressure altitude line, to above the 50°F value (C). The density-altitude is 5500 feet (D). This makes a difference of about 1650 feet.

3293. C. The indicated altitude is 1380 feet MSL and, looking at the pressure altitude conversion factor (right column), you see that at an altimeter setting of 28.2, the value is a *plus* 1630 feet. Adding the two gives you an answer of 3010 feet. (If you *interpolate* between 28.2 and 28.3 to get 28.22, the answer is approximately a *plus* 1610 feet conversion, which gives you 2990 feet. But using the closest one-tenth inch gives a reasonable answer.)

3294. C. First you would convert the altimeter setting to get the correct pressure altitude. At 29.25 inches (interpolating), the answer is +626 feet, which gives you a pressure altitude of 5250 + 626 = 5876 feet (call it 5900 feet). Moving along a pressure altitude value of 5900 approximately, stopping over 81°F, and then moving across to the density-altitude, you'll see an answer of 8500 feet.

3295. A. The altimeter is 0.04 (four-hundredths) of an inch *higher* than standard, and since one inch of barometric pressure is (approximately) 1000 feet, you could make a stab at this by subtracting 40 feet from the 3563 feet given, or come up with an answer of 3523 feet. Looking at figure 8 you get a value (interpolated) of 36 feet subtracted, for an answer of 3527 feet.

3296. C. This one is close to question 3292. Find the density-altitude at 3000 feet pressure altitude and 30°F, which is *1600 feet.* Looking at 3000 feet pressure altitude and 50°F, *2900 feet* density-altitude is found, or *1300 feet.*

3297. A. Interpolating between 29.92 (0 feet correction) and 30.0 (−73 feet correction) for the 29.97 value (−46 feet), you would subtract this from the 1386 given to get an answer of 1340 feet, which is closest to the choice of 1341.

3298. A. Using figure 8 and interpolating between the conversion factors for 30.3 and 30.4, you'll get a −392 feet, or the pressure altitude is 3894 − 392 = 3502 feet (call it 3500, what the heck). Moving down the 3500 foot pressure altitude line (exactly between the 3000 and 4000 foot lines) until +25°F is reached, you'd then move horizontally to the left to get a density-altitude of 2000 feet.

3299. B. Okay, this one is a little more complicated. If you read *SPFM*, Chapter 3, Simulated High-Altitude Takeoffs, you'll see by using that procedure in this case the pressure altitude increased by 500 feet *but* the *decreased* temperature of 30°F gives an advantage of a 2000 foot *lower* density-altitude (each −15° = 1000 feet lower) for a *lower* density-altitude of 1500 feet. But look at figure 8 and work out the details: At 90°F a pressure altitude of 1250 feet gives a density-altitude of 3600 feet. At 1750 feet pressure altitude and 60°F, the density-altitude is about 2200 feet, a difference (*decrease*) of approximately 1400 feet, so answer B is the best.

3300. B. It decreases performance. High, Hot, or Humid Hurts.

3301. A. The horizontal component of lift is what turns the airplane. See Fig. 8-7.
  B. *SPFM*, Fig. 8-7 and 9-2. The vertical component of lift is working to hold the airplane *up*.
  C. *Centrifugal force* is an *apparent* not a real force, and your neighborhood physics professor doesn't like this term.

3302. C. *SPFM*, Fig. 6-7, and also remember that when the wind is a strong quartering tailwind, you set the ailerons and elevator to "dive and turn away from the wind."

3303. A. *SPFM*, Fig. 6-7 and remember that you "climb and turn into the wind" with a quartering headwind.

3304. A. *SPFM*, Chapter 6, Flight Controls in Taxiing.

3305. A. *SPFM*, Fig. 6-7, but holding some back pressure is good, also.

3306. A. This is the best combination for a tail wheel airplane.

3307. C. This one *agrees* with *SPFM*, Fig. 6-7, and gives the best control positions for the left quartering tailwind.

3308. B. *SPFM*, Fig. 6-7. Dive and turn away from the wind.

3309. C. You can't spin without stalling. *SPFM*, Chapter 12, Spins.

3310. A. This is the best answer, but in the incipient phase (usually the first 2 turns of a *spin*), one wing may not be stalled. In some steep cyclic spins the conditions are such that both wings are stalled, and then only one wing is stalled, with the airplane moving back and forth between these two modes.

3311. C. An airplane always stalls at the *same angle of attack* for a given airplane configuration.
 A. *No.* The *stall speed* is variable and will increase as the CG is moved forward, requiring a heavier tail-down load for a standard-type airplane (tail in rear). *SPFM*, Chapter 23, Weight and Balance.
 B. *No.* See answer C above. The *stall speed* varies as the *square root* of the change in *weight* or load factor.

3312. A. *SPFM*, Chapter 17, Short-Field Landing.
 B. *No.* Induced drag *decreases* in ground effect. See the reference for answer A.
 C. *No.* In ground effect, the wings do an even *better* job of supporting the airplane.

3313. A. *SPFM*, Chapter 17, Short-Field Landing.
 B. *No.* See A.
 C. *No.* The airplane actually gets the same lift at a *lower* angle of attack.

3314. B. *SPFM*, Chapter 17, Short-Field Landing.
 A. *No.* Wingtip vortices decrease in ground effect, decreasing induced drag; the latter is strongest at low speeds such as during takeoffs or landings. *SPFM*, Chapter 17, Short-Field Landings.
 C. *SPFM*, Chapter 23, Running a Weight and Balance. A full stall landing will require *more* up elevator deflection because in ground effect, the airplane is more stable and resists the upward nose movement.

3315. B. The airplane may become airborne at a lower airspeed than is safe for continuing the climbout. *SPFM*, Chapter 17, Short-Field Landing, Chapter 23, Running a Weight and Balance.
 A. *No.* Ground effect has just the opposite effect. See the references for answer B.
 C. No. See the references above.

3316. A. The stall speed goes up as the square root of the load factor increases. *SPFM*, Chapter 9, Load Factors in the Turn.
 B. *No.* But it *could* be argued that since the airplane stalls at a higher airspeed, an unexpected spin might be induced at that higher airspeed, but that's not the correct answer.
 C. *No.* Not particularly.

3317. A. *SPFM*, Chapter 2, Angle of Attack.

* * *

Questions 3381–3524 pertain to weather.

3381. C. Heat exchange. To quote *Aviation Weather*, "Temperature variations create forces that drive the atmosphere in its endless motions."
 A. No. *SPFM*, Chapter 22, Fog, *Radiation fog*. Radiation (ground) for is a good example of

weather with heat exchange, but without movement of air.
 B. *No.* Heat exchange is still the answer.

3382. A. See answer C in question 3381.
 B. *No.* Barometric pressures are corrected to sea level for aviation purposes, so altimeter *settings* aren't a factor for this choice.
 C. *No.* Coriolis force is created by the earth's rotation and influences wind direction (*AW*).

3383. C. *AW.* The term "inversion" indicates that the temperature *increases* with altitude, unlike the normal condition.
 A. *No.* An inversion aloft permits warm rain to fall through cold air below. (Freezing rain may occur.)
 B. *No.* Just the opposite. The visibility can be bad because of trapped fog, smoke, and other restrictions.

3384. A. *AW.* Because of an inversion (the ground radiates and cools much faster than the air well above) ground fog or radiation fog is normally formed in conditions of clear sky, little or no wind, and small temperature or dewpoint spread. (If you are shooting landings at night and the runway lights start appearing "fuzzy," better make the next one a full stop and/or be prepared to fly to an alternate. Inversions may also occur with warm air aloft overrunning cold air near the surface (C) but A is the best answer.

3385. A. *AW.* Fog, haze, and other restrictions occur.
 B. *No.* There is no wind shear.
 C. *No.* Turbulent air and showers?

3386. A. *SPFM*, Chapter 3, Altimeter.
 B. *No.* A temperature of 59°C is warm indeed!
 C. *No.* The standard pressure at sea level is 29.92" Hg or *1013.2* millibars.

3387. C. Rolling the setting-window numbers down also decreases the hands' indications (and vice versa). Since 1 inch of window change would be 1000 feet indication change, a *decrease* of 0.15 inches, from 30.11 to 29.96, would indicate a decrease of 150 feet.

3388. B. Since altimeters are set in reference to sea level, standard atmospheric pressure (pressure of 29.92" Hg, 59°F or 15°C at sea level, and normal lapse rate), when set to 29.92 (pressure altitude), the altimeter would indicate true altitude, which is the height above sea level (Fig. 3-3).

3389. C. At standard temperature the density- and pressure altitudes are the same. *SPFM*, Fig. 3-4, and Chapter 3, Altimeter.
 A. *No.* The density-altitude will be nearly 2000 feet *below* sea level at that temperature. See the thumb rule, *SPFM*, p. 156, Simulated High-Altitude Takeoff.
 B. *No.* What about temperature?

3390. C. *SPFM*, Chapter 3, Altimeter. Look back at Fig. 27-4.
 A. *No.* See the answer to question 3388.
 B. *No. SPFM*, Chapter 3, Altimeter.

3391. B. See the explanation for question 3390. Also see *SPFM*, Chapter 3, Altimeter.
  A. *No.* Remember H to L = H or HLH (from a High to a Low the altimeter reads High, and vice versa).
  C. *No.* See the answer to 3390 again.

3392. A. Remember HALT (High Altimeter because of Low Temperature). The altimeter will read higher than true altitude. (*SPFM*, Chapter 3, Altimeter.)
  B. *No.* Just the opposite.
  C. *No.* *True* altitude is the actual height above sea level. A seaplane can be sitting at sea level and, on a hot day, have a density-altitude of, say, 2000 feet.

3393. C. Look at the answer to question 3392 and reverse the HALT situation. Low Altimeter because of a High Temperature. (In this case LAHT would be the reference, but remember HALT to keep from hitting the terrain and reverse it if necessary.)
  A. *No.* The opposite would occur. See answer above.
  B. *No.* Density-altitude and *true* altitude would be the same only in standard atmospheric conditions. Remember that a high density-altitude means "thin air." This book hyphenates density-altitude to avoid confusion.

3394. B. *SPFM*, Chapter 17, Simulated High-Altitude Take-offs.
  A. *No.* An increase in barometric pressure would *decrease* the density-altitude if the temperature remained the same. As indicated in *SPFM*, p. 17-9, pressure changes aren't as big a factor as temperature changes.
  C. *No.* This would help matters slightly as far as performance is concerned.

3395. B. *AW*, see the section on Friction. Surface friction affects the Coriolis force and the wind direction will be different between the surface and 5000 feet AGL.
  A. *No.* This would affect *velocity*, not direction.
  C. *No.* See the explanation above.

* * *

3397. C. *AW.* *Dewpoint* is the temperature to which air must be cooled to become saturated. Be alert for fog formation when the temperature and dewpoint are within 4°F of each other.
  A. *No.* See the reference for C.
  B. *No.* Sounds logical, but C is the answer.

3398. B. *AW.* An old rule of thumb is that for every 20°F increase in its temperature, the ability of a parcel of air to hold moisture is doubled.

3399. A. *AW.* Condensation is the process of changing water vapor to liquid water. As water vapor condenses on condensation nuclei, drops of liquid or ice particles begin to grow.
  B. *No.* Water vapor is always present but may not have condensed to the point for visible moisture.
  C. *No.* The relative humidity may be 100 percent without visible moisture occurring.

3400. A. *AW.* Glossary of Weather Terms (change of state). By evaporation, liquid water is converted to water vapor. Sublimation is the process by which ice is converted to water vapor. Either process can add moisture to unsaturated air.

3401. B. *AW.* *Dew* forms on those still, clear nights when the temperature of the air surrounding an object cools by radiation to or below the dewpoint. *Frost forms when the dewpoint of the surrounding air is below freezing.*
  A. *No.* Dew and frost *form* from the moisture in the surrounding air. No small droplets "*fall*" on the surface and freeze. (That could be freezing drizzle or rain.)
  C. *No.* Sorry. Again, no droplets are falling on the collective surface.

3402. C. *AW.* "Ice pellets always indicate freezing rain at a higher altitude."
  A. *No.* Hail and/or rain are the forms of precipitation normally associated with thunderstorms.
  B. *No.* Precipitation from a cold front is normally freezing rain, hail, or rain.

3403. B. *AW.* "Stability runs the gamut from absolutely stable to absolutely unstable and the atmosphere usually is in a delicate balance somewhere in between. *A change in ambient temperature lapse rate of an air mass can tip this balance.*" (Emphasis added.)
  A. *No.* There can be stable air or unstable air at various atmospheric pressures.
  C. *No.* See the correct answer above.

3404. A. *AW.* "Surface heating or cooling aloft can make the air more unstable."
  B. *No.* *AW.* "Surface cooling or warming aloft often tips the balance toward greater stability."
  C. *No.* A decrease in water vapor alone would not affect stability.

3405. A. *AW.* "Thus within a *stable* layer, clouds are stratiform."
  B. *No.* *AW.* "Restricted visibility at or near the surface over large areas usually indicates stable air."
  C. *No.* *AW.* "Unstable air favors convection. . . . Thus within an unstable layer, clouds are cumuliform (cumulus)."

3406. A. *AW.* Mechanical (upslope) lifting of stable air produces stratiform (stratus-type) clouds.
  B. *No.* That's *unstable* air being mechanically lifted.
  C. *No.* Again (see question 3405), unstable air favors convective action and cumulus-type clouds.

3407. C. Unstable air favors convective action with vertical development and turbulence.
  A. *No.* See 3406.
  B. *No.* Stratus clouds normally are associated with smooth air.

3408. A. *AW.* "If the temperature increases with altitude through a layer—an inversion—the layer is stable and convection is suppressed. Air may be unstable beneath the inversion." Inversion is smooth air in that layer.
  B. *No.* See A.
  C. *No.* Chinook winds raise temperatures because air is heated as it moves mechanically down a slope.

3409. B. Using a 4.4°F convergence of temperature and dewpoint: 22 ÷ 4.4 = 5000 feet. *SPFM*, p. 202, col. 1, para. 2.
   A. *No.* This doesn't compute with either *AW* or *SPFM*. Other meteorological texts agree with *SPFM* on the 4.4 value.
   C. *No. AW.* The difference in temperature and dewpoint is 22°F. Using a 4.4°F convergence of temperature and dewpoint, a value of 22 ÷ 4.4 = 5000 not 6000.

3410. B. *AW.* Dividing by 4.4°F, a height of 10,000 feet is obtained.
   A. *No.* It doesn't compute with a 4.4°F lapse rate.
   C. *No.* The difference in temperature and dewpoint is 82°F − 38°F = 44°F. Dividing that by 4.4°F results in a base of 10,000 feet, not 11,000 feet.

* * *

3412. A. *AW.* "Unstable air favors convection and cumuliform clouds." Cumuliform clouds produce showery precipitation.
   B. *No.* Those are the characteristics of *stable* air.
   C. *No. AW.* "Precipitation from stratiform clouds is usually steady."

3413. A. Turbulence is associated with unstable air and convection. Surface visibility is usually good unless there are showers.
   B. *No.* Turbulence, yes, but visibility is usually good.
   C. *No.* Nimbostratus (rain-stratus) clouds have little turbulence and the visibility is poor in precipitation.

3414. C. *AW.* Smooth air is a typical result of stable air.
   A. *No. AW.* The precipitation in stable air is more apt to be steady. Unstable air produces convective clouds and showery precipitation.
   B. *No.* This is a characteristic of *unstable* air.

3415. B. *AW.* Nimbostratus clouds are named "rain clouds" but may produce continuous rain, snow, or ice pellets. "Nimbus" means rain.
   A. *No.* That's a CB (cumulonimbus) or towering cumulus.
   C. *No.* Nimbus means "rain."

3416. B. *AW.* Also see *SPFM*, Chapter 22, Cloud Types.
   A. *No.* Some clouds are shaped like animals (temporarily) or some odd shapes at the beach.
   C. *No.* Clouds are composed of water droplets or ice crystals and aren't divided this way.

3417. C. *AW.* Lenticular clouds (they're shaped like spectacles or other lenses) are usually found over mountain barriers. The presence of these clouds indicates strong winds and turbulence.
   A. *No.* A cloud with winds of 50 knots or more is not apt to be considered an "inactive" frontal cloud.
   B. *No.* A funnel cloud is a tornado-type that hasn't touched the ground. Head for the storm cellar, Maude!

3418. B. *AW.* See the answer to 3417.
   A. *No. AW.* Mammatocumulus clouds are so named because of the hanging protuberances on the underside of the cloud. Further description is unnecessary.
   C. *No. AW.* This is the troublemaker located on the leading edge of a *cumulonimbus* or, less often, on a rapidly developing cumulus.

3419. B. *AW.* Cumulonimbus (CB), the worst type for turbulence, hail, icing, and heavy rain. The accompanying illlustration shows how hail develops in a CB.
   A. *No.* These are bad but are *still developing* into real trouble as they become CBs.
   C. *No. AW.* Nimbostratus clouds have little turbulence.

3420. C. *AW.* Towering cumulus, by their very name, indicate convection, unstable air, and *turbulence*.
   A. *No. AW.* Cirrus clouds are very high clouds composed of ice crystals. Wispy cirrus has little turbulence; dense, banded cirrus may have turbulence, but not like towering Cu. Stop worrying, you can't get your general aviation trainer up to these clouds anyway.
   B. *No.* As noted in answer B in 3419, these stratus clouds contain precipitation and usually have smooth air.

3421. C. *AW.* "The boundary between two different air masses." Also see *SPFM*, p. 197, col. 2, para. 3.
   A. *No. AW.* Frontolysis is the dissipation of a *front.*
   B. *No. AW.* Frontogenesis is the initial formation of a *front* or *frontal zone.*

3422. A. *AW.* These are changes in temperature, humidity, and wind, and *temperature* is the only choice that really fits as an answer to this question.
   B. *No.* There are dry fronts with little or no clouds.
   C. *No.* There usually will be a *decrease* in relative humidity when a *cold* front passes.

3423. A. *AW.* "The wind always changes across a front."
   B. *No.* The clouds will usually be dissipating after a cold front passage.
   C. *No.* The air may be turbulent (unstable) after a cold front passage, but there may be little change in stability in a warm front passage.

3424. C. *AW.* Steady precipitation is a sign of stable air with stratiform clouds and little or no turbulence.
   A. *No.* The "moderate turbulence" shoots this choice down.
   B. *No.* The "cumuloform clouds" eliminates this as an answer. Remember that this type of cloud is associated with showery precipitation.

3425. A. *AW.* Be prepared for strong turbulence under the conditions cited.
   B. *No.* By definition, *mountain wave* describes the turbulence when strong winds cross a *barrier.*
   C. *No.* The *parallel* to a mountain *peak* shoots this one down.

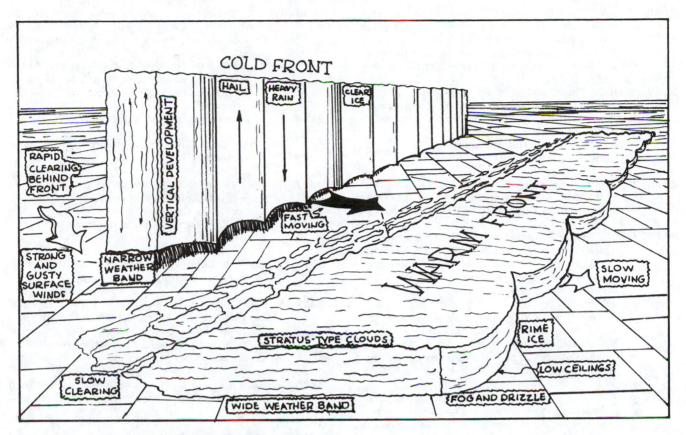

Fig. 27-5. Characteristics of "typical" warm and cold fronts. The vertically developed clouds may produce clear ice, whereas the status type (stable air) clouds usually associated with warm fronts may produce rime icing at or above the freezing level.

3426. C. *AW.* "Wind shear may be associated with either a wind shift or wind speed gradient *at any level in the atmosphere.*" (Emphasis added.)
  A. *No.* You can get wind shear on an approach.
  B. *No.* You can get wind shear at higher altitudes.

3427. B. *AW.* "Three conditions [for wind shear] are of interest—(1) wind shear with low-level temperature inversion, (2) wind shear in a frontal zone, and (3) clear air turbulence."
  A. *No.* That's a mountain wave.
  C. *No.* See answer B above.

3428. C. *AW.* "You can be relatively certain of a shear zone in the inversion if you know the wind at 2000 to 4000 feet is *25 knots or more.*" (Emphasis added.)

3429. C. *AW.* "Two conditions are necessary for structural icing in flight: (1) The aircraft must be flying through visible water such as rain or cloud droplets, and (2) the temperature at the point where the moisture strikes the aircraft must be 0°C or colder."
  A. *No.* This could be referring to fog. (Note that the temperature isn't mentioned.)
  B. *No.* You get rime ice in stratiform clouds and then again you don't—depending on the temperature.

3430. A. *AW.* "The condition most favorable for very hazardous icing is the presence of many *large, super-cooled* water drops. This could be found in cumulus cloud formations." (Emphasis added.)
  B. *No.* Freezing drizzle is bad but the accretion is not as high as A.
  C. *No.* This one, however, gets this writer's personal opinion that *freezing rain* can be extremely hazardous for the VFR pilot (since he or she may not be able to climb into the above-freezing air level without getting into IMC).

3431. C. *SPFM*, Chapter 22, Frost. *AW* indicates that frost may spoil the smooth flow of air and cause early airflow separation. A heavy layer of *frost* may increase stall speed by 5 to 10 percent.
  A. *No.* It would have to be an extremely thick and rough layer of frost in various places to change the *shape* of the airfoil. (That's impossible.)
  B. *No.* Frost does *not* increase control effectiveness.

3432. C. *SPFM*, Chapter 22, Frost, and *AW.*
  B. *No.* As noted in the explanation in 3431A.
  C. *No.* Just the opposite.

3433. B. *AW.* "For a thunderstorm to form, the air must have (1) sufficient water vapor, (2) an unstable lapse rate, and (3) an initial upward boost (lifting). . . ."

3434. B. *AW.* "The key feature of the cumulus stage [of a thunderstorm] is an updraft. . . ."
  A. *No.* The roll cloud is associated with the mature stage of a thunderstorm.
  C. *No.* Lightning is associated with the mature stage.

3435. B. *AW.* "Precipitation beginning to fall from the cloud base is your signal that a downdraft has developed and that the cell has entered the mature stage."
  A. *No.* See B. The anvil is seen primarily in the dissipating stage.
  C. *No.* That's the cumulus stage.

3436. A. *AW.* See the answer for question 3433.
  B. See 3433.
  C. See 3433. The term "cloud cover" eliminates this one.

3437. B. *AW.* "Downdrafts characterize the dissipating stage of the thunderstorm cell. . . ."
  A. *No.* This is the stage for *updrafts*.
  C. *No.* There are up *and* downdrafts in the mature stage.

3438. A. *AW.* "All thunderstorm hazards reach their greatest intensity during the mature stage."

3439. A. *SPFM*, Chapter 25, Weather, and *AW.* "The squall line presents the single most intense weather hazard to aircraft."
  B. *No.* But these are bad enough.
  C. *No.* See (A) above.

3440. B. *AW.* "A *squall line* is a non-frontal, narrow band of thunderstorms. Often it develops ahead of a cold front. . . ."

3441. B. *SPFM*, Chapter 25, Weather.
  A. *No.* Precipitation static is not the big problem here.
  C. *No.* Steady *heavy* rain could be hazardous in reducing visibility but B is the big problem.

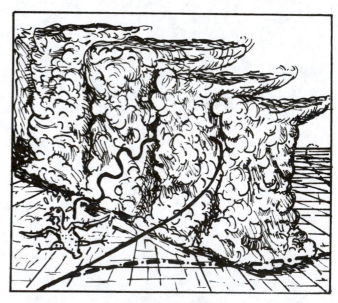

Fig. 27-6. An airplane should fly the dashed-line route, staying safely away (at least 5 miles) from the cloud and overhang, or complete the 180° turn well *before* getting to the squall line. Note that the turkey who tried to go through lost some feathers.

3442. C. *AW.* "Maintain a constant *attitude.* Maneuvers performed in an attempt to maintain constant *altitude* place additional stresses on the aircraft."
A. *No.* You cannot do this anyway, so don't even try. See C above.
B. *No.* This can't be done either.

3443. A. *AW.* "Conditions favorable for radiation fog are clear sky, little or no wind, and small temperature-dewpoint spread.
B. *No.* That's advection fog.
C. *No.* See A.

3444. C. That's a good combination for fog or low clouds.
A. *No.* The temperature is too high for that.
B. *No.* The temperature is slightly too low to expect thunderstorms to develop.

3445. B. *AW.* "Advection fog forms when moist air moves over colder ground or water." (This choice did say that it is winter.)

3446. C. *AW.* Advection and upslope fog depend on air mass movement (wind).
A. *No.* Radiation fog is formed in conditions of little or no wind.
B. *No.* Same as A for radiation (ground) fog.

3447. A. *AW.* Fog induced by precipitation (rain) is "especially critical because it occurs in the proximity of precipitation and other hazards such as icing, turbulence, and thunderstorms."

3448. C. *AW.* ". . . thermals depend on solar heating. . . ."

3449. C. Wind shear and turbulence are the most hazardous conditions for *any* kind of flying.
A. *No.* Static electricity is not a big problem.
B. *No.* Lightning is not the *big* factor.

Fig. 27-7. Convection currents are good for sailplanes, but on summer days they can cause some too-high approaches.

3450. C. *AW.* "At night cool air from the land flows toward warmer water—the land breeze. During the day, wind blows from the water to the warmer land— *the sea breeze.*"
A. *No.* See C.
B. *No.* Land is normally warmer than water during the day.

* * *

3452. A. *AW.* Lightning *always* occurs with thunderstorms.
B. Heavy rain is *often* associated with a thunderstorm because the thickness (height) of the cloud is a factor in producing rain and other precipitation.
C. *No.* Hail *often* occurs with thunderstorms.

3453. A. "Generally the TWEB contains route oriented data. . . ." Aeronautical Information Manual (AIM).
B. *No.* Terminal Forecasts are for *terminal* points, not routes.
C. *No.* Area Forecasts are for areas of states or parts of states, *not* routes.

3454. B. AIM.
A. *No.* That's primarily for local traffic advisories.
C. *No.* That's local airport recorded information.

3455. A. AIM.
B. *No.* The weather briefer doesn't need that.
C. *No.* That's not required on the preflight *weather* briefing.

3456. C. AIM. The standard briefing should be requested if you have not received a previous briefing or gotten the information through a Transcribed Weather Broadcast (TWEB), etc.
A. *No.* There's no "general briefing."
B. *No.* This is more for an update of a previous briefing.

3457. C. AIM. This is the thorough briefing.
A. *No.* The outlook briefing is for a proposed time of departure 6 hours or more from the time of the briefing.
B. *No.* The abbreviated briefing is for an update on earlier information.

3458. C. AIM. The abbreviated briefing is the update for earlier briefings or mass-disseminated data (TWEB, etc.)

3459. A. AIM. See 3458 C.
B. *No.* See 3456 C.
C. *No.* The outlook briefing is more of a forecast-type briefing (6 hours or more before the proposed departure time) for planning purposes.

3460. A. AIM. Also see the explanations in 3457 A and 3459 C.

3461. A. AIM. See the explanations in 3457 A and 3459 C.

3462. C. KINK (Wink, TX) has 15 statute miles (SM) visibility and clear sky (SKC). KBOI (Boise, ID) has a 30 SM visibility and scattered clouds at 15,000 feet. KLAX has 6 SM visibility and scattered at 700 feet (and not a ceiling).
  A. and B. No. KMDW (Chicago Midway) and KJFK (Kennedy International) have issued Specials and have visibilities of ½ SM and 1½ SM plus overcast clouds at 700 and 500 feet respectively.

3463. B. Review Chapter 22, METARs and TAFs.

3464. A. The wind at KJFK is 18004KT or from 180° *true* at 4 knots. METAR winds and winds aloft are measured from true north. ATC towers, ATIS, and Airport Advisory Service report wind as *magnetic*.

3465. B. Winds at KINK (Wink) are 11012G18KT or from 110° (true) at 12 knots, gusting to 18 knots.
  A. No. The winds aren't calm. The symbol for that would be 00000KT.
  C. No. Not a bad stretch, though.

3466. B. The rain began at KMDW at 1835Z.

3467. A. There's a Special (SPECI) for KMDW which means a significant change in the weather and the ceiling is 700 feet (007) overcast and visibility is 1½ SM in rain (RA).
  B. No. The OVC is at 007, not 070 (7000 feet).
  C. No. The visibility is 1½ miles in rain.

* * *

3472. C. *AWS.* The sky report (SK) indicates that the pilot was reporting *two* layers: a *broken* (BKN) layer from 1800 feet (018) to 5500 feet (055) *MSL* and an *overcast* (OVC) layer from 7200 (072) to 8900 (089) *MSL.* So, the overcast layer was from 7200 feet to 8900 feet.

3473. C. *AWS.* The airplane is at a flight level of 12,000 feet. The temperature (TA) is −9° *Celsius* and the wind (WV) is from 090° (true) at 21 knots (0921).
  A. *No.* For one thing, temperatures are always given as *Celsius* on a pilot report.
  B. *No.* Again, temperatures on a UA are given as Celsius.

3474. A. The terrain elevation is 1295 MSL and the base of the first layer is 1800 feet MSL (the base of the *broken* layer, and broken clouds constitute a ceiling). The ceiling is 505 feet AGL. (A ceiling is always measured AGL.)

3475. B. Turbulence (TB) is moderate (MDT) between 5500 feet MSL (055) and 7200 feet MSL (072). No other altitudes are mentioned in connection with turbulence.

3476. B. The pilot reports light to moderate clear icing (IC) between 7200 and 8900 feet MSL.
  A. *No.* The report includes the type of icing (clear).
  C. *No.* Not rime ice, but clear ice is reported.

* * *

3478. C. The Area Forecast will give the expected weather at the estimated time of arrival in the *area* of the destination, if a Terminal Forecast is not available. (See *SPFM*, Chapter 22, Area Forecast.)
  A. *No.* The Low Level Prog Chart is too general to get weather for specific points, although it does show areas of expected MVFR and IFR weather.
  B. *No.* The Weather Depiction Chart gives a good view of areas of MVFR and IFR weather but does not describe the ceilings and visibilities expected for an area as well as the Area Forecast.

3479. C. The KMEM valid period is 121818, or from the 12th of the month at 1800Z to 1800Z on the 13th.

3480. A. At KMEM (and anywhere else) "SHRA" is the short-hand for "Showers of Rain." (OK, rain showers.)

3481. B. The situation at KMEM is BECMG (Becoming) between 1012 (1000Z to 1200Z) wind calm (00000KT) with *3 SM* in Mist (BR) with a clear sky (SKC). So 3 statute miles is the answer.
  A. The ½ SM value is found in the next period (TEMPO) and is forecast to occur temporarily between 1200Z and 1400Z.
  C. No. The *last period* (from 1600Z) is forecast to be greater than 6 SM (P6SM).

3482. C. The KMEM forecast wind from 1600Z to the end of the forecast is directionally variable wind at 04 knots (VRB04KT).

3483. A. From (FM) 1600Z the wind is forecast to be from 160° at 10 knots. At 2224 (2200Z to 2400Z) the wind is forecast to be from 200° at 13, gusting to 20 knots, visibility 4 SM in rain showers (SHRA), and an overcast at 2000 feet (OVC020).

3484. A. The forecast is for the weather becoming (BECMG) from 2200Z to 2400Z, wind from 200° at 13, gusts to 20 knots, 4 statute miles, in rain showers (SHRA) with an overcast at 2000 feet (OVC020).
B. The 40% probability is for 0006 (0000Z–0600Z).
C. Nope, that period 2200Z to 2400Z indicates an overcast at 2000 feet (OVC020).

3485. C. At KOKC from 0600Z to 0800Z (0608) the note is NSW (No Significant Weather).
A. No. The wind is no particular problem.
B. The visibility is better than 6 SM (P6SM) with scattered clouds at 4000 feet, which is not significant weather.

3486. B. The cumulonimbus (CB) is the only cloud type forecast in a TAF. (See the end of the first line in the KMEM TAF (OVC008*CB*).

3487. A. *SPFM*, Chapter 22, Area Forecast.
B. *No. AWS.* "Weather Depiction Charts are prepared from surface aviation [SA—hourly sequence] reports and give a broad overview of *observed* flying category *conditions as of the valid time of the chart.*" (Emphasis added.) It's not a forecast.
C. *No.* Satellite maps show the *cloud cover* at the time the picture was taken. (No information on forecast ceilings or visibilities.)

3488. C. *SPFM*, Chapter 22, Area Forecast. The time period given (2300–0500Z) for TN is the outlook period of VFR. This means ceilings of 3000 feet or higher and visibilities of 5 miles or greater.

3489. C. See the Area Forecast (figure 16) and look at the 4th paragraph (DFWI . . . ICING AND FREEZING LEVEL VALID UNTIL 042300).

3490. A. See figure 16, section DFWC FA 041040. It forecasts the weather by states or other geographical areas, not along specific routes.

3491. C. Under DFWT (Turbulence)—In figure 16, TN AL and CSTL WTRS can expect moderate turbulence from 25,000 to 38,000 feet MSL due to jetstream. This turbulence hazard was noted for this area under Flight Precautions under Hazards at the beginning of the forecast.

3492. C. IFR Texas (visibility less than 3 miles) from San Antonio (SAT) to Palacios, Texas (PSX—probably a misprint) to Mobile (MOB—another misprint; MOV is Monument Valley, Utah) to San Antonio, the visibility will be below 3 miles in fog until 1500Z. The less than 3 miles in fog is also stated in the 5th line from the bottom in figure 16.

3493. B. Check the 6th and 5th lines from the bottom in figure 16. This tells it all.

3494. A. AIM. TWEBs are route-oriented and picked up on a VOR receiver or an ADF.

3495. C. *SPFM*, Chapter 22, In-Flight Weather Advisories. CONVECTIVE SIGMETs forecast thunderstorms that are *imbedded* (obscured by massive cloud layers).
A. *No.* CONVECTIVE SIGMETs are warnings about thunderstorms that cover ⅘ of the area (or more).
B. *No.* Moderate turbulence is covered in AIRMETS. Severe and extreme turbulence are covered in SIGMETs.

3496. A. *SPFM*, Chapter 22, In-Flight Weather Advisories.

3497. C. *SPFM*, Chapter 22, In-Flight Weather Advisories. SIGMETs are for *all* aircraft.

3498. B. *SPFM*, Chapter 22, In-Flight Weather Advisories.

3499. A. The AIM defines the AIRMET as being potentially hazardous to aircraft that have limited capability because of lack of equipment, instrumentation, or pilot qualifications. (In earlier times AIRMETs were specifically aimed at "small single-engine aircraft.")

3500. B. *SPFM*, Chapter 22, Wind Information. The wind at STL at 6000 is from 230° *true* at 25 knots (2325). *Published winds are always in true directions to the nearest 10°, and in knots.*

3501. A. At STL at 18,000 feet the wind is from 230° *true* at 56 knots. The temperature is a cool −16°C.

3502. B. Denver is forecasting the wind at 30,000 feet to be 230° true at 53 knots with the temperature a minus 47°C. (There's a note that temperatures are negative above 24,000 feet, which saves putting in all those minus signs.)
A. *No.* Dadgummit, see the explanations in questions 3500 and 3501.
C. *No.* Winds are to the nearest 10° true. Here the wind speed and the temperature got mixed up.

3503. A. The wind at Kansas City (MKC) is from 050° true at 7 knots. No temperature is forecast for the 3000-foot (MSL) level or for a level within 2500 feet of station elevation. Where did this direction (360°) for B and C come from?

3504. C. *SPFM*, Chapter 22, Wind Information. Subtract 50 from 73 to get 23 (230° true) and add 100 to 06 to get 106 knots. The temperature is minus 49°C.

3505. C. *SPFM*, Chapter 22, Wind Information. See the explanations for questions 3500 and 3501.

3506. B. *SPFM*, Chapter 22, Wind Information. Light and variable winds (less than 5 knots) are noted by 9900.
A. *No.* 0000 is for *surface winds.* See *SPFM*, Fig. 22-11.
C. *No.* 9999 isn't used in Winds Aloft Forecasts.

3507. A. *SPFM*, Chapter 22, Weather Information, Fig 22-2. Think of it as the warm front and cold front symbols pulling against each other so that no front movement occurs.

B. *No. SPFM*, Chapter 22, Weather Information, and Fig. 22-2 and Fig. 22-7. The occluded front symbols show that the cold front has caught up with the warm front and the barbs have "stuck through" the warm front.

C. *No.* That's not the case here. When a front is reversing its movement, it's normally referred to as "backing up," though "retreating" may be a classier term.

3508. B. *AWS.* In the Weather Depiction Chart (figure 18), the symbol for fog is = , and ⊗ means sky partially obscured; 2 = ⊗ means 2 miles visibility in fog, sky partially obscured.

3509. A. *AWS.* "The Weather Depiction Chart is the choice place to do your weather briefing and flight planning. From it, you can determine general weather conditions more readily than any other source." (It's a *now* condition for the valid time of the chart and is *not* a forecast.)

3510. A. *AWS.* The symbol ⌐⟨. is for thunderstorms and rain showers. That's the problem in southeast New Mexico.

B. *No.* The ceiling is *6000* feet.

C. *No.* The visibility is 4 miles; that's VFR.

3511. B. *AWS.* Low ceilings (and fog) are the problem along the coast of Oregon. Conditions are 300 overcast and 200 feet, sky obscured. Stratus clouds and fog are the problem. *SPFM*, Fig. 22-26, and accompanying material on *Weather Depiction Chart* in figure 18.

3512. C. *AWS.* Looking at figure 18, you see the symbols ◖ and ◗ which indicate scattered and broken clouds respectively at 25,000 feet. (Always add 2 zeros to the ceiling values.)

3513. B. *AWS.* Radar Weather Reports (RAREPS) "include the location of the precipitation along with type, intensity, and intensity trend." Radar only is able to get echoes of precipitation and so would be unable to detect large areas of low ceilings and fog (if there's no rain, etc., to provide echoes).

3514. A. *SPFM*, Chapter 22, Radar Summary Chart. Radar doesn't pick up *ceilings* and can't tell stratus from cirrus.

3515. B. Checking figure 19, area A, you see that the indication (right by the "A") is an arrow pointing eastward with the "feather" being one and one-half barbs, indicating a 15-knot movement. (A whole barb is 10 knots; a half barb is 5 knots.)

A. *No.* The "220" is the highest precipitation top at "A."

C. *No.* The "220" has to do with the highest precipitation top, although it could be confused with direction and velocity of the area movement.

3516. C. *SPFM*, Chapter 22, Radar Summary Chart. The symbol RW+ shows rain showers (RW) increasing in intensity (+) in area "C." This is the symbol used for increasing intensity in Radar Summary Charts but might be misunderstood since "RW+" is the notation for *heavy* rain showers used in hourly reports and Terminal Forecasts.

3517. B. *AWS.* The arrow at "D" in figure 19 indicates that a TRW of increasing intensity (+) is moving northeast at 20 knots (20).

3518. C. *AWS.* This one could be a little confusing, but there is a symbol for a precipitation top of "300" (30,000 feet) and this is the only choice that fits. (There's a "220" too, but it's not one of the choices.)

3519. B. *AWS.* "Severe weather watch areas are outlined by heavy dashed lines, usually in the form of a large rectangular box." (The area of a severe weather watch on your commercial TV stations—which includes severe thunderstorm and tornado watches—is outlined by a solid line.)

3520. B. *AWS. SPFM*, Chapter 22, Weather Prognostic Charts. Checking figure 20, you see that freezing levels and turbulence areas are shown (see the top two panels). These could help identify areas to avoid.

A. *No.* "All altitudes" covers a multitude of choices and figure 20 is for low levels (SFC–400 MB, or surface to 24,000 feet).

C. *No.* The chart is a *prognostic* (forecast) chart for the valid times given, not *current* information.

3521. A. (Check out figure 20.) *AWS.* "Forecast areas of moderate or greater turbulence are inclosed by long-dashed lines." The numbers below and above the short line show expected base and top of the turbulent layer in hundreds of feet MSL. Absence of a number below the line indicates turbulence from the surface upward.

3522. B. *AWS. SPFM*, Chapter 22, Weather Prognostic Charts. In figure 20, the 12-hour Surface Prog Chart shows an area outlined by dot-dash lines (which indicate showery precipitation). Inside the boundary are symbols for thunderstorms and rain showers (though the rain shower symbol is somewhat hard to read). As the chart legend indicates, thunderstorms imply moderate or greater turbulence.

A. *No.* That area outlined on the Significant Weather Prog Chart (figure 20) by the smooth line would have ceilings lower than 1000 feet and/or visibility less than 3 miles.

C. *No.* Rain and thundershowers are not shown moving northeasterly; that line is a slash (/) separating the two symbols discussed in A.

3523. A. *AWS. SPFM,* Chapter 22, Weather Prognostic Charts. The direction and speed of a pressure area are indicated by an arrow and number. In the 12-hour Surface Prog Chart (figure 20) the Low in the corner of Utah and Wyoming is forecast to move due east at 30 knots. The High in Oregon is forecast to move southeast at 30 knots. Notice on the 24-hour Surface Prog that there is *no* Low shown in the western states.

3524. B. *AWS.* Looking at the upper right panel (the 24-hour Significant Weather Prog Chart) you'll see that the dashed line (freezing level above mean sea level) indicated over northeastern Oklahoma can be followed out to 80 (8000 feet MSL) at the upper right corner of the chart.

* * *

Questions 3526–3650 pertain to navigation and chart reading.

3526. A. AIM. Tell them you're a pilot or the folks at the FSS or Weather Service Office might think you want the forecast for the office picnic. (*This year,* try to behave yourself at the picnic.)
  B. *No.* While this would be information that would keep the briefers on the edge of their seats, it's not an important factor for a pre-flight weather briefing.
  C. *No.* Suppose they think you're carrying too many people or don't like the color of your aircraft—what then?

3527. B. AIM. They want to know your intended route, destination, and type of aircraft, plus other information. 5-1f. Your name and address have no bearing on the weather briefing. Save this for the filing of a flight plan.

3528. C. AIM. The limitations of the pilot and the equipment are of importance to the briefer. His or her briefing and suggestions will be based on whether you are going strictly VFR or can go IFR if necessary.
  A. *No.* The briefer doesn't need your name at this point (maybe later if you file the flight plan by phone). Also, what if you are flying alone and not in a formation?
  B. *No.* It's none of their business, but if you're going flying you'd save yourself a lot of trouble if you possess a current pilot certificate.

3529. C. See *SPFM,* Chapter 19, Getting the Groundspeed, for the idea. Okay you're over Hampton Roads Airport (upper left corner of the page) at 1456 and over Chesapeake Municipal at 1501, 5 minutes later. The distance is 10 nautical miles, working out to a groundspeed of 120 knots. Chesapeake Municipal to First Flight is a distance of 50 NM. At 2 NM per minute it should take another 25 minutes; 1501 + 25 = 1526.

3530. A. In figure 21, draw a horizontal line through Currituck County Airport parallel to the 36°30′ parallel. The airport is 6 minutes south of that, or 36°30′ − 6′ = 36°24′N (all latitudes are *North, and increase northward* in the United States). The 76° meridian runs through Virginia Beach (upper right corner of the chart). Draw a vertical line through the airport parallel to the meridian. (Are you confused yet?) The longitude is about 1 minute west of that meridian, or 76°01′W. (All longitude values in the United States are *West and increase westward.*)
  B. *No.* That's NAS Oceana. For Pete's sake, don't land *there.*
  C. *No.* That's roughly 120 NM north of *Ottawa, Canada.* Better get that compass in your airplane checked.

3531. C. *SPFM,* Chapter 19, Plotter. Use the meridian (*SPFM,* Fig. 19-1) that runs right by the Currituck County Airport as the center meridian. Using the plotter you should get 321° true. The variation (9°W) is *added* to 321° to get 330° magnetic.

3532. B. *SPFM,* Fig. 25-4 and explanation. Be sure that you get the 340 radial and *don't* use that "340" on the 325 radial. *That* "340" is the MSL height of that antenna.

3533. C. Looking at figure 21 you see that Shawboro is on the 030 radial, which means that if you center the needle, the course or omnibearing selector (OBS) will indicate 030 or 210. Of the three choices in figure 29 (5, 6, and 8), 8 is the one because it indicates 210° TO, which shows that at the current position of the airplane (over Shawboro) it is 210° TO the VOR.
  A. *No.* The indication of 5 shows that the airplane would fly 030° TO the station, so its position is southwest of the VOR, *not* over Shawboro.
  B. *No.* The OBS is set at 030° and the airplane is FROM the station, but it's not on a line to Shawboro.

3534. B. Using an E-6B–type computer or an electronic computer, you find the following: (1) True heading is 002° and (2) groundspeed is 80 knots. The measured distance between Mercer County Regional Airport to Minot International is 59 miles. Using your computer you'll get an answer of 44 + minutes en route. Add 3½ minutes for climb and 48 minutes is obtained.

3535. B. Find 47°39′30″N and draw a horizontal line. Find 100°53′00″W and draw a vertical. Lo and behold, where these lines cross is the airport in question (Crooked Lake). Note that there are two Johnson airports in the lower right corner of the chart. One Johnson is at 2000-feet elevation and 2000 feet long. The other Johnson is at 1920-feet elevation and 1700 feet long.

3536. C. Find 47°21′N (it's very near the bottom of the chart) and draw a horizontal line. Do the same for 101°01′W and draw a vertical line. Where the two lines cross is *Washburn.*

3537. B. The airship flew 8 miles in 12 minutes (1056 to 1108), so the groundspeed is a rip-roaring 40 knots. Measure from the Minot VOR. (This is more accurate than finding the creek.) At 40 knots and a time of 1 hour and 15 minutes from Minot VOR-TAC, a distance of 50 NM has been covered and the airship should be crossing the road east of the town of Underwood.

(Question: If airplanes do wingovers, do airships do bagovers?)

3538. C. Okay, we can hie back to the work you should have done with question 3534. The true heading with a course of 012° true, TAS of 100 knots, and winds at 330/25 was 002°. If the magnetic variation is 10° E, you'd *subtract* this from 002° to get 352°. See *SPFM*, p. 163, Variation.

3539. A. Aha! You'll have to know that VORs and VORTACs are oriented on *Magnetic North*. The Minot omnirose is offset so that the 000° value is pointed about 11° east of True North, as indicated by the meridians. See where the drawn course line crosses the south boundary of the VORTAC rose and take the reciprocal of that value. *SPFM*, p. 181, VOR, and Fig. 21-4.

B. *No.* That is the *true* course and you'd have to subtract the variation of 11°E to get the course that the VOR is based on.

C. *No.* That's 180° out of phase, and if flying north to Minot, the course deviation indicator (needle) would have reverse sensing. (Turn *away* from the needle to get back on course.)

3540. C. Working out the problem on both an electronic computer and an E-6B–type gave you a time of 34 minutes. You may find *minor* discrepancies between the two types. (Groundspeed 104, distance 58 miles = 34 minutes.)

3541. C. The true course is 144° (measured), wind 030° at 12 knots, TAS 95 knots. Using the two types of computers, the true heading was found to be 137° and groundspeed to be 99 knots. At 99 knots and (measured) distance of 48 NM, a time of 29 minutes was required. There's no 29 minutes choice. But wait! There's the 2 minutes for climbout. Adding this 2 minutes for climbout, 31 minutes is the answer.

3542. B. 43 minutes. The true course (measured) is 345°, wind 300° at 14 knots. The computers give a true heading of 339° and a groundspeed of 80 knots. The distance is 53 NM, for a time of 40 minutes. Add the 3 minutes for climbout to get the answer.

3543. B. Drawing vertical and horizontal lines through Shoshone County Airport, as we did for location problems earlier, you find that the answer is 47°33′N. (You can count *down* from 48°N to confirm this. Also, Shoshone is 3′ north of the 47°30′ line.) Looking at the longitude, you'll see that the airport is 11′ *west* of the 116° meridian. The closest answer is 47°32′N, 116°11′W.

3544. A. This one was included, even though it was for a balloon, because it's a navigation problem. This is relatively simple since it can be assumed that with the wind from 220° at 5 knots the balloon will drift 040° *true* at 5 knots, and after 2 hours it should be 10 NM from the airport on a 040° bearing. The balloon is about 1½ NM east of Hackney (PVT) Airport.

3545. B. The computed true heading is 187° (groundspeed 104 knots), variation (see the isogonic line running from upper right to lower left on the chart) is 19°E. This is subtracted from the 187° to get an answer of 168° *magnetic heading*, which is within 1° of answer B.

Note: In measuring the true course you may be 1° off and a 1- to 2-knot or 1- to 2-degree difference was sometimes found between the answers given by the two computer types. But there is enough difference between the choices to keep this from being a problem.

3546. A. Computation shows a true heading of 136° (with groundspeed 99 knots). Note that the 19°E isogonic line crosses in the middle of the course line. Subtracting 19° from 136° gives a *magnetic heading of 117°*, the closest to the 116° choice. Again, *your* course measurement and the question-writer's measurement may be slightly different, but this is taken into account, and the *closest* answer to yours will be the *right* answer if you did the problem right.

3547. A. True course 345°, wind 300° at 14 knots, TH is 339° (GS is 80 knots). Variation is 19°E so 339° − 19° = 320°, which is closest to 319°. (Again, probably due to a difference in TC measurements.)

3548. B. The TC (measured) is 212°, distance 57 NM, TAS 90 knots. With a wind of 090/16, TH is 203° and GS is 97 knots. At 97 knots, to fly 57 NM takes 35 minutes, plus 2 minutes for the climbout to give 37 minutes. When you are working a problem such as this, write down other information (such as TH, etc.) because it may be used in a following question. *Be sure, however,* that wind and TAS haven't been changed in a new question pertaining to *that same route.*

3549. B. TC from Claxton-Evans County to Hampton Varnville is 044°, distance 57 NM. With a wind of 290° at 18 knots and a TAS of 85 knots, the answer is a TH of 033° and a GS of 91 knots. To go 57 NM at 91 knots takes 37+ minutes, plus 2 minutes for climbout, giving a total of 39+ minutes.

3550. C. *SPFM*, Chapter 19, Variation, and Deviation. TH is found to be 203°. The closest isogonic line (4°W) is in the upper right corner of the chart. Magnetic heading is 203° + 4° = 207°. Looking at the compass card in the upper left corner of figure 24, you see that 207° is close to the 210° value where 4° is added (compass heading is 214° there), so add 4° to 207° to get *211°*.

3551. A. After plugging the *given* numbers into your computer you find TH to be 033° (with GS of 91). Finding CH requires getting the magnetic heading (variation is 4°W, so MH is 037°) and using this in the compass card at the upper left corner of the chart. There it indicates that 3° should be subtracted in the 030° magnetic heading area. CH is 037° − 3° = 034°.

3552. A. *SPFM*, Chapter 25, Problems and Emergencies, and Fig. 25-4. The lines intersect at Guyton. Note that the Allendale VOR is *not* on the airport but is exactly on the top margin of figure 24.

3553. B. *SPFM*, Fig. 25-6. One way to do it would be to draw a line from the Savannah VORTAC to Hampton Varnville Airport (003°) and add 180° (the reciprocal) to get 183°. On a full chart you could draw a line form Hampton Varnville on *through* the Savannah VORTAC rose and read the number off directly.

3554. C. This one is pretty simple in that the 248 radial makes up the MILEN intersection on V-185 and 216R is DOVER intersection. The distance between these is 10 NM and the time en route is 9 minutes (0951 to 1000) for a groundspeed of 67 knots. (The aircraft isn't exactly making sonic booms.) The distance from DOVER to the Savannah VORTAC is a measured 40.5 NM, so the ETA is 36 minutes later; that is, 1000 + 36 = 1036.

3555. B. This question is included because it uses basic navigation skills, even though 36 knots TAS means a pretty slow-moving object. Anyway, a groundspeed of 41 knots is found and it takes 59 minutes to travel 40 miles. The pilot's log for *this* trip might read: "Thursday, May 4th. Cruising as before. Vicinity of Miller Grove. Water running low. Crew mutinous. Scurvy aboard."

3556. A. *SPFM*, Chapter 19, Variation. The true course (TC) is 082°. With a magnetic variation of 6°30′E you would subtract this (East is *least* when going from *true* to magnetic) to get 075° MC. Since VOR roses are based on Magnetic North, you could lay a straight edge through the Quitman VORTAC parallel to the course line and get 075°.

* * *

3558. A. Working out the wind triangle for a TC of 102°, we find a true heading of 084°. Subtracting 6°30′ (call it 7°) from this gives an answer of 077°, which is close enough to the 078° of answer A. Remember again that your answer may be off by a degree or a nautical mile or so from the right choice.

3559. B. Drawing a 245° line from Sulphur Springs VORTAC and a 130° one form Blue Ridge VORTAC results in an intersection right over Meadowview Airport.

3560. A. Okay, this is a *little* tricky. Remember that radials *radiate* from the center of the VOR/VORTAC. For instance, if you were on the 240 radial, the proper thing to do would be to set up 060-TO on your VOR receiver in order *to* fly to the VOR. Here, I'd set my VOR to 101-TO, even though I'm on the 281 radial.

3561. C. The VOR receiver is tuned to Blue Ridge VORTAC and the Omni bearing selector (OBS) is set to 210°. This is 90° from your present bearing of 121°, so the aircraft is in an "area of confusion" as far as your setting is concerned. You are neither TO or FROM the VORTAC as shown in illustration 7 in figure 29.
  A. *No.* Fig. 29, item 1, shows that the airplane is approximately *south-southwest* of the station (030-TO) and needs to fly left to intersect the 030 bearing TO the VORTAC. (It can't be over Lone Oak.)
  B. *No.* Illustration 4 in figure 29 shows that the airplane is close to being on a 030° course TO the VORTAC but would have to fly right to get on that 030-TO (and a bearing of 210° FROM or south-southwest of) the VORTAC. It cannot be over Lone Oak, which is *southeast* of the VORTAC.

3562. A. Using the numbers given, the answers are TH = 138°, GS = 102K. The distance is 22 NM and requires 13 minutes. You don't need magnetic variation for this problem.

3563. A. True heading is 192°, GS = 126. To go 17 miles from Addison to Redbird takes 8 minutes.

3564. B. Great Scott! Another one of those slow aircraft (35K TAS). If you had wrongly *added* 7°E variation you would have gotten answer C.

3565. A. One computer got a TH of 010°, the other 011°. Subtract 7° (east) from the 010° to get an MH of 003°. If you'd *added east* variation, you would have gotten 017° (answer B).

3566. A. All right, if the OBS setting is on 253° and TO, the airplane would be on a bearing of 073° FROM the station if the needle was centered. The CDI needle deflection is to the right, and that means you'd have to fly to the right (more north) to center it; you are east-northeast of the VORTAC.

3567. A. Draw a horizontal line through Cooperstown Airport and note that the latitude is 47°25′N. It's 5′ below the 47°30′N parallel. Draw a vertical line through it and see that the longitude is 98°06′W as it is 6′ west of the 98°W meridian.
  B. *No.* Remember that in the United States the numbers on the chart *increase* as you go up (north) and left (west) on the chart. The latitude (47°25′N) is okay but the longitude 99°54′W is wrong.
  C. *No.* The longitude is right (98°06′W), but there's been too much latitude given here. (Sorry.)

3568. C. You don't even have to use your protractor on this one since the VOR/DME is located on the airport and *oriented to Magnetic North* (as all VOR/VOR-TACS are).

3569. B. *Watch this one*—it's in MPH, not knots, and gives a magnetic course to begin with. First convert 282° magnetic to *true* course (got to work backward here and *add* the 8°E variation because we're going from magnetic to true for a change). The true course from Eckelson is 290°, and using your protractor, you would draw a vertical line through Eckelson to help with this. Put the protractor hole over the town and rotate the protractor until the drawn line intercepts 290° on the protractor. You'd then draw the 290° line and measure 25 *statute* miles (2.5 hours at 10 mph) along that line. It's right over the town of Buchanan. Although this is a balloon problem, it was included to show that all questions should be read *carefully*.

3570. C. The indicator is set at 030° FROM (Wimbledon is about 025° from the Jamestown VOR). The indicator says that the airplane must fly about 5 degrees to the right to get the 030 radial and so that's the place.

3571. C. The airplane departs at 0945 EDT for a 2-hour flight, so the landing is made at 1145 EDT (the pilot forgot to change his watch). UTC requires the addition of 4 hours (figure 28) to EDT, so 1145 + 4 = 1545Z.

3572. B. To keep down confusion, let the pilot keep CST time until after the landing. Okay, he leaves at 0930 CST and arrives at 1130 CST. Setting his watch back an hour to MST gives a time of 1030 MST.

3573. C. The airplane departs at 0845 CST and flies for 2 hours, arriving at its destination at 1045 CST. Fig. 28 shows that 6 hours is to be added to CST to get UTC (Zulu) time. 1045 + 6 = 1645Z. (In other words, converting to MST might confuse the issue.)

3574. B. The aircraft departs at 1615 MST and, after a flight of 2 hours and 15 minutes, arrives at the PST airport at 1830 MST. Then the pilot's watch must be set back 1 hour to 1730 PST.

3575. C. The aircraft departs at 1030 PST, so the plane would land in the central time zone at 1430 PST. Eight hours must be added to 1430 PST to get 2230Z and then the pilot's watch must be set 2 hours ahead for the local time of 1630 CST.

3576. A. The airplane departs at 1515 MST and flies for 2 hours and 30 minutes, arriving at 1745 MST. The pilot's watch then must be set back one hour because the local time is 1645 PST.

3577. C. The OBS is set on 030-TO, so the aircraft needs to fly left (or more westerly) to get on that radial. The airplane is south of the VOR.

3578. B. The OBS is set on 030 and there is no TO-FROM indication, meaning that the aircraft's position is approximately on a 90° bearing from that 030 radial. The needle indicates that the selected radial is to the left so the aircraft is southeast of the VOR.

3579. A. The OBS is set to 210° and the needle is centered with a TO indication. At that instant of time the airplane would have to fly a *course* of 210° to get to the station, so it's on the 030 radial. Radials are spokes radiating from the VOR.

3580. C. Assuming that the movable card has been set to the aircraft magnetic heading, you'd read the magnetic bearing to the NDB (nondirectional beacon), which is read directly and is 210°. *SPFM*, Fig. 21-2.

3581. C. Again, assuming that the card has been set to the aircraft's magnetic heading, read 190° directly from the face.

3582. C. In illustration 2 the aircraft bearing TO the NDB is 190° (010° FROM) or the aircraft is north-north-east of the station (draw it). Draw a line running north from the station (this would be the track for 180° TO). You'll see that the aircraft will have to be turned more westerly to get over to that 180°-TO line and 220° is the only one that fits.

3583. B. Again assuming that the card has been set to the magnetic heading of the aircraft, just read the FROM bearing on the *tail* of the arrow.

3584. C. The bearing to the station in illustration 4 is 200°, but the aircraft is flying on a magnetic heading of 220°, using a +20° correction angle for a right crosswind.

3585. A. In illustration 1 it's 210° TO the station as it must be in a position 030° FROM (outbound bearing) the NDB.

3586. C. *Relative* bearing is required here. The station is 60° to the left of the tail, or, counting clockwise, it's 60° past the 180° position, making a *relative* bearing of 240°. *SPFM*, Fig. 21-2 and explanation.

3587. B. Watch that *relative* bearing again. The needle is 55° past the 180° relative bearing or the relative bearing is 235°. Another way would be to subtract the angle (left) of 125° from the 360° or 000° relative position to get 235°.

3588. C. The needle is pointing 20° to the *left* of straight ahead (360° or 000° position), so 360° − 20° = 340° relative bearing TO the station.

Note questions 3589–3597 refer to a Fixed Card ADF. (SPFM, Fig. 21-2, top A, B, and C.)

3589. C. The *relative* bearing on a fixed card ADF is what it shows in *relation* to 000° (or 360°-straight ahead), so 315° is the answer, since the arrow *head* is at that value.

3590. A. Again, the *relative* bearing TO the station on a fixed card ADF is what the arrow *head* says. Without knowing the magnetic heading, it would be impossible to find the magnetic bearing or course to the NDB.

3591. B. Look at the head of the arrow for TO information and 180° is the relative position TO the station.

3592. B. *SPFM,* Chapter 21, Radio Beacon, and Fig. 21-2. On a magnetic heading of 320° and a relative bearing of 225°, the magnetic bearing TO the station is 545°. Hold it! Subtract 360° from that for an answer of 185° TO the station. You can also note that the needle is 135° from the nose (left) so, by subtracting this from the reading of 320°, an answer of 185° bearing TO the station is obtained.

3593. A. *SPFM,* Chapter 21, Radio Beacon. On a magnetic heading of 035° and a relative bearing of 000°, the magnetic bearing TO the station is, of course, 035°—it's right on the nose.

3594. B. Magnetic heading of 120° and a relative bearing of 045° is 120° + 45° = 165°.

3595. C. Magnetic *bearing* TO the station is 240°, so the magnetic *heading* is 195°. The aircraft heading is 45° to the left of that magnetic bearing of 240° (240° − 45° = 195°).

3596. B. The magnetic *bearing* TO the station is 030° and the relative bearing is 270° (or 90° to the left). It means that the needle is pointing along a line of 030°, and the aircraft is headed 90° to the *right* of that, or 120°.

3597. C. Magnetic *bearing* TO the station is 135° so the only way to get this combination is to be headed 360°. With a fixed card ADF, the only time relative bearings and magnetic bearings TO the station are the same is on a heading of 360°.

3598. C. AIM. Remember the Cessna 182 when using the VOT (180°-TO) and, of course, the opposite is 000° FROM.

3599. C. Check the segment of the Class B airspace over Fort Worth Meacham Field and you'll note at a point of about 155° and 1¾ inches (on the chart) that the information given is that the segment extends from 4000 to 10,000 feet MSL ($\frac{100}{40}$).

3600. B. The Class B airspace is centered at Dallas–Ft. Worth International Airport. Looking at Addison Airport (area B), you'll see that the floor of the Class B airspace is 3000 feet MSL. (That segment extends from 3000 to 10,000 feet MSL as you can check by looking almost due east at about 1 inch on the chart.)

3601. A. AIM. Warning Areas may contain the same dangers to aircraft as Restricted Areas, which have unusual, often invisible hazards, such as aerial gunnery or guided missiles. Warning Areas are basically Restricted Areas over *international* waters.

3602. B. AIM. You could expect to encounter military flight training, including aerobatics.

3603. A. AIM. Military Training Routes such as IR 644 will have some, but not necessarily all, of their operations above 1500 AGL. These will be high-speed operations (250 knots or more).

3604. A. AIM. The frequency of 122.9 MHz is used for airports without communications, and you should broadcast in the blind, as indicated in answer A.

3605. B. That's what it says. The (Ⓒ) by a frequency indicates that it be used as a Common Traffic Advisory Frequency (CTAF).

3606. C. Note in App. 2, Legend 1, that the "C" symbol denotes the CTAF and in that example is the tower frequency (118.3). The AIM Glossary states that CTAF is "a frequency designed for the purpose of carrying out airport advisory practices while operating to or from an uncontrolled airport. The CTAF may be Unicom, Multicom or tower frequency . . ." In figure 23 the (Ⓒ) symbol for CTAF accompanies the Unicom frequency of 122.8. Fig. 32 indicates in the Communications section that the CTAF is 122.8 (MHz), the Unicom frequency.

3607. C. See the answer for question 3606. When the tower is not in operation, it is an uncontrolled airport and figures 23 and 32 indicate that 122.8 (MHz) is the CTAF.

3608. C. The sectional chart and *A/FD* both indicate this frequency as Unicom. (*Don't* call the tower for fuel information.)

3609. A. The tower frequency for Redbird is 120.3 MHz and the symbol (Ⓒ) for the CTAF is adjacent to that number. See the explanation for answer 3606.

3610. A. The CTAF at Cooperstown is listed as 122.9 MHz. There is no Unicom. Circling the airport in a left turn is one form of communications, showing that you plan to (perhaps) land, but it's not the best answer.

3611. B. The number just before the (Ⓒ) CTAF symbol is 123.0 MHz. (See App. 2, Legend 1, Airport Data.)

3612. B. The CTAF/Unicom frequency at Barnes County is 122.8 MHz.

3613. A. AIM, says that on initial contact the full aircraft identification should be given using the phonetic alphabet for the letters in the identification.

3614. A. AIM. "Four thousand, five hundred."

3615. C. AIM. "One zero thousand, five hundred" is the correct usage. (You don't have to indicate feet; it's understood that it isn't yards, miles, or fathoms.)

3616. C. AIM, and *SPFM*, Chapter 22, Other Services. Call "Flight Watch" on 122.0 MHz.

3617. A. AIM, and *SPFM*, Chapter 22, Other Services. EFAS is designed to communicate such information as turbulence, winds, and icing between pilots and Flight Watch specialists. No flight plan or preferential routing service is done on Flight Watch.

3618. A. AIM. "All aircraft are requested to maintain a minimum altitude of 2000 feet above the surface of the following: "National Parks . . . Wildlife Reserves . . ." This information will also be noted on the border of the sectional chart.

3619. B. See the *A/FD* for information on the permanent type of operations as indicated on the sectional chart. (See *A/FD*, Airport Remarks, and the section Parachute Jumping Areas, which give the location as a distance and radial from the nearest VOR/VORTAC with maximum altitudes and times.) When such operations first started, it's likely that a NOTAM was published. Notes on the border of the chart pertain to tower frequencies, restricted and prohibited areas, etc.

3620. A. AIM, and *SPFM*, Fig. 21-20. The *less* than 1200 feet AGL and daylight gives a good clue here. Wall is not within controlled airspce (it's in Class G airspace).

3621. B. AIM. It's in Class G airspace. Cooperstown is outside of controlled airspace, so in the daylight hours to operate VFR above 1200 feet AGL, but at less than 10,000 MSL, a minimum of 1 mile visibility with a cloud clearance of 500 feet below, 1000 feet above, and 2000 feet horizontal is required.

3622. A. Lowe Airport is in Class G airspace from the surface to 14,500 feet MSL (at present).

3623. C. The color magenta designates Class E airspace that exists at *700 feet AGL upward* and is used for instrument approaches. The airspace below 700 feet in that area is Class G (uncontrolled).

3624. A. The magenta boundary around McKinney Municipal indicates Class E airspace, with uncontrolled airspace from the surface to 700 feet AGL. The airport *is* in the DFW Terminal Control Area *veil*, which extends from 20 to 30 miles out, but the only requirement to operate in a TCA veil is to have an operating transponder and encoder so that the airplane can be better seen on radar. *No* communication or control is required up to 700 feet, with certain conditions.

3625. C. Note that Meacham has a tower and lies *within* Class D airspace, as indicted by the segmented blue circle. The box in that circle indicates that the top of the Class D airspace is at 3200 feet MSL.

3626. B. Class C airspace is usually from the surface to 4000 feet above the surface within the NM radius of the inner circle and some higher base or bases to 4000 feet AGL in the outer circle. The entire outer circle at Savannah has a base of 1300 MSL ($\frac{41}{13}$). When set to the nearest 100 feet above the airport, 1200 feet AGL (A) is also correct.

3627. B. FARs 91.130 and 91.215. Norfolk International is in Class C airspace and a two-way radio and a Mode C transponder are required.

3628. B. Airports that prohibit special VFR operations have a notation "NO SVFR" directly above the airport name box. Looking carefully at figure 26, you'll see that only Dallas/Ft. Worth and Dallas Love Field have that notation.

3629. C. If the *floor* is *other* than 1200 feet above the surface, it will be noted with an altitude (7500 MSL). See App. 2, Legend 1, and FAR 71.6(b)(6). The low-altitude airway system, as shown here, extends up to 17,999 feet.

3630. A. The small black square in the lower right hand corner of the frequency box for Minot VORTAC indicates that Hazardous In-Flight Weather Advisory Service is available. (See the Sectional Chart Legend following page 360.) The "normal" communication on that VORTAC frequency is not available (frequency underlined).

3631. C. The caution box doesn't indicate that the cable is radar-outfitted, but there is a notice that there's a balloon on a cable to 3000 feet MSL.

3632. C. That's a visual checkpoint. See App. 2, Legend 1.

3633. A. Chesapeake Municipal Airport has an elevation of *20 feet*, lights, a hard-surface runway 4200 feet long, with 122.7 MHz being the CTAF. There's a seaplane base there, also.

3634. B. This one requires some thought. In App. 2, Legend 1, at the bottom center, there is a note indicating that contour intervals are drawn every 500 feet. This would mean that if the terrain in the light-tan area (2000–3000 feet MSL) exceeded 2500 feet, a contour interval line would be shown. No contour line is shown so the terrain is between 2000 and 2500 feet MSL.

3635. A. Minot and Mercer County Regional have the "cogs" indicating that fuel is available (see App. 2, Legend 1). Garrison has neither of these, shooting down B and C.

3636. C. These are visual checkpoints. See App. 2, Legend 1.

3637. B. The lighted obstacle, approximately 7 NM southwest of Savannah International, is 1549 MSL and (1532) AGL. See App. 2, Legend 1.

3638. B. App. 2, Legend 1. The larger (top) number on the sectional chart is the MSL value of the obstacle with the AGL figure in parentheses below. The AGL height is (400) feet.

3639. B. The top of the tower is 773 MSL (remember the larger numbers are MSL and the smaller numbers in parentheses are height above ground). To clear it by 500 feet would require an altitude of 1273 feet MSL.

3640. C. The top of the obstacle is 903 feet MSL, so an altitude of 903 + 500 = 1403 feet MSL is required to clear it by 500 feet.

3641. B. The control tower (CT) frequency is 126.0 MHz (given in bold numbers with the other airport information).

3642. C. SPFM, FAR 91.119. (A clue is given when "congested area" is mentioned, and you can see this by looking at figure 26, area H.) Minimum Safe Altitude: General states: "(b) Over congested areas. Over any congested area of a city, town or settlement, or over any open-air assembly of persons, an altitude of 1000 feet above the highest obstacle within a horizontal radius of 2000 feet of the aircraft." The Cedar Hill towers are 2349 feet MSL so 3349 feet MSL is required by FAR.

3643. B. Look at Legend 1, the Sectional Chart Legend (Radio Aids to Navigation) immediately following page 360 in this book. The symbol at DFW is that for a VORTAC.

* * *

Questions 3651–3856 pertain to operations.

3651. A. The engine needs to be producing less power and getting more air across the cylinders, and this is the best combination given.

3652. A. SPFM, App. D, Engine Cooling. Enrich the mixture. The other choices go the wrong way.

3653. A. SPFM, Chapter 23, Notes on Constant-Speed Propeller Operation.
   B. No. The propeller control regulates a constant RPM; the prop governor changes the blade angles as necessary to do this.
   C. No. The throttle for a constant-speed propeller controls manifold pressure.

3654. B. SPFM, Chapter 23, Notes on Constant-Speed Propeller Operation. A fixed-pitch propeller is only efficient in a comparatively small range of airspeeds, these airspeeds being in a low range ("climb"-type, low pitch) or higher range ("cruise"-type, higher pitch). The constant-speed propeller is efficient through a wide range of airspeeds. The pilot sets the RPM, which in turn gives a more efficient performance because of the governor-controlled blade pitch angles.

3655. B. SPFM, Chapter 23, Notes on Constant-Speed Propeller Operation. This is the "proper" answer, but there's a lot of misinformation about manifold pressure and RPM. Some people believe that the power setting should be "squared," that is, that the manifold pressure (inches Hg) should never be higher than the RPM (hundreds), as in 23″ Hg and 2300 RPM. You certainly can damage an engine by a very high manifold pressure and extra-low RPM, but use the manufacturer's figures for a given power setting. Many times that recommendation is for a higher manifold pressure (inches) than RPM (hundreds).
   A. No. See B.
   C. No. You may be at cruise at a high RPM and not need to use a rich mixture.

3656. A. Set the RPM for proper operation and check the engine gauges (particularly the oil pressure) for desired indications.
   B. No. SPFM, Chapter 7, Pretakeoff or Cockpit Check. It's best to do this at the end of the flight.
   C. No. That should be done before the start and the regular brakes checked again as the airplane starts to move for taxi.

3657. B. SPFM, Chapter 5, Propping the Plane.
   A. No. That would tell the guy inside to turn the ignition ON.
   C. No. The man doing the propping calls out all commands.

3658. B. SPFM, Fig. 4-1. While there is no FAR requirement for draining the fuel or a walk-around, you should do a careful walk-around inspection of the aircraft, including visually checking the level and draining quick drains before every flight (SPFM, Chap. 4). This is the best answer.

3659. A. The checklist ensures that all items are checked in a logical sequence. You should always use a checklist, even if you are flying an "uncomplicated" Cessna 152.
   B. No. Don't memorize. Distractions could make you forget something. (Distractions can occur when using a checklist; if so, start over.)
   C. No. Well, there's two sides to this. Most passengers will be impressed with your professionalism, while others may wonder why you have to read how to operate the airplane. (Jeez, you don't know how to fly this thing yet?) To heck with 'em, use a checklist.

3660. C. The hint here is "extended period of time." There are real dangers of mud daubers, birds, and mice, etc. building nests and creating serious problems with pitot-static systems, engine/carburetor obstruction, etc.
   A. No. If you're doing flight training within 50 NM of the airport, an ELT is not required anyway (SPFM, FAR 91.207).
   B. No. Condensation in the fuel tanks can be a problem, but it's more a normal, not a special, check. (See the explanation for answer 3659.)

Questions 3661–3677 pertain to weight and balance.
(Some moments have been rounded off.)

3661. A. *SPFM*, Chapter 23, Running a Weight and Balance. A is the closest answer, but airplanes certified in 1976 and after include *full* oil and unusable fuel in the empty weight. *Standard Empty Weight* is the weight of an "average" airplane of that model (including full oil and unusable fuel). *Basic Empty Weight* is the Standard Empty Weight plus the added equipment installed in a *particular airplane* (for instance, C-152 N7557L). *Older airplanes* have a *Licensed Empty Weight,* which includes unusable fuel and undrainable oil, which makes choice A correct.

3662. C. *SPFM*, Chapter 23, Running a Weight and Balance. Aviation gasoline weighs 6 pounds per gallon, so 18.4 gallons (rounded off) must be drained to drop 110 pounds.

3663. C. Aviation gasoline still weighs 6 pounds per gallon, so 15 gallons must be drained to reduce the weight by 90 pounds.

3664. B. *SPFM*, Chapter 23, Running a Weight and Balance. Adding all the weights (180 pounds for fuel), a total of 2055 pounds is found. To find the pilot and passenger moments, multiply 380.0 pounds by 64.0 inches to get 24,320 pound-inches. The fuel weight (180 pounds) is multiplied by 96.0 inches to get 17,280 pound-inches. Adding everything:

Weight = 2055
Moment = 193,193 pound-inches (lb-in.)
$$CG = \frac{193,193}{2055} = 94.01 \text{ inches aft of datum}$$

3665. B. *SPFM*, Chapter 23, Running a Weight and Balance. The weight is 2970 pounds, 20 pounds over the max. Note that the arms are given for each position (front seats, rear seats, fuel, baggage, etc.) and you would use these. The moment was found (with interpolation) to be 247,640 pound-inches, within the allowable CG range. A further check shows that the CG position is 247,640/2970 = 83.4 inches aft of datum and within the 2970 pound CG limits (see question 3667).

Note: For different airplanes' weight-and-balance papers a moment of, say, 280,000 pound-inches might be noted as moment/100 = 2800 (figures 33 and 34), or moment/1000 = 280 (figure 35).

3666. A. The weight without baggage adds up to 2905 pounds, and the moment without baggage is approximately 239,600 pound-inches (with some interpolation in figure 33). This leaves 45 pounds in weight allowed, but a check is made to see if this affects the moment (figure 34). Finding the moment for the 45 pounds of baggage, the total is approximately 2460, well within the 2422 to 2499 allowed at 2950 pounds. See *SPFM*, p. 233. Use the arms as given in figure 33 for passengers, fuel, and baggage.

3667. B. Calculating:

	WEIGHT (LB)	MOMENT (LB-IN.)
Empty weight	2015	155,400
Front seat	350	29,750
Rear seat	325	39,325
Fuel (6 × 35)	210	15,750
Baggage	27	3,780
TOTAL	2927	244,005

$CG = \frac{Moment}{Weight} = \frac{244,005}{2927} = 83.4$ inches—closest answer. Check figure 34 and note that it is within limits.

3668. C. Calculating (and rounding off):

	WEIGHT (LB)	MOMENT (LB-IN.)
Empty weight	2015	155,400
Front seat	415	35,300
Rear seat	110	13,300
Fuel	378	
(Main tanks)	264	19,800
(Aux. tank)	114	10,700
Baggage	32	4,500
TOTAL	2950	239,000

The weight is within limits, but the CG is too far forward. The total moment is too small (for the weight of 2950 pounds: 242,200 to 249,900 are the limits), and the CG is too far forward at *239,100 pound-inches;* CG position = 81.0 inches aft of datum.

3669. A. Calculating:

	WEIGHT (LB)	MOMENT/1000
Empty weight	1350	51.5
Pilot and front passenger	250	9.5
Rear passengers	400	29.2
Fuel (30 gals)	180	9.0
Oil (8 qt @ 7.5 lb/gal)	15	−0.2
TOTAL	2195	99.0

Note: Add up all of the positive moments (99.2); then subtract any negative moments (−0.2 here) to get a final total moment of 99.0 (or 99,000 pound-inches). Add all weights together.

The gross weight (Normal Category) is 2300 lb., so 105 pounds may be carried *if* the CG limit is *not* exceeded, so

	WEIGHT (LB)	MOMENT (LB-IN.)
	105 lb	10,000
	2195 lb	99,000
TOTAL	2300 lb	109,000 (right on the rearward limit)

3670. B. Get a separate piece of paper and lay it out as was done for questions 3667 and 3668. Adding up the weights gives a value of 1999 pounds. Adding up the positive moments (81.0 pound-inches) and subtracting the negative moment (0.2) from that total gives a total of 80.8 (or 80,800) pound-inches. Looking at the Center of Gravity Moment Envelope in figure 35, you see that the moment of 80.8 *and* the weight of 1999 pounds put it in the Utility Category Envelope (barely).

3671. C. Lay it out on paper, as in questions 3667 and 3668: Total weight = 2060 pounds; total positive moment = 91,000 pound-inches (91.2 − 0.2 = 91.0). There are 240 pounds left in weight for fuel, or 40 gallons (2300 − 2060 = 240 pounds). (It's assumed that the airplane has long-range tanks because 38 gallons is the maximum for standard tanks.) A quick check of this added weight of the fuel on the moment shows a total of 102,500 pound-inches, within the Normal Category Envelope. *Incidentally, manufacturers design their aircraft so that the CG moves as little as possible as fuel is burned.*

3672. B. Back to the paper: Lay things out in order, now. (Full standard tanks have a weight of 228 pounds and a moment of 11.0—see figure 35.) Your number may vary slightly from the 74.9 (or 74,900) pound-inches according to how you read the graph, but the choices are far enough apart to get the correct answer.

3673. B. The airplane weighs 2033 pounds, which is out of the Utility Category and into the Normal Category Envelope. The total *positive* moment is 79.4 (or 79,400/1000) pound-inches and 0.2 is taken from this for the oil to give a moment of 79.2.

3674. A. Okay, the total weight was 2690 pounds, and the moment/100 was 2260 before the passengers started playing musical chairs. Looking at it logically, you know the weight will be *180 pounds* less after the front seat passenger gets out. The 204-pound passenger moving forward from the rear seat (there's nobody there now) will then move the CG forward a noticeable amount. The moment before the transfer (working with full numbers) was 226,000 pound-inches. The CG then was 226,000/2690 = 84.01 inches aft of datum.

Looking at the moment change, assume that the front seat and rear seat passengers both got out. The loss of moment is (figure 33):

Front seat passenger	=	− 15,300 lb-in.
Rear seat passenger	=	− 24,700 lb-in.
TOTAL LOSS		− 40,000 lb-in.
NEW TOTAL MOMENT		186,000 lb-in.

The former rear seat passenger (204 pounds) returns from the bathroom and gets into the front seat for the next leg of the trip. The added moment (figure 33) is 17,300 pound-inches (rounded off).

New total moment = 186,000 + 17,300 = 203,300

$$\text{New CG} = \frac{\text{New total moment}}{\text{New weight}} = \frac{203,300}{2510}$$
$$= 81.03 \text{ inches aft of datum}$$

or *the CG is about 3 inches farther forward* (84.01 − 80.99 = 3.02 inches)

There are equations that can be used for this, but the longer steps were given here to show the principle involved.

3675. B. The airplane is 54 pounds heavier and 9 gallons of fuel will take care of that. The fuel draining will reduce the moment by 4050 from 247,600 to 243,550 pound-inches, within the envelope. *Weight* was the problem, not CG.

3676. A. The old weight was 2890 pounds and the moment was 245,200 pound-inches. (Sometimes it seems easier to use the full number, not 2452). The CG was 245,200/2890 = 84.8 inches. Subtracting the weight (210 pounds) and the moment decrease caused by the 35-gallon fuel burn (15,800 pound-inches), the new CG is found: 229,400/2680 = 85.6 inches. The CG is aft.

Looking at the limits in figure 34, you see that the aft (or maximum) moment at 2680 pounds is 228,700; the CG is aft of limits.

3677. B. In figure 34, the safe range for 2790 lb is 224,300 to 237,400 pound-inches. Working out the moment at the weights and moments given in the question, the moment is 222,200 pound-inches and out forward of the envelope. By adding 100 pounds to the baggage compartment (weight *now* 2890 lb), the moment is 236,200 pound-inches. Looking at 2890 lb in figure 34, it's seen that the safe range is 235,400–245,200 pound-inches. (Adding weight doesn't help performance.)

3678. C. Some interpolation is involved here. Note that *all* of the settings in figure 36 are for 65% maximum continuous power. (See the note just under the heading "Cruise Power Settings.") The *left column* is the −20°C (−36°F) or below standard. There's a question here: Apparently the *standard sea level temperature* used in this chart (ISA) is 63°F and 17°C instead of 59°F and 15°C. *So be it.* On with the problem . . .

So, using the left column, at 8000 feet the TAS is 181 MPH and at 10,000 feet it is 184 MPH. Since 9500 feet MSL is three-fourths of the difference between the altitudes and the difference is 3 MPH (not easy to do, if you remember your fractions), add 2 MPH to 181 to get 183 MPH.

3679. B. Okay, the temperature of 22°C is found in the *right* column. Note that the full consumption is 11.5 gallons per hour at a TAS of 164 knots. The wind is calm so that's no factor. So, dividing the distance (1000 NM) by the TAS of 164 knots (you recall that a "knot" is 1 NM per hour), an answer of 1000/164 = 6.1 hours at 11.5 GPH = 6.1 × 11.5 = 70.1 gallons is obtained.

If you'd used MPH for the 1000-NM trip, you might have come close to A. Don't mix apples and oranges. (That's an original expression—tell 'em, you first read it here.)

3680. B. Look at the right column, since +29°C is a well above average temperature for 4000 feet. The TAS is 159 knots. (Remember this is a 500-NM flight). Dividing, 550 NM by 159 K gives 3.14 hours at a fuel consumption of 11.5 GPH, or 3.14 × 11.5 = *36.1 gallons.*

3681. B. This calls for interpolation. As indicated earlier, everything in figure 36 is for 65% max continuous power or full throttle operation. On a standard day at 10,000 feet the fuel flow is 11.5 GPH; at 12,000 feet it's 10.9 GPH. At 11,000 feet the fuel flow is exactly between these, or 11.2 GPH.

3682. C. Okay, look at the right column for the ISA + 36°F information. For 6000 feet the manifold pressure required is 21.0″ Hg; at 8000 feet it's 20.8″ Hg, a decrease of 0.2 inch. Moving *one-quarter* of the way from 21.0 inches at 6000 feet to the 20.8 inches at 8000 feet (or at 6500 feet), you'd subtract 0.05 from 21.0 to get 20.95″ Hg. This value, of course, is easily read on any manifold pressure gauge. Maybe you'd better use 21.0, or answer C.

3683. B. *SPFM*, Chapter 17, Takeoff and Landing Performance. Assume that Runway 18 is exactly 180° magnetic. A wind from 220° at 30 knots would be off the nose 40°. Looking at figure 37, move out the 40° line (toward A) until you intersect the 30-knot ring. Make a dot there. Then move over to the *Headwind Component* axis and get an answer of 23 knots.

3684. C. Start at the 25-knot crosswind component and move up until you intercept the 45° line. Read the value wind velocity on the circle to get 35 knots (halfway between the 30- and 40-knot rings).

3685. C. Find the crosswind component of 12 knots and move up to the 30° line to get an answer *almost halfway* between the 20- and 30-knot rings. The choice of 24 knots is closest to the mark.

3686. C. Assume that the wind is from exactly 360° at 20 knots and the Runways 6, 29, or 32 are even numbers (060°, 290°, and 320°). Take 3 steps: Runway 6—the wind is 60° off the nose. Follow the 20-knot arc to 60°. The crosswind component for this value is 17 knots (nope). Do the same for 290°: 20-knot arc to the 70° angle. This gives a 19-knot crosswind component. Well, maybe the remaining choice of Runway 32 is the answer. The angle is 40°, and moving around the 20-knot ring to the 40° angle, you find a value of 13 knots. Eureka!

3687. B. The wind is from 180° at 20 knots, and again it's assumed that the runways are exactly lined up with their cardinal numbers. Max crosswind component = 13 knots. Runway 14 (140° magnetic). The wind is 40° off the nose at 20 knots. Moving around the 20-knot ring to 40°, you see that 13 knots is the crosswind component.
  A. Runway 10 (100° magnetic). The wind is 80° off the nose. Using figure 37 and moving around the 20-knot ring until the 70° angle is intercepted, you find that the crosswind component is about 19 knots. Not this one.
  C. Runway 24 is 60° off and the crosswind component is about 17 knots.

3688. A. The wind is 40° off the nose at 30 knots. The crosswind component is 19 knots. *Don't* inadvertently read the *headwind* component of 23 knots, which is given as answer B.

3689. B. It's pretty tough to read that graph. (Try reading a pressure altitude of 3965 feet and a weight of 2814, as given in the sample in figure 38.) Remember the airplane is landing over a 50-foot obstacle.

3690. A. The combination works out to 850 feet total distance. Assume no obstacle, since it wasn't mentioned.

3691. C. This is the closest number, based on trying to read the graph.

3692. C. Again, this is a tough one to call, but it looks like 1950 feet is the most reasonable figure for ground roll with a *tailwind component* of 10 knots. (Figure 38 has probably gotten well marked up by the time you finished the last four problems.) You aren't likely to get more than one or two of them on the actual test.

3693. B. This one is pretty straightforward after figure 38. At sea level and standard temperature, the ground roll is 445 feet. Note 1 says to decrease the distances shown by 10 percent for each 4 knots of headwind, so subtract 44.5 feet from 445 feet to get 400.5 feet. (Okay, call it 401 feet like answer B says.)

3694. B. *SPFM*, Chapter 17, Takeoff and Landing Performance. At 7500 feet (standard temperature) the total distance to land over a 50-foot obstacle is 1255 feet. The 8-knot headwind cuts this by 20 percent (10 percent for each 10 knots) to 1004 feet, *but* the airplane is rolling out on dry grass, which gives poorer braking than dry concrete, so 20 percent is added to get a figure of *1205 feet*.

3695. B. Total distance is 1195 feet (standard temperature). The headwind of 8 knots knocks off 20 percent (see note 1) so the answer is *956 feet* (hard-surface runway).

3696. C. "Great balls of sheet iron!" as Grandpaw Pettibone (mythical old Navy pilot) would say. No wind and 101°F! The standard temperature at 5000 feet is 41°F so it's 60°F high or 10 percent will be added, or 1.1 × 1195 = *1314 feet*. It's hoped that the pilot doesn't turn around and *takeoff* with a heavy load under these conditions.

3697. A. A little interpolation is needed. The ground roll would be a distance of 483 feet. That is halfway between 2500 feet (470) and 5000 feet altitude (495). The 12 knots of headwind then reduces this 483 feet by 30 percent, and the answer is *338 feet*.

3698. B. Again, interpolation is needed. The landing at standard sea level is 445 feet; at 2500 feet it is 470 feet. The roll at 1250 feet altitude is 458 feet, but the 8-knot headwind reduces this by 20% (see notes 1), and *366 feet* is the answer.

* * *

3705. B. Under the conditions cited, a distance of 1750 feet is required to clear a 50-foot obstacle.

3706. B. A total distance of 1400 feet is required under the conditions cited.

3707. A. The graph is so small that the 1150 and 1300 might be mixed up, but we'll opt for 1150 feet ground roll.

3708. A. Another question with the graph of figure 41. Well, this is the last one, with an answer of *650 feet*.
SPFM-WRITER'S NOTE: Each of the questions 3689–3698 and 3705–3708 was worked at least 3 times with the graphs of figures 38 and 41. The last check was with the figures enlarged by 41 percent to get more accuracy.

3709. B. Some FAA Flight Standards District Offices (FS-DOs) have *some* of the free material (answer A), but the GPO has *all* of the advisory circulars. The material available at the GPO is listed in Advisory Circular AC 00-2.5 (and later replacements). The Federal Register doesn't list the AC.

3710. B. Look around!
A. Check the instruments, but look around.
C. Sure, call the Unicom, FSS, or tower; hey, they'd be interested in hearing about that stall you're about to do.

3711. A. *SPFM*, Chapter 13, Emergencies on Takeoff and in the Pattern.

3712. B. *SPFM*, Chapter 26, Cones and Rods. You see better at night by *not* looking directly at an object.

3713. A. *SPFM*, Chapter 26, Cones and Rods. Look to the side and scan slowly.

3714. C. *SPFM*, Fig. 26-2.

3715. A. *SPFM*, Fig. 26-6. When you can see the red "rotating" beacon *and* a red light, you're looking at the left side of an airplane, so it's crossing to the left.

3716. A. *SPFM*, Fig. 26-6. When directly behind another airplane, you'll only see the flashing beacon and the white taillight.

3717. C. *SPFM*, Fig. 26-6. When you can see both the red and green wing lights, and they are at your altitude, better give way to the right quickly.

3718. B. *SPFM*, Fig. 26-5. Taxiway lights are blue.

3719. C. *FTH*, Chap. 14. It has been recommended that the night approach should be made the same as during the day, and if a VASI (Visual Approach Slope Indicator) is available, it should be used. At a strange airport (or at a familiar airport for that matter), do *not* get low on the approach because of possible obstacle encounters. Many pilots tend to make a *slightly* steeper approach at night, thinking of the trees, etc., reaching up for them.

* * *

3759. A. *SPFM*, Chapter 25, Getting Lost. *Communicate*. All you need for help from VHF/DF facilities is the ability to communicate verbally with them.

* * *

3760. B. AIM. If all 4 lights are white the airplane is *high*; while 3 whites and 1 red is *slightly high*; 2 white and 2 red, *on glide path*; 1 white and 3 red, *slightly low*; 4 red is *low*.

3761. A. AIM. Red means danger. You are *low*.

3762. C. AIM. The amber is a *caution* that you are above the glide slope and need to correct it.

3763. B. AIM. Green is good.

3764. C. Pulsating red is a danger signal that you are *low* on the approach slope.

3765. B. Right on the glide slope. (Red over white is a pilot's delight.) The airplane is too low for the far lights (red) and too high for the near lights (white), so it's right where it should be.

3766. B. The airplane is too high for *all* the lights.

3767. B. The airplane is below the glide slope. All red is danger. (Red over red could make some folks dead.)

3768. C. AIM suggests that the mike be keyed 7 times in 5 seconds for highest intensity at first, then 5 times in 5 seconds for *medium intensity*, and then 3 times for lowest.

3769. B. FAR 91.155 and the AIM. When the rotating beacon is in operation, weather in the Class D airspace is below basic minimums.

* * *

3771. B. *SPFM*, Chapter 26, Airport Lighting. Don't land at an airport *with these beacon "flashes" unless you have previous permission, or it is an emergency.*

3772. B. See explanation for 3771.

3773. B. AIM. That's a *displaced* threshold, normally set up because of required obstacle clearance on approach. That portion can be used for taxi, landing rollouts, and takeoffs, but *not for landing*.

3774. B. AIM. As noted in 3773, that portion (displaced threshold) may be used for takeoffs, but airplanes must touch down past B.

UH, HARRY, UH, IT <u>COULD</u> TURN OUT THAT THIS <u>ISN'T</u> JONESVILLE MUNICIPAL..."

Fig. 27-8. Know the difference between military and civilian rotating beacons.

3775. A. AIM. "A" is a displaced threshold, but "E" is a stopway and is an unusable runway because of the nature of its structure.

3776. C. AIM. That's a closed runway; it's been X'd out.

3777. C. AIM. That's a displaced threshold. Obstacles are usually the problem.

3778. C. *SPFM*, Chapter 6, Wind Indicators.

3779. C. The vertical limit of Class C airspace is 4000 feet above the primary airport.

3780. C. Class C outer areas have a radius of 20 NM.

3781. C. AIM; FAR 91.215.

3782. C. AIM. The pilot must contact ATC as soon as practicable after taking off from a satellite airport within Class C airspace.

3783. B. AIM.

* * *

3785. C. AIM. "Pilots operating under VFR should exercise extreme caution while flying within an MOA when military activity is being conducted."

3786. B. AIM. ". . . pilots of participating aircraft as well as pilots transiting the area shall be equally responsible for collision avoidance."

3787. C. AIM.

3788. C. You'd better talk to the control tower of the primary airport to avoid "foreign entanglements."

3789. C. AIM, Airport Advisory Areas. "At such locations, the FSS provides advisory service to arriving and departing aircraft."

* * *

3791. C. AIM. ATIS is the continuous broadcast of recorded non-control information in selected high activity terminal areas.
  A. *No.* ATIS is *prerecorded* information and won't help pilots in dangerous proximity to terrain, etc.
  B. *No.* The ATIS does reduce frequency congestion, but the information is *not* non-essential.

3792. B. The airplane is on a heading of due east; 3 o'clock is 90° to the right, so the pilot should look south.

3793. A. The heading is due north; 10 o'clock is about 60° to the left, or northwest.

3794. C. AIM. Radar only tracks the airplane's *path,* and the controller does not know its exact heading. Since 2 o'clock is in the *right front quadrant,* it would be wise to check between directly ahead and 90° to the right.

3795. C. Okay, there's no wind and the pilot is flying north (and it's assumed that the airplane is doing the same). The position of the traffic is at 9 o'clock and so is 90° to the left, or west of the airplane.

3796. A. AIM. In addition to the use of radar for the control of IFR aircraft, all commissioned radar facilities provide safety alerts, traffic advisories, and limited vectoring (on a workload permitting basis) to VFR aircraft.

3797. C. AIM. Pilots of departing VFR aircraft are encouraged to request radar traffic information by notifying ground control on initial contact with their request and proposed direction of flight.

3798. C. AIM. The purpose of this service is to provide separation between all participating VFR aircraft and all IFR aircraft operating in the TRSA.

3799. A. Call approach control before entering Class C airspace.

3800. C. AIM. The transponder code 7500 is squawked if the airplane is being hijacked. 7600 is for loss of communications, and 7700 is the emergency code. You'd get the attention of the folks in ATC if you selected those.

3801. A. AIM. Use code 1200 for VFR operations, regardless of altitude, unless otherwise instructed by ATC.

3802. A. AIM. The pilot's certificate has nothing to do with the required code. It's still 1200 for VFR operations.

3803. B. You're back on your own when radar service is terminated, so it's back to code 1200.

3804. A. AIM.

3805. B. *SPFM,* Chapter 6, Segmented Circle. Runway 22 is closed. *SPFM,* Fig. 25-3.

3806. A. Land on Runway 18 and expect a right crosswind. *Don't land on 22*—it's closed. Don't land on 36, or you'll have a left, tail crosswind.

3807. A. *SPFM,* Fig. 6-2. Extend the drawn traffic patterns for all runways as is done in Fig. 6-2.

3808. C. Note that turns are made *away* from the southeast quadrant.

3809. A. *SPFM,* Fig. 6-1. Check the windsock or wind cone. The airplane flies *out* of the big end of the trumpet, into the wind.

3810. C. *SPFM,* Figs. 6-1 and 6-2. Use left-hand traffic on Runway 35 and expect a left crosswind. Landing on Runway 8 or 17 could, depending on wind speed, result in some excitement for both you and any spectators present.

3811. A. AIM. A pilot who has just landed should not change from the tower frequency to the ground control frequency until he is directed to do so by the (tower) controller.

3812. A. AIM.

3813. B. AIM. The Air Traffic Control Tower is the controlling agency of the choices here. Center and the Automated FSS *don't* control the traffic and special VFR clearances.

3814. A. AIM. Who's to say that there aren't airplanes flying off the center line of the airway? Clear the area by gentle turns and scanning.

3815. A. Circumstances (weather) may warrant a change of highest or lowest cruising altitudes that may not be valid. The highest cruising altitude would *perhaps* be of some value to check to see how far you may have glided off the proposed course, but AIM, para. 5-4 says to enter the appropriate VFR cruising altitude.

3816. A. AIM. Suppose you filed a *4-hour* flight plan with interim stops but only listed the destination airport. You could go down 15 minutes after takeoff, but no form of search would begin until 4 hours and *30 minutes* after the departure time. Because of this, the FAA encourages separate (short) VFR flight plans, in this instance, for quick rescue efforts if necessary.

3817. C. AIM. This will give time to decide when the airplane has to be down (if overdue) and also to calculate how far it could have traveled.

3818. B. It's up to the pilot to close the VFR flight plan with an FSS or other FAA facility. The tower doesn't know or care if you are on a VFR flight plan. They have problems of their own controlling local traffic. *Don't* take up the controller's time by asking for that service.

3819. B. AIM. The frequencies in A and C are tower and Unicom frequencies.

3820. A. FAR 91.207.
   B. *No.* If it was required on every 100-hour inspection, the battery might have to be replaced every month or 6 weeks for some training airplanes.
   C. *No.* Suppose the battery has been used up during that period? Otherwise it would be replaced every 24 calendar months.

3821. C. AIM.

3822. C. AIM. Some pilots turn to 121.5 MHz for a few seconds on climbout from uncontrolled airports to catch any ELT signals early. (Don't switch if you are flying out of a controlled airport and should be listening to departure control.)

3823. A. AIM. "FSS's are allocated frequencies for different functions; for example, 122.0 MHz is assigned as the En Route Flight Advisory Service frequency at selected FSS's."
  B. *No.* The FSS frequency 122.1 MHz is a one-way frequency from *you* to *them.* (You'd likely listen on the VOR frequency.)
  C. *No.* That's the airport advisory frequency at non-tower FSS locations.

3824. C. AIM. "... trailing vortices are a by-product of wing lift."

3825. C. The heavy and slow makes sense, but why clean? When the airplane is dirty (flaps and gear), the concentrated wingtip vortices are broken up, as is the case for turbulent air. The worst situation for wake turbulence is a perfectly calm day or night; wind and thermals tend to break it up.

3826. A. *SPFM,* App. A.

3827. C. *SPFM,* App. A.

3828. B. *SPFM,* App. A.

3829. A. *SPFM,* App. A.

3830. B. *SPFM,* App. A.

3831. B. AIM.

Questions 3832–3836 and 3844–3853 pertain to aeromedical factors.

3832. C. *SPFM,* App. C.
  A. *No.* Could a tightness across the forehead be construed as a headache? Or is it because your halo is too tight?
  B. *No.* Enough carbon monoxide can cause loss of muscular power, but C is the answer because one of the symptoms of CO poisoning is that of hypoxia, which gives (among other things) a sense of well-being. (This question could be a little confusing.)

3833. C. Based on 3500 hours as an aerobatic instructor, I've found that haze tends to make the eyes overwork, and the trainee has problems in detecting relative movement of the ground references. Because the eyes are a strong factor in the onset of nausea, a hazy day can cause an aerobatic trainee, who had no problem yesterday in a clear atmosphere, to get queasy in the haze of today.

3834. B. *SPFM,* App. C.

3835. A. *SPFM,* App. C.

3836. C. *SPFM,* App. C.

3837. C. AIM, Glossary. "Air Traffic Clearance—an authorization by air traffic control, for the purpose of preventing collision between known aircraft, for an aircraft to proceed under specified traffic conditions within *controlled airspace.*" (Emphasis added.)
  A. *No.* What about the other aircraft that have a clearance, also?
  B. *No.* If you are in VFR conditions, even if in an IFR clearance, it's *your* responsibility to ensure separation from other traffic.

3838. A. Since the aircraft is approaching from the west (270°), it would fall into the sector of 170°–349°, so Lincoln approach control is on 124.0 MHz. The tower and approach control operate between 1200Z (0600CST) and (0000CST); you'd be able to talk to them at noon local time. (See figure 28 for the Lincoln time zone also.)

3839. C. Looking at figure 53, you see that Lincoln Municipal is in Class C airspace. ATC provides sequencing, traffic advisories, and as appropriate, safety alerts in Class C airspace.

3840. A. AIM. A CTAF (Common Traffic Advisory Frequency) is a frequency designated for the purpose of carrying out airport advisory practices while operating to or from an airport without an operating control tower. The CTAF for Lincoln is the tower frequency of 118.5 MHz, and you would do as choice A says.

3841. B. Loup City Municipal Airport is 1 mile northwest of the *center of the city.* See App. 2, Legend 3, Item 4.

3842. B. If there's no traffic direction given it will be left. Note that Runways 17L and 35L have no direction given, while 17R and 35R have "Rgt Tfc" (right traffic) indicated.

3843. B. *SPFM.* VHF/DF is "VHF Direction Finding," and information on it will be found in the *Airport/Facility Directory,* Radio Aids to Navigation.

3844. A. *SPFM,* App. C.

3845. A. *SPFM,* App. C.

3846. A. *SPFM,* App. C.

3847. B. *SPFM,* App. C.

3848. A. *SPFM,* App. C. Carbon monoxide may cause the symptoms and effects of hypoxia, and the higher the altitude, the more the additive effect.

3849. C. *SPFM,* Chapter 26, Cones and Rods, and *SPFM,* App. C, para. 605b(2).

3850. B. *SPFM,* App. C.

3851. A. *SPFM*, App. C.

3852. B. *SPFM*, App. C.
 A. *No.* You *must* ignore the sensations of muscles and inner ear when visual references are lost.
 C. *No.* This can lead to the problem.

3853. A. *SPFM*, App. C.
 B. *No.* This makes the problem worse.
 C. *No.* You'd do this to decrease hyperventilation symptoms.

3854. A. FAR 61. The full FAR 60s relate to airmen (not just to pilots, but mechanics, etc.). However FAR 61 pertains to pilots and instructors.

3855. B. The FAR 70s pertain to airspace.

3856. C. FARs Part 91, for instance, covers General Operating and Flight Rules. Other 90s pertain to other areas of ATC and General Operations.

* * *

Comments and suggestions about these answers (keep 'em clean) should be sent to:

William K. Kershner
P.O. Box 3266
Sewanee, TN 37375

28

THE PRACTICAL (FLIGHT) TEST

INTRODUCTION

After you've passed the written test, you'll fly more dual and solo flights to review any particular problems and bring your skill up to private pilot level.

```
┌─────────────────────────────────────────────┐
│     APPLICANT'S PRACTICAL TEST CHECKLIST      │
│                                               │
│         APPOINTMENT WITH EXAMINER:            │
│                                               │
│  EXAMINER'S NAME_____   │
│                                               │
│  LOCATION _____   │
│                                               │
│  DATE/TIME _____   │
│                                               │
│  ACCEPTABLE AIRCRAFT                          │
│                                               │
│    ☐  Aircraft Documents:                     │
│         Airworthiness Certificate             │
│         Registration Certificate             │
│         Operating Limitations                 │
│    ☐  Aircraft Maintenance Records:           │
│         Logbook Record of Airworthiness Inspections │
│         and AD Compliance                     │
│    ☐  Pilot's Operating Handbook, FAA-Approved │
│         Airplane Flight Manual                 │
│    ☐  FCC Station License                     │
│                                               │
│  PERSONAL EQUIPMENT                           │
│                                               │
│    ☐  View-Limiting Device                    │
│    ☐  Current Aeronautical Charts             │
│    ☐  Computer and Plotter                    │
│    ☐  Flight Plan Form                        │
│    ☐  Flight Logs                             │
│    ☐  Current AIM, Airport Facility Directory, and Appropriate │
│         Publications                          │
│                                               │
│  PERSONAL RECORDS                             │
│                                               │
│    ☐  Identification - Photo/Signature ID     │
│    ☐  Pilot Certificate                       │
│    ☐  Current and appropriate Medical Certificate │
│    ☐  Completed FAA Form 8710-1, Airman Certificate and/or │
│         Rating Application with Instructor's Signature (if │
│         applicable)                           │
│    ☐  AC Form 8080-2, Airman Written Test Report, or │
│         Computer Test Report                  │
│    ☐  Pilot Logbook with Appropriate Instructor Endorsements │
│    ☐  FAA Form 8060-5, Notice of Disapproval (if applicable) │
│    ☐  Approved School Graduation Certificate (if applicable) │
│    ☐  Examiner's Fee (if applicable)          │
└─────────────────────────────────────────────┘
```

Fig. 28-1. Practical test checklist.

As noted in Chapter 27, you should take the written test sometime after the first solo cross-country so that you'll have the procedures well in mind; but allow a few days and some flying time after getting your passing grade back before taking the practical test. (This chapter is for the Airplane, Single-Engine Land test.)

Figure 28-1 is a recommended checklist to use when going to take the practical test; it wouldn't look too good, for instance, if the examiner (check pilot) found that the airplane was lacking required papers.

Following are areas of knowledge necessary for you to safely perform the privileges of a private pilot. The chapter numbers in parentheses after the various procedures/maneuvers are from this book, *The Student Pilot's Flight Manual* (*SPFM*), that apply to the private pilot practical test. There has been some personal advice added, plus some paraphrasing of the various elements of the test. *For full details get the latest issue of the FAA Practical Test Standards*, currently FAA-S-8081-14(1995).

References (in addition to *this* book) upon which this practical test is based include

FAR Part 43 Maintenance, Preventive Maintenance, Rebuilding, and Alteration
FAR Part 61 Certification: Pilots and Flight Instructors
FAR Part 91 General Operating and Flight Rules
FAR Part 97 Standard Instrument Approach Procedures
NTSB Part 830 Notification and Reporting of Aircraft Accidents and Incidents
AC 00-2 Advisory Circular Checklist
AC 00-6 Aviation Weather
AC 00-45 Aviation Weather Services
AC 61-21 Flight Training Handbook
AC 61-23 Pilot's Handbook of Aeronautical Knowledge
AC 61-27 Instrument Flying Handbook
AC 61-65 Certification: Pilots and Flight Instructors
AC 61-67 Stall/Spin Awareness Training
AC 61-84 Role of Preflight Preparation
AC 67-2 Medical Handbook for Pilots
AC 90-48 Pilot's Role in Collision Avoidance
AC 91-23 Pilot's Weight and Balance Handbook
AC 91-69 Seaplane Safety for FAR Part 91 Operations
AC 120-51 Crew Resource Management Training
AIM Aeronautical Information Manual
AFD Airport Facility Directory

NOTAMs Notices to Airmen
Pilot Operating Handbooks
FAA-Approved Flight Manuals

SOME MORE IDEAS TO CONSIDER

You may encounter the term AREA OF OPERA-TION in the practical test. Areas of operation are phases of flight arranged in a logical sequence within the guide. They begin with the flight's preparation and end with its conclusion. The examiner, however, may conduct the practical test in any sequence that results in a complete and efficient test. Certain demonstrations may be impractical (for example, night flying), so your knowledge may be checked by oral testing.

TASKS are procedures and maneuvers appropriate to an area of operation. PREFLIGHT PREPARATION, for example is an area of operation; one task in that area would be an explanation of the certificates and documents pertaining to you *and* the airplane.

Each task is broken down into OBJECTIVES. The objectives of all tasks must be demonstrated and evaluated sometime during the practical tests.

As an example of an area of operation on the practical test:

VIII. SLOW FLIGHT AND STALLS (AREA OF OPERATION)
A. Maneuvering During Slow Flight (Task)
Objectives: (Gives expected performance criteria for 8 listed requirements)
B. Power-Off Stalls (Task)
(Objectives follow)
C. Power-On Stalls (Task)
(Objectives follow)
D. Spin Awareness (Task)
(Objectives follow)

Consistently exceeding the tolerances noted before taking corrective action will be considered unsatisfactory. Any procedures, or lack thereof, that require the examiner to take over will be disqualifying. If you don't look around and clear the flight area adequately during the practical test, you could be failed.

The examiner will be looking for safe performance as a pilot; you won't be expected to fly like an ace, but the main thing is to use good clearing procedures, checklists, and other techniques and aids in avoiding accidents and incidents. There also will be emphasis on coping with spatial disorientation and wake turbulence hazards.

You'd better know the meaning and significance of airplane performance speeds, including:

V_{so}—the stalling speed or minimum steady flight speed in landing configuration.

V_Y—the speed for the best rate of climb.

V_x—the speed for the best angle of climb.

V_A—the design and maneuvering speed.

V_{NE}—the never exceed speed.

V_{NO}—max normal operating speed (max structural cruising speed).

V_{FE}—max flaps extended speed.

V_{S1}—the stalling speed, or the minimum steady flight speed obtained in a specific configuration (normally with flaps and landing gear retracted).

The examiner will emphasize stall/spin awareness, spatial disorientation, wake turbulence avoidance, runway incursion avoidance, and as noted earlier, checklist usage.

You'll hear the term "stabilized approach," which is not to be construed in the same context as the term used in large aircraft. As used for the practical test guide, it means that the airplane is in a position where minimum output of all controls will result in a safe landing. Excessive control input at any point could be an indication of improper planning.

When you're doing simulated emergency procedures, consideration will always be given to local conditions, including weather and terrain. If safety will be jeopardized, the examiner will simulate that part of the task.

GENERAL PROCEDURES FOR PRACTICAL TESTS

■ Satisfactory Performance

You'll be judged on your ability to safely:

1. Perform the approved *areas of operation* for the certificate or rating sought within the approved standards.

2. Demonstrate mastery of the aircraft, with the successful outcome of each *task* performed never seriously in doubt.

3. Demonstrate satisfactory proficiency and competency within the approved standards.

4. Demonstrate sound judgment.

5. Demonstrate single-pilot competence if the aircraft is certificated for single-pilot operations.

■ Unsatisfactory Performance

If you fail any of the required *tasks*, you fail the practical test. The examiner or you may discontinue the test at any time when the failure of a required task makes you ineligible for the certificate or rating sought. If the test is discontinued, you are entitled to credit for only those pilot operations, or tasks, that you have successfully performed. At the discretion of the examiner any task may be reevaluated, including those previously passed.

Typical areas of unsatisfactory performance and grounds for disqualification are:

1. Any action or lack of action by you that requires intervention by the examiner to maintain safe flight.

2. Failure to use proper and effective visual scanning techniques to clear the area before and while performing maneuvers.

3. Consistently exceeding tolerances stated in the *objectives.*

4. Failure to take prompt corrective action when tolerances are exceeded.

■ Private Pilot Practical Test Prerequisites

As an applicant for the private pilot practical test you are required by the Federal Aviation Regulations to have (1) passed the appropriate private pilot written test within 24 calendar months before the date you take the practical test; (2) completed the applicable instruction and aeronautical experience prescribed for a private pilot certificate; (3) obtained a first-, second-, or third-class medical certificate issued within the past 24 months; (4) reached at least 17 years of age; and (5) have a written statement from an appropriately certificated flight instructor certifying that you have been given flight instruction in preparation for the practical test within 60 days preceding the date of application, that you are competent to pass the test, and that you have a satisfactory knowledge of the subject areas in which you were shown to be deficient by the Airman Knowledge Test Report.

■ Aircraft and Equipment Required

You're required to provide an appropriate, airworthy, certificated aircraft for the practical test. The aircraft must be equipped for, and its operating limitations must not prohibit, the performance of all tasks required on the test. As a practical suggestion aircraft should have the following:

1. A two-way radio suitable for voice communications with aeronautical ground stations.

2. A radio receiver that can be utilized for available radio navigation facilities (it may be the same radio used for communications).

3. Appropriate flight instruments for the control of the airplane during instrument conditions. Appropriate flight instruments are considered to be those required by FAR 91 for flight under instrument flight rules.

4. Engine and flight controls that are easily reached and operated in a normal manner by both pilots.

5. A suitable view-limiting device that is easy to install and remove in flight, for simulating instrument flight conditions.

6. Operating instructions and limitations. You should have an appropriate checklist, a *Pilot's Operating Handbook (POH)*, or, if required for the airplane used, an FAA-approved airplane flight manual. Any operating limitations or other published recommendations of the manufacturer that are applicable to the specific airplane will be observed.

A good question on the oral part of the test is: Does this particular airplane require a propeller spinner to be legal to fly? In some cases the spinner is required to smooth airflow into the cowling for cooling, and the airplane has been flight tested and is certificated *on that basis.* It's possible that without the spinner, the engine could run hot. Look at the equipment list in the *POH* or pertinent airplane papers to see if the letter designating required equipment is by the spinner listing.

Take time to go through the *POH* well beforehand so that you'll at least know where to look for information.

■ Use of Distractions during Practical Tests

It's been noted that many accidents occur when the pilot's attention is distracted during various phases of flight. Many accidents have resulted from engine failure during takeoffs and landings under conditions in which safe flight would have been possible had the distracted pilot not used improper techniques.

Distractions that have been found to cause problems are:

1. Preoccupation with situations inside or outside the cockpit.

2. Maneuvering to avoid other traffic.

3. Maneuvering to clear obstacles during takeoffs, climbs, approaches, or landings.

The examiner may provide realistic distractions during the practical test, which could include:

1. Simulating engine failure.

2. Simulating radio tuning and transmissions.

3. Identifying a field suitable for emergency landings.

4. Identifying features or objects on the ground.

5. Reading the outside air temperature gauge.

6. Removing objects from the glove compartment or map case. (This could start an avalanche of objects you didn't even know were in there, including the ham sandwich you misplaced while on that cross-country flight last month.)

7. Questioning by the examiner.

PREFLIGHT PREPARATIONS

Certificates and documents—You'll be required to present your pilot and medical certificates and locate and explain the airplane's papers (including airworthiness directives) and other records including your and the airplane's logbooks. Review the airworthiness and registration certificates and weight and balance documentation (Chapter 3).

Obtaining weather information—You'll need to obtain and be able to read and analyze weather charts and forecasts, surface weather reports, wind aloft forecasts and reports, pilot weather reports, SIGMETs, AIRMETs, wind shear reports, and NOTAMs. Check the available weather information and understand its significance to a particular flight and make a sound GO/NO-GO decision based on that information (Chapter 22).

Cross-country flight planning—You'll be expected to plan a VFR cross-country flight near the maximum range of the airplane. You'd better be sure that your sectional chart(s) and *Airport/Facility Directory* are up to date and pertain to the route and airports. Check NOTAMs. Plot your course(s) and double-check for fuel stops, if necessary; be sure that these stops have the proper fuel for your airplane, considering the maximum allowable passenger and baggage loads. Your checkpoints should be prominent and useful in navigating the trip. Compute the flight time, headings, and fuel requirements; don't forget a *reserve!* Check for navigation and communications facilities that can be used on the flight, and check the route carefully so that you don't fly through a prohibited area or a less-restrictive but still hazardous one. Check obstructions enroute. Choose airports than can be used as alternates and check their facilities also. You'll be expected to complete a navigation log and simulate filing a VFR flight plan.

National airspace system—Know the basic VFR weather minimums for all classes of airspace. Be sure that you also know the airspace classes—their boundaries, pilot certification, and airplane equipment requirements for Classes A, B, C, D, E, and G. How's your knowledge of special-use airspace and other airspace areas? (See Chapter 21.)

Performance and limitations—You may be quizzed on performance capabilities, approved operating procedures, and limitations of your airplane. Know the power settings, placarded speeds, and fuel and oil requirements. You may be asked to demonstrate the application of the approved weight and balance data for your airplane to determine if the weight and CG locations are within limits for various passenger, fuel, and/or baggage combinations. What are the effects of adding, removing, or shifting weight? Use the manufacturer's charts and graphs for performance figures. (Review Chapter 23.)

Know what could happen if you exceed various limitations, such as loading, so that the CG is out of the envelope. What are the effects of frost or ice on the airplane's takeoff performance? (See Chapter 22; review Chapters 17 and 23 also.)

Operations of airplane systems—You'll be expected to be knowledgeable about the following items; some of the factors are discussed in various parts of the book, but the chapters (or other references) cited are the major sources of information:

1. Primary flight controls and trim (Chapters 8 and 9).
2. Wing flaps and leading edge devices (Chapters 12, 13, and 17).
3. Flight instruments, pitot-static and vacuum/pressure systems (Chapters 3 and 15).
4. Landing gear (Chapter 13, *The Advanced Pilot's Flight Manual*).

5. Engine (Chapters 4, 5, and 7).
6. Propeller (Chapters 2, 4, 5, 7, and 23).
7. Fuel system (Chapters 3, 4, 5, 7, and 23).
8. Hydraulic system (Chapter 14, *The Advanced Pilot's Flight Manual*).
9. Electrical system (Chapters 3 and 25).
10. Environmental system (Chapter 19, *The Advanced Pilot's Flight Manual*).
11. Oil system (Chapters 4, 7, and 23).
12. De-ice and anti-ice systems (Chapter 7, *The Instrument Flight Manual*).
13. Avionics (Chapter 21).

Minimum equipment list—Be able to answer questions from FAR 91.205 concerning required instruments and equipment for day VFR and night VFR flight. Know the procedures for operating the airplane with inoperative instruments and equipment.

Aeromedical factors—You'll be expected to have knowledge of the aeromedical factors, including symptoms, and the effects of and corrective action for hypoxia, hyperventilation, middle ear and sinus infection, spatial disorientation, motion sickness, and carbon monoxide poisoning. Review the effects of alcohol and over-the-counter drugs, plus scuba diving as factors in creating flying hazards. (Check Appendix C.)

PREFLIGHT PROCEDURES

Preflight inspection—Be able to explain *why* you are checking various items on the airplane (and use the checklist). You'll be expected to determine if the airplane is safe for flight; if you try to take the airplane when one wing is lying flat on the ground, certain questions (and eyebrows) might be raised. Know the colors of the various fuels and the type of fuel and oil used (and also be able to find all fuel and oil filler points). Know where the battery and other components are located. Check for fuel contamination and be able to explain what you are looking for and when it is most likely to occur. Know capacities of fuel and oil tanks and where to look for leaks in these systems. (This also goes for the hydraulic and brake systems.) Check and explain the flight controls and look at the balance weights on the rudder, elevator, and ailerons, as applicable. (You should know by now that in normal operations when one aileron goes up, the other goes down.) Check for structural damage (hangar rash) as you go around, and for Pete's sake, don't try to taxi off with the airplane tied down, control lock(s) in place, or chocks under the wheels. Again, you should know that ice and frost are detrimental to takeoff operations. See to the security of baggage, cargo, and equipment. You are acting as pilot-in-command and will be checked on your judgment of the airplane's readiness to fly safely (Chapters 4 and 23).

Cockpit management—You are expected to have an adequate knowledge of cockpit management and be able to explain the related safety factors. In other words, you'll be expected to organize your material and equipment in a manner that makes them readily available. (Taking off with the charts and other navigation equipment well out of reach in the far corner of the baggage compartment could cost you several points.) Make sure that the safety belts and harnesses are fastened (with you and the examiner *inside* them). Your seat should be adjusted—and locked—in the proper position. Be sure that the controls are free and have full and proper movement. Since the examiner is a "passenger," you should brief him or her (and any others on board) on the use of the belts and any emergency procedures. Use an appropriate checklist (Chapter 4).

Engine starting—You'll be expected to explain the starting procedures, including starting under various conditions. *Use the checklist—all items*—for prestart, start, and poststart procedures. You'll be expected to be safe, with emphasis on (1) positioning the airplane to avoid creating hazards (don't blast people or airplanes behind you), (2) making sure that the propeller area is clear, and (3) ensuring that the engine controls are properly adjusted (for instance, a wide-open throttle with a cold engine, a mixture left in idle cutoff, or a fuel selector OFF could cause problems). Set (or better yet, *hold*) the brakes before, during, and after start, and don't let the airplane *inadvertently* move after the start. *Don't* let the engine roar up after the start; keep the rpm under control. Check the engine instruments after start. (Read Chapter 5—now—not while starting the engine on the check flight.)

Check the *Pilot's Operating Handbook* (POH) of your airplane to confirm the proper procedures for using an external power source. (Some airplanes require the battery master switch to be OFF and others require it to be ON during a start using an external power source.)

Taxiing—You'll be expected to explain and use safe taxiing procedures. Check the brakes for effectiveness as you move out of the parking spot. You'll be checked on safe ground operations, your use of brakes and power to control taxiing speeds, and proper positioning of the flight controls for the existing wind conditions. You'll be expected to be aware of ground hazards and comply with taxiing procedures and instructions. (Review Chapter 6.) (As one applicant said to the check pilot after sideswiping two airplanes, nearly running over a flight instructor, and ending up with the airplane's nose sticking into the flight office, "I suppose this means that I've flunked the surface operations part of the test.")

Don't abuse the brakes by carrying too much power and using them to keep the taxi speed down. This is a common error.

Before takeoff check—Review Chapter 7. During the flight test, as anytime, you should be able to explain the reasons for checking the items. Don't create a hazard during the runup. *Don't* stick your head down in the cockpit and keep it there; divide your attention. Touch each control or switch as you *follow*

the checklist and call out each item as you check it. Give the instrument reading, if applicable, after checking it on the list. Make sure, again as always, that the airplane is in a safe operating condition, emphasizing (1) flight controls and instruments, (2) engine and propeller operation, (3) seats adjusted and locked, (4) belts and harnesses fastened, and (5) doors and windows secured. If there is any discrepancy, you'll be expected to determine if the airplane is safe for flight or requires maintenance. Know the takeoff performance—airspeeds and takeoff distances (Chapter 17)—and be able to describe takeoff emergency procedures (Chapter 13). You'll handle the takeoff and departure clearances and will note the takeoff time. Make sure that there's no traffic conflict when you taxi into takeoff position.

AIRPORT OPERATIONS

Radio communication and ATC light signals—Be at ease using the radio, including knowing pertinent frequencies. You are required to interpret and comply with ATC light signals (see Fig. 21-21). Use prescribed procedures following radio communications failure. Make sure you comply with traffic pattern procedures instructions.

Traffic pattern—You'll be expected to explain traffic pattern procedures at controlled and noncontrolled airports. You'll be expected to follow prescribed traffic patterns, using proper arrival and departure procedures. Watch your wind drift corrections, spacing with other aircraft and altitudes, airspeeds, and ground traffic. Know the factors of runway incursion, collision avoidance, wake turbulence, and wind shear. Better hold altitudes within ±100 feet and airspeeds within ±10 K of those required. Prompt corrections will be expected. Complete the prelanding checklist.

You are expected to maintain orientation with the runway in use, perform the prelanding cockpit check on the downwind leg, and have a final approach leg at a stabilized airspeed (Chapter 13).

Airport and runway marking and lighting—Review Chapters 25 and 26.

TAKEOFFS, LANDINGS, AND GO-AROUNDS

Normal takeoff and climb—Be able to describe the elements of a normal takeoff and climb, including airspeeds, configurations, and emergency procedures. The examiner will watch for safe operations in normally anticipated conditions. Align the airplane on the runway center line and open the throttle *smoothly* to max power—don't ram it open. Check the engine instruments as the run commences and watch your directional control. Rotate at the POH-recommended speed and establish the pitch attitude for V_Y +10/ −5 K during the climb. Retract the flaps (and gear, if applicable) at a safe point and altitude. (Keep the landing gear extended until you can no longer land on the runway.) *Use* the takeoff power until at a safe altitude and don't forget to maintain a straight track on the climbout. Comply with noise abatement procedures. Use the checklist (Chapter 13).

Crosswind takeoff and climb—Know and be able to explain the crosswind takeoff and climb, including airspeeds, configurations, and emergency procedures. For normal takeoffs, apply power smoothly after you are well lined up on the center line. Use full aileron deflection (in the proper direction) and check the engine instruments as the takeoff run starts. Stay on the center line and adjust the aileron deflection pressure as the run progresses. (Note: For some runways with center-line lighting you may want to keep the nosewheel slightly to the left or right of the line.) Again, you'll rotate at the *POH*-recommended speed and establish and maintain V_Y +10/−5 K. Get the gear and flaps up as applicable at a safe point and altitude during the climb. Maintain the takeoff power to a safe maneuvering altitude and don't drift over into the next county during the straight portion of the climbout. You'll be judged on power application, smoothness, and wind drift correction (make sure the airplane doesn't skip and drift during the takeoff and climb). Comply with noise abatement procedures and use a checklist. (Review Chapter 13.)

Normal approach and landing—Know what elements affect a normal approach and landing, including airspeed and configuration. Be aware of wind conditions, landing surface, and obstructions. Keep a good straight track on final at the proper approach and landing configuration and power, maintaining the recommended airspeed but not more than 1.3 V_{so} +10/−5 K with a gust factor applied. You'll be expected to make smooth, timely, and correct control applications during the final approach and the transition to landing. You should touch down smoothly on the main gear (tricycle gear) or three-point gear (tail-wheel landing gear) at the approximate stalling speed, beyond and within 400 feet of a specified spot, with no drift and the longitudinal axis aligned on the runway center line. Maintain directional control during the roll-out. You could be asked to make landings at different flap configurations (Chapter 13).

Crosswind approach and landing—You'll be expected to show adequate knowledge by explaining crosswind approach and landing techniques, limitations, and recognition of excessive crosswind. You'll be expected to establish the approach and landing configuration and power and to maintain a straight track on final. Keep the airspeed as recommended but not more than 1.3 V_{so} +10/−5 K and make smooth, timely, and correct control application through the approach to landing attitude. The airplane is to touch down smoothly (it says here) on the upwind main gear (tricycle gear) or upwind main gear and tailwheel at approximately the stalling speed, beyond and within 400 feet of a specified point, with no drift and the longitudinal axis aligned along the runway center line.

Maintain directional control after touchdown, increasing aileron deflection into the wind as necessary (Chapter 13).

Soft-field takeoff and climb—You should be able to describe the elements and possible emergency situations of the soft-field takeoff and climb, noting that attempting to climb at airspeeds of less than V_x can be fraught with peril. Select the proper flap setting and keep rolling as you turn onto the takeoff surface, but don't make a full-power, hairy, tire-ruining turn onto the runway. Set the pitch attitude to get the weight off the wheels as soon as possible and maintain directional control. (Here's where problems might occur at the slow speed and blocked forward view.) If obstructions are a problem, maintain V_x +10/−5 K until you've cleared them, then V_Y +10/−5 K. Lift off at the lowest possible airspeed and use ground effect to help acceleration. The same rules apply in this maneuver concerning when to retract flaps (and gear) and tracking straight after takeoff (Chapter 17).

Soft-field approach and landing—Know airspeeds, configurations, and operations on various surfaces. Check out the obstructions, landing surface, and wind condition; establish the recommended soft-field approach and landing configuration and recommended airspeed. If no approach speed is recommended, make it at not more than 1.3 V_{so} +10/−5 K with a gust factor applied. Keep it lined up on final; touch down with a minimum descent rate and groundspeed, with no drift and the airplane lined up with the landing area. Maintain directional control after touchdown and maintain sufficient speed to taxi on a soft surface (Chapter 17).

Short-field takeoff and climb—Know and be able to describe the procedure and the expected performance for the existing conditions. Review Chapter 17 and the *POH* for your airplane and be able to use the takeoff performance charts. Select the proper wing flap setting, line up with the center line, and smoothly apply power. Keep your directional control. (Taking off *across* the runway won't cut it.) Set up the proper pitch attitude to get the best acceleration and rotate the airplane at the recommended airspeed. Accelerate to V_x and maintain within +10/−5 K until at least 50 feet above the surface or until the "obstacle" is cleared; then assume V_Y +10/−5 K. Don't suddenly pull up the flaps and settle back in; retract the flaps and the landing gear at a safe point and altitude, as noted in the earlier takeoff procedures. Keep full power on until the "obstacle" is cleared and check back occasionally during the straight part of the climbout to see that you are tracking properly. Don't forget to comply with noise abatement procedures on all takeoffs cited here. Airspeed control is a big factor for best climb performance. (Chapter 17.)

Short-field approach and landing—You'll be expected to explain short-field approach and landing procedures, including airspeeds, configurations, and performance data. You'll be asked to evaluate obstructions, landing surface, and wind condition and to select the most-suitable touchdown point. Be prepared for a go-around if necessary. You are to establish the recommended configuration and airspeed or use a speed of not more than 1.3 V_{so} +10/−5 K, if no other information is available. Maintain precise control of the descent rate along the extended runway center line. There should be little float, and you should touch down beyond and within 200 feet of a specified point, lined up with the runway center line, with no side drift. Keep the airplane directionally straight after landing. Brake as necessary to stop in the shortest distance consistent with safety (no squared-off tires, please). Holding the control wheel or stick full back while braking and upping the flaps (not the *gear*) will give more-efficient braking (Chapter 17).

Forward slip to landing—You should have a good grasp of the principles involved and be aware of technique, limitations, and effect on the airspeed indicator. Be aware of any manufacturer's limitations on slips with flaps. You are to demonstrate a forward slip at a point from which a landing can be made in a desired area, using the recommended configuration. Remember that if there's a crosswind, you drop the wing that's into the wind (Fig. 13-23). Keep the airplane's track straight with the runway and maintain an airspeed that results in little floating during the landing. The requirements for touchdown point (at or within 400 feet beyond a specified point), with no side drift and good directional control, are the same as for the normal landing just cited (Chapter 13).

Go-around—You may have to demonstrate the procedure for go-around on an approach (balked landing). You'll be expected to make a timely decision, fly the recommended airspeeds, and cope with any undesirable pitch and yaw tendency (keep up with the trim requirements). Apply takeoff power, attain the proper pitch attitude (airspeed), and retract the flaps as recommended at a safe altitude. If the airplane has retractable gear, pull it up after a rate of climb has been established and you can't land on the remaining runway if the engine fails. Trim the airplane, climb at V_Y +10/−5 K, and track a good traffic pattern for the next landing (Chapter 13).

PERFORMANCE MANEUVER

Steep turns—Review this maneuver in Chapter 10. Be able to explain the requirements involved, including increased load factors, additional power, and other elements. Stay above 1500 AGL. Establish the airspeed below the airplane's maneuvering speed (V_A) and set up a bank angle of 45° ±5°, maintaining coordinated flight for 360° of turn, as directed, then roll out and do a 360° turn in the opposite direction. Divide your attention between airplane control and orientation, keep the bank ±10° of that required, and rollout with ±10° of the desired heading. Keep the altitude within ±100 feet and the airspeed within ±10 K.

GROUND REFERENCE MANEUVERS

Rectangular course—Know the principles of wind drift correction in straight and turning flight and the tie-in between the rectangular course and airport traffic patterns. You'll be expected to pick the field or fields, set up the maneuver, and enter at the proper distance from the boundary at a 45° angle to the downwind leg at traffic pattern altitude (600-1000 feet AGL). (No, you aren't allowed to fly up and down between those altitudes; keep the altitude within ±100 feet of that selected.) Keep your flight path parallel to the sides and be prepared to fly it with both left or right turns. You'll be expected to be able to divide your attention between airplane control and the ground track. Check your coordination and maintain airspeed within ±10 K.

Avoid excessive maneuvering (keep the bank angle to a maximum of 45° and airspeed to ±10 K) and stay above the minimum legal altitude for local obstructions. You'll reverse course as directed by the examiner. Exit at the point of entry at the same airspeed and altitude the maneuver was begun. Watch for other airplanes (Chapter 10).

S-turns across a road—Be able to expound on the principles of this maneuver. Select a long, straight reference perpendicular to the wind (know its approximate direction and speed) and fly it at a selected altitude of 600-1000 feet AGL. Keep the turns with a constant radius (see Figures 10-6 and 10-7), divide your attention between airplane control and the ground, and keep the maneuver coordinated. Reverse the turns directly over the reference line and keep the altitude within ±100 feet of that selected. The airspeed should stay within ±10 K of that desired; it is suggested that you don't exceed a bank of 45° at the steepest point of the turn. Keep it legal as far as obstructions are concerned and don't have a midair collision. Also, don't take the examiner on a cross-country but reverse the turns as directed.

Turns about a point—Pick the point promptly and enter, checking your heading by the heading indicator or a geographic reference. The practical test guide requires two turns and the examiner will ask for such, so you'd better get in the habit of knowing where you are during practice sessions. As Chapter 16 notes, be prepared to do the turns in either direction, but most folks start out with left turns. (The examiner will probably call for reversal of the turns after the first two turns.) The altitude, airspeed, and max bank (45°) limitations are the same as for the S-turns just covered (Chapter 16).

NAVIGATION

Pilotage and dead reckoning—You should be able to give the examiner the straight information on piloting and dead reckoning techniques and procedures. You'll demonstrate that you are able to follow the preplanned course solely by reference to landmarks and identify these landmarks by relating the surface features to chart symbols.

You also must be able to navigate by means of precomputed headings, groundspeed, and elapsed time and *combine* pilotage and dead reckoning with a verification of the airplane's position within 3 NM of the flight-plan route at all times. You're expected to hit enroute checkpoints within 5 minutes of the initial or revised ETA and make your destination within 5 minutes of the ETA. Be aware of the remaining fuel at all times. Make corrections for differences between preflight groundspeed and heading calculations and those determined enroute (and record them). Keep your altitude within ±200 feet. You'd better maintain the desired heading within ±15°. Use those climb, cruise, and descent checklists. Don't get lost (Chapters 24 and 25).

Navigation systems and radar services—Be able to explain radio navigation equipment, procedures, and limitations. You'll be required to select, tune, and identify the desired radio facility and locate the airplane's position relative to that facility. Better practice using VOR and NDB facilities and be able to intercept and track a given radial or bearing. Know how to locate the airplane's position using cross bearings (see Fig. 25-4). Recognize station passage and/or be able to describe it. Check with your flight instructor to find out the best method of recognizing signal loss and how best to regain the signal, as applied to the equipment you're using. While you're doing the radio navigation, stay within ±200 feet of the altitude chosen. Use proper communications procedures when dealing with ATC radar services. (Be aware of the radar services available.) Check on your local sectional chart legend (control tower frequencies) to see what facilities have ASR (airport surveillance radar) for approaches. You'll be expected to follow verbal or radio navigation guidance and determine the minimum safe altitude. Maintain altitude within ±200 feet (Chapters 21 and 25).

Diversion—The examiner will want to see how you cope with having to change your plans during flight. You'll be expected to recognize the conditions that require a diversion (such as unexpected headwinds, lowering ceilings and visibilities, or other factors). At the examiner's prompting, you'll select an alternate airport and route and proceed to it without delay. Make an accurate estimate of heading, groundspeed, arrival time, and fuel required to get there. Use pilotage, dead reckoning, or radio navigation aids and ATC assistance. Keep the altitude within ±200 feet during the sashay off the preplanned course. Keep the heading within ±15° once you're established on course (Chapter 25).

Lost procedures—Make sure that you know and can explain lost procedures, including the following items:

1. Maintaining the original or appropriate head-

ing, identifying landmarks, and climbing if necessary (Chapter 25).

2. Rechecking the calculations. ("Ah, Igor, it's really 110 K groundspeed not 1100!")

3. Proceeding to and identifying the nearest concentration of prominent landmarks. ("Is that the Atlantic or the Pacific Ocean over there?")

4. Using available communications and navigation aids for contacting the appropriate facility for assistance. ("Jonesville approach control, I'd like a *practice* vector to the airport.")

5. Planning a precautionary landing if deteriorating visibility and/or fuel exhaustion is imminent. Use your skills at dragging the area and short- and soft-field landings. (One pilot inadvertently made a precautionary landing in the middle of a nudist colony in 1985 and has not yet gotten over it.)

You'll be expected to *select the best course of action* when given a "lost" situation. (See Chapter 17, and take another good look at Chapter 25.)

SLOW FLIGHT AND STALLS

Maneuvering during slow flight—Know the background of this requirement and *why* it is an important part of flight training. Stay above 1500 feet AGL. Establish and maintain an airspeed at 1.2 V_{s1} +10/−5 K while (1) in coordinated straight and turning flight in various configurations and bank angles as established by the examiner and (2) in coordinated climbs and descents in various configurations as specified by the examiner. Maintain the specified angle of bank, not to exceed 30° +0/−10° in level flight and maintain the specified angle of bank, not to exceed 20° +0/−10° in climbing or descending flight. Roll out on the specified heading ±10°. You'll be expected to maintain the desired altitude within ±100 feet, when a constant altitude is specified, and level off from climbs and descents within ±100 feet. Your heading during straight flight must be maintained within ±10° of that specified in coordinated flight (Chapter 12).

Power-off stalls—These are basically the *approach to landing stalls* covered in Chapter 12, but you might want to review Chapter 14 also. You'll be expected to explain the aerodynamics of the stall and know the flight situations that result in power-off stalls, including the proper recovery procedures. Know the hazards of stalling during slips and skids. Recover above 1500 feet AGL or the recommended altitude, whichever is higher. Set up the normal approach or landing configuration and airspeed with the throttle closed or at a reduced power setting. Establish a straight glide (±10° heading) or gliding turn of (suggested) 30° (+0/−10°) in coordinated flight as required. Maintain good coordination, establish a landing pitch attitude, and keep it there to induce a full stall. You'll be expected to recognize the indications of a stall and promptly recover by decreasing the angle of attack, simultaneously leveling the wings, and advancing the power to regain the normal flight attitude. Retract the flaps (if extended) and landing gear (if retractable) and establish a climb. Accelerate to V$_Y$ before the final flap retraction. Return to the altitude, heading, and airspeed specified by the examiner. Don't get a secondary stall, build up excessive airspeed, lose excessive altitude, spin, or get below 1500 AGL at any time.

Power-on stalls—The examiner will want to know that you understand the principles of these stalls and what can contribute to aggravating them (slips and skids) and will look at your recovery effectiveness. Again, don't get below 1500 AGL at any time during the demonstration. Basically, these stalls are the *takeoff and departure stalls* covered in Chapter 12. Set the airplane up in the takeoff or normal climb (departure) configuration and establish the takeoff or climb airspeed before applying takeoff or climb power. (You may have to use reduced power to avoid excessive pitch-up during the entry, but then apply more power as things progress.) You'll smoothly set up and maintain a pitch attitude that will induce a stall in both wings when they are level (±10° constant heading) and in 20° (+0/−10°) banks. "Torque" may tend to skid and turn the airplane to the left; watch your coordination during the entry. Increase elevator (stabilator) back pressure smoothly until the stall occurs, as indicated by (1) a sudden loss of control effectiveness, (2) buffeting, or (3) uncontrollable pitching.

The required recovery will be to promptly and positively reduce the angle of attack, simultaneously level the wings, and regain normal flight attitude with coordinated use of flight and power controls. (*Always* use full power when recovering from stalls unless the examiner asks for a power-off recovery as a demonstration of your knowledge that reducing the angle of attack is the main action in stall recovery; remember that the use of power reduces the altitude loss.) After recovery, clean up the airplane and establish a climb. Accelerate to V$_Y$ before the final flap retraction and return to the altitude, heading, and airspeed specified by the examiner. Again, if the examiner has to take over to keep you from spinning or having other problems, such as going below 1500 feet AGL, or if you get a secondary stall during recovery, you will be disqualified (Chapters 12 and 14).

Spin awareness—You'll be expected to explain flight situations where unintentional spins may occur. (A good all-round answer is "when the pilot is distracted.") Review Chapter 12 and your *POH*, and discuss with your instructor ways to recover from unintentional spins, with particular emphasis on the spin recovery procedure for the airplane you're using on the practical test.

BASIC INSTRUMENT MANEUVERS

Straight and level flight—Know why you are looking at the various instruments in your scan (cross-check) and how you control the airplane when flying by reference to instruments. (It's exactly the same as when using outside references.) Your control application should be smooth, timely, and coordinated; if the ball in the turn indicator leaves the cockpit, you may be in a little trouble. You are to maintain straight and level flight keeping the heading within ±20°, the assigned altitude within ±200 feet, and the desired airspeed within ±10 K. Use all flight instruments available (Chapter 15).

Straight constant-airspeed climbs—You should be able to explain what you are doing and what makes the airplane climb. Establish the climb pitch attitude and power setting on an assigned heading. Fly smoothly and with coordination. The heading must be maintained within ±20° (watch that "torque" effect) and the airspeed maintained within ±10 K; you are to level off and maintain flight within ±200 feet of the desired altitude (Chapter 15.)

Constant-airspeed descents—Again, know what you're doing and be able to talk about it. Set up the airspeed and power (see Chapter 15 for some suggestions) for about a 500-fpm descent and be smooth and coordinated. The heading should be kept within ±20° and airspeed limits within ±10 K; be sure to level off and maintain flight within ±200 feet of the desired altitude.

Turns to headings—Know the background of what you propose to do. Use standard-rate turns and don't make the examiner sick with your rough control. Keep the altitude within ±200 feet and airspeed within ±10 K, maintain a standard-rate turn, and roll out within ±20° of the assigned heading (Chapter 15).

Recovery from unusual flight attitudes—You'll be expected to demonstrate that you can cope with the situation when it's necessary to recover from a "nonstandard" flight situation. You should recognize the critical attitudes promptly; interpret the instruments; and recover by prompt, smooth, coordinated controls applied in the proper sequence. If you bend something, go too fast, spin the airplane, and/or make the examiner's voice rise a couple of octaves while he's saying "I've got it," you aren't doing a satisfactory job of flying (Chapter 15).

Radio communications, navigation systems/ facilities, and radar services—Know how to use the NAV/COMM system to help yourself if you get caught in instrument conditions. Follow verbal instructions and/or navigation systems/facilities for guidance. Determine the minimum safe altitude. Maintain altitude within ±200 feet, heading within ±20°, and airspeed within ±10 K.

EMERGENCY OPERATIONS

Emergency descent—An emergency descent is a matter for urgency, and you should have checked out the procedure in the *POH*. Know the emergency descent configuration and airspeed (and establish it within ±5 K). Keep oriented, use proper planning, and divide your attention between the proper procedure and flying the airplane. Don't forget to use the emergency checklist (Chapter 18).

Emergency approach and landing—Get to that best-glide attitude configuration and airspeed (± 10 K). Be sure to pick a suitable emergency landing area *within gliding distance*. (This is important!) You should set up and follow a flight pattern to that area, considering altitude, wind, terrain, and obstructions. When you're set up, try to figure out the reason for the malfunction and correct it, if possible. (But *fly the airplane*, don't spin it while distracted—or anytime for that matter.) Don't forget the appropriate emergency checklist (Chapter 18).

Systems and equipment malfunctions—You better know what problems are possible with your airplane. The examiner will set up simulated malfunctions and/or emergencies such as partial or complete power loss, engine roughness or overheat, carburetor or induction icing, loss of oil pressure, fuel starvation, electrical or flight instruments malfunction, landing gear or flap troubles, inoperative trim, inadvertent door or window opening, structural icing, smoke/fire engine compartment fire, and any other emergency appropriate to the airplane being used for the practical test. Use the emergency checklist.

Emergency equipment and survival gear—Know the location, method of operation, or use for all emergency equipment and survival gear. Check the servicing requirements and method of safe storage; be aware of the equipment and survival gear appropriate for operation in various climates and topographical environments. Use that emergency checklist.

NIGHT FLIGHT OPERATIONS

Preparation and equipment—You may be asked how you would go about preparing for a local cross-country night flight. Be able to answer questions on airport lighting and the airplane lights and their operation. Know what equipment you'd need; a flashlight or another kind of backup would be essential. Weather factors are particularly pertinent to night flight (such as the relationship of temperature and dew point and the possibility of ground fog forming during night flights). Be aware of the possible need for reference to the flight instruments (1) over unlighted areas and (2) if actual instrument conditions are unexpectedly encountered. Understand the factors of night vision, including the effects of changing light conditions, coping with illusions, and how the pilot's physical condition affects visual acuity. If you don't meet night flying experience requirements, this area won't be evaluated and your certificate will bear the limitation, "Night Flying Prohibited" (Chapter 26).

Night flight—If the demonstration of night flight is impracticable, your competency may be determined by oral testing. If night navigation is required by the examiner, you will follow procedures similar to those described earlier under "Cross-Country Flying."

Be aware of night flying procedures, including safety precautions and emergency action. *Use a checklist.* You should be sharp at explaining (and demonstrating, if required) the starting, taxiing, and pretakeoff check. Use good operating practices for night takeoff, climb, approach, and landing. Know the factors in navigation and orientation for night flight under VFR conditions. Don't forget your flashlight and checklist (Chapter 26).

POSTFLIGHT PROCEDURES

After landing—Know what should be done after the landing and taxi to the parking/refueling area, using the proper wind control technique and obstacle avoidance procedures. Complete the appropriate checklist.

Parking and securing—You'll be expected to know the parking hand signals and understand the factors for deplaning passengers. (Shutting the engine(s) down is *always* the safest procedure before people enplane or deplane.) Park the airplane *properly*, considering the other aircraft and the safety of nearby persons and property on the ramp. Use the recommended procedure for engine shutdown and secure the cockpit and the airplane. Perform a satisfactory postflight inspection and complete the appropriate checklist.

AERONAUTICAL DECISION MAKING (ADM)

Analysis of aviation accidents over the past years shows that a large percentage are caused by a lack of judgment or the pilot's lack of aeronautical decision making. Airplanes land short of destination airports, killing or injuring occupants because pilots decide to skip that interim refueling stop, figuring they "can make it." Or perhaps the weather was not checked or weather warnings were ignored by the pilot, resulting in a fatal accident.

More and more emphasis is being placed on aeronautical decision making (ADM) during the practical tests by examiners. Review each TASK indicated in this chapter and see how *your* good or bad ADM could affect the outcome. For instance, a superficial preflight check (or *no* preflight check), not using a checklist, failing to clear the area before starting a maneuver, or not continuing to look around as you proceed are some instances of poor judgment or poor ADM. A very large portion of flying is headwork, and the examiner will be interested in your grasp of this. If you are careless on a check ride that you've been preparing for, the examiner will wonder how your mind will work after you get out on your own and meet unforeseen situations head-on.

Fig. 28-2. As shown here, the check pilot is just an ordinary human being like yourself—with no ax to grind.

CHECKITIS

Checkitis is a mysterious disease that strikes just before a practical test. The victims are usually of at least average intelligence, but as soon as this malady hits, they are lucky to remember their own names. It is characterized by shaking knees, a knotted stomach, sweating, and loss of memory. You may not get checkitis—if this is the case, you're one of the chosen few.

The examiner probably has also had it and will make allowances. The good thing about checkitis is that as soon as you get busy in the practical test, its symptoms lessen or fade completely. The best insurance against it is to know what you're doing beforehand. If you don't do too well on the oral, it may hurt the entire flight because of added nervousness. Learn all you can about the plane and flying in general and then don't worry about checkitis.

You may feel that the check pilot is indeed a sinister figure (Fig. 28-2).

THE AIRPLANE NUMBERS

AIRPLANE PAPERS AND LOGBOOKS

1. Logbooks

   Date of last annual inspection _____

   Date of last 100-hour inspection _____

   Next 100-hour inspection due (tachometer reading) _____ hours

2. Airworthiness Certificate—Standard

   Issued to (number of airplane) _____

3. Certificate of Registration

   Issued to (name) _____ on (date) _____

4. Operational limitations

   Up-to-date? _____

5. Weight and balance form

   Is it up-to-date, or has a major repair and alteration form superseded it?

   Latest information on equipment list and empty weight (date) _____

AIRPLANE STATISTICS

1. Basic Empty Weight

   _____ pounds

2. Empty Weight Center of Gravity

   _____ inches aft of datum; or empty weight moment _____ pound-inches

3. Maximum Certificated (Gross) Weight

   Normal category _____ pounds

   Utility category _____ pounds. (either or both may apply)

4. Fuel

   Total capacity _____ gallons

   Usable _____ gallons (fuel weights 6 pounds per gallon)

5. Oil

   Total capacity _____ quarts

   Minimum for flight _____ quarts (oil weight 7½ pounds per gallon)

   Brand name and viscosity—winter _____; summer _____

6. Baggage

   Maximum weight _____ pounds

PERFORMANCE

(All figures are for maximum certificated weight)

1. Range

   At 65% power, _____ NM, _____ gallons per hour

   Max range: At _____ % power, _____ K, _____ NM

2. Rate of Climb at Sea Level

   _____ feet per minute

   Best rate-of-climb speed: _____ K at sea level

   Max angle-of-climb speed: _____ K at sea level, _____° flaps

3. Takeoff Distance at Sea Level (run:

   _____ feet

   Total distance over 50-foot obstacle: _____ feet

4. Landing Roll

   _____ feet

   Total distance over 50-foot obstacle: _____ feet (sea level)

5. Service Ceiling

   _____ feet. (The service ceiling is the altitude at which the airplane's maximum rate of climb is 100 feet

   per minute.)

MANEUVERS

1. Slow Flight

   _____ K (clean). 20°–30° banked turns, climbs and glides, cruise and landing configurations

2. Takeoff and Departure Stalls

   Reduce power and slow to _____ K, then apply climb power (a) straight ahead and (b) 10°–20°

   banked turns

3. Approach to Landing Stalls

   Start at _____ K and full flaps. Slow to stall. Gliding turns (20°–30° banks) and straight ahead (trim set)

4. Accelerated Stalls

   Cruise configuration: No flaps, 45° bank, _____ rpm, slow to _____ K (Bottom of green arc plus 10 K)

CLEAR THE AREA BEFORE DOING STALLS

5. Emergency Flying by Reference to Instruments

   Climbs: Full power and best rate of climb speed _____ K

   Descents: At _____ K and _____ rpm (to give approximately 500 fpm descent)

6. Turns Around a Point and Other Ground Reference Maneuvers

   Note heading of entry

   Altitude for maneuver _____ feet MSL, or _____ feet above the local surface

7. Short-Field Takeoff

   Flap setting _____°

   Climb speed to clear obstacle _____ K.

8. Short-Field Landing

   Flap setting (usually full flaps) _____°

   Approach speed _____ K

9. Soft- or Rough-Field Takeoff

   Flap setting _____°

10. Soft-field Landings

    Flap setting _____°

    Approach speed _____ K

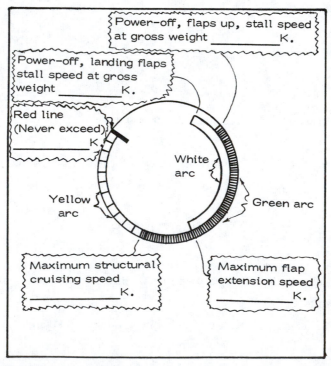

Fig. 28-3.　Important airspeeds.

AFTER THE PRACTICAL TEST

It seems that immediately after getting the certificate and getting out on his or her own, the average private pilot sets about developing as many bad habits as possible. Actually, it's not a deliberate move, but sometimes it seems so. More about this later.

The training, while intended to be realistic, couldn't cover all of the problems that can be encountered in a long-term flying career.

■ High, Hot, Humid, and Heavy

For instance, the chances are that you haven't flown a loaded four-place airplane (with people *and* baggage) from a high density-altitude airport. Probably you took your flight training in that same four-place airplane with only you and the instructor (and then only you, solo). And maybe your flying was in low density-altitude conditions. Be prepare for a loss of performance when you take those three friends and their baggage out of a high, hot, and humid airport. Give yourself plenty of time and distance. You may have to (1) wait overnight for a cooler, early morning takeoff, (2) fly your passengers one or two at a time to a lower elevation airport, or (3) have somebody drive them to that airport.

■ Preflight Check, Slackening of

The preflight check you so carefully learned and used during the flight training (and particularly during the practical test!) is oftentimes like memory as a person gets older, the first or second thing to go. This is particularly a problem in extreme temperature conditions. It's too hot on the ramp, and the pilot moves around the airplane at a rate closely approximating Indians riding around a circled wagon train. When the temperatures are nudging zero, it's tough to get a good fuel check (gasoline is *cold*), shortcuts are established, and the overall preflight check suffers as well. *Your habit pattern should be so established for a thorough preflight check that you wouldn't take off without one* because you'd feel uncomfortable during the entire flight. You'll have to stop or slow down from time to time, and *really* look at the airplane as you preflight it.

■ Shortcuts

One shortcut is the one just mentioned about the preflight check. Another is the lack of use of a checklist for starting and the pretakeoff and prelanding checks. There are a couple of ways to use checklists: (1) Take it step by step, item by item as it appears on the list or (2) use the checklist after the fact to confirm that the times are covered. The first choice seems to work better for most people, but be sure that you don't skip an item as you go through it. (This is the most common mistake.)

Another common shortcut is to use less than the full runway even when there is no pressure for a quick takeoff. Plenty of pilots whip out on the runway, leaving a couple of hundred feet (or more) behind them. You might need that unused runway to brake to a stop if a takeoff has to be aborted for one reason or another.

It's been noted that many accidents and incidents aren't caused by a catastrophic occurrence but a series of "minor" upsets and delays.

You're leaving with your family on that long-planned flying vacation:

If you don't get off by a certain time the weather will preclude going VFR. (You don't have an instrument rating yet.) The car has a flat tire. This takes time to fix. There's more luggage than you'd counted on. Traffic is heavy (and slow) on the way to the airport. The airplane's master switch was left ON by the other airplane owner yesterday, requiring the use of an external power unit before *you* can fly today.

The weather has moved in, but after all this trouble you've decided that you can be out of it with only a "little" scud running.

So you take off.

You're nervous and upset and this detracts from your judgment of the weather problem and also in flying the airplane. If you make it, you'll swear you'll never do *that* again.

If there is one thing that could sum up your flying attitude after you get out on your own, it is to never be pressured into flying when *you* don't want to. (Many pilots have sat a number of days on the ground at remote airports because the weather was worse than they or the airplane's equipment could cope with.) On every flight (local or otherwise) *always,* repeat, *always* have an alternative plan of action.

There's an old, old truism, "If you have time to spare, go by air." There'll be times when you have to rent a car to get home and then have the hassle of going back and getting the airplane later. *So welcome to the club.*

Another shortcut is to *not* oversee the fueling of your airplane at a strange airport. (Sure, you're in a hurry to get to the motel.)

■ Currency

After a layoff for several weeks (or months) you'll feel unsure of yourself about flying again. The longer the layoff, the more unsure you'll be about even going out to the airport.

Even experienced pilots don't feel quite as at home with the airplane, and it may take them 30 minutes or an hour to get back to where "things feel right," even in an airplane they have flown for many hours.

Get some dual at selected intervals. Discuss this with your instructor. Too many private pilots think that dual is now a check ride. Of course, you'll get a Flight Review as necessary but additional dual as you (or your instructor) feel necessary doesn't hurt. For instance, you might get instruction on some of the commercial maneuvers such as chandelles, lazy eights, and eights-on-pylons. You'll be learning new items. Who knows, you may want to go on to the commercial anyway. When done in a properly certificated airplane and with an experienced instructor, aerobatics and spin training is a great confidence builder. (Like everybody else, you'll approach such training with trepidation that first time but will find that your other areas of flying will profit after you've completed it.)

Work on a higher certificate or rating. Just flying around the local area is expensive and can get boring after a while. Working on a new certificate or rating doesn't mean that you have to fly *every* flight with that in mind, but at least you have a long-term goal. For instance, the instrument rating would be a good goal to work on right after the private certificate. Probably by the time you get the private you'll have around 66 hours total (this is the national average). Fifty hours of cross-country experience is required including the 10 hours solo you got as a student pilot. You can be flying some long cross-countries with family and friends, interspersed with instrument instruction and practice—with a qualified instructor or safety pilot, of course.

Another way of keeping flying fresh for you is to get a good checkout in an airplane type or model you haven't flown before. Perhaps you would like to fly a tailwheel airplane and so would contact a school that specializes in this. Or maybe you have an eye on that four-place with the constant-speed propeller, which would be good for those long cross-country trips with the family.

It's been said that a long cross-country flight is no more than a series of short cross-country segments stuck together. That's true to a certain extent, but fatigue and weather become more important factors on the long ones. After flying 7 or 8 hours and several hundred miles you'll find that many times both have put a certain amount of added pressure on you. Your weather judgments may be decayed by fatigue; "get-homeitis" has caused more than one bad accident near the end of a long flight home. Don't push weather beyond your capabilities; as said earlier, there will be times when you'll have to leave the airplane at a distant airport and find other means to get home. Remember that there is weather in which *nobody* flies (airlines, corporate or military), and your minimums should be appreciably higher at this early stage of your flying career.

As time goes on, get a night checkout and some local solo night time. This is a good confidence builder. Later you'll want to extend your night flights to cross-countries to nearby airports and then farther. (A tip for night cross-countries—keep up with the lighted towns on the sectional chart as you come to them; it's easier than goofing off and trying to figure if that's Smithfield down there, or whether you've passed it, or not.) Of course, airport rotating beacons and their relative position to a town are a good aid in identifying a particular town or city, but it's best to keep up with things as you go rather than, for instance, needing VOR cross bearings to finally locate yourself.

■ Bad Habits

Earlier it was mentioned that (probably inadvertently) you'll start to acquire some bad habits, maybe subtly starting on the day after the private pilot check ride.

For instance, your approaches and landings start getting faster and faster. You'll start landing that tricycle-gear airplane in a flatter attitude until you'll touch down with all three wheels at once, using much more runway than necessary and finding that the brakes are often abused.

Another bad habit is to let your traffic pattern flying deteriorate, with too-far or too-close downwind legs, imprecise turns at all "four corners" of the pattern, and poor wind drift correction on the legs.

Some private pilots still are leery of crosswind landings; if this is the situation for you, get some dual in crosswind conditions a little stronger (but still safe) than you had during training. That's a very good confidence builder and breaker of bad habits.

If your time and money are limited, you'll get the best recurrent training per dollar or per hour if you shoot an hour or so of landings every so often. Don't shoot all touch-and-go's, but break it up with full stops and taxiing back for another go. If you don't want to get dual instruction during some of these sessions, you might have a more experienced friend or a flight instructor, who's sitting out in front of the airport office, critique your takeoffs, patterns, and landings (from the ground).

SUMMED UP

Don't let yourself grow stale; regularly scheduled dual both on the basics and new areas of flying will increase your safety and hold your interest as you gain experience.

No matter how skilled or experienced you are in *your* line of work, flying is comparatively new to you at this point. Stay within your own limitations and never let yourself get pressured to fly.

APPENDIX A

WAKE TURBULENCE

(From *Aeronautical Information Manual*)

7–3–1. GENERAL

a. Every aircraft generates a wake while in flight. Initially, when pilots encountered this wake in flight, the disturbance was attributed to "prop wash." It is known, however, that this disturbance is caused by a pair of counter rotating vortices trailing from the wing tips. The vortices from larger aircraft pose problems to encountering aircraft. For instance, the wake of these aircraft can impose rolling moments exceeding the roll-control authority of the encountering aircraft. Further, turbulence generated within the vortices can damage aircraft components and equipment if encountered at close range. The pilot must learn to envision the location of the vortex wake generated by larger (transport category) aircraft and adjust the flight path accordingly.

b. During ground operations and during takeoff, jet engine blast (thrust stream turbulence) can cause damage and upsets if encountered at close range. Exhaust velocity versus distance studies at various thrust levels have shown a need for light aircraft to maintain an adequate separation behind large turbojet aircraft. Pilots of larger aircraft should be particularly careful to consider the effects of their "jet blast" on other aircraft, vehicles, and maintenance equipment during ground operations.

7–3–2. VORTEX GENERATION

Lift is generated by the creation of a pressure differential over the wing surface. The lowest pressure occurs over the upper wing surface and the highest pressure under the wing. This pressure differential triggers the roll up of the airflow aft of the wing resulting in swirling air masses trailing downstream of the wing tips. After the roll up is completed, the wake consists of two counter rotating cylindrical vortices. (See FIG 7–3–1). Most of the energy is within a few feet of the center of each vortex, but pilots should avoid a region within about 100 feet of the vortex core.

7–3–3. VORTEX STRENGTH

a. The strength of the vortex is governed by the weight, speed, and shape of the wing of the generating aircraft. The vortex characteristics of any given aircraft can also be changed by extension of flaps or other wing configuring devices as well as by change in speed. However, as the basic factor is weight, the vortex strength increases proportionately. Peak vortex tangential speeds exceeding 300 feet per second have

Wake Vortex Generation

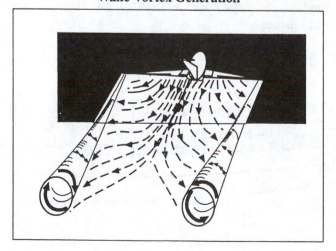

FIG 7–3–1

Wake Encounter Counter Control

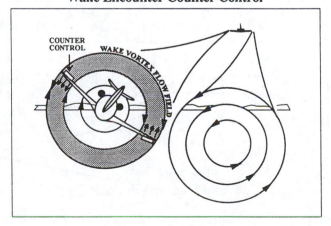

FIG 7–3–2

been recorded. The greatest vortex strength occurs when the generating aircraft is HEAVY, CLEAN, and SLOW.

b. INDUCED ROLL

1. In rare instances a wake encounter could cause inflight structural damage of catastrophic proportions. However, the usual hazard is associated with induced rolling moments which can exceed the roll-control authority of the encountering aircraft. In flight experiments, aircraft have been intentionally flown directly up trailing vortex cores of larger aircraft. It was shown that the capability of an aircraft to counteract the roll imposed by the wake vortex primarily depends on the wingspan and counter-control responsiveness of the encountering aircraft.

2. Counter control is usually effective and induced roll minimal in cases where the wingspan and ailerons of the encountering aircraft extend beyond the rotational flow field of the vortex. It is more difficult for aircraft with short wingspan (relative to the generating aircraft) to counter the imposed roll induced by vortex flow. Pilots of short span aircraft, even of the high performance type, must be especially alert to vortex encounters. (See FIG 7–3–2).

3. The wake of larger aircraft requires the respect of all pilots.

7–3–4. VORTEX BEHAVIOR

a. Trailing vortices have certain behavioral characteristics which can help a pilot visualize the wake location and thereby take avoidance precautions.

Wake Ends/Wake Begins

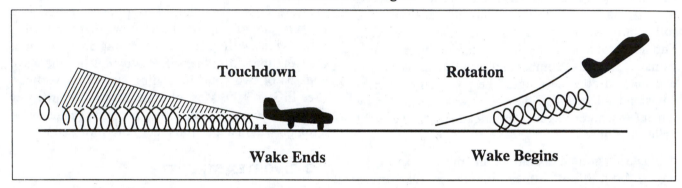

FIG 7–3–3

1. Vortices are generated from the moment aircraft leave the ground, since trailing vortices are a by-product of wing lift. Prior to takeoff or touchdown pilots should note the rotation or touchdown point of the preceding aircraft. (See FIG 7–3–3).

2. The vortex circulation is outward, upward and around the wing tips when viewed from either ahead or behind the aircraft. Tests with large aircraft have shown that the vortices remain spaced a bit less than a wingspan apart, drifting with the wind, at altitudes greater than a wingspan from the ground. In view of this, if persistent vortex turbulence is encountered, a slight change of altitude and lateral position (preferably upwind) will provide a flight path clear of the turbulence.

Vortex Flow Field

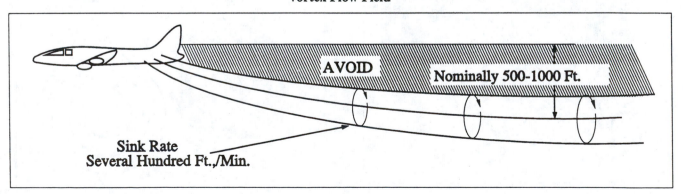

FIG 7–3–4

3. Flight tests have shown that the vortices from larger (transport category) aircraft sink at a rate of several hundred feet per minute, slowing their descent and diminishing in strength with time and distance behind the generating aircraft. Atmospheric turbulence hastens breakup. Pilots should fly at or above the preceding aircraft's flight path, altering course as necessary to avoid the area behind and below the generating aircraft. (See FIG 7-3-4). However vertical separation of 1,000 feet may be considered safe.

4. When the vortices of larger aircraft sink close to the ground (within 100 to 200 feet), they tend to move laterally over the ground at a speed of 2 or 3 knots. (See FIG 7-3-5).

Vortex Movement Near Ground – No Wind

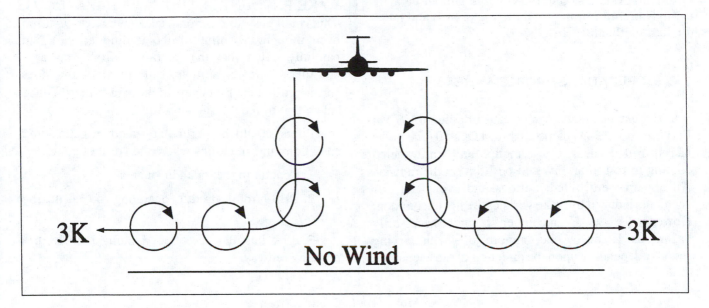

FIG 7-3-5

Vortex Movement Near Ground – with Cross Winds

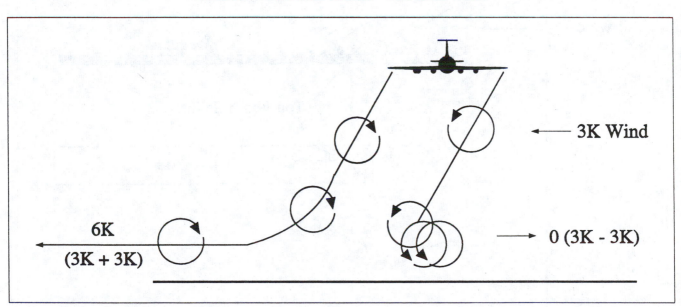

FIG 7-3-6

b. A crosswind will decrease the lateral movement of the upwind vortex and increase the movement of the downwind vortex. Thus a light wind with a cross runway component of 1 to 5 knots could result in the upwind vortex remaining in the touchdown zone for a period of time and hasten the drift of the downwind vortex toward another runway. (See FIG 7–3–6). Similarly, a tailwind condition can move the vortices of the preceding aircraft forward into the touchdown zone. THE LIGHT QUARTERING TAILWIND RE-QUIRES MAXIMUM CAUTION. Pilots should be alert to large aircraft upwind from their approach and take-off flight paths. (See FIG 7–3–7).

7–3–5. OPERATIONS PROBLEM AREAS

a. A wake encounter can be catastrophic. In 1972 at Fort Worth a DC–9 got too close to a DC–10 (two miles back), rolled, caught a wingtip, and cartwheeled coming to rest in an inverted position on the runway. All aboard were killed. Serious and even fatal GA accidents induced by wake vortices are not uncommon. However, a wake encounter is not necessarily hazardous. It can be one or more jolts with varying severity depending upon the direction of the encounter,

weight of the generating aircraft, size of the encountering aircraft, distance from the generating aircraft, and point of vortex encounter. The probability of induced roll increases when the encountering aircraft's heading is generally aligned with the flight path of the generating aircraft.

b. AVOID THE AREA BELOW AND BEHIND THE GENERATING AIRCRAFT, ESPECIALLY AT LOW ALTITUDE WHERE EVEN A MOMENTARY WAKE ENCOUNTER COULD BE HAZARDOUS. This is not easy to do. Some accidents have occurred even though the pilot of the trailing aircraft had carefully noted that the aircraft in front was at a considerably lower altitude. Unfortunately, this does not ensure that the flight path of the lead aircraft will be below that of the trailing aircraft.

c. Pilots should be particularly alert in calm wind conditions and situations where the vortices could:

1. Remain in the touchdown area.

2. Drift from aircraft operating on a nearby runway.

3. Sink into the takeoff or landing path from a crossing runway.

Vortex Movement in Ground Effect
Tailwind

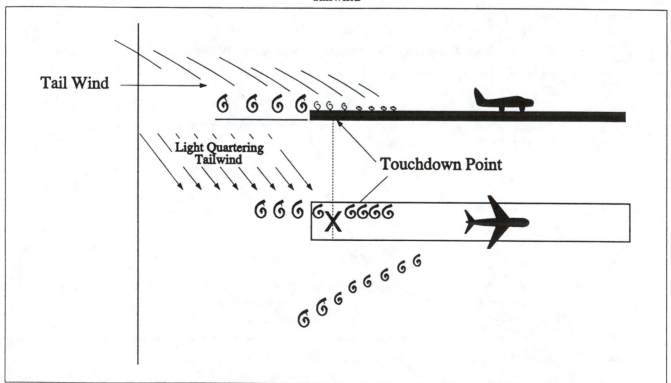

FIG 7–3–7

4. Sink into the traffic pattern from other airport operations.

5. Sink into the flight path of VFR aircraft operating on the hemispheric altitude 500 feet below.

d. Pilots of all aircraft should visualize the location of the vortex trail behind larger aircraft and use proper vortex avoidance procedures to achieve safe operation. It is equally important that pilots of larger aircraft plan or adjust their flight paths to minimize vortex exposure to other aircraft.

7–3–6. VORTEX AVOIDANCE PROCEDURES

a. Under certain conditions, airport traffic controllers apply procedures for separating IFR aircraft. The controllers will also provide to VFR aircraft, with whom they are in communication and which in the tower's opinion may be adversely affected by wake turbulence from a larger aircraft, the position, altitude and direction of flight of larger aircraft followed by the phrase "CAUTION – WAKE TURBULENCE." WHETHER OR NOT A WARNING HAS BEEN GIVEN, PILOTS ARE EXPECTED TO ADJUST THEIR OPERATIONS AND FLIGHT PATH(S) AS NECESSARY TO PRECLUDE SERIOUS WAKE ENCOUNTERS.

b. The following vortex avoidance procedures are recommended for the various situations:

1. *Landing behind a larger aircraft– same runway:* Stay at or above the larger aircraft's final approach flight path-note its touchdown point-land beyond it.

2. *Landing behind a larger aircraft– when parallel runway is closer than 2,500 feet:* Consider possible drift to your runway. Stay at or above the larger aircraft's final approach flight path– note its touchdown point.

3. *Landing behind a larger aircraft– crossing runway:* Cross above the larger aircraft's flight path.

4. *Landing behind a departing larger aircraft– same runway:* Note the larger aircraft's rotation point– land well prior to rotation point.

5. *Landing behind a departing larger aircraft– crossing runway:* Note the larger aircraft's rotation point– if past the intersection– continue the approach– land prior to the intersection. If larger aircraft rotates prior to the intersection, avoid flight below the larger aircraft's flight path. Abandon the approach unless a

landing is ensured well before reaching the intersection.

6. *Departing behind a larger aircraft:* Note the larger aircraft's rotation point and rotate prior to the larger aircraft's rotation point. Continue climbing above the larger aircraft's climb path until turning clear of the larger aircraft's wake. Avoid subsequent headings which will cross below and behind a larger aircraft. Be alert for any critical takeoff situation which could lead to a vortex encounter.

7. *Intersection takeoffs– same runway:* Be alert to adjacent larger aircraft operations, particularly upwind of your runway. If intersection takeoff clearance is received, avoid subsequent heading which will cross below a larger aircraft's path.

8. *Departing or landing after a larger aircraft executing a low approach, missed approach or touch-and-go landing:* Because vortices settle and move laterally near the ground, the vortex hazard may exist along the runway and in your flight path after a larger aircraft has executed a low approach, missed approach or a touch-and-go landing, particular in light quartering wind conditions. You should ensure that an interval of at least 2 minutes has elapsed before your takeoff or landing.

9. *En route VFR (thousand-foot altitude plus 500 feet):* Avoid flight below and behind a large aircraft's path. If a larger aircraft is observed above on the same track (meeting or overtaking) adjust your position laterally, preferably upwind.

7–3–7. HELICOPTERS

In a slow hover taxi or stationary hover near the surface, helicopter main rotor(s) generate downwash producing high velocity outwash vortices to a distance approximately three times the diameter of the rotor. When rotor downwash hits the surface, the resulting outwash vortices have behavioral characteristics similar to wing tip vortices produced by fixed wing aircraft. However, the vortex circulation is outward, upward, around, and away from the main rotor(s) in all directions. Pilots of small aircraft should avoid operating within three rotor diameters of any helicopter in a slow hover taxi or stationary hover. In forward flight, departing or landing helicopters produce a pair of strong, high-speed trailing vortices similar to wing tip vortices of larger fixed wing aircraft. Pilots of small aircraft should use caution when operating behind or crossing behind landing and departing helicopters.

7–3–8. PILOT RESPONSIBILITY

a. Government and industry groups are making concerted efforts to minimize or eliminate the hazards of trailing vortices. However, the flight disciplines necessary to ensure vortex avoidance during VFR operations must be exercised by the pilot. Vortex visualization and avoidance procedures should be exercised by the pilot using the same degree of concern as in collision avoidance.

b. Wake turbulence may be encountered by aircraft in flight as well as when operating on the airport movement area.

REFERENCE–
PILOT/CONTROLLER GLOSSARY TERM– WAKE TURBULENCE.

c. Pilots are reminded that in operations conducted behind all aircraft, acceptance of instructions from ATC in the following situations is an acknowledgment that the pilot will ensure safe takeoff and landing intervals and accepts the responsibility for providing wake turbulence separation.

1. Traffic information,

2. Instructions to follow an aircraft, and

3. The acceptance of a visual approach clearance.

d. For operations conducted behind **heavy** aircraft, ATC will specify the word "heavy" when this information is known. Pilots of **heavy** aircraft should always use the word "heavy" in radio communications.

e. Heavy and large jet aircraft operators should use the following procedures during an approach to landing. These procedures establish a dependable baseline from which pilots of in–trail, lighter aircraft may reasonably expect to make effective flight path adjustments to avoid serious wake vortex turbulence.

1. Pilots of aircraft that produce strong wake vortices should make every attempt to fly on the established glidepath, not above it; or, if glidepath guidance is not available, to fly as closely as possible to a "3–1" glidepath, not above it.

EXAMPLE–
FLY 3,000 FEET AT 10 MILES FROM TOUCHDOWN, 1,500 FEET AT 5 MILES, 1,200 FEET AT 4 MILES, AND SO ON TO TOUCHDOWN.

2. Pilots of aircraft that produce strong wake vortices should fly as closely as possible to the approach course centerline or to the extended centerline of the runway of intended landing as appropriate to conditions.

f. Pilots operating lighter aircraft on visual approaches in–trail to aircraft producing strong wake vortices should use the following procedures to assist in avoiding wake turbulence. These procedures apply only to those aircraft that are on visual approaches.

1. Pilots of lighter aircraft should fly on or above the glidepath. Glidepath reference may be furnished by an ILS, by a visual approach slope system, by other ground–based approach slope guidance systems, or by other means. In the absence of visible glidepath guidance, pilots may very nearly duplicate a 3–degree glideslope by adhering to the "3 to 1" glidepath principle.

EXAMPLE–
FLY 3,000 FEET AT 10 MILES FROM TOUCHDOWN, 1,500 FEET AT 5 MILES, 1,200 FEET AT 4 MILES, AND SO ON TO TOUCHDOWN.

2. If the pilot of the lighter following aircraft has visual contact with the preceding heavier aircraft and also with the runway, the pilot may further adjust for possible wake vortex turbulence by the following practices:

(a) Pick a point of landing no less than 1,000 feet from the arrival end of the runway.

(b) Establish a line–of–sight to that landing point that is above and in front of the heavier preceding aircraft.

(c) When possible, note the point of landing of the heavier preceding aircraft and adjust point of intended landing as necessary.

EXAMPLE–
A PUFF OF SMOKE MAY APPEAR AT THE 1,000–FOOT MARKINGS OF THE RUNWAY, SHOWING THAT TOUCHDOWN WAS THAT POINT; THEREFORE, ADJUST POINT OF INTENDED LANDING TO THE 1,500–FOOT MARKINGS.

(d) Maintain the line–of–sight to the point of intended landing above and ahead of the heavier preceding aircraft; maintain it to touchdown.

(e) Land beyond the point of landing of the preceding heavier aircraft.

3. During visual approaches pilots may ask ATC for updates on separation and groundspeed with respect to heavier preceding aircraft, especially when there is any question of safe separation from wake turbulence.

7–3–9. AIR TRAFFIC WAKE TURBULENCE SEPARATIONS

a. Because of the possible effects of wake turbulence, controllers are required to apply no less than specified minimum separation for aircraft operating behind a **heavy** jet and, in certain instances, behind large **non–heavy** aircraft (i.e., B757 aircraft).

1. Separation is applied to aircraft operating directly behind a **heavy/B757** jet at the same altitude or less than 1,000 feet below:

(a) **Heavy** jet behind **heavy** jet–4 miles.

(b) **Large/heavy** behind **B757** – 4 miles.

(c) **Small** behind **B757** – 5 miles.

(d) **Small/large** aircraft behind **heavy** jet – 5 miles.

2. Also, separation, measured at the time the preceding aircraft is over the landing threshold, is provided to small aircraft:

(a) **Small** aircraft landing behind **heavy** jet– 6 miles.

(b) **Small** aircraft landing behind **B757** – 5 miles.

(c) **Small** aircraft landing behind **large** aircraft– 4 miles.

REFERENCE–
PILOT/CONTROLLER GLOSSARY TERM– AIRCRAFT CLASSES.

3. Additionally, appropriate time or distance intervals are provided to departing aircraft:

(a) Two minutes or the appropriate 4 or 5 mile radar separation when takeoff behind a **heavy/B757** jet will be:

(b) from the same threshold

(c) on a crossing runway and projected flight paths will cross

(d) from the threshold of a parallel runway when staggered ahead of that of the adjacent runway by less than 500 feet and when the runways are separated by less than 2,500 feet.

NOTE–
CONTROLLERS MAY NOT REDUCE OR WAIVE THESE INTERVALS.

b. A 3–minute interval will be provided when a **small** aircraft will takeoff:

1. From an intersection on the same runway (same or opposite direction) behind a departing **large** aircraft,

2. In the opposite direction on the same runway behind a large aircraft takeoff or low/missed approach.

NOTE–
THIS 3–MINUTE INTERVAL MAY BE WAIVED UPON SPECIFIC PILOT REQUEST.

c. A 3–minute interval will be provided for all aircraft taking off when the operations are as described in subparagraph b. 1. and 2. above, the preceding aircraft is a **heavy/B757** jet, and the operations are on either the same runway or parallel runways separated by less than 2,500 feet. Controllers may not reduce or waive this interval.

d. Pilots may request additional separation i.e., 2 minutes instead of 4 or 5 miles for wake turbulence avoidance. This request should be made as soon as practical on ground control and at least before taxiing onto the runway.

NOTE–
FAR PART 91.3(A) STATES: "THE PILOT IN COMMAND OF AN AIRCRAFT IS DIRECTLY RESPONSIBLE FOR AND IS THE FINAL AUTHORITY AS TO THE OPERATION OF THAT AIRCRAFT."

e. Controllers may anticipate separation and need not withhold a takeoff clearance for an aircraft departing behind a **large/heavy** aircraft if there is reasonable assurance the required separation will exist when the departing aircraft starts takeoff roll.

APPENDIX B

(From Airport/Facility Directory)

DIRECTORY LEGEND SAMPLE

① ③ ④ ⑤ ⑥ ⑦

CITY NAME

AIRPORT NAME (ORL) 4 E GMT-5(-4DT) 28°32'43"N 81°20'10"W   JACKSONVILLE
200 B S4 FUEL 100, JET A OX 1, 2, 3 TPA—1000(800) AOE CFR Index A Not insp.   H-4G, L-19C

② ⑨ ⑩ ⑪ ⑫ ⑬ ⑭ ⑮ ⑯ ⑰ ⑧
                                      IAP

⑱ — RWY 07-25: H6000X150 (ASPH-PFC)   S-90, D-160, DT-300   HIRL CL
     RWY 07: ALSF1. Trees.   RWY 25: REIL. Rgt tfc.
   RWY 13-31: H4620X100 (ASPH)   HIRL
     RWY 13: VASI(V2L)—GA 3.3° TCH 89' Pole.   RWY 31: VASI(V2L)—GA 3.1° TCH 36'. Tree. Rgt tfc.
⑲ — AIRPORT REMARKS: Special Air Traffic Rules—Part 93, see Regulatory Notices. Attended 1200-0300Z‡. Parachute
     Jumping. CAUTION cattle and deer on arpt. Acft 100,000 lbs or over ctc Director of Aviation for approval (305) 894-
     9831. Fee for all airline charters, travel clubs and certain revenue producing acft. Flight Notification Service
     (ADCUS) available. Control Zone effective 1500-0700Z‡.   ⑤
⑳ — WEATHER DATA SOURCES: AWOS-1 120.3 (202) 426-8000. LLWAS.
㉑ — COMMUNICATIONS: ATIS 127.25   UNICOM 122.95
     NAME FSS (ORL) on fld. 123.65 122.65 122.2 LD305-894-0861. NOTAM FILE ORL.   ⑤
  Ⓡ NAME APP/DEP CON 128.35 (1200-0400Z‡)
     TOWER 118.7   GND CON 121.7   CLNC DEL 125.55   PRE TAXI CLNC 125.5
     TCA GROUP II: See VFR Terminal Area Chart.
㉒ — RADIO AIDS TO NAVIGATION: NOTAM FILE ORL. VHF/DF ctc FSS.
   (H) ABVORTAC 112.2   ▪ MCO   Chan 59   28°32'33"N 81°20'07"W   at fld.   1110/8E.
     NOTAM FILE MCO.   TWEB avbl 1300-0100Z‡.   ⑤
      VOR unusable 050-060° beyond 15 NM below 5000'
     HERNY NDB (LOM) 221   OR   28°30'24"N 81°26'03"W   067° 5.4 NM to fld.
     ILS 109 I-ORL Rwy 07. LOM HERNY NDB
     ASR/PAR
㉓ — COMM/NAVAID REMARKS: Emerg frequency 121.5 not available at tower.

TENNESSEE

SEWANEE

FRANKLIN CO (UOS) 1 E UTC-6(-5DT) N35°12.24' W85°53.92'   ATLANTA
1950   B   FUEL 80, 100LL   S-20   LIRL.   L-14H
RWY 06-24: H3300X50 (ASPH)   S-20   LIRL.
     RWY 06: SAVASI(S2L)—GA 3.5°TCH 28'. Trees.   RWY 24: VASI(V2L)—GA 4.0° TCH 31'. Trees.
AIRPORT REMARKS: Attended dalgt hours. For svc after hours call 615-598-0745. Sporadic crosswinds and
     turbulence. Deer on and in vicinity of rwy.
COMMUNICATIONS: CTAF/UNICOM 122.8
     NASHVILLE FSS (BNA) TF 1-800-WX-BRIEF. NOTAM FILE BNA.
RADIO AIDS TO NAVIGATION: NOTAM FILE BNA.
     SHELBYVILLE (L) VOR/DME 109.0   SYI   Chan 27   N35°33.72' W86°26.35'   130° 34.1 NM to fld.
     810/01W.
     SEWANEE NDB (MHW) 275   UOS   N35°12.26' W85°53.75'   at fld. (VFR only).

COLUMBIA/MOUNT PLEASANT

MAURY CO (MRC) 2 NE UTC-6(-5DT) N35°33.26' W87°10.82'   ATLANTA
677   B   S4   FUEL 100LL, JET A   H-4H, L-14G
RWY 05-23: H5003X75 (ASPH)   S-28, D-43, DT-70   MIRL   IAP
     RWY 05: PAPI(P4R). Trees.   RWY 23: ODALS(NSTD). VASI(V2L). Trees.
RWY 17-35: 2500X300 (TURF)
     RWY 17: Trees.   RWY 35: Thld dsplcd 600'. Trees.
AIRPORT REMARKS: Attended 1300Z‡-dark. ACTIVATE ODALS Rwy 23—CTAF.
WEATHER DATA SOURCES: AWOS-3   128.625   (615) 379-0844.
COMMUNICATIONS: CTAF/UNICOM 122.8
     JACKSON FSS (MKL) TF 1-800-WX-BRIEF. NOTAM FILE MKL.
  Ⓡ MEMPHIS CENTER APP/DEP CON 125.85
RADIO AIDS TO NAVIGATION: NOTAM FILE MKL.
     GRAHAM (L) VORTAC 111.6   GHM   Chan 53   N35°50.04' W87°27.11'   139° 21.4 NM to fld. 770/03E.
     NDB (MHW) 365   PBC   N35°36.49' W87°05.48'   235° 5.4 NM to fld.
     SDF 108.7 I MRC Rwy 23.

ALABAMA

HUNTSVILLE INTL-CARL T JONES FLD (HSV) 9 SW UTC-6(-5DT)   ATLANTA
     N34°38.47' W86°46.45'   H-4H, L-14H
630   B   S4   FUEL 100LL, JET A   OX 1, 2, 3, 4   LRA   ARFF Index C   IAP
RWY 18R-36L: H8000X150 (ASPH-GRVD)   S-75, D-200, DT-350, DDT-850   HIRL CL
     RWY 18R: ALSF2 TDZ Rgt tfc.   RWY 36L: MALSR
RWY 18L-36R: H8000X150 (ASPH-GRVD)   S-90, D-108, DT-170   HIRL
     RWY 18L: MALSR. PAPI(P4L)—GA 3.0° TCH 70'.   RWY 36R: REIL. VASI(V4L)—GA 3.0° TCH 55'. Rgt tfc.
AIRPORT REMARKS: Attended continuously. Rwy 18L-36R first 1,000' CLOSED both ends. Migratory birds Oct 1–Mar
     15, wildlife refuge S and W of arpt. When twr clsd ACTIVATE MALSR Rwy 18L; REIL Rwy 36R; VASI Rwy 36R;
     and PAPI Rwy 18L—CTAF. PPR 24 hours for unscheduled air carrier ops with more than 30 passenger seats call
     arpt manager 205-772-8728. Rwy 18L MALSR out of svc indefinitely. Rwy 18L and 36R VASI out of svc
     indefinitely. Rwy 36R REIL out of svc indefinitely. Flight Notification Service (ADCUS) available. Control Zone
     effective continuously.
WEATHER DATA SOURCES: LLWAS.
COMMUNICATIONS: CTAF 127.6   UNICOM 122.95   ATIS 121.25
     ANNISTON FSS (ANB) TF 1-800-WX-BRIEF. NOTAM FILE HSV.
  Ⓡ APP CON 125.6 (360°-179°) 118.05 (180°-359°)118.75 (1200-0600Z‡)
     MEMPHIS CENTER APP CON 120.8 (0600-1200Z‡)
     TOWER 127.6 (1200-0600Z‡)   GND CON 121.9   CLNC DEL 120.35
  Ⓡ DEP CON 125.6 (359°-179°) 118.05 (180°-358°) 120.35 (1200-0600Z‡)
     MEMPHIS CENTER DEP CON 120.8 (0600-1200Z‡)
     ARSA ctc APP CON
RADIO AIDS TO NAVIGATION: NOTAM FILE ANB.
     DECATUR (L) VORW/DME 112.8   DCU   Chan 75   N34°38.90' W86°56.37'   094° 8.2 NM to fld. 590/01W.
     CAPSHAW NDB (MHW) 350   CWH   N34°46.42' W86°46.74'   179° 7.9 NM to fld. NOTAM FILE HSV.
     ILS 109.3 I-HSV Rwy 18R. (Unmonitored when twr closed).
     ILS 108.5 I-ELL Rwy 36L. (Unmonitored when twr closed)
     ILS 111.9 I-TVN Rwy 18L. ILS out of svc indefinitely.
     ASR

DIRECTORY LEGEND
LEGEND

This Directory is an alphabetical listing of data on record with the FAA on all airports that are open to the public, associated terminal control facilities, air route traffic control centers and radio aids to navigation within the conterminous United States, Puerto Rico and the Virgin Islands. Airports are listed alphabetically by associated city name and cross referenced by airport name. Facilities associated with an airport, but with a different name, are listed individually under their own name, as well as under the airport with which they are associated.

The listing of an airport in this directory merely indicates the airport operator's willingness to accommodate transient aircraft, and does not represent that the facility conforms with any Federal or local standards, or that it has been approved for use on the part of the general public.

The information on obstructions is taken from reports submitted to the FAA. It has not been verified in all cases. Pilots are cautioned that objects not indicated in this tabulation (or on charts) may exist which can create a hazard to flight operation.

Detailed specifics concerning services and facilities tabulated within this directory are contained in Airman's Information Manual, Basic Flight Information and ATC Procedures.

The legend items that follow explain in detail the contents of this Directory and are keyed to the circled numbers on the sample on the preceding page.

① CITY/AIRPORT NAME

Airports and facilities in this directory are listed alphabetically by associated city and state. Where the city name is different from the airport name the city name will appear on the line above the airport name. Airports with the same associated city name will be listed alphabetically by airport name and will be separated by a dashed rule line. All others will be separated by a solid rule line.

② NOTAM SERVICE

§—NOTAM "D" (Distance teletype dissemination) and NOTAM "L" (local dissemination) service is provided for airport. Absence of annotation § indicates NOTAM "L" (local dissemination) only is provided for airport. Airport NOTAM file identifier will be shown as "NOTAM FILE IAD" for all public-use airports. See AIM, Basic Flight Information and ATC Procedures for detailed descriptions of NOTAM.

③ LOCATION IDENTIFIER

A three or four character code assigned to airports. These identifiers are used by ATC in lieu of the airport name in flight plans, flight strips and other written records and computer operations.

④ AIRPORT LOCATION

Airport location is expressed as distance and direction from the center of the associated city in nautical miles and cardinal points, i.e., 4 NE.

⑤ TIME CONVERSION

Hours of operation of all facilities are expressed in Coordinated Universal Time (UTC) and shown as "Z" time. The directory indicates the number of hours to be subtracted from UTC to obtain local standard time and local daylight saving time UTC-5(-4DT). The symbol ‡ indicates that during periods of Daylight Saving Time effective hours will be one hour earlier than shown. In those areas where daylight saving time is not observed (-4DT) and ‡ will not be shown. All states observe daylight savings time except Arizona and that portion of Indiana in the Eastern Time Zone and Puerto Rico and the Virgin Islands.

⑥ GEOGRAPHIC POSITION OF AIRPORT

⑦ CHARTS

The Sectional Chart and Low and High Altitude Enroute Chart and panel on which the airport or facility is located.

⑧ INSTRUMENT APPROACH PROCEDURES

IAP indicates an airport for which a prescribed (Public Use) FAA Instrument Approach Procedure has been published.

⑨ ELEVATION

Elevation is given in feet above mean sea level and is the highest point on the landing surface. When elevation is sea level it will be indicated as (00). When elevation is below sea level a minus (–) sign will precede the figure.

⑩ ROTATING LIGHT BEACON

B indicates rotating beacon is available. Rotating beacons operate dusk to dawn unless otherwise indicated in AIRPORT REMARKS.

⑪ SERVICING

S1:	Minor airframe repairs.
S2:	Minor airframe and minor powerplant repairs.
S3:	Major airframe and minor powerplant repairs.
S4:	Major airframe and major powerplant repairs.

⑫ FUEL

CODE	FUEL
80	Grade 80 gasoline (Red)
100	Grade 100 gasoline (Green)
100LL	Grade 100LL gasoline (low lead) (Blue)
115	Grade 115 gasoline
A	Jet A—Kerosene freeze point-40° C.
A1	Jet A-1—Kerosene, freeze point-50° C.
A1+	Jet A-1—Kerosene with icing inhibitor, freeze point-50° C.
B	Jet B—Wide-cut turbine fuel, freeze point-50° C.
B+	Jet B—Wide-cut turbine fuel with icing inhibitor, freeze point-50° C.

⑬ OXYGEN

OX 1	High Pressure
OX 2	Low Pressure
OX 3	High Pressure—Replacement Bottles
OX 4	Low Pressure—Replacement Bottles

⑭ TRAFFIC PATTERN ALTITUDE

Traffic Pattern Altitude (TPA)—The first figure shown is TPA above mean sea level. The second figure in parentheses is TPA above airport elevation.

⑮ AIRPORT OF ENTRY AND LANDING RIGHTS AIRPORTS

AOE—Airport of Entry—A customs Airport of Entry where permission from U.S. Customs is not required, however, at least one hour advance notice of arrival must be furnished.

LRA—Landing Rights Airport—Application for permission to land must be submitted in advance to U.S. Customs. At least one hour advance notice of arrival must be furnished.

NOTE: Advance notice of arrival at both an AOE and LRA airport may be included in the flight plan when filed in Canada or Mexico, where Flight Notification Service (ADCUS) is available the airport remark will indicate this service. This notice will also be treated as an application for permission to land in the case of an LRA. Although advance notice of arrival may be relayed to Customs through Mexico, Canadian, and U.S. Communications facilities by flight plan, the aircraft operator is solely responsible for insuring that Customs receives the notification. (See Customs, Immigration and Naturalization, Public Health and Agriculture Department requirements in the International Flight Information Manual for further details.)

DIRECTORY LEGEND

⑯ CERTIFICATED AIRPORT (FAR 139)

Airports serving Department of Transportation certified carriers and certified under FAR, Part 139, are indicated by the CFR index; i.e., CFR Index A, which relates to the availability of crash, fire, rescue equipment.

FAR–PART 139 CERTIFICATED AIRPORTS

INDICES AND FIRE FIGHTING AND RESCUE EQUIPMENT REQUIREMENTS

Airport Index	Required No. Vehicles	Aircraft Length	Scheduled Departures	Agent + Water for Foam
A	1	≤90′	≥1	500#DC or 450#DC + 50 gal H_2O
AA	1	[illegible] ≤126′	< 5	300#DC + 500 gal H_2O
B	[illegible]	[illegible]126′	[illegible]	[illegible]

⑰ FAA INSPECTION

All airports not inspected by FAA will be identified by the note: Not insp. This indicates that the airport information has been provided by the owner or operator of the field.

⑱ RUNWAY DATA

Runway information is shown on two lines. That information common to the entire runway is shown on the first line while information concerning the runway ends are shown on the second or following line. Lengthy information will be placed in the Airport Remarks.

Runway direction, surface, length, width, weight bearing capacity, lighting, gradient (when gradient exceeds 0.3 percent) and appropriate remarks are shown for each runway. Direction, length, width, lighting and remarks are shown for sealanes. The full dimensions of helipads are shown, i.e., 50X150.

RUNWAY SURFACE AND LENGTH

Runway lengths prefixed by the letter "H" indicate that the runways are hard surfaced (concrete, asphalt). If the runway length is not prefixed, the surface is sod, clay, etc. The runway surface composition is indicated in parentheses after runway length as follows:

(AFSC)—Aggregate friction seal coat
(ASPH)—Asphalt
(CONC)—Concrete
(DIRT)—Dirt
(GRVD)—Grooved
(GRVL)—Gravel, or cinders
(PFC)—Porous friction courses
(RFSC)—Rubberized friction seal coat
(TURF)—Turf
(TRTD)—Treated
(WC)—Wire combed

RUNWAY WEIGHT BEARING CAPACITY

Runway strength data shown in this publication is derived from available information and is a realistic estimate of capability at an average level of activity. It is not intended as a maximum allowable weight or as an operating limitation. Many airport pavements are capable of supporting limited operations with gross weights of 25-50% in excess of the published figures. Permissible operating weights, insofar as runway strengths are concerned, are a matter of agreement between the owner and user. When desiring to operate into any airport at weights in excess of those published in the publication, users should contact the airport management for permission. Add 000 to figure following S, D, DT, DDT and MAX for gross weight capacity:

S—Runway weight bearing capacity for aircraft with single- wheel type landing gear, (DC-3), etc.
D—Runway weight bearing capacity for aircraft with dual-wheel type landing gear, (DC-6), etc.
DT—Runway weight bearing capacity for aircraft with dual-tandem type landing gear, (707), etc.
DDT—Runway weight bearing capacity for aircraft with double dual- tandem type landing gear, (747), etc.

Quadricycle and dual-tandem are considered virtually equal for runway weight bearing consideration, as are single-tandem and dual-wheel.

Omission of weight bearing capacity indicates information unknown.

RUNWAY LIGHTING

Lights are in operation sunset to sunrise. Lighting available by prior arrangement only or operating part of the night only and/or pilot controlled and with specific operating hours are indicated under airport remarks. Since obstructions are usually lighted, obstruction lighting is not included in this code. Unlighted obstructions on or surrounding an airport will be noted in airport remarks.

Temporary, emergency or limited runway edge lighting such as flares, smudge pots, lanterns or portable runway lights will also be shown in airport remarks.

Types of lighting are shown with the runway or runway end they serve.

LIRL—Low Intensity Runway Lights
MIRL—Medium Intensity Runway Lights
HIRL—High Intensity Runway Lights
REIL—Runway End Identifier Lights
CL—Centerline Lights
TDZ—Touchdown Zone Lights
ODALS—Omni Directional Approach Lighting System.
AF OVRN—Air Force Overrun 1000′ Standard Approach Lighting System.
LDIN—Lead-In Lighting System.
MALS—Medium Intensity Approach Lighting System.
MALSF—Medium Intensity Approach Lighting System with Sequenced Flashing Lights.
MALSR—Medium Intensity Approach Lighting System with Runway Alignment Indicator Lights.
SALS—Short Approach Lighting System.
SALSF—Short Approach Lighting System with Sequenced Flashing Lights.
SSALS—Simplified Short Approach Lighting System.
SSALF—Simplified Short Approach Lighting System with Sequenced Flashing Lights.
SSALR—Simplified Short Approach Lighting System with Runway Alignment Indicator Lights.
ALSAF—High Intensity Approach Lighting System with Sequenced Flashing Lights
ALSF1—High Intensity Approach Lighting System with Sequenced Flashing Lights, Category I, Configuration.
ALSF2—High Intensity Approach Lighting System with Sequenced Flashing Lights, Category II, Configuration.
VASI—Visual Approach Slope Indicator System.

VISUAL APPROACH SLOPE INDICATOR SYSTEMS

VASI—Visual Approach Slope Indicator
SAVASI—Simplified Abbreviated Visual Approach Slope Indicator

S2L	2-box SAVASI on left side of runway
S2R	2-box SAVASI on right side of runway
V2R	2-box VASI on right side of runway
V2L	2-box VASI on left side of runway
V4R	4-box VASI on right side of runway
V4L	4-box VASI on left side of runway
V6R	6-box VASI on right side of runway
V6L	6-box VASI on left side of runway
V12	12-box VASI on both sides of runway
V16	16-box VASI on both sides of runway
*NSTD	Nonstandard VASI, VAPI, or any other system not listed above

VASI approach slope angle and threshold crossing height will be shown when available; i.e., GA 3.5° TCH 37.0′.

PILOT CONTROL OF AIRPORT LIGHTING

Key Mike	Function
7 times within 5 seconds	Highest intensity available
5 times within 5 seconds	Medium or lower intensity (Lower REIL or REIL-Off)
3 times within 5 seconds	Lowest intensity available (Lower REIL or REIL-Off)

Available systems will be indicated in the Airport Remarks, as follows:

ACTIVATE MALSR Rwy 7, HIRL Rwy 7-25-122.8.
or
ACTIVATE MIRL Rwy 18-36-122.8.
or
ACTIVATE VASI and REIL, Rwy 7-122.8.

Where the airport is not served by an instrument approach procedure and/or has an independent type system of different specification installed by the airport sponsor, descriptions of the type lights, method of control, and operating frequency will be explained in clear text. See AIM, "Basic Flight Information and ATC Procedures," for detailed description of pilot control of airport lighting.

DIRECTORY LEGEND

RUNWAY GRADIENT

Runway gradient will be shown only when it is 0.3 percent or more. When available the direction of slope upward will be indicated, i.e., 0.5% up NW.

RUNWAY END DATA

Lighting systems such as VASI, MALSR, REIL; obstructions; displaced thresholds will be shown on the specific runway end. "Rgt tfc"—Right traffic indicates right turns should be made on landing and takeoff for specified runway end.

⑲ AIRPORT REMARKS

Landing Fee indicates landing charges for private or non-revenue producing aircraft, in addition, fees may be charged for planes that remain over a couple of hours and buy no services, or at major airline terminals for all aircraft.

Remarks—Data is confined to operational items affecting the status and usability of the airport.

⑳ WEATHER DATA SOURCES

AWOS—Automated Weather Observing System

AWOS-1—reports altimeter setting, wind data and usually temperature, dewpoint and density altitude.
AWOS-2—reports the same as AWOS-1 plus visibility.
AWOS-3—reports the same as AWOS-1 plus visibility and cloud/ceiling data.
See AIM, Basic Flight Information and ATC Procedures for detailed description of AWOS.

SAWRS—identifies airports that have a Supplemental Aviation Weather Reporting Station available to pilots for current weather information.

LAWRS—Limited Aviation Weather Reporting Station where observers report cloud height, weather, obstructions to vision, temperature and dewpoint (in most cases), surface wind, altimeter and pertinent remarks.

LLWAS—indicates a Low Level Wind Shear Alert System consisting of a center field and several field perimeter anemometers.

HIWAS—See RADIO AIDS TO NAVIGATION

㉑ COMMUNICATIONS

Communications will be listed in sequence in the order shown below:

Common Traffic Advisory Frequency (CTAF), Automatic Terminal Information Service (ATIS) and Aeronautical Advisory Stations (UNICOM) along with their frequency is shown, where available, on the line following the heading "COMMUNICATIONS." When the CTAF and UNICOM is the same frequency, the frequency will be shown as CTAF/UNICOM freq.

Flight Service Station (FSS) information. The associated FSS will be shown followed by the identifier and information concerning availability of telephone service, e.g., Direct Line (DL), Local Call (LC-384-2341), Long Distance (LD 202-426-8800 or LD 1-202-555-1212) etc. The airport NOTAM file identifier will be shown as "NOTAM FILE IAD." Where the FSS is located on the field it will be indicated as "on arpt" following the identifier. Frequencies available will follow. The FSS telephone number will follow along with any significant operational information. FSS's whose name is not the same as the airport on which located will also be listed in the normal alphabetical name listing for the state in which located. Remote Communications Outlet (RCO) providing service to the airport followed by the frequency and name of the Controlling FSS.

FSS's provide information on airport conditions, radio aids and other facilities, and process flight plans. Airport Advisory Service is provided on the CTAF by FSS's located at non-tower airports or airports where the tower is not in operation.

(See AIM, Par. 157/158 Traffic Advisory Practices at airports where a tower is not in operation or AC 90 - 42C.)

Aviation weather briefing service is provided by FSS specialists. Flight and weather briefing services are also available by calling the telephone numbers listed.

Remote Communications Outlet (RCO)—An unmanned air/ ground communications facility, remotely controlled and providing UHF or VHF communications capability to extend the service range of an FSS.

Civil Communications Frequencies—Civil communications frequencies used in the FSS air/ground system are now operated simplex on 122.0, 122.2, 122.3, 122.4, 122.6, 123.6; emergency 121.5; plus receive-only on 122.05, 122.1, 122.15, and 123.6.

a. 122.0 is assigned as the Enroute Flight Advisory Service channel at selected FSS's.
b. 122.2 is assigned to all FSS's as a common enroute simplex service.
c. 123.6 is assigned as the airport advisory channel at non-tower FSS locations, however, it is still in commission at some FSS's collocated with towers to provide part time Airport Advisory Service.
d. 122.1 is the primary receive-only frequency at VOR's. 122.05, 122.15 and 123.6 are assigned at selected VOR's meeting certain criteria.
e. Some FSS's are assigned 50 kHz channels for simplex operation in the 122-123 MHz band (e.g. 122.35). Pilots using the FSS A/G system should refer to this directory or appropriate charts to determine frequencies available at the FSS or remoted facility through which they wish to communicate.

Part time FSS hours of operation are shown in remarks under facility name.

Emergency frequency 121.5 is available at all Flight Service Stations, Towers, Approach Control and RADAR facilities, unless indicated as not available.

Frequencies published followed by the letter "T" or "R", indicate that the facility will only transmit or receive respectively on that frequency. All radio aids to navigation frequencies are transmit only.

TERMINAL SERVICES

CTAF—A program designed to get all vehicles and aircraft at uncontrolled airports on a common frequency.
ATIS—A continuous broadcast of recorded non-control information in selected areas of high activity.
UNICOM—A non-government air/ground radio communications facility utilized to provide general airport advisory service.
APP CON—Approach Control. The symbol Ⓡ indicates radar approach control.
TOWER—Control tower
GND CON—Ground Control
DEP CON—Departure Control. The symbol Ⓡ indicates radar departure control.
CLNC DEL—Clearance Delivery.
PRE TAXI CLNC—Pre taxi clearance
VFR ADVSY SVC—VFR Advisory Service. Service provided by Non-Radar Approach Control. Advisory Service for VFR aircraft (upon a workload basis) ctc APP CON.
STAGE II SVC—Radar Advisory and Sequencing Service for VFR aircraft
STAGE III SVC—Radar Sequencing and Separation Service for participating VFR Aircraft within a Terminal Radar Service Area (TRSA)
ARSA—Airport Radar Service Area
TCA—Radar Sequencing and Separation Service for all aircraft in a Terminal Control Area (TCA)
TOWER, APP CON and DEP CON RADIO CALL will be the same as the airport name unless indicated otherwise.

㉒ RADIO AIDS TO NAVIGATION

The Airport Facility Directory lists by facility name all Radio Aids to Navigation, except Military TACANS, that appear on National Ocean Service Visual or IFR Aeronautical Charts and those upon which the FAA has approved an Instrument Approach Procedure.

All VOR, VORTAC ILS and MLS equipment in the National Airspace System has an automatic monitoring and shutdown feature in the event of malfunction. Unmonitored, as used in this publication for any navigational aid, means that FSS or tower personnel cannot observe the malfunction or shutdown signal. The NAVAID NOTAM file identifier will be shown as "NOTAM FILE IAD" and will be listed on the Radio Aids to Navigation line. When two or more NAVAIDS are listed and the NOTAM file identifier is different than shown on the Radio Aids to Navigation line, then it will be shown with the NAVAID listing. Hazardous Inflight Weather Advisory Service (HIWAS) will be shown where this service is broadcast over selected VOR's.

NAVAID information is tabulated as indicated in the following sample:

TWEB — TACAN/DME Channel — Geographical Position — Site Elevation

NAME (L) ABVORTAC 117.55 ■ ABE Chan 122(Y) 40°43′36″N 75°27′18″W 180° 4.1 NM to fld. 1110/8E. HIWAS.

Class — Frequency — Identifier — Bearing and distance facility to airport — Magnetic Variation — Hazardous Inflight Weather Advisory Service

VOR unusable 020°-060° beyond 26 NM below 3500′

Restriction within the normal altitude/range of the navigational aid (See primary alphabetical listing for restrictions on VORTAC and VOR/DME).

Note: Those DME channel numbers with a (Y) suffix require TACAN to be placed in the "Y" mode to receive distance information.

HIWAS—Hazardous Inflight Weather Advisory Service is a continuous broadcast of inflight weather advisories including summarized SIGMETs, convective SIGMETs, AIRMETs and urgent PIREPs. HIWAS is presently broadcast over selected VOR's and will be implemented throughout the conterminous U.S.

ASR/PAR—Indicates that Surveillance (ASR) or Precision (PAR) radar instrument approach minimums are published in U.S. Government Instrument Approach Procedures.

DIRECTORY LEGEND

RADIO CLASS DESIGNATIONS

Identification of VOR/VORTAC/TACAN Stations by Class (Operational Limitations):

Normal Usable Altitudes and Radius Distances

Class	Altitudes	Distance (miles)
(T)	12,000′ and below	25
(L)	Below 18,000′	40
(H)	Below 18,000′	40
(H)	Within the Conterminous 48 States only, between 14,500′ and 17,999′	100
(H)	18,000′ FL 450	130
(H)	Above FL 450	100

(H) = High (L) = Low (T) = Terminal

NOTE: An (H) facility is capable of providing (L) and (T) service volume and an (L) facility additionally provides (T) service volume.

The term VOR is, operationally, a general term covering the VHF omnidirectional bearing type of facility without regard to the fact that the power, the frequency protected service volume, the equipment configuration, and operational requirements may vary between facilities at different locations.

AB	Automatic Weather Broadcast (also shown with ⬛ following frequency.)
DF	Direction Finding Service.
DME	UHF standard (TACAN compatible) distance measuring equipment.
DME(Y)	UHF standard (TACAN compatible) distance measuring equipment that require TACAN to be placed in the "Y" mode to receive DME.
H	Non-directional radio beacon (homing), power 50 watts to less than 2,000 watts (50 NM at all altitudes).
HH	Non-directional radio beacon (homing), power 2,000 watts or more (75 NM at all altitudes).
H-SAB	Non-directional radio beacons providing automatic transcribed weather service.
ILS	Instrument Landing System (voice, where available, on localizer channel).
ISMLS	Interim Standard Microwave Landing System.
LDA	Localizer Directional Aid.
LMM	Compass locator station when installed at middle marker site (15 NM at all altitudes).
LOM	Compass locator station when installed at outer marker site (15 NM at all altitudes).
MH	Non-directional radio beacon (homing) power less than 50 watts (25 NM at all altitudes).
MLS	Microwave Landing System
S	Simultaneous range homing signal and/or voice.
SABH	Non-directional radio beacon not authorized for IFR or ATC. Provides automatic weather broadcasts.
SDF	Simplified Direction Facility.
TACAN	UHF navigational facility-omnidirectional course and distance information.
VOR	VHF navigational facility-omnidirectional course only.
VOR/DME	Collocated VOR navigational facility and UHF standard distance measuring equipment.
VORTAC	Collocated VOR and TACAN navigational facilities.
W	Without voice on radio facility frequency.
Z	VHF station location marker at a LF radio facility.

FREQUENCY PAIRING PLAN AND MLS CHANNELING

The following is a list of paired VOR/ILS VHF frequencies with TACAN channels and MLS channels.

TACAN CHANNEL	VHF FREQUENCY	MLS CHANNEL	TACAN CHANNEL	VHF FREQUENCY	MLS CHANNEL	TACAN CHANNEL	VHF FREQUENCY	MLS CHANNEL
17X	108.00	.	50Y	111.35	606	94X	114.70	.
17Y	108.05	540	51X	111.40	.	94Y	114.75	648
18X	108.10	500	51Y	111.45	608	95X	114.80	.
18Y	108.15	542	52X	111.50	534	95Y	114.85	650
19X	108.20	.	52Y	111.55	610	96X	114.90	.
19Y	108.25	544	53X	111.60	.	96Y	114.95	652
20X	108.30	502	53Y	111.65	612	97X	115.00	.
20Y	108.35	546	54X	111.70	536	97Y	115.05	654
21X	108.40	.	54Y	111.75	614	98X	115.10	.
21Y	108.45	548	55X	111.80	.	98Y	115.15	656
22X	108.50	504	55Y	111.85	616	99X	115.20	.
22Y	108.55	550	56X	111.90	538	99Y	115.25	658
23X	108.60	.	56Y	111.95	618	100X	115.30	.
23Y	108.65	552	57X	112.00	.	100Y	115.35	660
24X	108.70	506	57Y	112.05	.	101X	115.40	.
24Y	108.75	554	58X	112.10	.	101Y	115.45	662
25X	108.80	.	58Y	112.15	.	102X	115.50	.
25Y	108.85	556	59X	112.20	.	102Y	115.55	664
26X	108.90	508	59Y	112.25	.	103X	115.60	.
26Y	108.95	558	70X	112.30	.	103Y	115.65	666
27X	109.00	.	70Y	112.35	.	104X	115.70	.
27Y	109.05	560	71X	112.40	.	104Y	115.75	668
28X	109.10	510	71Y	112.45	.	105X	115.80	.
28Y	109.15	562	72X	112.50	.	105Y	115.85	670
29X	109.20	.	72Y	112.55	.	106X	115.90	.
29Y	109.25	564	73X	112.60	.	106Y	115.95	672
30X	109.30	512	73Y	112.65	.	107X	116.00	.
30Y	109.35	566	74X	112.70	.	107Y	116.05	674
31X	109.40	.	74Y	112.75	.	108X	116.10	.
31Y	109.45	568	75X	112.80	.	108Y	116.15	676
32X	109.50	514	75Y	112.85	.	109X	116.20	.
32Y	109.55	570	76X	112.90	.	109Y	116.25	678
33X	109.60	.	76Y	112.95	.	110X	116.30	.
33Y	109.65	572	77X	113.00	.	110Y	116.35	680
34X	109.70	516	77Y	113.05	.	111X	116.40	.
34Y	109.75	574	78X	113.10	.	111Y	116.45	682
35X	109.80	.	78Y	113.15	.	112X	116.50	.
35Y	109.85	576	79X	113.20	.	112Y	116.55	684
36X	109.90	518	79Y	113.25	.	113X	116.60	.
36Y	109.95	578	80X	113.30	.	113Y	116.65	686
37X	110.00	.	80Y	113.35	620	114X	116.70	.
37Y	110.05	580	81X	113.40	.	114Y	116.75	688
38X	110.10	520	81Y	113.45	622	115X	116.80	.
38Y	110.15	582	82X	113.50	.	115Y	116.85	690
39X	110.20	.	82Y	113.55	624	116X	116.90	.
39Y	110.25	584	83X	113.60	.	116Y	116.95	692
40X	110.30	522	83Y	113.65	626	117X	117.00	.
40Y	110.35	586	84X	113.70	.	117Y	117.05	694
41X	110.40	.	84Y	113.75	628	118X	117.10	.
41Y	110.45	588	85X	113.80	.	118Y	117.15	696
42X	110.50	524	85Y	113.85	630	119X	117.20	.
42Y	110.55	590	86X	113.90	.	119Y	117.25	698
43X	110.60	.	86Y	113.95	632	120X	117.30	.
43Y	110.65	592	87X	114.00	.	120Y	117.35	.
44X	110.70	526	87Y	114.05	634	121X	117.40	.
44Y	110.75	594	88X	114.10	.	121Y	117.45	.
45X	110.80	.	88Y	114.15	636	122X	117.50	.
45Y	110.85	596	89X	114.20	.	122Y	117.55	.
46X	110.90	528	89Y	114.25	638	123X	117.60	.
46Y	110.95	598	90X	114.30	.	123Y	117.65	.
47X	111.00	.	90Y	114.35	640	124X	117.70	.
47Y	111.05	600	91X	114.40	.	124Y	117.75	.
48X	111.10	530	91Y	114.45	642	125X	117.80	.
48Y	111.15	602	92X	114.50	.	125Y	117.85	.
49X	111.20	.	92Y	114.55	644	126X	117.90	.
49Y	111.25	604	93X	114.60	.	126Y	117.95	.
50X	111.30	532	93Y	114.65	646			

㉓ COMM/NAVAID REMARKS:
Pertinent remarks concerning communications and NAVAIDS.

GENERAL INFORMATION

This Airport/Facility Directory is a Civil Flight Information Publication published and distributed every eight weeks by the National Ocean Service, NOAA, Department of Commerce, Rockville, Maryland 20852. It is designed for use with Aeronautical Charts covering the conterminous United States, Puerto Rico and the Virgin Islands.

It contains an Airport/Facility Directory of all airports, seaplane bases and heliports open to the public, communications data, navigational facilities and certain special notices and procedures.

PROCUREMENT

Subscriptions to this publication are for sale by the:
National Ocean Service
NOAA Distribution Branch, N/CG33
Riverdale, Maryland 20737
Telephone (301) 436-6990

Change of address notices, or inquiries concerning your subscription, should be directed to the National Ocean Service, at the above address.

Purchases on a nonsubscription basis may be obtained through the above address, or any NOS authorized aeronautical chart agent.

ERRORS, OMISSIONS OR SUGGESTED CHANGES AND COMMENTS

"CRITICAL information such as equipment malfunction, abnormal field conditions, hazards to flight, etc., should be reported as soon as possible to the nearest FAA facility, either in person or by reverse charge telephone call"

Data of a non-critical nature should be submitted to:

Federal Aviation Administration
National Flight Data Center, ATO-250
800 Independence Avenue, SW
Washington, D.C. 20591

APPENDIX C

(From *Aeronautical Information Manual*)

MEDICAL FACTS FOR PILOTS

Section 1. FITNESS FOR FLIGHT

8–1–1. FITNESS FOR FLIGHT

a. Medical Certification:

1. All pilots except those flying gliders and free air balloons must possess valid medical certificates in order to exercise the privileges of their airman certificates. The periodic medical examinations required for medical certification are conducted by designated Aviation Medical Examiners, who are physicians with a special interest in aviation safety and training in aviation medicine.

2. The standards for medical certification are contained in FAR Part 67. Pilots who have a history of certain medical conditions described in these standards are mandatorily disqualified from flying. These medical conditions include a personality disorder manifested by overt acts, a psychosis, alcoholism, drug dependence, epilepsy, an unexplained disturbance of consciousness, myocardial infarction, angina pectoris and diabetes requiring medication for its control. Other medical conditions may be temporarily disqualifying, such as acute infections, anemia, and peptic ulcer. Pilots who do not meet medical standards may still be qualified under special issuance provisions or the exemption process. This may require that either additional medical information be provided or practical flight tests be conducted.

3. Student pilots should visit an Aviation Medical Examiner as soon as possible in their flight training in order to avoid unnecessary training expenses should they not meet the medical standards. For the same reason, the student pilot who plans to enter commercial aviation should apply for the highest class of medical certificate that might be necessary in the pilot's career.

CAUTION–
THE FAR'S PROHIBIT A PILOT WHO POSSESSES A CURRENT MEDICAL CERTIFICATE FROM PERFORMING CREWMEMBER DUTIES WHILE THE PILOT HAS A KNOWN MEDICAL CONDITION OR INCREASE OF A KNOWN MEDICAL CONDITION THAT WOULD MAKE THE PILOT UNABLE TO MEET THE STANDARDS FOR THE MEDICAL CERTIFICATE.

b. Illness:

1. Even a minor illness suffered in day-to-day living can seriously degrade performance of many piloting tasks vital to safe flight. Illness can produce fever and distracting symptoms that can impair judgment, memory, alertness, and the ability to make calculations. Although symptoms from an illness may be under adequate control with a medication, the medication itself may decrease pilot performance.

2. The safest rule is not to fly while suffering from any illness. If this rule is considered too stringent for a particular illness, the pilot should contact an Aviation Medical Examiner for advice.

c. Medication:

1. Pilot performance can be seriously degraded by both prescribed and over-the-counter medications, as well as by the medical conditions for which they are taken. Many medications, such as tranquilizers, sedatives, strong pain relievers, and cough-suppressant preparations, have primary effects that may impair judgment, memory, alertness, coordination, vision, and the ability to make calculations. Others, such as antihistamines, blood pressure drugs, muscle relaxants, and agents to control diarrhea and motion sickness, have side effects that may impair the same critical functions. Any medication that depresses the nervous system, such as a sedative, tranquilizer or antihistamine, can make a pilot much more susceptible to hypoxia.

2. The FAR's prohibit pilots from performing crewmember duties while using any medication that affects the faculties in any way contrary to safety. The safest rule is not to fly as a crewmember while taking any medication, unless approved to do so by the FAA.

d. Alcohol:

1. Extensive research has provided a number of facts about the hazards of alcohol consumption and flying. As little as one ounce of liquor, one bottle of beer

or four ounces of wine can impair flying skills, with the alcohol consumed in these drinks being detectable in the breath and blood for at least 3 hours. Even after the body completely destroys a moderate amount of alcohol, a pilot can still be severely impaired for many hours by hangover. There is simply no way of increasing the destruction of alcohol or alleviating a hangover. Alcohol also renders a pilot much more susceptible to disorientation and hypoxia.

2. A consistently high alcohol related fatal aircraft accident rate serves to emphasize that alcohol and flying are a potentially lethal combination. The FAR's prohibit pilots from performing crewmember duties within 8 hours after drinking any alcoholic beverage or while under the influence of alcohol. However, due to the slow destruction of alcohol, a pilot may still be under influence 8 hours after drinking a moderate amount of alcohol. Therefore, an excellent rule is to allow at least 12 to 24 hours between "bottle and throttle," depending on the amount of alcoholic beverage consumed.

e. Fatigue:

1. Fatigue continues to be one of the most treacherous hazards to flight safety, as it may not be apparent to a pilot until serious errors are made. Fatigue is best described as either acute (short-term) or chronic (long-term).

2. A normal occurrence of everyday living, acute fatigue is the tiredness felt after long periods of physical and mental strain, including strenuous muscular effort, immobility, heavy mental workload, strong emotional pressure, monotony, and lack of sleep. Consequently, coordination and alertness, so vital to safe pilot performance, can be reduced. Acute fatigue is prevented by adequate rest and sleep, as well as by regular exercise and proper nutrition.

3. Chronic fatigue occurs when there is not enough time for full recovery between episodes of acute fatigue. Performance continues to fall off, and judgment becomes impaired so that unwarranted risks may be taken. Recovery from chronic fatigue requires a prolonged period of rest.

f. Stress:

1. Stress from the pressures of everyday living can impair pilot performance, often in very subtle ways. Difficulties, particularly at work, can occupy thought processes enough to markedly decrease alertness. Distraction can so interfere with judgment that unwarranted risks are taken, such as flying into deteriorating weather conditions to keep on schedule. Stress and fatigue (see above) can be an extremely hazardous combination.

2. Most pilots do not leave stress "on the ground." Therefore, when more than usual difficulties are being experienced, a pilot should consider delaying flight until these difficulties are satisfactorily resolved.

g. Emotion:

Certain emotionally upsetting events, including a serious argument, death of a family member, separation or divorce, loss of job, and financial catastrophe, can render a pilot unable to fly an aircraft safely. The emotions of anger, depression, and anxiety from such events not only decrease alertness but also may lead to taking risks that border on self-destruction. Any pilot who experiences an emotionally upsetting event should not fly until satisfactorily recovered from it.

h. Personal Checklist: Aircraft accident statistics show that pilots should be conducting preflight checklists on themselves as well as their aircraft for pilot impairment contributes to many more accidents than failures of aircraft systems. A personal checklist, which includes all of the categories of pilot impairment as discussed in this section, that can be easily committed to memory is being distributed by the FAA in the form of a wallet-sized card.

i. PERSONAL CHECKLIST: *I'm physically and mentally safe to fly; not being impaired by:*

Illness

Medication

Stress

Alcohol

Fatigue

Emotion

8–1–2. EFFECTS OF ALTITUDE

a. Hypoxia:

1. Hypoxia is a state of oxygen deficiency in the body sufficient to impair functions of the brain and other organs. Hypoxia from exposure to altitude is due only to the reduced barometric pressures encountered at

altitude, for the concentration of oxygen in the atmosphere remains about 21 percent from the ground out to space.

2. Although a deterioration in night vision occurs at a cabin pressure altitude as low as 5,000 feet, other significant effects of altitude hypoxia usually do not occur in the normal healthy pilot below 12,000 feet. From 12,000 to 15,000 feet of altitude, judgment, memory, alertness, coordination and ability to make calculations are impaired, and headache, drowsiness, dizziness and either a sense of well-being (euphoria) or belligerence occur. The effects appear following increasingly shorter periods of exposure to increasing altitude. In fact, pilot performance can seriously deteriorate within 15 minutes at 15,000 feet.

3. At cabin pressure altitudes above 15,000 feet, the periphery of the visual field grays out to a point where only central vision remains (tunnel vision). A blue coloration (cyanosis) of the fingernails and lips develops. The ability to take corrective and protective action is lost in 20 to 30 minutes at 18,000 feet and 5 to 12 minutes at 20,000 feet, followed soon thereafter by unconsciousness.

4. The altitude at which significant effects of hypoxia occur can be lowered by a number of factors. Carbon monoxide inhaled in smoking or from exhaust fumes, lowered hemoglobin (anemia), and certain medications can reduce the oxygen-carrying capacity of the blood to the degree that the amount of oxygen provided to body tissues will already be equivalent to the oxygen provided to the tissues when exposed to a cabin pressure altitude of several thousand feet. Small amounts of alcohol and low doses of certain drugs, such as antihistamines, tranquilizers, sedatives and analgesics can, through their depressant action, render the brain much more susceptible to hypoxia. Extreme heat and cold, fever, and anxiety increase the body's demand for oxygen, and hence its susceptibility to hypoxia.

5. The effects of hypoxia are usually quite difficult to recognize, especially when they occur gradually. Since symptoms of hypoxia do not vary in an individual, the ability to recognize hypoxia can be greatly improved by experiencing and witnessing the effects of hypoxia during an altitude chamber "flight." The FAA provides this opportunity through aviation physiology training, which is conducted at the FAA Civil Aeromedical Institute and at many military facilities across the United States, to attend the Physiological Training Program at the Civil

Aeromedical Institute, Mike Monroney Aeronautical Center, Oklahoma City, OK, contact by telephone (405) 954–6212, or by writing Airmen Education Program Branch, AAM–420, CAMI, Mike Monroney Aeronautical Center, P.O. Box 25082, Oklahoma City, OK 73125.

NOTE–
TO ATTEND THE PHYSIOLOGICAL TRAINING PROGRAM AT ONE OF THE MILITARY INSTALLATIONS HAVING THE TRAINING CAPABILITY, AN APPLICATION FORM AND A FEE MUST BE SUBMITTED. FULL PARTICULARS ABOUT LOCATION, FEES, SCHEDULING PROCEDURES, COURSE CONTENT, INDIVIDUAL REQUIREMENTS, ETC. ARE CONTAINED IN THE PHYSIOLOGICAL TRAINING APPLICATION, FORM NUMBER AC 3150–7, WHICH IS OBTAINED BY CONTACTING THE ACCIDENT PREVENTION SPECIALIST OR THE OFFICE FORMS MANAGER IN THE NEAREST FAA OFFICE.

6. Hypoxia is prevented by heeding factors that reduce tolerance to altitude, by enriching the inspired air with oxygen from an appropriate oxygen system, and by maintaining a comfortable, safe cabin pressure altitude. For optimum protection, pilots are encouraged to use supplemental oxygen above 10,000 feet during the day, and above 5,000 feet at night. The FAR's require that at the minimum, flight crew be provided with and use supplemental oxygen after 30 minutes of exposure to cabin pressure altitudes between 12,500 and 14,000 feet and immediately on exposure to cabin pressure altitudes above 14,000 feet. Every occupant of the aircraft must be provided with supplemental oxygen at cabin pressure altitudes above 15,000 feet.

b. Ear Block:

1. As the aircraft cabin pressure decreases during ascent, the expanding air in the middle ear pushes the eustachian tube open, and by escaping down it to the nasal passages, equalizes in pressure with the cabin pressure. But during descent, the pilot must periodically open the eustachian tube to equalize pressure. This can be accomplished by swallowing, yawning, tensing muscles in the throat, or if these do not work, by a combination of closing the mouth, pinching the nose closed, and attempting to blow through the nostrils (Valsalva maneuver).

2. Either an upper respiratory infection, such as a cold or sore throat, or a nasal allergic condition can produce enough congestion around the eustachian tube to make equalization difficult. Consequently, the difference in pressure between the middle ear and aircraft cabin can build up to a level that will hold the eustachian tube closed, making equalization difficult if not impossible. The problem is commonly referred to as an "ear block."

3. An ear block produces severe ear pain and loss of hearing that can last from several hours to several days. Rupture of the ear drum can occur in flight or after landing. Fluid can accumulate in the middle ear and become infected.

4. An ear block is prevented by not flying with an upper respiratory infection or nasal allergic condition. Adequate protection is usually not provided by decongestant sprays or drops to reduce congestion around the eustachian tubes. Oral decongestants have side effects that can significantly impair pilot performance.

5. If an ear block does not clear shortly after landing, a physician should be consulted.

c. Sinus Block:

1. During ascent and descent, air pressure in the sinuses equalizes with the aircraft cabin pressure through small openings that connect the sinuses to the nasal passages. Either an upper respiratory infection, such as a cold or sinusitis, or a nasal allergic condition can produce enough congestion around an opening to slow equalization, and as the difference in pressure between the sinus and cabin mounts, eventually plug the opening. This "sinus block" occurs most frequently during descent.

2. A sinus block can occur in the frontal sinuses, located above each eyebrow, or in the maxillary sinuses, located in each upper cheek. It will usually produce excruciating pain over the sinus area. A maxillary sinus block can also make the upper teeth ache. Bloody mucus may discharge from the nasal passages.

3. A sinus block is prevented by not flying with an upper respiratory infection or nasal allergic condition. Adequate protection is usually not provided by decongestant sprays or drops to reduce congestion around the sinus openings. Oral decongestants have side effects that can impair pilot performance.

4. If a sinus block does not clear shortly after landing, a physician should be consulted.

d. Decompression Sickness After Scuba Diving:

1. A pilot or passenger who intends to fly after scuba diving should allow the body sufficient time to rid itself of excess nitrogen absorbed during diving. If not, decompression sickness due to evolved gas can occur during exposure to low altitude and create a serious inflight emergency.

2. The recommended waiting time before going to flight altitudes of up to 8,000 feet is at least 12 hours after diving which has not required controlled ascent (nondecompression stop diving), and at least 24 hours after diving which has required controlled ascent (decompression stop diving). The waiting time before going to flight altitudes above 8,000 feet should be at least 24 hours after any SCUBA dive. These recommended altitudes are actual flight altitudes above mean sea level (AMSL) and not pressurized cabin altitudes. This takes into consideration the risk of decompression of the aircraft during flight.

8–1–3. HYPERVENTILATION IN FLIGHT

a. Hyperventilation, or an abnormal increase in the volume of air breathed in and out of the lungs, can occur subconsciously when a stressful situation is encountered in flight. As hyperventilation "blows off" excessive carbon dioxide from the body, a pilot can experience symptoms of lightheadedness, suffocation, drowsiness, tingling in the extremities, and coolness and react to them with even greater hyperventilation. Incapacitation can eventually result from incoordination, disorientation, and painful muscle spasms. Finally, unconsciousness can occur.

b. The symptoms of hyperventilation subside within a few minutes after the rate and depth of breathing are consciously brought back under control. The buildup of carbon dioxide in the body can be hastened by controlled breathing in and out of a paper bag held over the nose and mouth.

c. Early symptoms of hyperventilation and hypoxia are similar. Moreover, hyperventilation and hypoxia can occur at the same time. Therefore, if a pilot is using an oxygen system when symptoms are experienced, the oxygen regulator should immediately be set to deliver 100 percent oxygen, and then the system checked to assure that it has been functioning effectively before giving attention to rate and depth of breathing.

8–1–4. CARBON MONOXIDE POISONING IN FLIGHT

a. Carbon monoxide is a colorless, odorless, and tasteless gas contained in exhaust fumes. When breathed even in minute quantities over a period of time, it can significantly reduce the ability of the blood to carry oxygen. Consequently, effects of hypoxia occur.

b. Most heaters in light aircraft work by air flowing over the manifold. Use of these heaters while exhaust fumes are escaping through manifold cracks and seals is responsible every year for several nonfatal and fatal aircraft accidents from carbon monoxide poisoning.

c. A pilot who detects the odor of exhaust or experiences symptoms of headache, drowsiness, or dizziness while using the heater should suspect carbon monoxide poisoning, and immediately shut off the heater and open air vents. If symptoms are severe or continue after landing, medical treatment should be sought.

8–1–5. ILLUSIONS IN FLIGHT

a. Introduction: Many different illusions can be experienced in flight. Some can lead to spatial disorientation. Others can lead to landing errors. Illusions rank among the most common factors cited as contributing to fatal aircraft accidents.

b. Illusions Leading to Spatial Disorientation:

1. Various complex motions and forces and certain visual scenes encountered in flight can create illusions of motion and position. Spatial disorientation from these illusions can be prevented only by visual reference to reliable, fixed points on the ground or to flight instruments.

2. *The leans:* An abrupt correction of a banked attitude, which has been entered too slowly to stimulate the motion sensing system in the inner ear, can create the illusion of banking in the opposite direction. The disoriented pilot will roll the aircraft back into its original dangerous attitude, or if level flight is maintained, will feel compelled to lean in the perceived vertical plane until this illusion subsides.

(a) *Coriolis illusion:* An abrupt head movement in a prolonged constant-rate turn that has ceased stimulating the motion sensing system can create the illusion of rotation or movement in an entirely different axis. The disoriented pilot will maneuver the aircraft into a dangerous attitude in an attempt to stop rotation. This most overwhelming of all illusions in flight may be prevented by not making sudden, extreme head movements, particularly while making prolonged constant-rate turns under IFR conditions.

(b) *Graveyard spin:* A proper recovery from a spin that has ceased stimulating the motion sensing system can create the illusion of spinning in the opposite direction. The disoriented pilot will return the aircraft to its original spin.

(c) *Graveyard spiral:* An observed loss of altitude during a coordinated constant-rate turn that has ceased stimulating the motion sensing system can create the illusion of being in a descent with the wings level. The disoriented pilot will pull back on the controls, tightening the spiral and increasing the loss of altitude.

(d) *Somatogravic illusion:* A rapid acceleration during takeoff can create the illusion of being in a nose up attitude. The disoriented pilot will push the aircraft into a nose low, or dive attitude. A rapid deceleration by a quick reduction of the throttles can have the opposite effect, with the disoriented pilot pulling the aircraft into a nose up, or stall attitude.

(e) *Inversion illusion:* An abrupt change from climb to straight and level flight can create the illusion of tumbling backwards. The disoriented pilot will push the aircraft abruptly into a nose low attitude, possibly intensifying this illusion.

(f) *Elevator illusion:* An abrupt upward vertical acceleration, usually by an updraft, can create the illusion of being in a climb. The disoriented pilot will push the aircraft into a nose low attitude. An abrupt downward vertical acceleration, usually by a downdraft, has the opposite effect, with the disoriented pilot pulling the aircraft into a nose up attitude.

(g) *False horizon:* Sloping cloud formations, an obscured horizon, a dark scene spread with ground lights and stars, and certain geometric patterns of ground light can create illusions of not being aligned correctly with the actual horizon. The disoriented pilot will place the aircraft in a dangerous attitude.

(h) *Autokinesis:* In the dark, a static light will appear to move about when stared at for many seconds. The disoriented pilot will lose control of the aircraft in attempting to align it with the light.

3. *Illusions Leading to Landing Errors:*

(a) Various surface features and atmospheric conditions encountered in landing can create illusions of incorrect height above and distance from the runway threshold. Landing errors from these illusions can be prevented by anticipating them during approaches, aerial visual inspection of unfamiliar airports before landing, using electronic glide slope or VASI systems when available, and maintaining optimum proficiency in landing procedures.

(b) *Runway width illusion:*
A narrower-than-usual runway can create the illusion that the aircraft is at a higher altitude than it actually is. The pilot who does not recognize this illusion will fly a lower approach, with the risk of striking objects along the approach path or landing short. A wider-than-usual runway can have the opposite effect, with the risk of leveling out high and landing hard or overshooting the runway.

(c) *Runway and terrain slopes illusion:* An upsloping runway, upsloping terrain, or both, can create the illusion that the aircraft is at a higher altitude than it actually is. The pilot who does not recognize this illusion will fly a lower approach. A downsloping runway, downsloping approach terrain, or both, can have the opposite effect.

(d) *Featureless terrain illusion:* An absence of ground features, as when landing over water, darkened areas, and terrain made featureless by snow, can create the illusion that the aircraft is at a higher altitude than it actually is. The pilot who does not recognize this illusion will fly a lower approach

(e) *Atmospheric illusions:* Rain on the windscreen can create the illusion of greater height, and atmospheric haze the illusion of being at a greater distance from the runway. The pilot who does not recognize these illusions will fly a lower approach. Penetration of fog can create the illusion of pitching up. The pilot who does not recognize this illusion will steepen the approach, often quite abruptly.

(f) *Ground lighting illusions:* Lights along a straight path, such as a road, and even lights on moving trains can be mistaken for runway and approach lights. Bright runway and approach lighting systems, especially where few lights illuminate the surrounding terrain, may create the illusion of less distance to the runway. The pilot who does not recognize this illusion will fly a higher approach. Conversely, the pilot overflying terrain which has few lights to provide height cues may make a lower than normal approach.

8–1–6. VISION IN FLIGHT

a. Introduction: Of the body senses, vision is the most important for safe flight. Major factors that determine how effectively vision can be used are the level of illumination and the technique of scanning the sky for other aircraft.

b. Vision Under Dim and Bright Illumination:

1. Under conditions of dim illumination, small print and colors on aeronautical charts and aircraft instruments become unreadable unless adequate cockpit lighting is available. Moreover, another aircraft must be much closer to be seen unless its navigation lights are on.

2. In darkness, vision becomes more sensitive to light, a process called dark adaptation. Although exposure to total darkness for at least 30 minutes is required for complete dark adaptation, a pilot can achieve a moderate degree of dark adaptation within 20 minutes under dim red cockpit lighting. Since red light severely distorts colors, especially on aeronautical charts, and can cause serious difficulty in focusing the eyes on objects inside the aircraft, its use is advisable only where optimum outside night vision capability is necessary. Even so, white cockpit lighting must be available when needed for map and instrument reading, especially under IFR conditions. Dark adaptation is impaired by exposure to cabin pressure altitudes above 5,000 feet, carbon monoxide inhaled in smoking and from exhaust fumes, deficiency of Vitamin A in the diet, and by prolonged exposure to bright sunlight. Since any degree of dark adaptation is lost within a few seconds of viewing a bright light, a pilot should close one eye when using a light to preserve some degree of night vision.

3. Excessive illumination, especially from light reflected off the canopy, surfaces inside the aircraft, clouds, water, snow, and desert terrain, can produce glare, with uncomfortable squinting, watering of the eyes, and even temporary blindness. Sunglasses for protection from glare should absorb at least 85 percent of visible light (15 percent transmittance) and all colors equally (neutral transmittance), with negligible image distortion from refractive and prismatic errors.

c. Scanning for Other Aircraft:

1. Scanning the sky for other aircraft is a key factor in collision avoidance. It should be used continuously by the pilot and copilot (or right seat passenger) to cover all areas of the sky visible from the cockpit. Although pilots must meet specific visual acuity requirements, the ability to read an eye chart does not ensure that one will be able to efficiently spot other aircraft. Pilots must develop an effective scanning technique which maximizes one's visual capabilities. The probability of spotting a potential collision threat obviously increases with the time spent looking outside the cockpit. Thus, one must use timesharing techniques to efficiently scan the surrounding airspace while monitoring instruments as well.

2. While the eyes can observe an approximate 200 degree arc of the horizon at one glance, only a very small center area called the fovea, in the rear of the eye, has the ability to send clear, sharply focused messages to the brain. All other visual information that is not processed directly through the fovea will be of less detail. An aircraft at a distance of 7 miles which appears in sharp focus within the foveal center of vision would have to be as close as $7/10$ of a mile in order to be recognized if it were outside of foveal vision. Because the eyes can focus only on this narrow viewing area, effective scanning is accomplished with a series of short, regularly spaced eye movements that bring successive areas of the sky into the central visual field. Each movement should not exceed 10 degrees, and each area should be observed for at least 1 second to enable detection. Although horizontal back-and-forth eye movements seem preferred by most pilots, each pilot should develop a scanning pattern that is most comfortable and then adhere to it to assure optimum scanning.

3. Studies show that the time a pilot spends on visual tasks inside the cabin should represent no more that $1/4$ to $1/3$ of the scan time outside, or no more than 4 to 5 seconds on the instrument panel for every 16 seconds outside. Since the brain is already trained to process sight information that is presented from left to right, one may find it easier to start scanning over the left shoulder and proceed across the windshield to the right.

4. Pilots should realize that their eyes may require several seconds to refocus when switching views between items in the cockpit and distant objects. The eyes will also tire more quickly when forced to adjust to distances immediately after close-up focus, as required for scanning the instrument panel. Eye fatigue can be reduced by looking from the instrument panel to the left wing past the wing tip to the center of the first scan quadrant when beginning the exterior scan. After having scanned from left to right, allow the eyes to return to the cabin along the right wing from its tip inward. Once back inside, one should automatically commence the panel scan.

5. Effective scanning also helps avoid "empty-field myopia." This condition usually occurs when flying above the clouds or in a haze layer that provides nothing specific to focus on outside the aircraft. This causes the eyes to relax and seek a comfortable focal distance which may range from 10 to 30 feet. For the pilot, this means looking without seeing, which is dangerous.

8–1–7. AEROBATIC FLIGHT

a. Pilots planning to engage in aerobatics should be aware of the physiological stresses associated with accelerative forces during aerobatic maneuvers. Many prospective aerobatic trainees enthusiastically enter aerobatic instruction but find their first experiences with G forces to be unanticipated and very uncomfortable. To minimize or avoid potential adverse effects, the aerobatic instructor and trainee must have a basic understanding of the physiology of G force adaptation.

b. Forces experienced with a rapid push-over maneuver result in the blood and body organs being displaced toward the head. Depending on forces involved and individual tolerance, a pilot may experience discomfort, headache, "red-out," and even unconsciousness.

c. Forces experienced with a rapid pull-up maneuver result in the blood and body organ displacement toward the lower part of the body away from the head. Since the brain requires continuous blood circulation for an adequate oxygen supply, there is a physiologic limit to the time the pilot can tolerate higher forces before losing consciousness. As the blood circulation to the brain decreases as a result of forces involved, a pilot will experience "narrowing" of visual fields, "gray-out," "black-out," and unconsciousness. Even a brief loss of consciousness in a maneuver can lead to improper control movement causing structural failure of the aircraft or collision with another object or terrain.

d. In steep turns, the centrifugal forces tend to push the pilot into the seat, thereby resulting in blood and body organ displacement toward the lower part of the body as in the case of rapid pull-up maneuvers and with the same physiologic effects and symptoms.

e. Physiologically, humans progressively adapt to imposed strains and stress, and with practice, any maneuver will have decreasing effect. Tolerance to G forces is dependent on human physiology and the individual pilot. These factors include the skeletal anatomy, the cardiovascular architecture, the nervous system, the quality of the blood, the general physical state, and experience and recency of exposure. The pilot should consult an Aviation Medical Examiner prior to aerobatic training and be aware that poor physical condition can reduce tolerance to accelerative forces.

f. The above information provides pilots with a brief summary of the physiologic effects of G forces. It does not address methods of "counteracting" these effects. There are numerous references on the subject of G forces during aerobatics available to pilots. Among these are "G Effects on the Pilot During Aerobatics," FAA–AM–72–28, and "G Incapacitation in Aerobatic Pilots: A Flight Hazard" FAA–AM–82–13. These are available from the National Technical Information Service, Springfield, Virginia 22161.

REFERENCE–
FAA AC 91–61, A HAZARD IN AEROBATICS: EFFECTS OF G–FORCES ON PILOTS.

8–1–8. JUDGMENT ASPECTS OF COLLISION AVOIDANCE

a. Introduction: The most important aspects of vision and the techniques to scan for other aircraft are described in AIM, Vision in Flight, paragraph 8–1–6. Pilots should also be familiar with the following information to reduce the possibility of mid-air collisions.

b. Determining Relative Altitude: Use the horizon as a reference point. If the other aircraft is above the horizon, it is probably on a higher flight path. If the aircraft appears to be below the horizon, it is probably flying at a lower altitude.

c. Taking Appropriate Action: Pilots should be familiar with rules on right-of-way, so if an aircraft is on an obvious collision course, one can take immediate evasive action, preferably in compliance with applicable Federal Aviation Regulations.

d. Consider Multiple Threats: The decision to climb, descend, or turn is a matter of personal judgment, but one should anticipate that the other pilot may also be making a quick maneuver. Watch the other aircraft during the maneuver and begin your scanning again immediately since there may be other aircraft in the area.

e. Collision Course Targets: Any aircraft that appears to have no relative motion and stays in one scan quadrant is likely to be on a collision course. Also, if a target shows no lateral or vertical motion, but increases in size, *take evasive action.*

f. Recognize High Hazard Areas:

1. Airways, especially near VOR's, and Class B, Class C, Class D, and Class E surface areas are places where aircraft tend to cluster.

2. Remember, most collisions occur during days when the weather is good. Being in a "radar environment" still requires vigilance to avoid collisions.

g. Cockpit Management: Studying maps, checklists, and manuals before flight, with other proper preflight planning; e.g., noting necessary radio frequencies and organizing cockpit materials, can reduce the amount of time required to look at these items during flight, permitting more scan time.

h. Windshield Conditions: Dirty or bug-smeared windshields can greatly reduce the ability of pilots to see other aircraft. Keep a clean windshield.

i. Visibility Conditions: Smoke, haze, dust, rain, and flying towards the sun can also greatly reduce the ability to detect targets.

j. Visual Obstructions in the Cockpit:

1. Pilots need to move their heads to see around blind spots caused by fixed aircraft structures, such as door posts, wings, etc. It will be necessary at times to maneuver the aircraft; e.g., lift a wing, to facilitate seeing.

2. Pilots must insure curtains and other cockpit objects; e.g., maps on glare shield, are removed and stowed during flight.

k. Lights On:

1. Day or night, use of exterior lights can greatly increase the conspicuity of any aircraft.

2. Keep interior lights low at night.

l. ATC support: ATC facilities often provide radar traffic advisories on a workload-permitting basis. Flight through Class C and Class D airspace requires communication with ATC. Use this support whenever possible or when required.

APPENDIX D

ADDED NOTES ON ENGINES AND OTHER SYSTEMS

ADDED NOTES ON ENGINE THEORY, CONSTRUCTION, AND OPERATIONS

There have been cases of certificated private and commercial pilots who have never fully seen the engine of the airplane they've been flying for many hours. (They open the small access door to check the oil and only get a glimpse of the top of the engine crankcase.) The cowlings of many airplanes are designed so that it is impractical for the pilot to open them before every flight; probably the only time the engine is fully seen is when a 100-hour or annual inspection is done.

The point is that you should have a basic understanding of how various things work, so that a lot of the mystery disappears and you aren't sweating out some unknown and unexplained failure. You might avoid a bad situation such as loss of electrical power at night, for instance, by noting that the belt that drives the alternator is getting frayed and should be replaced before it breaks. If you didn't know the function of the belt, you might not be so concerned about replacing it before that long night flight over the Louisiana swamps.

■ Engine Operation

In order to know what to look for in the preflight check or how to analyze engine problems in flight, you should have some basic knowledge of how an engine works. Most general aviation airplane engines are horizontally opposed and air cooled. The horizontally opposed engine is "flat" and has an equal number of cylinders on each side of the crankcase (Fig. D-1).

PROPELLER

Fig. D-1. The Lycoming 0-235 L2C four-cylinder, horizontally opposed, air-cooled engine. This is used in current two-place trainers. (*Avco Lycoming*)

The engine uses the "Otto cycle" in operation and works this way: The pistons, enclosed in the cylinders, move *in* and *out* within a limited distance. The burning of the fuel-air mixture pushes the piston in, and it is attached by a connecting rod to an offset section of the crankshaft. Because of that offset section, the back and forth motion of the piston is translated to a rotary motion of the crankshaft, which in turn rotates the propeller either directly or through a gearing system. Most lighter airplane engines have direct drive; that is, the propeller is directly attached to and rotates at the same speed as the engine crankshaft. Higher horsepower engines may have gears that reduce the propeller rotation speed. The engine has four strokes (Fig. D-2):

1. *Intake stroke*—The piston moves toward the base of the cylinder with the intake valve open (exhaust

valve closed), pulling the fuel-air mixture into the cylinder.
2. *Compression*—As the crankshaft rotates, the piston moves away from the base of the cylinder with *both* valves closed so that the fuel-air mixture is compressed.
3. *Power stroke*—The fuel-air mixture is *ignited* in its compressed state, which gives greater power than if it was not compressed, and the piston is moved downward (or inward), rotating the crankshaft and thus turning the propeller.
4. *Exhaust stroke*—As the crankshaft turns, the piston moves up (out) with the exhaust valve open, purging the burned gases out through the exhaust system.

The crankshaft has to make two full revolutions for each cylinder's power stroke, as you can see by analyzing Figure D-2. This sequence is happening in other cylinders at different times, and each power stroke turns the crankshaft and provides for the intake, compression, and exhaust strokes as needed for the other cylinders.

■ Engine Components

The pistons, which in most cases are made from aluminum alloy forgings, have rings that fit into grooves cut in them. The rings are made from a good stock of cast iron and have a slightly larger outside diameter than the piston (Fig. D-3).

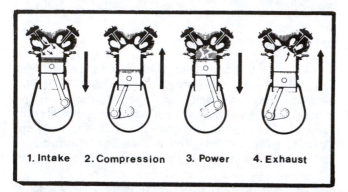

1. Intake 2. Compression 3. Power 4. Exhaust

Fig. D-2. The four strokes. The cylinder is shown as upright rather than horizontal here. (*FAA AC 61-32A*)

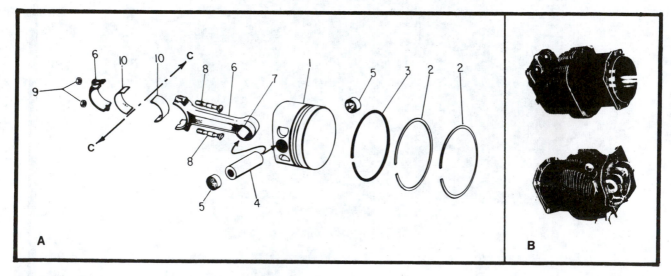

Fig. D-3. A—Piston and components: (1) piston, (2) compression rings, (3) oil regulating or oil scraper ring, (4) piston pin (sometimes called "wrist pin"), (5) piston pin plugs, (6) connecting rod (from piston to crankshaft), (7) bushing, (8) connecting rod bolts, (9) connecting rod nuts, (10) connecting rod bearings (replaceable). The line C-C shows where the connecting rod joins the crankshaft. B—Two views of a cylinder. Note the cooling fins. Look at Figure D-2 again.

Compression rings are designed to prevent the escape of the fuel-air mixture or exhaust gases into the crankcase. There are usually two or more of these per piston, and they are replaceable if worn or broken. The barrel of the cylinder is forged steel; for a certain number of hours from the time the new engine is first started, the rings will be considered to be "seating," or fitting to the cylinder walls. (An operating time of 50 hours is considered a reasonable period for this.) The manufacturer usually recommends that straight mineral oil, which has no additives, be used because this allows more friction so that the slight wear needed for seating is speeded up. Once the rings are seated, ashless dispersant oils are normally used to keep friction to a minimum.

One possible later problem, as the hours pile up on the engine, is that these piston rings get worn or may break, allowing oil from the crankcase to get into the combustion chamber and burn with the fuel-air mixture. That's what was happening in that car in front of you the other day; the driver was laying a smoke screen because oil as well as gasoline was burning. Not only was too much oil being burned but the residue was fouling the spark plugs and gumming up the valves so that they didn't close properly.

Mechanics run compression checks on the 100-hour and annual inspections to check for valve and ring problems. If the airplane you're flying isn't developing the power it should or the oil consumption has gone up radically, rings may be the problem. This can also be checked by an experienced pilot or the mechanic, who (after making sure the ignition is off, the mixture is at idle cutoff, and the airplane is chocked or tied down) pulls the prop through and checks for a cylinder (or cylinders) that doesn't resist the pulling

through like the others. Extreme caution is suggested when pulling the prop through, even though the mag switch "says" OFF.

The oil control rings are below the compression rings and regulate the thickness of the oil coating on the cylinder walls.

The camshaft is geared to the crankshaft at the rear of the engine. The camshaft has nodes or cams that are offset around the shaft; as it rotates, the cams move up against pushrods that open the spring-loaded exhaust or intake valves for the various cylinders in the proper sequence. A pushrod extends from the camshaft to the rocker arm, which pushes the valve (exhaust or intake) down and opens it (Fig. D-4).

Engines are classified by the cubic-inch displacement in the cylinders or the volume of air in all cylinders if the pistons are in the bottom dead center or "withdrawn" position. For example, suppose the *diameter* of each piston (with rings) of a four-cylinder engine is 5.4 inches (its *bore*) and it *travels* 3 inches (its *stroke*) from full up to full down (or in a horizontally opposed engine, full out to full back) (Fig. D-5). One cylinder's displacement is the volume of that travel. Volume (V) or displacement = area × distance. (Area = πr^2.)

Radius (r) = 2.7 inches; r^2 = (2.7)^2 = 7.3 square inches; πr^2 = 22.9 square inches = area

Stroke = 3 inches; displacement = 3 × 22.9 = 68.7 cubic inches

There are four cylinders, so the total displacement is 4 × 68.7 = 274.8 cubic inches, and this fictitious engine would have a designation of O-275 because it is a horizontally opposed type. If it had fuel injection, the designation would be IO-275; if it was also turbo-

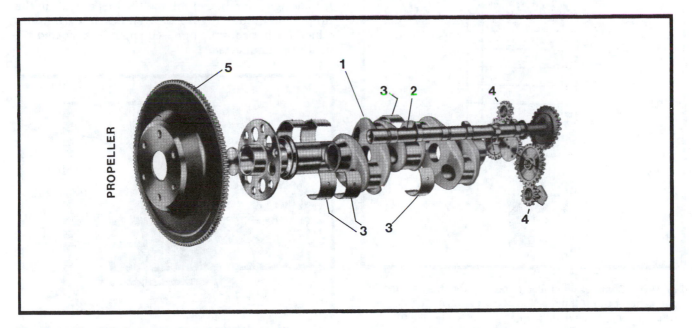

Fig. D-4. Crankshaft, camshaft, and other parts: (1) crankshaft, (2) camshaft, (3) crankshaft bearing (same as item 10 in Fig. D-3), (4) magneto gears, (5) starter ring gear.

charged, the designation would be TIO-275. Another letter G would be included (TGIO-275) if the engine was geared and not a direct drive. It's likely that an engine with this small amount of displacement would not have all the refinements mentioned, but it's an example. The term "compression ratio" is used for reciprocating engines to indicate the ratio of the amount of volume of the mixture when the piston is at the bottom dead center position (fully withdrawn down or back in the cylinder) to top dead center (at the greatest compression). Figure D-6 gives an idea of the relative volumes for a compression ratio of 7 to 1.

TIME BETWEEN OVERHAULS. You'll hear the expression time between overhauls (TBO) as you progress as a pilot, and the tendency is to think that this is a definite time required by regulation until a complete major overhaul of the engine. Also, you might think there would be an immediate failure if the engine was secretly operated past that limit. Not so. The manufacturer sets a *recommended* limit based on inputs of users from around the world and its own experience in overhauling engines returned for that purpose. An engine having a recommended TBO of 2000 hours, for instance, may need overhauling at 1200 hours (if abused and the manufacturer's recommendations for operations neglected), or it may go 2500 hours or more if properly operated and maintained.

Another term you'll hear is *top overhaul,* and this usually consists of removing only the cylinders so that piston rings may be replaced and valves may be replaced or reseated. This is done by some operators between *major* overhauls (in which the engine is totally disassembled). Some manufacturers feel that if the engine is properly maintained and operated, it should go to at least the recommended TBO without the need for a top overhaul. However, local conditions, need for special operations, and use of higher lead content fuels may necessitate a top overhaul.

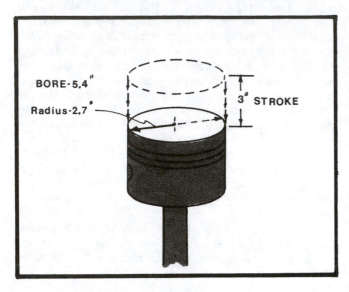

Fig. D-5. Piston displacement.

■ Carburetion

The engine has a way to get the fuel-air mixture in, compress it, fire it, and exhaust it, so the next step is to design a method of mixing the fuel and air in the proper amounts. Figure D-7 shows a carburetor, which for horizontally opposed engines is located beneath the crankcase.

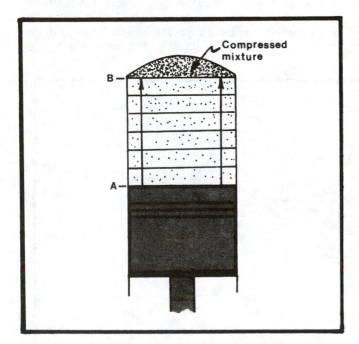

Fig. D-6. Compression ratio. In moving from the bottom dead center point (A), the piston compresses the fuel-air mixture to one-seventh of its original volume at (B); the compression ratio is 7 to 1. Low-compression engines are considered to be those from 6.5 to 1 to 7.9 to 1. Above that they are considered high-compression types.

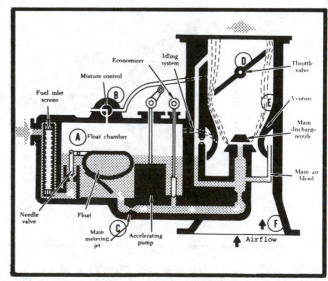

Fig. D-7. Float-type carburetor. (*FAA AC 61-32C*)

The carburetor has a float chamber that stores a small amount of fuel for use in mixing with the incoming air. As the fuel in the chamber is depleted, the float moves down, opening a valve to let in more fuel. This is a repeating process as the engine runs.

The fuel is vaporized by the main discharge nozzle and mixes with the incoming air (as shown in Figure D-7) just past the venturi (which speeds up the airflow like the upper camber of an airfoil). As you move the throttle forward, the throttle valve opens more and allows a greater amount of fuel and air to enter the engine and provide more power. The temperature is decreased in the venturi discharge nozzle and throttle valve area by two events: (1) the pressure of the air moving through the venturi is lowered, lowering the temperature, and (2) the vaporization of the fuel lowers the temperature so that a drop of up to 60°F is possible. If the incoming air contains moisture, it may freeze in that area and eventually shut off the air (and fuel) flow, with engine stoppage resulting. Look again at Chapter 4 for the background of carburetor heat and for diagrams of how it works.

FUEL INJECTION. The fuel injection system is normally used for higher-powered engines, and later you will get a chance to fly an airplane so equipped. The advantage of this system is that the fuel is injected into the air just as it enters the cylinder, resulting in less chance of carburetor ice and a much better fuel distribution.

When you lean a carburetor-equipped engine, it will become rough (you can feel it), since the air and fuel are mixed at the carburetor and the mixture will vary between cylinders because of the distances involved. One or more cylinders may be leaner than the others, and as you move the mixture control out to lean the engine, those cylinders start losing power earlier. The engine runs rough, and you move the mixture control forward just enough to smooth it out.

Because of the better fuel distribution, the fuel injection normally can't be leaned by feel and you'll use a fuel-flow gauge or an EGT (exhaust-gas temperature) gauge. Fuel-injected engines don't use carburetor heat as such but have an alternate air source (usually from inside the cowling) if induction icing (icing in the air intake area) occurs. Fuel injection is best used in an engine of over 200 horsepower with a constant-speed propeller.

If a tank is run dry in a carburetor system, it normally takes 1–2 seconds for power to be restored after the tanks are switched; a fuel injection system may require 8–10 seconds and special techniques to get power back. It's best never to deliberately run a tank dry with *any* system. Check your *Pilot's Operating Handbook* now for the procedure if you should inadvertently run a tank dry.

■ Ignition

MAGNETOS. There must be a system to ignite the mixture when it's been compressed in the cylinder. The magneto is used in an airplane because it is self-sufficient, as noted in Chapter 4. Two magnetos are used for two reasons: (1) there is one remaining to keep the engine going if one fails, and (2) two spark plugs provide a smoother burning of the mixture in the cylinder. With one magneto and one spark plug the ignition would start in one corner of the combustion chamber and proceed across, taking some time (milliseconds), whereas *two* separate sparks can each set off a portion of the fuel-air mixture and make for smoother and more efficient combustion. That's one reason why you get a drop in rpm in the mag check when going from BOTH to just LEFT or RIGHT.

The magneto is "timed" so that the spark plug gives ignition to its cylinder at a time that would result in the most efficient burning and best power. The initial thought would be that the spark should occur just at the instant the piston reaches its farthest point up (or out) in the cylinder when the compression is greatest (top dead center, or TDC). The problem in that case is that by the time the burning reached its peak, even though it's burning fast, the piston would have already started down and the peak power wouldn't result. The idea is to time the spark so that the maximum effect of combustion results. Since it takes time for the combustion to propagate, the ignition should be started *before* the piston reaches TDC so that the best "push" is obtained. The mechanic "times" the magneto by setting it to fire at some point, say 25° of crankshaft rotation before TDC, as recommended by the manufacturer. Magnetos get out of time and mechanics have to check them. Timing is critical to smooth operation; for instance, you can imagine what would happen if the timing was accidentally set too early (for instance, 90° before TDC) so that the piston was still moving up when the full force of combustion occurred. There would be pandemonium in the powerhouse.

It's possible that in flight one of the magnetos might slip out of timing and the engine would run very rough. You might go from BOTH to LEFT *or* RIGHT to find the magneto that is acting properly. Since the airplane is now running on one mag, it would be best to get on to the nearest safe airport for repairs.

In some cases, the magnetos may be timed slightly apart. One is set to give the word to the spark plug at 25° before TDC, and the other fires its plug in the same cylinder slightly later, at 23° before TDC. This may be done to get an even smoother operation; it depends on the engine.

If you get a chance, go out into the hangar or shop some day between flights and observe some of the engine work being done. Most mechanics don't mind good questions as long as they don't interfere with the work. There's a sign that mechanics show to airplane owners:

MAINTENANCE CHARGES

1. Our mechanic alone $30.00/hr
2. If you stand around and
                 offer suggestions   35.00/hr
3. If you help   40.00/hr

The pilots come back with a sign that says:

IT TAKES SIX PEOPLE TO WORK

ON AIRPLANES HERE:

ONE MECHANIC

AND FIVE MORE TO HOLD THE PILOT BACK.

Most engines have each magneto separately geared to the engine accessory section, but some newer models use a common engine drive with both magnetos in the same housing. Advantages of this are a decrease in weight and freeing of one magneto drive that might be used for other accessories. Some argue against having a *single* drive for *both* mags and prefer a separate one for each.

The magnetos and plug leads ("leeds") must be shielded to prevent ignition noise from coming through the radio receivers. If you hear a repetitive crackling noise coming through the radio, you might try checking to see which mag and its ignition components (plugs, leads, etc.) are causing the problem. You can tell the mechanic which is the guilty system and save time in analysis.

SPARK PLUGS. Spark plugs are seemingly simple pieces of equipment, but they can affect the performance of the engine to a great extent. They are subjected to extremely high temperatures when the engine is running.

A great deal depends on the plug "gap" or gaps, that is, how far apart the electrodes are. The spark has to "jump" a certain length to get the optimum ignition—the wider or narrower the gap from that recommended by the manufacturer, the less efficient the result.

Spark plug electrodes are subjected to temperatures of over 3000°F, gas pressures of up to 2000 pounds per square inch, and electrical pressures of up to 15,000 volts.

It's noted in Chapters 4 and 7 that each magneto furnishes ignition (through the spark plugs) to each cylinder, and the wiring route may vary between makes and models. The ignition wiring on one four-cylinder engine used in a current trainer is established so that the right magneto fires the two lower right and two upper left plugs and the left magneto fires the lower left and upper right (Fig. D-8). Usually though not always, the lower plugs foul more easily, so if you had a bad permanent drop on the left mag, under the arrangement just described you'd probably start a maintenance check, looking at those bottom left plugs first. Of course, the mechanic would know the wiring arrangement, but you should have an idea of the general layout of your airplane ignition system.

Fig. D-8. In this particular installation the left magneto (1) fires the bottom plugs on the left two cylinders (2) and the top plugs on the right cylinders (not shown). The right mag (not shown) fires the bottom right (not shown) and top left (3). While it's not a part of the ignition system, (4) shows the ground-service plug receptacle for use of an external electrical power source.

Problems occurring in the combustion chamber are *preignition* and *detonation*, and sometimes people get confused between the two.

Preignition, as defined by the *Lycoming Flyer*, is a situation in which the combustion takes place within the cylinder before the timed spark occurs. Usually it's caused by combustion deposits, such as lead, which cause local hot spots. It can be caused by a cracked valve or piston or a broken spark plug insulation that sets up a local glow spot. The mixture is not fired at the proper time, the engine runs rough and backfires, and a sudden increase in cylinder-head temperature is noted if you have a CHT (cylinder head temperature) gauge. Older airplanes without mixture control had to be shut down by turning off the ignition, and if the engine was hot it continued running. Also, some model-airplane engines use the principle of preignition by electrically preheating a certain type plug, which is then kept hot by combustion.

Preignition is a serious condition that can cause

burned pistons and damaged valves and is similar to early timing of the spark.

Detonation results when limits of compression and temperature are exceeded; this is most likely at higher power settings. Leaning the mixture at high power settings and abrupt opening of the throttle can cause detonation.

Unless it is heavy, the pilot doesn't know it's happening. Light to medium detonation may *not* cause noticeable roughness, observable oil temperature or CHT increase, or loss of power. But after an engine has been detonating, a teardown shows damage to valves, pistons, and cylinder heads.

In *normal combustion* the piston is pushed smoothly down (or in, in a horizontally opposed engine); with *detonation* the piston acts as if it was being hit by a sledgehammer.

The best temporary in-flight methods for correcting preignition and detonation are to retard the throttle, richen the mixture, open cowl flaps if available, or do all three.

You might remember the causes of detonation by the following three "too's":

Too lean—mixture too lean for higher power settings.

Too hot—climbing at too slow an airspeed and/or too lean a mixture.

Too cheap—lower grade fuel than recommended.

■ **Oil System**

The engine operation would be short-lived without lubrication. The oil used in the engine has three major purposes: (1) to maintain a film between moving parts so that they do not actually make metal-to-metal contact, (2) to carry off heat from the various internal parts as it circulates, and (3) to pick up foreign particles for depositing in a filter.

The oil used in the airplanes you are flying is designed to take care of the three items just mentioned. Without going into laborious details about the theory of oil viscosity (or flow rate ability), the higher the commercial aviation number or SAE (Society of Automotive Engineers) number, the higher the viscosity (thickness). The viscosity of oil varies inversely with heat—the colder the temperature, the thicker the oil. If you choose a low-viscosity (thin oil) for use in hot weather, the viscosity gets even lower as the engine heats up, and metal-to-metal contact results. The use of high-viscosity oil in cold climates would mean that it would be like grease, making the engine difficult to start, and the oil circulation would be very poor. Your *Pilot's Operating Handbook* will note the oil grades to be used for various ambient temperature ranges and, as indicated earlier, will probably note that after the first 50 hours of operation, ashless dispersant oil meeting certain criteria is to be used. It may recommend use of multiviscosity oils.

Some typical light-trainer oil system components

are shown in Figure D-9. This is a wet-sump system, meaning that the oil is contained in the lower part of the engine (the sump), as in most automobiles.

The carburetor for this engine is located as shown in Figure D-9 and gets warmth from the warm oil in the sump. Because of this, the engine is less susceptible to carburetor ice than the dry-sump type, which has the oil in a container separate from the engine and not close to the carburetor. The wet-sump engine *can* have carb ice troubles, and carb heat should be used to check for the presence of ice before the throttle is retarded to lower power settings, such as when starting an approach.

With less availability of 80-octane fuel and more use of grade 100 and 100 Low-Lead, it's important that the oil be changed more often when using these higher-grade fuels with their attendant higher lead contents. The old oil, when drained, will carry off impurities, thus lengthening the engine life.

Your instructor may show you the engine oil breather, which is an attachment and hose designed to release pressures that build up in the crankcase because of the high temperature there. The breather usually comes out the top of the crankcase, with the hose running down and out the bottom of the cowling (item 9, Fig. D-9). During the preflight check in the winter, be sure that the breather is not closed by ice (the warm water vapor from the engine will freeze at the breather outlet).

Since lighter airplanes don't have a cylinder-head temperature gauge, the oil temperature gauge becomes very important in indicating whether the engine is being abused in a too-slow climb or a too-lean setting for the flight condition. As noted at the beginning of this section, part of the job of the oil is to remove heat from the engine components. The *Airframe and Powerplant Mechanics Powerplant Handbook* (DOT-FAA) indicates that the thermal distribution of the heat released by combustion in an engine is approximately as follows:

1. 25–30 percent is converted into useful power.
2. 40–45 percent is carried out with the exhaust.
3. 5–10 percent is *removed by the oil*.
4. 15–20 percent is removed by the cylinder cooling fins.

■ **Engine Cooling**

The principle of the cylinder cooling fins is radiation of heat from the cylinder into the air moving past the engine. The fins present a large surface area and are oriented so that the air moves around and between them, thus removing heat. In a preflight inspection you may note that some of the fins are broken off; this should call for a discussion with your instructor or mechanic because if enough cooling fin area is missing serious heat problems could arise and the cylinder should be changed.

In the spring, birds may build a nest in the cowling, with the result that the cooling flow is shut off or redirected and a hot spot exists, leading to serious damage. Baffles are shaped aluminum plates attached to the engine or cowling to help direct the airflow for best engine cooling. These should be checked for security and proper alignment during the preflight if possible.

Your current light trainer may not have cowl flaps (which are normally located at the bottom rear of the cowling), but probably one or more of the airplanes you fly later will have them.

Engine cooling, from a pilot's standpoint, results from two main sources: (1) rich mixture and (2) airflow past the cylinders. Looking at the airflow, there's usually no problem at cruise, and a small opening at

the bottom rear of the cowling is sufficient for the warmed ram air to escape; there is a large volume of air moving over the engine each second, carrying off engine heat. Also, the power is back to cruise, so there's a comparatively low heat problem and plenty of air passing by to take care of it.

When the airplane is sitting on the ground with the engine running, there is comparatively little air moving across the cylinders, and the cowl flaps should be fully open to allow the greatest amount of heat escape possible. Some airplanes require that the cowl flaps be set fully open for takeoff and not quite fully open for a cruise climb. After cruise is established, the cowl flaps are closed to cut down on aerodynamic drag and also, particularly in cold weather,

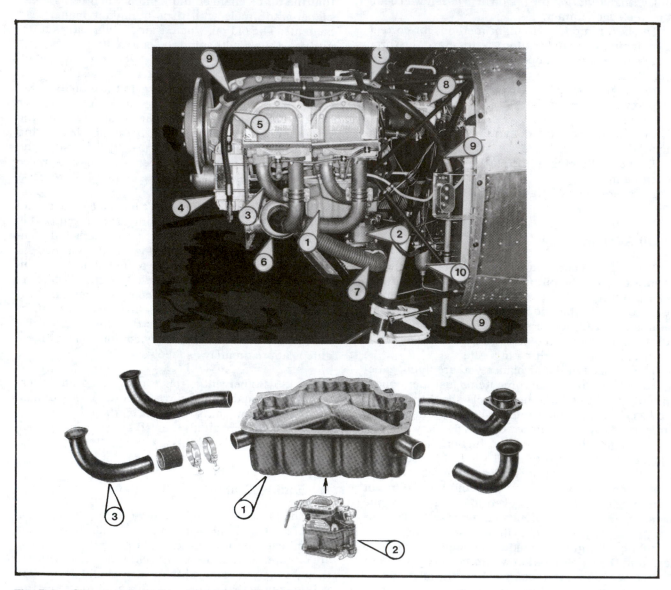

Fig. D-9. Oil system and other components: (1) oil sump (the bottom picture shows a front view of the sump), (2) carburetor (note its position in relation to the oil sump), (3) intake pipe, (4) oil cooler (radiator), (5) oil line, (6) carburetor heat/cabin heat muff, (7) carburetor heat hose, (8) oil filter, (9) breather pipe, (10) fuel strainer.

to keep the engine from running *too* cool.

You may also fly an airplane using exhaust augmenters, which use exhaust gas velocity to draw air over the engine. The simpler types do not require cowl flaps or other controls to be adjusted by the pilot. The system is automatic; as the exhaust gases are released through the ejector, the hot air is pulled from the engine compartment and sent out with the exhaust.

■ Exhaust Systems

The exhaust system should seem simple enough; all that is apparently needed is a pipe to let the burned gases out. The most efficient (and noisiest) system was that of the short-stacks type used earlier on radial engines; it consisted of a short, curved exhaust pipe for each cylinder. This meant that there would be little back pressure in the ''system.'' When the exhaust system requires long pipes or collector rings (as for bigger, cowled radial engines), it is less efficient because of the resistance of the burned gases in the longer system. The exhaust gas just emerging from the exhaust port has to push against this mass to get out. The exhaust system must provide a method of getting the hot waste safely out of the cowling and away from the airframe; this may require a long crossover exhaust pipe system in some airplanes.

The metal of the exhaust system must withstand very high temperatures without burning through at a critical place in the engine compartment. Exhaust systems on each side of horizontally opposed engines may be joined to a single final exhaust stack (Fig. D-10).

Fig. D-10. Exhaust system. The exhaust stacks (1) join to one outlet (2). The other end of the carburetor heat/cabin heat muff and the other two stacks can be seen in Figure D-9. The cabin heat hose is shown at (3). Item (4) is the oil filler cap, admittedly not a part of the exhaust system, but there it is anyway.

EXHAUST GAS TEMPERATURE GAUGE. The exhaust gas temperature (EGT) gauge measures the relative change in the exhaust gas temperature with change in mixture and gives an indication of the best economy and best power mixtures.

One simple system consists of the gauge, wiring, and a probe into one of the exhaust stacks at a point 1.5 inches, or slightly more, from the cylinder head. The gauge may be calibrated in 25°F increments (with no fixed numbers shown) or in a range of 1200–1750°F. One manufacturer's engines normally operate in the 1200–1600°F EGT range.

Peak EGT—As you slowly lean the mixture from full rich, the fuel-air ratio decreases and the cooling effect of the fuel is lessened. Watching the EGT gauge, you can see the gradual increase in the temperature until at some point it peaks and further leaning will cause a decrease in temperature. The peak EGT is where the maximum utilization of the fuel and air occurs; it's a chemically correct mixture. As the leaning continues, the ratio of air to fuel increases and power will be lost, and the exhaust temperature will decrease for that one examined cylinder. If the mixture control is moved forward, richening the mixture from peak, the EGT will drop again as fuel cooling takes place. More expensive EGT systems have a probe for each cylinder, with a selector to check each.

Operating at peak EGT is only practical in *fuel-injected* engines of over 250 horsepower.

Remember the earlier discussions pointing out that the fuel distribution was more even in fuel-injected engines because it was injected at the intake port of each cylinder? It was also noted that the carburetor did not allow this good distribution, and an engine so equipped can be leaned by ear or feel, since the leanest cylinder would start complaining first as the mixture control was moved out and the rough running engine tells you to richen it forward just a tad. Some pilots are afraid that this roughness means detonation and so hesitate to lean. If leaning is done at the manufacturer's recommended top cruise power setting or below, no detonation can occur. Leaning in climb or other high power settings *can* cause detonation. Peak EGT operations are limited to 75 percent rated power or below in some direct drive, normally aspirated (nonsupercharged) engines.

Best economy mixture—For practical purposes, the best economy for some engines starts at the peak temperature. Expect a slight loss of horsepower and a loss of 2–3 knots at cruise if you use this. Other engines may use a temperature drop of 125°F on the rich side of peak.

Best power mixture—This is sometimes referred to as *maximum power range*, which is in the range of a temperature drop of 100–150°F from the peak *on the rich side*. (Remember that the temperature will drop from the peak when the mixture is either leaned or richened.) This provides a safe mixture for higher than cruise power settings (except for takeoff). A rule of thumb compromise between best economy and best power mixtures has been a mixture setting of

50°F on the rich side of the peak. The *Pilot's Operating Handbook* for a particular airplane is the final word on this as well as for other operations.

ELECTRICAL SYSTEM

As a pilot, the electrical system may be a mystery to you (that is, how the electrons move through it), but you should know how to check it in the preflight and know what procedures to use in case of problems. Note the electrical system in Figure 3-26. Maybe it doesn't make sense with all the symbols, but at least you can tell what components of the airplane are electrically powered by looking at the circuit breakers. Maybe you didn't know that the oil temperature gauge depended on electrical power for indication (the oil *pressure* gauge does not). The oil pressure switch seen at the left of the electrical system diagram energizes the flight recorder when the engine starts, and oil pressure is then available.

Note that the ignition switch is in the electrical system, not because the magnetos are battery powered, but because the spring-loaded starter position is part of that switch.

■ The Alternator

The airplane's alternator is like that of your automobile. It is engine-driven by gears or a belt, and it has more than enough capacity to keep the battery charged when working properly. If an overvoltage occurs, the system will automatically turn off the alternator and you are warned by a red light or other indication. Check with your instructor concerning procedures in event of overloading and alternator loss. Procedures will differ among airplane types.

The alternator has replaced the generator in airplanes and cars because it develops electrical power at lower engine rpm and is more efficient in the idle power range. (While it's working at lower rpm or down to idle, it still may not be keeping up with the electrical requirements, so run the engine up a little if in doubt.) The older generators required 1200 rpm or more even to "cut in," and the battery could be discharged when the situation called for low rpm and large electrical requirements (such as taxiing at night with taxi and navigation lights and radios on). Figure D-11 shows components of the electrical system for a light trainer. This system is protected by circuit breakers (which can be *reset* in the event of an electrical overload) except for the clock, flight hour recorder, and external power plug, which are protected by fuses that are replaced if an overload burns them out. In other words, the circuit breakers and fuses are deliberately established "weaker" spots so that if an electrical overload in the system occurs, the circuit breaker will "pop" or an inexpensive fuse will burn out rather than damaging such components as radios or other

expensive electrically operated items. Of course, you should have all unnecessary electrical items off for starting, particularly radios that could be damaged by surges of transient voltages during the start.

Fig. D-11. Electrical components: (1) alternator, (2) belt drive of alternator, (3) battery box (the battery is inside), (4) fuses for the clock and flight hour recorder and external power receptacle. (See item 4, Fig. D-8.)

AMMETER. The ammeter indicates the current flow in amperes from the alternator through the battery to the system. If the ammeter shows a discharge, it means that either the alternator is not keeping up with the load being placed on the system or it's out of action and the battery is being discharged at the rate shown (amperes). For instance, if the battery in your airplane is rated at 32 amp-hours, it means that the battery, when fully charged without alternator input, should produce an electric current flow of 16 amps for 2 hours, or 8 amps for 4 hours, etc. However, at high usage the battery loses power at a higher rate; if 32 amps were being used with a 32 amp-hour battery, it might last only 30 minutes or less rather than the hour indicated by the earlier arithmetic. If you have an alternator overvoltage problem and have to pull it from the system (turn that ALT half of the switch OFF), you'd lengthen the battery power availability by turning off unnecessary electrical components. For instance, landing and taxi lights are usually the biggest users of current. The cigarette lighter takes a great deal of current, and you shouldn't be smoking in the airplane anyway; it's bad for the vacuum-driven gyro instruments (attitude indicator and heading indicator), as will be shown shortly.

VACUUM SYSTEM

Older airplanes used a venturi, shown in Figure 3-17, but the drawback with that system is that the airplane had to be moving for some time before the suction got the instruments up to speed. Now, even the smaller trainers have an engine-driven vacuum pump to operate the gyro instruments. As noted in Chapter 3, the usual procedure is to have the attitude and heading indicators run by the vacuum system and the turn and slip or turn coordinator electrically driven. Figure D-12 shows a vacuum system installation for a light trainer.

The vacuum pump is at the rear of the engine (at the accessory section) and pulls air from the cockpit area through the gyro instruments and discharges it into the engine compartment.

Note in Figure D-12 that a suction gauge is a part

of the system, and for this airplane the normal range is 4.6–5.4 inches of mercury. These figures indicate the drop in pressure in the system as compared with atmospheric pressure and give the suction available to the instruments.

The air filter is located on the aft side of the firewall and, as just noted, takes air from the cockpit area. This is a good reason for not smoking in the airplane, since tars and other contaminants will clog the filter and cut down the accuracy and dependability of the gyro instruments.

The more you know about the systems, the better able you will be to avoid or correct problems in flight. Most systems are based on simple principles and are presented here from a pilot's standpoint. The Bibliography contains references for further reading.

REVIEW QUESTIONS

Okay, answer the following questions on this section of Appendix D, and after finding out those you missed (if any), read Chapter 4 and this section again.

01. The four strokes of an Otto-cycle airplane engine are

1—thrust, rotary, power, and exhaust.
2—intake, power, thrust, and exhaust.
3—intake, compression, power, and exhaust.
4—intake, thrust, compression, and power.

02. Which of the following would most likely result in a sudden increase in oil consumption in the engine of an airplane? (There are no outside leaks in evidence.)

1—Eroded spark plug electrodes.
2—Worn or broken piston rings.
3—A poorly adjusted carburetor.
4—A coagulated frammis.

03. A horizontally opposed six-cylinder, direct drive, nonturbocharged engine with fuel injection has a bore of 5 inches and a stroke of 4 inches. Its designation by the manufacturer would be

1—GO-470.
2—IO-470.
3—IO-540.
4—TIO-470.

04. The addition of carburetor heat

1—leans the mixture.
2—has no effect on the mixture.
3—richens the mixture.
4—causes oil pressure fluctuations.

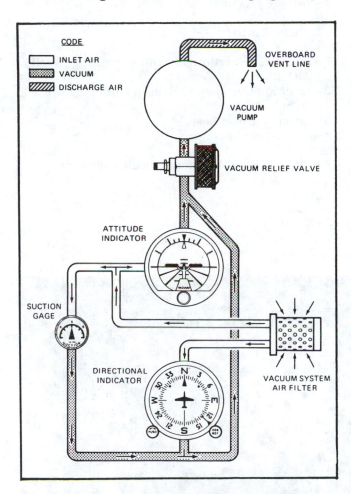

Fig. D-12. Vacuum system schematic. The overboard vent line shown here can be seen in Figure D-11 behind and above the oil filler cap; it's that aluminum tube with two 90° bends. If you look closely, you can see that the vacuum pump, which is behind the oil filler, is a black drum with a light metallic band around it. You also might look just above item 8 in Figure D-9 to see the pump and overboard vent line. (The pump is almost hidden by part of the engine mount.)

05. An advantage in the fuel injection system over the carburetor is that it

1—has better fuel distribution to the cylinders.
2—does not require an alternate air system.
3—can be used best with fixed-pitch propellers.
4—has a simpler oil system.

06. As an all-around procedure you should

1—deliberately run a tank dry to use all the fuel only if it is a fuel injection system.
2—expect a delay of 1–2 seconds in power after the selector is switched from an empty to a usable tank with the fuel injection system.
3—deliberately run a tank dry only at altitudes above 1500 feet MSL.
4—never deliberately run a tank dry.

07. During the pretakeoff check after checking the mags, you pull the carburetor heat ON at 1800 rpm (or the manufacturer's recommended number). There is *no* change of rpm (plus or minus) when you pull it ON or for 30 seconds thereafter. This is a sign that

1—carburetor ice was present before the heat was applied.
2—there was no carburetor ice present before the heat was applied.
3—the carburetor is not getting heat from the system.
4—you did not leave the heat on long enough to get an effect.

08. Smoking in an airplane

1—is useful in finding out the wind direction by blowing the smoke out a window.

2—may cause you to be thrown overboard by a non-smoker.
3—can cause clogging of the vacuum system air filter.
4—can clog the carburetor heat.

Questions 09–17 are based on Figure D-13.

09. Item A is the

1—carburetor air filter.
2—oil cooler.
3—alternator.
4—carburetor heat muff.

10. If item B is ruptured, you would

1—expect a rapidly rising oil temperature.
2—lose use of the carburetor heat.
3—have ignition problems on the vacuum pump.
4—lose tire pressure.

11. Item C is the

1—magneto overflow simplex routing system (MOSRS).
2—air line to the nosewheel, used for filling the tire in flight.
3—carburetor heat hose.
4—oil breather line.

12. Item D is a component of which system?

1—Oil.
2—Fuel.
3—Electrical.
4—Ignition.

Fig. D-13. Engine and system components for questions 09–17.

13. Item G is important in the event of

1—low oil pressure.
2—electrical overload.
3—carburetor ice.
4—alternator failure.

14. Item H is

1—a magneto.
2—the alternator.
3—the carburetor.
4—the oil reservoir.

15. Item I is

1—an intake manifold.
2—a fuel line.
3—an oil line.
4—an exhaust manifold.

16. If item J was severed, the result would be

1—a drop in oil pressure.
2—an ignition problem.
3—a fuel leak.
4—an alternator failure.

17. Items E and F, respectively, are part of which systems?

1—Oil and fuel.
2—Electrical and fuel.
3—Oil and electrical.
4—Oil and brake.

■ **Answers**

01.   3.   See the text at the beginning of Appendix D.
02.   2.   As indicated in this chapter, worn or broken piston rings may allow oil to get into the combustion chamber and be burned. A coagulated frammis (4) is not the answer because it is a factor that makes dentures slip and also may cause the heartbreak of psoriasis.
03.   2.   IO-470. The engine is opposed (O) and has fuel injection (I). The bore of 5 inches would mean a piston surface area of πr^2 or $(2.5)^2 \times \pi$. Using 3.1416 for π, the area is 19.635 square inches. The stroke (piston movement) is 4 inches so that each cylinder displaces $4 \times 19.635 = 78.54$. There are six cylinders, so that the total displacement is 471.24 cubic inches. The engine is direct drive (not geared), so a "G" wouldn't be part of the numbers. Your numbers may differ slightly from the above answers, depending on how far you take the decimal points.
04.   3.   Chapter 4. The warmer air is less dense, so for the same weight of fuel being metered the mixture is richened. If, say at cruise, you need to use carburetor heat (and in trainers you should use *full* heat if you use *any*), you'll have to re-lean

the mixture to get smooth running.

05.   1.   The fuel injection system has better fuel distribution because the mixing is done just before it enters the cylinders. Because of this, the system is less susceptible to carburetor ice. Induction icing (caused by visible moisture freezing in the intake area) can still occur however.
06.   4.   Because of possible restart problems, a good all-around philosophy is *never* to deliberately run a tank dry, particularly with a fuel injection system.
07.   3.   The carburetor is not getting heat. In this book it is noted that the warmer air is less dense and a slight power loss would result. (A figure of 100 rpm is given as an average, but it could vary among airplanes.) If no ice was present, the rpm would stay at the lower value as long as the heat was ON. If ice had been present when the heat was applied, the rpm would drop initially because of the warmer air and then pick up *above* the first setting as the ice cleared out. This question ties in with question 13 in that there is a good possibility that the heat hose is detached at one or both ends. (See item G in Fig. D-13 and also see Fig. 4-2.) You may have missed this in the preflight check but might catch it here, since that's the purpose of pulling the heat ON. *Don't* pull the carburetor heat ON and quickly shove it OFF. Leave it ON for 10 seconds to check an initial drop in rpm and to see if it has cleared out ice. As the answer to question 13 notes, without an operating carburetor heat you might get into icing conditions that could cause a forced landing. Taxi back to have the problem checked if the heat isn't working properly during the pretakeoff check.
08.   3.   Students fall into categories. Some students have fallen out of airplanes (or were pushed) when they smoked and contributed to vacuum-system air-filter problems.
09.   2.   That's the oil cooler.
10.   1.   If item B ruptures, you can expect to have a large oil leak, a rapidly rising oil temperature, and possible engine stoppage.
11.   4.   That's the oil breather line. If you picked the first choice, it's possible that you are overwhelmed by technical gobbledygook. (A magneto overflow simplex routing system?) As far as filling the nosewheel tire in flight is concerned, when you get the private certicate you can send out that passenger who is always making sneering comments about your landings.
12.   1.   Item D is the oil sump where the engine oil is stored.
13.   3.   That's the carburetor heat hose, and during your preflight you should check it for attachment at the heat muff *and* at the carburetor. If it is off at either point, you could find yourself in a bad icing situation with no heat available.
14.   1.   It's the right magneto.
15.   4.   Exhaust manifold. In the picture, you might initially mistake it for an intake manifold, but look at it disappearing into the carburetor heat muff. When looking at an actual engine, you could readily see evidence of prior heating, whereas the intake manifold usually stands (or sits) in unheated glory.

16.   2.   That's an ignition lead going to the bottom plug of the right front cylinder. The engine would still run on BOTH with that lead cut, but a mag check would show a problem. This particular installation has the right mag firing the lower right and upper left plugs. Selecting the right mag would show a prodigious drop, since on that mag one cylinder would be totally out of action. Remember that different engines have different ignition wire routings.

17.   2.   E is the alternator and F is the carburetor.

MORE NOTES ON BRAKES

The brakes in smaller airplanes are an independent hydraulic system; that is, they have their own reservoir, separate from a main hydraulic system that might actuate the landing gear and flaps. On some airplanes the brakes are the *only* hydraulic system on board.

Hydraulics for brakes, flaps, or landing gear work because liquid, unlike gas, is incompressible. Any force transmitted at one end goes directly to the other end of the system. (Purists will argue that liquids are compressible, but compared with air, not much.) If the lines had been filled with air, there would be little action at the far end because the air would allow itself to be compressed before passing on some of the forces. An analogy might be pushing against the back of a toy wagon with an iron bar (liquid) and a rubber hose (air). The iron bar will send your force directly and get the wagon rolling, but the rubber hose may bend and flex before transmitting part of your force to get the same (or lesser) result.

You may run into the problem of air in the brake lines in the airplane. The air is compressible, so as you press the pedal(s) there is no immediate response to pedal pressure because the air compresses rather than sending a direct message as the liquid brake fluid would do. A small amount of air in the system would mean a relatively small amount of "sponginess" of the pedals, but if the pedals start feeling this way, you'd better check with the mechanic who will "bleed" the system of air and check for loose connections or tighten up the system to stop the further encroachment of air. More about this later.

One type of light-airplane brake system consists of a master cylinder for each brake pedal, with its own reservoir of hydraulic fluid. Additional components consist of a mechanical linkage from each pedal to its cylinder, with required fluid lines to a brake assembly on each main wheel. In other words, each wheel has its own separate brake system.

Airplanes with a brake handle (no toe brakes) have a single master cylinder that transfers pressure to both main wheels at the same time.

Other systems may have a separate reservoir in the engine compartment, and the fluid level should be checked during each preflight inspection (if you can get to it).

During the preflight check you should look at the brake lines and at the surface under each main wheel to check for signs of brake fluid leaks. The wheel covers (speed fairings) on some airplanes make it difficult to check the condition of the brake disks and cylinders. Incidentally, if your airplane is such that the disks and brake can be seen—and touched—don't put your hand on the disk if the airplane has just come in from a flight. A painful burn could result because the disks will remain hot for a while.

The parking brake is essentially a valve that locks pressure in the wheel cylinders. You'll press on the pedals and then set the parking brake, which (it says here) will hold the pressure exerted until released.

The parking brake generally should not be trusted. The pressure may bleed off, and the brakes would no longer hold. More than one pilot has set the parking brake for the engine run-up and looked up from a checklist to find that the airplane was moving rapidly toward some expensive object.

If the parking brake is set in the morning, rising temperatures during the day can cause fluid expansion and destroy seals in the system. Chocks or tie-downs are more dependable for an airplane that is to be left sitting for more than a few minutes.

■ Brake Problems

Probably one of the most common problems you'll run into is "soft brakes" or air in the system. The pedal or pedals may go easily to the limit with little or no brake effectiveness. If you unexpectedly encounter such a problem (you've taxied into the ramp too fast and suddenly realize that a very expensive mistake is about to be made), *pump* the brake pedals. This will build up temporary pressure for braking (Fig. D-14). It's a good idea to press the brakes—and release them—before coming in to land, so that you'll know whether the pedals feel normal. If they are soft, you might plan on pumping the brakes during roll-out or consider that you may not have *any* braking effectiveness and should make a short-field approach or fly to another airport with longer runways.

If on roll-out you discover that one brake of your tricycle-gear airplane isn't working, press the other brake gently and use rudder pedal pressure on the failed brake side to help keep the airplane straight.

People have opened doors on the roll-out after brake failure; the air drag of the door(s) can definitely help slow a smaller airplane.

Another problem is that of "hard" brakes. The brake pedals feel like they are solid and not hinged, and braking effectiveness is very low. You should get a mechanic on this also, as soon as practicable.

Dragging brakes (you can hear them sometimes) can cause tire wear on touchdown, brake heating problems, and longer takeoff runs. You can sometimes see the effects of a brake dragging by looking at a main wheel (high-wing airplane) as the airplane is taxied on a smooth surface; one landing gear shakes while the nondragging side moves smoothly.

Older airplanes have mechanical brakes in which a cable is used instead of the hydraulic line. When the pedal (usually a heel brake) was pressed, the cable moved the brake pad against the disk. A disadvantage of this type of system was that one of the cables would wear and break, usually at an "inopportune time." (See Fig. 6-10 to check their use.) Press the brake pedals *before* starting the engine and also check their effectiveness just as the airplane starts to move for taxi.

You aren't expected to be a mechanic, and these problem descriptions are presented so you'll know that people in maintenance should be consulted and you can give them an idea of what to look for, thus saving time (and money).

Fig. D-14.

BIBLIOGRAPHY

AND RECOMMENDED READING

Aeronautical Information Manual. Washington, D.C., USGPO.

Aviation Weather. AC 00-6A. Washington, D.C., USGPO, 1975.

Aviation Weather Services. AC 00-45D. Washington, D.C., USGPO, 1985

Kershner, William K., *The Advanced Pilot's Flight Manual,* 6th ed. Ames, Iowa State University Press, 1994.

Langewiesche, Wolfgang, *Stick and Rudder.* New York, McGraw-Hill, 1944.

Private Pilot Practical Test Standards (Airplane SEL) FAA-S-8081-14, Washington, D.C., 1995.

Recreational Pilot and Private Pilot Written Test Book, FAA-T-8080-15B, Washington, D.C., USGPO, 1993.

Information received from companies and *Pilot's Operating Handbooks* referred to:

Bendix/King, General Aviation Avionics Division, Olathe, KS 66082.

Castleberry Instruments and Avionics, Inc. Austin, TX Aircraft Instruments.

Cessna Aircraft Co., Wichita, KS Pilot Operating Handbooks.

INDEX